HANDBOOK OF CHEMICAL AND ENVIRONMENTAL ENGINEERING CALCULATIONS

HANDBOOK OF CHEMICAL AND ENVIRONMENTAL ENGINEERING CALCULATIONS

Joseph P. Reynolds
John S. Jeris
Louis Theodore

WILEY-
INTERSCIENCE

A John Wiley & Sons, Inc., Publication

Chemistry Library

Copyright © 2002 by John Wiley & Sons, Inc., New York. All rights reserved.

Published simultaneously in Canada.

For ordering and customer service information please call 1-800-CALL-WILEY.

Library of Congress Cataloging-in-Publication Data is available.

ISBN 0-471-40228-1

Printed in the United States of America.
10 9 8 7 6 5 4 3 2 1

To Barbara, Megan, and Marybeth
for their unrelenting support, good-natured criticism,
and putting up with their husband and father [J.P.R.]

To my lovely wife Helen,
who for years endured neglect as I followed the path of
professional growth but still found it possible to
offer her strong support and love [J.S.J]

To Brother Conrad Timothy Burris, F.S.C.,
Professor Emeritus of Chemical Engineering and
Former Dean of Engineering at Manhattan College
for having the foresight to allow the School of Engineering to
achieve its potential during his tenure as dean,
for providing the leadership necessary for the school to reach its
potential, and for hiring me in 1960 [L.T.]

CONTENTS

PREFACE

Because of the pervasive nature of environmental problems, the overlap and interrelationship between the chemical and environmental engineering disciplines have become unavoidable. Further, many have agreed that environmental engineering involves the application of chemical engineering fundamentals and principles to the environment.

From an academic perspective, environmental engineering programs have traditionally been offered at the graduate level. More recently, formal environmental engineering programs—with the accompanying degree—are being offered at the undergraduate level. As a result, courses similar in content are often offered in both programs. This recent phenomenon has created a need for the development of material concerned with both chemical and environmental engineering calculations. The resulting end product is the *Handbook of Chemical and Environmental Engineering Calculations*.

As is usually the case in preparing any text, the question of what to include and what to omit has been particularly difficult. However, the problems and solutions in the *Handbook* attempt to address calculations common to both chemical and environmental engineering. This *Handbook* provides the reader with nearly 600 solved problems in the chemical and environmental engineering fields. Of the eight parts, two are concerned with chemical engineering and six with environmental engineering. The interrelationship between both fields is emphasized in all parts. Each part is divided into a number of problem sets, each set containing anywhere from 8 to 12 problems and solutions.

This project was a unique undertaking. Rather than prepare a textbook in the usual format—essay material, illustrative examples, nomenclature, bibliography, problems, etc.,—the authors considered writing a calculations handbook that could be used as a self-teaching aid. One of the key features of this book is that the solutions to the problems are presented in a stand-alone manner. Throughout the book, the problems are laid out in such a way as to develop the reader's technical understanding of the subject in question. Each problem contains a title, problem statement, and data and solution, with the more difficult problems located at or near the end of each problem set. Thus, this *Handbook* offers material not only to individuals with limited technical background but also to those with extensive industrial experience. As such, this *Handbook* can be used as a text in either a general chemical engineering or environmental engineering course and (perhaps primarily) as a training tool for industry.

The authors cannot claim sole authorship to all the problems and material in this *Handbook*. The present book has evolved from a host of sources including: exam problem prepared by Dr. Sum Marié Flynn for her undergraduate Process Control course; notes, homework problems and exam problems prepared by J. Jeris for graduate environmental engineering courses; notes, homework problems, and exam problems prepared by L. Theodore for several chemical and environmental engineering graduate and undergraduate courses; problems and solutions drawn (with permission) from numerous Theodore Tutorials; and problems and solutions developed by faculty participants during National Science Foundation (NSF) Undergraduate Faculty Enhancement Program (UFEP) workshops.

One of the objectives of the NSF workshops included the development of illustrative examples by the faculty. Approximately 40 out of the nearly 600 problems provided in this *Handbook* were drawn, in part, from the original work of these faculty. We would like to acknowledge the following professors whose problems, in original or edited form, are included on this *Handbook*. (The problem numbers are noted in parenthesis alongside each name.)

Prof. William Auberle; Civil and Environmental Engineering, Northern Arizona University (ENC.1, ULT.4)

Dr. Howard Bein; Chemistry, U.S. Merchant Marine Academy, (ISO.2, ISO.3)

Dr. Seymour Block; Chemical Engineering, University of Florida (MED.8)

Dr. Ihab Farag; Chemical Engineering, University of New Hampshire (MED.9)

Dr. Kumar Ganesan; Environmental Engineering, Montana Tech of the University of Montana (ISO.6, IAQ.4, IAQ.5, IAQ.6)

Dr. David James; Civil and Environmental Engineering, University of Nevada at Las Vegas (HZA.2, ENC.4, ULT.5)

Dr. Christopher Koroneos; Chemical Engineering, Columbia University (ECO.2, ECO.4, ECO.5, ECO.8)

Dr. SoonSik Lim; Chemical Engineering, Youngstown State University (CHR.7)

Dr. Sean X. Liu; Civil and Environmental Engineering, University of California at Berkley (ULT.6, ECO.6, MUN.5, MUN.6, MED.6)

Dr. P.M. Lutchmansingh; Petroleum Engineering, Montana Tech of the University of Montana (ECO.7)

Dr. Suwanchai Nitisoravut; Civil Engineering, University of North Carolina at Charlotte (ULT.7, ISO.5, CHR.5)

Dr. Holly Peterson; Environmental Engineering, Montana Tech of the University of Montana (IAQ.1)

Dr. Lisa Reidl; Civil Engineering, University of Wisconsin at Platteville (RCY.7)

Dr. Carol Reifschneider; Science and Math, Montana State University (ISO.4)

Dr. Dennis Ryan; Chemistry, Hofstra University, (CHR.6)

Dr. Dilip K. Singh; Chemical Engineering, Youngstown University (ENC.10)

Dr. David Stevens, Civil and Environmental Engineering, Utah State University (HRA.4, WQA.10)

Dr. Bruce Thomson; Civil Engineering, University of New Mexico (CHR.8)

Dr. Frank Worley; Chemical Engineering, University of Houston (MED.7)

Dr. Ronald Wukash; Civil Engineering, Purdue University (MED.10)

Dr. Poa-Chiang (PC) Yuan; Civil Engineering, Jackson State University (ISO.1, MUN.7, MUN.8, HRA.1, HRA.2)

During the preparation of this *Handbook*, the authors were ably assisted in many ways by a number of graduate students in Manhattan College's Chemical Engineering Master's Program. These students contributed much time and energy researching and classroom testing various problems in the book.

Two other sources that were employed in preparing the problems included numerous Theodore Tutorials (plus those concerned with the professional engineering exam) and the Wilcox and Theodore 1999 Wiley-Interscience text, *Engineering and Environmental Ethics*. Finally, the authors wish to acknowledge the National Science Foundation for supporting several faculty workshops (described above) that produced a number of problems appearing in this *Handbook*.

Somehow the editor usually escapes acknowledgment. We were particularly fortunate to have Bob Esposito ("Espo" to us) of John Wiley & Sons serve as our editor. He had the vision early on to realize the present need and timeliness for a handbook of this nature.

Joseph P. Reynolds
John S. Jeris
Louis Theodore

HANDBOOK OF CHEMICAL AND ENVIRONMENTAL ENGINEERING CALCULATIONS

PART I
Chemical Engineering Fundamentals

Robert Ambrosini

1   Units and Dimensions (UAD)

UAD.1   UNIT CONVERSION FACTORS

Convert the following:

1. 8.03 yr to seconds
2. 150 mile/h to yard/h
3. 100.0 m/s^2 to ft/min^2
4. 0.03 g/cm^3 to lb/ft^3

Solution

1. The following conversion factors are needed:

   365 day/yr
   24 h/day
   60 min/h
   60 s/min

   Arranging the conversion factors so that units cancel to leave only the desired units, the following is obtained:

$$(8.03 \text{ yr})\left(\frac{365 \text{ day}}{\text{yr}}\right)\left(\frac{24 \text{ h}}{\text{day}}\right)\left(\frac{60 \text{ min}}{\text{h}}\right)\left(\frac{60 \text{ s}}{\text{min}}\right) = 2.53 \times 10^8 \text{ s}$$

2. In similar fashion, $\left(\frac{150 \text{ mile}}{\text{h}}\right)\left(\frac{5280 \text{ ft}}{\text{mile}}\right)\left(\frac{\text{yd}}{3 \text{ ft}}\right) = 2.6 \times 10^5 \text{ yd/h}$

3. $(100.0 \text{ m/s}^2)\left(\frac{100 \text{ cm}}{\text{m}}\right)\left(\frac{\text{ft}}{30.48 \text{ cm}}\right)\left(\frac{60 \text{ s}}{\text{min}}\right)^2 = 1.181 \times 10^6 \text{ ft/min}^2$

4. $(0.03 \text{ g/cm}^3)\left(\frac{\text{lb}}{454 \text{ g}}\right)\left(\frac{30.48 \text{ cm}}{\text{ft}}\right)^3 = 2 \text{ lb/ft}^3$

UAD.2   CHEMICAL CONVERSIONS

Answer the following:

1. What is the molecular weight of nitrobenzene ($C_6H_5O_2N$)?
2. How many moles are there in 50.0 g of nitrobenzene?

3. If the specific gravity of a substance is 1.203, what is the density in g/cm^3?

4. What is the volume occupied by 50.0 g of nitrobenzene in cm^3, in ft^3, and in $in.^3$?

5. If the nitrobenzene is held in a cylindrical container with a base of 1 in. in diameter, what is the pressure at the base? What is it in gauge pressure?

6. How many molecules are contained in 50.0 g of nitrobenzene?

Solution

1. Pertinent atomic weights are listed below:

Carbon $= 12$
Hydrogen $= 1$
Oxygen $= 16$
Nitrogen $= 14$

The molecular weight of nitrobenzene is then

$$(6)(12) + (5)(1) + (2)(16) + (1)(14)$$
$$= 123 \text{ g/gmol}$$

2. To convert a mass to moles, divide by the molecular weight:

$$(50.0 \text{ g})\left(\frac{\text{gmol}}{123 \text{ g}}\right) = 0.407 \text{ gmol}$$

3. Since specific gravity is a ratio of density to the density of water $(1.000 \text{ g/cm}^3$ at $4°C)$,

$$\rho = (\text{SG})(\rho_{H_2O} \text{ at } 4°C)$$
$$= (1.203)(1.000 \text{ g/cm}^3) = 1.203 \text{ g/cm}^3$$

4. The results of part 3 may be employed to calculate the volume:

$$V = \left(\frac{50.0 \text{ g}}{1.203 \text{ g/cm}^3}\right) = 41.6 \text{ cm}^3$$

Applying conversion factors,

$$V = (41.6\,\text{cm}^3)\left(\frac{\text{ft}}{30.48\,\text{cm}}\right)^3 = 1.46 \times 10^{-3}\,\text{ft}^3$$

$$= (1.46 \times 10^{-3}\,\text{ft}^3)\left(\frac{12\,\text{in.}}{\text{ft}}\right)^3 = 2.52\,\text{in.}^3$$

5. The cross-sectional area of the base is calculated as:

$$A = \left(\frac{\pi D^2}{4}\right) = \left[\frac{\pi(1\,\text{in.})^2}{4}\right] = 0.78\,\text{in.}^2$$

Since there are 454 g/lb,

$$50.0\,\text{g} = 0.110\,\text{lb}$$

Using Newton's law, the force exerted by the mass is

$$F = m\left(\frac{g}{g_c}\right) = (0.110\,\text{lb})\left(1\,\frac{\text{lb}_\text{f}}{\text{lb}}\right) = 0.110\,\text{lb}_\text{f}$$

Note that, in the English system,

$$g = 32.2\,\text{ft/s}^2$$

$$g_c = 32.2\,\frac{\text{lb} \cdot \text{ft}}{\text{lb}_\text{f} \cdot \text{s}^2}$$

Since the force is exerted over an area of $0.78\,\text{in.}^2$, the pressure at the base is

$$P = \frac{F}{A} = \frac{0.110\,\text{lb}_\text{f}}{0.78\,\text{in.}^2} = 0.14\,\text{psi}$$

By definition:

$$P_\text{gauge} = P_\text{absolute} - P_\text{atmospheric}$$

Thus,

$$P_\text{gauge} = 0.14\,\text{psi} - 14.7\,\text{psi} = -14.6\,\text{psig}$$

6. There are 6.02×10^{23} (Avogadro's number) molecules/gmol. Therefore,

$$(0.406 \text{ gmol})(6.02 \times 10^{23} \text{ molecules/gmol}) = 2.44 \times 10^{23} \text{ molecules}$$

UAD.3   TEMPERATURE CONVERSIONS

Convert the following temperatures:

1. 20°C to °F, K, and °R
2. 20°F to °C, K, and °R

Solution

The following key equations are employed:

$T \, (°F) = 1.8T \, (°C) + 32$
$T \, (K) = T \, (°C) + 273$
$T \, (°R) = T \, (°F) + 460$
$T \, (°R) = 1.8T \, (K)$

Thus,

1. $T \, (°F) = 1.8(20°C) + 32 = 68°F$

    $T \, (K) = (20°C) + 273 = 293K$
    $T \, (°R) = 1.8(293K) = 527°R$
2. $T \, (°C) = (20°F - 32)/1.8 = -6.7°C$

    $T \, (K) = -6.7°C + 273 = 266K$
    $T \, (°R) = 20°F + 460 = 480°R$

UAD.4   PRESSURE CALCULATIONS

The height of a liquid column of mercury is 2.493 ft. Assume the density of mercury is $848.7 \, \text{lb/ft}^3$ and atmospheric pressure is $2116 \, \text{lb}_f/\text{ft}^2$ absolute. Calculate the gauge pressure in lb_f/ft^2 and the absolute pressure in lb_f/ft^2, psia, mm Hg, and in. H_2O.

Solution

Expressed in various units, the standard atmosphere is equal to:

1.0	Atmospheres (atm)
33.91	Feet of water (ft H_2O)
14.7	Pounds-force per square inch absolute (psia)
2116	Pounds-force per square foot absolute (psfa)
29.92	Inches of mercury (in. Hg)
760.0	Millimeters of mercury (mm Hg)
1.013×10^5	Newtons per square meter (N/m^2)

The equation describing the gauge pressure in terms of the column height and liquid density is

$$P_g = \rho g h / g_c$$

where P_g = gauge pressure
ρ = liquid density
h = height of column
g = acceleration of gravity
g_c = conversion constant

Thus,

$$P_g = (848.7 \, \text{lb/ft}^3)\left(1\frac{\text{lb}_f}{\text{lb}}\right)(2.493 \, \text{ft})$$

$$= 2116 \, \text{lb}_f/\text{ft}^2 \text{ gauge}$$

The pressure in lb$_f$/ ft^2 absolute is

$$P_{\text{absolute}} = P_g + P_{\text{atmospheric}}$$

$$= 2116 \, \text{lb}_f/\text{ft}^2 + 2116 \, \text{lb}_f/\text{ft}^2$$

$$= 4232 \, \text{lb}_f/\text{ft}^2 \text{ absolute}$$

The pressure in psia is

$$P(\text{psia}) = (4232 \, \text{psfa})\left(\frac{1 \, \text{ft}^2}{144 \, \text{in.}^2}\right) = 29.4 \, \text{psia}$$

The corresponding gauge pressure in psi is

$$P(\text{psig}) = 29.4 - 14.7 = 14.7 \, \text{psig}$$

The pressure in mm Hg is

$$P(\text{mm Hg}) = (29.4\,\text{psia})\left(\frac{760\,\text{mm Hg}}{14.7\,\text{psia}}\right) = 1520\,\text{mm Hg}$$

Note that 760 mm Hg is equal to 14.7 psia, which in turn is equal to 1.0 atm. Finally, the pressure in in. H_2O is

$$P_{(\text{in. } H_2O)} = \left(\frac{29.4\,\text{psia}}{14.7\,\text{psia/atm}}\right)\left(\frac{33.91\,\text{ft } H_2O}{\text{atm}}\right)\left(\frac{12\,\text{in.}}{\text{ft}}\right)$$

$$= 813.8 \text{ in. } H_2O$$

The reader should note that absolute and gauge pressures are usually expressed with units of atm, psi, or mm Hg. This statement also applies to partial pressures. One of the most common units employed to describe pressure drop is inches of H_2O, with the notation in. H_2O or IWC (inches of water column).

UAD.5   ENGINEERING CONVERSION FACTORS

Given the following data for liquid methanol, determine its density in lb/ft^3 and convert heat capacity, thermal conductivity, and viscosity from the International System of Units (SI) to English units:

Specific gravity = 0.92 (at 60°F)
Density of reference substance (water) = 62.4 lb/ft^3 (at 60°F)
Heat capacity = 0.61 cal/(g · °C) (at 60°F)
Thermal conductivity = 0.0512 cal/(m · s · °C) (at 60°F)
Viscosity = 0.64 cP (at 60°F)

Solution

The definition of specific gravity for liquids and solids is

$$\text{Specific gravity} = \frac{\text{Density}}{\text{Density of water at } 4°C}$$

Note that the density of water at 4°C is 62.4 lb/ft^3 in English engineering units or 1.0 g/cm^3.

Calculate the density of methanol in English units by multiplying the specific gravity by the density of water.

$$\text{Density of methanol} = (\text{Specific gravity})(\text{Density of water})$$

$$= (0.92)(62.4) = 57.4 \, \text{lb/ft}^3$$

The procedure is reversed if one is interested in calculating specific gravity from density data. The notation for density is usually ρ.

Convert the heat capacity from units of cal/(g · °C) to Btu/(lb · °F).

$$\left(\frac{0.61 \, \text{cal}}{\text{g} \cdot °\text{C}}\right)\left(\frac{454 \, \text{g}}{\text{lb}}\right)\left(\frac{\text{Btu}}{252 \, \text{cal}}\right)\left(\frac{°\text{C}}{1.8°\text{F}}\right) = 0.61 \, \text{Btu/(lb} \cdot °\text{F})$$

Note that $1.0 \, \text{Btu/(lb} \cdot °\text{F})$ is equivalent to $1.0 \, \text{cal/(g} \cdot °\text{C})$. This also applies on a mole basis, i.e.,

$$1 \, \text{Btu/(lbmol} \cdot °\text{F}) = 1 \, \text{cal/(gmol} \cdot °\text{C})$$

The usual notation for heat capacity is C_p. In this book, C_p represents the heat capacity on a mole basis, while c_p indicates a mass basis.

Convert the thermal conductivity of methanol from cal/(m · s · °C) to Btu/(ft · h · °F).

$$\left(\frac{0.0512 \, \text{cal}}{\text{m} \cdot \text{s} \cdot °\text{C}}\right)\left(\frac{\text{Btu}}{252 \, \text{cal}}\right)\left(\frac{0.3048 \, \text{m}}{\text{ft}}\right)\left(\frac{3600 \, \text{s}}{\text{h}}\right)\left(\frac{°\text{C}}{1.8°\text{F}}\right) = 0.124 \, \text{Btu/(ft} \cdot \text{h} \cdot °\text{F})$$

The usual engineering notation for thermal conductivity is k.

Convert viscosity from centipoise to lb/(ft · s):

$$(0.64 \, \text{cP})\left(\frac{6.72 \times 10^{-4} \, \text{lb}}{\text{ft} \cdot \text{s} \cdot \text{cP}}\right) = 4.3 \times 10^{-4} \, \text{lb/(ft} \cdot \text{s})$$

The notation for viscosity is typically μ. The kinematic viscosity, v, is defined by the ratio of viscosity to density, i.e., $v = \mu/\rho$ with units of length^2/time.

Finally, note that the above physical properties are strong functions of the temperature but weak functions of the pressure. Interestingly, the viscosity of a gas increases with increasing temperature, while the viscosity of a liquid decreases with an increase in temperature.

UAD.6 MOLAR RELATIONSHIPS

A mixture contains 20 lb of O_2, 2 lb of SO_2, and 3 lb of SO_3. Determine the weight fraction and mole fraction of each component.

Solution

By definition:

Weight fraction = weight of A/total weight
Moles of A = weight of A/molecular weight of A
Mole fraction = moles of A/total moles

First, calculate the weight fraction of each component:

Compound	Weight (lb)	Weight Fraction
O_2	20	$20/25 = 0.8$
SO_2	2	0.08
SO_3	3	0.12
Total	25	1.00

Calculate the mole fraction of each component, noting that moles = weight/molecular weight.

The molecular weights of O_2, SO_2 and SO_3 are 32, 64 and 80, respectively. The following table can be completed:

Compound	Weight	Molecular Weight	Moles	Mole Fraction
O_2	20	32	$20/32 = 0.6250$	0.901
SO_2	2	64	0.0301	0.045
SO_3	3	80	0.0375	0.054
Total			0.6938	1.000

The reader should note that, in general, weight fraction (or percent) is <u>not</u> equal to mole fraction (or percent).

UAD.7 FLUE GAS ANALYSIS

The mole percent (gas analysis) of a flue gas is given below:

$N_2 = 79\%$
$O_2 = 5\%$
$CO_2 = 10\%$
$CO = 6\%$

Calculate the average molecular weight of the mixture.

Solution

First write the molecular weight of each component:

$$MW(N_2) = 28$$
$$MW(O_2) = 32$$
$$MW(CO_2) = 44$$
$$MW(CO) = 28$$

By multiplying the molecular weight of each component by its mole percent the following table can be completed:

Compound	Molecular Weight	Mole Fraction	Weight (lb)
N_2	28	0.79	22.1
O_2	32	0.05	1.6
CO_2	44	0.10	4.4
CO	28	0.06	1.7
Total		1.00	

Finally, calculate the average molecular weight of the gas mixture:

$$\text{Average molecular weight} = 22.1 + 1.6 + 4.4 + 1.7 = 29.8$$

The sum of the weights in pounds represents the average molecular weight because the calculation above is based on 1.0 mol of the gas mixture.

The reader should also note that in a gas, molar percent equals volume percent and vice versa. Therefore, a volume percent can be used to determine weight fraction as illustrated in the table. The term y is used in engineering practice to represent mole (or volume) fraction of gases; the term x is often used for liquids and solids.

UAD.8   PARTIAL PRESSURE

The exhaust to the atmosphere from an incinerator has a SO_2 concentration of 0.12 mm Hg partial pressure. Calculate the parts per million of SO_2 in the exhaust.

Solution

First calculate the mole fraction, y. By Dalton's law,

$$y = p_{SO_2}/P$$

Since the exhaust is discharged to the atmosphere, the atmospheric pressure, 760 mm Hg, is the total pressure, P:

$$y = (0.12 \, \text{mm Hg})/(760 \, \text{mm Hg}) = 1.58 \times 10^{-4}$$

$$\text{ppm} = (y) \, (10^6) = (1.58 \times 10^{-4}) \, (10^6)$$

$$= 158 \, \text{ppm}$$

UAD.9   CONCENTRATION CONVERSION

Express the concentration 72 g of HCl in 128 cm^3 of water into terms of fraction and percent by weight, parts per million, and molarity.

Solution

The fraction by weight can be calculated as follows:

$$72 \, \text{g}/200 \, \text{g} = 0.36$$

The percent by weight can be calculated from the fraction by weight.

$$(0.36)(100\%) = 36\%$$

The ppm (parts per million) can be calculated as follows:

$$(72 \, \text{g}/128 \, \text{cm}^3)(10^6) = 562,500 \, \text{ppm}$$

The molarity (M) is defined as follows:

$$M = \text{moles of solute/volume of solution (L)}$$

Using atomic weights,

$$\text{MW of HCl} = 1.0079 + 35.453 = 36.4609$$

$$M = \left[(72 \, \text{g HCl}) \left(\frac{1 \, \text{mol HCl}}{36.4609 \, \text{g HCl}} \right) \right] \Big/ \left(\frac{128 \, \text{cm}^3}{1000 \, \text{cm}^3/\text{L}} \right) = 15.43 \, \text{mol/L}$$

UAD.10   FILTER PRESS APPLICATION

A plate and frame filter press is to be employed to filter a slurry containing 10% by mass of solids. If 1 ft^2 of filter cloth area is required to treat 5 lb/h of solids, what cloth area, in ft^2, is required for a slurry flowrate of 6000 lb/min?

Solution

Convert the slurry flowrate, $\dot{m}$, to lb/h:

$$\dot{m}\ (\text{slurry}) = (6000\ \text{lb/min})(60\ \text{min /h}) = 360{,}000\ \text{lb/h}$$

Calculate the solids flowrate in the slurry:

$$\dot{m}\ (\text{solids}) = (0.1)(360{,}000\ \text{lb/h}) = 36{,}000\ \text{lb/h}$$

Calculate the filter cloth area, A, requirement:

$$A = (36{,}000\ \text{lb/h})\left(\frac{1}{5}\frac{\text{h} \cdot \text{ft}^2}{\text{lb}}\right) = 7200\ \text{ft}^2$$

2 Conservation Law for Mass (CMA)

CMA.1 PROCESS CALCULATION

An external gas stream is fed into an air pollution control device at a rate of 10,000 lb/h in the presence of 20,000 lb/h of air. Due to the energy requirements of the unit, 1250 lb/h of a vapor conditioning agent is added to assist the treatment of the stream. Determine the rate of product gases exiting the unit in pounds per hour (lb/h). Assume steady-state conditions.

Solution

The conservation law for mass can be applied to any process or system. The general form of this law is given by:

$$\text{Mass accumulated} = \text{Mass in} - \text{Mass out} + \text{Mass generated}$$

Apply the conservation law for mass to the control device on a rate basis:

$$\text{Rate of mass in} - \text{Rate of mass out} + \text{Rate of mass generated}$$
$$= \text{Rate of mass accumulated}$$

Rewrite this equation subject to the conditions in the problem statement:

$$\text{Rate of mass in} = \text{Rate of mass out}$$

or

$$\dot{m}_{in} = \dot{m}_{out}$$

Note that mass is not generated and steady conditions (no accumulation) apply. Refer to the problem statement for the three inlet flows:

$$\dot{m}_{in} = 10{,}000 + 20{,}000 + 1250 = 31{,}250 \, \text{lb/h}$$

Determine $\dot{m}_{out}$, the product gas flowrate.

14

Since $m_{in} = m_{out}$,

$$\dot{m}_{out} = 31{,}250\,\text{lb/h}$$

Finally, the conservation law for mass may be written for any compound whose quantity is not changed by chemical reaction and for any chemical element whether or not it has participated in a chemical reaction. It may be written for one piece of equipment, around several pieces of equipment, or around an entire process. It may be used to calculate an unknown quantity directly, to check the validity of experimental data, or to express one or more of the independent relationships among the unknown quantities in a particular problem situation.

CMA.2   COLLECTION EFFICIENCY

Given the following inlet loading and outlet loading of an air pollution particulate control unit, determine the collection efficiency of the unit.

Inlet loading $= 2\,\text{gr/ft}^3$
Outlet loading $= 0.1\,\text{gr/ft}^3$

Solution

Collection efficiency is a measure of the degree of performance of a control device; it specifically refers to the degree of removal of a pollutant and may be calculated through the application of the conservation law for mass. *Loading* refers to the concentration of pollutant, usually in grains (gr) of pollutant per cubic feet of contaminated gas stream.

The equation describing collection efficiency (fractional), E, in terms of inlet and outlet loading is

$$E = \frac{\text{Inlet loading} - \text{Outlet loading}}{\text{Inlet loading}}$$

Calculate the collection efficiency of the control unit in percent for the rates provided.

$$E = \frac{2 - 0.1}{2}\,100 = 95\%$$

The term η is also used as a symbol for efficiency E.

The reader should also note that the collected amount of pollutant by the control unit is the product of E and the inlet loading. The amount discharged to the atmosphere is given by the inlet loading minus the amount collected.

CMA.3 OVERALL COLLECTION EFFICIENCY

A cyclone is used to collect particulates with an efficiency of 60%. A venturi scrubber is used as a second control device. If the required overall efficiency is 99.0%, determine the minimum operating efficiency of the venturi scrubber.

Solution

Many process systems require more than one piece of equipment to accomplish a given task, e.g., removal of a gaseous or particulate pollutant from a flow stream. The efficiency of each individual collector or equipment may be calculated using the procedure set forth in Problem CMA.2. The overall efficiency of multiple collectors may be calculated from the inlet stream to the first unit and the outlet stream from the last unit. It may also be calculated by proceeding sequentially through the series of collectors.

Calculate the mass of particulate leaving the cyclone using a basis of 100 lb of particulate entering the unit.

Use the efficiency equation:

$$E = (W_{in} - W_{out})/(W_{in})$$

where E = fraction efficiency
W = loading

Rearranging the above gives:

$$W_{out} = (1 - E)(W_{in}) = (1 - 0.6)(100) = 40\,lb$$

Calculate the mass of particulate leaving the venturi scrubber using an overall efficiency of 99.0%:

$$W_{out} = (1 - E)(W_{in}) = (1 - 0.99)(100) = 1.0\,lb$$

Calculate the efficiency of the venturi scrubber using W_{out} from the cyclone as W_{in} for the venturi scrubber. Use the same efficiency equation above and convert to percent efficiency:

$$E = (W_{in} - W_{out})/(W_{in}) = (40 - 1.0)/(40) = 0.975 = 97.5\%$$

An extremely convenient efficiency-related term employed in pollution control calculations is the penetration, P. By definition:

$$P = 100 - E \quad \text{(percent basis)}$$
$$P = 1 - E \quad \text{(fractional basis)}$$

Note that there is a 10-fold increase in P as E goes from 99.9 to 99%. For a multiple series of n collectors, the overall penetration is simply given by:

$$P = P_1 P_2, \ldots, P_{n-1} P_n$$

For particulate control in air pollution units, penetrations and/or efficiencies can be related to individual size ranges. The overall efficiency (or penetration) is then given by the contribution from each size range, i.e., the summation of the product of mass fraction and efficiency for each size range. This is treated in more detail in Part III, Air Pollution Control Equipment.

CMA.4 SPRAY TOWER APPLICATION

A proposed incineration facility design requires that a packed column and a spray tower be used in series for the removal of HCl from the flue gas. The spray tower is to operate at an efficiency of 65% and the packed column at an efficiency of 98%. Calculate the mass flowrate of HCl leaving the spray tower, the mass flowrate of HCl entering the packed tower, and the overall efficiency of the removal system if 76.0 lb of HCl enters the system every hour.

Solution

As defined in Problem CMA.3:

$$E = (W_{in} - W_{out})/W_{in}$$

Then,

$$W_{out} = (1 - E)(W_{in})$$

For the spray tower,

$$W_{out} = (1 - 0.65)(76.0) = 26.6 \text{ lb/h HCl}$$

The mass flowrate of HCl leaving the spray tower is equal to the mass flowrate of HCl entering the packed column. For the packed column,

$$W_{out} = (1 - 0.98)(26.6) = 0.532 \text{ lb/h HCl}$$

The overall efficiency can now be calculated:

$$E = (W_{in} - W_{out})/W_{in} = (76.0 - 0.532)/(76.0)$$
$$= 0.993$$
$$= 99.3\%$$

CMA.5   COMPLIANCE DETERMINATION

A proposed incinerator is designed to destroy a hazardous waste at 2100°F and 1 atm. Current regulations dictate that a minimum *destruction and removal efficiency (DRE)* of 99.99% must be achieved. The waste flowrate into the unit is 960 lb/h while that flowing out of the unit is measured as 0.08 lb/h. Is the unit in compliance?

Solution

Select as a basis 1 h of operation. The mass equation employed for efficiency may also be used to calculate the minimum destruction and removal efficiency:

$$DRE = [(w_{in} - w_{out})/w_{in}](100) = [(960 - 0.08)/960](100)$$
$$= 99.992\%$$

Thus, the unit is operating in compliance with present regulations. The answer is *yes*.

CMA.6   COAL COMBUSTION

A power plant is burning anthracite coal containing 7.1% ash to provide the necessary energy for steam generation. If 300 ft³ of total flue gas are produced for every pound of coal burned, what is the maximum effluent particulate loading in gr/ft³? Assume no contribution to the particulates from the waste. The secondary ambient air quality standard for particulates is 75 µg/m³. What dilution factor and particulate collection efficiency are required to achieve this standard?

Solution

Select 1.0 lb of coal as a basis. Calculate the mass of particulates (ash), M:

$$M = (1.0)(0.071) = 0.071 \text{ lb}$$

The maximum particulate loading, w, is then obtained by dividing by the volume of the flue gas, V:

$$w = M/V = 0.071/300 = 2.367 \times 10^{-4} \text{ lb/ft}^3 = 1.66 \text{ gr/ft}^3$$

Note: 1 lb = 7000 gr.
  Converting to $\mu g/m^3$:

$$w = \left(2.367 \times 10^{-4} \frac{\text{lb}}{\text{ft}^3}\right)\left(\frac{\text{ft}}{0.3048 \text{ m}}\right)^3 \left(\frac{454 \text{ g}}{\text{lb}}\right)\left(\frac{10^6 \text{ } \mu g}{\text{g}}\right) = 3.795 \times 10^6 \frac{\mu g}{m^3}$$

The dilution factor (DF) is required to achieve the ambient quality standard. Then

$$DF = w_{in}/w_{std} = 3.795 \times 10^6/75 = 5.04 \times 10^4$$

Alternately, the required fractional collection efficiency is

$$E = (3.795 \times 10^6 - 75)/ \ 3.795 \times 10^6 = 0.99998 = 99.998\%$$

CMA.7   VELOCITY DETERMINATION

If 20,000 ft³/min of water exits a system through a pipe whose cross-sectional area is 4 ft², determine the mass flowrate in lb/min and the exit velocity in ft/s.

Solution

The continuity equation is given by

$$\dot{m} = \rho A v$$

where ρ = liquid density
        A = cross-sectional area
        v = velocity

Since 20,000 ft³/min of water enters the system, then

$$A_1 v_1 = 20,000 \text{ ft}^3/\text{min}$$

where subscript 1 refers to inlet conditions.
  Therefore,

$$\dot{m} = \rho_1 A_1 v_1 = (62.4)(20,000) = 1,248,000 \text{ lb/min}$$

Under steady-state conditions, this same mass flow must also leave the system. Thus, in accordance with the conservation law for mass

$$\dot{m} = \rho_2 A_2 v_2$$

where subscript 2 refers to exit conditions.
  Therefore,

$$v_2 = \left[\frac{1{,}248{,}000 \text{ lb/min}}{(62.4 \text{ lb/ft}^3)(4 \text{ ft}^2)}\right]\left(\frac{\min}{60 \text{ s}}\right) = 83.3 \text{ ft/s}$$

CMA.8 CONVERGING PIPE

Water (density $= 1000 \text{ kg/m}^3$) flows in a converging circular pipe (see Figure 1). It enters at point 1 and leaves at point 2. At point 1, the inside diameter d is 14 cm and the velocity is 2 m/s. At point 2, the inside diameter is 7 cm. Determine:

1. The mass and volumetric flowrates
2. The mass flux of water
3. The velocity at point 2

Solution

The conservation law for mass may be applied to a fluid device with one input and one output. For a fluid of constant density ρ and a uniform velocity v at each cross section A, the mass rate $\dot{m}$, the volumetric rate q, and the mass flux G, are

$$\dot{m} = \rho A v; \quad \text{mass/time}$$

$$q = Av; \quad \text{volume/time}$$

$$G = \dot{m}/A = \rho v; \quad \text{mass/time} \cdot \text{area}$$

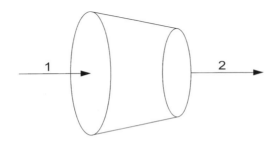

Converging pipe

Figure 1. Diagram for Problem CMA.8.

First calculate the flowrates, q and $\dot{m}$, based on the information at point 1.

$$A_1 = \frac{\pi d_1^{\,2}}{4} = \frac{\pi (0.14)^2}{4} = 0.0154 \text{ m}^2$$

$$q_1 = A_1 v_1 = (0.0154)(2) = 0.031 \text{ m}^3/\text{s} = 488 \text{ gpm}$$

$$\dot{m}_1 = \rho q_1 = (1000)(0.031) = 31 \text{ kg/s}$$

Note the use of the following conversion factor for q_1:

$$1 \text{ gpm} = 6.309 \times 10^{-5} \text{ m}^3/\text{s}$$

Obtain the mass flux G:

$$G = 31/0.0154 = 2013 \text{ kg}/(\text{m}^{-2} \cdot \text{s})$$

Noting that

$$1 \text{ ft/s} = 0.3048 \text{ m/s}$$

the velocity at point 2, v_2, may now be calculated:

$$q_2 = q_1 = 0.031 \text{ m}^3/\text{s}$$

$$v_2 A_2 = v_1 A_1$$

$$v_2 = v_1 (A_1/A_2) = v_1 (d_1^{\,2}/d_2^{\,2}) = 2(14/7)^2 = 8 \text{ m/s} = 26.25 \text{ ft/s}$$

As expected, for steady-state flow of an incompressible fluid, a decrease in cross-sectional area results in an increase in the flow velocity.

CMA.9 HUMIDITY EFFECT

A flue gas [molecular weight (MW) = 30, dry basis] is being discharged from a scrubber at 180°F (dry bulb) and 125°F (wet bulb). The gas flowrate on a dry basis is 10,000 lb/h. The absolute humidity at the dry-bulb temperature of 180°F and wet-bulb temperature of 125°F is 0.0805 lb H_2O/lb dry air.

1. What is the mass flowrate of the wet gas?
2. What is the actual volumetric flowrate of the wet gas?

Solution

Curves showing the relative humidity (ratio of the mass of the water vapor in the air to the maximum mass of water vapor the air can hold at that temperature, i.e., if the air were saturated) of humid air appear on the psychrometric chart. (See Figure 2.) The curve for 100% relative humidity is also referred to as the saturation curve. The abscissa of the humidity chart is air temperature, also known as the dry-bulb temperature (T_{DB}). The wet-bulb temperature (T_{WB}) is another measure of humidity; it is the temperature at which a thermometer with a wet wick wrapped around the bulb stabilizes. As water evaporates from the wick to the ambient air, the bulb is cooled; the rate of cooling depends on how humid the air is. No evaporation occurs if the air is saturated with water; hence T_{WB} and T_{DB} are the same. The lower the humidity, the greater the difference between these two temperatures. On a psychrometric chart, constant wet-bulb temperature lines are straight with negative slopes. The value of T_{WB} corresponds to the value of the abscissa at the point of intersection of this line with the saturation curve.

Calculate the flowrate of water in the air. Note that both the given flowrate and humidity are on a dry basis.

$$\text{Water flowrate} = (0.0805)(10{,}000) = 805\,\text{lb/h}$$

Calculate the total flowrate by adding the dry gas and water flowrates:

$$\text{Total flowrate} = 10{,}000 + 805 = 10{,}805\,\text{lb/h}$$

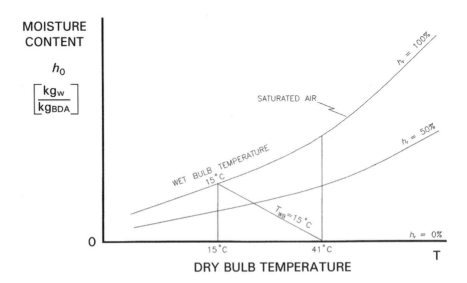

Figure 2. Diagram of a psychrometric chart.

The moles of water and dry gas are thus

$$\text{Moles gas} = 10,000/30 = 333.3 \text{ lbmol/h}$$
$$\text{Moles water} = 805/18 = 44.7 \text{ lbmol/h}$$

Calculate the mole fraction of water vapor using the above units:

$$y_{water} = 44.7/(44.7 + 333.3) = 0.12$$

The average molecular weight of the mixture becomes

$$MW = (1.0 - 0.12)(30) + (0.12)(18)$$
$$= 28.6 \text{ lb/lbmol}$$

The molar flowrate of the wet gas may now be determined:

$$\dot{n} = 10,805/28.6 = 378 \text{ lbmol/h}$$

The ideal gas law may be applied to calculate the volumetric flowrate of the wet gas:

$$q = \dot{n} RT/P$$
$$= (378)(0.73)(460 + 180)/1.0$$
$$= 1.77 \times 10^5 \text{ ft}^3/\text{h}$$

The following are some helpful points on the use of psychrometric charts.

1. In problems involving the use of the humidity chart, it is convenient to choose a mass of dry air as a basis, since the chart uses this basis.
2. Heating or cooling at temperatures above the dew point (temperature at which the vapor begins to condense) corresponds to a horizontal movement on the chart. As long as no condensation occurs, the absolute humidity stays constant.
3. If the air is cooled, the system follows the appropriate horizontal line to the left until it reaches the saturation curve and follows this curve thereafter.

CMA.10   RESIDENTIAL WATER CONSERVATION

Assume that the average water usage of a community is approximately 130 gal per person per day. After implementing water saving practices, the average water usage drops to 87 gal per person per day. A 10 million gallon per day (MGD) wastewater treatment plant is used for which 60% of the flow is residential wastewater. Calculate the water savings in gallons per day (gal/day) and the percent reduction in water usage. Also calculate the current residential wastewater flow in gal/day, millions of gal (Mgal)/yr, lb/day and kg/day. Assume the specific gravity of the water to be 1.0.

Solution

First calculate the water savings, W_s, in gal/day:

$$W_s = (130 - 87) = 43 \, \text{gal/day}$$

The percent reduction is

$$\%_{red} = (43/130)(100)$$
$$= 33.1\%$$

Finally, the current residential wastewater flow, q, is

$$q = (0.60)(10\text{MGD})$$
$$= 6.0 \, \text{MGD} = 6 \times 10^6 \, \text{gal/day}$$

Converting to Mgal/yr,

$$q = (6 \times 10^6 \, \text{gal/day})(365 \, \text{day/yr})(\text{Mgal}/10^6 \, \text{gal}) = 2190 \, \text{Mgal/yr}$$

Converting to ft^3/day,

$$q = (6 \times 10^6 \, \text{gal/day})(1 \, \text{ft}^3/7.48 \, \text{gal}) = 8.02 \times 10^5 \, \text{ft}^3/\text{day}$$

Converting to lb/day,

$$\dot{m} = (8.02 \times 10^5 \, \text{ft}^3/\text{day})(62.4 \, \text{lb/ft}^3) = 5.00 \times 10^7 \, \text{lb/day}$$

Converting to kg/day,

$$\dot{m} = (5.00 \times 10^7 \, \text{lb/day})/(2.2 \, \text{lb/kg}) = 2.28 \times 10^7 \, \text{kg/day}$$

CMA.11   FLOW DIAGRAM

As part of a pollution prevention program, flue gas from a process is mixed with recycled gas from an absorber (A), and the mixture passes through a waste heat boiler (H), which uses water as the heat transfer medium. It then passes through a water spray quencher (Q) in which the temperature of the mixture is further decreased and, finally, through an absorber (A) in which water is the absorbing

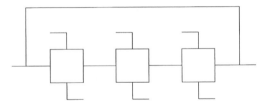

Figure 3. Line diagram.

agent (solvent) for one of the species in the flue gas stream. Prepare a simplified flow diagram for this process.

Solution

Before attempting to calculate the raw material or energy requirements of a process, it is desirable to attain a clear picture of the process. The best way to do this is to draw a flow diagram. A flow diagram is a line diagram showing the successive steps of a process by indicating the pieces of equipment in which the steps occur and the material streams entering and leaving each piece of equipment. Flow diagrams are used to conceptually represent the overall process.

Lines are usually used to represent streams, and boxes may be used to represent equipment. A line diagram of the process is first prepared (Figure 3).

The equipment in Figure 3 may now be labeled as in Figure 4. Label the flow streams as shown in Figure 5.

The reader should note that four important processing concepts are *bypass*, *recycle*, *purge*, and *makeup*. With bypass, part of the inlet stream is diverted around

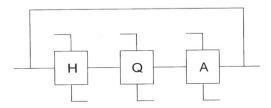

Figure 4. Line diagram with equipment labels.

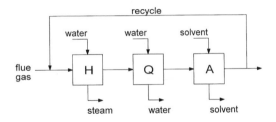

Figure 5. Line diagram with stream labels.

the equipment to rejoin the main stream after the unit. This stream effectively moves in parallel with the stream passing through the equipment. In recycle, part of the product stream is sent back to mix with the feed. If a small quantity of nonreactive material is present in the feed to a process that includes recycle, it may be necessary to remove the nonreactive material in a purge stream to prevent its building up above a maximum tolerable value. This can also occur in a process without recycle; if a nonreactive material is added in the feed and not totally removed in the products, it will accumulate until purged. The purging process is sometimes referred to as *blowdown*. *Makeup*, as its name implies, involves adding or making up part of a stream that has been removed from a process. Makeup may be thought of, in a final sense, as the opposite of purge and/or blowdown.

CMA.12   VELOCITY MAGNITUDE AND DIRECTION

A fluid device has four openings, as shown in Figure 6. The fluid has a constant density of $800\,kg/m^3$. Steady-state information on the system is also provided in the following table:

Section	Flow area (m^2)	Velocity (m/s)	Direction (relative to the device)
1	0.2	5	In
2	0.3	7	In
3	0.25	12	Out
4	0.15	?	?

Determine the magnitude and direction of the velocity v_4. What is the mass flowrate at section 4?

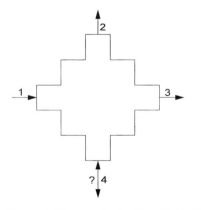

Figure 6. Diagram for Problem CMA.12.

Solution

Calculate the volumetric flowrate through each section:

$$q_1 = (0.2)(5) = 1 \, \text{m}^3/\text{s}$$
$$q_2 = (0.3)(7) = 2.1 \, \text{m}^3/\text{s}$$
$$q_3 = (0.25)(12) = 3 \, \text{m}^3/\text{s}$$

Since the fluid is of constant density, the continuity equation may be applied on a volume rate basis:

$$q_1 + q_2 = q_3 + q_4$$
$$1 + 2.1 = 3 + q_4$$
$$q_4 = 0.1 \, \text{m}^3/\text{s (exiting)}$$

Therefore, flow is out of the control volume at section 4.
   Calculate the velocity v_4:

$$v_4 = 0.1/0.15$$
$$= 0.667 \, \text{m/s}$$

Calculate the mass flowrate across section 4, $\dot{m}_4$:

$$\dot{m}_4 = (800)(0.1)$$
$$= 80 \, \text{kg/s}$$

3 Conservation Law for Energy (CLE)

CLE.1 OUTLET TEMPERATURE

Heat at 18.7×10^6 Btu/h is transferred from the flue gas of an incinerator. Calculate the outlet temperature of the gas stream using the following information:

Average heat capacity, c_p, of gas $= 0.26$ Btu/(lb $\cdot^\circ$ F)
Gas mass flowrate, $\dot{m} = 72,000$ lb/h
Gas inlet temperature, $T_1 = 1200^\circ$F

Solution

The first law of thermodynamics states that energy is conserved. For a flow system, neglecting kinetic and potential effects, the energy transferred, Q, to or from the flowing medium is given by the enthalpy change, ΔH, of the medium. The enthalpy of an ideal gas is solely a function of temperature; enthalpies of liquids and most real gases are almost always assumed to depend on temperature alone. Changes in enthalpy resulting from a temperature change for a single phase material may be calculated from the equation

$$\Delta H = m c_p \, \Delta T$$

or

$$\Delta \dot{H} = \dot{m} c_p \, \Delta T$$

where ΔH = enthalpy change
$\quad$ m = mass of flowing medium
$\quad$ c_p = average heat capacity per unit mass of flowing medium across the temperature range of ΔT
$\quad$ $\Delta \dot{H}$ = enthalpy change per unit time
$\quad$ $\dot{m}$ = mass flowrate of flowing medium

Note: The symbol Δ means "change in."

Solve the conservation law for energy for the gas outlet temperature T_2:

$$\dot{Q} = \Delta \dot{H} = \dot{m}c_p \, \Delta T = \dot{m}c_p(T_2 - T_1)$$

where $\dot{Q}$ is the rate of energy transfer.

$$T_2 = [\dot{Q}/(\dot{m}C_p)] + T_1$$

The gas outlet temperature is therefore

$$T_2 = [-18.7 \times 10^6/\{(72,000)(0.26)\}] + 1200$$
$$= 200^\circ F$$

The above equation is based on adiabatic conditions, i.e., the entire heat load is transferred from the flowing gas. The unit is assumed to be perfectly insulated so that no heat is transferred to the surroundings. However, this is not the case in a real-world application.

As with mass balances (see Chapter 2), an enthalpy balance may be performed within any properly defined boundary, whether real or imaginary. For example, an enthalpy balance can be applied across the entire unit or process. The enthalpy of the feed stream(s) is equated with the enthalpy of the product stream(s) plus the heat loss from the process. All the enthalpy terms must be based on the same reference temperature.

Finally, the enthalpy has two key properties that should be kept in mind:

1. Enthalpy is a point function, i.e., the enthalpy change from one state (say 200°F, 1 atm) to another state (say 400°F, 1 atm) is a function only of the two states and not the path of the process associated with the change.
2. Absolute values of enthalpy are not important. The enthalpy of water at 60°F, 1 atm, as recorded in some steam tables is 0 Btu/lbmol. This choice of zero is arbitrary however. Another table may indicate a different value. Both are correct! Note that changing the temperature of water from 60 to 100°F results in the same change in enthalpy using either table.

Enthalpy changes may be obtained with units (English) of Btu, Btu/lb, Btu/lbmol, Btu/scf, or Btu/time depending on the available data and calculation required.

CLE.2   POTENTIAL ENERGY CALCULATION

A process plant pumps 2000 lb of water to an elevation of 1200 ft above the turbogenerators. Determine the change in potential energy in Btu and ft · lb$_f$.

Solution

The potential energy change (ΔPE) is given by

$$\Delta \text{PE} = \frac{mg}{g_c}(\Delta Z)$$

where m = mass, lb
$\quad g$ = acceleration due to gravity, 32.2 ft/s^2 at sea level
$\quad g_c$ = gravitational constant, 32.2 lb $\cdot$ ft/(lb$_f$ $\cdot$ s^2)
$\quad \Delta Z$ = change in height

Substituting the data yields

$$\text{PE} = (2000\,\text{lb})\left[\frac{32.2\,\text{ft/s}^2}{32.2\,\text{lb} \cdot \text{ft/(lb}_f \cdot \text{s}^2)}\right]600\,\text{ft} = 1.2 \times 10^6\,\text{ft} \cdot \text{lb}_f$$

Since 1 Btu = 778.17 ft $\cdot$ lb$_f$,

$$\text{PE} = (1.2 \times 10^6\,\text{ft} \cdot \text{lb}_f)(1\,\text{Btu}/778.17\,\text{ft} \cdot \text{lb}_f) = 1543\,\text{Btu}$$

CLE.3   KINETIC ENERGY CALCULATION

If 2000 lb of water has its velocity increased from 8 to 30 ft/s, calculate the change in kinetic energy of the water in Btu and ft $\cdot$ lb$_f$ and the minimum energy required to accomplish this change.

Solution

By definition, the kinetic energy (KE) is given by

$$\text{KE} = \frac{mv^2}{2g_c}$$

where m = mass, lb
$\quad v$ = velocity of water flow
$\quad g_c$ = gravitational constant, lb $\cdot$ ft/lb$_f$ $\cdot$ s^2

This equation permits one to evaluate the energy possessed by a body of mass m, and having a velocity v, relative to a stationary reference; it is customary to use the earth as the reference.

The kinetic energy of the body initially is

$$\text{KE} = \frac{(2000\,\text{lb})(8\,\text{ft/s})^2}{2[32.174\,\text{ft} \cdot \text{lb/(s}^2 \cdot \text{lb}_f)]} = 1989\,\text{ft} \cdot \text{lb}_f$$

The kinetic energy at its terminal velocity of 30 ft/s is

$$KE = \frac{(2000 \text{ lb})(30 \text{ ft/s})^2}{2[31.174 \text{ ft} \cdot \text{lb}/(\text{s} \cdot \text{lb}_f)]} = 27,972 \text{ ft} \cdot \text{lb}_f$$

The kinetic energy change or difference, ΔKE, is then

$$\Delta KE = 1989 - 27,972 = -25,983 \text{ ft} \cdot \text{lb}_f$$

Converting the answer to Btu yields

$$\Delta KE = (-25,983 \text{ ft} \cdot \text{lb}_f)(1 \text{ Btu}/778.17 \text{ ft} \cdot \text{lb}_f) = -33.390 \text{ Btu}$$

CLE.4   SPHERE VELOCITY

A 1-kg steel sphere falls 100 m. If the sphere was initially at rest, determine its kinetic energy in $N \cdot m$ and velocity in ft/s at the end of its fall (and prior to any impact).

Solution

The law of conservation of energy, which like the law of conservation of mass applies for all processes that do not involve nuclear reactions, states that energy can neither be created nor destroyed. As a result, the energy level of a system can change only when energy crosses the system boundary, i.e.,

$$\Delta(\text{Energy level of system}) = \text{Energy crossing boundary}$$

For a closed system, i.e., one in which there is no mass transfer between system and surroundings, energy crossing the boundary can be classified in one of two different ways: heat, Q, or work, W. Heat is energy moving between the system and the surroundings by virtue of a temperature driving force. Heat flows from high to low temperature. The temperature in a system can vary; the same can be said of the surroundings. If a portion of the system is at a higher temperature than a portion of the surroundings and, as a result, energy is transferred from the system to the surroundings, that energy is classified as heat. If part of the system is at a higher temperature than another part of the system and energy is transferred between the two parts, that energy is not classified as heat because it is not crossing the boundary. Work is also energy moving between the system and the surroundings. Here, the driving force can be anything but a temperature difference, e.g., a mechanical force, a pressure difference, gravity, a voltage difference, a magnetic field, etc. Note also that the definition of work is a force acting through a distance. All of the examples of driving forces just cited can be shown to provide a force capable of acting through a distance.

The energy level of the system has three principal contributions: kinetic energy (KE), potential energy (PE), and internal energy (U). Any body in motion possesses kinetic energy. If the system is moving as a whole, its kinetic energy is proportional to the mass of the system and the square of the velocity of its center of gravity. The phrase "as a whole" indicates that motion inside the system relative to the system's center of gravity does not contribute to the KE term, but rather to the internal energy or U term. The terms *external* kinetic energy and *internal* kinetic energy are sometimes used here. An example would be a moving railroad tank car carrying propane gas. (The propane gas is the system.) The center of gravity of the propane gas is moving at the velocity of the train—this constitutes the system's external kinetic energy. The gas molecules are also moving in random directions relative to the center of gravity—this constitutes the system's internal kinetic energy due to motion inside the system. The potential energy (PE) involves any energy the system as a whole possesses by virtue of its position (more precisely, the position of its center of gravity) in some force field, e.g., gravity, centrifugal, electrical, etc., that gives the system the potential for accomplishing work. Again the phrase "as a whole" is used to differentiate between external potential energy and internal potential energy. Internal potential energy refers to potential energy due to force fields inside the system. For example, the electrostatic force fields (bonding) between atoms and molecules provide those particles with the potential for work. The internal energy U is the sum of all internal kinetic and internal potential energy contributions.

The law of conservation of energy, which is also called the first law of thermodynamics, may now be written as

$$\Delta(U + \text{KE} + \text{PE}) = Q + W$$

or equivalently as

$$\Delta U + \Delta \text{KE} + \Delta \text{PE} = Q + W$$

It is important to note the sign convention for Q and W defined by the above equation. Since any Δ term is always defined as the final minus the initial state, both the heat and work terms must be positive when they cause the system to gain energy, i.e., when they represent energy being transferred from the surroundings to system. Conversely, when the heat and work terms cause the system to lose energy, i.e, when energy is transferred from the system to the surroundings, they are negative in sign.

The above sign convention is not universal, and the reader must exercise caution and check what sign convention is being used by a particular author when studying the literature. For example, work is often defined in some texts as positive when the system does work on the surroundings (cf. the THEODORE TUTORIAL, *Thermodynamics*, ETS International, Roanoke, VA, 1995.).

Thus, regarding the problem statement, only kinetic and potential effects are present. In line with the conservation law for energy, the initial energy must equal the final energy:

$$PE_1 + KE_1 = PE_2 + KE_2$$

where subscripts 1 and 2 refer to initial and final states, respectively.

If the final reference state with respect to position is chosen as $Z = 0$, then $PE_2 = 0$. Further, since the velocity at state 1 may be assigned to be zero, $KE_1 = 0$. Thus,

$$PE_1 = KE_2$$

Substituting yields

$$(1\,\text{kg})(100\,\text{m})(9.81\,\text{m/s}^2) = \frac{(1\,\text{kg})v^2(\text{m/s}^2)}{2}$$

$$v^2 = (2)(9.81)(100)$$

$$v = 44.3\,\text{m/s}$$

Interestingly, the final result is independent of the mass and density of the body. The reader is left the exercise of showing that the final KE (at state 2) is $KE_2 = 981\,\text{N} \cdot \text{m}$.

CLE.5   INTERNAL ENERGY CALCULATION

A gas at 200°C and 10 atm is contained in a cylinder with a movable piston. As the gas is cooled slowly to room temperature, it is allowed to expand. If the amount of work due to the expansion is 2780 J and the amount of heat given off during the cooling is 3942 cal, calculate the change in internal energy (in kcal) of the gas during this process.

Solution

For a closed system, the only type of work, outside of shaft work, that can be performed is work done by virtue of an expansion or contraction of the system. This is called pressure–volume work, W_{PV}, and is given by

$$W_{PV} = -\int_{v_1}^{v_2} P_{\text{surroundings}}\, dV$$

The minus sign is required because of the sign convention employed. Work done by the system, e.g., during expansion must be negative. For a constant pressure process, this becomes

$$W_{PV} = -\int_{v_1}^{v_2} P \, dV = -P \, \Delta V$$

$$= -\Delta(PV)$$

The law of conservation of energy for a closed system is

$$\Delta U + \Delta KE + \Delta PE = Q + W$$

In most chemical engineering applications, the ΔKE and ΔPE terms are negligible, and many textbooks present the energy balance equations with these two terms omitted.

Therefore

$$\Delta KE = 0$$
$$\Delta PE = 0$$

Substituting yields

$$Q = (-3942 \, \text{cal})(1 \, \text{kcal}/1000 \, \text{cal}) = -3.942 \, \text{kcal}$$
$$= -3.942 \, \text{kcal}$$
$$W = (-2780 \, \text{J})(2.39 \times 10^{-4} \, \text{kcal}/\text{J}) = -0.664 \, \text{kcal}$$
$$= -0.664 \, \text{kcal}$$

The change in internal energy of the gas is therefore

$$\Delta U + 0 + 0 = -3.942 - 0.664$$
$$\Delta U = -4.606 \, \text{kcal}$$

In older thermodynamics textbooks, work is defined as positive if the system does work on the surroundings. This means that a loss of energy in the form of work from the system is considered positive while a loss of energy in the form of heat from the system is considered negative. To the authors, this thermodynamic convention is inconsistent, and potentially confusing.

CLE.6   COOLING RIVER REQUIREMENT

Determine the percentage of a river stream's flow available to an industry for cooling such that the river temperature does not increase more than $10°F$. Fifty percent of the industrial withdrawal is lost by evaporation and the industrial water returned to the river is $60°F$ warmer than the river.

Note: This problem is a modified and edited version (with permission) of an illustrative example prepared by Ms. Marie Gillman, a graduate mechanical engineering student at Manhattan College.

Solution

Draw a flow diagram representing the process as shown in Figure 7. Express the volumetric flow lost by evaporation from the process in terms of that entering the process:

$$q_{lost} = 0.5q_{in}$$

Express the process outlet temperature and the maximum river temperature in terms of the upstream temperature:

$$T_{out} = T_{up} + 60°F$$

$$T_{mix} = T_{up} + 10°F$$

Using the conservation law for mass, express the process outlet flow in terms of the process inlet flow. Also express the flow bypassing the process in terms of the upstream and process inlet flows:

$$q_{out} = 0.5q_{in}$$

$$q_{byp} = q_{up} - q_{in}$$

$$q_{mix} = q_{up} - 0.5q_{in}$$

The flow diagram with the expressions developed above are shown in Figure 8.

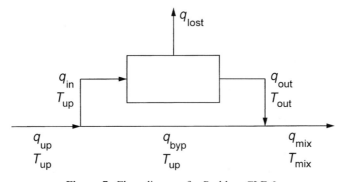

Figure 7. Flow diagram for Problem CLE.6.

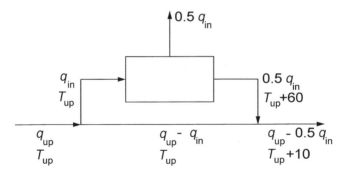

Figure 8. Flow diagram after applying mass balances.

Noting that the enthalpy of any stream can be represented by $qc_p\rho(T - T_{ref})$, an energy balance around the downstream mixing point leads to

$$(q_{up} - q_{in})c_p\rho(T_{up} - T_{ref}) + 0.5q_{in}c_p\rho(T_{up} + 60 - T_{ref})$$
$$= (q_{up} - 0.5q_{in})c_p\rho(T_{up} + 10 - T_{ref})$$

Note that T_{ref} in arbitrary and indirectly defines a basis for the enthalpy. Setting $T_{ref} = 0$ and assuming that density and heat capacity are constant yield

$$(q_{up} - q_{in})T_{up} + 0.5q_{in}(T_{up} + 60) = (q_{up} - 0.5q_{in})(T_{up} + 10)$$

The equation may now be solved for the inlet volumetric flow to the process in terms of the upstream flow:

$$q_{up}T_{up} - q_{in}T_{up} + 0.5q_{in}T_{up} + 30q_{in}$$
$$= q_{up}T_{up} + 10q_{up} - 0.5q_{in}T_{up} - 5q_{in}$$

Canceling terms produces

$$35q_{in} = 10q_{up}$$
$$q_{in} = 0.286q_{up}$$

Therefore, 28.6% of the original flow, q_{up}, is available for cooling.

Note that the problem can also be solved by setting $T_{ref} = T_{up}$. Since for this condition, $T_{ref} - T_{up} = 0$, the solution to the problem is greatly simplified.

CLE.7   STEAM REQUIREMENT

It is desired to evaporate 1000 lb/h of 60°F water at 1 atm at a power plant. Utility superheated steam at 40 atm and 1000°F is available, but since this stream is to be used elsewhere in the plant, it cannot drop below 20 atm and 600°F. What mass flowrate of the utility steam is required? Assume that there is no heat loss in the evaporator.

From the steam tables:

$$P = 40\,\text{atm} \qquad T = 1000°\text{F} \qquad H = 1572\,\text{Btu/lb}$$

$$P = 20\,\text{atm} \qquad T = 600°\text{F} \qquad H = 1316\,\text{Btu/lb}$$

For saturated steam:

$$P = 1\,\text{atm} \qquad H = 1151\,\text{Btu/lb}$$

For saturated water:

$$T = 60°\text{F} \qquad H = 28.1\,\text{Btu/lb}$$

Solution

A detailed flow diagram of the process is provided in Figure 9.

Assuming the process to be steady state and noting that there is no heat loss or shaft work, the energy balance on a rate basis is

$$\dot{Q} = \Delta\dot{H} = 0$$

This equation indicates that the sum of the enthalpy changes for the two streams must equal zero.

The change in enthalpy for the vaporization of the water stream is

$$\Delta\dot{H}_{\text{vaporization}} = (1000\,\text{lb/h})(1151 - 28.1\,\text{Btu/lb})$$

$$= 1.123 \times 10^6\,\text{Btu/h}$$

The change of enthalpy for the cooling of the superheated steam may now be determined. Since the mass flowrate of one stream is unknown, its ΔH must be expressed in terms of this mass flowrate, which is represented in Figure 9 as $\dot{m}$ lb/h.

$$\Delta\dot{H}_{\text{cooling}} = \dot{m}(1572 - 1316) = (256)\,\dot{m}\,\text{Btu/h}$$

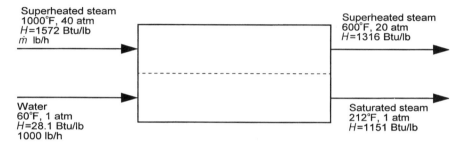

Figure 9. Flow diagram for Problem CLE.7.

Since the overall $\Delta \dot{H}$ is zero, the enthalpy changes of the two streams must total zero. Thus,

$$\Delta \dot{H}_{\text{vaporization}} + \Delta \dot{H}_{\text{cooling}} = 0$$

$$1.123 \times 10^6 = (256)\dot{m}$$

$$\dot{m} = 4387 \, \text{lb/h}$$

Tables of enthalpies and other state properties of many substances may be found in R. H. Perry and D. W. Green, Ed., *Perry's Chemical Engineers' Handbook*, 7th ed., McGraw-Hill, New York, 1996.

CLE.8   PROCESS COOLING WATER REQUIREMENT

Determine the total flowrate of cooling water required for the services listed below if a cooling tower system supplies the water at 90°F with a return temperature of 115°F. How much fresh water makeup is required if 5% of the return water is sent to "blowdown?" Note that the cooling water heat capacity is $1.00 \, \text{Btu/(lb} \cdot \text{°F)}$, the heat of vaporization at cooling tower operating conditions is 1030 Btu/lb, and the density of water at cooling tower operating conditions is $62.0 \, \text{lb/ft}^3$.

Process Unit	Heat Duty (Btu/h)	Required Temperature (°F)
1	12,000,000	250
2	6,000,000	200–276
3	23,500,000	130–175
4	17,000,000	300
5	31,500,000	150–225

Solution

The required cooling water flowrate, q_{CW}, is given by the following equation:

$$q_{\text{CW}} = \dot{Q}_{\text{HL}} / [(T)(c_p)(\rho)]$$

where $\dot{Q}_{\text{HL}}$ = heat load, Btu/min
T = change in temperature = $115°\text{F} - 90°\text{F} = 25°\text{F}$
c_p = heat capacity = $1.00 \, \text{Btu/(lb} \cdot \text{°F)}$
ρ = density of water = $(62.0 \, \text{lb/ft}^3)(0.1337 \, \text{ft}^3/\text{gal})$
    = $8.289 \, \text{lb/gal}$

The heat load is

$$\dot{Q}_{\text{HL}} = (12 + 6 + 23.5 + 17 + 31.5)(10^6 \, \text{Btu/h})/60 \, \text{min/h}$$

$$= 1,500,000 \, \text{Btu/ min}$$

Thus,

$$q_{CW} = \frac{1,500,000 \text{ Btu/min}}{(25°F)(1.00 \text{ Btu/lb} \cdot °F)(8.289 \text{ lb/gal})} = 7250 \text{ gpm}$$

The blow-down flow, q_{BD}, is given by the following:

$$q_{BD} = (BDR)(q_{CW})$$

where BDR is the blow-down rate $= 5\% = 0.05$.
    Thus,

$$q_{BD} = (0.05)(7250 \text{ gpm}) = 362.5 \text{ gpm}$$

The amount of water vaporized by the cooling tower, q_v, is given by:

$$q_v = (q_{HL})/[(\rho)(HV)]$$

where HV is the heat of vaporization $= 1030 \text{ Btu/lb}$.
    Substitution yields

$$q_v = (1,500,000 \text{ Btu/min})/[(8.289 \text{ lb/gal})(1030 \text{ Btu/lb})]$$

$$= 175.7 \text{ gpm}$$

CLE.9   STEAM REQUIREMENT OPTIONS

Determine how many pounds per hour of steam are required for the following two cases: (a) if steam is provided at 500 psig and (b) if steam is provided at both 500 and 75 psig pressures. The plant has the following heating requirements:

Process Unit	Unit Heat Duty (UHD) (Btu/h)	Required Temperature (°F)
1	10,000,000	250
2	8,000,000	450
3	12,000,000	400
4	20,000,000	300

Note the properties of saturated steam are as follows:

Pressure Provided (psig)	Saturation Temperature (°F)	Enthalpy of Vaporization (ΔH_v) (Btu/lb)
75	320	894
500	470	751

Solution

The total required flowrate of 500 psig steam, $\dot{m}_{BT}$, is given by:

$$\dot{m}_{BT} = \dot{m}_{B1} + \dot{m}_{B2} + \dot{m}_{B3} + \dot{m}_{B4}$$

For the above equation:

$\dot{m}_{B1}$(mass flowrate of 500 psig steam through unit 1)

$\qquad = \text{UHD}/\Delta H_v = 13{,}320 \text{ lb/h}$

$\dot{m}_{B2}$(mass flowrate of 500 psig steam through unit 2)

$\qquad = \text{UHD}/\Delta H_v = 10{,}655 \text{ lb/h}$

$\dot{m}_{B3}$(mass flowrate of 500 psig steam through unit 3)

$\qquad = \text{UHD}/\Delta H_v = 15{,}980 \text{ lb/h}$

$\dot{m}_{B4}$(mass flowrate of 500 psig steam through unit 4)

$\qquad = \text{UHD}/\Delta H_v = 26{,}635 \text{ lb/h}$

Thus,

$$\dot{m}_{BT} = 66{,}590 \text{ lb/h}$$

The required combined total flowrate of 500 and 75 psig steam, $\dot{m}_{CT}$, is given by

$$\dot{m}_{CT} = \dot{m}_{75,1} + \dot{m}_{B2} + \dot{m}_{B3} + \dot{m}_{75,4}$$

For this situation:

$\dot{m}_{75,1}$(mass flowrate of 75 psig steam through unit 1)

$\qquad = \text{UHD}/\Delta H_v = 11{,}185.7 \text{ lb/h}$

$\dot{m}_{75,4}$(mass flowrate of 75 psig steam through unit 4)

$\qquad = \text{UHD}/\Delta H_v = 22{,}371.4 \text{ lb/h}$

Thus,

$$\dot{m}_{CT} = 60{,}192 \text{ lb/h}$$

Note that since the saturation temperature of 75 psig steam is lower than two of the process units, the 500 psig steam must be used for process units 2 and 3.

CLE.10   POWER GENERATION

A power plant employs steam to generate power and operates with a steam flowrate of 450,000 lb/h. For the adiabatic conditions listed below, determine the power produced in horsepower, kilowatts, Btu/h, and Btu/lb of steam.

	Inlet	Outlet
Pressure, psia	100	1
Temperature, °F	1500	350
Steam velocity, ft/s	120	330
Steam vertical position, ft	0	−20

Solution

Apply the following energy equation:

$$\frac{Z_1}{J}\left(\frac{g}{g_c}\right) + \frac{v_1^2}{2g_cJ} + H_1 + Q = \frac{Z_2}{J}\left(\frac{g}{g_c}\right) + \frac{v_2^2}{2g_cJ} + H_2 + \frac{W}{J}$$

where Z_1, Z_2 = vertical position at inlet/outlet, respectively
  J = conversion factor from ft · lb_f to Btu
  v_1, v_2 = steam velocity at inlet/outlet, respectively
  H_1, H_2 = steam enthalpy at inlet/outlet, respectively
  W = work extracted from system

For adiabatic conditions, $Q = 0$. Substituting data from the problem statement yields

$$0 + \frac{(120)^2}{2(32.17)(778)} + 1505.4 + 0 = \frac{-20}{778} + \frac{(330)^2}{2(32.17)(778)} + 940.0 + \frac{W}{J}$$

$$0 + 0.288 + 1505.4 + 0 = -0.026 + 2.176 + 940.0 + \frac{W}{J}$$

$$\frac{W}{J} = 563.54\,\text{Btu/lb}$$

The total work of the turbine is equal to

$$(450,000\,\text{lb/h})(563.54\,\text{Btu/lb}) = 2.54 \times 10^8\,\text{Btu/h}$$

Converting to horsepower gives

$$(2.54 \times 10^8\,\text{Btu/h})(3.927 \times 10^{-4}\,\text{hp} \cdot \text{h/Btu}) = 9.98 \times 10^4\,\text{hp}$$

Converting to kilowatts gives

$$(9.98 \times 10^4 \, \text{hp})(0.746 \, \text{kW/hp}) = 7.45 \times 10^4 \, \text{kW}$$

The reader is left the exercise of showing that both kinetic and potential effects could have been neglected in the solution of the problem.

4 Conservation Law for Momentum (CLM)

CLM.1 REYNOLDS NUMBER

A liquid with a viscosity of 0.78 cP and a density of 1.50 g/cm³ flows through a 1-in. diameter tube at 20 cm/s. Calculate the Reynolds number. Is the flow laminar or turbulent?

Solution

By definition, the Reynolds number (Re) is equal to:

$$\text{Re} = \rho v d / \mu$$

where ρ = fluid density
v = fluid velocity
d = characteristic length, usually the conduit diameter
μ = fluid viscosity

Since

$$1\,\text{cP} = 10^{-2}\ \text{g/(cm} \cdot \text{s)}$$

$$\mu = 0.78 \times 10^{-2}\ \text{g/(cm} \cdot \text{s)}$$

$$1\,\text{in.} = 2.54\,\text{cm}$$

$$\text{Re} = (1.50)(20)(2.54)/(0.78 \times 10^{-2})$$

$$= 9769.23 \approx 9800$$

As noted below, the value of the Reynolds number indicates the nature of fluid flow in a duct or pipe:

Re < 2100; flow is streamline (laminar)
Re > 10,000; flow is turbulent
$2100 \leqslant \text{Re} \leqslant 10,000$; transition region

A more accurate classification between the latter two regions can be achieved if the relative pipe roughness, k/D, is known.

For this problem, the flow is turbulent.

CLM.2   SPRAY TOWER APPLICATION

The inlet gas to a spray tower is at 1600°F. It is piped through a 3.0-ft inside diameter duct at 25 ft/s to the spray tower. The scrubber cools the gas to 500°F. In order to maintain the velocity of 25 ft/s, what size duct would be required at the outlet of the unit? Neglect the pressure across the spray tower and any moisture considerations.

Solution

Applying the continuity equation, the volumetric flowrate into the scrubber is

$$q_1 = A_1 v_1$$

Since

$$A_1 = [\pi(3.0)^2]/4 = 7.07 \, \text{ft}^2$$

then

$$q_1 = (7.07)(25) = 176.6 \, \text{ft}^3/\text{s}$$

The volumetric flowrate out of the scrubber, using Charles' law, is

$$q_2 = q_1(T_2/T_1) \qquad (T \text{ in absolute units})$$
$$= (176.6)(960/2060)$$
$$= 82.31 \, \text{ft}^3/\text{s}$$

The cross-sectional area of outlet duct is given by

$$A_2 = q_2/v_2$$
$$= 82.31/25$$
$$= 3.29 \, \text{ft}^2$$

The diameter of outlet duct is then

$$D = [4(A_2)/\pi]^{1/2}$$
$$= [(4)(3.29)/\pi]^{1/2}$$
$$= 2.05 \text{ ft}$$

CLM.3   MERCURY COLUMN

A 3-in. column of mercury (specific gravity = 13.6) is open to the atmosphere. Determine the pressure at the base of the column in psia, psfa, and in. H_2O, if atmospheric pressure is 14.7 psia.

Solution

The density of mercury is

$$\rho = (13.6)(62.4) = 848.6 \text{ lb/ft}^3$$

The pressure difference across the column, ΔP, (of height L) is given by the equation

$$\Delta P = \rho(g/g_c)L$$
$$= (848.6 \text{ lb/ft}^3)\left(1\frac{\text{lb}_f}{\text{lb}}\right)(3 \text{ in.})(1 \text{ ft}/12 \text{ in.}) = 212.2 \text{ psf}$$

where g = acceleration due to gravity
g_c = units conversion factor [= 32.2 lb · ft/(lb$_f$ · s^2) in the American Engineering System and 1 (no units) in the International System (SI) system].

To find the absolute pressure at the base of the column, the atmospheric pressure must be added to ΔP.

$$P_{\text{base}} = (14.7)(144) + 212.2 = 2329 \text{ psfa}$$

To convert to the other units use the conversion factors

$$1 \text{ ft}^2/144 \text{ in.}^2$$

$$406.9 \text{ in. } H_2O/14.7 \text{ psia}$$

Thus,

$$P_{\text{base}} = (2329 \text{ psfa})(1 \text{ ft}^2/144 \text{ in.}^2)$$
$$= 16.2 \text{ psia}$$

and

$$P_{base} = (16.2\,\text{psia})(406.9\,\text{in.}\ H_2O/14.7\,\text{psia})$$
$$= 448\,\text{in.}\ H_2O\ (\text{absolute})$$

CLM.4 PRESSURE GAUGE READING

Calculate the pressure gauge reading in psia, psig, psfa and psfg in Figure 10.

Solution

As described in Problem CLM.3, the fundamental equation of fluid statics indicates that the rate of change of the pressure P is directly proportional to the rate of change of the depth Z, or

$$dP/dZ = -\rho(g/g_c)$$

where Z = vertical displacement, where upward is considered positive
  ρ = fluid density
  g = acceleration due to gravity
  g_c = units conversion factor (see Problem CLM.3)

For constant density, ρ is constant and the above equation may be integrated to give the hydrostatic equation

$$P_2 = P_1 + \rho g h/g_c$$

Here point 2 is located a distance h below point 1.

Manometers are often used to measure pressure differences. This is accomplished by a direct application of the above equation. Pressure differences in manometers may be computed by systematically applying the above equation to each leg of the manometer.

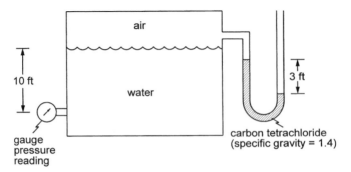

Figure 10. Diagram for Problem CLM.4.

Since the density of air is effectively zero, the contribution of the air to the 3-ft manometer reading can be neglected. The contribution to the pressure due to the carbon tetrachloride in the manometer is found by using the hydrostatic equation.

$$\Delta P = \rho g h / g_c$$
$$= (62.4)\,(1.4)\,(3) = 262.1\,\mathrm{psf}$$
$$= 1.82\,\mathrm{psi}$$

Since the right leg of the manometer is open to the atmosphere, the pressure at that point is atmospheric:

$$P = 14.7\,\mathrm{psia}$$

Note that this should be carried as a positive term.

The contribution to the pressure due to the height of water above the pressure gauge is similarly calculated using the hydrostatic equation.

$$\Delta P = (62.4)\,(3) = 187.2\,\mathrm{psf} = 1.3\,\mathrm{psi}$$

The pressure at the gauge is obtained by summing the results of the steps above, but exercising care with respect to the sign(s):

$$P = 14.7 - 1.82 + 1.3 = 14.18\,\mathrm{psia}$$
$$= 14.18 - 14.7 = -0.52\,\mathrm{psig}$$

The pressure may now be converted to psfa and psfg:

$$P = (14.18)\,(144) = 2042\,\mathrm{psfa}$$
$$= (-0.52)\,(144) = -75\,\mathrm{psfg}$$

Care should be exercised when providing pressure values in gauge and absolute pressure. The key equation is

$$P(\text{gauge}) = P(\text{absolute}) - P(\text{ambient}) \qquad (\text{consistent units})$$

CLM.5   STORAGE TANK CALCULATION

A cylindrical tank is 20 ft (6.1 m) in diameter and 45 ft high. It contains water (of density 1000 kg/m^3) to a depth of 9 ft and a 36 ft height of oil (of specific gravity 0.89) above the water. The tank is open to the atmosphere. Calculate:

1. The density of oil in kg/m^3 and N/m^3.
2. The gauge pressure at the oil–water interface.

3. The gauge pressure and pressure force at the bottom of the tank.
4. The pressure force in Newtons on the bottom 9 ft of the side of the tank (where the fluid is water).

Solution

The density of oil is

$$\rho_{oil} = (0.89)(1000) = 890 \, \text{kg/m}^3$$

Bernoulli's equation (which is presented in more detail in Chapter 6) is applied to calculate the gauge pressure at the water–oil interface, Z_2.

$$\frac{P_1 g_c}{\rho_{oil} g} + Z_1 = \frac{P_2 g_c}{\rho_{oil} g} + Z_2$$

Since

$Z_1 = 0, P_1 = 1 \, \text{atm}$
$Z_2 = -36 \, \text{ft} = -10.98 \, \text{m}$
$g_c = 1$ in the SI system

$$P_{2,g} = (\rho_{oil})(g/g_c)(Z_1 - Z_2)$$
$$= (890)(9.807)(10.98) = 95,771 \, \text{Pa (gauge)}$$
$$= 95,771/101,325 = 0.945 \, \text{atm (gauge)}$$
$$= (0.945)(14.69) = 13.9 \, \text{psig}$$

The gauge pressure at the bottom of the tank $(Z = Z_3)$ may be determined in a similar manner.

$$\frac{P_2 g_c}{\rho_w g} + Z_2 = \frac{P_3 g_c}{\rho_w g} + Z_3$$

$$P_3 = P_2 + (\rho_w)(g/g_c)(Z_2 - Z_3)$$
$$= 95,771 + (1000)(9.807)(9.0/3.281) = 122,673 \, \text{Pa (gauge)}$$
$$= 122,673/101,325 = 1.2 \, \text{atm (gauge)}$$

The pressure force at the bottom of the tank is then

$$F = (P_3)(S)$$
$$= (122,673 + 101,325)(\pi)(6.1)^2/4 = 6,537,684\,\text{N}$$
$$= (6,537,684)(0.2248) = 1,469,671\,\text{lb}_f$$

To calculate the force on the side of the tank within the water layer, one needs to first obtain an equation of the variation of pressure with height in the water layer:

$$\frac{P_2 g_c}{\rho_w g} + Z_2 = \frac{P g_c}{\rho_w g} + Z$$

$$P = P_2 + (\rho_w)(g/g_c)(Z_2 - Z)$$
$$= 95,771 + (1,000)(9.807)(-10.98 - Z)$$
$$= 95,771 - (9,807)(10.98 + Z)$$
$$= -11,910 - 9.807Z$$

The total force is then obtained by an integration since the pressure varies with height.

$$F = (\pi)(6.1)\int_{Z_3}^{Z_2}(-11,910 - 9.807Z)\,dZ$$
$$= 6.1\pi[-11,910(Z_2 - Z_3) - 4903.5(Z_2^2 - Z_3^2)]$$

Set

$$Z_3 = -13.72\,\text{m}\ (-45\,\text{ft}) \text{ and } Z_2 = -10.98\,\text{m}(-36\,\text{ft})$$
$$F = 6.1\pi[-11,910(-10.98 + 13.72) + 4903.5(10.98^2 - 13.72^2)]$$
$$= 5.73 \times 10^6\,\text{N}$$

Calculations of fluid pressure force on submerged surfaces is important in selecting the proper material and thickness. If the pressure on the submerged surface is not uniform, then the pressure force is calculated by integration. Also note that a nonmoving, simple fluid exerts only pressure forces; moving fluids exert both pressure and shear forces.

CLM.6 BALLOON VOLUME

A balloon containing air (assume $\rho_B = 0$) is attached to $3\,\text{ft}^3$ of concrete ($\rho_c = 150\,\text{lb/ft}^3$). What volume should the balloon be in ft^3 to float the concrete mass in water?

Solution

By definition, the buoyant force acting on a submerged body is given by the weight of the volume of fluid displaced. The resulting force acts through the center of geometry or centroid of the submerged body. The describing equation is given by

$$F_B = \rho(\text{volume displaced})(g/g_c)$$
$$= \rho V_B(g/g_c)$$

The net force acting on a submerged body in water can be obtained from

$$F_{NET} = (\rho_w - \rho_B)(\text{volume of body})(g/g_c)$$
$$= (\rho_w - \rho_B)V_B(g/g_c)$$

The gravity force acting on the concrete volume, V_C, is

$$F_G = (\rho_C - \rho_w)V_C(g/g_c)$$

The buoyant force associated with the balloon is

$$F_B = (\rho_w)V_B(g/g_c)$$

Equating the gravity and buoyant forces gives

$$(\rho_C - \rho_w)V_C = (\rho_w)V_B$$

Thus

$$V_B = V_C[(\rho_c - \rho_w)/\rho_w]$$
$$= 3.0[(150 - 62.4)/62.4]$$
$$= 4.21 \text{ ft}^3$$

The reader should note that the buoyant force is independent of depth.

CLM.7   SHEAR STRESS

Carbon tetrachloride at 1 atm and 20°C is sheared between two long horizontal parallel plates, 0.5 mm apart, with the bottom plate fixed and the top plate moving at a constant velocity v. Each plate is 2 m long and 0.8 m wide. The strain rate (or velocity gradient) is 5000 s^{-1}. Calculate the shear stress if the viscosity and density of the carbon tetrachloride are 0.97 cP and 1590 kg/m^3, respectively.

Solution

Forces that act on a fluid can be classified as either body forces or surface forces. Body forces are distributed throughout the material, e.g., gravitational, centrifugal, and electromagnetic forces. Surface forces are forces that act on the surface.

Stress is a force per unit area. If the force is parallel to the surface, the force per unit area is called a *shear stress*. When the force is perpendicular (normal) to a surface, the force per unit area is called a *normal stress*. By measuring the shear stress and the rate of change of velocity with height when a fluid is flowing between two parallel plates, it is possible to classify the flow behavior and define a fluid property called *viscosity*.

When a fluid flows past a stationary wall, the fluid adheres to the wall at the interface between the solid and the fluid. Therefore, the local velocity v of the fluid at the interface is zero. At some distance y normal to and displaced from the wall, the velocity of the fluid is finite. Therefore, there is a velocity variation from point to point in the flowing fluid. This causes a velocity field in which the velocity is a function of the normal distance from the wall, i.e., $v = f(y)$. If $y = 0$ at the wall, $v = 0$, and v increases with y. The rate of change of velocity with respect to distance is the velocity gradient:

$$\frac{dv}{dy} = \frac{\Delta v}{\Delta y}$$

This velocity derivative (or gradient) is also called the strain rate, shear, time rate of shear, or rate of deformation.

It has been shown that when a fluid is sheared with a shear stress τ, its strain rate (or deformation rate) is proportional (for most fluids) to the shear stress. The proportionality constant is termed the fluid viscosity μ. For a fluid sheared between two long parallel plates, the local velocity (at any height y) varies from zero at the fixed plate to v at the upper moving plate. As described above, the derivative of the local velocity (u) with respect to the height y (i.e., du/dy) is termed the *velocity gradient*, strain rate, or deformation rate. The shear stress is related to du/dy by the equation

$$\tau = \frac{\mu}{g_c} \frac{du}{dy}$$

where

$$g_c = 32.17 \left(\frac{\text{lb} \cdot \text{ft}}{\text{lb}_f \cdot \text{s}^2} \right) = 1 \left(\frac{\text{kg} \cdot \text{m}}{\text{N} \cdot \text{s}^2} \right)$$

This is Newton's law of viscosity. Fluids obeying Newton's law are termed *Newtonian fluids*. Note that since

$$1\text{N} = 1\,\text{kg} \cdot \text{m/s}^2$$

the last term in the equation for g_c is actually unitless.

Noting that $1\,cP = 0.001\,kg/(m \cdot s)$, the fluid dynamic (or absolute) viscosity, μ, can be converted to units of $kg/(m \cdot s)$:

$$\mu = 0.97\,cP = 0.0097P = 0.00097\,kg/(m \cdot s)$$

The shear stress is therefore

$$\tau = \frac{\mu}{g_c}\frac{du}{dy} = \frac{(0.00097)(5000)}{(1)} = 4.85\,kg/(m \cdot s^2)$$

$$= 4.85\,Pa = 4.85/6897.5 = 7.032 \times 10^{-4}\,psi$$

The reader is left the exercise of showing that the shear force on the plate is $7.76\,N$ ($1.74\,lb_f$) and that the velocity of the moving plate is $2.5\,m/s$ ($8.2\,ft/s$).

Fluids can be classified based on their viscosity. An imaginary fluid of zero viscosity is called a *Pascal fluid*. The flow of a Pascal fluid is termed inviscid (or nonviscous) flow.

Viscous fluids are classified based on their rheological (viscous) properties. These are detailed below:

1. *Newtonian fluids:* Fluids that obey Newton's law of viscosity, i.e., fluids in which the shear stress is linearly proportional to the velocity gradient. All gases are considered Newtonian fluids. Examples of Newtonian liquids are water, benzene, ethyl alcohol, hexane, and sugar solutions. All liquids of a simple chemical formula are considered Newtonian fluids.
2. *Non-Newtonian fluids:* Fluids that do not obey Newton's law of viscosity. Generally they are complex mixtures, e.g., polymer solutions, slurries, etc.

Non-Newtonian fluids are classified into three types:

1. *Time-independent fluids:* Fluids in which the viscous properties do not vary with time.
2. *Time-dependent fluids:* Fluids in which the viscous properties vary with time. These can be further classified.
3. *Visco-elastic or memory fluids:* Fluids with elastic properties that allow them to "spring back" after the release of a shear force. Examples include egg-white and rubber cement.

CLM.8   CAPILLARY RISE

The following data are given:

Liquid–gas system is water–air
Temperature is $30°C$, and pressure is 1 atm
Capillary tube diameter $= 8mm = 0.008\,m$

Water surface tension $= 0.0712\,\text{N/m}$
Water density $= 1000\,\text{kg/m}^3$
Contact angle, $\beta = 0°$

Calculate the capillary rise in millimeters.

Solution

A liquid forms an interface with another fluid. At the surface, the molecular layers are different in density than the bulk of the fluid. This results in surface tension and interfacial phenomena. The surface tension coefficient, σ, is the force per unit length of the circumference of the interface, i.e., N/m, or the energy per unit area of the interface area.

Surface tension causes a contact angle to appear when a liquid interface is in contact with a solid surface. If the contact angle, β, is $< 90°$, the liquid is termed *wetting*. If $\beta > 90°$, it is a *nonwetting* liquid.

Surface tension causes a fluid interface to rise, h (or fall), in a capillary of diameter d. The capillary rise is obtained by equating the vertical component of the surface tension force F_σ to the weight of the liquid (of height $= h$), F_g. Thus,

$$F_\sigma = \pi\, d\sigma(\cos\ \beta)$$

$$F_g = \frac{mg}{g_c} = \rho\ (\pi d^2/4)\ hg/g_c$$

where $\sigma = $ liquid–air surface tension
$\quad m = $ mass of liquid (of height $= h$)
$\quad \rho = $ density of liquid
$\quad g = $ acceleration due to gravity $(9.807\,\text{m/s}^2)$
$\quad g_c = $ conversion factor $(g_c = 1$ for the SI system)

Equating the two forces gives

$$\pi\, d\sigma(\cos\ \beta) = \rho(\pi d^2/4)hg/g_c$$

or

$$h = (4\sigma g_c/\rho\, dg)\cos\ \beta$$

Converting the water–air surface tension to units of kg/s^2,

$$\sigma = 0.0712\,\text{N/m} = 0.0712\,\text{kg/s}^2$$

The capillary rise, h, is then

$$h = \frac{(4)(1)(0.0712\,\text{kg/s}^2)(\cos\ 0°)}{(1000\,\text{kg/m}^3)(0.008\,\text{m})(9.807\,\text{m/s}^2)}$$

$$= 0.00363\,\text{m} = 3.63\,\text{mm}$$

CLM.9   AVERAGE VELOCITY

A liquid, with specific gravity (SG) 0.96, flows through a long circular tube of radius $R = 3$ cm. The liquid has the following linear distribution of the axial velocity v (the velocity in the direction of the flow):

$$v\ (\text{m/s}) = 6 - 200r$$

where r is the radial position (in meters) measured from tube centerline.
    Calculate the average velocity of the fluid in the tube.

Solution

As described earlier, a real fluid in contact with a nonmoving wall will have a velocity of zero at the wall. Similarly, the fluid in contact with a wall moving at a velocity v will move at the same velocity v. This is termed "the no-slip condition" of real fluids. Thus, for a fluid flowing in a duct, the fluid velocity at the walls of a duct is zero and increases as one moves further from the wall.
    To calculate the volumetric flowrate, q, of the fluid passing through a perpendicular surface, A, one must integrate the product of the component of the velocity that is normal to the area and the area over the whole cross-sectional area of the duct. This procedure is depicted in the equation

$$q = \int_A v\,dA$$

where q = volumetric flowrate (m^3/s, or ft^3/s, ...)
    v = absolute velocity, or the normal component of the velocity to the area dA
    A = surface area

The average velocity of fluid passing through the surface, A, is

$$v_{\text{av}} = q/A$$

On a differential level,

$$dq = v\,dA$$

and, in a cylindrical coordinate system

$$dA = 2\pi r\ dr$$

Thus,

$$dq = 2\pi r v\ dr$$
$$= 2\pi r(6 - 200r)\ dr = 2\pi(6r - 200r^2)\ dr$$

This equation can be integrated between the limits of $r = 0$ and $r = R$ to give

$$q = 2\pi \int_0^R (6r - 200r^2)\ dr = 2\pi\left[3R^2 - \left(\frac{200}{3}\right)R^3\right] = 2\pi R^2\left[3 - \left(\frac{200}{3}\right)R\right]$$

The volumetric flowrate is then

$$q = 2\pi(0.03)^2(3 - 2) = 0.00565\ \text{m}^3/\text{s} = 0.200\ \text{ft}^3/\text{s}$$

Since $\rho = (SG)(1000)\ \text{kg/m}^3$, the mass flowrate $\dot{m}$ is

$$\dot{m} = (0.96)(1000)(0.00565) = 5.42\ \text{kg/s}$$
$$= (5.42\ \text{kg/s})(1\ \text{lb}/0.454\ \text{kg}) = 11.95\ \text{lb/s}$$

The average velocity may now be calculated:

$$A = \pi R^2 = \pi(0.03)^2$$
$$v_{av} = \frac{q}{A} = \frac{2\pi R^2(3 - 2)}{\pi R^2} = 2\ \text{m/s}$$

The reader should note that average velocity is a useful value in scale-up and flow analysis.

CLM.10 PIPE FORCE REQUIREMENT

A 10-cm-diameter horizontal line carries saturated steam at 420 m/s. Water is entrained by the steam at the rate 0.15 kg/s. The line has a 90° bend. Calculate the force required to hold the bend in place due to the entrained water.

Solution

A line diagram of the system is provided in Figure 11. Select the fluid in the bend as the system and apply the conservation law for mass:

$$\dot{m}_1 = \dot{m}_2$$

Since the density and cross-sectional area are constant,

$$v_1 = v_2$$

where $\dot{m}_1, \dot{m}_2$ = mass flowrate at 1 and 2, respectively
$\quad\quad v_1, v_2$ = absolute velocity at 1 and 2, respectively

A linear momentum ($\dot{M}$) balance in the horizontal direction provides the force applied by the channel wall on the fluid in the x-direction, F_x:

$$F_x g_c = \dot{M}_{x,\text{out}} - \dot{M}_{x,\text{in}}$$
$$= \frac{d}{dt}(mv)_{x,\text{out}} - \frac{d}{dt}(mv)_{x,\text{in}}$$

Note that the equation assumes that the pressure drop across the bend is negligible. Since $v_{x,\text{out}} = 0$ and $dm/dt = \dot{m}$,

$$F_y g_c = 0 - \dot{m}v_{x,\text{in}} = -\frac{(0.15)(420)}{(1)} = -63\,\text{N} = -14.1\,\text{lb}_f$$

The x-direction supporting force acting on the 90° elbow is 14.1 lb$_f$ acting toward the left.

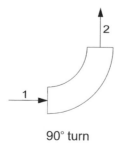

90° turn

Figure 11. Diagram for Problem CLM.10.

A linear momentum balance in the vertical direction results in

$$F_y g_c = \dot{M}_{y,\text{out}} - \dot{M}_{y,\text{in}}$$
$$= \dot{m}v_{y,\text{out}} - \dot{m}v_{y,\text{in}}$$
$$= \dot{m}v_2 - 0 = (0.15)(420) = 63 \text{ N} = 14.1 \text{ lb}_f$$

The y-direction supporting free force on the 90° elbow is 14.1 lb$_f$ acting upwards. The resultant supporting force is therefore

$$F_{\text{res}} = \sqrt{F_x^2 + F_y^2}$$
$$= \sqrt{(-63)^2 + 63^2} = 89.1 \text{ N} = 19.1 \text{ lb}_f$$

The direction is given by

$$\tan\theta = \frac{F_y}{F_x} = \frac{63}{-63} = -1$$
$$\theta = 135°$$

where θ is the angle between the positive x axis and the direction. The counterclockwise rotation of the direction from the x axis is defined as positive.

The supporting force is therefore 19.1 lb$_f$ acting in the "northwest" direction.

5 Stoichiometry (STC)

STC.1 REACTANTS/PRODUCTS RATIO

The reaction equation (not balanced) for the combustion of butane is shown below.

$$C_4H_{10} + O_2 \rightarrow CO_2 + H_2O$$

Determine the mole ratio of reactants to products.

Solution

A chemical equation provides a variety of qualitative and quantitative information essential for the calculation of the quantity of reactants reacted and products formed in a chemical process. The balanced chemical equation must have the same number of atoms of each type in the reactants and products. Thus the balanced equation for butane is

$$C_4H_{10} + (\tfrac{13}{2})O_2 \rightarrow 4CO_2 + 5H_2O$$

Note that:

Number of carbons in reactants = number of carbons in products = 4
Number of oxygens in reactants = number of oxygens in products = 13
Number of hydrogens in reactants = number of hydrogens in products = 10
Number of moles of reactants is 1 mol C_4H_{10} + 6.5 mol O_2 = 7.5 mol total
Number of moles of products is 4 mol CO_2 + 5 mol H_2O = 9 mol total

The reader should note that although the number of moles on both sides of the equation do not balance, the masses of reactants and products (in line with the conservation law for mass) must balance.

58

STC.2   STOICHIOMETRIC COMBUSTION

Consider the following equation:

$$C_3H_8 + 5O_2 \rightarrow 3CO_2 + 4H_2O$$

Determine the scf (standard cubic feet) of air required for stoichiometric combustion of 1.0 scf propane (C_3H_8).

Solution

Stoichiometric air is the air required to assure complete combustion of a fuel, organic, and/or waste. For complete combustion:

1. No fuel, organic, and/or waste remains
2. No oxygen is present in the flue gas
3. Carbon has combusted to CO_2, not CO
4. Sulfur has combusted to SO_2, not SO_3

Excess air (fractional basis) is defined by:

$$EA = \frac{\text{Air entering} - \text{Stoichiometric air}}{\text{Stoichiometric air}}$$

Noting that, for an ideal gas, the number of moles is proportional to the volume, the scf of O_2 required for the complete combustion of 1 scf of propane is 5 scf

The nitrogen-to-oxygen volume (or mole) ratio in air is $\frac{79}{21}$. Therefore the amount of N_2 in a quantity of air that contains 5.0 scf of O_2 is

$$\text{scf of } N_2 = (\tfrac{79}{21})(5)$$
$$= 18.81 \text{ scf}$$

The stoichiometric amount of air is then

$$\text{scf of air} = \text{scf of } N_2 + \text{scf of } O_2$$
$$= 18.81 + 5.0$$
$$= 23.81 \text{ scf}$$

Therefore the amount of flue gas produced is

$$\text{scf of flue gas} = \text{scf of } N_2 + \text{scf of } CO_2 + \text{scf of } H_2O$$
$$= 18.81 + 3.0 + 4.0$$
$$= 25.81 \text{ scf}$$

STC.3 LIMITING REACTANT

Complete combustion of carbon disulfide results in combustion products of CO_2 and SO_2 according to the reaction

$$CS_2 + O_2 \rightarrow CO_2 + SO_2$$

1. Balance this reaction equation.
2. If 500 lb of CS_2 is combusted with 225 lb of oxygen, which is the limiting reactant?
3. How much of each product is formed (lb)?

Data: MW of $CS_2 = 76.14$; MW of $SO_2 = 64.07$; MW of $CO_2 = 44$

Solution

1. The balanced equation is

$$CS_2 + 3O_2 \rightarrow CO_2 + 2SO_2$$

2. The initial molar amounts of each reactant is

$$(500 \text{ lb } CS_2)(1 \text{ lbmol } CS_2/76.14 \text{ lb } CS_2) = 6.57 \text{ lbmol } CS_2$$
$$(225 \text{ lb } O_2)(1 \text{ lbmol } O_2/32 \text{ lb } O_2) = 7.03 \text{ lbmol } O_2$$

The amount of O_2 needed to consume *all* the CS_2, i.e., the stoichiometric amount, is then

$$(6.57 \text{ lbmol})(3 \text{ lbmol}/1 \text{ lbmol}) = 19.71 \text{ lbmol } O_2$$

Therefore, O_2 is the limiting reactant since 19.7 mol of O_2 are required for complete combustion but only 7.03 mol of O_2 are available.

3. The limiting reactant is used to calculate the amount of product formed.

$$(7.03 \text{ lbmol } O_2)(1 \text{ lbmol } CS_2/3 \text{ lbmol } O_2) = 2.34 \text{ lbmol } CS_2$$
$$(2.34 \text{ lbmol } CS_2)(76.14 \text{ lb}/1 \text{ lbmol } CS_2) = 322 \text{ lb } CS_2 \text{ unreacted}$$

$$(7.03 \text{ lbmol } O_2)(1 \text{ lbmol } CO_2/3 \text{ lbmol } O_2) = 2.34 \text{ lbmol } CO_2$$
$$(2.34 \text{ lbmol } CO_2)(44 \text{ lb } CO_2/1 \text{ lbmol } CO_2) = 103 \text{ lb } CO_2 \text{ produced}$$

$$(7.03 \text{ lbmol } O_2)(2 \text{ lbmol } SO_2/3 \text{ lbmol } O_2) = 4.68 \text{ lbmol } SO_2$$
$$(4.68 \text{ lbmol } SO_2)(64.07 \text{ lb}/1 \text{ lbmol } SO_2) = 300 \text{ lb } SO_2 \text{ produced}$$

STC.4   EXCESS AIR COMBUSTION

Benzene is incinerated at 2100°F in the presence of 50% excess air (EA). Balance the combustion reaction for this process.

$$C_6H_6 + O_2 \rightarrow CO_2 + H_2O + O_2$$

Solution

Ignoring nitrogen, the balanced equation is

$$C_6H_6 + \tfrac{15}{2}(EA + 1)\, O_2 \rightarrow 6CO_2 + 3H_2O + \tfrac{15}{2}(EA)\, O_2$$

For 50% excess air,

$$EA = 50\% = 0.5$$

so that

$$O_2 \text{ (start)} = (\tfrac{15}{2})\,(1 + 0.5) = 11.25$$

and

$$O_2 \text{ (end)} = (\tfrac{15}{2})\,(0.5) = 3.75$$

The balanced equation now becomes

$$C_6H_6 + 11.25O_2 \rightarrow 6CO_2 + 3H_2O + 3.75O_2$$

To include N_2,

$$N_2 = 11.25\,(\tfrac{79}{21}) = 42.32$$

The balanced reaction equation, including the inert nitrogen, is now

$$C_6H_6 + 11.25O_2 + 42.32N_2 \rightarrow 6CO_2 + 3H_2O + 3.75O_2 + 42.32N_2$$

STC.5   INCINERATOR APPLICATION

C_6H_5Cl is fed into a hazardous waste incinerator at a rate of 5000 scfm (60°F, 1 atm) and is combusted in the presence of air fed at a rate of 3000 scfm (60°F, 1 atm). Both streams enter the incinerator at 70°F. Following combustion, the products are cooled from 2000°F and exit a cooler at 180°F. At what rate (lb/h) do the products exit the cooler? The molecular weight of C_6H_5Cl is 112.5; the molecular weight of air is 29.

Solution

Convert scfm to acfm using Charles' law for both chlorobenzene (q_1) and air (q_2).

$$q_1 = (5000 \text{ scfm})(460 + 70)/(460 + 60)$$
$$= 5096 \text{ acfm of } C_6H_5Cl$$
$$q_2 = (3000 \text{ scfm})(460 + 70)/(460 + 60)$$
$$= 3058 \text{ acfm of air}$$

Since 1 lbmol of any ideal gas occupies 379 ft^3 at 60°F and 1 atm, the molar flowrate, $\dot{n}$, may be calculated by dividing the volumetric flow rates at 60°F by 379.

$$\dot{n}_1 = 5000/379$$
$$= 13.2 \text{ lbmol/min}$$
$$= 792 \text{ lbmol/h}$$
$$\dot{n}_2 = 3000/379$$
$$= 7.92 \text{ lbmol/min}$$
$$= 475 \text{ lbmol/h}$$

The mass flowrate is found by multiplying the above by the molecular weight:

$$\dot{m}_1 = (792)(112.5)$$
$$= 89,100 \text{ lb/h}$$
$$\dot{m}_2 = (475)(29)$$
$$= 13,800 \text{ lb/h}$$

The mass rate exiting the cooler is then

$$\dot{m}_{out} = \dot{m}_{in}$$
$$= 89,100 + 13,800$$
$$= 102,900 \text{ lb/h}$$

STC.6   CHEMICAL FORMULA DETERMINATION

A separation unit produces a pure hydrocarbon compound with a concentration not high enough to economically justify recovering and recycling it. The engineering division of the company has made a decision that it would be worthwhile to combust the compound and recover the heat generated as a makeup heat source for the separation unit.

If this hydrocarbon compound contains three atoms of carbon, determine its chemical formula if the flue gas composition on a dry basis is:

CO_2: 7.5%   CO: 1.3% O_2: 8.1% N_2: 83.1%

Solution

Assume a basis 100 mol of dry flue gas. The moles of each component in the dry product gas is then

CO_2 7.5 mol
CO   1.3 mol
O_2 8.1 mol
N_2 83.1 mol

Determine the amount of oxygen fed for combustion.

Since nitrogen does not react, using the ratio of oxygen to nitrogen in air will provide the amount of oxygen fed:

$$O_{2,\text{fed}} = (\tfrac{21}{79})(83.1) = 22.1 \text{ mol}$$

A balanced equation for the combustion of the hydrocarbon in terms of N moles of hydrocarbon and n hydrogen atoms in the hydrocarbon yields

$$NC_3H_n + 22.1O_2 \rightarrow 7.5CO_2 + 1.3CO + 8.1O_2 + N(n/2)H_2O$$

The moles of hydrocarbon, N, is obtained by performing an elemental carbon balance:

$$3N = 7.5 + 1.3$$
$$N = 8.8/3 = 2.93$$

Similarly, the moles of water formed is obtained by performing an elemental oxygen balance:

$$2(22.1) = 2(7.5) + 1.3 + 2(8.1) + N(n/2)$$
$$N(n/2) = 44.2 - 15 - 1.3 - 16.2$$
$$= 11.7$$

The number of hydrogen atoms, n, in the hydrocarbon is then

$$n = 2(11.7)/N$$
$$= 23.4/2.93$$
$$= 7.99 \approx 8$$

Since $n = 8$, the hydrocarbon is C_3H_8, propane.

STC.7   ETHANOL COMBUSTION

When ethanol (C_2H_5OH) is completely combusted, the products are carbon dioxide and water.

1. Write the balanced reaction equation.
2. If 150 lbmol/h of water is produced, at what rate (molar) is the ethanol combusted?
3. If 2000 kg of the ethanol is combusted, what mass of oxygen is required?

The atomic weights of C, O, and H are 12, 16, and 1, respectively.

Solution

The balanced reaction equation is

$$C_2H_5OH + 3O_2 \rightarrow 2CO_2 + 3H_2O$$

The stoichiometric ratio of the C_2H_5OH consumed to the water produced may now be calculated.

$$\text{Ratio} = \frac{1 \text{ lbmol } C_2H_5OH}{3 \text{ lbmol } H_2O \text{ produced}}$$

The result may be used to calculate the amount of C_2H_5OH reacted in part 2.

$$(150 \text{ lbmol/h } H_2O \text{ produced})\left(\frac{1 \text{ lbmol } C_2H_5OH}{3 \text{ lbmol } H_2O \text{ produced}}\right)$$
$$= 50 \text{ lbmol/h } C_2H_5OH \text{ reacted}$$

The molar amount (in kgmol) of C_2H_5OH reacted in part 3 is

$$\frac{2000 \text{ kg } C_2H_5OH}{45 \text{ kg/kgmol } C_2H_5OH} = 43.48 \text{ kgmol of } C_2H_5OH$$

Using the stoichiometric ratio of oxygen to ethanol reacted, the number of kgmol of oxygen needed is then

$$(43.48\,\text{kgmol C}_2\text{H}_5\text{OH})\left(\frac{3\,\text{kgmol O}_2}{1\,\text{kgmol C}_2\text{H}_5\text{OH}}\right) = 130.4\,\text{kgmol O}_2 \text{ reacted}$$

The required mass of oxygen may now be calculated:

$$(130.4\,\text{kgmol O}_2)(32.0\,\text{kg/kgmol}) = 4174\,\text{kg O}_2 \text{ required}$$

STC.8 ISOTHERMAL EVAPORATION

Determine the minimum number of cubic feet of dry air required to evaporate 20 lb of ethanol if the total pressure is maintained at 740 mm Hg. The evaporation process is isothermal at 70°F and the vapor pressure of the alcohol is 0.9 psia at this temperature.

Solution

The number of lbmols of ethanol, n, is

$$n = 20/46$$

$$= 0.435\,\text{lbmol}$$

The volume V of ethanol is given by the ideal gas law

$$V = nRT/P$$

$$= (0.435)(0.73)(70 + 460)/(740/760)$$

$$= 173\,\text{ft}^3$$

The maximum (or saturation) mole fraction, y, of ethanol in the air is determined from the ethanol vapor pressure

$$y_{\text{max}} = (0.9/14.7)/(740/760)$$

$$= 0.0629$$

The minimum volume of dry air required to evaporate the alcohol may now be calculated:

$$y_{\text{air}} = 1 - y_{\text{max}}$$

$$= 0.937$$

$$V_{\text{air}} = V(y_{\text{air}}/y_{\text{max}})$$

$$= 173(0.937/0.0629)$$

$$= 2577\,\text{ft}^3$$

Calculating the total volume of the air–ethanol mixture is given by the sum of the ethanol and dry air volumes:

$$V_T = V + V_{air}$$
$$= 173 + 2577$$
$$= 2750 \text{ ft}^3$$

STC.9   EXTENT OF REACTION BALANCES

In one step of the Contact Process for the production of sulfuric acid, sulfur dioxide is oxidized to sulfur trioxide at high pressures in the presence of a catalyst:

$$2SO_2 + O_2 \rightarrow 2SO_3$$

A mixture of 300 lbmol/min of sulfur dioxide and 400 lbmol/min of oxygen are fed to a reactor. The flowrate of the unreacted oxygen leaving the reactor is 300 lbmol/min. Determine the composition (in mol%) of the exiting gas stream by three methods:

1. Molecular balances
2. Atomic balances
3. "Extent of reaction" balances

Solution

There are three different types of material balances that may be written when a chemical reaction is involved: the molecular balance, the atomic balance, and the "extent of reaction" balance. It is a matter of convenience which of the three types is used.

Assuming a steady-state continuous reaction, the accumulation term, A, is zero and, for all components involved in the reaction, the molecular balance equation becomes

$$I + G = O + C$$

where I = amount or rate of input
   G = amount or rate of material generated
   O = amount or rate of output
   C = amount or rate of material consumed

If a total material balance is performed, the above form of the balance equation must be used if the amounts or flowrates are expressed in terms of moles, e.g., lbmol or gmol/h, since the total number of moles can change during a chemical reaction. If, however, the amounts or flowrates are given in terms of mass, e.g., kg or lb/h, the G and C terms may be dropped, since mass cannot be lost or gained in a chemical reaction. Therefore,

$$I = O$$

In general, however, when a chemical reaction is involved, it is usually more convenient to express amounts and flowrates using moles rather than mass.

A material balance that is based not on the compounds (or molecules), but rather on the atoms that make up the molecules, is referred to as an atomic balance. Since atoms are neither created nor destroyed in a chemical reaction, the G and C terms equal zero and the balance again becomes

$$I = O$$

As an example, take once again the combination of hydrogen and oxygen to form water:

$$O_2 + 2H_2 \rightarrow 2H_2O$$

As the reaction progresses, the O_2 and H_2 molecules (or moles) are consumed while H_2O molecules (or moles) are generated. On the other hand, the number of oxygen atoms (or moles of oxygen atoms) and the number of hydrogen atoms (or moles of hydrogen atoms) do not change. Care must be taken to distinguish between molecular oxygen and atomic oxygen. If, in the above reaction, we start out with 1000 lbmol of O_2 (oxygen molecules), we are also starting out with 2000 lbmol of O (oxygen atoms).

The extent of reaction balance gets its name from the fact that the amounts of the chemicals involved in the reaction are described in terms of how much of a particular reactant has been consumed or how much of a particular product has been generated. As an example, take the formation of ammonia from hydrogen and nitrogen:

$$N_2 + 3H_2 \rightarrow 2NH_3$$

If one starts off with 10 mol of hydrogen and 10 mol of nitrogen, and lets z represent the amount of nitrogen consumed by the end of the reaction, the output amounts of all three components are $10 - z$ for the nitrogen, $10 - 3z$ for the hydrogen, and $2z$ for the ammonia. The following is a convenient way of representing the extent of reaction balance:

Input →	10	10	0	$\sum = 20$
	N_2 +	$3H_2$ →	$2NH_3$	
Output →	$10 - z$	$10 - 3z$	$2z$	$\sum = 20 - 2z$

In this problem, the reader is asked to employ all three types of material balances. A flow diagram of the process is given in Figure 12.

Using the molecular balance approach, one obtains

$$O_2: \quad I = O + C$$

$$400 = 300 + C$$

$$C = 100 \text{ lbmol/ min } O_2 \text{ consumed}$$

$$SO_2: \quad I = O + C$$

$$300 = \dot{n}_{SO_2} + \left(\tfrac{2}{1}\right)(100)$$

$$\dot{n}_{SO_2} = 100 \text{ lbmol/ min } SO_2 \text{ out}$$

$$SO_3: \quad G = O$$

$$\left(\tfrac{2}{1}\right)(100) = \dot{n}_{SO_3}$$

$$\dot{n}_{SO_3} = 300 \text{ lbmol/ min } SO_3 \text{ out}$$

Since the total flowrate of the exiting gas is 600 lbmol/ min,

$O_2:$ $(300/600)(100\%) = 50.0\%$
$SO_2:$ $(100/600)(100\%) = 16.7\%$
$SO_3:$ $(200/600)(100\%) = 33.3\%$

Using the atomic balance approach, one obtains

$$S: \quad I = O$$

$$\left(\tfrac{1}{1}\right)(300) = \left(\tfrac{1}{1}\right)\dot{n}_{SO_2} + \left(\tfrac{1}{1}\right)\dot{n}_{SO_3}$$

$$\dot{n}_{SO_2} + \dot{n}_{SO_3} = 300$$

$$O: \quad I = O$$

$$\left(\tfrac{2}{1}\right)(300) + \left(\tfrac{2}{1}\right)(400) = \left(\tfrac{2}{1}\right)(300) + \left(\tfrac{2}{1}\right)\dot{n}_{SO_2} + \left(\tfrac{3}{1}\right)\dot{n}_{SO_3}$$

$$2\dot{n}_{SO_2} + 3\dot{n}_{SO_3} = 800$$

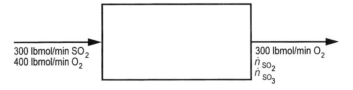

Figure 12. Flow diagram for Problem STC.9.

Solving these two equations simultaneously yields the flowrates of the SO_2 and SO_3 leaving the reactor.
As before,

$$\dot{n}_{SO_2} = 100 \text{ lbmol/ min SO}_2\text{out}$$

$$\dot{n}_{SO_3} = 200 \text{ lbmol/ min SO}_3\text{out}$$

Note: Since there are only two types of atoms, S and O, atomic balances can only provide 2 equations. This is not enough for a system with 3 unknowns. The oxygen output rate must be determined by one of the other methods.
Using the extent of reaction balances approach gives

Input →	300		400		0
	$2SO_2$	$+$	O_2	$\rightarrow$	$2SO_3$
Output →	$300 - z$		$400 - (\frac{1}{2})z$		$0 + z$

where z is the rate (lbmol/min) of SO_2 reacted.
Since the outlet O_2 flowrate is given as 300 lbmol/min,

$$300 = 400 - (\tfrac{1}{2})z$$

$$z = 200$$

SO_2 outlet flowrate $= 300 - z = 100$ lbmol/ min
SO_3 outlet flowrate $= z = 200$ lbmol/ min
O_2 outlet flowrate $= 400 - (\tfrac{1}{2})z = 300$ lbmol/ min

STC.10   ETHYLENE OXIDATION

Ethylene oxide is produced by the oxidation of ethylene with oxygen-enriched air:

$$C_2H_4 + \tfrac{1}{2}O_2 \rightarrow C_2H_4O$$

An undesired side reaction is the oxidation of ethylene to carbon dioxide:

$$C_2H_4 + 3O_2 \rightarrow 2CO_2 + 2H_2O$$

The feed stream to the ethylene oxide reactor consists of 45% (by mole) C_2H_4, 30% O_2, and 25% N_2. The amounts of ethylene oxide and carbon dioxide in the product stream are 20 gmol and 10 gmol per 100 gmol of feed stream, respectively. Determine the composition of the exiting gas stream.

Solution

Selecting a basis of 100 gmol feed stream, a flow diagram of the process may be generated as shown in Figure 13.

One may apply extent of reaction balances to determine expressions for each of the components leaving the reactor in terms of the amount of C_2H_4 converted to C_2H_4O and the amount converted to CO_2.

$$\text{Initial} \rightarrow \quad 45 \qquad\qquad 30 \qquad\qquad\qquad 0$$
$$C_2H_4 \quad + \quad \tfrac{1}{2}O_2 \qquad \rightarrow \qquad C_2H_4O$$
$$\text{Final} \rightarrow \quad 45 - y - z \quad 30 - (\tfrac{1}{2})y - (\tfrac{3}{1})z \qquad y$$

$$\text{Initial} \rightarrow \quad 45 \qquad\qquad 30 \qquad\qquad 0 \qquad 0$$
$$C_2H_4 \quad + \quad 3O_2 \qquad \rightarrow \qquad 2CO_2 \; + \; 2H_2O$$
$$\text{Final} \rightarrow \quad 45 - y - z \quad 30 - (\tfrac{1}{2})y - (\tfrac{3}{1})z \qquad 2z \qquad 2z$$

where y = amount of C_2H_4 converted to C_2H_4O
$\quad\quad z$ = amount of C_2H_4 converted to CO_2

The amounts of all components of the product gas stream may now be calculated:

$$\text{Amount } C_2H_4O = y = 20 \text{ gmol}$$
$$\text{Amount } CO_2 = 2z = 10 \text{ gmol} \rightarrow z = 5$$
$$\text{Amount } H_2O = 2z = 2(5) = 10 \text{ gmol}$$
$$\text{Amount } C_2H_4 = 45 - y - z = 45 - 20 - 5 = 20 \text{ gmol}$$
$$\text{Amount } O_2 = 30 - (\tfrac{1}{2})y - (\tfrac{3}{1})z$$
$$= 30 - (\tfrac{1}{2})20 - (\tfrac{3}{1})5 = 5 \text{ gmol}$$
$$\text{Amount } N_2 = 25 \text{ gmol}$$
$$\text{Total amount of product gas} = 90 \text{ gmol}$$

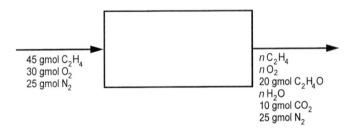

Figure 13. Flow diagram for Problem STC.10.

The mole fractions are therefore:

$$\text{Mole fraction } C_2H_4O = \tfrac{20}{90} = 0.2222$$
$$\text{Mole fraction } CO_2 = \tfrac{10}{90} = 0.1111$$
$$\text{Mole fraction } H_2O = \tfrac{10}{90} = 0.1111$$
$$\text{Mole fraction } C_2H_4 = \tfrac{20}{90} = 0.2222$$
$$\text{Mole fraction } O_2 = \tfrac{5}{90} = 0.0556$$
$$\text{Mole fraction } N_2 = \tfrac{25}{90} = 0.2778$$
$$\overline{1.0000}$$

PART II
Chemical Engineering Principles

6 Fluid Flow (FFL)

FFL.1   STACK VELOCITY

The exhaust gas flowrate from a facility is 1500 scfm. All of the gas is vented through a small stack, which has an inlet area of 1.3 ft^2. The exhaust gas temperature is 350°F. What is the velocity of the gas through the stack inlet in ft/s. Assume standard conditions to be 70°F and 1 atm. Neglect the pressure drop across the stack.

Solution

Applying Charles' law,

$$q_a = q_s \left(\frac{T_a}{T_s} \right)$$
$$= 1500 \left(\frac{350 + 460}{70 + 460} \right)$$
$$= 2292 \, \text{acfm}$$

The average velocity may now be calculated:

$$v = \frac{q_a}{A}$$
$$= \frac{2292}{(1.3)(60)}$$
$$= 29.4 \, \text{ft/s}$$

FFL.2   HORSEPOWER REQUIREMENT

Calculate the horsepower (HP) needed to process a 8500-acfm gas stream from an incinerator. The pressure drop across the various pieces of process equipment has been estimated to be 8.4 in. H$_2$O. The pressure loss for duct work, elbows, valves, and so on, and expansion-contraction losses are estimated at 5.8 in. H$_2$O. Assume an overall fan-motor efficiency of 58%.

Solution

The total pressure drop across the system is

$$\Delta P = (8.4 + 5.8) = 14.2 \text{ in. } H_2O$$

The describing equation for the brake horsepower (BHP) is

$$\text{BHP} = \frac{q_a \, \Delta P(1.575 \times 10^{-4})}{\text{Fractional fan efficiency}}$$

Substitution yields

$$\text{BHP} = \frac{(8500)(14.2)(1.575 \times 10^{-4})}{0.58}$$

$$= 32.78 \text{ bhp}$$

The reader should note that 1.575×10^{-4} is a conversion factor to obtain units of horsepower from $\text{ft} \cdot \text{lb}_f/\text{min}$.

FFL.3   FAN SPEED

A fan operating at a speed of 1750 rpm delivers 13,250 acfm at 5.5 in. H_2O static pressure and requires a BHP of 10.5. What will be the new operating conditions if the fan speed is increased to 2000 rpm?

Solution

Fan laws are equations that enable the results of a fan test (or operation) at one set of conditions to be used to calculate the performance at another set of conditions, including differently sized but geometrically similar models of the same fan design. The fan laws can be written in many different ways. The three key laws are provided in the following equations:

$$q_a = k_1(\text{RPM})D^3$$
$$P_s = k_2(\text{RPM})^2 D^2 \rho$$
$$\text{HP} = k_3(\text{RPM})^3 D^5 \rho$$

where q_a = actual volumetric flowrate
$\quad P_s$ = static pressure
$\quad \text{HP}$ = horsepower
$\quad \text{RPM}$ = revolution per minute
$\quad D$ = wheel diameter
$\quad \rho$ = gas density
$\quad k_1, k_2, k_3$ = proportionality constants

Thus, these three laws may be used to determine the effect of fan speed, fan size, and gas density on flowrate, developed static pressure head, and horsepower. For two conditions, where the constants k remain unchanged, these equations become

$$\left(\frac{q'_a}{q_a}\right) = \left(\frac{RPM'}{RPM}\right)\left(\frac{D'}{D}\right)^3$$

$$\left(\frac{P'_s}{P_s}\right) = \left(\frac{RPM'}{RPM}\right)^2\left(\frac{D'}{D}\right)^2\left(\frac{\rho'}{\rho}\right)$$

$$\left(\frac{HP'}{HP}\right) = \left(\frac{RPM'}{RPM}\right)^3\left(\frac{D'}{D}\right)^5\left(\frac{\rho'}{\rho}\right)$$

Note: The prime refers to the new condition. It is also important to note that the fan laws are approximations and should not be used over wide ranges or changes of flowrate, size, etc.

The new fan flowrate, q'_a, is

$$q'_a = q_a\left(\frac{RPM'}{RPM}\right)\left(\frac{D'}{D}\right)^3$$

$$= 13{,}250\left(\frac{2000}{1750}\right)(1)^3$$

$$= 15{,}143 \text{ acfm}$$

The new static pressure becomes

$$P'_s = P_s\left(\frac{RPM'}{RPM}\right)^2\left(\frac{D'}{D}\right)^2\left(\frac{\rho'}{\rho}\right)$$

$$= 5.5\left(\frac{2000}{1750}\right)^2(1)^2(1)$$

$$= 7.18 \text{ in. } H_2O$$

The required horsepower is

$$HP' = HP\left(\frac{RPM'}{RPM}\right)^3\left(\frac{D'}{D}\right)^5\left(\frac{\rho'}{\rho}\right)$$

$$= 10.5\left(\frac{2000}{1750}\right)^3(1)^5(1)$$

$$= 15.67 \text{ bhp}$$

Fans are usually classified as either the centrifugal or the axial-flow type. In centrifugal fans, the gas is introduced into the center of the revolving wheel (the eye) and discharges at right angles to the rotating blades. In axial-flow fans, the gas moves directly (forward) through the axis of rotation of the fan blades. Both are used in industry, but it is the centrifugal fan that is important at process facilities. Generally, centrifugal fans are easier to control, more robust in construction, and less noisy than axial units. They have a broader operating range at their highest efficiencies. Centrifugal fans are also better suited for operations in which there are flow variations, and they can handle dust and fumes better than axial fans.

FFL.4   PUMP CHARACTERISTICS

It is necessary to pump a constant-flow stream of liquid with a density and viscosity similar to that of water into a reactor at a rate of 70 gal/min. The pump must operate against a pressure of 200 psi. A pump with characteristics shown in Figure 14 is available with variable speed drive. At what speed should the pump be operated? What horsepower (hp) is needed to maintain this flow?

Solution

The information provided is superimposed on the curve in Figure 14. The speed is (see point A) located between the 400 rpm and 600 rpm line. Interpolation gives that speed to be approximately 380 rpm. Interpolation again on the horsepower curve (see point B) gives approximately 16 hp.

FFL.5   COMPRESSED AIR REQUIREMENTS

Compressed air is to be employed in the nozzle of a liquid injection incinerator to assist the atomization of a liquid hazardous waste. The air requirement for the nozzle is 0.5 lbmol/min at 50 psia. If atmospheric air is available at 50°F and 1.0 atm, calculate the power requirement. For this process, assume the polytropic constant, n, to be 1.3.

For an ideal gas, the compressor energy requirement (delivered to the air) is given by

$$W_s = -\frac{nRT_1}{n-1}\left[\left(\frac{P_2}{P_1}\right)^{(n-1)/n} - 1\right]$$

Solution

The ideal gas law may be assumed to apply at these conditions:

$$W_s = -\frac{(1.3)(1.987)(510)}{1.3-1}\left[\left(\frac{50}{14.7}\right)^{(1.3-1)/1.3} - 1\right]$$

$$= -1434\,\text{Btu/lbmol of air}$$

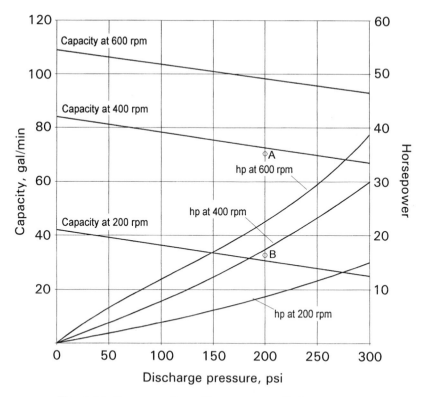

Figure 14. Pump capacity and horsepower vs. discharge pressure.

The horsepower, HP, is obtained by multiplying by the molar flowrate delivered to the air:

$$HP = -(1434)(0.5)(778)$$

$$= -557,634 \, \text{ft} \cdot \text{lb}_f / \min$$

$$= -16.9 \, \text{hp}$$

Note that there are $778 \, \text{ft} \cdot \text{lb}_f$ per Btu and $33,000 \, \text{ft} \cdot \text{lb}_f/\min$ per hp. The power *required* has the opposite sign, i.e., $+16.9 \, \text{hp}$. The reader should also note that this represents the *minimum* power required to accomplish this job.

FFL.6   GRAVITY DECANTER

1. In the system shown in Figure 15 calculate the gauge pressure at point 1 in psig (lb_f/in^2 gauge).

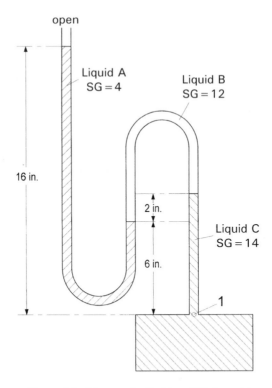

Figure 15. Diagram for Problem FFL.6, part 1.

2. The gravity decanter shown in Figure 16 is employed to separate three immiscible liquids. Calculate the heights of the two adjustable overflow legs and express the answer in feet.

Solution

Both of these problems involve the application of the hydrostatic pressure equation presented in Chapter 4.

For part 1 of this problem, three legs of fluid must be taken into account to calculate the pressure at point 1. Applying the hydrostatic pressure equation gives

$$P_1 = P_{atm} + (g/g_c)\rho_A h_A - (g/g_c)\rho_B h_B + (g/g_c)\rho_C h_C$$
$$= 0 + (1)(4)(62.4)[(16 - 6)/12] - (1)(12)(62.4)(2/12)$$
$$\quad + (1)(14)(62.4)([6 + 2]/12)$$
$$= 666 \, \text{lb}_f/\text{ft}^2 \, \text{(gauge)}$$
$$= 4.62 \, \text{psig}$$

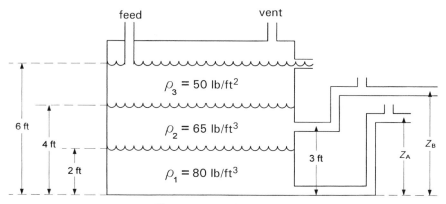

Figure 16. Gravity decanter.

For part 2, application of the hydrostatic equation to determine z_A gives

$$P_{atm} = P_{atm} + (g/g_c)\rho_3(6-4) + (g/g_c)\rho_2(4-2) + (g/g_c)\rho_1(2) - (g/g_c)\rho_1(z_A)$$
$$0 = 0 + (1)(50)(2) + (1)(65)(2) + (1)(80)(2) - (1)(80)(z_A)$$
$$z_A = 4.88\,\text{ft}$$

To determine z_B, the hydrostatic equation again is applied:

$$P_{atm} = P_{atm} + (g/g_c)\rho_3(6-4) + (g/g_c)\rho_2(4-3) - (g/g_c)\rho_2(z_B-3)$$
$$0 = 0 + (1)(50)(2) + (1)(65)(1) - (1)(65)(z_B-3)$$
$$z_B = 5.54\,\text{ft}$$

FFL.7 PIPING SYSTEM

Calculate P_3 in psig (lb_f/in^2) for the piping system shown in Figure 17.

Data
$\bar{v}_1 = 8\,\text{ft/s}$
$P_1 = 2\,\text{atm absolute}$
For pipe 1–2: 2 in. Sch 40; $h_f = 6\,\text{ft} \cdot \text{lb}_f/\text{lb}$
For pipe 2–3: 4 in. Sch 40; $h_f = 7\,\text{ft} \cdot \text{lb}_f/\text{lb}$
For pump: $\eta = 0.55$; HP $= 0.75\,\text{hp}$

Assume turbulent flow, i.e., $\alpha = 1$, and the Bernoulli equation to apply.

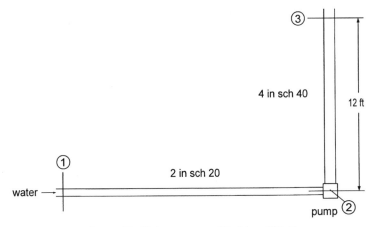

Figure 17. Piping system of Problem FFL.7.

Solution

The velocity in the vertical pipe can be found using the continuity equation and first determining the cross-sectional areas, S_1 and S_2, of the two pipes:

$$S_1 \text{ (2 in., Sch 40)} = 0.0233 \text{ ft}^2$$

$$S_2 \text{ (4 in., Sch 40)} = 0.0884 \text{ ft}^2$$

$$\bar{v} = \left(\frac{0.0233}{0.0884}\right)(8) = 2.11 \text{ ft/s}$$

$$P_1 = (2)(14.7) = 29.4 \text{ psia} = 14.7 \text{ psig}$$

Bernoulli's equation is given by

$$\frac{\Delta P}{\rho} + \frac{\Delta(\alpha \bar{v}^2)}{2g_c} + \Delta z\left(\frac{g}{g_c}\right) = \eta W_p - h_f$$

where P = pressure
     ρ = density of fluid
     α = kinetic energy correction factor
     $\bar{v}$ = velocity
     z = height
     g = gravitational acceleration
     g_c = gravitational conversion factor
     h_f = friction loss per unit mass of fluid
     η = pump efficiency
     W_p = work required per unit mass of fluid, $\dot{m}$

Before applying Bernoulli's equation between points 1 and 3, various terms in the equation are evaluated:

$$\eta W_p = (0.55)\,[(0.75)(550)/\dot{m}]$$

$$\dot{m} = (8)(0.0233)(62.4) = 11.63\ \text{lb/s}$$

$$\eta W_p = 19.51\ \text{ft}\cdot\text{lb}_f/\text{lb}$$

Substitution into Bernoulli's equation (with P in psig) gives

$$\frac{(P_2 - 14.7)(144)}{62.4} + \frac{(1)(2.11^2) - (1)(8^2)}{(2)(32.2)} + (12 - 0)(1) = 19.51 - 13$$

$$P_2 = 12.7\ \text{psig}$$

FFL.8   PUMP REQUIREMENTS

Water at 60°F is being pumped from a reservoir to a storage tank on top of a building through an open pipe. The reservoir's water level is 10 ft above the pipe inlet, and 200 ft below the water level in the tank. Both tanks are open to the atmosphere. The piping system has an inner diameter (ID) of 4 in., contains two gate valves, five 90° elbows, and is 525 ft long. A flowrate of 610 gal/min is desired. Calculate the pump requirement (in horsepower) if it is rated as 60% efficient. Also provide the pump requirements with units of kW, W, and N · m/s. Assume k (roughness parameter) $= 0.00015$ ft.

Solution

A diagram representing this system is provided in Figure 18. The average pipe velocity $\bar{v}$ is first calculated:

$$\bar{v} = \frac{q}{S}$$

$$q = \left(\frac{610\ \text{gal}}{\text{min}}\right)\left(\frac{1\ \text{min}}{60\ \text{s}}\right)\left(\frac{1\ \text{ft}^3}{7.481\ \text{gal}}\right) = 1.36\ \text{ft}^3/\text{s}$$

$$S = \frac{(\pi)(4/12)^2}{4} = 0.0873\ \text{ft}^2$$

$$\bar{v} = \frac{1.36}{0.0873} = 15.6\ \text{ft/s}$$

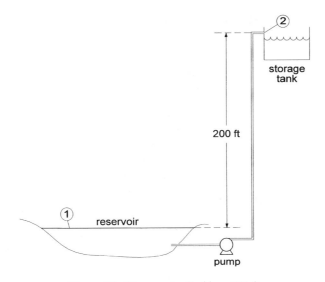

Figure 18. Diagram for Problem FFL.8.

The Reynolds number (Re) is calculated using the density and viscosity of water:

$$\rho = 62.37 \, \text{lb/ft}^3$$

$$\mu = 1.129 \, \text{cP}$$

$$\text{Re} = \frac{D\bar{v}\rho}{\mu}$$

$$= \frac{(4/12)(15.6)(62.37)}{(1.129)(6.72 \times 10^{-4})}$$

$$= 427{,}500 \qquad \text{(turbulent flow)}$$

Note that $1 \, \text{cP} = 6.72 \times 10^{-4} \, \text{lb/(ft} \cdot \text{s)}$.

To obtain the Fanning friction factor,

$$k/D = 0.00015/(\tfrac{4}{12}) = 0.00045$$

The Fanning friction factor f may be estimated from plots or correlations available in the literature. The Jain correlation is employed here:

$$\frac{1}{f^{0.5}} = 2.28 - 4 \, \log_{10}\left(\frac{k}{D} + \frac{21.25}{\text{Re}^{0.9}}\right)$$

$$= 2.28 - 4 \, \log_{10}\left(0.00045 + \frac{21.25}{427{,}500^{0.9}}\right)$$

$$f = 0.0044$$

There are various friction losses: skin friction, sudden contraction, sudden expansion, fittings, etc. These all contribute to the h_f term in Bernoulli's equation, as shown in the following expression.

$$h_f = h_{fs} + h_{fc} + h_{fe} + \sum_{\text{fittings}} h_{ff} = \left(4f\frac{L}{D} + K_c + K_e + \sum_{\text{fittings}} K_f\right)\frac{\bar{v}^2}{2g_c}$$

where h_f = total friction loss, energy per unit mass of fluid
  h_{fs} = skin friction loss
  h_{fc} = loss due to sudden contraction
  h_{fe} = loss due to sudden expansion
  h_{ff} = loss due to a fitting
  f = fanning friction factor
  K_c = contraction loss coefficient
  K_e = expansion loss coefficient
  K_f = fitting loss coefficient

The various coefficients are evaluated as follows:

1. For a sudden contraction between point 1 (upstream) and 2 (downstream):

$$K_c = 0.4\left(1 - \frac{S_2}{S_1}\right)$$

2. For a sudden expansion point 1 (upstream) and 2 (downstream):

$$K_e = \left(1 - \frac{S_1}{S_2}\right)^2$$

Note that, for both 1 and 2, $\bar{v}$ in the equation for h_f is the velocity in the *smaller* cross section.

3. For fittings, K is obtained from the following table:

Fitting	K_f
Globe valve, wide open	10.0
Angle valve, wide open	5.0
Gate valve, wide open	0.2
Gate valve, half open	5.6
Return bend	2.2
Tee	1.8
Elbow, 90°	0.9
Elbow, 45°	0.4

For the sudden contraction at the pipe entrance, note that the entrance cross section, S_1, is essentially infinity. Therefore,

$$K_c = 0.4\left(1 - \frac{S_2}{\infty}\right) = 0.4$$

The table above is used to calculate the losses due to the two gate valves and the five 90° elbows. The total loss is therefore given by

$$h_f = \left(4(0.0044)\frac{525}{4/12} + 0.4 + (2)(0.2) + 5\,(0.9)\right)\frac{(15.6)^2}{2(32.2)}$$

$$= 124.8\,\text{ft} \cdot 1\text{b}_f/\text{lb}$$

The Bernoulli equation may now be applied between points 1 and 2:

$$\frac{\Delta P}{\rho} + \frac{\Delta(\alpha \bar{v}^2)}{2g_c} + \Delta z\left(\frac{g}{g_c}\right) = \eta W_p - h_f$$

$$\frac{0-0}{62.37} + \frac{15.6^2 - 0^2}{2(32.2)} + (200)(1) = (0.60)\,W_p - 124.8$$

$$W_p = 547.6\,\text{ft} \cdot \text{lb}_f/\text{lb}$$

The mass flowrate, $\dot{m}$, which is required to determine the power requirement, is obtained from the continuity equation:

$$\dot{m} = \rho \bar{v} S = (62.37)(15.6)\,[\pi(4/12)^2/4] = 84.92\,\text{lb}/\text{s}$$

The power requirement to run the pump is then

$$HP = (547.6)(84.92)/550 = 84.5\,\text{hp}$$

Note that 1 hp = 550 ft · lb$_f$/s. The power requirement in the other units requested is

$$(84.5)(0.7456) = 63.0\,\text{kW}$$

$$= 63,000\,\text{W}$$

$$= 63,000\,\text{N} \cdot \text{m/s}$$

FFL.9   FLOW SYSTEM PRESSURE CALCULATION

Calculate the pressure P_1 (see Figure 19) such that 0.3 ft^3/s of water will be discharged to the atmosphere.

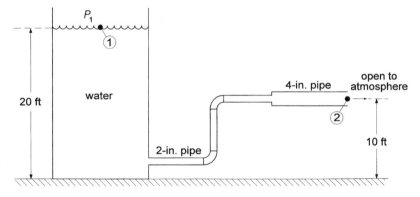

Figure 19. Diagram for Problem FFL.9.

Data

$k = 0.0004$ ft (roughness parameter)

2-in. pipe, 25 ft long with 2 90° elbows

4-in. pipe, 15 ft long

$\mu = 1.00$ cP

$\rho = 62.4$ lb/ft³

Solution

The solution to the problem once again requires the application of the Bernoulli equation:

$$\frac{\Delta P}{\rho} + \frac{\Delta(\alpha \bar{v}^2)}{2g_c} + \Delta z \left(\frac{g}{g_c} \right) = \eta W_p - h_f$$

For this problem,

$$\bar{v}_1 = 0$$

$$z_1 = 20 \text{ ft}$$

$$z_2 = 10 \text{ ft}$$

$$P_2 = 0 \text{ psig}$$

The velocity in the 4-in. pipe is given by

$$\bar{v}_{4 \text{ in.}} = \frac{0.3}{(\pi/4)(4/12)^2} = 3.44 \text{ ft/s}$$

and in the 2-in. pipe by

$$\bar{v}_{2\ in.} = \frac{0.3}{(\pi/4)(2/12)^2} = 13.8\ ft/s$$

The h_f term requires more detailed solution since there are several contributing factors. Here,

$$h_f = 4f_2 \frac{L_2}{D_2}\frac{(\bar{v}_{2\ in.})^2}{2g_c} + 4f_4 \frac{L_4}{D_4}\frac{(\bar{v}_{4\ in.})^2}{2g_c} + (K_c + 2K_{elb} + K_e)\frac{(\bar{v}_{2\ in.})^2}{2g_c}$$

where $K_c = 0.4$
    $K_{elb} = 0.9$

In addition,

$$K_e = \left(1 - \frac{S_{2\ in.}}{S_{4\ in.}}\right)^2 = 0.56$$

For the 2-in. pipe,

$$Re_{2\ in.} = \frac{D\bar{v}\rho}{\mu}$$

$$= \frac{(2/12)(13.8)(62.4)}{6.72 \times 10^{-4}}$$

$$= 214{,}000 \qquad \text{for turbulent flow, } \alpha = 1$$

$$k/D = (0.0004)/(2/12) = 0.0024$$

Using the Jain correlation (see Problem FFL.8), $f = 0.00640$.
    For the 4-in. pipe,

$$Re_{4\ in.} = \frac{D\bar{v}\rho}{\mu}$$

$$= \frac{(4/12)(3.44)(62.4)}{6.72 \times 10^{-4}}$$

$$= 106{,}000 \qquad \text{turbulent flow, } \alpha = 1$$

$$k/D = (0.0004)/(4/12) = 0.0012$$

Again, using the Jain correlation, $f = 0.00572$.

Therefore,

$$h_f = 4(0.00637)\frac{25}{2/12}\frac{(13.8)^2}{(2)(32.2)} + 4(0.00572)\frac{15}{4/12}\frac{(3.44)^2}{(2)(32.2)}$$

$$+ [0.4 + 2(0.9) + 0.56]\frac{(13.8)^2}{(2)(32.2)}$$

$$= 19.7 \, \text{ft} \cdot \text{lb}_f/\text{lb}$$

From the Bernoulli equation,

$$\frac{0 - P_1}{62.4} + \frac{3.44^2 - 0}{2(32.2)} + (10 - 20)(1) = 0 - 19.7$$

$$P_1 = 616.7 \, \text{lb}_f/\text{ft}^2 \, (\text{gauge}) = 616.7 \, \text{psfg}$$

$$= 4.3 \, \text{psig}$$

FFL.10   WELL WATER APPLICATION

Water at 20°C is pumped from a well to a reservoir 90 m above the well through 2 km of 300 mm ID commercial steel pipe ($k = 0.0046$ m).

1. Ignoring "minor" losses, find the pump power in kilowatts (or $N \cdot m/s$) required for a flowrate of 0.2 m³/s.
2. If the pump is 1.8 m above the water surface in the well, and the suction pipe is 6.5 m long with one 90° elbow, calculate the water pressure in kilopascals at the pump inlet (see Figure 20). Include minor losses in this part (2).

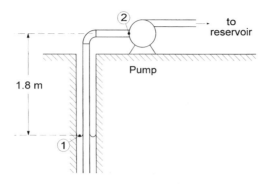

Figure 20. Diagram for problem FFL.10.

Solution

This problem is similar to Problem FFL.8, but in this case, SI units are employed. For part 1, the density and viscosity for water at 20°C are

$$\rho = 998 \text{ kg/m}^3$$

$$\mu = 1.00 \text{ cP} = 0.001 \text{ kg/(m} \cdot \text{s)}$$

$$\bar{v} = \frac{q}{S} = \frac{0.2}{(\pi/4)(0.30)^2} = 2.83 \text{ m/s}$$

$$\text{Re} = \frac{\rho \bar{v} D}{\mu} = \frac{(998)(2.83)(0.30)}{0.0010} = 847,000$$

Since $k = 0.0046$ m,

$$k/D = 0.0046/0.30 = 0.00015$$

Using the Jain correlation,

$$f = 0.00359$$

Thus,

$$h_f = 4f \frac{L}{D} \frac{\bar{v}^2}{2g_c}$$

$$= 4(0.00359) \frac{2000}{0.30} \frac{2.83^2}{2(1)}$$

$$= 383 \text{ m}^2/\text{s}^2$$

$$= 383 \text{ J/kg}$$

In applying the Bernoulli equation, point 1 is chosen at the liquid surface in the well and point 2 at the liquid level in the reservoir. (Note that these points are different from those in Figure 20.) For these conditions,

$$\frac{\Delta P}{\rho} + \frac{\Delta(\alpha \bar{v}^2)}{2g_c} + \Delta z \left(\frac{g}{g_c}\right) = \eta W_P - h_f$$

$$\frac{0-0}{998} + \frac{0^2 - 0^2}{2(1)} + (90 - 0) \frac{9.81}{1} = \eta W_P - 383$$

$$\eta W_p = 1266 \text{ J/kg}$$

Multiplying by the mass flowrate yields the brake power requirement, P_b

$$P_b = (1266 \text{ J/kg})(0.2 \text{ m}^3/\text{s})(998 \text{ kg/m}^3)$$
$$= 253{,}000 \text{ J/s}$$
$$= 253{,}000 \text{ W}$$
$$= 253 \text{ kW}$$

Note: The term $\dot{m}\eta W_p$ is the *fluid* power as opposed to the *brake* power ($\dot{m}W_p$). Fluid power is the power imparted to the fluid by the pump. Brake power is the power required to run the pump.

For part 2, the Bernoulli equation is now applied between points 1 and 2 in Figure 20. Contributions to the friction loss are now included. Besides skin friction, there are losses associated with the contraction at the pipe entrance and one 90° elbow:

$$h_f = h_{fs} + h_{fc} + h_{ff} = \left(4f\frac{L}{D} + K_c + K_f\right)\frac{\bar{v}^2}{2g_c}$$

$$h_f = \left[4(0.00359)\frac{6.5}{0.30} + 1.0 + 0.9\right]\frac{(2.83)^2}{2(1)}$$

$$= 8.85 \text{ J/kg}$$

Therefore,

$$\frac{P_2 - 1.013 \times 10^5}{998} + \frac{2.83^2 - 0^2}{2(1)} + (1.8 - 0)\frac{9.81}{1} = 0 - 8.85$$

Note: 1 atm $= 1.013 \times 10^5 \text{ N/m}^2 = 1.013 \times 10^5 \text{ Pa}$

$$P_2 = 70{,}800 \text{ Pa}$$
$$= 70.8 \text{ kPa (absolute)}$$

7 HEAT TRANSFER (HTR)

HTR.1   HEAT PUMP

A heat pump takes in 3500 gpm of water at a temperature of 38°F and discharges back to the lake at 36.2°F. How many Btu are removed from the water per day [C_p for $H_2O = 75.4\,J/(gmol \cdot °C)$, $\rho = 62.4\,lb/ft^3$]?

Solution

To calculate the heat load, the following equation is employed:

$$\dot{Q} = \dot{m}C_p(T_2 - T_1)$$

$$\dot{m} = \frac{(3500\,gal/\,min)(62.4\,lb/ft^3)(1440\,min/day)}{7.48\,gal/ft^3}$$

$$= 4.20 \times 10^7\,lb/day$$

The heat capacity is converted to consistent units and placed on a mass basis as follows:

$$c_p = \frac{[75.4\,J/(gmol \cdot °C)](454\,g/lb)}{(1054\,J/Btu)(18\,g/gmol)(1.8°F/°C)} = 1.00\,Btu/(lb \cdot °F)$$

Therefore,

$$\dot{Q} = (4.20 \times 10^7\,lb/day)[1.0\,Btu/(lb \cdot °F)](38 - 36.2\ °F)$$

$$= 1.36 \times 10^8\,Btu/day$$

HTR.2   THERMAL CONDUCTIVITY

Estimate the thermal conductivity of methane at 500°C. Assume the Lennard-Jones model to apply.

92

Solution

For polyatomic molecules such as methane, the Lennard-Jones equation may be written as

$$k = \frac{15}{4}\frac{R}{MW}\left(\frac{4}{15}\frac{C_v}{R} + \frac{3}{5}\right)\mu$$

where k = thermal conductivity, cal/(s · cm · °C)
   R = ideal gas constant, 1.987 cal/(gmol · °C)
   MW = molecular weight, g/gmol
   C_v = molar heat capacity at constant volume, cal/(gmol · °C)
   μ = absolute viscosity, g/(cm · s)

For methane,

$$c_p = 0.58\,\text{cal}/(\text{g} \cdot \text{°C})$$

$$c_p/c_v = 1.26$$

Therefore,

$$c_v = 0.58/1.26 = 0.46\,\text{cal}/(\text{g} \cdot \text{°C})$$

$$C_v = (0.46)(16) = 7.35\,\text{cal}/(\text{gmol} \cdot \text{°C})\,(\text{mole basis})$$

The viscosity of methane is approximately $2.26 \times 10^{-4}\text{g}/(\text{cm} \cdot \text{s})$ at these conditions. Therefore,

$$k = \left(\frac{15}{4}\right)\left(\frac{1.987}{16}\right)\left[\left(\frac{4}{15}\right)\left(\frac{7.35}{1.987}\right) + \frac{3}{5}\right](2.26 \times 10^{-4})$$

$$= 1.67 \times 10^{-4}\,\text{cal}/(\text{s} \cdot \text{cm} \cdot \text{°C})$$

$$= 0.040\,\text{Btu}/(\text{h} \cdot \text{ft} \cdot \text{°F})$$

HTR.3   OVEN INSULATION

One wall of an oven has a 3-in. insulation cover. The temperature on the inside of the wall is at 400°F; the temperature on the outside is at 25°C. What is the heat flux (heat flowrate per unit area) across the wall if the insulation is made of glass wool $[k = 0.022\,\text{Btu}/(\text{h} \cdot \text{ft} \cdot \text{°F})]$?

Solution

The thermal resistance associated with conduction is defined as

$$R = L/kA$$

where R = thermal resistance
$\quad k$ = thermal conductivity
$\quad A$ = area across which heat is conducted
$\quad L$ = length across which heat is conducted

The rate of heat transfer, $\dot{Q}$, is then

$$\dot{Q} = \Delta T / R$$

Thus,

$$\dot{Q} = kA \, \Delta T / L$$

Since 25°C is approximately 77°F and L is 0.25 ft,

$$\frac{\dot{Q}}{A} = \frac{0.022 \, (400 - 77)}{0.25}$$

$$= 28.4 \, \text{Btu}/(\text{h} \cdot \text{ft}^2)$$

HTR.4   OVERALL HEAT TRANSFER COEFFICIENT

A rectangular plane window glass panel is mounted on a house. The glass is $\frac{1}{8}$ in. thick has a surface area of $1.0 \, \text{m}^2$, and its thermal conductivity, k_2, is $1.4 \, \text{W}/(\text{m} \cdot {}^{\circ}\text{C})$. The inside house temperature, T_1, and the outside air temperature, T_4, are 25 and $-14{}^{\circ}\text{C}$, respectively. The heat transfer coefficient inside the room, h_1, is $11 \, \text{W}/(\text{m}^2 \cdot \text{K})$ and the heat transfer coefficient from the window to the surrounding cold air, h_3, is $9.0 \, \text{W}/(\text{m}^2 \cdot \text{K})$. Calculate the overall heat transfer coefficient in $\text{W}/(\text{m}^2 \cdot \text{K})$.

Solution

In most steady-state heat transfer problems, more than one heat transfer mode may be involved. The various thermal resistances due to thermal convection or conduction may be combined and described by an overall heat transfer coefficient, U. Using U, the heat transfer rate, $\dot{Q}$, can be calculated from the terminal and/or system temperatures. The analysis of this problem is simplified when the concepts of thermal circuit and thermal resistance are employed.

Consider heat transfer from one fluid to another by a three-step steady-state process: convection from a warmer fluid at T_1 to a solid (with a convection heat transfer coefficient, h_1), conduction through the wall (with a thickness, L_2, and a thermal conductivity, k_2), then convection to a colder fluid at T_4 (with a convection heat transfer coefficient, h_2). The heat transfer area is A. The thermal circuit consists of four nodes (T_1, T_2, T_3, and T_4) and three resistances (R_1, R_2, and R_3).

In Figure 21,

$$\dot{Q} = \text{rate of heat transfer}$$

$$R_1 = \text{inside convection resistance} = 1/h_1 A$$

$$R_2 = \text{conduction resistance through the wall} = L_2/k_2 A$$

$$R_3 = \text{outside convection resistance} = 1/h_3 A$$

$$T_1 = \text{temperature of the inside fluid}$$

$$T_2 = \text{temperature at the inside fluid–wall interface}$$

$$T_3 = \text{temperature at the wall–outside fluid interface}$$

$$T_4 = \text{temperature of the outside fluid}$$

$$h_1 = \text{heat transfer coefficient of inside fluid}$$

$$h_2 = \text{heat transfer coefficient of outside fluid}$$

Since all thermal resistances are in series, the total resistance is calculated as

$$R_{\text{tot}} = R_1 + R_2 + R_3$$

The heat transfer rate is calculated using the terminal temperatures, i.e.,

$$\dot{Q} = (T_1 - T_4)/R_{\text{tot}}$$

The overall heat transfer coefficient, U, based on the temperature difference between the two fluids, $T_1 - T_4$, is defined by

$$\dot{Q} = UA(T_1 - T_4)$$

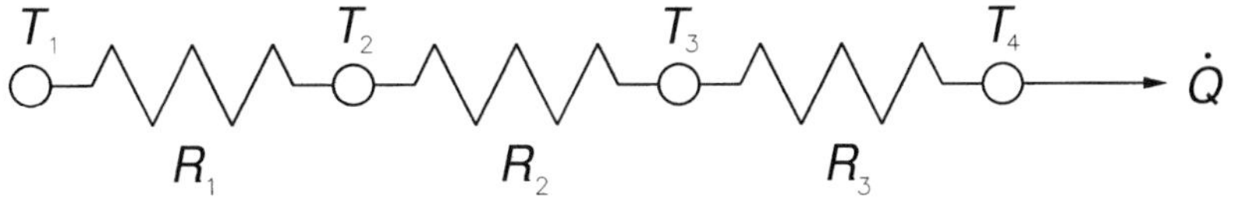

Figure 21. Thermal circuit for Problem HTR-4.

These equations are combined to give

$$R_{\text{tot}} = R_1 + R_2 + R_3 = 1/UA$$

or

$$\frac{1}{UA} = \frac{1}{h_1 A} + \frac{L_2}{k_2 A} + \frac{1}{h_3 A}$$

The overall heat transfer coefficient, U, has the same units as h.
The internal convection resistance is

$$R_1 = \frac{1}{(11)(1)}$$
$$= 0.0909°C/W$$

The conduction resistance through the glass panel is

$$L_2 = \frac{1}{8}\text{in.} = 0.00318\,\text{m}$$
$$R_2 = \frac{0.00318}{(1.4)(1)}$$
$$= 0.00227°C/W$$

The outside convection resistance is

$$R_3 = \frac{1}{(9)(1)}$$
$$= 0.111°C/W$$

The total thermal resistance is the sum of the above three resistances.

$$R_{\text{tot}} = 0.0909 + 0.00227 + 0.111$$
$$= 0.204°C/W$$

The overall heat transfer coefficient may now be calculated:

$$U = \frac{1}{(1)(0.204)}$$
$$= 4.9\,\text{W}/(\text{m}^2 \cdot °\text{C})$$

HTR.5   INCINERATOR WALL TEMPERATURE PROFILE

A flat incinerator wall with a surface area of 480 ft² consists of 6 in. of firebrick with a thermal conductivity of 0.61 Btu/(h · ft · °F) and an 8-in. outer layer of rock wool insulation with a thermal conductivity of 0.023 Btu/(h · ft · °F). If the temperature of the insulation of the inside face of the firebrick and the outside surface of the rock wool insulation are 1900 and 140°F, respectively, calculate the following:

1. The heat loss through the wall in Btu/h.
2. The temperature of the interface between the firebrick and the rock wool.

Solution

From the previous problem,

$$\dot{Q} = \frac{\Delta T}{\Sigma R}$$

The individual resistances are:

$$R_{\text{firebrick}} = \frac{L_f}{k_f A}$$

$$= \frac{0.5}{(0.61)(480)}$$

$$= 0.0017 \, \text{h} \cdot {}^\circ\text{F/Btu}$$

$$R_{\text{rock wool}} = \frac{L_{\text{rw}}}{k_{\text{rw}} A}$$

$$= \frac{0.67}{(0.023)(480)}$$

$$= 0.0604 \, \text{h} \cdot {}^\circ\text{F/Btu}$$

Thus,

$$\Sigma R = 0.0017 + 0.0604 = 0.0621 \, \text{h} \cdot {}^\circ\text{F/Btu}$$

The heat loss through the wall is then

$$\dot{Q} = \frac{1900 - 140}{0.0621}$$

$$= 28,341 \, \text{Btu/h}$$

The temperature at any interface can be calculated from the equation

$$\Delta T = \dot{Q} R_F$$

Applying this equation between the firebrick and rock wool gives

$$1900 - T_1 = (28,341)(0.0017)$$
$$T_1 = 1852°F$$

HTR.6   WALL RESISTANCE OPTION

Heat is flowing from steam on one side of a vertical steel sheet 0.375 in. thick to air on the other side. The steam heat-transfer coefficient is 1700 Btu/(h · ft² · °F) and that of the air is 2.0 Btu/(h · ft² · °F). The total temperature difference is 120°F. How would the rate of heat transfer be affected if the wall was copper rather than steel? By increasing the steam coefficient to 250? By increasing the air coefficient to 12.0? Note that the thermal conductivities, k, for steel and copper are 26 and 218 Btu/ (h · ft · °F), respectively.

Solution

The describing equation is once again

$$q = \frac{\sum \Delta T}{\sum \Delta R}$$

For the existing application, assume a basis of 1.0 ft², Therefore,

$$R_{steam} = \frac{1}{hA} = \frac{1}{(1700)(1)} = \frac{1}{1700}$$

$$R_{air} = \frac{1}{hA} = \frac{1}{(2)(1)} = \frac{1}{2}$$

$$R_{steel} = \frac{L}{kA} = \frac{0.375/12}{(26)(1)}; \quad k_{steel} = 26 \text{ Btu/h} \cdot \text{ft} \cdot °F$$

$$\therefore \quad \sum R = \frac{1}{1700} + \frac{1}{2} + \frac{0.375/12}{(26)}$$

$$= 0.502 \text{ h} \cdot °F/\text{Btu}$$

If copper [$k = 218$ Btu/(h · ft · °F)] is employed,

$$\Sigma R = \frac{1}{1700} + \frac{1}{2} + \frac{0.375/12}{(218)}$$

$$= 0.50 \, h \cdot °F/Btu$$

Thus, the rate of heat transfer is essentially unaffected. If h_{steam} is 2500 Btu/(h · ft^2 · °F),

$$\Sigma R = \frac{1}{2500} + \frac{1}{2} + \frac{0.375/12}{(26)}$$

$$= 0.50 \, h \cdot °F/Btu$$

The rate is again unaffected.
  However, if h_{air} is 12 Btu/(h · ft^2 · °F),

$$\Sigma R = \frac{1}{1700} + \frac{1}{12} + \frac{0.375/12}{(26)}$$

$$= 0.0852 \, h \cdot °F/Btu$$

The rate is affected for this case. Thus, it can be concluded that the air is the *controlling* resistance.

HTR.7   ROTARY KILN

A rotary kiln incinerator is 30 ft long, has a 12-ft ID and is constructed of $\frac{3}{4}$-in. carbon steel. The inside of the steel shell is protected by 10 in. of firebrick [$k = 0.608$ Btu/(h · ft ·° F)] and 5 in. of Sil-o-cel insulation [$k = 0.035$ Btu/(h · ft ·° F)] covers the outside. The ambient air temperature is 85°F and the average inside temperature is 1800°F. The present heat loss through the furnace wall is 6% of the heat generated by combustion of the waste. Calculate the thickness of Sil-o-cel insulation that must be added to cut the losses to 3%.

Solution

A diagram of this system is presented in Figure 22. Although the resistance of the steel can be neglected, the other two need to be considered. For cylindrical systems,

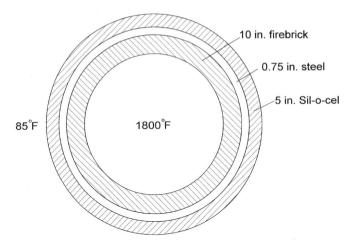

Figure 22. Diagram for Problem HTR.7.

the effect of the radius of curvature must be included in the resistance equations. These take the form presented below.

$$R_{\text{firebrick}} = \frac{\ln(r_{\text{fo}}/r_{\text{fi}})}{2\pi L k_{\text{f}}}$$

$$= \frac{\ln(6.000/5.167)}{2\pi(30)(0.608)}$$

$$= 1.304 \times 10^{-3}\, \text{h} \cdot ^\circ \text{F/Btu}$$

$$R_{\text{Sil-o-cel}} = \frac{\ln(r_{\text{so}}/r_{\text{si}})}{2\pi L k_{\text{s}}}$$

$$= \frac{\ln(6.479/6.063)}{2\pi(30)(0.035)}$$

$$= 10.059 \times 10^{-3}\, \text{h} \cdot ^\circ \text{F/Btu}$$

Thus,

$$\sum R = 11.363 \times 10^{-3}\, \text{h} \cdot ^\circ \text{F/Btu}$$

To cut the heat loss in half, R must be doubled. The additional Sil-o-cel resistance is therefore $11.363 \times 10^{-3}\, \text{h} \cdot ^\circ \text{F/Btu}$. The new outside radius, r_{o}, is calculated from

$$R_{\text{added Sil-o-cel}} = 11.363 \times 10^{-3} = \frac{\ln(r_{\text{o}}/6.479)}{2\pi(30)(0.035)}$$

$$r_{\text{o}} = 6.983\, \text{ft}$$

The extra thickness is $6.983 - 6.479 = 0.504\, \text{ft} = 6.05\, \text{in.}$

HTR.8 LOG-MEAN TEMPERATURE DIFFERENCE

A heavy hydrocarbon oil with heat capacity, $c_p = 0.55\,\mathrm{Btu}/(\mathrm{lb}\cdot{}^\circ\mathrm{F})$, is being cooled from $T_1 = 210^\circ\mathrm{F}$ to $T_2 = 170^\circ\mathrm{F}$. The oil flows inside a tube at the rate of 8000 lb/h and the inside tube surface temperature is maintained at $60^\circ\mathrm{F}$. The overall heat transfer coefficient, U, is $63\,\mathrm{Btu}/(\mathrm{h}\cdot\mathrm{ft}^2\cdot{}^\circ\mathrm{F})$. Calculate the log-mean temperature difference (LMTD) in degrees Fahrenheit.

Solution

When heat is exchanged between a surface and a fluid, or between two fluids flowing through a heat exchanger, the local temperature driving force, ΔT, varies with the flow path. For this condition, the concept of the log-mean temperature difference (LMTD) may be applied when the overall heat-transfer coefficient, U, is given. In this case,

$$\dot{Q} = UA\,\Delta T_{\mathrm{LM}}$$

Heat-transfer calculations using this equation and ΔT_{LM} are convenient when terminal temperatures are known. If the temperature of the fluid leaving the tube is not known, the procedure requires a trial-and-error calculation.

The heat transferred by the heavy oil is

$$\dot{Q} = (8000)(0.55)(170 - 210)$$
$$= -176{,}000\,\mathrm{Btu/h}$$

The thermal driving forces, or the approaches, at the pipe entrance and exit are given by

$$\Delta T_1 = 210 - 60 = 150^\circ\mathrm{F}$$
$$\Delta T_2 = 170 - 60 = 110^\circ\mathrm{F}$$

The LMTD by definition is then

$$\Delta T_{\mathrm{LM}} = \frac{\Delta T_1 - \Delta T_2}{\ln(\Delta T_1/\Delta T_2)} = \frac{150 - 110}{\ln(150/110)}$$
$$= 129^\circ\mathrm{F}$$

If something other than a single, single-pass (1–1) exchanger is enployed, the describing equation becomes

$$\dot{Q} = UAF\,\Delta T_{\mathrm{LM}}$$

where F is a function of the exchanger geometry and the fluid temperature. Many F curves (or F data) are provided in the literature.

HTR.9   ADIABATIC QUENCH TOWER

The composition of a combustion gas stream is given in the following table:

	MW	lb/h	lbmol/h	Mole Fraction
CO_2	44	12,023	273.3	0.0536
H_2O	18	9,092	505.1	0.0990
N_2	28	106,783	3,813.7	0.7476
HCl	36.5	193	5.3	0.0010
O_2	32	16,115	503.6	0.0988
Total		144,206	5,101.0	1.0

Also for the combustion gas, $c_p = 0.3\,\text{Btu}/(\text{lb}\cdot{}^\circ\text{F})$.

Calculate the mass of $0.91\,\text{Btu}/(\text{lb}\cdot{}^\circ\text{F})$ heat capacity solids required to cool the gas from 2050 to 180°F for one hour of operation. The initial temperature of the solids is 60°F, and an approach temperature of 20°F may be assumed.

Solution

Select as a basis one hour of operation. Applying an enthalpy balance,

$$\Delta H_{\text{flue}} = -\Delta H_{\text{solids}}$$

$$= m_{\text{flue}} c_p (T_2 - T_1)$$

$$= (144,206)(0.3)\,(180 - 2050)$$

$$= -8.09 \times 10^7\,\text{Btu}$$

$$\Delta H_{\text{solids}} = m_{\text{solids}} c_p (T_2 - T_1)$$

$$m_{\text{solids}} = \frac{\Delta H_{\text{solids}}}{c_p (T_2 - T_1)}$$

For a 20° approach,

$$T_2 = 180 - 20 = 160^\circ\text{F}$$

which is the final temperature of the solids. Thus,

$$m_{\text{solids}} = \frac{8.09 \times 10^7}{(0.91)\,(160 - 60)}$$

$$= 8.89 \times 10^5\,\text{lb solids needed}$$

HTR.10   DOUBLE-PIPE HEAT EXCHANGER

A heavy hydrocarbon oil with heat capacity, $c_p = 0.55\,\mathrm{Btu/(lb \cdot {}^\circ F)}$, is being cooled in a countercurrent double-pipe heat exchanger from $T_1 = 210^\circ F$ to $T_2 = 170^\circ F$. The oil flows inside a tube at the rate of 8000 lb/h and the inside tube surface temperature is maintained at $60^\circ F$. The overall heat-transfer coefficient, $U = 63\,\mathrm{Btu/(h \cdot ft^2 \cdot {}^\circ F)}$. Calculate the required inside heat-transfer area, A, in square feet.

Solution

In a countercurrent flow exchanger, the two fluids exchanging heat flow in opposite directions. The temperature approach at the tube entrance end, ΔT_1 or $(T_1 - T_4)$, and at the annulus entrance end, ΔT_2 or $(T_2 - T_3)$ are usually roughly the same. The thermal driving force is normally relatively constant over the length of the exchanger.

In designing a double-pipe heat exchanger, mass balance, heat balance, and heat-transfer equations are used. The steady-state heat balance equation is

$$\dot{Q} = -\dot{m}_1 \bar{c}_{p,1}(T_2 - T_1) = \dot{m}_2 \bar{c}_{p,2}(T_4 - T_3)$$

Besides steady state, this equation assumes no heat loss, no viscous dissipation, and no heat generation. The rate equation used to design an exchanger once again employs the log-mean temperature difference (LMTD),

$$\dot{Q} = UA\,\Delta T_{LM}$$

where $\dot{Q}$ is the heat load (transfer rate), U is the overall heat transfer coefficient, A is the heat-transfer area, and ΔT_{LM} is the log-mean temperature difference (LMTD). If the temperature difference driving forces are ΔT_1 and ΔT_2 at the entrance and exit of the heat exchanger, then

$$\begin{aligned}
\Delta T_{LM} &= \frac{(\Delta T_1 - \Delta T_2)}{\ln(\Delta T_1/\Delta T_2)} \\
&= \frac{(-150) - (-110)}{\ln(-150/-110)} \\
&= -129^\circ F
\end{aligned}$$

The describing equation for the area may be written as

$$\dot{Q} = UA\,\Delta T_{LM}$$

For this problem,

$$\dot{Q} = (8000)(0.55)(170 - 210)$$
$$= -176,000 \, \text{Btu/h}$$

$$A = \frac{\dot{Q}}{UA \, \Delta T_{\text{LM}}}$$
$$= \frac{-176,000}{(63)(-129)}$$
$$= 21.7 \, \text{ft}^2$$

The value of U was provided in the calculation. Typical values of U for new, clean exchangers are available in the literature and are here assigned the symbol U_{clean}. After several months of use, the tubes are fouled by scale or dirt. This scale causes a decrease in the overall heat-transfer coefficient, from U_{clean} to U_{dirty}. The relation between these two U's is given by

$$\frac{1}{U_{\text{dirty}}} = \frac{1}{U_{\text{clean}}} + R_{\text{f}}$$

where R_{f} is the fouling or dirt factor in typical units of $\text{m}^2 \cdot \text{K/W}$ or $\text{ft}^2 \cdot \text{h} \cdot {}^\circ\text{F/Btu}$.

HTR.11 SHELL-AND-TUBE HEAT EXCHANGER

A unique shell-and-tube heat exchanger has one pass on the shell side and two passes on the tube side (i.e., a 1–2 shell-and-tube heat exchanger). It is being used for oil cooling. The oil flows in the tube side. It enters at 110°C and leaves at 75°C. The shell-side fluid is water at a flowrate of 1.133 kg/s, entering at 35°C and leaving at 75°C. The heat capacity of the water is $4180 \, \text{J/(kg} \cdot \text{K)}$. The overall heat-transfer coefficient for the heat exchanger is $350 \, \text{W/(m}^2 \cdot \text{K)}$. The geometry factor F is 1.15. Calculate the heat-transfer area requirement for this unit in square meters.

Solution

For multipass and cross-flow exchangers, the log-mean temperature difference (LMTD) method is still applicable, i.e., the heat transfer rate is

$$\dot{Q} = UA \, \Delta T_{\text{LM}}$$

However, the ΔT_{LM} in the above equation must be corrected by a geometry factor, F. Therefore,

$$\Delta T_{LM} = F \, \Delta T_{LM,CF}$$

where $\Delta T_{LM,CF}$ = log-mean temperature difference from terminal ΔT's based on countercurrent operation

F = geometry factor, or correction factor, applied to a countercurrent flow with the same hot and cold fluid temperatures

The heat load is

$$\dot{Q} = \dot{m}\bar{c}_p \, (T_{out} - T_{in})$$
$$= (1.133)(4180)(75 - 35)$$
$$= 189{,}400 \, \text{W}$$

The countercurrent log-mean temperature difference is

$$\Delta T_1 = 110 - 75 = 35°C$$
$$\Delta T_2 = 75 - 35 = 40°C$$
$$\Delta T_{LM} = \frac{35 - 40}{\ln(35/40)}$$
$$= 37.4°C$$
$$= 310.6 \, \text{K}$$

The corrected log-mean temperature difference is calculated employing the correction factor, F:

$$\Delta T_{LM} = F \, \Delta T_{LM,CF}$$
$$= (1.15)(310.6)$$
$$= 357.2 \, \text{K}$$

The required heat-transfer area is then

$$A = \frac{\dot{Q}}{U \, \Delta T_{LM}}$$
$$= \frac{189{,}400}{(350)(357.2)}$$
$$= 1.5 \, \text{m}^2$$

The LMTD method is adequate when the terminal temperatures are known. If the heat-transfer area is given and the exit temperatures are unknown, the problem often requires solution by trial and error.

Shell-and-tube heat exchangers often provide a large heat-transfer area both economically and practically. The tubes are placed in a bundle and the ends of the tubes are mounted in tube sheets. The tube bundle is enclosed in a cylindrical shell through which the second fluid flows. The simplest shell-and-tube heat exchanger has a single pass through the shell and a single pass through the tubes. This is termed a *1–1 shell-and-tube* heat exchanger. Fluid flow through tubes at low velocity results in low heat-transfer coefficients and low pressure drops. To increase the heat-transfer rates, multipass operation may be used. Baffles are used to divert the tube fluid within the distribution header. An exchanger with one pass on the shell side and four tube passes is termed a *1–4 shell-and-tube* heat exchanger. It is also possible to increase the number of passes on the shell side by using dividers. A *2–8 shell-and-tube* heat exchanger has two passes on the shell side and eight passes on the tube side.

HTR.12   RADIATION HEAT-TRANSFER COEFFICIENT

An uninsulated steam pipe made of anodized aluminum has a diameter d of 0.06 m and a length L of 100m and is at a surface temperature $T_1 = 127°C$. The surface emissivity, ε, of anodized Al is 0.76. Calculate the emissive power of the pipe surface in watts per square meters.

Solution

Each surface emits radiation based on its temperature. The ideal radiator is called a *blackbody*. The rate of energy emission per unit area, q'', is also termed *blackbody emissive power*, E_b, and is given by the Stefan–Boltzmann law,

$$q'' = E_b = \sigma T^4$$

where σ = Stefan–Boltzmann constant
$= 5.669 \times 10^{-8} W/(m^2 \cdot K^4)$
$= 0.1713 \times 10^{-8} Btu/(ft^2 \cdot h \cdot °R^4)$
T = absolute temperature, K or °R
E_b = blackbody emissive power, W/m^2 or $Btu/(ft^2 \cdot h)$

Real surfaces emit less radiation than a blackbody surface. One may define

ε = surface emissivity

= emissive power of real surface/emissive power of blackbody

The emissive power, E, of a real surface is

$$E = q'' = \varepsilon\sigma T^4 = \varepsilon E_b \qquad 0 < \varepsilon < 1$$

Consider a small real surface of area A, surface emissivity ε, at an absolute temperature T_1, surrounded by a large enclosure at an absolute temperature T_2. The rate of heat transfer, $\dot{Q}$ can be written as

$$\dot{Q} = \varepsilon\sigma A(T_1{}^4 - T_2{}^4)$$
$$= \varepsilon A(E_{b1} - E_{b2})$$

where E_{b1} = blackbody emissive power of the surface = $\sigma T_1{}^4$
E_{b2} = blackbody emissive power of the surrounding = $\sigma T_2{}^4$

It is often convenient to express the radiation heat transfer in a Newton's law of cooling form. In this case, a radiation heat-transfer coefficient, h_r, is defined by equating the Newton's law of cooling rate to the radiation heat-transfer rate, i.e.,

$$\dot{Q} = h_r A(T_1 - T_2) = \varepsilon\sigma A(T_1{}^4 - T_2{}^4)$$
$$h_r = \varepsilon\sigma(T_1{}^4 - T_2{}^4)/(T_1 - T_2)$$

First convert the temperature to Kelvin:

$$T_1 = 127°C = 400\,K$$

Assuming it is a blackbody, the emissive power of the pipe surface is

$$E_b = (5.669 \times 10^{-8})(400^4)$$
$$= 1451\,W/m^2$$

The actual emissive power of the pipe surface is then

$$E = (0.76)(1451)$$
$$= 1103\,W/m^2$$

If the temperature difference, $|T_1 - T_2|$ in an application is $< 200°R$ (or $120\,K$), the radiative heat-transfer coefficient, h_r, is approximated as

$$h_r \approx 4\varepsilon\sigma T_{av}$$

where T_{av} is the average temperature of the two surfaces = $(T_1 + T_2)/2$.

8 Mass Transfer Operations (MTO)

MTO.1 *Txy* DIAGRAM FOR ETHANOL–WATER

Using the vapor–liquid equilibrium (VLE) data for the ethanol–water system provided in the following table, generate a *Txy* diagram. What are the liquid and vapor mole fractions of water if the liquid mixture is 30 mol % ethanol?

Vapor–Liquid Equilibrium at 1 atm (mole fractions)		
T (°C)	x_{ETOH}	y_{ETOH}
212	0.000	0.000
192	0.072	0.390
186	0.124	0.470
181	0.238	0.545
180	0.261	0.557
177	0.397	0.612
176	0.520	0.661
174	0.676	0.738
173	0.750	0.812
172	0.862	0.925
171	1.000	1

Solution

The *Txy* diagram is plotted in Figure 23. The top curve (y vs. T) represents saturated vapor and the bottom curve (x vs. T) is saturated liquid.

From the diagram, when $x = 0.3$, $y = 0.57$ (see the tie line at about 179°C); the ethanol vapor composition is 57%.

The water vapor mole fraction is therefore

$$y \text{ (water)} = 1 - 0.57 = 0.43$$

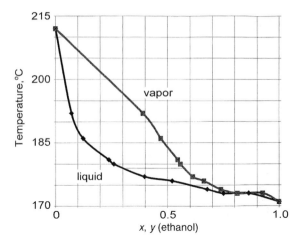

Figure 23. *Txy* diagram for the ethanol–water system.

MTO.2   HENRY'S LAW

Given Henry's law constant and the partial pressure of H_2S, determine the maximum mole fraction of H_2S that can be dissolved in solution at the given conditions. Data are provided below.

Partial pressure of H_2S = 0.01 atm
Total pressure = 1.0 atm
Temperature = 60°F
Henry's law constant, H_{H_2S} = 483 atm/mol fraction (1 atm, 60°F)

Solution

An important equilibrium phase relationship is that between liquid and vapor. Raoult's and Henry's laws theoretically describe liquid–vapor behavior and under certain conditions are applicable in practice. Raoult's law is sometimes useful for mixtures of components of similar structure. It states that the partial pressure of any component in the vapor is equal to the product of the vapor pressure of the pure component and the mole fraction of that component in the liquid, that is,

$$p_i = p'x_i$$

where p_i = partial pressure of component i in the vapor
p' = vapor pressure of pure i at the same temperature
x_i = mole fraction of component i in the liquid

This expression may be applied to all components. If the gas phase is ideal,

$$p_i = y_i P$$

and the first equation can be written as follows:

$$y_i = (p'/P) \, x_i$$

where y_i = mole fraction component i in the vapor
P = total system pressure

Thus, the mole fraction of water vapor in a gas that is saturated, that is, in equilibrium contact with pure water $(x = 1.0)$, is given simply by the ratio of the vapor pressure of water at the system temperature divided by the system pressure. These equations find application in distillation, absorption, and stripping calculations.

Unfortunately, relatively few mixtures follow Raoult's law. Henry's law is a more empirical relationship used for representing data on many systems. Here,

$$p_i = H_i x_i$$

where H_i is Henry's law constant for component i (in units of pressure).
If the gas behaves ideally, the above equation may be written as

$$y_i = m_i x_i$$

where m_i is a constant (dimensionless).

This is a more convenient form of Henry's law. The constant m_i (or H_i) has been determined experimentally for a large number of compounds and is usually valid at low concentrations. In word equation form, Henry's law states that the partial pressure of a solute in equilibrium in a solution is proportional to its mole fraction. The law is exact in the limit, as the concentration approaches zero.

The most appropriate form of Henry's law for this problem is the first, i.e.

$$p_i = H_i x_i$$

The maximum mole fraction of H_2S that can be dissolved in solution can now be calculated:

$$x_i = p_i / H_{H_2S}$$
$$= 0.01/483$$
$$= 0.0000207$$
$$= 20.7 \, \text{ppm}$$

Henry's law may be assumed to apply for most dilute solutions. This law finds widespread use in absorber calculations since the concentration of the solute in some process gas streams is often dilute. This greatly simplifies the study and design of absorbers. One should note, however, that Henry's law constant is a strong function of temperature.

MTO.3 PACKED COLUMN DESIGN

A packed column is designed to absorb ammonia from a gas stream. Given the operating conditions and type of packing below, calculate the height of packing and column diameter. The unit operates at 60% of the flooding gas mass velocity, the actual liquid flowrate is 25% more than the minimum, and 90% of the ammonia is to be collected.

Gas mass flowrate $= 5000\,lb/h$

NH_3 concentration in inlet gas stream $= 2.0\,mol\,\%$

Scrubbing liquid $=$ pure water

Packing type $=$ 1-in. Raschig rings; packing factor, $F = 160$

H_{OG} (height of a gas transfer unit) of the column $= 2.5\,ft$

Henry's law constant, $m = 1.2$

Density of gas (air) $= 0.075\,lb/ft^3$

Density of water $= 62.4\,lb/ft^3$

Viscosity of water $= 1.8\,cP$

The ordinate and abscissa of the graph in Figure 24 are dimensionless numbers where:

$G =$ mass flux (mass flowrate per unit cross-sectional area) of gas stream

$L =$ mass flux of liquid stream

$F =$ packing factor

$\psi =$ ratio of specific gravity of scrubbing liquid to that of water

$\mu_L =$ viscosity of liquid phase

$\rho_L =$ density of liquid phase

$\rho =$ density of gas phase

$g_c =$ Newton's law proportionality factor

Solution

To calculate the number of overall gas transfer units, N_{OG}, first calculate the equilibrium outlet concentration, x_1, at $y_1 = 0.02$.

$$x_1^* = y_1/m$$
$$= (0.02)/(1.20)$$
$$= 0.0167$$

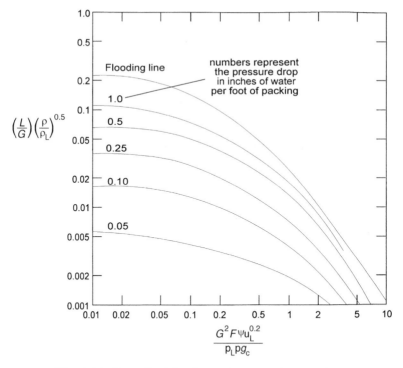

Figure 24. Generalized flooding and pressure drop correlation.

where x_1^* is the liquid mole fraction in equilibrium with the vapor of mole fraction, y_1.

Determine y_2 for 90% removal:

$$y_2 = \frac{(0.1)\,y_1}{(1-y_1)+(0.1)\,y_1}$$

$$= \frac{(0.1)(0.02)}{(1-0.02)+(0.1)(0.02)}$$

$$= 0.00204$$

The *minimum* ratio of molar liquid flowrate to molar gas flowrate, $(L_m/G_m)_{min}$, is determined by a material balance:

$$(L_m/G_m)_{min} = (y_1 - y_2)/(x_1^* - x_2)$$

$$= (0.02 - 0.00204)/(0.0167 - 0)$$

$$= 1.08$$

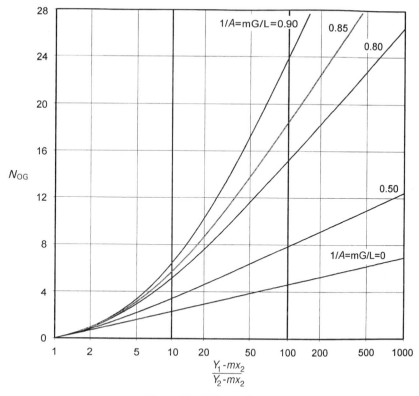

Figure 25. Colburn chart.

The *actual* ratio of molar liquid flowrate to molar gas flowrate, (L_m/G_m), is

$$L_m/G_m = (1.25)(L_m/G_m)_{min}$$
$$= (1.25)(1.08)$$
$$= 1.35$$

In addition,

$$(mG_m)/L_m = (1.2)/(1.35) = 0.889$$

The absorption factor, A, is defined as

$$A = L_m/(mG_m) = 1/0.889 = 1.125$$

The value of $(y_1 - mx_2)/(y_2 - mx_2)$ is

$$\frac{y_1 - mx_2}{y_2 - mx_2} = \frac{0.02 - (1.2)(0)}{0.00204 - (1.2)(0)} = 9.80$$

The term N_{OG} is calculated from Colburn's equation (or read from Colburn's chart, Figure 25),

$$N_{OG} = \frac{\ln([(y_1 - mx_2)/(y_2 - mx_2)][1 - (1/A)] + (1/A))}{1 - (1/A)}$$

$$= \frac{\ln[(9.80)\,[1 - (1/1.125)] + (1/1.125)]}{1 - (1/1.125)}$$

$$= 6.14$$

The height of packing, Z, is then

$$Z = N_{OG}H_{OG}$$

$$= (6.14)(2.5)$$

$$= 15.4 \text{ ft}$$

To determine the diameter of the packed column, the abscissa of Figure 24 is first calculated:

$$\left(\frac{L}{G}\right)\left(\frac{\rho}{\rho_L}\right)^{0.5} = \left(\frac{L_m}{G_m}\right)\left(\frac{18}{29}\right)\left(\frac{\rho}{\rho_L}\right)^{0.5} = (1.35)\left(\frac{18}{29}\right)\left(\frac{0.075}{62.4}\right)^{0.5}$$

$$= 0.0291$$

Note that the molecular weights of water and air are 18 and 29 lb/lbmol, respectively. The value of the ordinate at the flooding line is determined from Figure 24:

$$\frac{G^2 F \psi \mu_L^{0.2}}{\rho \rho_L g_c} = 0.21$$

The flooding gas mass velocity, G_f, in lb/ft$^2 \cdot$ s, is

$$G_f = \left(\frac{0.21 \rho_L \rho g_c}{F \psi \mu_L^{0.2}}\right)^{1/2} = \left(\frac{(0.21)(62.4)(0.075)(32.2)}{(160)(1)(1.8)^{0.2}}\right)^{1/2}$$

$$= 0.419 \text{ lb/(ft}^2 \cdot \text{s)}$$

The actual gas mass velocity, G_{act}, in $lb/(ft^2 \cdot s)$, is

$$G_{act} = (0.6)(0.419)$$
$$= 0.251 \, lb/(ft^2 \cdot s)$$
$$= 904 \, lb/(ft^2 \cdot h)$$

Calculate the diameter of the column, D, in feet:

$$D = [(4m)/(G_{act}\pi)]^{1/2}$$
$$= [(4)(5000)/(904)(\pi)]^{1/2}$$
$$= 2.65 \, ft$$

The column height (packing) and diameter are 15.4 and 2.65 ft, respectively.

MTO.4   SIZING OF A PACKED TOWER WITH NO DATA

Qualitatively outline how one can size (diameter, height) a packed tower to achieve a given degree of separation without any information on the physical and chemical properties of a gas to be cleaned.

To calculate the height, one needs both the height of a gas transfer unit, H_{OG}, and the number of gas transfer units, N_{OG}. Since equilibrium data are not available, assume m (slope of equilibrium curve) approaches zero. This is not an unreasonable assumption for most solvents that preferentially absorb (or react with) the solute. For this condition:

$$N_{OG} = \ln(y_1/y_2)$$

where y_1 and y_2 represent inlet and outlet concentrations, respectively.

Since it is reasonable to assume the scrubbing medium to be water or a solvent that effectively has the physical and chemical properties of water, H_{OG} can be assigned values usually encountered for water systems. These are given in the following table:

Packing Diameter (in.)	Plastic Packing H_{OG} (ft)	Ceramic Packing H_{OG} (ft)
1.0	1.0	2.0
1.5	1.25	2.5
2.0	1.5	3.0
3.0	2.25	4.5
3.5	2.75	5.5

For plastic packing, the liquid and gas flowrates are both typically in the range of 1500–2000 lb/(h · ft^2) of cross-sectional area. (*Note:* The "flow rates" are more properly termed "fluxes"; a *flux* is a flow rate per unit cross-sectional area with units of lb/h · ft^2 in this problem. However, it is common practice to use the term *rate*.) For ceramic packing, the range of flowrates is 500–1000 lb/(h · ft^2). For difficult-to-absorb gases, the gas flowrate is usually lower and the liquid flowrate higher. Superficial gas velocities (velocity of the gas if the column is empty) are in the 3–6 ft/s range. The height Z is then calculated from

$$Z = (H_{OG})(N_{OG})(SF)$$

where SF is a safety factor the value of which can range from 1.25 to 1.5. Pressure drops can vary from 0.15 to 0.40 in. H_2O/ft packing. Packing size increases with increasing tower diameter.

Note: This problem and design procedure were originally developed by Dr. Louis Theodore in 1985 and later published in 1988. These materials recently appeared in the 1992 Air and Waste Management Association text published by Van Nostrand Reinhold titled *Air Pollution Engineering Manual*. This was done without properly acknowledging the author, Dr. Louis Theodore, and without permission from the original publisher.

Solution

The reader is left the exercise of verifying the charts below. A sample calculation is presented after the first chart.

Packing Height, Z (ft)

Removal Efficiency (%)	Plastic Packing Size (in.)				
	1.0	1.5	2.0	3.0	3.5
63.2	1.0	1.25	1.5	2.25	2.75
77.7	1.5	1.9	2.25	3.4	4.1
86.5	2.0	2.5	3.0	4.5	5.5
90	2.3	3.0	3.45	5.25	6.25
95	3.0	3.75	4.5	6.75	8.2
98	3.9	4.9	5.9	8.8	10.75
99	4.6	5.75	6.9	10.4	12.7
99.5	5.3	6.6	8.0	11.9	14.6
99.9	6.9	8.6	10.4	15.5	19.0
99.99	9.2	11.5	13.8	20.7	25.3

Consider, as an example, a removal efficiency of 86.5% using 1.5-in. diameter plastic packing. Using a basis of 100 ppm for the inlet concentration, y_1, the outlet concentration, y_2, is obtained from

$$0.865 = \frac{100 - y_2}{100}$$

$$y_2 = 13.3 \text{ ppm}$$

The N_{OG} is then

$$N_{OG} = \ln(100/13.3)$$

$$= 2.02$$

From the chart in the problem statement, for 1.5-in. diameter plastic packing, H_{OG} is 2.5 ft. The required packing height is

$$Z = (2.5)(2.02)$$

$$= 5.0 \text{ ft}$$

For ceramic packing, one obtains

Packing Height, Z (ft)

Removal Efficiency (%)	Ceramic Packing Size (in.)				
	1.0	1.5	2	3	3.5
63.2	2.0	2.5	3.0	4.5	5.5
77.7	3.0	3.7	4.5	6.75	8.25
86.5	4.0	5.0	6.0	9.0	11.0
90	4.6	5.75	6.9	10.4	12.7
95	6.0	7.5	9.0	13.5	16.5
98	7.8	9.8	11.7	17.6	21.5
99	9.2	11.5	13.8	20.7	25.3
99.5	10.6	13.25	15.9	23.8	29.1
99.9	13.8	17.25	20.7	31.1	38.0
99.99	18.4	23.0	27.6	41.4	50.7

An equation for estimating the cross-sectional area of the tower, S, in terms of the gas volumetric flow rate, q, in acfs (actual cubic feet per second) is

$$S(\text{ft}^2) = q(\text{acfs})/4$$

An equation to estimate the tower packing pressure drop, ΔP, in terms of Z is

$$\Delta P(\text{in. } H_2O) = (0.2)Z \qquad Z = \text{ft}$$

The following packing size(s) is (are) recommended:

For $D \approx 3$ ft, use 1 in. packing
For $D < 3$ ft, use < 1 in. packing
For $D > 3$ ft, use > 1 in. packing

MTO.5   DESIGN OF A PACKED TOWER AIR STRIPPER

An atmospheric packed tower air stripper is used to clean contaminated groundwater with a concentration of 100 ppm trichloroethylene (TCE). The stripper was designed such that packing height is 13 ft, the diameter (D) is 5 ft, and the height of a transfer unit (HTU) is 3.25 ft. Assume Henry's law applies with a constant (H) of 324 atm at 68°F. Also, at these conditions the molar density of water is 3.47 lbmol/ft^3 and the air–water mole ratio (G/L) is related to the air–water volume ratio (G''/L'') through $G''/L'' = 130\,G/L$, where the units of G'' and L'' are ft^3/(s · ft^2).

1. If the stripping factor (R) used in the design is 5.0, what is the removal efficiency?
2. If the air blower produces a maximum air flow (q) of 106 acfm, what is the maximum water flow (in gpm) that can be treated by the stripper?

Solution

As described in Problem MTO.3, the height of a packed tower can be calculated by

$$Z = (N_{OG})(H_{OG}) = (\text{NTU})\,(\text{HTU})$$

In addition, the following equation has been developed for the calculation of the number of transfer units (NTU) for an air–water stripping system and is based on the stripping factor, R, and the inlet/outlet concentrations:

$$\text{NTU} = \frac{R}{R-1}\ln\left(\frac{(C_{in}/C_{out})(R-1)+1}{R}\right)$$

where C_{in} = inlet contaminant concentration, ppm
C_{out} = outlet contaminant concentration, ppm
R = stripping factor

For the purposes of this problem

$$R = \frac{H}{P}\frac{G}{L}$$

where H = Henry's law constant, atm
$\quad\ P$ = system pressure, atm
$\quad\ G$ = gas (air) loading rate (or flux) lbmol/(s · ft^2)
$\quad\ L$ = liquid loading rate (or flux) lbmol/(s · ft^2)

The number of transfer units (NTU) is first calculated:

$$Z = (N_{OG})(H_{OG}) = (\text{NTU})\ (\text{HTU})$$

$$\text{NTU} = Z/\text{HTU}$$

$$= 13/3.25$$

$$= 4$$

Rearranging the equation above,

$$C_{out} = \frac{C_{in}(R - 1)}{R\ \exp\ [(\text{NTU})(R - 1)/R] - 1}$$

$$= \frac{(100)(5.0 - 1)}{(5.0)\exp\ [(4.0)(5.0 - 1)/5.0] - 1}$$

$$= 3.3$$

The removal efficiency (RE) is then

$$\text{RE} = [(C_{in} - C_{out})/C_{out}]\ 100\%$$

$$= [(100 - 3.3)/100]\ 100\%$$

$$= 96.7\%$$

The air–water mole ratio, G/L, is

$$G/L = (P)(R)/H$$

$$= (1\ \text{atm})(5)/(324\ \text{atm})$$

$$= 0.0154$$

In addition,

$$G''/L'' = 130 \ G/L$$
$$= 130(0.0154)$$
$$= 2$$

Since the tower cross-sectional area, S, in ft^2, is

$$\text{Area} = S = \pi D^2/4$$
$$= \pi(5 \ \text{ft})^2/4$$
$$= 19.63 \ \text{ft}^2$$

the air volumetric loading rate, G'', in $\text{ft}^3/(\text{min} \cdot \text{ft}^2)$ is then

$$G'' = (106 \ \text{ft}^3/\text{min})/(19.63 \ \text{ft}^2)$$
$$= 5.4 \ \text{ft}^3/(\text{min} \cdot \text{ft}^2)$$

Also the water volumetric loading rate, L'', in $\text{ft}^3/(\text{min} \cdot \text{ft}^2)$ is

$$L'' = 2G''$$
$$= 2(5.4 \ \text{ft}^3/(\text{min} \cdot \text{ft}^2))$$
$$= 10.8 \ \text{ft}^3/(\text{min} \cdot \text{ft}^2)$$

This can be converted to gpm:

$$L = \frac{[10.8 \ \text{ft}^3/(\text{min} \cdot \text{ft}^2)](3.47 \ \text{lbmol/ft}^3)(18 \ \text{lb/lbmol})(19.63 \ \text{ft}^2)}{8.33 \ \text{lb/gal}}$$

$$= 1590 \ \text{gpm}$$

Once the volatile organic compounds (VOCs) have been recovered from a process wastewater or groundwater stream, the off-gas (air/VOC mixture) usually needs to be treated. This entails the installation of other equipment to handle the disposal of the VOCs. Some typical methods include flaring, carbon adsorption, and incineration. Flaring is potentially hazardous due to the oxygen that is allowed to enter the flare header. Carbon adsorption can be an efficient means of recovering the VOCs from the off-gas, but it can generate large quantities of solid hazardous waste. A utility boiler (incineration) is probably the best alternative since VOC destruction is typically more than 99%, and it is safer and more inexpensive. In addition, catalytic incinerators can achieve high VOC destruction efficiencies.

MTO.6 ADSORBER COLUMN HEIGHT

Determine the required height of adsorbent for an adsorber that treats a degreaser ventilation stream contaminated with trichloroethylene (TCE) given the following operating and design data:

Volumetric flowrate of contaminated air stream $= 10,000$ scfm
Standard conditions $= 60°F$, 1 atm
Operating temperature $= 70°F$
Operating pressure $= 20$ psia
Adsorbent $=$ activated carbon
Bulk density of activated carbon, $\rho_B = 36\,lb/ft^3$
Working capacity of activated carbon $= 28$ lb TCE/100 lb carbon
Inlet concentration of TCE $= 2000$ ppm (ppm by volume)
Molecular weight of TCE $= 131.5$

The adsorption column cycle is set at 4 h in the adsorption mode, 2 h in heating and desorbing, 1 h in cooling, and 1 h in standby. The adsorber recovers 99.5% by weight of TCE. A horizontal cylinder unit with an inside diameter of 6 ft and length of 15 ft is used.

Solution

The actual volumetric flowrate of the contaminated gas stream, q, in acfh (actual cubic feet per hour), is obtained using Charles' law.

$$q = 10,000 \left(\frac{70 + 460}{60 + 460} \right) \left(\frac{14.7}{20} \right)$$

$$= 7491\,\text{acfm}$$

$$= 4.5 \times 10^5\,\text{acfh}$$

The volumetric flowrate of TCE in acfh is

$$q_{TCE} = (y_{TCE})(q_a)$$

$$= (2000 \times 10^{-6})(4.5 \times 10^5)$$

$$= 900\,\text{acfh}$$

The mass flow rate of TCE, $\dot{m}$, in lb/h can be calculated:

$$\dot{m} = \dot{m}_{TCE} = q_{TCE}\frac{P(MW)}{RT}$$

$$= (900)\frac{(131.5)(20)}{(10.73)(70 + 460)}$$

$$= 416.2\,\text{lb/h}$$

The mass of TCE adsorbed during the 4-h period is

$$\text{TCE adsorbed} = (\dot{m})(0.995)(4)$$
$$= (416.2)(0.995)(4)$$
$$= 1656 \text{ lb}$$

The volume of activated carbon, V_{AC}, required is

$$V_{AC} = \frac{\text{TCE adsorbed}}{(28 \text{ lb TCE adsorbed}/100 \text{ lb carbon})(\text{bulk density})}$$
$$= \frac{1656}{(28/100)(36)}$$
$$= 164 \text{ ft}^3$$

The height of the adsorbent, Z, is

$$Z = \frac{\text{Activated carbon volume}}{\text{Cross-sectional area}}$$
$$= \frac{164}{(6)(15)}$$
$$= 1.83 \text{ ft}$$

MTO.7   ADSORBENT BREAKTHROUGH CALCULATION

The R&D group at a local adsorbent manufacturer has recently developed a new granulated activated column (GAC) adsorbent, JB26, for the removal of common water pollutants. Some data have been collected on the adsorption isotherm for a few solutes, but no extensive tests have been conducted as of yet. However, a major client is very interested in the new adsorbent and would like to know approximately how long one of its 65-ft^3 units could operate with JB26 before breakthrough would occur.

The following information was given to estimate how many days a 56,000-gal/day unit could run until breakthrough. From limited testing, the isotherm of interest is described by

$$Y_T = 0.002C^{3.11}$$

where Y_T = lb adsorbate/lb adsorbent
C = adsorbate concentration, mg/L

Currently, the client's unit operates 30 days until regeneration; the client would like to double that time if possible so as to limit down time and increase profits. The density of JB26 is 42 lb/ft³ and it will treat a stream with an inlet concentration (C_i) of 3.5 mg/L. In addition, the breakthrough concentration has been set at 0.5 mg/L.

In the design of the GAC it is important to be able to measure or predict the approximate time until an adsorbent will reach its maximum capacity for adsorption. The point at which this occurs is referred to as the *breakthrough adsorption capacity*, Y_B, and corresponds to an adsorbate (solute) concentration at breakthrough, C_B. Once breakthrough occurs, undesired solute will pass through the bed without being adsorbed, contaminating product quality. The breakthrough adsorption capacity typically ranges between 25 and 50% of the theoretical capacity, Y_T, which is determined from an adsorption isotherm, such as the Freundlich equation, evaluated at the initial solute concentration in solution, C_i. The time to breakthrough is then given by the following equation:

$$t_B = \frac{Y_B m_C}{8.34q[C_i - (C_B/2)]}$$

where t_B = time to breakthrough, days
$\quad Y_B$ = adsorption capacity at breakthrough, lb adsorbate/lb adsorbent
$\quad m_C$ = mass of carbon in column, lb
$\quad q$ = volumetric flow rate of solution, Mgal/day (millions of gallons per day)
$\quad C_i$ = adsorbate feed concentration, mg/L
$\quad C_B$ = adsorbate concentration at breakthrough, mg/L
Note: 8.34 lb = 1 Mgal · L/mg

Solution

The theoretical adsorption capacity, Y_T, is

$$Y_T = 0.002C^{3.11}$$

$$= 0.002(3.5)^{3.11}$$

$$= 0.09842 \text{ lb adsorbate/lb adsorbent}$$

Assume the actual value is 50% of the theoretical value (see comment above). Thus,

$$Y_B = (0.5)(0.09842) = 0.04921 \text{ lb adsorbate/lb adsorbent}$$

The mass of carbon in the unit is then

$$m_C = (65 \text{ ft}^3)(42\text{lb/ft}^3)$$

$$= 2730 \text{ lb carbon}$$

The breakthrough time can now be calculated:

$$t_B = \frac{(0.04921)(2730)}{(8.34)(0.56)[3.5 - (0.5/2)]}$$

$$= 88.5 \, \text{days}$$

MTO.8   FLASH DISTILLATION

An organic chemistry professor performed a flash distillation experiment in a laboratory for his students. A 10-kgmol/h liquid feed mixture consisted of 20 mol % ethanol and 80 mol % water at 1 atm. While the professor was able to determine that 30 mol % of the feed vaporized at 70°F, he lacked the necessary equipment to measure the liquid and vapor compositions of the more volatile component, ethanol. Determine the liquid and vapor compositions, as well as the percent ethanol recovery from the flash. Equilibrium data for the ethanol–water system at 1 atm are provided in MTO.1.

The separation of a volatile component from a liquid process stream can be achieved by way of *flash distillation*. It is referred to as a *flash* since the more volatile component of a gas mixture rapidly vaporizes upon entering a tank or drum that is at a lower pressure and/or a higher temperature than the incoming feed. If the feed is considered to be "cold," a pump and heater may be required to elevate the pressure and temperature, respectively, to achieve an effective flash (see Figure 26). As the feed enters the tank/drum, it may impinge against the wall or an internal deflector plate, which would promote liquid–vapor separation of the feed mixture.

As a result of the flash, the vapor phase will contain most of the more volatile component. Typically, flash distillation is not an efficient means of separation.

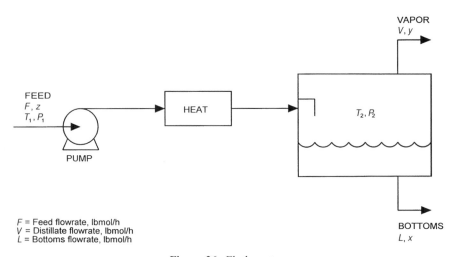

F = Feed flowrate, lbmol/h
V = Distillate flowrate, lbmol/h
L = Bottoms flowrate, lbmol/h

Figure 26. Flash system.

However, it can be a necessary and economical method of separating two or more components with different relative volatilities. A high relative volatility indicates the component more likely to vaporize from a mixture when the temperature is raised or the pressure is lowered. For example, the relative volatility (α) of component A in mixture AB is defined as:

$$\alpha_{AB} = K_A/K_B$$

where K = vapor-to-liquid mole fraction ratio at equilibrium for each component.

As with many process operations, an overall material balance can be written to describe the system illustrated in Figure 26 as follows:

$$F = L + V$$

Based on the above, a material balance can be written for component i as follows:

$$z_i F = x_i L + y_i V$$

where z_i = feed composition of component i
$\quad x_i$ = liquid composition of component i
$\quad y_i$ = vapor composition of component i

The overall balance equation can be rearranged to a linear form, i.e., $y = mx + b$,

$$y = -(L/V)x + (F/V)z$$

where $m = -(L/V)$ = slope of the operating line
$\quad b = (F/V)z = y$ intercept

Note that V/F is the fraction of feed vaporized.

The above equation defines an operating line for this system. Since this operation is assumed to occur at equilibrium, it is defined as an *equilibrium stage*. The equation therefore relates the liquid and vapor compositions leaving an equilibrium stage.

As shown in Figure 27, plots of the equilibrium data, the $y = x$ line, and the operating line allow a procedure to calculate the unknown variables for a system, typically, the outlet liquid and vapor compositions. The $y = x$ line simplifies the graphical solution method and intersects the operating line at the feed composition, z. Thus, at this point, $y = x = z$. The unknown compositions, in the vapor and liquid product streams, are determined by the intersection of the operating line and the equilibrium curve.

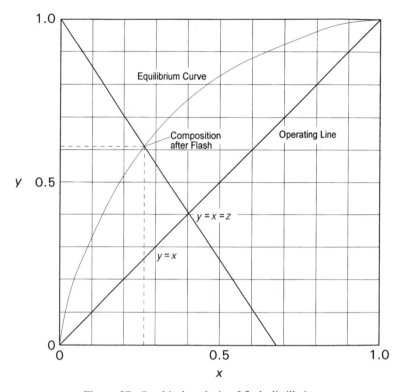

Figure 27. Graphical analysis of flash distillation.

Solution

The equilibirum data, are plotted in Figure 28. Superimposed on the graph is the $y = x$ line. Since the describing equation for the operating line is

$$y = -(L/V)x + (F/V)z$$

the slope of the operating line may be calculated:

$$V/F = 0.3$$
$$V = 0.3F$$
$$L = 0.7F$$
$$\text{Slope} = -L/V = 0.7F/0.3F = -2.33$$

The operating line also appears on the plot.

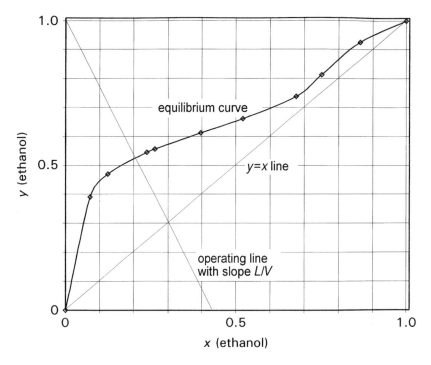

Figure 28. McCabe–Thiele diagram for flash calculation.

The liquid and vapor compositions as a result of the flash are found from the intersection of the operating line and the equilibrium curve:

$$x = 0.208$$

$$y = 0.52$$

These compositions can be used to check if a material balance is satisfied:

$$z_i F = x_i L + y_i V$$

$$(0.3)(250) = 75 = (0.208)(175) + (0.52)(75) = 75.4$$

$$< 1\% \text{ error}$$

MTO.9   DISTILLATION COLUMN DESIGN

A chemical manufacturer would like to evaluate the separation of ethanol from a 250-lbmol/h 20 mol % ethanol/80 mol % water stream at 300°F and 1 atm. The cost estimating group of the plant needs to know the height and diameter of a tray column required to obtain 70 mol % ethanol in the overhead distillate product and 5 mol % ethanol in the bottoms if the tray efficiency is 65% and a tray spacing of 24 in. is specified for this design. A total condenser and a partial reboiler, an external reflux ratio (L/D) of 1.25, and a feed quality, q, of 1.2 are all specified for the design. Equilibrium data are provided in the following table:

Ethanol–Water Vapor–Liquid Equilibrium at 1 atm (mole fractions)

T ($^{\circ}$C)	x_{ETOH}	y_{ETOH}
212	0.000	0.000
192	0.072	0.390
186	0.124	0.470
181	0.238	0.545
180	0.261	0.557
177	0.397	0.612
176	0.520	0.661
174	0.676	0.738
173	0.750	0.812
172	0.862	0.925
171	1.000	1

Columnwise distillation involves two sections; enriching (top) and stripping (bottom) (see Figure 29). In the stripping section, which lies below the feed, the more volatile components are stripped from the liquid. Above the feed, in the enriching or rectifying section, the concentration of the more volatile component is increased. A column may consist of one or more feeds and may produce two or more product streams. The product recovered at the top of a column is referred to as the "tops" or overheads, while the product at the bottom of the column is referred to as the "bottoms." Any products drawn at various stages between the top and bottom are referred to as "side streams." Multiple feeds and product streams do not alter the basic operation of a column, but they do complicate the analysis of the process to some extent. If the process requirement is to strip a volatile component from a relatively nonvolatile solvent, the rectifying (bottom) section may be omitted, and the unit is then called a *stripping column*. Virtually pure top and bottom products can be achieved by using many stages or sometimes additional columns; however, this is not usually economically feasible.

The top of the column is cooler than the bottom, so that the liquid stream becomes progressively hotter as it descends, and the vapor stream becomes progressively cooler as it rises. This heat transfer is accomplished by actual contact of liquid and vapor; and, for this purpose, effective contacting is desirable. Each plate in the column is assumed to approach equilibrium conditions. This type of plate is defined as a "theoretical plate," i.e., a plate on which the contact between the vapor and liquid is sufficiently good so that the vapor leaving the plate has the same

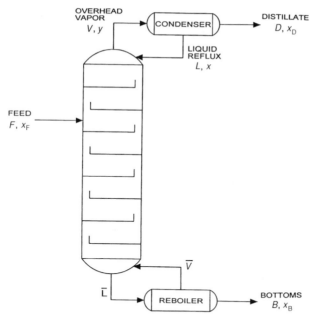

Figure 29. Schematic of a tray distillation column.

composition as the vapor in equilibrium with the liquid overflow from the plate. For such a plate the vapor and liquid leaving are related by the equilibrium curve. Distillation columns designed on this basis serve as a standard for comparison to actual columns. By such comparisons it is possible to determine the number of actual plates that are equivalent to a theoretical plate and then to reapply this factor when designing other columns for similar service.

In some operations where the top product is required as a vapor, the liquid condensed is sufficient only to provide reflux to the column, and the condenser is referred to as a *partial condenser*. In a partial condenser, the reflux will be in equilibrium with the vapor leaving the condenser and is considered to be an equilibrium stage in the development of the operating line and when estimating the column height. When the liquid is totally condensed, the liquid returned to the column will have the same composition as the top product and is not considered to be an equilibrium stage. A partial reboiler is utilized to generate vapor to operate the column and to produce a liquid product if necessary. Since both a liquid and vapor are in equilibrium, a partial reboiler is considered to be an equilibrium stage as well.

As described earlier, an operating line can be developed to describe the equilibrium relation between the liquid and vapor components. However, in staged column design, it is necessary to develop an operating line that relates the passing streams (liquid entering and vapor leaving) on each stage in the column. The following analysis will develop operating lines for the top, or enriching (rectifying), section and the bottom, or stripping, section of a column.

To accomplish this analysis, an overall material balance is written for the condenser as $V = L + D$, which relates the vapor (V) leaving the top stage, the liquid reflux returning (L) to the column from the condenser (reflux), and the distillate (D) collected. A material balance for component A is written as

$$Vy = Lx + DxD$$

This can be rearranged in the form of an equation for a straight line straight, $y = mx + b$ as

$$y = \frac{L}{D}x + \frac{D}{V}x_D$$

From the overall material balance, $D = V - L$ and

$$y = \frac{L}{V}x + \left(1 - \frac{L}{V}\right)x_D$$

where L/V is the slope. This is the internal reflux ratio (liquid reflux returned to the column/vapor from the top of the column). Also, the term $(1 - L/V)x_D$ represents the y intercept. Even though this equation has been developed around the condenser, it represents the equilibrium relationship of passing liquid/vapor streams and can be applied to the top of the column.

The corresponding operating line in the bottom, or stripping, section can be developed in a similar manner. The overall material balance around the reboiler is

$$\overline{L} = B + \overline{V}$$

and the component material balance is given by

$$\overline{L}x = \overline{V}y + Bx_B$$

Note that the terms with the bars over them represent the flow at the bottom of the column. Again, this material balance can be rearranged in the form of a straight line as

$$y = \frac{\overline{L}}{\overline{V}}x - \frac{B}{\overline{V}}x_B$$

But since

$$B = \overline{L} - \overline{V}$$

the equation becomes

$$y = \frac{\overline{L}}{\overline{V}}x - \left(\frac{\overline{L}}{\overline{V}} - 1\right)x_B$$

The term $\overline{L}/\overline{V}$ is the slope and $[(\overline{L}/\overline{V}) - 1]x_B$ represents the y-intercept. The slope can be calculated from the external reflux ratio by

$$\frac{\overline{L}}{\overline{V}} = \frac{(L/D)(z - x_B) + q\ (x_D - x_B)}{(L/D)(z - x_B) + q\ (x_D - x_B) - (x_D - z)}$$

where L/D = external reflux ratio; the ratio of liquid returned to the column to the distillate collected

q = feed quality; represents the fraction of feed remaining liquid below the feed stage

For further information on feed quality, the reader is referred to any chemical engineering text that addresses distillation. A procedure for designing a staged distillation column is provided below:

1. Plot the equilibrium data, the top and bottom operating lines, and the $y = x$ line on the same graph. This plot is called a McCabe–Thiele diagram (Figure 30).
2. Step off the number of stages required beginning at the top of the column. The point at the top of the curve represents the composition of the liquid entering (reflux) and gas leaving (distillate) the top of the column. The point at the bottom of the curve represents the composition of liquid leaving (bottoms) and the gas entering (vapor from reboiler) the bottom of the column. All stages are stepped off by drawing alternate vertical and horizontal lines in a stepwise manner between the top and bottom operating lines and the equilibrium curve. (See Figure 30.) The number of steps is the number of theoretical stages required. When a partial reboiler or partial condenser is employed, the number of equilibrium stages stepped off is $N + 1$ and the number of theoretical plates required would be N. Thus, if both are to be included, the number of equilibrium stages stepped off would be $N + 2$.
3. To determine the actual number of plates required, divide the result in step 2 by the overall plate fractional efficiency, typically denoted by E_0. Values can range from 0.4 to 0.8. The actual number of plates can be calculated from

$$N_{act} = N/E_0$$

4. The column height can be calculated by multiplying the result in step 3 by the tray spacing (9, 12, 15, 18, and 24 in. are typical tray spacings).

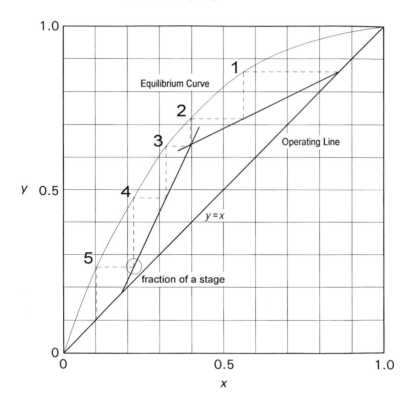

Figure 30. McCabe–Thiele diagram for a column distillation.

5. To determine the optimum feed plate location, draw a line from the feed composition on the $y = x$ line, through the intersection of the top/bottom operating lines, to the equilibrium curve. The step straddling the feed line is the correct feed-plate location.

6. Due to differences in vapor traffic throughout the column, the diameter will vary from the top to the bottom. Typical designs are swedged columns, where the diameter is larger at the bottom than at the top. A "conelike" connection links the bottom of the column with the top of the column. The diameter at the top of a column can be roughly sized by assuming a *superficial* vapor velocity of 5 ft/s and then using the following equation to calculate the diameter. (Note: A superficial velocity, sometimes referred to as an *empty tower velocity*, is the velocity the gas would have if only the vapor were flowing through the column.)

$$D = \left(\frac{4q}{\pi v}\right)^{0.5}$$

where D = diameter of the column, ft
q = vapor volumetric flowrate, ft^3/s
v = gas velocity in column, ft/s

A somewhat similar procedure is employed to calculate the diameter at the bottom of a column. The interested reader is referred to a mass transfer text that can provide a more detailed analysis.

Solution

The equilibrium data for the ethanol–water mixture and the $y = x$ line are plotted in Figure 31.

Since the describing equation for the operating line at the top of the column is

$$y = \frac{L}{V}\, x + \left(1 - \frac{L}{V}\right)$$

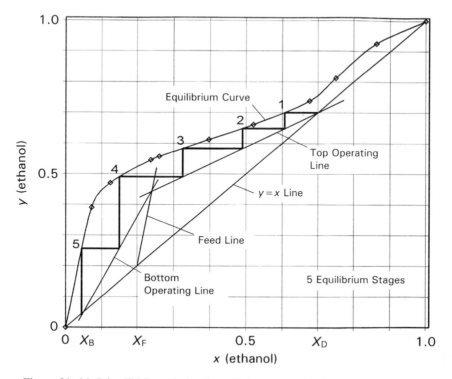

Figure 31. McCabe–Thiele analysis of equilibrium stages for the ethanol–water system.

the slope of this line may be determined:

$$V = D + L$$

$$\text{Slope} = \frac{L}{V} = \frac{L}{D + L} = \frac{L/D}{1 + (L/D)}$$

Thus, with $L/D = 1.25$,

$$L/V = 1.25/(1 + 1.25)$$
$$= 0.56$$

The operating line for the top of the column (see Figure 31) is then

$$y = (0.56)x + (1 - 0.56)(0.7)$$
$$= 0.56x + 0.308$$

The describing equation for the operating line at the bottom of the column is

$$y = \frac{\overline{L}}{\overline{V}} x - \left(\frac{\overline{L}}{\overline{V}} - 1 \right) x_B$$

where

$$\frac{\overline{L}}{\overline{V}} = \frac{(L/D)(z - x_B) + q\,(x_D - x_B)}{(L/D)(z - x_B) + q\,(x_D - x_B) - (x_D - z)}$$

Substitution yields

$$\frac{\overline{L}}{\overline{V}} = \frac{1.25\,(0.2 - 0.05) + 1.2\,(0.7 - 0.05)}{1.25\,(0.2 - 0.05) + 1.2\,(0.7 - 0.05) - (0.7 - 0.2)}$$
$$= 2.07$$

The operating line for the bottom of the column (see Figure 31) is

$$y = (2.07)x - (2.07 - 1)(0.2)$$
$$= 2.07x - 0.0535$$

By counting the number of equilibrium stages in Figure 31, one obtains five equilibrium stages. Note, however, that the partial reboiler counts as a stage. The number of equilibrium stages not counting the reboiler, N, is four. The stage efficiency E_0 is used to find the actual number of stages:

$$N_{act} = N/E_0 = 4/0.65$$

$$= 6.2$$

$$\approx 6 \text{ actual stages}$$

The column height is

$$6 \text{ stages } (2 \text{ ft}) = 12 \text{ ft}$$

The feed line is drawn from the $x = 0.2$ on the $y = x$ line through the intersection of the top and bottom operating lines to the equilibrium curve (see Figure 31). The optimum feed stage from the diagram is equilibrium stage 4, which is the lowest stage above the reboiler. This corresponds to the sixth actual stage. The optimum feed stage is stage 6.

Finally, the vapor volumetric flowrate is calculated:

$$q_a = \left(\frac{250 \text{ lbmol}}{h}\right)\left(\frac{1 \text{ h}}{3600 \text{ s}}\right)\left(\frac{379 \text{ ft}^3}{\text{lbmol}}\right)\left(\frac{300 + 460°R}{60 + 460°R}\right)$$

$$= 38.47 \text{ ft}^3/\text{s}$$

This enables one to estimate the diameter at the top of the column:

$$\text{Area} = \frac{q}{v} = \frac{38.47 \text{ ft}^3/\text{s}}{5 \text{ ft/s}} = 7.693 \text{ ft}^2$$

$$\text{Diameter} = D = \sqrt{\frac{4(7.693)}{\pi}}$$

$$= 3.13 \text{ ft}$$

$$\approx 3 \text{ ft}$$

MTO.10   EQUILIBRIUM STAGES OF A DISTILLATION COLUMN

J.B. Chemicals, Inc., has hired Theodore Distillation Consultants (TDC) to assist in the separation of 500 kmol/h of a saturated vapor feed (quality, $q = 0$) consisting of 40 mol % benzene (C), 30 mol % toluene (A), and 30 mol % cumene (B). It has been suggested that the distillation column include a partial reboiler and a total condenser. J.B. Chemicals requires 99% recovery of the toluene in the distillate $(FR_A)_{dist}$ and 95% recovery of the cumene in the bottoms $(FR_B)_{bot}$. As an engineer for TDC you need to determine the actual number of equilibrium stages required for the separation, as well as the fractional recovery (FR) of the benzene in the distillate. The column will operate at a reflux ratio of 2.0 times the minimum and will have relative volatilities of $\alpha_{CA} = 2$, $\alpha_{BA} = 0.25$, and $\alpha_{AA} = 1.0$.

A = toluene (light key)
B = cumene (heavy key)
C = benzene (light nonkey)

Solution

Fenske developed the following equation to calculate the minimum number of stages at total reflux for the separation of two key components (i.e., A and B in a mixture of A, B, and C). The original notation is employed here

$$N_{min} = \frac{\ln\left(\dfrac{FR_{A,dist}/(1 - FR_{A,dist})}{(1 - FR_{B,bot})/FR_{B,bot}}\right)}{\ln(1/\alpha_{BA})}$$

where N_{min} = minimum number of equilibrium stages
FR_i = fractional recovery in distillate (dist) or bottoms (bot) for component i
α_{BA} = relative volatility of component B versus that of component A

The following equation can be used to calculate the fractional recoveries of the nonkey components (i.e., of C) in the distillate and bottoms:

$$FR_{C,dist} = \frac{\alpha_{CB}^{N_{min}}}{\dfrac{FR_{B,bot}}{1 - FR_{B,bot}} + \alpha_{CB}^{N_{min}}}$$

The *Underwood* equation requires a trial-and-error solution and a subsequent material balance to estimate the minimum reflux ratio. First, the unknown, φ, is determined by trial and error, such that both sides of the following equation are equal. The unknown value of φ should lie between the relative volatilities of the light and heavy key components. The *key* components are those that have their fractional recoveries specified. The most volatile component of the keys is the light key and the least volatile is the heavy key. All other components are referred to as *nonkey* components. If a nonkey component is lighter than the light key component, it is a light nonkey; if it is heavier than the heavy key component it is a heavy nonkey component.

$$\Delta V_{feed} = F(1 - q) = \sum_{i=1}^{n} \frac{\alpha_i F z_i}{\alpha_i - \varphi}$$

where ΔV_{feed} = change in vapor flowrate at feed stage
F = feed rate
q = percentage of feed remaining liquid
z_i = feed composition of component i
n = number of components
φ = unknown root to be determined by trial and error

Next, calculate the minimum vapor flowrate, V_{min}, from the following equation:

$$V_{min} = \sum_{i=1}^{n} \frac{\alpha_i D_{xi,dist}}{\alpha_i - \varphi}$$

where $D_{xi,dist}$ = amount of component i in the distillate
$= F z_i (FR)_{i,dist}$

Once V_{min} is known, the minimum liquid flowrate is calculated from the following:

$$L_{min} = V_{min} - D$$

The minimum reflux ratio is then calculated by dividing the minimum liquid flowrate by the distillate flowrate as shown below:

$$R_{min} = (L/D)_{min}$$

The *Gilliland* correlation uses the results of the Fenske and Underwood equations to determine the actual number of equilibrium stages. The correlation has been fit to three equations:

$$\frac{N - N_{min}}{N + 1} = 1.0 - 18.5715x \qquad \text{for } 0 \leqslant x \leqslant 0.01$$

$$= 0.545827 - 0.591422x + 0.002743/x \qquad \text{for } 0.01 \leqslant x \leqslant 0.90$$

$$= 0.16595 - 0.16595x \qquad \text{for } 0.9 \leqslant x \leqslant 1.0$$

where N = actual number of equilibrium stages
$x = [(L/D) - (L/D)_{min}]/[(L/D) + 1]$

The values of N_{min} and $(L/D)_{min}$ have been previously defined as the minimum number of equilibrium stages (Fenske equation) and minimum reflux ratio (Underwood equation).

Thus, the calculational procedure is as follows:

1. Calculate the minimum number of equilibrium stages, N_{min}, from the Fenske equation.
2. Calculate the minimum reflux ratio, $(L/D)_{min}$, from the Underwood equation.
3. Select an actual reflux ratio, L/D, which usually ranges from 1 to 2.5 times the minimum ratio.

4. Calculate x.

5. Solve the Gilliland correlation for the actual number of equilbrium stages.

Following this procedure, the minimum number of equilibrium stages, N_{min}, at total reflux is

$$N_{min} = \frac{\ln\left(\dfrac{0.99/(1-0.99)}{(1-0.95)/0.95}\right)}{\ln(1/0.25)}$$

$$= 5.44$$

The fractional recovery of benzene, FR_C, in the distillate is

$$FR_{C,dist} = \frac{\left(\dfrac{2}{0.25}\right)^{0.44}}{\left(\dfrac{0.95}{1-0.95}\right) + \left(\dfrac{2}{0.25}\right)^{0.44}}$$

$$= 0.9998$$

Then φ is determined by trial and error using the Underwood equation:

$$\Delta V_{feed} = F(1-q) = \sum_{i=1}^{n} \frac{\alpha_i F z_i}{\alpha_i - \varphi}$$

$$= (500)(1-0)$$

$$= \frac{(1)(500)(0.3)}{(1)-\varphi} + \frac{(0.25)(500)(0.3)}{(0.25)-\varphi} + \frac{(2)(500)(0.4)}{(2)-\varphi}$$

$$\varphi = 0.561678$$

The recovery rate of each component in the distillate, D_{xi}, and the total distillate flowrate, D, are obtained from a simple material balance:

Toluene: $Z_A F(FR_{A,dist}) = D_A$ $(3)(500)(0.99) = 148.5\,\text{kmol/h}$

Cumene: $Z_B F(FR_{B,dist}) = D_B$ $(0.3)(500)(1-0.95) = 7.5\,\text{kmol/h}$

Benzene: $Z_C F(FR_{C,dist}) = D_C$ $(0.4)(500)(0.9998) = 199.95\,\text{kmol/h}$

The minimum overhead vapor flow rate, V_{min} is

$$V_{min} = \frac{1(148.5)}{1-0.561678} + \frac{0.25(7.5)}{0.25-0.561678} + \frac{2(199.95)}{2-0.561678}$$

$$= 610.81\,\text{kmol/h}$$

The minimum reflux ratio, $(L/D)_{min}$, is calculated from

$$L_{min} = V_{min} - D$$

$$= 610.81 - 355.95$$

$$= 254.86 \, \text{kmol/h}$$

$$R_{min} = (L/D)_{min}$$

$$= (254.86/355.95)$$

$$= 0.72$$

The actual reflux ratio is

$$R_{act} = 2(R_{min})$$

$$= 2(0.72)$$

$$= 1.44$$

The quantity x is required for the Guilliland correlation:

$$x = [(L/D) - (L/D)_{min}]/[(L/D) + 1]$$

$$= [(1.44) - (0.72)]/[(1.44) + 1]$$

$$= 0.294$$

Using the Guilliland equation,

$$\frac{N - N_{min}}{N + 1} = 0.545827 - 0.591422x + 0.002743/x \quad \text{(for } 0.01 \leqslant x \leqslant 0.90)$$

$$= 0.545827 - 0.91422(0.294) + 0.002743/(0.294)$$

$$N = 9.40 \sim 10 \, \text{stages}$$

Thus, one stage is assigned for the partial reboiler and nine stages for separation.

MTO.11   LIQUID–LIQUID EXTRACTION

A 200-lb/h process stream containing 20 wt % acetic acid (AA) in water is to be extracted with 400 lb/h of methyl isobutyl ketone (MIBK) ternary containing 0.05 wt % acetic acid and 0.005 wt % water. Determine the number of theoretical stages required to achieve an acetic acid concentration of 1 wt % in the process stream. The equilibrium data for the MIBK and acetic acid system, for acetic acid weight ratios between 0.01 and 0.25, can be represented by

$$Y = 0.48(X)^{1.5}$$

where X = weight ratio of solute to feed solvent in the raffinate phase
  Y = weight ratio of solute to extraction solvent in the extract phase

Solution

Liquid–liquid extraction (or liquid extraction) is a process for separating a solute from a solution based on the concentration driving force between two immiscible (nondissolving) liquid phases. Thus, liquid extraction involves the transfer of solute from one liquid phase into a second immiscible liquid phase. The simplest example involves the transfer of one component from a binary mixture into a second immiscible liquid phase such as is the case for the extraction of an impurity from wastewater into an organic solvent. Liquid extraction is usually selected when distillation or stripping is impractical or too costly (e.g., when the relative volatility for two components falls between 1.0 and 1.2).

The feed to an extractor consists of a liquid component, referred to as the *feed solvent*, and the solute(s) to be removed from the stream. The extraction solvent is usually an immiscible liquid that removes the solute from the feed without any appreciable dissolution of either solvent into the other phase. The extract phase is the solute-rich extraction solvent leaving the unit. The treated liquid feed phase remaining after contact with the extraction solvent is referred to as the *raffinate*.

A theoretical or equilibrium stage provides a mechanism by which two immiscible phases intimately mix until equilibrium concentrations are reached and then physically separated into clear layers. *Cross-current* extraction involves a series of stages in which the raffinate (R) from one extraction stage is contacted with additional fresh solvent (S) in a subsequent stage. *Cross-current* extraction is usually not economically attractive for large commercial processes because of the high solvent usage and low solute concentration in the extract. In *countercurrent extraction*, the extraction solvent enters a stage at the opposite end from where the feed (F) enters, and the two phases pass each other countercurrently. The purpose is to transfer one or more components from the feed solution to the extract (E).

The distribution coefficient, k, is defined as the ratio of the weight fraction of solute in the extract phase, y, to the weight fraction of solute in the raffinate phase, x, i.e.,

$$k = y/x$$

For shortcut methods the distribution coefficient is represented as the ratio of the weight ratio of solute to extraction solvent in the extract phase, Y, to the weight ratio of solute to feed solvent in the raffinate phase, X:

$$k' = Y/X$$

The cross-current extraction process is an ideal laboratory procedure since the extract and raffinate phases can be analyzed after each stage to generate equilibrium data as well as to achieve high solute removal. If the distribution coefficient, as well as the ratio of extraction solvent to feed solvent (S'/F'), are constant and the fresh extraction solvent is pure, then the number of cross-current stages (N) required to reach a specified raffinate composition can be estimated from

$$N = \frac{\log\,(X_f/X_r)}{\log\left(\dfrac{k'S'}{F'} + 1\right)}$$

where X_f = weight ratio of solute in feed
$\quad\;\; X_r$ = weight ratio of solute in raffinate
$\quad\;\; S'$ = mass flowrate of solute-free extraction solvent
$\quad\;\; F'$ = mass flowrate of solute-free feed solvent

Most liquid extraction systems can be treated as having either (a) immiscible (mutually non-dissolving) solvents, (b) partially miscible solvents with a low solute concentration in the extract, or (c) partially miscible solvents with a high solute concentration in the extract. Only case (a) will be addressed in this problem. For further information on cases (b) and (c), the reader is referred to *Perry's* (R. H. Perry and D. W. Green, Eds., *Perry's Chemical Engineers' Handbook*, 7th Edition, McGraw Hill, New York, 1996.) or any other chemical engineering book dealing with liquid extraction.

For case (a), since the solvents are immiscible, the rate of solvent in the feed stream (F') is the same as the rate of feed solvent in the raffinate stream (R'). Also, the rate of extraction solvent (S') entering the unit is the same as the extraction solvent leaving the unit in the extract phase (E'). However, the total flowrates entering and leaving the unit will be different since the extraction solvent is removing solute from the feed. Thus, the ratio of extraction-solvent to feed-solvent flowrates (S'/F') is equivalent to (E'/R').

By writing an overall material balance around the unit, a McCabe–Thiele type of operating line with a slope (F'/S') can be generated:

$$Y_e = \frac{F'X_f + S'Y_s - R'X_r}{E'}$$

where Y_e = weight ratio of solute to the extraction solvent at the feed stage
$\quad\;\; Y_s$ = weight ratio of solute to the extraction solvent at the raffinate stage

Note that the feed solvent enters at the "feed stage" and exits at the "raffinate stage."

If the equilibrium line is straight, its intercept is zero and the operating line is straight. The number of theoretical stages can then be calculated with one of the following equations, which are forms of the Kremser equation. When the intercept of the equilibrium line is greater than zero, Y_s/k'_s should be used instead of Y_s/m, where k'_s is the distribution coefficient at Y_s. Also, these equations contain an *extraction factor*, ε, which is calculated by dividing the slope of the equilibrium line, m, by the slope of the operating line, F'/S':

$$\varepsilon = mS'/F'$$

where m is the slope of equilibrium line.

If the equilibrium line is not straight, a geometric mean value of m should be used (Treybal, *Liquid Extraction*, 2nd ed., McGraw-Hill, 1963). This quantity is determined by the following equation:

$$m = \sqrt{m_1 m_2}$$

where m_1 = slope of the equilibrium line at the concentration leaving the feed stage
m_2 = slope of the equilibrium line at the concentration leaving the raffinate stage

When $\varepsilon \neq 1.0$

$$N = \frac{\ln\left[\frac{(X_f - Y_s)/m}{(X_r - Y_s)/m}\left(1 - \frac{1}{\varepsilon}\right) + \frac{1}{\varepsilon}\right]}{\ln \varepsilon}$$

When $\varepsilon = 1.0$,

$$N = \frac{(X_f - Y_s)/m}{(X_r - Y_s)/m} - 1$$

The feed solvent flow rate (F') in lb/h is

$$F' = (200 \text{ lb/h})(0.8) = 160 \text{ lb/h}$$

The extraction solvent flowrate (S') in lb/h is

$$S' = (400 \text{ lb/h})(1 - 0.00055) = 399.8 \text{ lb/h}$$

The weight ratios of the AA in the feed (X_f), raffinate (X_r), and extraction solvent (Y_s) are

$$X_f = (0.2)(200 \text{ lb/h})/160 = 0.25$$
$$X_r = 1/(100 - 1) = 0.0101$$
$$Y_s = 0.0005/(1 - 0.0005) = 0.0005$$

The weight ratio of the AA in the extract (Y_e) is

$$Y_e = \frac{F'X_f + S'Y_s - R'X_f}{E'}$$

$$= \frac{(160)(0.25) + (399.8)(0.0005) - (160)(0.010)}{399.8}$$

$$= 0.0965$$

The weight ratio of the liquid leaving the first stage (X_1), which is in equilibrium with the vapor leaving the same stage (Y_e), is given by

$$X_1 = (0.0965/1.23)^{(1/1.1)}$$

$$= 0.988$$

The slope of the equilibrium line is next calculated:

$$Y = 1.23 \ (X)^{1.1}$$

$$dY/dX = (1.1)(1.23)X^{0.1}$$

$$= 1.353X^{0.1}$$

The slope of the equilibrium line at the feed stage (m_1) is

$$m_1 = dY/dX = 1.353 \ (0.988)^{0.1}$$

$$= 1.351$$

The slope of the equilibrium line at the raffinate stage (m_r) is

$$m_r = dY/dX = 1.353 \ (0.0101)^{0.1}$$

$$= 0.8545$$

The geometric mean equilibrium slope (m) is

$$m = [(1.351)(0.8545)]^{0.5}$$

$$= 1.074$$

The extraction factor (ε) is

$$\varepsilon = 1.074 \ (399.8/160)$$

$$= 2.68$$

The number of theoretical stages (N) is, therefore,

$$N = \frac{\ln\left[\left(\dfrac{(X_f - Y_s)/m}{(X_r - Y_s)/m}\right)\left(1 - \dfrac{1}{\varepsilon}\right) + \dfrac{1}{\varepsilon}\right]}{\ln \ \varepsilon}$$

$$= \frac{\ln\left[\left(\dfrac{(0.25 - 0.0005)/1.074}{(0.0101 - 0.0005)/1.074}\right)\left(1 - \dfrac{1}{2.68}\right) + \dfrac{1}{2.68}\right]}{\ln \ 2.68}$$

$$= 2.85$$

MTO.12   LEACHING

A countercurrent leaching system (Figure 32) is to treat 10 kg/h of crushed sugar stalks with impurity-free water as the solvent. Analysis of the stalks is as follows:

Water: 38% (by mass)
Sugar: 10%
Pulp: 52%

If 95% sugar is to be recovered and the extract phase leaving the system is to contain 12% sugar, determine the number of actual stages required if each kilogram of the dry pulp retains 2.5 kg of solution. Assume an overall stage efficiency of 85%.
In Figure 32,

V = overflow solution (no inerts in overflow)

L = underflow solution exclusive of inerts

x = solute mass fraction in the underflow solution retained by the inerts

y = solute mass fraction in the extract overflow phase

The following notation is also employed: subscripts A, B, and I refer to solute, solvent, and inerts, respectively; and R represents the total underflow solution without inerts.

It is usually assumed that the inerts are constant from stage to stage and insoluble in the solvent. Since no inerts are usually present in the extract (overflow) solution,

$$y = y_A$$

When the solution retained by the inerts is approximately constant, both the underflow L_N and overflow V_N are constant, and the equation for the operating line approaches a straight line. Since the equilibrium line is also straight, the number of stages can be shown to be

$$N = \frac{\log\left[(y_{N+1} - x_N)/(y_1 - x_1)\right]}{\log\left[(y_{N+1} - y_1)/(x_N - x_1)\right]}$$

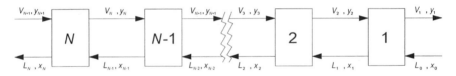

Figure 32. Countercurrent leaching system.

The above equation should not be used for the entire extraction cascade if L_0 differs from $L_1, L_2, \ldots, L_N$, i.e., the underflows vary within the system. For this case, the compositions of all the streams entering and leaving the first stage should first be calculated before applying this equation to the remaining cascade.

Solution

For a basis of 100 kg of sugar stalks,

   Sugar: 38 kg
   Water: 10 kg
   Pulp: 52 kg

For 95% sugar recovery, the extracted solution contains

$$0.95(10) = 9.5 \, \text{kg sugar}$$

and

$$[(1 - 0.12)/0.12](9.5) = 69.7 \, \text{kg water}$$

The total extract solution is then

$$V_1 = 9.5 + 69.7$$
$$= 79.2 \, \text{kg}$$

The solution is

$$L_1 = L_2 = \cdots = L_N = (2.5)(52) = 130 \, \text{kg}$$

Since $L_0 = 10 + 38 = 48 \, \text{kg}$, a material balance on the initial stage must be performed:

$$L_0 + V_2 = V_1 + L_1$$

or

$$48 + V_2 = 79.2 + 130$$
$$V_2 = 161.2 \, \text{kg}$$

Applying a componential solute (sugar) balance across the first stage,

$$161.2 \, (y_2) + 10 = 9.5 + 130 \, (0.12)$$
$$y_2 = 0.0937$$

As indicated above, the remaining $(N - 1)$ stages operate with the underflow and overflow solutions relatively constant. For this part of the system,

$$N - 1 = \frac{\log \left[(y_{N+1} - x_N)/(y_2 - x_1) \right]}{\log \left[(y_{N+1} - y_2)/(x_N - x_1) \right]}$$

For this equation,

$$y_{N+1} = 0.0$$

$$x_N = (0.05)(10)/(130) = 0.00385$$

Substitution of the values into the above equation gives

$$N - 1 = \frac{\log \left[(0 - 0.00385)/(0.0937 - 0.12) \right]}{\log \left[(0 - 0.0937)/(0.00385 - 0.12) \right]}$$

$$= 8.96$$

$$N = 9.96 \, \text{stages}$$

This represents the theoretical number of stages. The actual number of stages, N_{act}, is

$$N_{act} = 9.96/0.85 = 11.71$$

Twelve (12) stages are suggested.

9 Thermodynamics (THR)

THR.1 PARTIAL PRESSURE

A storage tank contains a gaseous mixture comprised of 30% CO_2, 5% CO, 5% H_2O, 50% N_2, and 10% O_2, by volume. What is the partial pressure of each component if the total pressure is 2 atm? What are their pure-component volumes if the total volume is 10 ft³? What are their concentrations in ppm (parts per million)?

Solution

Dalton's law states that the partial pressure, p_a, of an ideal gas is given by

$$p_a = y_a P$$

where y_a = mole fraction of component a
P = total pressure

Thus,

$$p_{CO_2} = 0.30(2) = 0.60 \text{ atm}$$

$$p_{CO} = 0.05(2) = 0.10 \text{ atm}$$

$$p_{H_2O} = 0.05(2) = 0.10 \text{ atm}$$

$$p_{N_2} = 0.50(2) = 1.00 \text{ atm}$$

$$p_{O_2} = 0.10(2) = \underline{0.20 \text{ atm}}$$

$$P \qquad\qquad = 2.00 \text{ atm}$$

Amagat's law states that the pure component volume, V_a, of an ideal gas is given by

$$V_a = y_a V$$

where V is the total volume.

Thus,

$$V_{CO_2} = 0.30(10) = 3.00 \, \text{ft}^3$$
$$V_{CO} = 0.05(10) = 0.50 \, \text{ft}^3$$
$$V_{H_2O} = 0.05(10) = 0.50 \, \text{ft}^3$$
$$V_{N_2} = 0.50(10) = 5.00 \, \text{ft}^3$$
$$V_{O_2} = 0.10(10) = \underline{1.00 \, \text{ft}^3}$$
$$V \qquad\qquad = 10.00 \, \text{ft}^3$$

By definition, the parts per million (ppm) is given by

$$\text{ppm}_a = y_a 10^6$$

Thus,

$$\text{ppm}_{CO_2} = 0.30(10^6) = 3.00 \times 10^5 \, \text{ppm}$$
$$\text{ppm}_{CO} = 0.05(10^6) = 0.50 \times 10^5 \, \text{ppm}$$
$$\text{ppm}_{H_2O} = 0.05(10^6) = 0.50 \times 10^5 \, \text{ppm}$$
$$\text{ppm}_{N_2} = 0.50(10^6) = 5.00 \times 10^5 \, \text{ppm}$$
$$\text{ppm}_{O_2} = 0.10(10^6) = 1.00 \times 10^5 \, \text{ppm}$$

Unless otherwise stated, ppm for gases is by mole or volume and is usually designated as ppmv. When applying the term to liquids or solids, the basis is almost always by mass; the notation may appear as ppmw or ppmm.

THR.2   CLAPEYRON EQUATION

For acetone at 0°C, the Clapeyron equation contains coefficients that have been experimentally determined to be

$$A = 15.03$$
$$B = 2817$$

for p' (vapor pressure) and T in mm Hg and K, respectively. Also for acetone at 0°C, the Antoine coefficients are

$$A = 16.65$$
$$B = 2940$$
$$C = -35.93$$

with p' and T in the same units.

(1) Use the Clapeyron equation to estimate the vapor pressure of acetone at 0°C. (2) Use the Antoine equation to estimate the vapor pressure of acetone at 0°C. (3) If the vapor pressure of acetone is 71 mm Hg, calculate the maximum concentration (mole fraction) of acetone at 1 atm total pressure.

Solution

The Clapeyron equation is given by

$$\ln p' = A - (B/T)$$

where p' and T are the vapor pressure and temperature, respectively. The Antoine equation is given by

$$\ln p' = A - B/(T + C)$$

The vapor pressure, p', of acetone at 0°C using the Clapeyron equation is

$$\ln p' = 15.03 - 2817/(0 + 273)$$
$$= 4.7113$$
$$p' = 111.2 \text{ mm Hg}$$

The vapor pressure of acetone at 0°C predicted by the Antoine equation is

$$\ln p' = 16.65 - 2940/(273 - 35.93)$$
$$= 4.2486$$
$$p' = 70.01 \text{ mm Hg}$$

The maximum concentration of a component in a noncondensable gas is given by its vapor pressure p' divided by the total pressure, P:

$$y_{max} = p'/P$$

Any increase in concentration will result in the condensation of the component in question. Thus, the maximum mole fraction of acetone in air at $0°C$ and 1 atm is

$$y_{max} = p'/P = 70.01/760$$

$$= 0.0921$$

The reader should note that the Clapeyron equation generally overpredicts the vapor pressure at or near ambient conditions. The Antoine equation is widely used in industry and usually provides excellent results. Also note that, contrary to statements appearing in the *Federal Register* and some *Environmental Protection Agency (EPA)* publications, vapor pressure is not a function of pressure.

THR.3   IDEAL GAS LAW

Given the following pressure, temperature, and molecular weight data of an ideal gas, determine its density:

Pressure $= 1.0$ atm
Temperature $= 60°F$
Molecular weight of gas $= 29$

Solution

An ideal gas is an imaginary gas that exactly obeys certain simple laws (e.g., Boyle's law, Charles' law, and the ideal gas law). No real gas obeys the ideal gas law exactly, although the "lighter" gases (hydrogen, oxygen, air, etc.) at ambient conditions approach ideal gas law behavior. The "heavier" gases such as sulfur dioxide and hydrocarbons, particularly at high pressures and low temperatures, deviate considerably from the ideal gas law. Despite these deviations, the ideal gas law is routinely used in engineering calculations. The ideal gas law equation takes the form

$$PV = nRT$$

where P = absolute pressure
$\quad V$ = volume
$\quad T$ = absolute temperature
$\quad n$ = number of moles
$\quad R$ = ideal gas law constant

The ideal gas law in terms of density, ρ, is

$$\rho = m/V = n(\text{MW})/V$$

$$= P(\text{MW})/RT$$

where MW = molecular weight
$\quad m$ = mass of gas
$\quad \rho$ = density of gas

Typical values of R are given below:

$$R = 10.73 \, \text{psia} \cdot \text{ft}^3/(\text{lbmol} \cdot {}^\circ\text{R})$$
$$= 1545 \, \text{psfa} \cdot \text{ft}^3/(\text{lbmol} \cdot {}^\circ\text{R})$$
$$= 0.73 \, \text{atm} \cdot \text{ft}^3/(\text{lbmol} \cdot {}^\circ\text{R})$$
$$= 555 \, \text{mm Hg} \cdot \text{ft}^3/(\text{lbmol} \cdot {}^\circ\text{R})$$
$$= 82.06 \, \text{atm} \cdot \text{cm}^3/(\text{gmol} \cdot \text{K})$$
$$= 8.314 \, \text{kPa} \cdot \text{m}^3/(\text{kgmol} \cdot \text{K})$$
$$= 1.986 \, \text{cal}/(\text{gmol} \cdot \text{K})$$
$$= 1.986 \, \text{Btu}/(\text{lbmol} \cdot {}^\circ\text{R})$$

The choice of R is arbitrary, provided consistent units are employed.
The density of the gas using the appropriate value of R may now be calculated:

$$\rho = P(\text{MW})/RT$$
$$= (1)(29)/(0.73)(60 + 460)$$
$$= 0.0764 \, \text{lb}/\text{ft}^3$$

Since the molecular weight of the given gas is 29, this calculated density may be assumed to apply to air.

Also note that the effect of pressure, temperature, and molecular weight on density can be obtained directly from the ideal gas law equation. Increasing the pressure and molecular weight increases the density; increasing the temperature decreases the density.

THR.4   ACTUAL VOLUMETRIC FLOWRATE

Given a standard volumetric flowrate, determine the actual volumetric flowrate. Data are provided below.

   Standard volumetric flowrate of a gas stream = 2000 scfm
   Standard conditions = 60°F and 1 atm
   Actual operating conditions = 700°F and 1 atm

Solution

The actual volumetric flowrate, usually in acfm (actual cubic feet per minute), is the volumetric flowrate based on actual operating conditions (temperature and pressure of the system). The standard volumetric flowrate, usually in scfm (standard cubic feet per minute), is the volumetric flowrate based on standard conditions. The standard conditions have to be specified, for there are different sets of standard conditions. For most engineering environmental applications, standard conditions are 60°F and 1 atm or 32°F and 1 atm.

Charles' law states that the volume of an ideal gas is directly proportional to the temperature at constant pressure. Boyle's law states that the volume of an ideal gas is inversely proportional to the pressure at constant temperature. One can combine Boyle's law and Charles' law to relate the actual volumetric flowrate to the standard volumetric flowrate:

$$q_a = q_s(T_a/T_s)(P_s/P_a)$$

where q_a = actual volumetric flowrate
q_s = standard volumetric flowrate
T_a = actual operating temperature, °R or K
T_s = standard temperature, °R or K
P_s = standard pressure, absolute
P_a = actual operating pressure, absolute

This equation may be used to calculate the actual volumetric flowrate in acfm:

$$q_a = q_s(T_a/T_s)$$
$$= 2000(700 + 460)/(60 + 460)$$
$$= 4462 \, \text{acfm}$$

If it is desired to convert from acfm to scfm, reverse the procedure and use the following equation:

$$q_s = q_a(T_s/T_a)(P_a/P_s)$$

The reader is cautioned on the use of acfm and/or scfm (see Problem THR.5). Predicting the performance and the design of equipment should always be based on actual conditions. Designs based on standard conditions can lead to disastrous results, with the unit usually underdesigned. For example, the ratio of acfm (2140°F) to scfm (60°F) for a particular application is 5.0. The reader is again reminded that absolute temperatures and pressures must be employed in all ideal gas law calculations.

THR.5   STANDARD VOLUMETRIC FLOWRATE

Given a mass flowrate, determine the standard volumetric flowrate. Data are provided below.

Mass flowrate of flue gas, $\dot{m} = 50\,\text{lb/min}$

Average molecular weight of flue gas, MW $= 29\,\text{lb/lbmol}$

Standard conditions $= 60°\text{F}$ and 1 atm

Solution

Another application of the ideal gas law arises when one is interested in converting a mass (or molar) flowrate to a volumetric flowrate (actual or standard), or vice versa. The ideal gas equation is rearranged and solved for one variable in terms of the others. For example, the volume of 1 lbmol of ideal gas is given by

$$PV = nRT \qquad R = 0.73\,\text{atm} \cdot \text{ft}^3/(\text{lbmol} \cdot °\text{R})$$

$$V/n = RT/P$$

$$= (0.73)(60 + 460)/1.0$$

$$= 379\,\text{scf/lbmol}$$

Thus, if the standard conditions are $60°\text{F}$ and 1 atm, there are 379 standard cubic feet of gas per lbmol for any ideal gas.

The standard volumetric flowrate, q_s, in scfm is

$$q_s = (m/\text{MW})(379\,\text{scf/lbmol})$$

$$= (50/29)(379)$$

$$= 653\,\text{scfm}$$

The previous result is an important number to remember in many engineering calculations: 1 lbmol of any ideal gas at $60°\text{F}$ and 1 atm occupies 379 ft^3; and, equally important, 1 lbmol of any ideal gas at $32°\text{F}$ and 1.0 atm occupies 359 ft^3. In SI units, 1 gmol of any ideal gas occupies 22.4 liters at $0°\text{C}$ and 1.0 atm.

THR.6 ENTHALPY OF COMBUSTION

Find the enthalpy of combustion of cyclohexane from enthalpy of formation data. Perform the calculation for both liquid and gaseous cyclohexane. The following enthalpy of formation ($25°\text{C}$) data are available:

Species	ΔH_f° (cal/gmol)
C_6H_{12} (l)	$-37,340$
C_6H_{12} (g)	$-29,430$
CO_2	$-94,052$
H_2O	$-68,317$

Solution

To simplify the solution that follows, examine the equation

$$a\mathrm{A} + b\mathrm{B} \rightarrow c\mathrm{C} + d\mathrm{D}$$

If the above reaction is assumed to occur in the standard state, the standard enthalpy of reaction, ΔH°, is given by

$$\Delta H^\circ = c\,(\Delta H_f^\circ)_\mathrm{C} + d\,(\Delta H_f^\circ)_\mathrm{D} - a\,(\Delta H_f^\circ)_\mathrm{A} - b\,(\Delta H_f^\circ)_\mathrm{B}$$

where $(\Delta H_f^\circ)_i$ is the standard enthalpy of formation of species i.

Thus, the (standard) enthalpy of a reaction is obtained by taking the difference between the (standard) enthalpy of formation of products and reactants. If the (standard) enthalpy of reaction or formation is negative (exothermic), as is the case with most combustion reactions, then energy is liberated due to the chemical reaction. Energy is absorbed if ΔH is positive (endothermic).

The balanced combustion equation for cyclohexane, C_6H_{12}, is

$$C_6H_{12}\ (l) + 9O_2 \rightarrow 6CO_2 + 6H_2O\ (l)$$

The enthalpy of combustion for the above reaction is

$$\Delta H_C^\circ = 6\,(\Delta H_f^\circ)_{CO_2} + 6\,(\Delta H_f^\circ)_{H_2O(l)} - (\Delta H_f^\circ)_{C_6H_{12}}$$

Using the data provided gives

$$\Delta H_C^\circ = 6\,(-94052) + 6\,(-68317) - (-37340)$$
$$= 936{,}874\ \mathrm{cal/gmol\ of\ liquid\ cyclohexane}$$

For gaseous cyclohexane,

$$\Delta H_C^\circ = 6\,(-94052) + 6\,(-68317) - (-29430)$$
$$= 944{,}784\ \mathrm{cal/gmol\ of\ gaseous\ cyclohexane}$$

Tables of enthalpies of formation, combustion, and reaction are available in the literature (particularly thermodynamics texts/reference books) for a wide variety of compounds. It is important to note that these are valueless unless the stoichiometric equation and the states of the reactants and products are included. However, enthalpy of reaction is not always employed in engineering reaction/combustion calculations. The two other terms that have been used are the gross (or higher) heating value and the net (or lower) heating value. These are discussed in the next problem.

THR.7   GROSS HEATING VALUE

Given the following gas mixture and combustion properties, determine the gross heating value, HV_G, in Btu/scf:

Species	Mole Fraction	Gross Heating Value (Btu/scf)
N_2	0.0515	0
CH_4	0.8111	1013
C_2H_6	0.0967	1792
C_3H_8	0.0351	2590
C_4H_{10}	0.0056	3370

Solution

The gross heating value (HV_G) represents the enthalpy change or heat released when a gas is stoichiometrically combusted at 60°F, with the final (flue) products at 60°F and any water present in the liquid state. Stoichiometric combustion requires that no oxygen be present in the flue gas following combustion of the hydrocarbons.

The gross heating value of the gas mixture, HV_G, may now be calculated using the equation

$$HV_G = \sum_{i=1}^{n} x_i \, HV_{G,i}$$

where x_i is the mole fraction of ith component.
    Thus,

$$HV_G = (0.0515)(0) + (0.8111)(1013) + (0.0967)(1792)$$
$$+ (0.0351)(2590) + (0.0056)(3370)$$
$$= 1105 \, \text{Btu/scf}$$

The *net heating value* (HV_N) is similar to HV_G except the water is in the vapor state. The net heating value is also known as the *lower heating value*, and the gross heating value is also known as the *higher heating value*.

THR.8   GROSS HEATING VALUE FOR A COMBUSTIBLE MIXTURE

Calculate the gross heating value in Btu/lbmol for a combustible gas mixture of 75 mol % methane, 10 mol % propane, and 15 mol % n-butane. The following gross heating value data are available:

Species	HV_G (Btu/lb)
CH_4	23,879
C_3H_8	21,661
C_4H_{10}	21,308

Solution

The HV_G data may be converted to a mole basis using molecular weights:

$$HV_{G,CH_4} = 23,879 \, \text{Btu/lb}(16.04 \, \text{lb/lbmol})$$

$$= 383,019 \, \text{Btu/lbmol}$$

$$HV_{G,C_3H_8} = 21,661 \, \text{Btu/lb}(44.09 \, \text{lb/lbmol})$$

$$= 955,033 \, \text{Btu/lbmol}$$

$$HV_{G,C_4H_{10}} = 21,308 \, \text{Btu/lb}(58.12 \, \text{lb/lbmol})$$

$$= 1,238,421 \, \text{Btu/lbmol}$$

Using the equation given in the previous problem gives

$$HV_G = 0.75(383,019) + 0.10(955,033) + 0.15(1,238,421)$$

$$= 568,531 \, \text{Btu/lbmol}$$

As demonstrated in this and the previous problem, the reader should note that various units can be employed when dealing with combustion properties such as $\Delta H, \Delta H^\circ, \Delta H_c^\circ, \Delta H_f^\circ, HV_G, HV_N$, etc. The two most common sets of units are Btu/scf and Btu/lbmol.

THR.9   THEORETICAL ADIABATIC FLAME TEMPERATURE

Calculate the theoretical adiabatic flame temperature of benzene (in air). The following standard heats of formation and heat capacity data are provided:

Species	ΔH_f° (cal/gmol)	a	$b \times 10^3$	$c \times 10^{-5}$
C_6H_6	11,720	—	—	—
CO_2	−94,052	10.57	2.10	−2.06
H_2O	−57,798	7.30	2.46	0
N_2	0	6.83	0.90	−0.12

The terms a, b, and c are constants for the heat capacity equation $C_P = a + bT + cT^{-2}$, where T is in K and C_P is in cal/(gmol $\cdot$ °C).

Solution

If all the heat liberated by a reaction goes into heating up the products of combustion (the flue gas), the temperature achieved is defined as the *flame temperature*. If the combustion process is conducted adiabatically, with no heat transfer loss to the surroundings, the final temperature achieved by the flue gas is defined as the *adiabatic flame temperature*. If the combustion process is conducted with theoretical or stoichiometric air (0% excess), the resulting temperature is defined as the *theoretical adiabatic flame temperature* (TAFT). TAFT represents the maximum temperature the products of combustion (flue) can achieve if the reaction is conducted both stoichiometrically and adiabatically. For this condition, all the energy liberated on combustion at or near standard conditions (ΔH_c° and/or ΔH_{298}°) appears as sensible heat in heating up the flue products, ΔH_p. This may be represented in equation form as

$$\Delta H_c^{\circ} + \Delta H_p = 0$$

where ΔH_c° = standard heat of combustion at 25°C
ΔH_p = enthalpy change of the products as the temperature increases from 25°C to the theoretical adiabatic flame temperature

If the heat capacity for each product (flue) gas is expressed as

$$C_P = a + bT + cT^{-2}$$

then the heat capacity for the flue gas may be represented by

$$\Sigma C_P = \Sigma a + (\Sigma b)T + (\Sigma c)T^{-2}$$

where

$$\Sigma a = \sum_{products} na$$

n = lbmol or stoichiometric coefficient (depending on basis of calculation) of individual product

with similar definitions for Σb and Σc. Since 25°C = 298 K, the enthalpy change associated with heating the flue products is given by

$$\Delta H_p = \int_{298}^{T_2} \Sigma C_P dT \qquad T_2 = \text{theoretical adiabatic flame temperature (K)}$$

Substituting ΣC_P obtained previously and integrating yields

$$\Delta H_p = \Sigma a(T_2 - 298) + \frac{\Sigma b}{2}(T_2^2 - 298^2) - \Sigma c(T_2^{-1} - 298^{-1})$$

This expression may now be equated with the negative of the enthalpy of reaction (or combustion) to calculate the theoretical adiabatic flame temperature. This procedure is illustrated in the solution to this problem.

The balanced combustion equation for benzene is

$$C_6H_6 + 7.5O_2 \rightarrow 6CO_2 + 3H_2O$$

The heat of combustion (using the data provided) is then

$$\Delta H_c^\circ = 6(-94,052) + 3(-57,798) - 11,720$$
$$= -749,426 \text{ cal/gmol}$$

Rewriting the above reaction equation for air (21 mol% O_2 and 79 mol % N_2),

$$C_6H_6 + 7.5O_2 + 28.2N_2 \rightarrow 6CO_2 + 3H_2O + 28.2N_2$$

Since

$$\Delta H_p = \int_{298}^{T_2} \Sigma C_P dT \qquad T_2 = \text{final temperature}$$

the term ΣC_p needs to be evaluated. By definition,

$$\Sigma C_P = 6C_{P,CO_2} + 3C_{P,H_2O} + 28.2C_{P,N_2}$$

Based on the heat capacity data given in the problem statement,

$$\Sigma a = 277.9$$
$$\Sigma b = 0.0454$$
$$\Sigma c = -1.57 \times 10^6$$

and

$$\Sigma C_P = \Sigma a + \Sigma bT + \Sigma cT^{-2}$$
$$= 277.9 + 0.0454T - 1.57 \times 10^{-6}T^{-2}$$

Substituting ΣC_P into the equation for ΔH_P and integrating yields

$$\Delta H_P = 277.9(T_2 - 298) + \left(\frac{0.0454}{2}\right)(T_2^2 - 298^2) + 1.57 \times 10^6 \left(\frac{1}{T_2} - \frac{1}{298}\right)$$

For adiabatic conditions,

$$\Delta H_P = -\Delta H_c^\circ$$
$$= 749,426 \, \text{cal/gmol}$$

Substitution gives

$$839,524 = 277.9T_2 + 0.0227T_2^2 + 1.57 \times 10^{-6}T_2^{-1}$$

This equation can be solved by trial-and-error. However, a simpler method can be used. As a first guess, neglect the last term on the right-hand side (RHS):

$$0.0227T_2^2 + 277.9T_2 - 839524 = 0$$

$$T_2 = \frac{-277.9 \pm \sqrt{(277.9)^2 + (4)(0.0227)(839524)}}{0.0454}$$

$$= 2507 \, \text{K}$$

Substituting this temperature into the RHS of the original equation and evaluating the RHS gives a value of 839,992. Since this compares very favorably with the left-hand side (LHS) value of 839,524 cal/gmol, the assumption is valid, i.e., the third term of the RHS is negligible. Thus,

$$T_2 = 2507 \, \text{K}$$
$$= 4513^\circ \text{R}$$
$$= 4053^\circ \text{F}$$

THR.10   EQUILIBRIUM CONSTANT

The equilibrium constant (K_p) for the reaction (at 1 atm pressure)

$$SO_2 + \tfrac{1}{2}O_2 \rightleftharpoons SO_3$$

may be expressed in the form (Theodore, personal notes)

$$Y = Ae^{BX}$$

where $A = 0.148 \times 10^{-4}$
$\quad B = 11{,}700$
$\quad Y = K_{\mathrm{p}}$
$\quad X = 1/T \qquad T = \text{kelvin (K)}$

Develop a more rigorous equation for K_{p} as a function of the absolute temperature (K). The following standard free energy of formation $(\Delta G_{\mathrm{f}}^{\circ})$, standard heat of formation, and heat capacity data are provided.

Species	$\Delta H_{\mathrm{f}}^{\circ}$ (cal/g)	$\Delta G_{\mathrm{f}}^{\circ}$ (cal/g)	a	$b \times 10^{3}$	$c \times 10^{-5}$
O_2	—	—	7.16	−1.0	−0.4
SO_2	−70,960	−71,790	11.04	1.88	−1.84
SO_3	−94,450	−88,520	13.90	6.10	−3.72

Once again, a, b, and c are constants for the equation $C_P = a + bT + cT^{-2}$, where T is in K and C_P is in cal/(gmol · °C).

Solution

Chemical reaction equilibrium calculations are structured around a thermodynamic term referred to as *free energy*. This so-called energy (G) is a thermodynamic property that cannot be easily defined without some basic grounding in thermodynamics. No attempt will be made to define it here, and the interested reader is directed to the literature for further development of this topic. Free energy has the same units as enthalpy and may be used on a mole or total mass basis. Consider the equilibrium reaction

$$a\mathrm{A} + b\mathrm{B} \rightleftharpoons c\mathrm{C} + d\mathrm{D}$$

For this reaction

$$\Delta G_{298}^{\circ} = c(\Delta G_{\mathrm{f}}^{\circ})_{\mathrm{C}} + d(\Delta G_{\mathrm{f}}^{\circ})_{\mathrm{D}} - a(\Delta G_{\mathrm{f}}^{\circ})_{\mathrm{A}} - b(\Delta G_{\mathrm{f}}^{\circ})_{\mathrm{B}}$$

The standard free energy of reaction ΔG° may be calculated from standard free energy of formation data in a manner similar to that for the standard enthalpy of reaction. The following equation is used to calculate the chemical reaction equilibrium constant K at a temperature T:

$$\Delta G_{\mathrm{T}}^{\circ} = -RT \ln K$$

The effect of temperature on $\Delta G_{\mathrm{T}}^{\circ}$ and K is provided below. If the molar heat capacity for each chemical species taking part in the reaction is known and can be expressed as a power series in T (kelvins), then

$$C_P = \alpha + \beta T + \gamma T^{2}$$

If $\Delta\alpha$, etc., represent the difference between the α values for the products and the reactants, then

$$\ln K_T = -\frac{\Delta H_0}{RT} + \frac{\Delta\alpha}{R}\ln T + \frac{\Delta\beta}{2R}T + \frac{\Delta\gamma}{6R}T^2 + I$$

and

$$\Delta G_T^\circ = \Delta H_0 - \Delta\alpha(T)\ln T - \frac{\Delta\beta}{2}T^2 - \frac{\Delta\gamma}{6}T^3 - I(R)(T)$$

Two unknowns, ΔH_0 and I, appear in the above two equations. The term ΔH_0 may be evaluated from the equation

$$\Delta H_T^\circ = \Delta H_0 + \Delta\alpha T + \frac{\Delta\beta}{2}T^2 + \frac{\Delta\gamma}{3}T^3$$

if ΔH_T° is given (or available) at a specified temperature (usually 25°C). The term I can be calculated if ΔG_T° or K is given at a specified temperature.

If the heat capacities for the gases are in the form (as with this problem)

$$C_P = a + bT + cT^{-2}$$

then

$$\ln K_T = -\frac{\Delta H_0}{RT} + \frac{\Delta a}{R}\ln T + \frac{\Delta b}{2R}T + \frac{\Delta c}{2R}T^{-2} + I$$

and

$$\Delta G_T^\circ = \Delta H_0 - \Delta a(T)\ln T - \frac{\Delta b}{2}T^2 - \frac{\Delta c}{2}T - I(R)(T)$$

with

$$\Delta H_T^\circ = \Delta H_0 + \Delta a T + \frac{\Delta b}{2}T^2 - \frac{\Delta c}{T}$$

Applying all of the above to the reaction equation,

$$\Delta G_{298}^{\circ} = -88,520 - (-71,790)$$

$$= -16,730 \, \text{cal/gmol}$$

$$\Delta H_{298}^{\circ} = -94,450 - (-70,960)$$

$$= -23,490 \, \text{cal/gmol}$$

$$\Delta a = 13.90 - [11.04 + 0.5(7.16)]$$

$$= -0.72$$

$$\Delta b = 6.10 \times 10^{-3} - [1.88 \times 10^{-3} + 0.5(-1.0 \times 10^{-3})]$$

$$= 3.72 \times 10^{-3}$$

$$\Delta c = -3.72 \times 10^{5} - [-1.84 \times 10^{5} + (0.5)(-0.4 \times 10^{5})]$$

$$= -1.18 \times 10^{5}$$

Thus,

$$\Delta H_0 = \Delta H_{298}^{\circ} - 298\Delta a - \frac{298^2}{2}\Delta b + \frac{1}{298}\Delta c$$

$$= -23,490 - 298(-0.72) - \frac{298^2}{2}(3.72 \times 10^{-3}) + \frac{1}{298}(-1.18 \times 10^{-5})$$

$$= -23,836 \, \text{cal/gmol}$$

In addition

$$\ln K = -\frac{\Delta G_{298}^{\circ}}{RT} = \frac{16,730}{(1.99)(298)} = 28.2$$

One may now solve for I using the above two values:

$$\ln K = -\frac{\Delta H_0}{RT} + \frac{\Delta a}{R}\ln T + \frac{\Delta b}{2R}T + \frac{\Delta c}{2R}T^{-2} + I$$

$$28.2 = \frac{23,836}{(1.99)(298)} - \frac{0.72}{1.99}(\ln 298) + \frac{3.72 \times 10^{-3}}{2(1.99)}(298) - \frac{1.18 \times 10^{5}}{2(1.99)}298^{-2} + I$$

$$I = -9.88$$

The equation for the equilibrium constant as a function of temperature is therefore

$$\ln K = \frac{11,978}{T} - 0.362 \ln T + 9.35 \times 10^{-4}T - 2.96 \times 10^{4}T^{-2} - 9.88$$

THR.11   EQUILIBRIUM PARTIAL PRESSURES

Refer to Problem THR. 10. The chemical reaction equilibrium constant based on partial pressures (K_P) was obtained as a function of temperature for the reaction

$$SO_2 + \tfrac{1}{2}O_2 \rightleftharpoons SO_3$$

The final result took the form

$$\ln K = \frac{11{,}978}{T} - 0.362 \ln T + 9.35 \times 10^{-4}T - 2.96 \times 10^4 T^{-2} - 9.88$$

If the initial partial pressures of SO_2, O_2, and SO_3 are 0.000257, 0.102, and 0.0 atm, respectively, estimate the equilibrium partial pressure of the SO_3. The operating conditions are 1.0 atm and 1250 K.

Solution

Once the chemical reaction equilibrium constant (for a particular reaction) has been determined, one can proceed to estimate the quantities of the participating species at equilibrium. The problem that remains is to relate K to understandable physical quantities. For gas-phase reactions, as in an incinerator operation, the term K may be approximately represented in terms of the partial pressures of the components involved. This functional relationship for the hypothetical reaction

$$aA + bB \rightleftharpoons cC + dD$$

is given by

$$K_P = \frac{p_C{}^c p_D{}^d}{p_A{}^a p_B{}^b}$$

where p_A is the partial pressure of component A, etc.

Assuming a K value is available or calculable, this equation may be used to determine the partial pressures of the participating components at equilibrium. To determine these quantities, one must account for the amounts reacted or produced. The following equations are valid if the partial pressures of the participating components are low. If x represents the increase in p_C due to the reaction, then

$$p_C = p_C(\text{initial}) + x$$
$$p_D = p_D(\text{initial}) + (d/c)x$$
$$p_A = p_A(\text{initial}) - (a/c)x$$
$$p_B = p_B(\text{initial}) - (b/c)x$$

The reaction equation

$$SO_2 + \tfrac{1}{2}O_2 \rightleftharpoons SO_3$$

is to be employed with the following initial partial pressures:

$$p_{SO_2} = 0.000257 \text{ atm}$$
$$p_{O_2} = 0.102 \text{ atm}$$
$$p_{SO_3} = 0.0 \text{ atm}$$

For this reaction

$$K_P = \frac{p_{SO_3}}{p_{SO_2} p_{O_2}^{0.5}}$$

The partial pressures may be approximated by

$$p_{SO_3} = x$$
$$p_{SO_2} = 0.000257 - x$$
$$p_{O_2} = 0.102 - 0.5x$$

For $T = 1250$ K,

$$\ln K = \frac{11,978}{1250} - 0.362(\ln 1250) + 9.35 \times 10^{-4}(1250) - \frac{2.96 \times 10^4}{1250^2} - 9.88$$
$$= -1.73$$
$$K = 0.1773$$

For ideal conditions, K_p may be assumed to be equal to K. Therefore,

$$0.1773 = \frac{x}{(0.000257 - x)(0.102 - 0.5x)^{0.5}}$$

Solving by trial and error,

$$x = 13.8 \times 10^{-6} \text{ atm}$$
$$= 13.8 \text{ppm}$$

THR.12   SPRAY CHAMBER APPLICATION

An air stream at 90°F, 14.7 psia, and a relative humidity of 60% is flowing into a water-spray chamber where enough water is added to reach saturation. The outlet temperature and pressure of the saturated stream are 80°F and 12.0 psia, respectively. After the spray chamber, the stream is heated and compressed to 30 psia and has a final relative humidity of 30%.

1. In the spray chamber, what is the ratio of the amount (moles) of water added (i.e., absorbed by the air stream) to the amount of incoming moist air?
2. What is the dew point of the moist air leaving the compressor/heater unit?
3. What is the final temperature of the moist air?

The following data are from steam tables and may be used for this problem:

Temperature (°F)	Vapor Pressure (psia)
32	0.08859
40	0.12163
50	0.17796
60	0.25611
70	0.36292
80	0.50683
90	0.69813
100	0.94924
110	1.2750
120	1.6927
130	2.2230
140	2.8892
150	3.7184
160	4.7414
170	5.9926
180	7.5110
190	9.340
200	11.526
212	14.696

Solution

The reader should become familiar with the following definitions, which are pertinent to vapor–liquid systems (some specifically to air–water systems).

- *Absolute saturation*: mass of vapor/mass of bone dry gas
- *Absolute humidity* (h_a): same definition as absolute saturation but applied to the air–water system, i.e.,

$$h_a = \text{mass of water vapor/mass of BDA}$$

where BDA is bone dry air.

- *Relative saturation*: (partial pressure vapor/vapor pressure) $\times 100\%$
- *Relative humidity* (h_r): same definition as relative saturation, but applied to the air–water system, i.e.,

$$\frac{\text{Partial pressure of water vapor}}{\text{Vapor pressure of water}} \times 100\%$$

or

$$h_r = (p_w/p'_w) \times 100\%$$

Note that since the vapor pressure is the maximum partial pressure that a gas can have at a given temperature, the relative humidity (or relative saturation) is a measure of how much vapor the gas is holding relative to the maximum (saturation) amount it can hold.

Another term that measures how concentrated the vapor is in the gas phase is the *dew point*. Since vapor pressure decreases with decreasing temperature, a gas holding a specific amount of vapor can be brought to the saturation point by lowering the temperature. The temperature at which the gas reaches saturation in this fashion is called the *dew point*. As an example, suppose the air in a room at 70°F is at 50% relative humidity. From the steam tables, the vapor pressure of water at 70°F is 0.3631 psi, which means that the air at 50% humidity holds water vapor with a partial pressure of (0.50) (0.3631) or 0.1812 psi. If the temperature is dropped to the point where 0.1812 psi equals the water vapor pressure (around 52°F), the air becomes saturated with water and any further drop in temperature will cause condensation. The dew point of this air mixture is then 52°F. Obviously, if the air were already saturated at 70°F, (i.e., 100% relative humidity), the dew point would also be 70°F.

With the aid of steam tables, the reader should now be able to find the mole fractions of air and water in an air–water system, given only the temperature, T, plus either

1. The dew point
2. Relative humidity or
3. Absolute humidity

These three cases are discussed below.

1. If the dew point temperature (DPT) is known, the vapor pressure at the dew point temperature should be found from the steam tables. The mole fraction of water, y_w, is then found using Dalton's law.

$$y_w = p'_{\text{DPT}}/P$$

2. If the relative humidity, h_r, is given, the vapor pressure at the gas temperature is first found from the steam tables. The partial pressure of the water is then

$$p_w = h_r p'_w$$

where h_r is expressed as a fraction, not as a percent. The water mole fraction is then obtained from Dalton's law:

$$y_w = p_w/P$$

3. If the absolute humidity, h_a, is known, the molecular weights of water (18) and bone dry air (29) must be used to convert from a mass ratio to a molar ratio. The molar ratio of water vapor to BDA is

$$(29/18)h_a = 1.611h_a$$

The mole fraction of water vapor is now

$$y_w = 1.611h_a/(1.611h_a + 1)$$

In all three of these examples, the mole fraction of air, y_a, is found by subtracting y_w from unity.

For the problem at hand, choose a basis of 100 lbmol feed. A flow diagram of the process as shown in Figure 33.

Dalton's law and the steam table data given in the problem statement may now be used to determine the compositions (mole fractions) of the moist air feed stream, the stream leaving the spray chamber, and the stream leaving the compressor/heater unit.

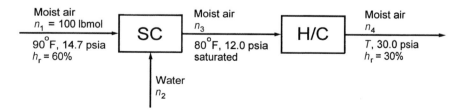

Figure 33. Flow diagram for THR.12.

Feed stream:

$$p_w = h_r p'_w \quad \text{(from the table provided, } p'_w \text{ at } 90°\text{F} = 0.69813 \text{ psia)}$$
$$= (0.60)(0.69813)$$
$$= 0.4189 \text{ psia}$$
$$y_w = p_w/P$$
$$= 0.4189/14.7$$
$$= 0.0285$$
$$y_a = 1 - 0.0285$$
$$= 0.9715$$

Stream exiting the spray chamber:

$$p_w = h_r p'_w \quad \text{(from the table, } p'_w \text{ at } 80°\text{F} = 0.50683 \text{ psia)}$$
$$= (1.00)(0.50683)$$
$$= 0.50683 \text{ psia}$$
$$y_w = p_w/P$$
$$= 0.50683/12.0$$
$$= 0.0422$$
$$y_a = 1 - 0.0422$$
$$= 0.9578$$

Stream exiting the compressor/heater unit:
Since no water is added to or removed from the air stream in the compressor/heater unit, the composition must remain the same. Therefore,

$$y_w = 0.0422$$
$$y_a = 0.9578$$

A material balance on the air around the water spray unit is used to find the amount of moist air leaving this unit:

$$(0.9715)100 = (0.9578)n_3$$
$$n_3 = 101.43 \text{ lbmol}$$

A total material balance around the water-spray unit provides the amount of water added:

$$100 + n_2 = n_3$$

$$n_2 = 101.43 - 100$$

$$= 1.43 \, \text{lbmol water added}$$

Thus the ratio of the amount of water added to the amount of incoming moist air is

$$\text{Ratio} = n_2/n_1 = 1.43/100$$

$$= 0.0143$$

The steam table can again be used to determine the dew point of the moist air leaving the compressor/heater unit.

In this stream,

$$p_w = y_w P$$

$$= (0.0422)(30 \, \text{psia})$$

$$= 1.266 \, \text{psia}$$

From the table, the temperature at which the air stream would be saturated (dew point) is somewhere between $100°F$ ($0.94924 \, \text{psia}$) and $110°F$ ($1.2750 \, \text{psia}$). Using linear interpolation,

$$\frac{1.2750 - 0.94924}{1.2750 - 1.266} = \frac{110 - 100}{110 - T}$$

$$T = 109.7°F$$

$$\approx 110°F$$

The vapor pressure of the water in the air stream leaving the compressor/heater unit is given by

$$p_w = y_w P$$

$$= (0.0422)(30 \, \text{psia})$$

$$= 1.266 \, \text{psia}$$

$$p'_w = p_w/h_r$$

$$= 1.266/0.30$$

$$= 4.22 \, \text{psia}$$

From the table, the temperature of moist air with a p'_w of 4.22 psia is somewhere between 150°F (3.7184 psia) and 160°F (4.7414 psia). Using linear interpolation,

$$\frac{4.7414 - 4.22}{4.7414 - 3.7184} = \frac{160 - T}{160 - 150}$$

$$T = 154.9°F$$

$$\approx 155°F$$

10 Chemical Kinetics (KIN)

KIN.1   REACTION MECHANISMS

Assuming power law (elementary reaction rate) kinetics to apply, write the rate equation for the following reactions:

1. First-order, irreversible reaction

$$A \rightarrow products$$

2. Second-order, irreversible reaction

$$2A \rightarrow products$$

3. Second-order, irreversible reaction

$$A + B \rightarrow products$$

4. Third-order, irreversible reaction

$$3A \rightarrow products$$

5. Third-order, irreversible reaction

$$2A + B \rightarrow products$$

6. Fractional or higher order, irreversible reaction

$$nA \rightarrow products$$

7. First-order reversible reaction

$$A \rightleftharpoons B$$

8. First-/second-order, reversible reaction

$$A \rightleftharpoons B + C$$

9. Simultaneous irreversible reactions

$$A \xrightarrow{k_{A1}} \text{products}$$
$$A + B \xrightarrow{k_{A2}} \text{products}$$
$$3A \xrightarrow{k_{A3}} \text{products}$$

10. Consecutive irreversible reactions

$$A \xrightarrow{k_A} B$$
$$B \xrightarrow{k_B} C$$

Solution

Based on experimental evidence, the rate of reaction is a function of:

1. Concentration of components existing in the reaction mixture (this includes reacting and inert species)
2. Temperature
3. Pressure
4. Catalyst variables

This may be put in equation form:

$$r_A = r_A(C_i, P, T, \text{catalyst variables})$$

or simply

$$r_A = \pm k_A f(C_i)$$

where k_A incorporates all the variables other than concentration. This $\pm$ notation is included to account for the reaction or formation of A. One may think of k_A as a constant of proportionality. It is defined as the *specific reaction rate* or more commonly the *reaction velocity constant*. It is a "constant" that is *independent* of concentration but *dependent* on the other variables. This approach has, in a sense, isolated one of the variables. The reaction velocity constant, like the rate of reaction, *must* refer to one of the species in the reacting system. However, K almost always is based on the same species as the rate of reaction. Consider, for example, the reaction

$$aA + bB \rightarrow cC + dD$$

where the notation $\rightarrow$ represents an irreversible reaction, i.e., if stoichiometric amounts of A and B are initially present, the reaction will proceed to the right until all the A and B have reacted (disappeared) and C and D have been formed. If the reaction is *elementary*, the rate of the above reaction is given by

$$r_A = -k_A C_A{}^a C_B{}^b$$

where the negative sign is introduced to account for the disappearance of A and the product concentrations do not affect the rate. For elementary reactions, the reaction mechanism for r_A is simply obtained by multiplying the molar concentrations of the reactants raised to powers of their respective stoichiometric coefficients (power law kinetics). For nonelementary reactions, the mechanism can take any form.

The order of the above reaction with respect to a particular species is given by the exponent of that concentration term appearing in the rate expression. The above reaction is, therefore, of order a with respect to A, and of order b with respect to B. The overall order n, usually referred to as *the order*, is the sum of the individual orders, i.e.,

$$n = a + b$$

All real and naturally occurring reactions are reversible. A reversible reaction is one in which products react to form reactants. Unlike irreversible reactions that proceed to the right until completion, reversible reactions achieve an equilibrium state after an infinite period of time. Reactants and products are still present in the system. At this (equilibrium) state, the reaction rate is zero. For example, consider the following reversible reaction:

$$aA + bB \rightleftharpoons cC + dD$$

where the notation $\rightleftharpoons$ is a reminder that the reaction is reversible; $\rightarrow$ represents the forward reaction contribution to the total or net rate, while $\leftarrow$ represents the contribution of the reverse reaction. The notation $=$ is employed if the reaction system is at equilibrium. The rate of this reaction is given by

$$r_A = \underset{\text{forward reaction}}{-k_A C_A{}^a C_B{}^b} + \underset{\text{reverse reaction}}{k_A' C_C{}^c C_D{}^d}$$

and

$$K_A = k_A / k_A'$$

With this as an introduction, the reader is now provided with the answers to this 10-part problem.

Since these are elementary reactions, power law kinetics apply, so that the solutions become

1. $r_A = -k_A C_A$
2. $r_A = -k_A C_A^2$
3. $r_A = -k_A C_A C_B$
4. $r_A = -k_A C_A^3$
5. $r_A = -k_A C_A^2 C_B$
6. $r_A = -k_A C_A^n$
7. $r_A = -k_A C_A + k_A' C_B$
   $\quad = -k_A(C_A - K_A C_B)]$

where K_A is the chemical reaction equilibrium constant for A based on concentration:

$$K_A = k_A'/k_A$$

Note that the contribution from the reverse reaction needs to be included in the rate term for A. The reaction rate constant for the reversible reaction is normally designated by k'.

8. $r_A = -k_A C_A + k_A' C_B C_C$
9. $r_A = -k_{A1} C_A - k_{A2} C_A C_B - k_{A3} C_A^3$
   $\quad r_B = -k_{A2} C_A C_B$

Note that the contribution from each of the three reactions needs to be included in the rate term.

10. $r_A = -k_A C_A$
    $\quad r_B = k_A C_A - k_B C_B$
    $\quad r_C = k_B C_B$

Note that the rate term for C is positive since it is being formed.

Before leaving this problem, the reader should note that a *consistent* and *correct* definition of the rate of a reaction is essential before meaningful kinetic and reactor applications can be discussed. The rate of a chemical reaction is defined as the time rate of change in the quantity of a particular species (say A) participating in a reaction divided by a factor that characterizes the reacting system's geometry. The choice of this factor is also a matter of convenience. For homogeneous media, the factor is almost always the volume of the reacting system. In equation form, then

$$r_A = \frac{1}{V}\frac{dn_A}{dt}$$

where r_A = rate of reaction for A
n_A = moles of A at time t
t = time
V = volume of reacting system

If the volume term is constant, one may write the above equation as

$$r_A = \frac{d(n_A/V)}{dt} = \frac{dC_A}{dt}$$

where C_A is the molar concentration of A.

This equation states that the reaction rate is equal to the rate of change of concentration with respect to time—all in consistent units. Rates expressed using concentration changes require the assumption of constant volume. The units of the rate become lbmol/(h · ft^3) and gmol/(s · L) in the engineering and metric (SI) systems, respectively. For fluid–solid reaction systems, the factor is often the mass of the solid. For example, in gas-phase catalytic reactions, the factor is the mass of catalyst (cat). The units of the rate are then lbmol/(h · lb) cat. In line with this definition for the rate of reaction, the rate is positive if species A is being formed or produced since C_A increases with time. The rate is negative if A is reacting or disappearing due to the reaction because C_A decreases with time. The rate is zero if the system is at equilibrium.

KIN.2 ARRHENIUS EQUATION

The following reaction velocity constant data were obtained for the reaction between two inorganic chemicals:

T (°C)	k [L/(gmol · h)]
0	5.20
20	12.0
40	21.0
60	39.0
80	60.0
100	83.0

Using the values of k at 40 and 100°C, calculate the constants of the Arrhenius equation. Use these to obtain k at 75°C.

Equations to describe the rate of reaction at the macroscopic level have been developed in terms of meaningful and measurable quantities. The reaction rate is affected not only by the concentration of species in the reacting system but also by the temperature. An increase in temperature will almost always result in an increase in the rate of reaction; in fact, the literature states that, as a general rule, a 10°C increase in reaction temperature will double the reaction velocity constant. However, this is generally no longer regarded as a truism, particularly at elevated temperatures.

The Arrhenius equation relates the reaction velocity constant with temperature. It is given by

$$k = Ae^{-E_a/RT}$$

where A = frequency factor constant and is usually assumed to be independent of temperature

R = universal gas constant

E_a = activation energy and is also usually assumed independent of temperature

Solution

The Arrhenius equation

$$k = Ae^{-E_a/RT}$$

may also be written as

$$\ln k = \ln A - E_a/RT$$

When the data are plotted on $\ln k$ vs. $1/T(K)$ coordinates, an approximate straight line passing through the points ($T = 283$ K, $k = 8.5$ L/gmol · h) and ($T = 363$ K, $k = 68.3$ L/gmol · h) results. For 40°C, substitution yields

$$\ln 8.5 = \ln A - E_a/(1.987)(283)$$

At 100°C,

$$\ln 68.3 = \ln A - E_a/(1.987)(363)$$

The two equations given above may be solved simultaneously. Subtracting the first equation from the second gives

$$\ln 8.5 - \ln 68.3 = -E_a/562.3 + E_a/721.3$$

Solving for E_a,

$$E_a = 5316 \, \text{cal/gmol}$$

Solving for A yields

$$A = 108,400 \, \text{L/(gmol} \cdot \text{h)}$$

Thus, k at 75°C is

$$k = 108,400 \exp[-5316/(1.987)(348)]$$
$$= 49.7 \, \text{L/(gmol} \cdot \text{h)}$$

Note that linearly interpolating between 60 and 80°C gives a value of 54.8 L/(gmol · h).

Accordingly, the Arrhenius equation should yield a straight line of slope $-E_a/R$ and intercept A if $\ln k$ is plotted against $1/T$. Implicit in this statement is the assumption that E_a is constant over the temperature range in question. Despite the fact that E_a generally varies significantly with temperature, the Arrhenius equation has wide applicability in industry. This method of analysis can be used to test the rate law, describe the variation of k with T, and/or evaluate E_a. The numerical value of E_a will depend on the choice and units of the reaction velocity constant.

KIN.3   LINEAR REGRESSION ANALYSIS

The following data have been obtained for the rate $-r_A$ versus concentration C_A.

$-r_A$ [lbmol/(ft^3 · s)]	C_A (lbmol/ft^3)
48	8
27	6
12	4
3	2

Using the above data, estimate the coefficient k_A and α in

$$-r_A = k_A C_A{}^{\alpha}$$

Solution

If the functional relationship between one variable and another is linear, a straight-line plot would be obtained on arithmetic-coordinate graph paper. If the relationship approaches a linear one, the "best" method of fitting the data to a linear model would be through the *method of least squares*. The resulting linear equation (or line) would have the properties of lying as "close" as possible to the data. For statistical purposes, "close" and/or "best" fit is defined as that linear equation or line for which the sum of the squared vertical distances between the data (values of Y or independent variable) and line is minimized. These distances are called *residuals*. This approach is employed in the solution below.

Linearizing the above equation may be accomplished by taking the natural logarithm (ln) of both sides of the equation:

$$\ln(-r_A) = \ln k_A + \alpha \ln C_A$$

Setting $\ln(-r_A)$ equal to Y and $\ln C_A$ equal to X produces

$$Y = A + BX$$

where $A = \ln k_A$
$B = \alpha$

The above data (four data points), when substituted into the equation, yield:

$$\ln(3) = A + B \, \ln(2)$$
$$\ln(12) = A + B \, \ln(4)$$
$$\ln(27) = A + B \, \ln(6)$$
$$\ln(48) = A + B \, \ln(8)$$

As indicated earlier, the method of least squares requires that the sum of the squared residuals be minimized. The resulting coefficients are

$$A = -0.2878$$
$$B = 2.0$$

These calculations may be done through a long-hand calculation. However, they are more often accomplished with the aid of computer software.

Taking the inverse natural logarithm of A to obtain k_A leads to

$$k_A = 0.75$$
$$\alpha = 2.0$$

The equation for the rate of reaction is therefore

$$-r_A = 0.75 C_A^{2.0}$$

KIN.4 TIME REQUIREMENT—FIRST-ORDER REACTION

Calculate the time required to achieve a 99.99% conversion of benzene at a temperature of 980°C. The arrhenius constants are

Frequency factor, $A = 3.3 \times 10^{10} \, s^{-1}$
Activation energy, $E = 35,900 \, cal/gmol$

Assume benzene undergoes a first-order irreversible reaction.

Solution

The kinetic equation for a first-order reaction is

$$\frac{dC}{dt} = -kC$$

where C = concentration of the compound undergoing reaction
k = reaction constant
t = time

The solution to the above differential equation becomes

$$\ln \frac{C}{C_0} = -kt \qquad C_0 \text{ is the initial concentration}$$

For a 99.99% conversion, the ratio C/C_0 is

$$C/C_0 = (0.0001/1.0)$$
$$= 0.0001$$

The reaction rate constant, k, is calculated directly from the Arrhenius equation:

$$k = A e^{-E_a/RT}$$
$$= 3.3 \times 10^{10} \exp[(-35,900 \, cal/gmol)/[1.98 \, cal/(gmol \cdot K)](980 + 273)K]$$
$$= 1.804 \times 10^4 \, s^{-1}$$

Therefore, the required residence time t is

$$t = -\ln(C/C_0)/k$$
$$= -\ln(0.0001)/1.804 \times 10^4$$
$$= 5.106 \times 10^{-4}\,\mathrm{s}$$

Note that all real and naturally occurring reactions are reversible. A reversible reaction is one in which the products also react to form the reactants. Unlike irreversible reactions that proceed to the right until completion, reversible reactions achieve an equilibrium state after an infinite period of time. Reactants and products are still present in the system. At this equilibrium state the net reaction rate is zero.

KIN.5   LIMITING AND EXCESS REACTANTS

When ammonia is burned, the products are nitric oxide and water.

$$4NH_3 + 5O_2 \rightarrow 4NO + 6H_2O$$

Ammonia is fed to a combustion device at the rate of 150 kgmol/h and oxygen at a rate of 200 kgmol/h.

1. Which is the *limiting* reactant?
2. What is the percent excess of the *excess* reactant?
3. If 85 kgmol/h of nitric oxide is produced, at what rate in kg/h is water produced, what is the *fractional conversion* of the ammonia, and what is the *fractional conversion* of the oxygen?

Solution

When methane is burned completely, the stoichiometric equation for the reaction is

$$CH_4 + 2O_2 \rightarrow CO_2 + 2H_2O$$

The stoichiometric ratio of the oxygen to the methane is

$$0.5\,\text{mol methane consumed/mol oxygen consumed}$$

If one starts out with 1 mol of methane and 3 mol of oxygen in a reaction vessel, only 2 mol of oxygen would be used up, leaving an excess of 1 mol. In this case, the oxygen is called the excess reactant and methane is the limiting reactant. The *limiting reactant* is defined as the reactant that would be completely consumed if the reaction went to completion. All other reactants are *excess* reactants. The amount by which a reactant is present in excess is defined as the *percent excess* and is given by

$$\% \text{ excess} = \left(\frac{n - n_s}{n_s} \right) \times 100\%$$

where n = number of moles of the excess reactant at the start of the reaction
n_s = stoichiometric number of moles of the excess reactant (i.e., the exact number of moles needed to react completely with the limiting reactant)

In the example above, the *stoichiometric* amount of oxygen is 2 mol, since that is the amount that would react with the 1 mol of methane. The *excess* amount of oxygen is 1 mol, which is a *percentage excess* of 50% or a *fractional excess* of 0.50. These definitions are employed in the solution below.

The *extent of reaction* balance on this reaction is

Input $\rightarrow$	150		200		0		0
	$4NH_3$	$+$	$5O_2$	$\rightarrow$	$4NO$	$+$	$6H_2O$
Output $\rightarrow$	$150 - z$		$200 - (5/4)z$		z		$(6/4)z$

where z is the rate of NH_3 consumed.

To determine which is the limiting reactant, allow the output rate of NH_3 to go to zero, i.e.,

$$150 - z = 0$$
$$z = 150 \text{ kgmol/h}$$

The output rate of O_2 is then

$$200 - [(\tfrac{5}{4})(150)] = +12.5 \text{ kgmol/h}$$

Since the output rate of O_2 is still positive when the NH_3 is all consumed, NH_3 is the limiting reactant. The stoichiometric O_2 rate is $(\tfrac{5}{4})$ (150) or 187.5 kgmol/h.

Since 12.5 kgmol/h is the "excess" rate of O_2 over and above that needed to react with all of NH_3,

$$\text{Percent excess } O_2 = \frac{12.5}{187.5} (100\%)$$
$$= 6.67\%$$

Based on the problem statement, NO is produced at the rate of 85 kgmol/h. Therefore, the reaction does not go to completion and

$$z = 85 \, \text{kgmol/h}$$

$$\text{Rate } H_2O \text{ produced} = (\tfrac{6}{4})z = (1.5)(85) = 127.5 \, \text{kgmol/h}$$

$$\text{Mass flowrate} = (127.5 \, \text{kgmol/h})(18 \, \text{kg/kgmol})$$

$$= 2295 \, \text{kg/h}$$

The fractional conversion of the ammonia, f_{NH_3}, becomes

$$\text{Fractional conversion of } NH_3 = \text{rate consumed/input rate}$$

$$f_{NH_3} = z/150$$

$$= 85/150$$

$$= 0.567$$

The fractional conversion, f_{O_2}, of the oxygen is

$$f_{O_2} = \tfrac{5}{4}z/200$$

$$= \tfrac{5}{4}(85)/200$$

$$= 0.531$$

KIN.6   CONVERSION, YIELD, AND SELECTIVITY

Acetaldehyde can be formed by oxidizing ethane:

$$C_2H_6 + O_2 \rightarrow CH_3CHO + H_2O$$

However, some carbon dioxide is usually an unwelcome by-product:

$$C_2H_6 + \tfrac{7}{2}O_2 \rightarrow 2CO_2 + 3H_2O$$

If 65% (by mole) of the ethane feed stream forms acetaldehyde, 15% forms CO_2 and the remainder is unreacted, calculate

1. The fractional conversion of ethane
2. The fractional yield of acetaldehyde
3. The selectivity of acetaldehyde over carbon dioxide production

Solution

When multiple reactions (either parallel or consecutive) occur, the side or undesired reactions compete with the main or desired reaction; the less predominant the main reaction is, the smaller will be the amount of desired product for a given amount of reactant. Suppose the following hypothetical parallel reactions occur:

$$A \rightarrow B$$
$$A \rightarrow C$$
$$A \rightarrow D$$

Suppose also that, of a starting amount of 10 mol of A, 4 form the desired product B, 2 form the undesired product C, 1 forms the undesired product D, and 3 remain unreacted. While this process produces 4 mol of valuable product, it could have produced 10 if everything went the way we wanted it to, i.e., if all 10 mol of A reacted to form B. The ratio of the 4 mol of B actually produced to the 10 it potentially could have produced is called the *yield* (0.40 or 40%). By definition, the *yield* of a reaction is a measure of how much of the desired product is produced relative to how much would have been produced if only the desired reaction occurred and if that reaction went to completion. Obviously, the fractional yield must be a number between zero and unity. Another term used in conjunction with multiple reactions is selectivity. *Selectivity* is a measure of how predominant the desired reaction is relative to one of the side reactions. The value of the selectivity is obtained by dividing the number of moles of a desired product actually generated by the number of moles of one of the undesired products produced by a side reaction. In the example above, the selectivity of B over C is 2.0 and that of B over D is 4.0. These definitions are employed in the solution below.

Choose a basis of 100 lbmol/h C_2H_6 input. Perform extent of reaction balances on the two reactions. Only the C_2H_6, CH_3CHO, and CO_2 need be considered since information on the other components is not required in the solution:

Input $\rightarrow$	100			0		
	C_2H_6	$+$	O_2	$\rightarrow$ CH_3CHO	$+$	H_2O
Output $\rightarrow$	$100 - z - y$		$(1/1)z$	z		
Input $\rightarrow$	100			0		
	C_2H_6	$+$ $(7/2)O_2$	$\rightarrow$	$2CO_2$	$+$	$3H_2O$
Output $\rightarrow$	$100 - z - y$			$2y$		

where $z =$ amount of C_2H_6 that forms CH_3CHO
$y =$ amount of C_2H_6 that forms CO_2

The fractional conversion, f, of the C_2H_6 is

$$f = \text{rate } C_2H_6 \text{ consumed/rate } C_2H_6 \text{ input}$$
$$= (z + y)/100$$
$$= (65 + 15)/100$$
$$= 0.80$$

If all of the C_2H_6 had reacted to form CH_3CHO, 100 lbmol/h of the CH_3CHO would have been produced. Thus, the fractional yield Y is

$$Y = 65/100$$
$$= 0.65$$

The selectivity S of the CH_3CHO over CO_2 production may now be calculated:

$$S = \text{rate of } CH_3CHO \text{ produced/Rate of } CO_2 \text{ produced}$$
$$= z/2y$$
$$= 65/[(2)(15)]$$
$$= 2.17$$

The reader should note that the terms *conversion, yield,* and *selectivity* are sometimes confusing. Some people in the field use them interchangeably (which is poor practice). Note that *conversion* is measured solely from the disappearance of a particular reactant; *yield* is measured in terms of the generation of a particular product; and *selectivity* is a comparison of the amounts or flowrates of two generated products.

KIN.7   LIQUID-PHASE REVERSIBLE REACTION

Compound B is prepared from compound A according to the following liquid-phase elementary reversible reaction:

$$A \underset{k_2}{\overset{k_1}{\rightleftarrows}} B$$

A 10-liter batch reactor will be used to perform the reaction. Determine the amount of time necessary to achieve a conversion of 40%.
    Additional information:

$C_{A0} = 3.0 \text{ lbmol/L}$
$k_1(\text{forward}) = 6.0 \text{ min}^{-1}$
$k_2(\text{reverse}) = 0.53 \text{ min}^{-1}$

Solution

A batch reactor is a solid vessel or container. It may be open or closed. Reactants are usually added to the reactor simultaneously. The contents are then mixed (if necessary) to ensure no spatial variations in the concentration of the species present. The reaction then proceeds. There is no transfer of mass into or out of the reactor during this period. The concentration of reactants and products change with time; thus, this is a transient or unsteady-state operation. The reaction is terminated when the desired chemical change has been achieved. The contents are then discharged and sent elsewhere, usually for further processing.

The describing equation for chemical reaction mass transfer is obtained by applying the conservation law for either mass or moles on a time rate basis to the contents of a batch reactor. It is best to work with moles rather than mass since the rate of reaction is most conveniently described in terms of molar concentrations.

The describing equation for species A in a batch reactor takes the form

$$\frac{dn_A}{dt} = r_A V$$

where n_A = moles A at time t
$\quad\quad r_A$ = rate of reaction of A; change in moles A/time-volume
$\quad\quad V$ = reactor volume contents

The above equation may also be written in terms of the conversion variable X since

$$n_A = n_{A0} - n_{A0}X$$

where n_{A0} = initial moles of A
$\quad\quad X = X_A$ = moles of A reacted/moles of A at the start of the reaction.

Thus,

$$n_{A0}\frac{dX}{dt} = -r_A V$$

The integral form of this equation is

$$t = n_{A0}\int_0^X \left(-\frac{1}{r_A V}\right)dX$$

If V is constant (as with most liquid-phase reactions),

$$t = \frac{n_{A0}}{V}\int_0^X \left(-\frac{1}{r_A}\right)dX$$

$$= C_{A0}\int_0^X \left(-\frac{1}{r_A}\right)dX$$

The general rate expression for the reaction, r_A, in terms of the concentrations of the participating species is

$$-r_A = k_1 C_A - k_2 C_B$$

The expressions for the concentrations of A and B in terms of the conversion variable X_A can also be written:

$$C_A = C_{A0}(1 - X_A)$$
$$C_B = C_{A0}(\theta_B + X_A)$$
$$= C_{A0} X_A$$

where θ_B = initial moles of B/initial moles of A

Note that $\theta_B = 0$ since the initial amount of B present is zero. The reaction rate equation may be rewritten in terms of the conversion:

$$-r_A = 6C_{A0} - 6C_{A0}X_A - 0.53C_{A0}X_A \qquad C_{A0} = 3$$
$$= 18(1 - 1.0883X_A)$$

This expression may be inserted into the characteristic batch reactor design equation:

$$t = C_{A0} \int_0^{X_A} \frac{1}{(18)(1 - 1.0883X_A)} dX_A$$

$$= \frac{C_{A0}}{18} \int_0^{X_A} \frac{1}{(1 - 1.0883X_A)} dX_A$$

The integrated form of this equation is

$$t = (C_{A0}/18)[1/(-1.0883)] \ln[1 - 1.0883X_A])$$
$$= (3/18)[1/(-1.0883)] \ln[1 - (1.0883)(0.4)]$$
$$= 0.0875 \text{ min}$$
$$= 5.25 \text{ s}$$

KIN.8   VOLUME FOR TWO CSTRs IN SERIES

Consider the elementary liquid-phase reaction between benzoquinone (B) and cyclopentadiene (C):

$$B + C \rightarrow \text{product}$$

If one employs a feed containing equimolal concentrations of reactants, the reaction rate expression can be written as

$$-r_B = kC_C C_B = kC_B^2$$

Calculate the reactor size requirements for one continuously stirred tank reactor (CSTR). Also calculate the volume requirements for a cascade composed of two identical CSTRs. Assume isothermal operation at $25°C$ where the reaction rate constant is equal to $9.92 \, m^3/(kgmol \cdot ks)$. Reactant concentrations in the feed are each equal to $0.08 \, kgmol/m^3$, and the liquid feed rate is equal to $0.278 \, m^3/ks$. The desired degree of conversion is 87.5%.

Solution

Another reactor where mixing is important is the tank flow or CSTR. This type of reactor, like the batch reactor, also consists of a tank or kettle equipped with an agitator. It may be operated under steady or transient conditions. Reactant(s) are fed continuously, and the product(s) are withdrawn continuously. The reactant(s) and product(s) may be liquid, gas, or solid or a combination of these. If the contents are perfectly mixed, the reactor design problem is greatly simplified for steady conditions because the mixing results in uniform concentration, temperature, etc. throughout the reactor. This means that the rate of reaction is constant. The describing equations are therefore not differential and do not require integration. In addition, since the reactor contents are perfectly mixed, the concentration and/or conversion in the CSTR is exactly equal to the concentration and/or conversion leaving the reactor. The describing equation for the CSTR can then be shown to be

$$V = \frac{F_{A0} X_A}{-r_A}$$

where V = volume of reacting mixture
F_{A0} = inlet molar feed rate of A
X_A = conversion of A; moles of A reactant/moles of A entering the system
$-r_A$ = rate of reaction of A

If the volumetric flowrates entering and leaving the CSTR are constant (this is equivalent to a constant density system), the above equation becomes

$$\frac{V}{q} = \frac{C_{A1} - C_{A0}}{r_A} = \frac{C_{A0} - C_{A1}}{-r_A}$$

where q = total volumetric flowrate through the CSTR
C_{A0} = inlet molar concentration of A
C_{A1} = exit molar concentration of A

The rate equation, $-r_B$, in terms of the conversion variable X is

$$-r_B = k_B C_B^2$$
$$C_B = C_{B0}(1 - X)$$
$$-r_B = k_B C_{B0}^2 (1 - X)^2$$

If only one reactor is employed,

$$V = \frac{F_{B0} X}{-r_B}$$
$$= \frac{F_{B0} X}{k_B C_{B0}^2 (1 - X)^2}$$

Since

$$F_{B0} = C_{B0} q$$
$$= (0.08)(0.278)$$
$$= 0.02224 \, \text{kgmol/ks}$$
$$= 0.02224 \, \text{gmol/s}$$

The volume may be calculated:

$$V = (0.02224)(0.875)/[(9.92)(0.08)^2(1 - 0.875)^2]$$
$$= 19.6 \, \text{m}^3$$

The design equation above may also be applied to two reactors in series:

$$V = \frac{F_{B0}}{k_B C_{B0}^2} \frac{X_1}{(1 - X_1)^2}$$
$$= \frac{F_{B0}}{k_B C_{B0}^2} \frac{\Delta X}{(1 - X_2)^2}$$

For reactor 1,

$$\frac{k C_{B0}^2 V}{F_{B0}} = \frac{X_1}{(1 - X_1)^2}$$

For reactor 2,

$$\frac{kC_{B0}^2 V}{F_{B0}} = \frac{X_2 - X_1}{(1 - X_2)^2}$$

Since the LHSs (left-hand sides) of both of the above equations are equal,

$$\frac{X_1}{(1 - X_1)^2} = \frac{X_2 - X_1}{(1 - X_2)^2}$$

Solving with $X_2 = 0.875$ yields

$$X_1 = 0.7251$$

Thus,

$$V_1 = \frac{(0.278)(0.08)(0.7251)}{(9.92)(0.08)^2(1 - 0.7251)^2}$$

$$= 3.36 \, \text{m}^3$$

$$V_T = V_1 + V_2$$

$$= (2)(3.36)$$

$$= 6.72 \, \text{m}^3$$

KIN.9   GAS-PHASE REACTION

The following elementary gaseous reaction

$$A \rightarrow 2B \qquad k_A = 0.7 \, \text{s}^{-1}$$

is conducted in a tubular flow reactor. The volumetric flowrate and inlet concentration of A are 2.0 L/s and 0.1 gmol/L, respectively. Determine the volume of reactor required to achieve a conversion of 90%. Perform the calculations with and without the volume effects included in the analysis.

Solution

This problem is somewhat complicated because of the gaseous nature of the reacting medium. Before proceding to the actual solution, consider the reaction

$$aA + bB \rightarrow cC + dD$$

The volume of an ideal gas in a reactor at any time can be related to the initial conditions by the following equation:

$$V = V_0 \left(\frac{P_0}{P}\right)\left(\frac{T}{T_0}\right)\left(\frac{n_T}{n_{T0}}\right)$$

where P = absolute pressure

T = absolute temperature

V = volume

n_T = total number of moles

0 = subscript denotes initial conditions

If δ is defined as the increase in the total number of moles per moles of A reacted, then

$$\delta = (d/a) + (c/a) - (b/a) - 1$$

The term ε is defined as the change in the total number of moles when the reaction is complete divided by the total number of moles initially present in (or fed to) the reactor.

This may be shown to be

$$\varepsilon = y_{A0}\delta$$

where y_{A0} is the initial mole fraction of A.

The gas volume can now be expressed as follows:

$$V = V_0 \left(\frac{P_0}{P}\right)\left(\frac{T}{T_0}\right)(1 + \varepsilon X)$$

If the pressure does not change significantly, the equation becomes

$$V = V_0 \left(\frac{T}{T_0}\right)(1 + \varepsilon X)$$

For isothermal operation, one obtains

$$V = V_0(1 + \varepsilon X)$$

The most common type of tubular flow reactor is the single-pass cylindrical tube. Another type is one that consists of a number of tubes in parallel. The reactor(s) may be vertical or horizontal. The feed is charged continuously at the inlet of the tube, and the products are continuously removed at the outlet. If heat exchange with surroundings is required, the reactor setup includes a jacketed tube. If the reactor is empty, a homogeneous reaction—one phase present—usually occurs. If the reactor contains catalyst particles, the reaction is said to be heterogeneous.

Tubular flow reactors are usually operated under steady conditions so that, at any point, physical and chemical properties do not vary with time. Unlike the batch and tank flow reactors, there is no mechanical mixing. Thus, the state of the reacting fluid will vary from point to point in the system, and this variation may be in both the radial and axial direction. The describing equations are then differential, with position as the independent variable.

For the describing equations presented below, the reacting system is assumed to move through the reactor in plug flow (no velocity variation through the cross section of the reactor). It is further assumed that there is no mixing in the axial direction, and complete mixing occurs in the radial direction so that the concentration, temperature, etc., do not vary through the cross section of the tube. Thus, the reacting fluid flows through the reactor in an undisturbed plug of mass. For these conditions, the describing equation for a tubular flow reactor is

$$V = F_{A0} \int -\frac{dX_A}{r_A}$$

Since $F_{A0} = C_{A0} q_0$,

$$\frac{V}{q_0} = C_{A0} \int -\frac{dX_A}{r_A}$$

The RHS (right-hand side) of the above represents the residence time in the reactor based on inlet conditions. If q does not vary through the reactor, then

$$V/q = \theta$$

where θ is the residence time in the reactor.

The rate equation for $-r_A$ is

$$-r_A = k_A C_A$$

If volume effect changes are included,

$$C_A = C_{A0}\left(\frac{1-X}{1+\varepsilon X}\right)$$

so that

$$-r_A = k_A C_{A0}\left(\frac{1-X}{1+\varepsilon X}\right)$$

Without volume effect changes,

$$C_A = C_{A0}(1-X)$$

and

$$-r_A = k_A C_{A0}(1 - X)$$

The tubular flow reactor design equation is

$$V = F_{A0} \int_0^X -\frac{dX}{r_A}$$

Since $F_{A0} = C_{A0}q_0$,

$$V = C_{A0}q_0 \int_0^X -\frac{dX}{r_A}$$

The volume of the reactor, neglecting any volume effect changes, may now be calculated. Since

$$-r_A = k_A C_{A0}(1 - X)$$

$$V = -(q_0/k_A) \int_0^{0.9} -\frac{dX}{1 - X}$$

$$= -(q_0/k_A) \ \ln(1 - X) \big]_0^{0.9}$$

$$= -(2.0/0.7) \ \ln(0.1)$$

$$= 6.57 \text{ liters}$$

In order to include volume effect changes, the terms y_{A0}, δ, and ε must be determined:

$$y_{A0} = 1.0$$
$$\delta = 2 - 1$$
$$= 1.0$$
$$\varepsilon = (1)(1)$$
$$= 1.0$$

The volume of the reactor with volume effect changes becomes

$$V = \frac{q_0}{k_A} \int \left(\frac{1 + \varepsilon X}{1 - X} \right) dX \qquad \varepsilon = 1.0$$

$$= \frac{q_0}{k_A} \int_0^{0.9} \left(\frac{1 + X}{1 - X} \right) dX$$

Integration yields

$$V = \frac{q_0}{k_A}[-X - 2\ \ln(1-X)]_0^{0.9}$$
$$= (2.0/0.7)[-0.9 - (2)\ \ln(0.1) + 0 - (2)\ \ln(1)]$$
$$= 10.59\ \text{liters}$$

The reader should note the difference in the two calculations for the volume. Some researchers have incorrectly neglected this effect. This effect becomes more pronounced for reactions involving large changes in the moles of the reacting system (e.g., petrochemical reactions) and where there are significant changes in the temperature and/or pressure.

KIN.10   CSTR VS. TUBULAR FLOW REACTOR

The liquid reaction

$$A \rightarrow \text{products}$$

follows the rate law

$$-r_A = \frac{k_1 C_A^{1/2}}{1 + k_2 C_A^2}$$

where $k_1 = 5(\text{gmol/L})^{1/2}/\text{h}$ and $k_2 = 10(\text{L/gmol})^2$. The initial concentration of A is 0.5 gmol/L, and the feed rate is 200 gmol/h of A.

1. Find the volume necessary to achieve 60% conversion in a CSTR.
2. Would a larger or smaller tubular flow reaction be required to achieve the same degree of conversion? Explain the difference in results.

Solution

The rate of reaction, $-r_A$, in terms of the conversion variable X is

$$C_A = C_{A0}(1 - X)$$
$$-r_A = \frac{k_1 C_{A0}^{1/2}(1-X)^{1/2}}{1 + k_2 C_{A0}^2(1-X)^2}$$

The design equation if the reaction is conducted in a CSTR is

$$V = \frac{F_{A0}X}{-r_A}$$

$$= F_{A0}X \frac{1 + k_2 C_{A0}^2(1 - X)^2}{k_1 C_{A0}^{1/2}(1 - X)^{1/2}} \qquad X = 0.6$$

Substitution yields

$$V = (200)(0.6)\frac{1 + (10)(0.5)^2(1 - 0.6)^2}{(5)(0.5)^{1/2}(1 - 0.6)^{1/2}} \qquad X = 0.6$$

$$= 75.1 \text{ liters}$$

The design equation if the reaction is conducted in a tubular flow reactor is

$$V = F_{A0}\int_0^x -\frac{dX}{r_A}$$

$$= \frac{F_{A0}}{k_1 C_{A0}^{1/2}}\int_0^x \frac{1 + k_2 C_{A0}^2(1 - X)^2}{(1 - X)^{1/2}}dX$$

The integral I may be defined as

$$I = \int_0^{0.6} \frac{1 + k_2 C_{A0}^2(1 - X)^2}{(1 - X)^{1/2}}dX$$

$$= \int_0^{0.6} f(X)dX$$

Simpson's three-point rule is used to evaluate the integral:

$$I = (h/3)[f(0.0) + (4)f(0.3) + f(0.6)] \qquad h = 0.3$$
$$= (0.3/3)[(3.5) + (4)(2.659) + 2.214)]$$
$$= 1.635$$

The volume is then

$$V = \frac{F_{A0}I}{k_1 C_{A0}^{1/2}}$$

$$= 92.5 \text{ liters}$$

Surprisingly, the CSTR requires a smaller volume. This result arises because of the unique nature of the rate expression. Also note that the integral could have been evaluated by any one of several methods.

KIN.11 BATCH REACTOR

The elementary liquid-phase reaction

$$A + B \rightarrow P$$

is carried out to a conversion of A of 0.5 in a constant volume, insulated batch reactor. The reactor is charged with equimolar concentrations of A and B at 27°C.
Additional information:

Rate constant $k = 1.0 \times 10^{-4} \, \text{L/min} \cdot \text{gmol}$ at 27°C

Activation energy $E_a = 3000 \, \text{cal/gmol}$

Heat capacities $C_{PA} = C_{PB} = 10 \, \text{cal/(gmol} \cdot \text{K)}$
    $C_{PP} = 20 \, \text{cal/(gmol} \cdot \text{K)}$

Heat of reaction $\Delta H_R = -5000 \, \text{cal/gmol}$ of A at 27°C

Initial concentrations $C_{A0} = C_{B0} = 5 \, \text{gmol/L}$

Calculate the reaction temperature at $X_A = 0.5$, and outline a method to calculate the time required to achieve this conversion.

Solution

The equation describing temperature variations in batch reactors is obtained by applying the conservation law for energy on a time–rate basis to the reactor contents. Since batch reactors are stationary (fixed in space), kinetic and potential effects can be neglected. The equation describing the temperature variation in reactors due to energy transfer, subject to the assumptions in its development, is

$$nC_P(dT/dt) = \dot{Q} + V(-\Delta H_A)| - r_A|$$

where n = number of moles of the reactor contents at time t

$\quad C_P$ = heat capacity of the reactor contents

$\quad \dot{Q}$ = heat transfer rate across the walls of the reactor

$\quad V$ = reactor volume

$\quad -\Delta H_A$ = enthalpy of reaction of species A

$\quad |-r_A|$ = absolute value of the rate of reaction of A

$\quad T$ = reactor temperature

In addition,

$$\dot{Q} = UA(T - T_a)$$

where U = overall heat transfer coefficient

A = area available for heat transfer

T_a = temperature surrounding reactor walls

For nonisothermal reactors, one of the reactor design equations, the energy transfer equation (see above), and an expression for the rate in terms of concentration and temperature must be solved simultaneously to give the conversion as a function of time. Note that the equations may be interdependent; each can contain terms that depend on the other equation(s). These equations, except for simple systems, are usually too complex for analytical treatment.

Regarding the energy equation, note that for an adiabatic system the above equation reduces to

$$nC_p \frac{dT}{dt} = V(-\Delta H_A)| - r_A|$$

Since, by definition,

$$-r_A = \frac{1}{V} \frac{dn_A}{dT} \qquad n_A = n_{A0}(1 - X)$$

$$= \frac{1}{V} \frac{d}{dt}[n_{A0}(1 - X)]$$

Thus,

$$V|-r_A| = n_{A0} \frac{dX}{dT}$$

and the above equation becomes

$$nC_p dT = (-\Delta H_A)n_{A0} \, dX$$

This equation may be integrated directly to give the temperature as a function of conversion,

$$nC_p(T - T_0) = (-\Delta H_A)n_{A0}(X - X_0)$$

where the subscript 0 refers to initial conditions. This equation directly relates the conversion with the temperature. The need for the simultaneous solution of the mass and energy equations is, therefore, removed for this case. Note that all the terms in the above equation must be dimensionally consistent. The reader should also note that both the heat capacity and enthalpy of reaction have been assumed to be constant.

Fogler, in his Prentice-Hall text *Elements of Chemical Reaction Engineering*, 1999, has derived a somewhat similar equation relating the conversion to temperature:

$$T = T_0 - \frac{X \, \Delta H^\circ(T_0)}{\Sigma \theta_i C_{Pi} + X \Delta C_P}$$

where X = conversion of A
$\Delta H^\circ(T_0)$ = standard enthalpy of reaction at inlet temperature T_0
ΔC_P = heat capacity difference between products and reactants
$\theta_i C_{Pi}$ = applies only to the initial reaction feed mixture (including inerts); θ_i = moles i/moles of A; C_{Pi} = heat capacity of i.

For the reaction

$$a\mathrm{A} + b\mathrm{B} \rightarrow c\mathrm{C} + d\mathrm{D}$$

one may write

$$\Delta C_P = cC_{PC} + dC_{PD} - aC_{PA} - bC_{PB}$$

Note that this term accounts for enthalpy of reaction variation with temperature. Folger's approach is employed in the solution to this problem.

Calculate ΔC_P for this reaction:

$$\begin{aligned}\Delta C_P &= C_{PP} - C_{PA} - C_{PB} \\ &= 20 - 10 - 10 \\ &= 0\end{aligned}$$

Since the enthalpy of reaction is not a function of temperature,

$$\begin{aligned}\Delta H(T) &= \Delta H^\circ(T_R) + \Delta C_P(T - T_R) \\ &= \Delta H^\circ(T_R) \\ &= -5000 \, \text{cal/gmol A}\end{aligned}$$

Thus, $\Delta H(T)$ is a constant.
By definition,

$$\theta_\mathrm{A} = 1.0$$

and for an equimolar feed,

$$\theta_B = 1.0$$

Using Fogler's approach, the temperature in the reactor may be obtained as a function of the conversion:

$$T = T_0 - \frac{X \, \Delta H(T)}{\Sigma \theta_i C_{Pi}}$$

$$\Sigma \theta_i C_{Pi} = (1)(10) + (1)(10)$$

$$= 20$$

$$T = 300 - [(X)(-5000)/(20)]$$

$$= 300 + 250X$$

The temperature when the conversion is 0.5 becomes

$$T = 300 + 125$$

$$= 425 \text{ K}$$

$$= 152°C$$

The rate of reaction, $-r_A$, in terms of the conversion variable X is

$$-r_A = k_A C_A C_B$$

Since

$$C_A = C_{A0}(1 - X)$$
$$C_B = C_{A0}(\theta_B - X)$$
$$= C_{A0}(1 - X)$$
$$-r_A = k C_{A0}^2 (1 - X)^2$$

The design equation for this batch reactor becomes

$$\frac{dC_A}{dT} = r_A \qquad C_A = C_{A0}(1 - X)$$

$$C_{A0} \frac{dX}{dt} = -r_A$$

$$\frac{dX}{dt} = k C_{A0}(1 - X)^2$$

The reaction velocity constant, k, may be expressed as a function of the conversion X through the Arrhenius equation:

$$k = k_0 \exp\left(-\frac{E}{R}\left\{\frac{1}{T} - \frac{1}{T_0}\right\}\right) \qquad T_0 = 300\,\text{K}$$

$$T = 300 + 250X$$

$$k = k_0 \exp\left(-\frac{E}{R}\left\{\frac{1}{300 + 250X} - \frac{1}{T_0}\right\}\right)$$

$$= (10^{-4})\exp\left(-\frac{3000}{1.987}\left\{\frac{1}{300 + 250X} - \frac{1}{300}\right\}\right)$$

The term k in the design equation can now be replaced by the expression obtained above. Thus,

$$\frac{dX}{dt} = (5)(10^{-4})(1 - X)^2 \exp\left(-\frac{1510}{300 + 250X} + 5.03\right) = I$$

Therefore,

$$dt = (1.0/I)\,dX$$

This resulting integral may be evaluated to obtain the time required to achieve an $X = 0.5$. One procedure would be to employ the trapezoidal rule. However, the Runge–Kutta method would produce more accurate results.

The reader is left the exercise of showing that the required time is approximately 820 min.

KIN.12 COMPLEX SYSTEMS

The following liquid-phase reaction takes place in a constant volume, isothermal batch reactor:

$$A + B + C \rightarrow P$$

The *initial* reaction rates were measured at various *initial* concentrations at $0°C$. The following data were obtained (Castellan, *Physical Chemistry*, Addison-Wesley, Boston, 1964):

Run	$C_{A0} \times 10^4$ (gmol/L)	$C_{B0} \times 10^2$ (gmol/L)	$C_{C0} \times 10^2$ (gmol/L)	Initial Reaction Rate $\times 10^7$ [gmol/(L·min)]
1	0.712	2.06	3.0	4.05
2	2.40	2.06	3.0	14.60
3	7.20	2.06	3.0	44.60
4	0.712	2.06	1.8	0.93
5	0.712	2.06	3.0	4.05
6	0.712	2.06	9.0	102.00
7	0.712	2.06	15.0	508.00
8	0.712	2.06	3.0	4.05
9	0.712	5.18	3.0	28.0
10	0.712	12.50	3.0	173.0

Assuming a reaction mechanism of the form

$$-r_A = k_A C_A{}^\alpha C_B{}^\beta C_C{}^\gamma$$

1. Estimate the order of the reaction with respect to A, α.
2. Estimate the order of the reaction with respect to B, β.
3. Estimate the order of the reaction with respect to C, γ.
4. Estimate the overall order of the reaction.
5. Outline how to calculate the reaction velocity constant, k_A.
6. Outline how to calculate the above (parts 1–5) numerically by regressing the data.

Solution

The application of any interpretation of kinetic data to reaction rate equations containing more than one concentration term can be somewhat complex. Consider the reaction

$$iA + jB \rightarrow \text{products}$$

where

$$\frac{dC_A}{dt} = -k_A C_A^i C_B^j$$

For the differentiation technique, this equation is written

$$\log\left(-\frac{dC_A}{dt}\right) = \log k_A + i \log C_A + j \log C_B$$

Three unknowns appear in this equation and can be solved by regressing the data; a trial-and-error graphical calculation is required. Using the integration technique is also complex. If one assumes $i = 1$ and $j = 1$, then

$$\frac{dC_A}{dt} = -k_A C_A C_B$$

which may be integrated to give

$$k_A = \frac{1}{t(C_{A0} - C_{B0})} \ln\left(\frac{C_{B0} C_a}{C_{A0} C_B}\right)$$

A constant value of k_A for the data would indicate that the assumed order $(i = 1, j = 1)$ is correct.

The isolation method has often been applied to the above mixed reaction system. The procedure employed here is to set the initial concentration of one of the reactants, say B, so large that the change in concentration of B during the reaction is vanishingly small. The rate equation may then be approximated by

$$\frac{dC_A}{dt} \approx -k_A^* C_A^i$$

where

$$k_A^* = k_A C_B^j \approx \text{constant}$$

Either the differentiation or integration method may then be used. For example, in the differentiation method, a log-log plot of $(-dC_A/dt)$ vs. C_A will give a straight line of slope i. A similar procedure is employed to obtain j.

The initial rate of reaction, $-r_{A0}$, in terms of the initial concentrations C_{A0}, C_{B0}, and C_{C0} for the problem at hand is

$$-r_{A0} = kC_{A0}^\alpha C_{B0}^\beta C_{C0}^\gamma$$

Linearizing the equation by taking logs,

$$\log(-r_{A0}) = \log(k) + \alpha \log(C_{A0}) + \beta \log(C_{B0}) + \gamma \log(C_{C0})$$
$$\quad (1) \qquad\qquad (2) \qquad\qquad (3) \qquad\qquad (4)$$

One may also use *ln* instead of *log* in the above equation.

For runs (1)–(3), the concentrations of B and C are constant (for each run at the initial conditions specified). Therefore, terms (1), (3), and (4) above may be treated as constants. Thus,

$$\log(-r_{A0}) = \log K + \alpha \log(C_{A0})$$

where

$$K = kC_{B0}^{\beta}C_{C0}^{\gamma}$$

To obtain α graphically, plot

$\log(-r_{A0})$	$\log C_{A0}$
−6.3925	−4.1475
−5.8356	−3.6197
−5.3506	−3.1426

or note that $\alpha \approx 1.0$.

Use runs (8)–(10) in order to establish the order of the reaction with respect to B:

$\log(-r_{A0})$	$\log C_{B0}$
−6.3925	−1.6861
−5.5528	−1.2856
−4.7619	−0.9039

For this set of runs, $\beta \approx 2.0$.

Use runs (4)–(7) to estimate the order the reaction with respect to C:

$\log(-r_{A0})$	$\log C_{C0}$
−7.0315	−1.7440
−6.3925	−1.5228
−4.9913	−1.0457
−4.2941	−0.8239

For this case, $\gamma \approx 3.0$.

The overall order of the reaction, n, is therefore

$$n = \alpha + \beta + \gamma$$
$$\approx 1.0 + 2.0 + 3.0 = 6.0$$

To calculate k_A, use the α, β, and γ values above and calculate 10 values of k_A. Sum the results and average. The units are $L^5/(gmol^5 \cdot min)$.

In order to calculate α, β, γ, and k by regressing the data, the equation (with logs) is of the form

$$Y = A_0 + A_1 X_1 + A_2 X_2 + A_3 X_3$$

where $Y = \log(-r_{A0})$
$A_0 = \log k$
$A_1 = \alpha$
$A_2 = \beta$
$A_3 = \gamma$
$X_1 = \log(C_{A0})$
$X_2 = \log(C_{B0})$
$X_3 = \log(C_{C0})$

The method of least squares may now be used to generate the A's.

11 Process Control (CTR)

CTR.1 FEEDBACK CONTROL

A heavy oil is heated by condensing steam in a shell-and-tube heat exchanger. Which sketch in Figure 34 contains a feedback control loop that will deliver the oil at the proper temperature by adjusting the amount of steam going to the exchanger? Note that TRC = temperature recorder and controller.

Solution

A control system or scheme is characterized by an output variable (e.g., temperature, pressure, liquid level, etc.) that is automatically controlled through the manipulation of inputs (input variables). Suppressing the influence of external disturbances on a process is the most common objective of a controller in a chemical plant. Such disturbances, which denote the effect that the external world has on a process, are usually out of reach of the human operator. Consequently, a control mechanism must be introduced that will make the proper changes on the process to cancel the negative impact that such disturbances may have on the desired operation of the process. Control engineers usually refer to the combination of a sensing element and a control device with a set point as a "control loop."

The central element in any control loop is the process to be controlled. Therefore, the control objectives must be defined (e.g., maintain a desired outlet temperature and/or composition, maintain the level in a tank at a certain height, etc.). Once the control objective is specified, variables are measured in order to monitor the operational performance of the process (sensing element). Next, the input variables that are to be manipulated are determined. Finally, after the control objectives, the possible measurements, and the available manipulated variables have been identified, the control configuration is defined. The control configuration is the information structure used to connect the available measurements to the available manipulated variables. The two general types of control configurations are feedback control and feedforward control. Details on feedback control are discussed below in this problem. Feedforward control is discussed in the next problem, CTR.2.

Feedback control, the general structure of which is illustrated in Figure 35, uses direct measurements of the controlled variable to adjust the values of the manipulated variable. The objective is to keep the controlled variable at a desired level (set point). In a feedback control loop, a measurement is made downstream of the

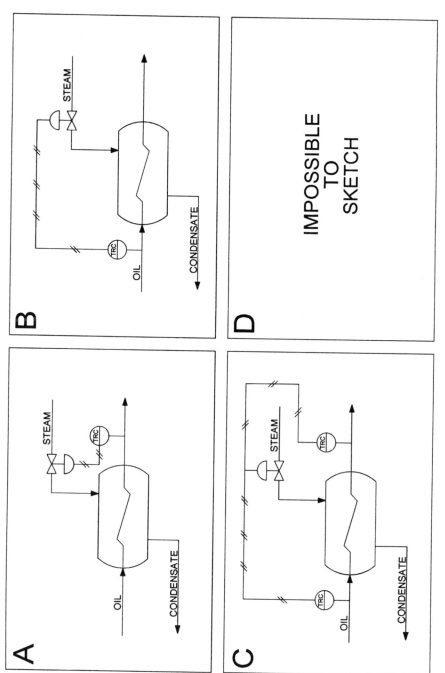

Figure 34. Answers to Problem CTR.1.

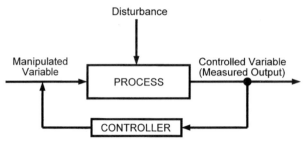

Figure 35. Feedback control.

process being controlled in order to manipulate a variable and achieve the desired set point.

Since the *controlled variable* (the outlet oil temperature) in Figure 34A is being measured, this is a feedback control system. Since the *disturbance* (the inlet oil temperature) in Figure 34B is being measured, this is a feedforward control system and *not* a feedback control system. Since both the outlet and inlet oil temperatures in Figure 34C are being measured, this is *not* a feedback control system. Therefore, the correct answer is Figure 34A.

CTR.2   FEEDFORWARD CONTROL

Consider a tank's contents that are heated by steam in a coil. Which sketch in Figure 36 contains a feedforward control loop that will deliver the fluid at the proper temperature by adjusting the flowrate of the steam in the heating coil? Note once again that TRC = temperature recorder and controller.

Solution

Feedforward control, the general structure of which is illustrated in Figure 37, uses direct measurement of the disturbances to adjust the values of the manipulated variables. The objective is to keep the values of the controlled output variables at desired levels.

Feedforward control does not wait until the effect of the disturbances has been felt by the system but acts appropriately before the external disturbance affects the system by anticipating what its effect will be.

Since the controlled variable (the outlet fluid temperature) in Figure 36A is being measured, this is a feedback control system and not a feedforward control system. Since the disturbance (the inlet fluid temperature) in Figure 36B is being measured, this is a feedforward control system. Since both the outlet and inlet fluid temperatures in Figure 36C are being measured, this is not a feedforward control system. Therefore, the correct answer is Figure 36B.

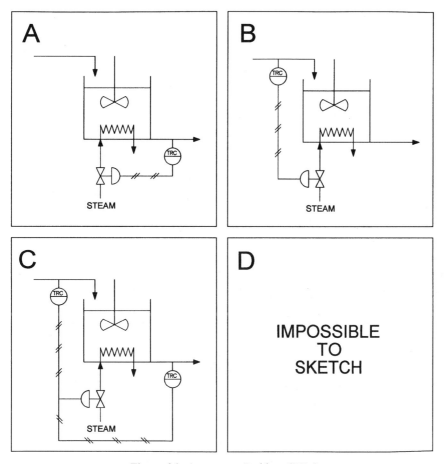

Figure 36. Answers to Problem CTR.2.

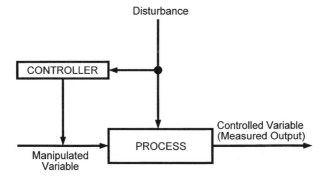

Figure 37. Feedforward control.

CTR.3   BLOCK DIAGRAMS

Consider the block diagram in Figure 38. Calculate the output z if $a = 14$, $b = 4$, and $c = 3$.

Solution

A block diagram is a line diagram representation of the cause-and-effect relationship between the input and output of a physical system. As such, it provides a convenient and useful method for characterizing the functional relationships among the various components of a control system. System components are alternatively called *elements* of the system.

The simplest form of the block diagram is a single block, with one input and one output (see Figure 39). The interior of the block usually contains a description of or the name of the element or symbol for the mathematical operation to be performed on the input (I) to yield the output (O). The arrows represent the direction of flow of information or *signal*. Two such operations are presented in Figure 40.

One can think of this operation in either of two ways. On one hand, one could say that O is the result of a control element operating on I. On the other hand, one can simply say that O is the product of a control element and I. Each of these descriptions is correct. Thus, the blocks representing the various components of a control system can be connected in a fashion that characterizes their functional relationship within the system.

For operations of addition and subtraction, the block is replaced by a small circle, called a summing point, with the appropriate plus or minus sign associated with the

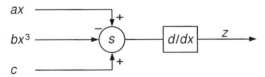

Figure 38. Block diagram for Problem CTR.3.

Figure 39. Single block with one input and one output.

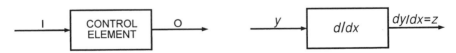

Figure 40. Symbols representing mathematical operations.

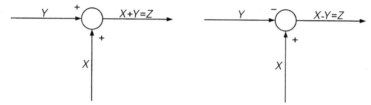

Figure 41. Summing points.

arrows entering the circle. The output is simply the algebraic sum of the inputs, and any number of inputs may enter a summing point (see Figure 41).

If the same signal or variable is an input to more than one block or summing point, a point is used. This allows the signal to proceed unaltered along several different paths to several destinations. For example, see Figure 42.

In this problem (see initial Figure 38), the output from the summing point, S, is

$$S = ax - bx^3 + c$$

The output from the control element, dS/dx, is

$$dS/dx = d(ax - bx^3 + c)/dx$$

$$= d(ax)/dx - d(bx^3)/dx + dc/dx$$

$$= a - 3bx^2 + 0$$

Since

$$dS/dx = z$$

then

$$z = a - 3bx^2$$

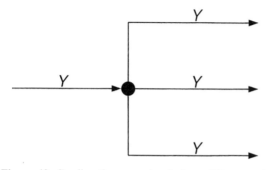

Figure 42. Sending the same signal along different paths.

Substituting $a = 14$ and $b = 4$ yields

$$z = 14 - 12x^2$$

In control system analysis, the block diagrams are used to show the functional relationship between the various parts of the system. As described earlier, each part is represented by a rectangle or box with one input and one output. The "transfer function" (or its symbol), which is the mathematical relationship between output (response function) and input (forcing function), is written inside the box, and the blocks are connected by arrows that indicate the flow of the information in the system. The outputs and inputs are considered as signals.

CTR.4   FORCING FUNCTIONS

Obtain the function shown in Figure 43 in the time domain, t.

Solution

Commonly encountered forcing functions (or input variables) in process control are step inputs (positive or negative), pulse functions, impulse functions, and ramp functions (refer to Figure 44).

Consider a step input of magnitude 1. The step function may be represented by $u(t)$, where $u(t) = 1$ for $t > 0$ and $u(t) = 0$ for $t < 0$. Substituting the value of this function into the definition of the Laplace transform and solving results in $\mathscr{L}[u(t)] = 1/s$.

Details on Laplace transforms are beyond the scope of this Handbook. The reader is referred to the literature for more information. However, the Laplace transforms of the above functions are now discussed.

Now consider a rectangular pulse whose magnitude is H and duration is T units of time. In the time interval $0 < t < T$, the function $f(t) = H$. When $t < 0$ or when $t > T$, the function $f(t) = 0$. Substituting the value of the pulse function into the

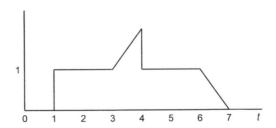

Figure 43. Forcing function for Problem CTR.4.

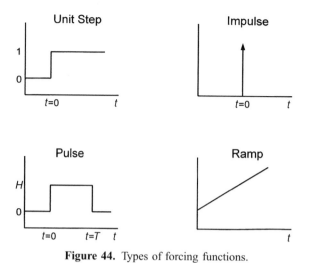

Figure 44. Types of forcing functions.

definition of the Laplace transform, and integrating between the limits 0 to T, ultimately yields

$$\mathscr{L}[f(t)] = F(s) = (H/s)(1 - e^{-Ts})$$

Consider an impulse function. The impulse function is also known as a Dirac delta function and is represented by $\delta(t)$. The function has a magnitude ∞ and an area equal to unity at time $t = 0$. The Laplace transform of an impulse function is obtained by taking the limit of a pulse function of unit area as $t \to 0$. Thus, the area of pulse function $HT = 1$. The Laplace transform is given by

$$\mathscr{L}[\delta(t)] = \lim_{T \to 0} \mathscr{L}f(t)$$

$$= \lim_{T \to 0} \frac{H}{S}(1 - e^{-Ts})$$

Using L'Hopital's rule, the limit is found to be 1, i.e., $\mathscr{L}[\delta(t)] = 1$.

Now consider a ramp function described by $f(t) = t[u(t)]$. The Laplace transform of this function is $1/s^2$. Finally, the Laplace transform of a sine-wave disturbance, $\sin(\omega t)$, is $\omega/(s^2 + \omega^2)$.

The second-shifting theorem, which permits the mathematical treatment of "lag" or "dead time," is defined in the following manner:

If

$$\mathscr{L}[f(t)] = F(s)$$

then,

$$\mathscr{L}(t - t_0) = e^{-st_0} F(S)$$

Conversely,

$$\mathscr{L}^{-1} e^{-st_0} = f(t - t_0)$$

Figure 43 is a combination of a positive step input at $t = 1$, a ramp function at $t = 3$, a negative step input at $t = 4$, and a negative ramp at $t = 6$. Thus, the time-domain equation is

$$f(t) = u(t - 1) + (t - 3)u(t - 3) - u(t - 4) - (t - 4)u(t - 4) - (t - 6)u(t - 6)$$

The reader is left the exercise of showing that the equation in the Laplace domain is

$$F(s) = \frac{\exp(-s)}{s} + \frac{\exp(-3s)}{s^2} - \frac{\exp(-4s)}{s} - \frac{\exp(-6s)}{s^2}$$

CTR.5   TIME TO REACH A SET TEMPERATURE

The transfer function relating the thermometer temperature (in a simple mercury thermometer), T_{th}, to the fluid temperature, T_f, is described by a first-order transfer function. (The term *first order* refers to the fact that the differential equation describing the transient process is first order). The transfer function is

$$\frac{T_{th}}{T_f} = \frac{K}{\tau s + 1}$$

where K and τ are the gain and time constant, respectively, and the temperatures are deviation variables.

The following data are provided:

Time constant, $\tau = 0.2 \, \text{min}$
Initial temperature of fluid $= 20°C$
Magnitude of step change in temperature of fluid $= 10°C$
Thermometer is at steady state at time $t = 0$ at $20°C$

At what time (in minutes) will the thermometer read $25°C$?

Solution

The transfer function relates two variables in a process; one of these is the forcing function or input variable, and the other is the response or output variable. The transfer function completely describes the dynamic characteristics of a system. The input and output variables are usually expressed in the Laplace domain and are written as deviations from the set-point values.

The process time constant, τ, provides an indication of the speed of the response of the process, i.e., the speed of the output variable to a forcing function or change in the input variable. The slower the response, the larger is the time constant, and vice versa. The time constant is usually composed of the different physical properties and operating parameters of the process. Constant τ has the units of time.

The steady-state gain gives an indication of how much the output variable changes (in the time domain) for a given change in the input variable (also in the time domain). The gain is also dependent on the physical properties and operating parameters of the process. The term "steady state" is applied if a step change in the input variable results in a change in the output variable, which reaches a new steady state that can be predicted by application of the final-value theorem. Gain may or may not be dimensionless.

Process dead time refers to the delay in time before the process starts responding to a disturbance in an input variable. It is sometimes referred to as "transportation lag" or "time delay". Dead time or delays can also be encountered in measurement sensors such as thermocouples, pressure transducers, and in transmission of information from one point to another. In these cases, it is referred to as *measurement lag*.

For the problem at hand, the response in the time domain is first determined. The forcing function is a step change of magnitude 10. Hence,

$$T_f(s) = 10/s$$

Therefore,

$$T_{th}(t) = 10(1 - e^{-t/0.2})$$

Note: For more in-depth examples of this type of calculation, see Problems CTR.10 and CTR.12.

When the thermometer reads 25°C, $T_{th}(t) = 5°$. Hence,

$$5 = 10(1 - e^{-t/0.2})$$

Solving yields

$$t = 0.139 \, \text{min}$$

CTR.6   STABILITY IN A FEEDBACK CONTROL SYSTEM

A feedback control system is represented in Figure 45. In this diagram

$$G_p = \frac{1}{3s^3 + 2s^2 + s - 5}$$

What value(s) of the proportional gain, K_c, will produce stable closed-loop responses?

Solution

A dynamic system is considered to be stable if for every bounded input, the output is also bounded, irrespective of its initial state. The transfer function $G(s)$ relates an output variable to an input variable or forcing function and is defined as:

$$G(s) = \frac{Y(s)}{X(s)}$$

The location of the roots of a transfer function of a dynamic system is important, since even if one root has a positive real part (i.e., lies on the right side of the complex plane), the system is unstable.

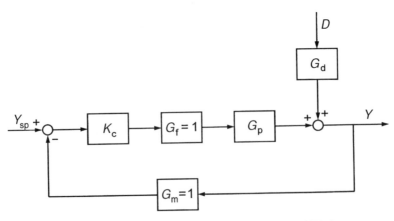

Figure 45. Feedback control system for Problem CTR.6.

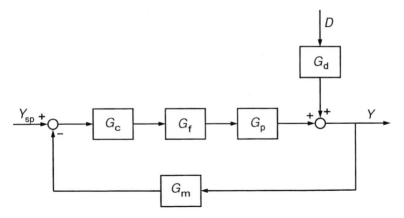

Figure 46. Generalized feedback control system.

The generalized feedback control system is shown in Figure 46 where

$$G_c = \text{transfer function of the controller}$$

$$G_f = \text{transfer function of the final control element}$$

$$G_p = \text{transfer function of the process}$$

$$G_d = \text{transfer function of the disturbance}$$

$$G_m = \text{transfer function of the measuring device}$$

$$Y = \text{output}$$

$$Y_{sp} = \text{set point}$$

$$D = \text{disturbance}$$

The general closed-loop response, which gives the transfer functions relating $Y(s)$, the output, to $Y_{sp}(s)$, the set point, and $D(s)$, the load disturbance, is given by

$$Y(s) = \frac{G_p G_f G_c}{1 + G_p G_f G_c G_m} Y_{sp}(s) + \frac{G_d}{1 + G_p G_f G_c G_m} D(s)$$

or equivalently,

$$Y(s) = G_{sp} Y_{sp}(s) + G_{load} D(s)$$

It is clear that the denominator of the overall transfer function for both load and set-point changes is the same. This denominator is 1 plus (+) the product of the transfer functions in the feedback loop $(1 + G_p G_f G_c G_m)$. The numerator of an overall closed-loop transfer function is the product of the transfer functions on the forward path between the set point or the load disturbance and the controlled variable.

The stability characteristics of the closed-loop response is determined by the poles of the transfer functions G_{sp} and G_{load}. These poles are common for both transfer functions (because they have a common denominator) and are given by the solution of the equation

$$1 + G_p G_f G_c G_m = 0$$

This equation is called the characteristic equation for the generalized feedback system in Figure 46.

Let $r_1, r_2, \ldots, r_n$ be the n roots of the characteristic equation:

$$1 + G_p G_f G_c G_m = (s - r_1)(s - r_2), \ldots, (s - r_n)$$

Then one can state the following criterion for the stability of a closed-loop system:

A feedback control system is stable if all the roots of its characteristic equation have negative real parts (i.e., are to the left of the imaginary axis). If any root of the characteristic equation has a real positive part (i.e., is on or to the right of the imaginary axis), the feedback system is unstable.

Examination of the characteristic equation indicates that it is not necessary to compute the actual values of the roots. All that is required is a knowledge of the location of the roots, i.e., if the roots lie on the right- or left-hand side of the imaginary axis. A simple test known as the *Routh–Hurwitz* test allows one to determine if any root is located on the right side of the imaginary axis, therefore rendering the system unstable.

The Routh–Hurwitz criteria for stability are:

1. Expand the characteristic equation into a polynomial form, which gives

$$1 + G_p G_f G_c G_m = a_0 s^n + a_1 s^{n-1} + \cdots + a_{n-1} s + a_n = 0$$

2. If a_0 is negative, multiply both sides of the equation by -1.
3. Place coefficients of polynomial into a Routh array as follows:

Row 1	a_0	a_2	a_4	...
Row 2	a_1	a_3	a_5	...
Row 3	A_1	A_2	A_3	...
Row 4	B_1	B_2	B_3	...
Row 5	C_1	C_2	C_3	...
Row $(n+1)$	Z_1	...	...	...

where

$$A_1 = \frac{a_1 a_2 - a_0 a_3}{a_1} \qquad B_1 = \frac{A_1 a_3 - a_1 A_2}{A_1} \qquad C_1 = \frac{B_1 A_2 - A_1 B_2}{B_1}$$

$$A_2 = \frac{a_1 a_4 - a_0 a_5}{a_1} \qquad B_2 = \frac{A_1 a_5 - a_1 A_3}{A_1} \qquad C_2 = \frac{B_1 A_3 - A_1 B_3}{B_1}$$

etc.

4. *First test for stability:* If any of the coefficients $a_1, a_2, \ldots, a_n$ is negative, then no further analysis is necessary since there is at least one root on the right-hand side of the imaginary axis, and the system is therefore unstable.

5. *Second test for stability:* If all the coefficients are positive, then it is difficult to conclude anything about the location of the roots and the first column must be examined. If the elements of the first column of the array a_0, a_1, A_1, B_1, C_1, etc., are all positive, then the system is stable. If any element in the first column is negative, there is at least one root to the right of the imaginary axis and the system is unstable. The number of sign changes in the elements of the first column is equal to the number of roots to the right of the imaginary axis.

The characteristic equation for the system in the problem statement is

$$1 + G_p G_c G_m G_f = 1 + \frac{1}{3s^3 + 2s^2 + s - 5}(K_c)(1)(1) = 0$$

This equation may be expanded into nth-order polynomial form.

$$3s^3 + 2s^2 + s + (K_c - 5) = 0$$

The first test for stability is performed by examining the coefficients of the nth-order polynomial.

$$\text{First test } K_c - 5 > 0 \qquad K_c > 5$$

The second test for stability is performed by examining column 1 of the array.

	Routh Array		
Row 1	3	1	0
Row 2	2	$K_c - 5$	0
Row 3	$\frac{1}{2}[2 - 3K_c - 5)]$	0	0
Row 4	$K_c - 5$	0	0

$$\text{Second test } \tfrac{1}{2}[2 - 3(K_c - 5) > 0 \qquad K_c < 5\tfrac{2}{3}$$

Therefore, stability occurs at

$$5 < K_c < 5\tfrac{2}{3}$$

CTR.7   MIXING PROCESS IN A FOOD PROCESSING PLANT

Consider the thermoneutral ($\Delta H_{rxn} = 0$) mixing process that takes place in a food processing plant. Pure water (W), pure oil (O), and pure grain (G) are combined to create a livestock feed mixture (see Figure 47). While the product flowrate F is not critical, the temperature T and composition of the product mixture, P (i.e., C_W, C_O, C_G), must be precisely maintained.

Develop a control loop to monitor each of the exit stream variables. At least one loop must be a feedforward loop. For each loop list:

1. Control objective
2. Measured variable(s)
3. Manipulated variable(s)

The following additional information is available:

1. The pure water reservoir is supplied by two (2) sources.
2. The pure oil reservoir is supplied by three (3) sources.
3. The holding tanks must be constantly stirred to maintain uniform temperature since the temperatures of the source streams are not equal and may vary.

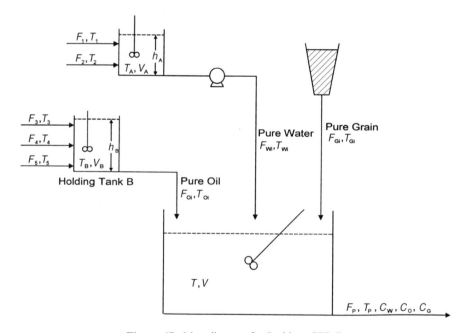

Figure 47. Line diagram for Problem CTR.7.

4. The temperature of the grain does not vary significantly with external conditions and may be considered a constant.

5. An adjustable pump regulates the flowrate of the water feed, F_{Wi}. However, the viscosity of the oil is such that an in-line pump would be prohibitively expensive. Therefore, the flowrate of the oil feed, F_{Oi}, is directly dependent on the height in the holding tanks, h_B, and an in-line adjustable valve on this stream is not recommended for process control.

6. The exit product flowrate is a function of the height in the mixer.

7. Assume the density and heat capacities of all feed streams are constant and equal to 1.0.

Solution

The line diagram in Figure 48 is employed to maintain the appropriate flowrate. Feedforward control is applied.

Note: The only manipulated variables available for any loop are the inlet feed flowrates.

1. Control objective: F_p
2. Measured variables: $F_3, F_4, F_5, F_{Wi}, F_{Gi}$
3. Manipulated variables: $F_3, F_4, F_5, F_{Wi}, F_{Gi}$

The line diagram in Figure 49 is employed for temperature (feedback) control.

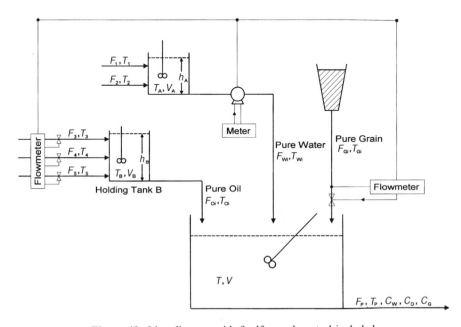

Figure 48. Line diagram with feedforward control included.

The following applies for temperature control.

1. Control objective: T_P
2. Measured variables: T_P (plus $T_{Wi}, T_{Gi}, T_3, T_4, T_5$)
3. Manipulated variables: $F_{Wi}, F_{Gi}, F_3, F_4, F_5$

The line diagram in Figure 50 is employed for concentration (feedback) control. The following applies for concentration control.

1. Control objectives: C_W, C_O, C_G
2. Measured variables: C_W, C_O, C_G
3. Manipulated variables: $F_{Wi}, F_3, F_4, F_5, F_{Gi}$

CTR.8 TWO CSTRs IN SERIES

Reactant A forms product B quite readily. However, this reaction will only take place in the presence of a preheated catalyst:

$$A + \text{catalyst} \longrightarrow B + \text{catalyst}$$

The above endothermic reaction takes place in two continuous stirred tank reactors (CSTRs) in series, and the heat required by the reaction is supplied by steam (see

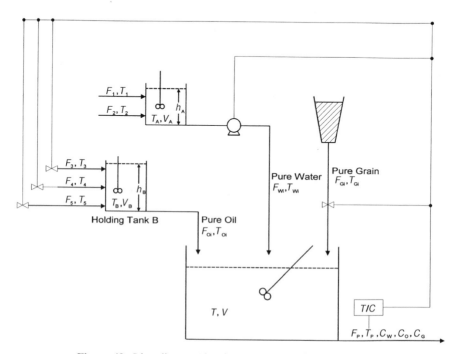

Figure 49. Line diagram showing temperature feedback control.

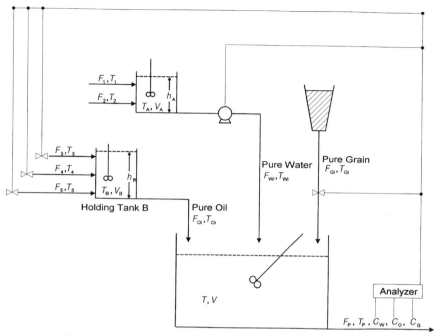

Figure 50. Line diagram showing concentration feedback control.

Figure 51). The temperature of reactant A and the temperature of the catalyst are set by an upstream process.

1. The primary control objective is to maintain the product flowrate F_4 at a desired level. Identify five additional control objectives.

2. Identify all of the variables in Figure 51 as either input or output variables. (Do not include the intermediate variables F_3, C_{A3}, and C_{B3} associated with stream 3.)

3. Develop six independent feedback control loops to satisfy each of the six control objectives (i.e., one loop for each objective). For each loop, the control objective and the manipulated variable must be identified.

4. Develop three independent feedforward loops to satisfy only three of the control objectives. Again, for each loop, identify the control objective and the manipulated variable as well as the variable being measured.

Hint: For parts 3 and 4 consider a composition analyzer on either the second reactor or the exit stream of the second reactor.

Solution

1. Additional control objectives are: h_1, h_2, C_{B4}, T_3, T_4.

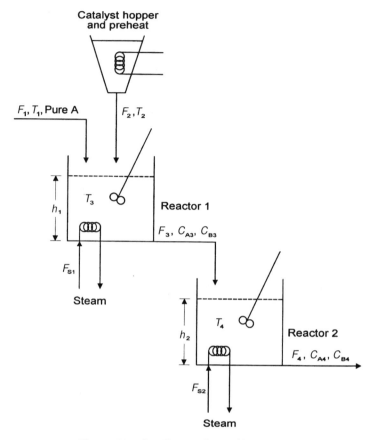

Figure 51. Line diagram for Problem CTR.8.

2. Input variables are: $F_1, T_1, F_2, T_2, F_{S1}, F_{S2}$.
   Output variables are: $h_1, h_2, T_3, T_4, F_4, C_{B4}$.
3. Six possible feedback loops are developed to satisfy each of the six control objectives. The control objective and the manipulated variable are identified for each one. Control objectives and manipulated variables for feedback control are provided in tabular format:

Loop	Control Objective	Manipulated Variable
1	h_1	F_1
2	h_2	F_3
3	C_{B4}	F_2
4	T_3	F_{S1}
5	T_4	F_{S2}
6	F_4	F_4

Note: For loop 6, it is permissible to go back and manipulate the flow if too much product is being made.

The system is pictured in Figure 52 for feedback control. Note that the loops are numbered according to the answer to part 1.

4. Control objectives and manipulated variables for feedforward control are provided in tabular format:

Loop	Control Objective	Measured Variable	Manipulated Variable
1	h_1	F_1	F_3
2	h_2	F_3	F_4
3	C_{B4}	F_2	F_2
4	T_3	T_1	T_{S1}
3	C_{B4}	C_{A3}	F_{S2}
5	T_4	T_3	F_{S2}

The system is pictured in Figure 53 for feedforward control.

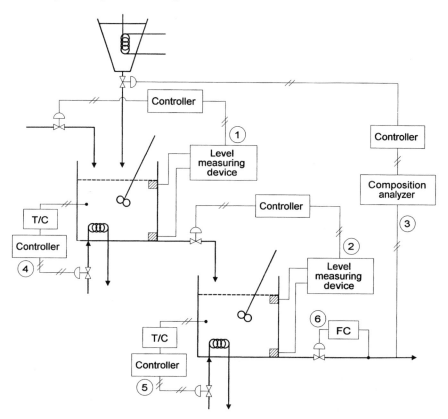

Figure 52. Line diagram for feedback control.

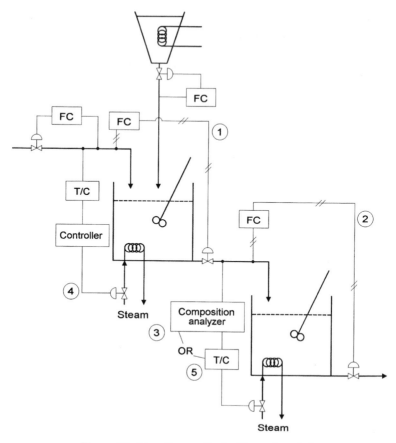

Figure 53. Line diagram for feedforward control.

CTR.9   DILUTION OF A NONREACTING IMPURITY

Consider the slightly endothermic reaction

$$A + 2B \xrightarrow{\text{Heat}} P$$

Pure A reacts with B to form the product P. Unfortunately, the stream supplying reactant B also contains some nonreacting impurity C. Therefore, it is often necessary to dilute the impurity in that stream with pure B supplied from a small holding tank (see Figure 54).

1. Develop a mathematical model for this system. The following information/ directives may be useful:
   a. Include all units of the system, including the mixing junction. *Hint:* the mixing junction has no volume.

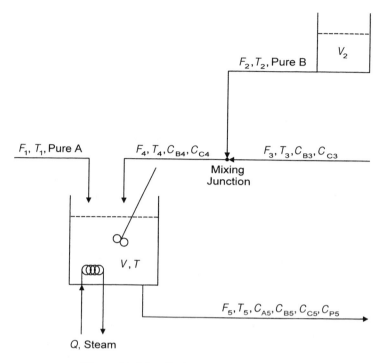

Figure 54. Line diagram for Problem CTR. 9.

b. It is not necessary to include the relationship between F and h in the solution.

c. Include the heat of solution (ΔH_s) where applicable.

d. Neglect any additional downstream effects.

2. Perform a degree of freedom analysis on the CSTR.

Solution

1. The following models (equations) are provided for each unit in the system: For the holding tank, the total mass balance is

$$\frac{dV_2}{dt} = -F_2$$

There is no component balance since only pure B is involved. There is no energy balance since there is no temperature change and no mixing.

For the mixing junction, there is no accumulation. The total mass balance is

$$\frac{dV}{dt} = 0 = F_2 + F_3 - F_4$$

The component balance on B is

$$0 = C_{B2}F_2 + C_{B3}F_3 - C_{B4}F_4$$

A component balance on C may be employed in place of that on B:

$$0 = C_{C3}F_3 - C_{C4}F_4$$

An energy balance yields

$$0 = \rho c_P F_2(T_2 - T_R) + [\rho c_P F_3(T_3 - T_R) + C_{B3}F_3 \Delta H_{S3}]$$
$$- [\rho c_P F_4(T_4 - T_R) + C_{B4}F_4 \Delta H_{S4}]$$

If T_R is arbitrarily set equal to zero, and ρ and c_P are assumed constant, the above equation reduces to

$$0 = F_2 T_2 + F_3 T_3 + \frac{1}{\rho c_P} C_{B3}F_3 \Delta H_{S3} - F_4 T_4 - \frac{1}{\rho c_P} C_{B4}F_4 \Delta H_{S4}$$

For the mixer (CSTR), the total mass balance is

$$\frac{dV}{dT} = F_1 + F_4 - F_5$$

There are three component balances. The following kinetics are employed in these balances:

$$r_A = k_1 C_A C_B{}^2$$
$$r_B = 2r_A$$
$$r_P = -r_A$$

For component A,

$$\frac{d}{dt}(C_A V) = C_{A1}F_1 - C_{A5}F_5 - r_A V$$

Also note that

$$C_{A5} = C_A$$

Expanding the derivative yields

$$\frac{d}{dt}(C_A V) = C_A \frac{dV}{dt} + V \frac{dC_A}{dt} = C_{A1}F_1 - C_A F_5 - r_A V$$

$$V \frac{dC_A}{dt} = C_{A1}F_1 - C_A F_5 - r_A V - C_A F_1 - C_A F_4 + C_A F_5$$

$$\frac{dC_A}{dt} = \frac{C_{A1}F_1}{V} - \frac{C_A}{V}(F_1 + F_4) - r_A$$

For component B,

$$\frac{d}{dt}(C_B V) = C_B \frac{dV}{dt} + V \frac{dC_B}{dt} = C_{B4}F_4 - C_{B5}F_5 - r_B V$$

$$V \frac{dC_B}{dt} = C_{B4}F_4 - C_B F_5 - r_B V - C_B F_1 - C_B F_4 + C_B F_5$$

$$\frac{dC_B}{dt} = \frac{C_{B4}F_4}{V} - \frac{C_B}{V}(F_1 + F_4) - r_B$$

For component C,

$$\frac{d}{dt}(C_c V) = C_c \frac{dV}{dt} + V \frac{dC_c}{dt} = C_{c4}F_4 - C_{c5}F_5$$

$$V \frac{dC_c}{dt} = C_{c4}F_4 - C_c F_5 - C_c F_1 - C_c F_4 + C_c F_5$$

$$\frac{dC_c}{dt} = C_{c4}F_4 - C_c F_1 - C_c F_4$$

Note: $C_c = C_{c5}$

For component P,

$$\frac{d}{dt}(C_P V) = C_P \frac{dV}{dt} + V \frac{dC_P}{dt} = -C_{P5}F_5 - r_P V$$

$$V \frac{dC_P}{dt} = -C_P F_5 + r_P V - C_P F_1 - C_P F_4 + C_P F_5$$

$$\frac{dC_P}{dt} = r_P - \frac{C_P}{V}(F_1 + F_4)$$

The energy balance is

$$\frac{d}{dt}[\rho V c_P T + C_A V \, \Delta H_s] = \rho F_1 c_P T_1 + [\rho F_4 c_P T_4 + C_{B4} F_4 \, \Delta H_{s4}]$$
$$- [\rho F_5 c_P T_5 + C_{A5} F_5 \Delta H_{s5}]$$
$$+ Q - \Delta H_{rxn}(r_A V)$$

Assume that $\Delta H_s \approx \Delta H_{s4} \approx \Delta H_{s5} = \Delta H_s$. Also set

$$\frac{\Delta H_s}{\rho c_P} = \Delta H'_s$$

$$\frac{\Delta H_{rxn}}{\rho c_P}(r_A) = \Delta H'_r$$

$$\frac{Q}{\rho c_P} = Q'$$

The equation above becomes

$$\frac{d}{dt}(VT) + \Delta H'_s \frac{d}{dt}(C_A V) = F_1 T_1 + F_4 T_4 + C_{B4} F_4 \, \Delta H'_s - F_5 T_5 + C_{A5} F_5 \, \Delta H'_s$$
$$+ Q' - \Delta H'_r(V)$$

The left-hand side of the above equation may be expanded to

$$\frac{d}{dt}(VT) + \Delta H'_s \frac{d}{dt}(C_A V) = V\frac{dT}{dt} + T\frac{dV}{dt} + \Delta H'_s C_A \frac{dV}{dt} + \Delta H'_s V \frac{dC_A}{dt}$$

Substituting for dV/dt and dC_A/dt yields

$$V\frac{dT}{dt} + (T + \Delta H'_s C_A)(F_1 + F_4 - F_5) + \Delta H'_s V\left(\frac{C_{A1} F_1}{V} - \frac{C_A}{V}(F_1 + F_4) - r_A\right)$$

Thus,

$$V\frac{dT}{dt} + (T)(F_1 + F_4 - F_5) + \Delta H'_s(C_{A1} F_1 - r_A V - C_A F_5)$$
$$= F_1 T_1 + F_4 T_4 + C_{B4} F_4 \, \Delta H'_s - F_5 T_5 + C_{A5} F_5 \, \Delta H'_s$$
$$+ Q' - \Delta H'_{rxn}(V)$$

$$V\frac{dT}{dt} = F_1(T_1 - T) + F_4(T_4 - T) + Q' + \Delta H'_{rxn}(V)$$
$$+ \Delta H'_s(C_{B4} F_4 + 2C_{A5} F_5 - C_{A1} F_1 + r_A V)$$

2. A degree of freedom (DOF) analysis around the CSTR (total moles not specified) gives

Number of equations = 5 (total mass, 3 component, energy)
Number of variables = $13(F_1, T_1, F_4, T_4, C_{B4}, C_{C4}, F_5, T_5, C_{A5}, C_{B5}, C_{C5}, C_{P5}, Q)$
Number of degrees of freedom = $13 - 5 = 8$
Number of control objectives = $4(T, F_5, C_{P5}, C_{A5})$
Number of disturbances = $4(F_1, T_1, F_4, T_4)$

CTR.10   HEIGHTS OF TWO MIXING TANKS

Two constantly stirred mixing tanks with a recycle stream are being used to dilute a very concentrated $NaNO_3$ solution by addition of pure water (see Figure 55).

1. Assume that cross-sectional areas of the tanks A_1 and A_2 are constant.
2. Assume that the inlet flowrate F_1 is held constant by a downstream process.
3. Assume that F_2 is a function of the height in tank 1 *only* and that the following relationship exists:

$$F_2 = k\sqrt{h_1}$$

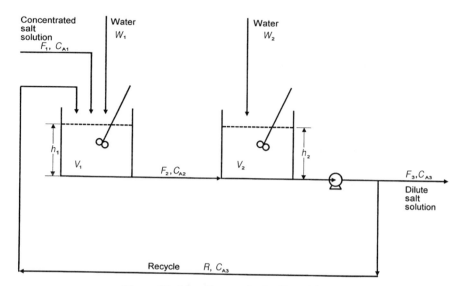

Figure 55. Line diagram for Problem CTR.10.

Determine the height in each tank after 1 h given the fact that each of the pure water streams experiences a unit impulse disturbance.

The following information is also available:

$$k = 2 \text{ ft}^3/\text{h}^{3/2}; \qquad A_1 = 1 \text{ ft}^2$$

$$h_{1s} = h_{2s} = 1 \text{ ft}; \qquad A_2 = 1 \text{ ft}^2$$

Solution

1. A mass balance around tank 1 gives

$$A_1 \frac{dh_1}{dt} = R + F_1 + W_1 - F_2 \qquad F_2 = k\sqrt{h_1}$$

$$A_1 \frac{dh_1}{dt} = R + F_1 + W_1 - k\sqrt{h_1}$$

2. This equation may be linearized by noting that

$$kh_1^{1/2} \approx kh_{1s}^{1/2} + \frac{1}{2} kh_{1s}^{-1/2}(h_1 - h_{1s})$$

Thus,

$$A_1 \frac{dh_1}{dt} = R + F_1 - kh_{1s}^{1/2} - \frac{1}{2} kh_{1s}^{-1/2}(h_1 - h_{1s})$$

3. To obtain the deviation form of the equation, note that

$$A_1 \frac{dh_{1s}}{dt} = R + F_1 + W_{1s} - kh_{1s}^{1/2}$$

Subtracting this equation from the last equation in step 2 yields

$$A_1 \frac{dh_1'}{dt} = W_1' - \frac{1}{2} kh_{1s}^{-1/2} h_1'$$

4. The Laplace transform of the above equation may now be obtained:

$$\frac{dh_1'}{dt} = \frac{W_1'}{A_1} - \frac{kh_{1s}^{-1/2}}{2A_1} h_1'$$

Letting

$$\alpha = \frac{kh_{1s}^{-1/2}}{2A_1}$$

yields

$$\frac{dh_1'}{dt} = \frac{W_1'}{A_1} - \alpha h_1'$$

$$\frac{dh_1'}{dt} + \alpha h_1' = \frac{W_1'}{A_1}$$

$$s\overline{h}_1' - h_1'(0) + \alpha \overline{h}_1' = \frac{\overline{W}_1'}{A_1}$$

Noting that $h_1'(0) = 0$, then

$$s\overline{h}_1' + \alpha \overline{h}_1' = \frac{\overline{W}_1'}{A_1}$$

$$\overline{h}_1' = \frac{\overline{W}_1'}{A_1(s + \alpha)}$$

Note that since $\overline{W}_1' \Rightarrow 1$ and $A_1 = 1$,

$$\overline{h}_1' = \frac{1}{s + \alpha}$$

5. Heaviside expansion is not applicable here.
6. Generating the inverse Laplace transform yields

$$h_1' = e^{-\alpha t}$$

$$\alpha = \frac{(1)(2)(1)}{(2)(1)} = 1$$

$$h_1' = e^{-t}$$

7. The height can now be calculated:

$$h_1 - h_{1s} = e^{-1}$$

$$h_1 = e^{-1} + 1 = 0.368 + 1$$

$$= 1.368 \text{ ft}$$

The same seven steps are now applied to tank 2.

1. A mass balance around tank 2 gives

$$A_2 \frac{dh_2}{dt} = +F_2 + W_2 - F_3 - R$$

$$A_2 \frac{dh_2}{dt} = k\sqrt{h_1} + W_2 - F_3 - R$$

2. The linearized version of $k\sqrt{h_1}$ is substituted:

$$A_2 \frac{dh_2}{dt} = kh_{1s}^{1/2} + \frac{1}{2}kh_{1s}^{-1/2}(h_1 - h_{1s}) + W_2 - F_3 - R$$

3. The deviation form of the equation is now obtained:

$$A_2 \frac{dh_{2s}}{dt} = kh_{1s}^{1/2} + W_{2s} - F_3 - R$$

$$A_2 \frac{dh'_2}{dt} = \frac{1}{2}kh_{1s}^{-1/2}h'_1 + W'_2$$

4. The Laplace transform is:

$$\frac{dh'_2}{dt} = +\frac{kh_{1s}^{-1/2}}{2A_2}h'_1 + \frac{W'_2}{A_2}$$

Letting

$$\beta = \frac{kh_{1s}^{-1/2}}{2A_2}$$

yields

$$\frac{dh'_2}{dt} = \beta h'_1 + \frac{W'_2}{A_2}$$

$$s\bar{h}'_2 - h'_2(0) = \beta \bar{h}'_1 + \frac{\overline{W}'_2}{A_2}$$

Noting that $h'_2(0) = 0$, then

$$s\bar{h}'_2 = \frac{\beta \bar{h}'_1}{s} + \frac{\overline{W}'_2}{A_1 s}$$

Since

$$\bar{h}'_1 = \frac{\overline{W}'_1}{A_1(s+\alpha)}$$

then

$$\bar{h}'_2 = \frac{\beta\overline{W}'_1}{A_1 s(s+\alpha)} + \frac{\overline{W}'_2}{A_2 s}$$

5. Heaviside expansion gives

$$\frac{1}{s(s+\alpha)} = \frac{A}{s} + \frac{B}{s+\alpha}$$

$$A = \frac{1}{\alpha}$$

$$B = -\frac{1}{\alpha}$$

$$\frac{1}{s(s+\alpha)} = \frac{1/\alpha}{s} - \frac{1/\alpha}{s+\alpha}$$

6. The inverse Laplace transform is now generated:

$$\bar{h}'_2 = \frac{\beta\overline{W}'_1}{A_1}\left(\frac{1/\alpha}{s} - \frac{1/\alpha}{s+\alpha}\right) + \frac{\overline{W}'_2}{A_2 s}$$

Since

$$\alpha = \frac{(1)(2)(1)}{(2)(1)} = 1$$

$$\beta = \frac{(1)(2)(1)}{(2)(1)} = 1$$

$$A_1 = 1$$

$$A_2 = 1$$

$$\overline{W}'_2 = 1$$

and

$$\bar{h}'_2 = \left(\frac{1}{s} - \frac{1}{s+1}\right) + \frac{1}{s}$$

the inverse Laplace is

$$h_2' = 2 - e^{-t}$$

7. The height is

$$h_2 - h_{2s} = 2 - e^{-1}$$
$$h_2 = 2 - e^{-1} + 1 = 2 - 0.368 + 1$$
$$= 2.632 \text{ ft}$$

CTR.11   CSTR WITH PROPORTIONAL FEEDBACK CONTROL

A unimolecular, first-order reaction takes place in the CSTR pictured in Figure 56. The inlet stream to the reactor, F_1, experiences disturbances from a downstream process. A proportional feedback control system has been installed on the exit stream, F_2, in order to monitor the height in the reactor and prevent it from overflowing. The mathematical model that determines the height of liquid in the CSTR has been defined as follows:

$$\overline{h}_1' = \frac{6}{(s+1)^3}\overline{F}_1' + \frac{0.5}{(s+1)(0.5s+1)}\overline{F}_2'$$

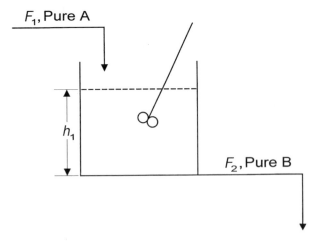

Figure 56. Flow diagram for Problem CTR.11.

1. Draw the block diagram for the above closed-loop process. Make sure that each block and the value of each block is labeled clearly. Use the following additional transfer functions to complete the diagram:

$$G_m = \frac{1}{0.5s + 1} \quad \text{and} \quad G_f = 2$$

2. Determine the acceptable values for the proportional gain, K_c.

Solution

1. The disturbance is F_1, where

$$G_d = \frac{6}{(s+1)^3}$$

The manipulated variable is F_2 where

$$G_p = \frac{0.5}{(s + 1)(0.5s + 1)}$$

For proportional control,

$$G_c = k_c$$

and

$$G_m = \frac{1}{0.5s + 1} \quad\quad G_f = 2$$

The block diagram is provided in Figure 57.

2. The characteristic equation is

$$1 + G_p G_f G_c G_m = 0$$

$$1 + \frac{0.5}{(s + 1)(0.5s + 1)} 2K_c \frac{1}{0.5s + 1} = 0$$

$$(s + 1)(0.5s + 1)^2 + K_c = 0$$

$$(s + 1)(0.25s^2 + s + 1) + K_c = 0$$

$$0.25s^3 + 1.25s^2 + 2s + (K_c + 1) = 0$$

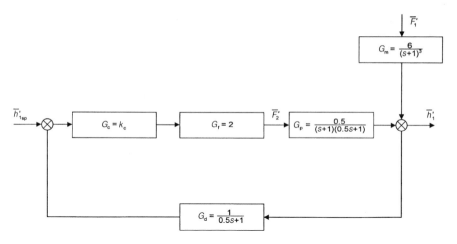

Figure 57. Block diagram for Problem CTR.11.

To determine acceptable values, employ the Routh–Hurwitz approach:

Row 1	0.25	2	0
Row 2	1.25	$K_c + 1$	0
Row 3	$\dfrac{(1.25)(2) - (K_c + 1)(0.25)}{1.25}$	0	0
Row 4	$K_c + 1$	0	0

First test: Since K_c is always positive, all coefficients are positive. Therefore, proceed to the second test.

Second test: Since $K_c + 1$ is always positive, one may write

$$\frac{2.5 - (K_c + 1)(0.25)}{1.25} > 0$$

$$\rightarrow \frac{2.5 - (K_c + 1)(0.25)}{1.25} > 0$$

$$\rightarrow 2.25 > 0.25 K_c \quad \text{and} \quad 9 > K_c$$

Thus,

$$0 < K_c < 9$$

CTR.12   HEIGHT OF LIQUID IN A TRAP

The second-order reaction

$$2A \rightarrow B$$

takes place in a perfectly mixed, isothermal CSTR with an overflow weir to maintain constant volume. The flowrate over the weir can be described by the following relationship:

$$F = k_F h_F^{2/3}$$

A trap is used to collect any overflow, which is then recycled back to the CSTR (see Figure 58). Two pumps are used to maintain the product flow, F_P, and the recycle flow, F_R, at a constant rate. Initially, h_F and h_2 are at steady state and the height of liquid in the CSTR never falls below h_w.

Determine $h_2(1)$, the height of liquid in the trap at $t = 1$ h, given the fact that there has been an impulse disturbance of 5 ft^3/h to the feed stream, F_0.

The following information may be useful:

$$A_1 = 3 \text{ ft}^2 \qquad h_W = 4 \text{ ft} \qquad F_O = 10 \text{ ft}^3/\text{h}$$
$$A_2 = 1.5 \text{ ft}^2 \qquad h_{Fs} = 0.1 \text{ ft} \qquad F_P = 10 \text{ ft}^3/\text{h}$$
$$k_F = 1 \text{ ft}^{3/2}/\text{h} \qquad h_{2s} = 0.2 \text{ ft} \qquad F_R = 4 \text{ ft}^3/\text{h}$$

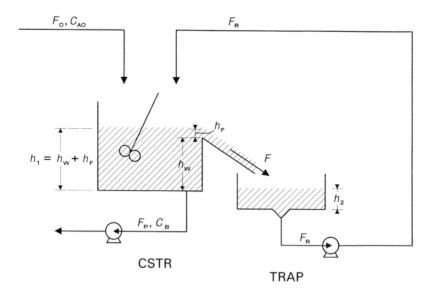

Figure 58. Diagram for Problem CTR.12.

Solution

Since the fluid is a liquid, the density may be assumed constant. The describing equation for tank 1 is

$$\text{Accumulation} = \text{In} - \text{Out}$$

$$\frac{dV_1}{dt} = F_0 + F_R - F_P - F$$

Since $V_1 = A_1 h_1 = A_1(h_W + h_F)$,

$$A_1 \frac{d(h_W + h_F)}{dt} = F_0 + F_R - F_P - k_F h_F^{2/3} \qquad h_W = \text{constant}$$

$$A_1 \frac{dh_F}{dt} = F_0 + F_R - F_P - k_F h_F^{2/3}$$

The describing equation for tank 2 is

$$\frac{dV_2}{dt} = F - F_R$$

$$A_2 \frac{dh_2}{dt} = k_F h_F^{2/3} - F_R$$

The two ordinary differential equations (ODE) above may be linearized by setting

$$h_F^{2/3} \approx h_{Fs}^{2/3} + \frac{2}{3} h_{Fs}^{-1/3}(h_F - h_{Fs})$$

The two tank balances become

$$A_1 \frac{dh_F}{dt} = F_0 + F_R - F_P - k_F h_{Fs}^{2/3} - \frac{2}{3} k_F h_{Fs}^{-1/3}(h_F - h_{Fs})$$

$$A_2 \frac{dh_2}{dt} = k_F h_{Fs}^{2/3} + \frac{2}{3} k_F h_{Fs}^{-1/3}(h_F - h_{Fs}) - F_R$$

The steady-state form of the mass balance for tank 1 is

$$A_1 \frac{dh_{Fs}}{dt} = F_0 + F_R - F_P - k_F h_{Fs}^{2/3}$$

The deviation form is found by subtracting this equation from the linearized mass balance and letting

$$h_F' = h_F - h_{Fs}$$

and

$$F_0' = F_0 - F_{0s}$$

The result is

$$A_1 \frac{dh_F'}{dt} = F_0' - \frac{2}{3} k_F h_{Fs}^{-1/3} h_F'$$

For tank 2, the linearized steady-state form of the mass balance is

$$A_2 \frac{dh_{2s}}{dt} = k_F h_{Fs}^{2/3} - F_R$$

Letting

$$h_2' = h_2 - h_{2s}$$

leads to

$$A_2 \frac{dh_2'}{dt} = \frac{2}{3} k_F h_{Fs}^{-1/3} h_F'$$

Taking the Laplace transforms of both equations,

$$\frac{dh_F'}{dt} = \frac{F_0'}{A_1} - \frac{2}{3} \frac{k_F h_{Fs}^{-1/3} h_F'}{A_1}$$

Letting $\alpha = \frac{2}{3} (k_F h_{Fs}^{-1/3})/A_1$,

$$\frac{dh_F'}{dt} + \alpha h_F' = \frac{F_0'}{A_1}$$

$$s\overline{h}_F' - h_F'(0) + \alpha \overline{h} F' = \frac{\overline{F}_0'}{A_1} \qquad h_F'(0) = 0$$

$$(s + \alpha)\overline{h}_F' = \frac{\overline{F}_0'}{A_1}$$

$$\overline{h}_F' = \frac{1}{A_1} \frac{\overline{F}_0'}{s + \alpha}$$

$$\frac{dh_2'}{dt} = \frac{2}{3} \frac{k_F h_{Fs}^{-1/3} h_F'}{A_2}$$

Letting

$$\beta = \frac{2}{3} \frac{k_F h_{Fs}^{-1/3}}{A_2}$$

$$\frac{dh_2'}{dt} = \beta h_F'$$

$$\overline{h}_2' = \beta \frac{\overline{h}_F'}{s}$$

Combining two of the equations above yields

$$\overline{h}_2' = \frac{\beta}{A_1} \frac{\overline{F}_0'}{s(s+\alpha)}$$

Since $\overline{F}_0' = 5$,

$$\overline{h}_2' = 5 \frac{\beta}{A_1} \frac{1}{s(s+\alpha)}$$

The Heaviside expansion yields

$$\frac{1}{s(s+\alpha)} = \frac{A}{s} + \frac{B}{s+\alpha}$$

$$A = \frac{1}{s+\alpha} \text{ at } s = 0 \qquad A = \frac{1}{\alpha}$$

$$B = \frac{1}{s} \text{ at } s = -\alpha \qquad B = -\frac{1}{\alpha}$$

Therefore,

$$\overline{h}_2' = \frac{5\beta}{A_1} \left(\frac{A}{s} + \frac{B}{s+\alpha} \right)$$

$$= \frac{5\beta}{A_1} \left(\frac{1}{\alpha s} - \frac{1}{\alpha(s+\alpha)} \right)$$

Taking the inverse Laplace transform yields

$$\mathcal{L}^{-1}(\overline{h}_2') = \mathcal{L}^{-1} \left[\frac{5\beta}{A_1} \left(\frac{1}{\alpha s} - \frac{1}{\alpha(s+\alpha)} \right) \right]$$

$$= \frac{5\beta}{A_1 \alpha} \left[\mathcal{L}^{-1} \left(\frac{1}{s} \right) - \mathcal{L}^{-1} \left(\frac{1}{s+\alpha} \right) \right]$$

$$h_2' = \frac{5\beta}{A_1 \alpha} (1 - e^{-\alpha t})$$

The term h_2 at $t = 1$ h may now be calculated.

$$\alpha = \frac{2}{3}\frac{k_F h_{Fs}^{-1/3}}{A_1} = \frac{(2)(1)(0.1)^{-1/3}}{(3)(3)} = 0.4788$$

$$\beta = \frac{2}{3}\frac{k_F h_{Fs}^{-1/3}}{A_2} = \frac{(2)(1)(0.1)^{-1/3}}{(3)(1.5)} = 0.9575$$

$$h_2 - h_{2s} = \frac{(5)(0.9575)}{(3)(0.4788)}(1 - e^{-(0.4788)(1)}) = 1.268$$

$$h_2 = 1.268 + 0.2$$

$$= 1.468 \text{ ft}$$

12 Process Design (PRD)

PRD.1 DILUTION AIR

Using ambient air at 60°F, calculate the quantity of dilution air required to cool 166,500 lb/h of gas from 2050 to 520°F. Design an air quencher to accomplish this if a 1.2-s residence time is recommended.

Solution

Under adiabatic conditions, the heat transferred from the flue gas is equal to that received by the dilution air. Thus,

$$q_{\text{flue}} = -q_{\text{air}}$$

For the flue gas,

$$q_{\text{flue}} = \dot{m}_{\text{flue}} C_P \Delta T$$
$$= (166,500)(0.3)(520 - 2050)$$
$$= -76.4 \times 10^6 \, \text{Btu/h}$$

Therefore,

$$\dot{m}_{\text{air}} = \frac{76.4 \times 10^6}{(0.26)(520 - 60)}$$
$$= 639,000 \, \text{lb/h}$$
$$\dot{n}_{\text{air}} = 639,000/28.8$$
$$= 22,200 \, \text{lbmol/h}$$

At 60°F, 1.0 lbmol of an ideal gas occupies 379 ft³. Therefore,

$$q_{\text{air}} = 84.1 \times 10^5 \, \text{ft}^3/\text{h} = 140,000 \, \text{acfm}$$

A more rigorous approach may be employed if detailed polynomial expressions for the heat capacities of the flue gas components are available (see Chapter 9).

242

The total mass flowrate, (including dilution air) is

$$\dot{m}_T = 166{,}500 + 639{,}000$$
$$= 805{,}500 \text{ lb/h}$$

The total volumetric flow, q_T, at 520°F is then

$$q_T = \frac{\dot{m}_T RT}{P(MW)}$$
$$= \frac{(805{,}500)(0.73)(520 + 460)}{(1.0)(28.8)}$$
$$= 2.00 \times 10^7 \text{ ft}^3/\text{h} = 5558 \text{ ft}^3/\text{s}$$

The volume of the tank, V_t, can now be calculated. Using a residence time t of 1.2 s,

$$V_t = q_T/t$$
$$= 5558/1.2$$
$$= 4630 \text{ ft}^3$$

Assume $L = D$ for minimum surface area. Then

$$V_t = \frac{\pi D^2}{4} D$$
$$D = \left(\frac{(4630)(4)}{\pi} \right)^{1/3} = 18.1 \text{ ft}$$
$$L = 18.1 \text{ ft}$$

PRD.2   MINIMUM VOLUME FOR TWO CSTRs IN SERIES

Consider the elementary liquid-phase reaction between benzoquinone (B) and cyclopentadiene (C):

$$B + C \rightarrow \text{product}$$

If one employs a feed containing equimolal concentrations of reactants, the reaction rate expression can be written as

$$-r_B = kC_C C_B = kC_B{}^2$$

Determine the effect of using a cascade of two CSTRs that differ in size on the volume requirements for the reactor network. For both reactors, assume isothermal operation at 25°C where the reaction rate constant is equal to $9.92 \, m^3/(kgmol \cdot ks)$. Reactant concentrations in the feed are each equal to $0.08 \, kgmol/m^3$ and the liquid feed rate is equal to $0.278m^3/ks$. Determine the minimum total volume required and the manner in which the volume should be distributed between the two reactors. A fractional overall conversion of 0.875 is to be achieved.

Solution

Set up an equation to determine the total volume requirement if the two reactor volumes are not equal:

$$V_T = V_1 + V_2$$

$$= \left(\frac{F_{B0}}{kC_{B0}^2} \right) \left(\frac{X_1}{(1 - X_1)^2} + \frac{X_2 - X_1}{(1 - X_2)^2} \right) \qquad X_2 = 0.875$$

Details of this equation are provided in Problem KIN.8.

Solve this equation for the intermediate conversion X_1 under minimum volume conditions. To minimize V_T, set

$$\frac{dV_T}{dX_1} = 0$$

and solve for X_1. Analytically differentiating yields

$$\frac{dV_T}{dX_1} = \frac{1}{(1 - X_1)^2} + \frac{2X_1}{(1 - X_1)^3} - 64 = 0$$

Solving by trial and error gives

$$X_1 = 0.702$$

The volumes of the two individual reactors and the total minimum volume requirement can now be calculated:

$$V_1 = \frac{(0.08)(0.278)(0.702)}{(9.92)(0.08)^2(1 - 0.702)^2}$$

$$= 2.77m^3$$

$$V_2 = \frac{(0.02224)(0.875 - 0.702)}{(9.92)(0.08)^2(1 - 0.875)^2}$$

$$= 3.88m^3$$

$$V_T = V_1 + V_2$$

$$= 2.77 + 3.88$$

$$= 6.65m^3$$

If the two reactor volumes are equal (see Problem KIN.8),

$$V_T = 6.72 \, \text{m}^3$$

As expected, the volume calculated in this problem is lower.

The reader is left the exercise of calculating the volume requirements for three equal-volume CSTRs in series. For this condition, $V_T = 4.77 \, \text{m}^3$ with $V_1 = V_2 = V_3 = 1.59 \, \text{m}^3$. This is left as an exercise only for those with a strong analytical mathematics background.

PRD.3   AMMONIA FROM NATURAL GAS

Figure 59 shows a simplified process flow diagram for the manufacture of ammonia from natural gas, which can be considered to be pure methane. In the first stage of the process, natural gas is partially oxidized in the presence of steam to produce a synthesis gas containing hydrogen, nitrogen, and carbon dioxide. The carbon dioxide is removed. The remaining mixture of hydrogen and nitrogen is reacted over a catalyst at high pressure to form ammonia.

The two reactions for the process, neglecting side reactions and impurities, are

$$7CH_4 + 10H_2O + 8N_2 + 2O_2 \rightarrow 7CO_2 + 8N_2 + 24H_2$$
$$N_2 + 3H_2 \rightarrow 2NH_3$$

Calculate the amount of methane gas, both in pounds per day and in standard cubic feet (60°F, 1 atm) per day, required for a world-scale ammonia plant producing 1200 tons of ammonia per day. The standard volume normally used in the natural gas industry is 379.5 cubic feet per lb · mol. It is based on conditions of 14.7 psia (one atmosphere) and 60°F.

Solution

The number of pound moles per day, $\dot{n}_a$, of ammonia formed is

$\dot{n}_a = (1200 \, \text{tons/day})(2000 \, \text{lb/ton})/(17.03 \, \text{lb ammonia/lbmol of ammonia})$

$\quad = 140{,}900 \, \text{lbmol/day}$

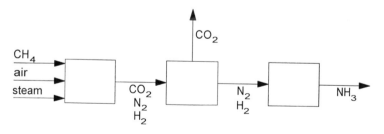

Figure 59. Flow diagram for Problem PRD.3.

The number of lbmol of methane required per lbmol of ammonia is obtained from the stoichiometry of both reactions. From the second reaction, 1 lbmol of ammonia requires $\frac{3}{2}$ lbmol of hydrogen. From the first, 1 lbmol of hydrogen requires $\frac{7}{24}$ lbmol of methane. Therefore, the overall requirement is

$$\left(\frac{3}{2}\right)\left(\frac{7}{24}\right) = \frac{7}{16}$$

The number of lbmol of methane, $\dot{n}_m$ required per day is

$$\dot{n}_m = (140,900 \text{ lbmol/day})(7/16)$$

$$= 61,600 \text{ lbmol/day}$$

PRD.4 COOLING WATER REQUIREMENTS

Determine the total flowrate of cooling water, in gallons per minute, required for the services listed below. If 5% of the return water is sent to "blowdown," how much fresh water makeup is required? The cooling tower system supplies water to the units at 90°F with a return temperature of 115°F.

Process Unit	Heat Duty (Btu/h)	Temperature (°F)
1	12,000,000	250
2	6,000,000	276–200
3	23,000,000	175–130
4	17,000,000	300
5	31,500,000	225–150

The heat capacity of the cooling water is 1.00 Btu/(lb · °F). The heat of vaporization at cooling tower operating conditions is 1030 Btu/lb.

Solution

The total cooling load is simply the sum of the cooling loads for the five process units:

$$\dot{Q} = (12.0 + 6.0 + 23.5 + 17.0 + 31.5)(10^6)$$

$$= 90,000,000 \text{ Btu/h}$$

The cooling water flow required, F, is the total heat duty divided by the enthalpy change per pound of water, which is the product of its heat capacity and temperature rise:

$$F = \frac{90.0 \times 10^6 \text{ Btu/h}}{(115 - 90°\text{F})[1.00 \text{ Btu/(lb} \cdot °\text{F)}]}$$

$$= 3,600,000 \text{ lb/h}$$

Since the density of water in lb/gal is 8.32,

$$F = \frac{3.6 \times 10^6 \, \text{lb/h}}{(60 \, \text{min/h})(8.32 \, \text{lb/gal})}$$

$$= 7200 \, \text{gal/min}$$

Note that the cooling water flowrate is independent of the temperatures of the process units. The sizes of the heat exchangers depend on all the temperatures, but the flowrate of cooling water does not. The heat discharge temperatures must, of course, all be higher than the cooling water temperature. Otherwise the cooling water would heat, not cool, the process.

The blowdown rate, B, in this case, is 5% of the circulating water:

$$B = (0.05)(3,600,000)$$

$$= 180,000 \, \text{lb/h}$$

$$= 360 \, \text{gal/min}$$

The amount of water vaporized, V, can be calculated from the total cooling system heat load. Ultimately, all the heat is dissipated to the atmosphere by the evaporating water. Therefore, the evaporation rate is the total heat load divided by the heat of vaporization (latent heat) of the cooling water:

$$V = (90,000,000 \, \text{Btu/h})/(1030 \, \text{Btu/lb})$$

$$= 87,400 \, \text{lb/h}$$

$$= 175 \, \text{gal/min}$$

The total freshwater makeup, M, is the sum of blowdown and the amount evaporated:

$$M = 180,000 + 87,400$$

$$= 267,400 \, \text{lb/h}$$

$$= 535 \, \text{gal/min}$$

Chemical additives are often added to the circulating cooling water to reduce corrosion rates, reduce scaling and fouling, and prevent growth of bacteria and algae. Plants often contract with specialty service companies to provide these additives and supervise their use.

PRD.5   NEW PROCESS EVALUATION

Provide a brief description of how "new" processes can be simply evaluated.

Solution

In comparing alternate processes or different options of a particular process from an economic point of view, it is recommended that the total capital cost be converted to an annual basis by distributing it over the projected lifetime of the facility. The sum of both the annualized capital costs (ACC) and the annual operating costs (AOC) is known as the total annualized cost (TAC) for the facility. The economic merit of the proposed facility, process, or scheme can be examined once the total annual cost is available. Alternate facilities or options (e.g., a distillation column versus an adsorber for solvent recovery, or two different processes for accomplishing the same degree of waste destruction) may also be compared. Note, a small flaw in this procedure is the assumption that the operating costs remain constant throughout the lifetime of the facility. However, since the analysis is geared to comparing different alternatives, the changes with time should be somewhat uniform among the various alternatives, resulting in little loss of accuracy. More details on economic factors can be found in Chapter 50.

PRD.6   ECONOMIC PIPE DIAMETER

Outline how to determine the optimum economic pipe diameter for a flow system.

Solution

The investment for piping can amount to an important part of the total cost for a chemical process. It is usually necessary to select pipe sizes that provide the minimum total cost of both capital and operating charges. For any given set of flow conditions, the use of an increased pipe diameter will result in an increase in the capital cost for the piping system and a decrease in the operating costs. (The operating cost is generally the energy costs associated with moving, i.e., pumping, the fluid of concern.) Thus, an optimum economic pipe diameter can be found by minimizing the sum of pumping (or energy) costs and capital charges of the piping system.

The usual calculational procedure is as follows:

1. Select a pipe diameter.
2. Obtain the annual operating cost.
3. Obtain the capital equipment cost.
4. Convert the capital cost to an annual basis.
5. Sum the two annual costs in steps 2 and 4.
6. Return to step 1.

The only variable that will appear in the resulting total-cost expressions is the pipe diameter. The optimum economic pipe diameter can be generated by taking the derivative of the total annual cost with respect to pipe diameter, setting the result equal to zero, and solving for the diameter. The derivative operation can be replaced by a trial-and-error procedure that involves calculating the total cost for various diameters and simply selecting the minimum.

PRD.7   FLUIDIZED-BED REACTOR

A fluidized-bed reactor is to be designed to destroy 99.99% of a unique liquid hazardous waste. Based on laboratory and pilot plant studies, researchers have described the waste reaction by a first-order reversible mechanism. Their preliminary findings are given below.

$$A \underset{k'}{\overset{k}{\rightleftharpoons}} R \quad A = \text{waste}$$

$$k = 1.0 \; e^{-10,000/T} \quad T = {}^{\circ}\text{R}$$

$$k' = 9.89 \; e^{-35,000/T} \quad T = {}^{\circ}\text{R}$$

Calculate a bed reaction operating temperature that will minimize the volume of the reactor and achieve the desired degree of waste conversion.

Solution

A fluid-bed reactor is best described by a continuously stirred tank reactor (CSTR) model. For more details on this, refer to Chapter 10, Problem KIN.8.
The rate of reaction, $-r_A$, in terms of the concentration of A, C_A, is

$$-r_A = kC_{A1} - k'C_{R1}$$

Since a fluidized-bed incinerator is best described by a CSTR,

$$\frac{V}{q} = \frac{C_{A0} - C_{A1}}{-r_A}$$

$$= \frac{C_{A0} - C_{A1}}{kC_{A1} - k'C_{R1}}$$

where V = reactor volume
q = volumetric flowrate of the waste

C_{A1} and C_{R1}, the outlet concentrations of A and R, respectively, can now be expressed in terms of C_{A0}. Note that for 99.99% destruction of the waste, A, the conversion variable X_A becomes

$$X_A = 0.9999$$

Thus

$$C_{A1} = 0.0001 C_{A0}$$

$$C_{R1} = 0.9999 C_{A0}$$

The design equation for V may be rewritten in terms of C_{A0} and the two k's:

$$\frac{V}{q} = \frac{C_{A0} - 0.0001(C_{A0})}{(k)(0.0001)C_{A0} - (k')(0.9999)C_{A0}}$$

$$= \frac{0.9999(C_{A0})}{(C_{A0})(0.0001\,k - 0.9999\,k')}$$

The reaction velocity constants, k and k', are described by the Arrhenius equation given in the problem statement. Thus,

$$k = Ae^{-E/RT}$$

$$k' = A'e^{-E'/RT}$$

$$A = 1.0$$

$$A' = 9.89$$

$$E/R = 10{,}000$$

$$E'/R = -35{,}000$$

To calculate the operating temperature that will require the minimum volume to accomplish a conversion of 99.99%, minimize the volume by setting $dV/dt = 0$ and solving for the temperature:

$$\frac{d(V/q)}{dT} = (0.9999)\frac{(dk/dT) - (9999)(dk'/dT)}{(k - 9999k')^2}$$

Setting

$$\frac{d(V/q)}{dT} = 0$$

$$\frac{dk}{dT} = (9999)\frac{dk'}{dT}$$

$$\approx (10^4)\frac{dk'}{dT}$$

Since

$$\frac{dk}{dT} = \frac{AE}{RT^2}e^{-E/RT}$$

$$\frac{dk'}{dT} = \frac{A'E'}{RT^2}e^{-E'/RT}$$

rearranging the above equation yields

$$10^4 = \left(\frac{A}{A'}\right)\left(\frac{E}{E'}\right)e^{(E'-E)/RT}$$

$$10^4 = \left(\frac{1.0}{9.89}\right)\left(\frac{-10,000}{-35,000}\right)e^{[(-35,000)'-(-10,000)]/T}$$

$$346,150 = e^{-25,000/T}$$

$$T = 1960°R$$

$$= 1500°F$$

PRD.8   HEAT EXCHANGER ECONOMICS

In a new process, it is necessary to heat 50,000 lb/h of an organic liquid from 150 to 330°F. The liquid is at a pressure of 135 psia. A simple steam-heated shell-and-tube floating-head carbon steel exchanger is the preferred equipment choice. Steam is available at 150 psia (135 psig) and 300 psia (285 psig). The higher pressure steam should result in a smaller heat exchanger, but the steam will cost more. Which steam choice would be better?

The heat capacity of the organic liquid is 0.6 Btu/(lb°F).

The plant stream factor is expected to be 90%. (The "stream factor" is the fraction of time that the plant operates over the course of a year.)

Steam Properties	150 psia	300 psia
Saturation temperature, °F	358	417
Latent heat, Btu/lb	863.6	809.0
Cost, $/1000 lb	5.20	5.75

Heat exchanger cost correlation (1979 basis) data are provided below.

Base cost = $117A^{0.65}$ (PF)
Installation factor = 3.29

Pressure factors (PF) affecting capital cost:
  100 to 200 psig = 1.15
  200 to 300 psig = 1.20
Cost indexes
  1979 = 230
  1998 = 360

Solution

The overall heat duty is

$$\dot{Q} = F c_P (T_2 - T_1) = (50{,}000 \text{ lb/h})[0.6 \text{ Btu/(lb·°F)}](330 - 150°F)$$
$$= 5{,}400{,}000 \text{ Btu/h}$$

The log-mean temperature difference (LMTD) for each case is

$$\text{LMTD}_{150} = \frac{(358 - 150) - (358 - 330)}{\ln\left(\dfrac{358 - 150}{358 - 330}\right)}$$
$$= 89.8°F$$

$$\text{LMTD}_{300} = \frac{(417 - 150) - (417 - 330)}{\ln\left(\dfrac{417 - 150}{417 - 330}\right)}$$
$$= 160.5°F$$

The required heat exchanger area is calculated from the standard heat transfer design equation:

$$A = \frac{Q}{(U)(\text{LMTD})}$$

Substitution gives

$$A_{150} = \frac{5{,}400{,}000}{(138)(89.8)} = 436 \text{ ft}^2$$

$$A_{300} = \frac{5{,}400{,}000}{(138)(160.5)} = 244 \text{ ft}^2$$

The capital cost for each case is

$$Cost_{150} = (117)(436)^{0.65}(360/230)(3.29)(1.15) = \$36,000$$
$$Cost_{300} = (117)(244)^{0.65}(360/230)(3.29)(1.20) = \$25,800$$

The steam requirement in pounds per year for each case is

$$St_{150} = (5,400,000\,Btu/h)(8760 \times 0.9\,h/yr)/(863.6\,Btu/lb)$$
$$= 49.3\,million\ lb/yr$$
$$St_{300} = (5,400,000\,Btu/h)(8760 \times 0.9\,h/yr)/(809.0\,Btu/lb)$$
$$= 52.6\,million\ lb/yr$$

The annual steam cost for each case is then

$$StCost_{150} = (49.3 \times 10^6 lb/yr)(0.00520\,\$/lb) = \$256,000/yr$$
$$StCost_{300} = (52.6 \times 10^6\ lb/yr)(0.00575\,\$/lb = \$303,000/yr$$

The 300-psia exchanger costs $10,200 less to purchase and install, but it costs $47,000 per year more to operate. The 150-psia exchanger is the obvious choice.

PRD.9   PUMP SELECTION

Two centrifugal pumps are being considered for purchase to handle a flow of 140 gpm with a required pressure increase of 120 psi. Both are nominal 20-HP pumps, which should be adequate for this service. Pump A costs less, but pump B operates more efficiently. Piping, installation, and foundation costs will be the same for either pump. Which pump should be selected?

Data: The pump is expected to operate with a 90% stream factor. (The "stream factor" is the fraction of time that the plant operates over the course of a year.) The cost of electricity at the plant is $0.085/kWh.

Pump Specifications	Pump A	Pump B
Cost delivered to plant	$4500	$6000
Efficiency at desired flow conditions	55%	60%

Plant economic criterion: simple 3-year payout

Solution

The hydraulic horsepower (HHP) required is

$$\text{HHP} = (0.000583)(q)(\Delta P) \qquad q[=]\text{gpm} \qquad \Delta P[=]\text{psi}$$
$$= (0.000583)(140)(120)$$
$$= 9.79 \text{ HP}$$

The brake horsepower (BHP) for each pump is

$$\text{BHP}_A = 9.79/0.55 = 17.8 \text{ HP}$$
$$\text{BHP}_B = 9.79/0.60 = 16.3 \text{ HP}$$

Converting to kW,

$$\text{kW}_A = (17.8)(0.745 \text{ kW/HP}) = 13.3 \text{ kW}$$
$$\text{kW}_B = (16.3)(0.745 \text{ kW/HP}) = 12.2 \text{ kW}$$

The total electrical energy in kWh required for 3 years becomes

$$\text{kWh}_A = (13.3)(3 \text{ yr})(8760 \text{ h/yr})(0.90) = 314{,}000 \text{ kWh}$$
$$\text{kWh}_B = (12.2)(3 \text{ yr})(8760 \text{ h/yr})(0.90) = 288{,}000 \text{ kWh}$$

The cost of electricity for 3 years (the payout period) is

$$\text{Cost}_A = (314{,}000 \text{ kWh})(\$0.085/\text{kWh}) = \$26{,}700$$
$$\text{Cost}_B = (288{,}000 \text{ kWh})(\$0.085/\text{kWh}) = \$24{,}400$$

The total 3-year cost for each pump can now be calculated:

Cost	Pump A	Pump B
Purchase cost	$4,500	$6,000
Electricity cost	$26,700	$24,400
Total	$31,200	$30,400

Pump B is the more economical choice. Note, however, that the difference is only about 2%. Small changes in the cost of electricity, the actual pump efficiencies, or the stream factor could shift the choice to pump A.

In the final design and purchase of equipment for a new plant, hundreds of decisions such as this are made after the overall process flow design is completed.

Note that a simple 3-year payout corresponds to a return on investment of about 14%, once taxes, depreciation, maintenance and other costs are accounted for.

PRD.10   HEAT EXCHANGER NETWORK

A plant has three streams to be heated and three streams to be cooled. Cooling water (90°F supply, 155°F return) and steam (saturated at 250 psia) are available. Devise a network of heat exchangers that will make full use of heating and cooling streams against each other, using utilities only if necessary.

The three streams to be heated are

Stream	Flowrate (lb/h)	c_P [Btu/(lb · °F)]	T_{in} (°F)	T_{out} (°F)
1	50,000	0.65	70	300
2	60,000	0.58	120	310
3	80,000	0.78	90	250

The three streams to be cooled are.

Stream	Flowrate (lb/h)	c_P [Btu/(lb · °F)]	T_{in} (°F)	T_{out} (°F)
4	60,000	0.70	420	120
5	40,000	0.52	300	100
6	35,000	0.60	240	90

Saturated steam at 250 psia has a temperature of 401°F.

Solution

The heating duties for all streams are first calculated. The results are tabulated below.

	Stream	Duty
1	Heating	7,745,000
2	Heating	6,612,000
3	Heating	9,984,000
4	Cooling	12,600,000
5	Cooling	4,160,000
6	Cooling	3,150,000

The total heating and cooling duties are next compared:

$$\text{Heating}: 7,745,000 + 6,612,000 + 9,984,000 = 24,341,000 \, \text{Btu/h}$$
$$\text{Cooling}: 12,600,000 + 4,160,000 + 3,150,000 = 19,910,000 \, \text{Btu/h}$$

As a minimum, 4,431,000 Btu/h will have to be supplied by steam.

Figure 60 represents a system of heat exchangers that will transfer heat from the hot streams to the cold ones in the amounts desired. It is important to note that this is but one of many possible schemes. The optimum system would require a trial-and-error procedure that would examine a host of different schemes. Obviously, the economics would come into play.

It should also be noted that in many chemical and petrochemical plants there are cold streams that must be heated and hot steams that must be cooled. Rather than use steam to do all the heating and cooling water to do all the cooling, it is often advantageous, as demonstrated in this problem, to have some of the hot streams heat the cold ones. The problem of optimum heat exchanger networks has been extensively studied and is available in the literature. This problem gives one simple illustration.

Finally, highly interconnected networks of exchangers can save a great deal of energy in a chemical plant. The more interconnected they are, however, the harder the plant is to control, start-up, and shut down. Often auxiliary heat sources and cooling sources must be included in the plant design in order to ensure that the plant can operate smoothly.

PRD.11   STORAGE REQUIREMENTS FOR A CHEMICAL PLANT

A chemical plant uses two liquid feeds and produces four different liquid chemical products. The flowrates and densities of the streams are given below.

Stream	Flowrate (lb/day)	Density (lb/ft^3)
Feed 1	110,000	49
Feed 2	50,000	68
Product A	40,000	52
Product B	25,000	62
Product C	10,000	52
Product D	95,000	47

The plant's storage requirements call for maintaining 4–5 weeks supply of each feed, 4–6 weeks supply of product A, B, and C, and 1–2 weeks supply of product D. The plant operates year round, but each tank must be emptied once a year for a week for maintenance. Tanks are normally "dedicated" to one feed or product, but one or two

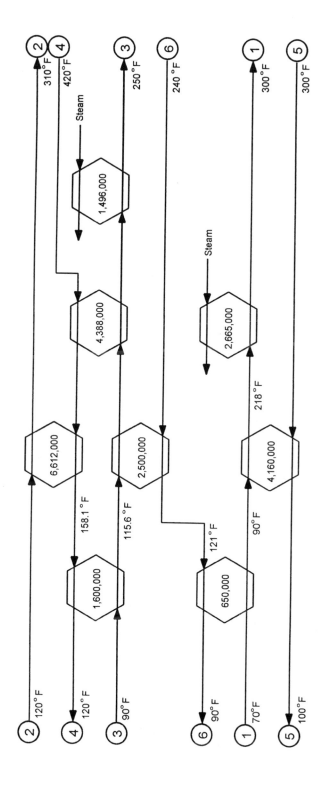

Numbers in heat exchanger boxes have units of Btu/h

Figure 60. Flow diagram for Problem PRD.10.

257

could be used as "swing" tanks, with one day of cleaning required between use with different liquids.

Specify an efficient set of tanks from the "standard" sizes given below to meet this plant's needs.

Standard Tank Sizes (gal)

2,800	16,800	281,000
5,600	28,100	561,000
8,400	56,100	1,123,000
11,200	140,000	

Solution

There is no single, simple method for determining the optimum mix of storage tanks for a chemical plant. Most often, estimates are made of the minimum and maximum amounts of feeds, intermediates, and products that must be kept on hand. Then some additional allowance is made to permit periodic cleaning and maintenance of the tanks.

The maximum and minimum amounts of each material to be stored is first obtained.

Stream	lb/day	gal/day	days	Gallons of Storage Required
Feed 1	110,000	16,800	28 to 35	470,200 to 587,800
Feed 2	50,000	5,500	28 to 35	154,000 to 192,500
Product A	40,000	5,750	28 to 42	161,000 to 241,700
Product B	25,000	3,020	28 to 42	84,500 to 126,700
Product C	10,000	14,390	28 to 42	40,300 to 60,400
Product D	95,000	15,120	7 to 14	105,800 to 211,700

The conversion to gallons requires dividing the rate in lb/day by the density in lb/ft^3, then multiplying by 7.481 gal/ft^3.

Select a set of tanks for each stream.

Feed 1: Use 2 × 281,000-gal tanks
Feed 2: Use 4 × 56,100-gal tanks
Product A: Use 2 × 140,000-gal tanks
Product B: Use 2 × 56,100-gal tanks
Product C: Use 1 × 56,100-gal tank
Product D: Use 2 × 140,000-gal tanks

Note that this set is 5% short on the maximum of feed 1, 11% short on product B, and 7% short on product C. For most situations, this would be acceptable.

Spare and/or swing tanks also need to be provided for maintenance.

Feed 1: Use an additional 281,000-gal tank.
Feed 2 and all products: Use one 56,100-gal swing tank.

This combination provides adequate auxiliary storage for maintenance periods.
   Thus, the total number of tanks required are three 281,000-gal, four 140,000-gal, and eight 56,100-gal tanks.
   The minimum number of tanks may not always be optimum if the tanks are extremely large. Several smaller tanks may cost somewhat more initially, but they offer more flexibility in use.

PDR.12   TROMBONE EXCHANGER

Design the trombone heat exchanger shown in Figure 61 using the data provided. This design employs a double-pipe heat exchanger for heating 2500 lb/h benzene from 60 to 120°F. Hot water at 200°F is available for heating purposes in amounts up to 4000 lb/h. Use schedule 40 brass pipe and an *integral* number of 15-ft-long sections.
   The design is to be based on achieving a Reynolds number of approximately 13,000 in both the inner pipe and annular region. As an additional constraint, the width of the annular region must be at least equal to $\frac{1}{4}$ the outside diameter of the inner pipe.
   For benzene at 90°F:

$$\mu = 0.55\,\text{cP} = 3.70 \times 10^{-4}\text{lb}/(\text{ft} \cdot \text{s})$$
$$\rho = 54.8\,\text{lb}/\text{ft}^3$$
$$c = c_p = 0.415\,\text{Btu}/(\text{lb} \cdot {}^\circ\text{F})$$
$$k = 0.092\,\text{Btu}/(\text{h} \cdot \text{ft} \cdot {}^\circ\text{F})\ \text{at}\ 86^\circ\text{F}$$

For water at 200°F:

$$\mu = 0.305\,\text{cP} = 2.05 \times 10^{-4}\,\text{lb}/(\text{ft} \cdot \text{s})$$
$$\rho = 60.13\,\text{lb}/\text{ft}^3$$
$$c = 1.0\,\text{Btu}/(\text{lb} \cdot {}^\circ\text{F})$$
$$k = 0.392\,\text{Btu}/(\text{h} \cdot \text{ft} \cdot {}^\circ\text{F})$$

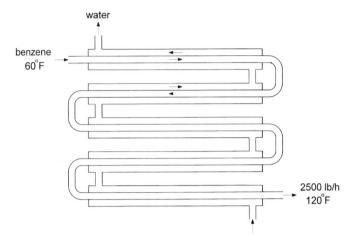

actual number of sections not shown

Figure 61. Trombone heat exchanger.

Solution

For flow inside tubes (pipes),

$$\text{Re} = \frac{D_i \bar{v} \rho}{\mu} = \frac{4\dot{m}}{\pi D_i \mu}$$

Solving for the inside diameter, D_i, yields

$$D_i = \frac{4\dot{m}}{\pi (\text{Re}) \mu}$$

$$= \frac{4(2500/3600)}{(\pi)(13,000)(3.70 \times 10^{-4})}$$

$$= 0.184 \, \text{ft}$$

$$= 2.20 \, \text{in}$$

The closest pipe with $D_i = 2.20$ in. is 2-in. schedule 40 pipe, which has a $D_i = 2.067$ in. This results in

$$\text{Re} = 13,874$$

For 2-in. schedule 40 pipe:

$D_{rmo}^{(i)}$	outside diameter of pipe	2.375 in.
$D_i^{(i)}$	inside diameter of pipe	2.067 in.
S_o	outside pipe area per unit length	0.622 ft^2/ft
S_i	inside pipe area per unit length	0.541 ft^2/ft
A_i	inside cross sectional area	0.0233 ft^2
x_w	pipe wall thickness	0.154 in.

The superscripts (i) and (o) represent "inside" and "outside," respectively

For flow in the annular region,

$$\mathrm{Re} = \frac{D_e \bar{v} \rho}{\mu} = \frac{4\dot{m}}{\pi D_e \mu}$$

The equivalent diameter, D_e, is given by

$$D_e = 4r_H = \frac{4S}{L_p} = \frac{4S}{\pi(D_o^{(i)} + D_i^{(i)})}$$

where r_H = hydraulic radius
L_p = wetted perimeter

Therefore

$$\mathrm{Re} = \frac{4\dot{m}}{\pi(D_o^{(i)} + D_i^{(i)})\mu}$$

However, one of the system constraints is that

$$1/2(D_i^{(o)} + D_i^{(i)}) \geqslant (1/4)D_o^{(i)}$$

Therefore,

$$D_i^{(o)} \geqslant (3/2)D_o^{(i)}$$

$$\geqslant 3.562 \text{ in.}$$

The smallest pipe that satisfies this constraint is 4 in., schedule 40 pipe. The outer pipe is therefore 4 in. schedule 40 pipe.

For Re $\geqslant 13,000$

$$\dot{m} = \frac{\pi(\text{Re})(D_o^{(i)} + D_i^{(i)})\mu}{4}$$

$$= \frac{\pi(13,000)(2.375 + 4.026)(2.05 \times 10^{-4})(3600)}{(4)(12)}$$

$$= 4019\,\text{lb/h}$$

This choice of outer pipe size is acceptable since the flowrate is approximately 4000 lb/h.

The outlet water temperature is obtained from a heat (enthalpy) balance:

$$\dot{Q} = \dot{m}_c c_c \Delta T_c = \dot{m}_h c_h \Delta T_h$$

$$= (2500)(0.415)(60) = (4000)(1.0)\Delta T_h$$

$$\Delta T_h = 62,250/4000 = 15.6°\text{F}$$

$$T_{h,\text{out}} = 200 - 15.6 = 184.4°\text{F}$$

$$\overline{T}_h = (200 + 184.4)/2 = 192.2°\text{F}$$

For water at 192°F:

$$\mu = 0.322\,\text{cP} = 2.16 \times 10^{-4}\,\text{lb/(ft} \cdot \text{s)}$$

$$\rho = 60.31\,\text{lb/ft}^3$$

$$c = 1.0\,\text{Btu/(lb} \cdot °\text{F)}$$

$$k = 0.390\,\text{Btu/(h} \cdot \text{ft} \cdot °\text{F)}$$

To determine the heat-transfer coefficients, the following equation is used. (See Chapter 7 for additional details.)

$$\text{St} = 0.023\,(\text{Re})^{-0.2}(\text{Pr})^{-2/3}$$

where St = Stanton number
      Pr = Prandtl number

For the inside pipe (benzene),

$$\text{Re} = 13,000$$

$$\text{Pr} = \frac{c\mu}{k} = \frac{(0.415)(3.70 \times 10^{-4})}{0.092/3600}$$

$$= 6.01$$

$$\text{St} = 0.023(13,000)^{-0.2}(6.01)^{-2/3}$$

$$= 0.00105$$

By definition,

$$St = \frac{h}{cG}$$

where G is the mass flux $= \dot{m}/S$.

Therefore,

$$h_i = (St)cG = \frac{(St)c\dot{m}}{S}$$

$$= \frac{(0.00105)(0.415)(2500)}{0.0233}$$

$$= 46.7 \, \text{Btu}/(\text{h} \cdot \text{ft}^2 \cdot °\text{F})$$

For the annular region (water),

$$Re = \frac{4\dot{m}}{\pi(D_0^{(i)} + D_i^{(o)})\mu}$$

$$= \frac{(4)(4000/3600)}{\pi\left(\dfrac{2.375 + 4.026}{12}\right)(2.16 \times 10^{-4})}$$

$$= 12,278$$

$$Pr = \frac{c\mu}{k} = \frac{(1.0)(2.16 \times 10^{-4})}{0.390/3600}$$

$$= 1.99$$

$$St = 0.023(12,278)^{-0.2}(1.99)^{-2/3}$$

$$= 0.00221$$

$$S_{\text{annular}} = \frac{\pi}{4}(D_i^{(o)2} - D_0^{(i)2})$$

$$= 0.0576 \, \text{ft}^2$$

$$h_0 = (St)cG = \frac{(St)c\dot{m}}{S}$$

$$= \frac{(0.00221)(1.0)(4000)}{0.0576}$$

$$= 154 \, \text{Btu}/(\text{h} \cdot \text{ft}^2 \cdot °\text{F})$$

Neglecting fouling,

$$\frac{1}{U_o} = \frac{D_o}{D_i h_i} + \frac{x_w D_o}{k_w D_{LM}} + \frac{1}{h_o}$$

$$D_{LM} = \frac{2.375 - 2.067}{\ln(2.375/2.067)} = 2.217 \text{ in.}$$

$$\frac{1}{U_o} = \frac{2.375}{(2.067)(46.7)} + \frac{(0.154)(2.375)}{(26)(2.217)} + \frac{1}{154}$$

$$= 0.0374$$

$$U_o = 26.7 \, \text{Btu}/(\text{h} \cdot \text{ft}^2 \cdot {}^\circ\text{F})$$

The length is now obtained by

$$\dot{Q} = U_o A_o \Delta T_{LM}$$

$$\Delta T_{LM} = \frac{124.4 - 80}{\ln(124.4/80)}$$

$$= 100.6^\circ \text{F}$$

$$A_o = 0.622L$$

$$62,250 = (26.7)(0.622)(L)(100.6)$$

$$L = 37.26 \, \text{ft}$$

Two sections might do the job; however, three sections are recommended, although the unit would be significantly overdesigned with three sections.

PART III
Air Pollution Control Equipment

13 Fluid Particle Dynamics (FPD)

FPD.1 CHECK FOR EMISSION STANDARDS COMPLIANCE

As a consulting engineer, you have been contracted to modify an existing control device used in fly ash removal. The federal standards for emissions have been changed to a total numbers basis. Determine if the unit will meet an effluent standard of $10^{5.7}$ particles/acf. Data for the unit are given below.

Average particle size, $d_p = 10\,\mu m$; assume constant
Particle specific gravity $= 2.3$
Inlet loading $= 3.0\,gr/ft^3$
Efficiency (mass basis), $E = 99\%$

Solution

The outlet loading, (OL) is

$$OL = (1.0 - 0.99)3.0$$
$$= 0.030\,gr/ft^3$$

Assume a basis of $1.0\,ft^3$:

$$Particle\ Mass = \rho_p V_p = \rho_p \frac{\pi d_p^3}{6}$$

$$= (7000\,gr/lb)\left(\frac{\pi[(10\,\mu m)(0.328 \times 10^{-5}\,ft/\mu m)]^3(2.33)(62.4\,lb/ft^3)}{6}\right)$$

$$= 1.881 \times 10^{-8}\,gr$$

$$Number\ of\ particles = (0.03\,gr)/(1.881 \times 10^{-8}\,gr/particle)$$
$$= 1.595 \times 10^6\ particles\ in\ 1\,ft^3$$

$$Allowable\ number\ of\ particles/ft^3 = 10^{5.7}$$
$$= 5.01 \times 10^5$$

Therefore, the unit will not meet a numbers standard.

FPD.2 LOG-NORMAL DISTRIBUTION

You have been requested to determine if a particle size distribution is log-normal. Data are provided below.

Particle Size Range, d_p μm	Distribution ($\mu g/m^3$)
< 0.62	25.5
0.62–1.0	33.15
1.0–1.2	17.85
1.2–3.0	102.0
3.0–8.0	63.75
8.0–10.0	5.1
> 10.0	7.65
Total	255.0

Solution

Particulates discharged from an operation consist of a size distribution ranging anywhere from extremely small particles (less than 1 μm) to very large particles (greater than 100 μm). Particle size distributions are usually represented by a cumulative weight fraction curve in which the fraction of particles less than or greater than a certain size is plotted against the dimension of the particle.

To facilitate recognition of the size distribution, it is useful to plot a size–frequency curve. The size–frequency curve shows the number (or weight) of particles present for any specified diameter. Since most dusts are comprised of an infinite range of particle sizes, it is first necessary to classify particles according to some consistent pattern. The number or weight of particles may then be defined as that quantity within a specified size range having finite boundaries and typified by some average diameter.

The shapes of the curves obtained to describe the particle size distribution generally follow a well-defined form. If the data include a wide range of sizes, it is often better to plot the frequency (i.e., number of particles of a specified size) against the logarithm of the size. In most cases an asymmetrical or "skewed" distribution exists; normal probability equations do not apply to this distribution. Fortunately, in most instances the symmetry can be restored if the logarithms of the sizes are substituted for the sizes. The curve is then said to be logarithmic normal (or log-normal) in distribution. Cumulative distribution plots described above are therefore generated by plotting particle diameter versus cumulative percent. For log-normal distributions, plots of particle diameter versus either percent less than stated size (% LTSS) or percent greater than stated size (% GTSS) produce straight lines on log-probability coordinates.

Cumulative distribution information can be obtained from the calculation results provided below.

d_p µm	% Total	Cumulative %GTSS
< 0.62	10	90
0.62–1.0	13	77
1.0–1.2	7	70
1.2–3.0	40	30
3.0–8.0	25	5
8.0–10.0	2	3
> 10.0	3	0

The cumulative distribution above can be plotted on log-probability graph paper. The cumulative distribution curve is shown in Figure 62. Since a straight line is obtained on log-normal coordinates, the particle size distribution is log-normal.

FPD.3   MEAN AND STANDARD DEVIATION OF A PARTICLE SIZE DISTRIBUTION

With reference to the previous log-normal distribution problem (FPD.2), estimate the mean and standard deviation from the size distribution information available.

The use of probability plots is of value when the arithmetic or geometric mean is required, since these values may be read directly from the 50% point on a logarithmic probability plot. By definition, the size corresponding to the 50% point on the probability scale is the *geometric mean diameter*. The geometric standard deviation is given (for % LTSS) by:

$$\sigma = 84.13\% \text{ size}/50\% \text{ size}$$

or

$$\sigma = 50\% \text{ size}/15.87\% \text{ size}$$

For % GTSS,

$$\sigma = 50\% \text{ size}/84.13\% \text{ size}$$

or

$$\sigma = 15.87\% \text{ size}/50\% \text{ size}$$

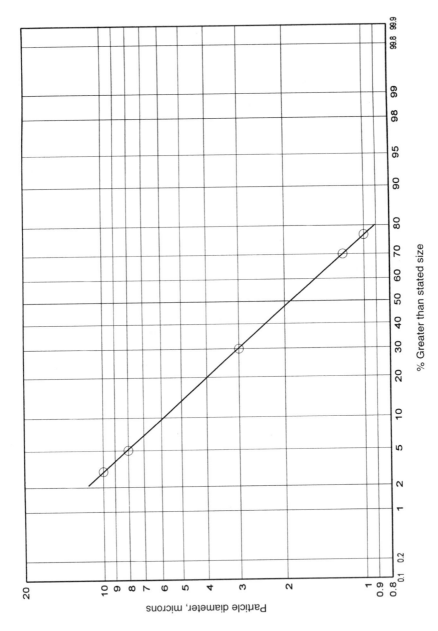

Figure 62. Cumulative distribution curve.

% Greater than stated size

Particle diameter, microns

270

Solution

The mean, as read from the 50% GTSS point on the graph in Figure 62 (and duplicated in Figure 63), is around 1.9 µm. A value of 1.91 µm is obtained from an expanded plot.

From the diagram, the particle size corresponding to the 15.87% point is

$$d_p(15.87\%) = 4.66 \, \mu m$$

The standard deviation may now be calculated. By definition,

$$\sigma = d_p(15.87\%)/d_p(50\%)$$

$$= 4.66/1.91$$

$$= 2.44$$

FPD.4   PARTICLE SETTLING VELOCITY

Three different sized fly ash particles settle through air. You are asked to calculate the particle terminal velocity and determine how far each will fall in 30 s. Also calculate the size of a fly ash particle that will settle with a velocity of 1.384 ft/s.

Assume the particles are spherical. Data are provided below.

Fly ash particle diameters = 0.4, 40, 400 µm
Air temperature and pressure = 238°F, 1 atm
Specific gravity of fly ash = 2.31

Solution

The terminal velocity of a particle is a constant value of velocity reached when all forces (gravity, drag, buoyancy, etc.) acting on the particle are balanced. The sum of all the forces is then equal to zero (no acceleration). To calculate this velocity, a dimensionless constant K determines the appropriate range of the fluid-particle dynamic laws that apply:

$$K = d_p \left(\frac{g(\rho_p - \rho)\rho}{\mu^2} \right)^{1/3}$$

where K = dimensionless constant that determines the range of the fluid-particle dynamic laws.
d_p = particle diameter
g = gravity force
ρ_p = particle density
ρ = fluid (gas) density
μ = fluid (gas) viscosity

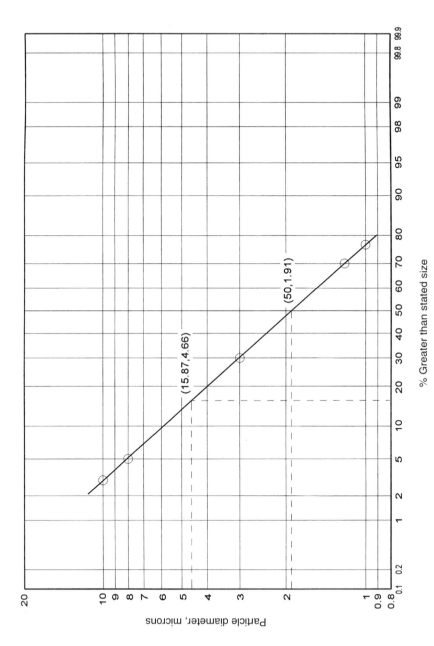

Figure 63. Cumulative distribution curve for Problem FPD.3.

272

A consistent set of units that will yield a dimensionless K is: d_p in ft, g in ft/s^2, ρ_p in lb/ft^3, and μ in lb/(ft · s). The numerical value of K determines the appropriate law:

$K < 3.3$; Stokes' law range

$3.3 < K < 43.6$; intermediate law range

$43.6 < K < 2360$; Newton's law range

For the Stokes' law range:

$$v = \frac{gd_p^2\rho_p}{18\mu}$$

For the intermediate law range:

$$v = 0.153 \frac{g^{0.71}d_p^{1.14}\rho_p^{0.71}}{\mu^{0.43}\rho^{0.29}}$$

For Newton's law range:

$$v = 1.74\left(\frac{gd_p\rho_p}{\rho}\right)^{0.5}$$

Larocca (*Chem. Eng.*, April 2, 1987 p. 73) and Theodore (personal notes), using the same approach employed above, defined a dimensionless term W that would enable one to calculate the diameter of a particle if the terminal velocity is known (or given). This particular approach has found application in catalytic reactor particle size calculations. The term W, which does not depend on the particle diameter, is given by

$$W = \frac{v^3\rho^2}{g\mu\rho_p}$$

The two key values of W that are employed in a manner similar to that for K are 0.2222 and 1514, i.e., for

$W < 0.2222$; Stokes' law range

$0.2222 < W < 1514$; intermediate law range

$1514 < W$; Newton's law range

When particles approach sizes comparable to the mean free path of other fluid molecules, the medium can no longer be regarded as continuous since particles can fall between the molecules at a faster rate than predicted by aerodynamic theory. To allow for this "slip", Cunningham's correction factor is introduced to Stokes' law.

$$v = \frac{gd_p^2 \rho_p}{18\mu} C_f$$

where C_f = Cunningham correction factor
$$= 1 + (2A\lambda/d_p)$$
$$A = 1.257 + 0.40e^{-1.10d_p/2\lambda}$$
λ = mean free path of the fluid molecules (6.53×10^{-6} cm for ambient air)

The Cunningham correction factor is usually applied to particle diameters equal to or smaller than 1 µm.

For the problem at hand, the particle density is calculated using the specific gravity given.

$$\rho_p = (2.31)(62.4)$$
$$= 144.14 \, lb/ft^3$$

The density of air is

$$\rho = P(MW)/RT$$
$$= (1)(29)/(0.7302)(238 + 460)$$
$$= 0.0569 \, lb/ft^3$$

The viscosity of air is

$$\mu = 0.021 \, cP$$
$$= 1.41 \times 10^{-5} \, lb/(ft \cdot s)$$

The value of K for each fly ash particle size settling in air may be calculated. Note that $\rho_p - \rho \approx \rho_p$.
For a d_p of 0.4 µm:

$$K = \frac{0.4}{(25,000)(12)} \left(\frac{(32.2)(144.1)(0.0569)}{(1.41 \times 10^{-5})^2} \right)^{1/3} = 0.0144$$

For a d_p of 40 µm:

$$K = \frac{40}{(25,400)(12)} \left(\frac{(32.2)(144.1)(0.0569)}{(1.41 \times 10^{-5})^2} \right)^{1/3} = 1.44$$

For a d_p of 400 µm:

$$K = \frac{400}{(25,400)(12)} \left(\frac{(32.2)(144.1)(0.0569)}{(1.41 \times 10^{-5})^2} \right)^{1/3} = 14.4$$

Therefore,

For $d_p = 0.4$ µm; Stokes' law range
For $d_p = 40$ µm; Stokes' law range
For $d_p = 400$ µm; intermediate law range

For a d_p of 0.4 µm (without the Cunningham correcton factor):

$$v = \frac{g d_p^2 \rho_p}{18\mu} = \frac{(32.2)[(0.4)/(25,400)(12)]^2(144)}{(18)(1.41 \times 10^{-5})}$$
$$= 3.15 \times 10^{-5} \text{ ft/s}$$

For a d_p of 40 µm:

$$v = \frac{g d_p^2 \rho_p}{18\mu} = \frac{(32.2)[(40)/(25,400)(12)]^2(144)}{(18)(1.41 \times 10^{-5})}$$
$$= 0.315 \text{ ft/s}$$

For a d_p of 400 µm:

$$v = 0.153 \frac{g^{0.71} d_p^{1.14} \rho_p^{0.71}}{\mu^{0.43} \rho^{0.29}}$$
$$= 0.153 \frac{(32.2)^{0.71}[(400)/(25,400)(12)]^{1.14}(144.1)^{0.71}}{(1.41 \times 10^{-5})^{0.43}(0.0569)^{0.29}}$$
$$= 8.90 \text{ ft/s}$$

The distance that the fly ash particles will fall in 30 s may also be calculated.

For a d_p of 40 µm:

$$\text{Distance} = (30)(0.315)$$
$$= 9.45 \text{ ft}$$

For a d_p of 400 µm:

$$\text{Distance} = (30)(8.90)$$
$$= 267 \text{ ft}$$

For a d_p of 0.4 µm, the Cunningham Correction factor should be included:

$$A = 1.257 + 0.40e^{-1.10d_p/2\lambda}$$
$$= 1.257 + 0.40 \ \exp -\left(\frac{1.10(0.4)}{(2)(6.53 \times 10^{-2})}\right)$$
$$= 1.2708$$
$$C_f = 1 + \frac{2A\lambda}{d_p}$$
$$= 1 + \frac{(2)(1.2708)(6.53 \times 10^{-2})}{(0.4)}$$
$$= 1.415$$
$$\text{Corrected } v = (3.15 \times 10^{-5})(1.415)$$
$$= 4.45 \times 10^{-5} \text{ ft/s}$$
$$\text{Distance} = (30)(4.45 \times 10^{-5})$$
$$= 1.335 \times 10^{-3} \text{ ft}$$

For the particle traveling with a velocity of 1.384 ft/s, first calculate the dimensionless number, W:

$$W = \frac{v^3\rho^2}{g\mu\rho_p}$$
$$= \frac{(1.384)^3(0.0569)^2}{(32.2)(144.1)(1.41 \times 10^{-5})}$$
$$= 0.1312$$

Since $W < 0.222$, Stokes's law applies, and

$$d_p = \left(\frac{18 \mu v}{g \rho_p} \right)^{0.5}$$

$$= \left(\frac{(18)(1.41 \times 10^{-5})(1.384)}{(32.2)(144.1)} \right)^{0.5}$$

$$= 2.751 \times 10^{-4} \text{ ft}$$

$$= 83.9 \text{ μm}$$

FPD.5   ATMOSPHERIC DISCHARGE CALCULATION

ETS engineers have been requested to determine the minimum distance downstream from a cement source emitting dust that will be free of cement deposit. The source is equipped with a cyclone. The cyclone is located 150 ft above ground level. Assume ambient conditions are at 60°F and 1 atm and neglect meteorological aspects. Additional data are given below.

Particle size range of cement dust $= 2.5–50.0$ μm
Specific gravity of the cement dust $= 1.96$
Wind speed $= 3.0$ miles/h

Solution

A particle diameter of 2.5 μm is used to calculate the minimum distance downstream free of dust since the smallest particle will travel the greatest horizontal distance. To determine the value of K for the appropriate size of the dust, first calculate the particle density using the specific gravity given and determine the properties of the gas (assume air).

$$\rho_p = (1.96)(62.4)$$

$$= 122.3 \text{ lb/ft}^3$$

$$\rho_a = P(MW)/RT$$

$$= (1)(29)/[(0.73)(60 + 460)]$$

$$= 0.0764 \text{ lb/ft}^3$$

Viscosity of air, μ, at 60°F $= 1.22 \times 10^{-5}$ lb/(ft · s)

The value of K is

$$K = d_p \left(\frac{g(\rho_p - \rho)\rho}{\mu^2} \right)^{1/3}$$

$$= \frac{2.5}{(25,400)(12)} \left(\frac{(32.2)(122.3 - 0.0764)(0.0764)}{(1.22 \times 10^{-5})^2} \right)^{1/3} = 0.104$$

The velocity is therefore in the Stokes' law range. Using the appropriate terminal settling velocity equation, the terminal settling velocity in feet per second is

$$v = \frac{gd_p^2 \rho_p}{18\mu} = \frac{(32.2)[(2.5)/(25,400)(12)]^2(122.3)}{(18)(1.22 \times 10^{-5})}$$

$$= 1.21 \times 10^{-3} \text{ ft/s}$$

The approximate time for descent is

$$t = h/v$$

$$= 150/1.21 \times 10^{-3}$$

$$= 1.24 \times 10^5 \text{ s}$$

Thus, the distance traveled, d, is

$$d = tu$$

$$= (1.24 \times 10^5)(3.0/3600)$$

$$= 103.3 \text{ miles}$$

FPD.6   PARTICLE DRAG

A spherical particle having a diameter of 0.0093 in. and a specific gravity of 1.85 is placed on a horizontal screen. Air is blown through the screen vertically at a temperature of 20°C and a pressure of 1.0 atm. Calculate the following:

1. The velocity required to just lift the particle
2. The particle Reynolds number at this condition
3. The drag force in both engineering and cgs units
4. The drag coefficient C_D

Solution

At 20°C or 68°F,

$$\rho = P(MW)/RT$$
$$= (1)(29)/[(0.73)(68 + 460)]$$
$$= 0.0752 \text{ lb/ft}^3$$
$$\mu = 1.23 \times 10^{-5} \text{ lb/(ft} \cdot \text{s)}$$

1. K is given by

$$K = d_p \left(\frac{g(\rho_p - \rho)\rho}{\mu^2} \right)^{1/3}$$
$$= \frac{9.3 \times 10^{-3}}{12} \left(\frac{(32.2)[(1.85)(62.4) - 0.0752](0.0752)}{(1.23 \times 10^{-5})^2} \right)^{1/3}$$
$$= 9.51$$

The intermediate range equation for v should be used:

$$v = 0.153 \frac{g^{0.71} d_p^{1.14} \rho_p^{0.71}}{\mu^{0.43} \rho^{0.29}}$$
$$= 0.153 \frac{(32.2)^{0.71}[(9.3 \times 10^{-3})/(12)]^{1.14}[(1.85)(62.4)]^{0.71}}{(1.23 \times 10^{-5})^{0.43}(0.0752)^{0.29}}$$
$$= 4.08 \text{ ft/s}$$

2. The Reynolds number is

$$\text{Re} = \frac{d_p v \rho}{\mu}$$
$$= \frac{(9.3 \times 10^{-3}/12)(4.08)(0.0752)}{1.23 \times 10^{-5}}$$
$$= 19.33$$

3. The drag force is calculated from the following equation:

$$F_D = \frac{2.31\pi(vd_p)^{1.4}\mu^{0.6}\rho^{0.4}}{g_c}$$
$$= \frac{2.31\pi[(4.08)(9.3 \times 10^{-3})/12]^{1.4}(1.23 \times 10^{-5})^{0.6}(0.0752)^{0.4}}{32.2}$$
$$= 2.866 \times 10^{-8} \text{ lb}_f$$

Using the conversion factor, 4.448×10^5 dyne/lb$_f$, yields

$$F_D = 0.0127 \, \text{dyn}$$

4. The value of C_D may be calculated from the following equation:

$$C_D = \frac{F_D/A_p}{\rho v^2/2g_c}$$

where A_p, the projected area, is given by

$$A_p = \pi d_p^2/4 = \pi(9.3 \times 10^{-3}/12)^2/4 = 4.72 \times 10^{-7} \, \text{ft}^2$$

Therefore,

$$C_D = \frac{(2.866 \times 10^{-8})/(4.72 \times 10^{-7})}{(0.0752)(4.08)^2/[(2)(32.2)]}$$
$$= 3.12$$

FPD.7   COLLECTION EFFICIENCY FOR PARTICLES SMALLER THAN 1 MICRON

Explain why nearly all particle size collection efficiency curves for high-efficiency control devices take the form shown in Figure 64.

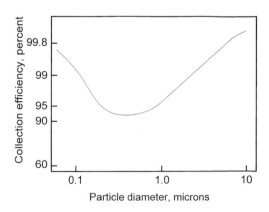

Figure 64. Effect of particle size on performance.

Solution

As illustrated in Figure 64, the collection efficiency for particulate control increases with increasing particle size over nearly the entire particle size range. However, for particles less than approximately 1.0 µm, the trend reverses and efficiency increases with decreasing size. This phenomena is experienced by almost all efficiency control devices, e.g., baghouses, venturi scrubbers, electrostatic precipitators, etc., and arises primarily because of molecular diffusion effects. The random, chaotic motion of submicron particles, similar to that predicted by the kinetic theory of gases, becomes more pronounced as the particle size decreases and approaches the molecular diameter of gases, resulting in higher efficiencies. This becomes an important consideration for systems requiring extremely high efficiencies, e.g., in excess of 99.5%.

FPD.8   DISPERSION OF SOAP PARTICLES

A plant manufacturing Ivory Soap detergent explodes one windy day. It disperses 100 tons of soap particles (specific gravity $= 0.8$) into the atmosphere (70°F, $\rho = 0.0752\,\text{lb/ft}^3$). If the wind is blowing 20 miles/h from the west and the particles range in diameter of 2.1–1000 µm, calculate the distances from the plant where the soap particles will start to deposit and where they will cease to deposit. Assume the particles are blown vertically 400 ft in the air before they start to settle. Also, assuming even ground-level distribution through an average 100-ft-wide path of settling, calculate the average height of the soap particles on the ground in the settling area. Assume the bulk density of the particles equal to half of the actual density.

Solution

The smallest particle will travel the greatest distance while the largest will travel the least distance. For the minimum distance, use the largest particle:

$$d_{\text{p}} = 1000 \ \mu\text{m} = 3.28 \times 10^{-3} \ \text{ft}$$

$$K = d_{\text{p}} \left(\frac{g(\rho_{\text{p}} - \rho)\rho}{\mu^2} \right)^{1/3}$$

$$= (3.28 \times 10^{-3}) \left(\frac{(32.2)[(0.8)(62.4) - 0.752](0.0752)}{(1.18 \times 10^{-5})^2} \right)^{1/3}$$

$$= 31.3$$

Using the intermediate range equation,

$$v = 0.153 \frac{g^{0.71} d_p^{1.14} \rho_p^{0.71}}{\mu^{0.43} \rho^{0.29}}$$

$$= 0.153 \frac{(32.2)^{0.71} (3.28 \times 10^{-3})^{1.14} [(0.8)(62.4)]^{0.71}}{(1.18 \times 10^{-5})^{0.43} (0.0752)^{0.29}}$$

$$= 11.9 \, \text{ft/s}$$

The descent time t is

$$t = H/v$$
$$= 400/11.9 = 33.6 \, \text{s}$$

The horizontal distance traveled, L, is

$$L = (33.6)\left(\frac{20}{(60)(60)}\right)(5280)$$
$$= 986 \, \text{ft}$$

For the maximum distance, use the smallest particle:

$$d_p = 2.1 \, \mu\text{m} = 6.89 \times 10^{-6} \, \text{ft}$$

$$K = (6.89 \times 10^{-6})\left(\frac{(32.2)[(0.8)(62.4) - 0.0752](0.0752)}{(1.18 \times 10^{-5})^2}\right)^{1/3}$$

$$= 0.066$$

The velocity v is in the Stokes regime and is given by

$$v = \frac{dg_p^2 \rho_p}{18\mu}$$

$$= \frac{(32.2)(6.89 \times 10^{-6})^2 (0.8)(62.4)}{(18)(1.18 \times 10^{-5})}$$

$$= 3.59 \times 10^{-4} \, \text{ft/s}$$

The descent time t is

$$t = H/v = 400/3.59 \times 10^{-4}$$
$$= 1.11 \times 10^6 \, \text{s}$$

The horizontal distance traveled, L, is

$$L = (1.11 \times 10^6)\left(\frac{20}{(60)(60)}\right)(5280)$$

$$= 3.26 \times 10^7 \text{ ft}$$

To calculate the depth D, the volume of particles (actual), V_{act}, is first determined:

$$V_{act} = (100)(2000)/[(0.8)(62.4)]$$

$$= 4006 \text{ ft}^3$$

The bulk volume V_B is

$$V_B = 4006/0.5 = 8012 \text{ ft}^3$$

The length of the drop area, L_D, is

$$L_D = 3.2 \times 10^7 - 994$$

$$= 3.2 \times 10^7 \text{ ft}$$

Since the width is 100 ft, the deposition area A is

$$A = (3.2 \times 10^7)(100) = 3.2 \times 10^9 \text{ ft}^2$$

$$V_B = AD$$

$$8012 = (3.2 \times 10^9)D$$

Therefore,

$$D = 2.5 \times 10^{-6} \text{ ft}$$

$$= 0.76 \text{ μm}$$

The deposition can be, at best, described as a "sprinkling."

14 Mechanical Collectors (MCC)

MCC.1 MINIMUM PARTICLE SIZE

A hydrochloric acid mist in air at 25°C is to be collected in a gravity settler. You are requested to calculate the smallest mist droplet (spherical in shape) that will always be collected by the settler. Assume the acid concentration to be uniform through the inlet cross section of the unit and Stokes' law applies. Operating data and information on the gravity settler are given below.

Dimensions of gravity settler $= 30$ ft wide, 20 ft high, 50 ft long
Actual volumetric flowrate of acidic gas $= 50$ ft^3/s
Specific gravity of acid $= 1.6$
Viscosity of air $= 0.0185$ cP $= 1.243 \times 10^{-5}$ lb/(ft · s)
Density of air $= 0.076$ lb/ft^3

Solution

Gravity settlers, or gravity settling chambers, have long been utilized industrially for the removal of solid and liquid waste materials from gaseous streams. Advantages accounting for their use are simple construction, low initial cost and maintenance, low pressure losses, and simple disposal of waste materials. Gravity settlers are usually constructed in the form of a long, horizontal parallelepipeds with suitable inlet and outlet ports. In its simplest form the settler is an enlargement (large box) in the duct carrying the particle-laden gases; the contaminated gas stream enters at one end, the cleaned gas exits from the other end. The particles settle toward the collection surface at the bottom of the unit with a velocity at or near their terminal settling velocity.

Gravity settlers can be designed to ensure capture of some minimum particle size. If Stokes' law applies, one can show that this minimum particle size is given by

$$d_{\mathrm{p}} = \left(\frac{18\mu g}{g\rho_{\mathrm{p}} BL} \right)^{1/2}$$

where $\mu =$ gas (fluid) viscosity
$q =$ volumetric flowrate
$g =$ gravity force
$\rho_p =$ particle or mist density
$B =$ width of settler
$L =$ length of settler

For the intermediate range,

$$d_p = \frac{q^{0.88}\rho^{0.524}\mu^{0.377}}{0.193(g\rho_p)^{0.623}(BL)^{0.88}}$$

Finally, for Newton's law range,

$$d_p = 0.547\left(\frac{\rho}{g\rho_p}\right)\left(\frac{q}{BL}\right)^2$$

For the problem at hand, first determine the density of the acid mist:

$$\rho_p = (62.4)(1.6)$$
$$= 99.84 \text{ lb/ft}^3$$

Calculate the minimum particle diameter both in feet and microns assuming Stokes's law applies:

$$d_p = \left(\frac{(18)(1.243 \times 10^{-5})(50)}{(32.2)(99.84)(30)(50)}\right)^{1/2}$$
$$= 4.82 \times 10^{-5} \text{ ft}$$

There are $3.048 \times 10^5 \,\mu\text{m}$ in 1 ft. Therefore,

$$d_p = 14.7\,\mu\text{m}$$

The particle diameter calculated above represents a limiting value since particles with diameters equal to or greater than this value will reach the settler collection surface and particles with diameters less than this value may escape from the unit. This limiting particle diameter can ideally be thought of as the minimum diameter of a particle that will automatically be captured for the above conditions. This diameter is denoted by d_p^* or $d_p(\min)$.

MCC.2   GRAVITY SETTLER DESIGN

As a recently hired engineer for an equipment vending company, you have been requested to design a gravity settler to remove all the iron particulates from a dust-laden gas stream. The following information is given:

$d_p = 35\ \mu m$; uniform, i.e., no distribution
gas = air at ambient conditions
$q = 130\ ft^3/s$; throughput velocity, $v = 10\ ft/s$
$\rho_p = 7.62\ g/cm^3$

Solution

First convert d_p and ρ_p to engineering units:

$$d_p = (35\ \mu m)(3.281 \times 10^{-6}\ ft/\mu m) = 11.48 \times 10^{-5}\ ft$$

$$\rho_p = (7.62\ g/\ cm^3)(1\ lb/453.6\ g)(28{,}316\ cm^3/ft^3) = 475.7\ lb/ft^3$$

The K value (see Problem, FPD.4) is calculated by

$$K = d_p \left(\frac{g(\rho_p - \rho)\rho}{\mu^2} \right)^{1/3}$$

$$= (11.48 \times 10^{-5}) \left(\frac{(32.2)(475.7 - 0.0775)(0.0775)}{(1.23 \times 10^{-5})^2} \right)^{1/3}$$

$$= 2.28 < 3.3$$

Stokes' law applies and the collection area required can be calculated from the equation

$$d_p = \left(\frac{18\mu g}{g\rho_p BL} \right)^{0.5}$$

Solving for BL,

$$BL = \frac{18\mu g}{g\rho_p d_p^2}$$

$$= \frac{18(1.23 \times 10^{-5})(130)}{(32.2)(475.5)(11.48 \times 10^{-5})^2}$$

$$= 142.5\ ft^2$$

The cross-sectional area for $v = 10 \, \text{ft/s}$ is

$$BH = q/v = 130/10$$
$$= 13 \, \text{ft}^2$$

Based on the minimum required for cleaning purposes, H is usually 3 ft. Then,

$$B = (13)/H = 13/3$$
$$= 4.33 \, \text{ft}$$

and

$$L = (142.5)/B = 142.5/4.33$$
$$= 32.9 \, \text{ft}$$

and the total volume of the settler is

$$V = (BL)H = (142.5)(3)$$
$$= 427.5 \, \text{ft}^3$$

The key process design variable for gravity settlers is the capture area, A, which is given by

$$A = BL$$

Once the capture area is calculated, the cost of the gravity settler internals may be assumed fixed. Because of material costs, however, the larger the outer casing of the physical system, the higher will be the total cost, all other factors being equal. These material costs generally constitute a significant fraction of the total cost of the settler. If the thickness of the outer casing is the same for alternate physical designs and if labor costs are linearly related to the surface area, then the equipment cost will roughly be a linear function of the outer surface area of the structural shell of the unit. This essentially means that the cost is approximately linearly related to the perimeter, P, where

$$P = 2L + 2B$$

To help minimize the cost (by minimizing the perimeter) one can equate the derivative of the perimeter to zero. Thus setting

$$B = A/L$$

gives

$$P = 2L + 2A/L$$

and

$$\frac{dP}{dL} = 2 - \frac{2A}{L^2}$$

Setting the above derivative equal to zero leads to

$$A = L^2$$

so that

$$L = B$$

since

$$A = BL$$

Interestingly, most gravity settlers (as well as electrostatic precipitators) are often designed physically in a form approaching a square box.

Thus

$$BL = L^2 = 142.5\,\text{ft}^2$$
$$B = L = 11.94\,\text{ft}$$
$$H = 3\,\text{ft}$$

and, in this case, the velocity of the gas would be

$$v = \frac{q}{BH} = \frac{(130)}{(11.94)(3)}$$
$$= 3.63\,\text{ft/s}$$

MCC.3   OVERALL COLLECTION EFFICIENCY

A settling chamber is installed in a small heating plant that uses a traveling grate stoker. You are requested to determine the overall collection efficiency of the settling chamber given the following operating conditions, chamber dimensions, and particle size distribution data.

Chamber width $= 10.8\,\text{ft}$
Chamber height $= 2.46\,\text{ft}$

Chamber length $= 15.0$ ft

Volumetric flowrate of contaminated air stream $= 70.6$ scfs

Flue gas temperature $= 446°F$

Flue gas pressure $= 1$ atm

Particle concentration $= 0.23$ gr/scf

Particle specific gravity $= 2.65$

Standard conditions $= 32°F$, 1 atm

Particle size distribution data of the inlet dust for the traveling grate stoker are given in the following table:

Particle Size Range (μm)	Average Particle diameter (μm)	C_i (gr/scf)	w_i (wt %)
0–20	10	0.0062	2.7
20–30	25	0.0159	6.9
30–40	35	0.0216	9.4
40–50	45	0.0242	10.5
50–60	55	0.0242	10.5
60–70	65	0.0218	9.5
70–80	75	0.0161	7.0
80–94	87	0.0218	9.5
+ 94	+94	0.0782	34.0
		0.2300	100.0

Assume that the actual terminal settling velocity is one-half of the Stokes law velocity.

Solution

The collection efficiency, E, for a monodispersed aerosol (particulates of one size) in laminar flow can be shown to be

$$E = \frac{v_t BL}{q} \ (100\%)$$

where v_t is the terminal settling velocity of the particle.

The validity of this equation is observed by noting that $v_t BL$ represents the hypothetical volume rate of flow of gas passing the collection area, while q is the volumetric flowrate of gas entering the unit to be treated. An alternate but equivalent form of the above equation is

$$E = \left(\frac{v_t}{v}\right)\left(\frac{L}{H}\right)(100\%)$$

where H is the height of the settling chamber and v is the gas velocity.

If the gas stream entering the unit consists of a distribution of particles of various sizes, then frequently a fractional or grade efficiency curve is specified for the settler. This is simply a curve describing the collection efficiency for particles of various sizes. The dependency of E on d_p arises because of the v_t term in the above equations.

Since the actual terminal settling velocity is assumed to be one-half of the Stokes' law velocity (according to the problem statement),

$$v = \frac{1}{2} \frac{g d_p^2 \rho_p}{18\mu}$$

and therefore

$$E = \frac{g d_p^2 \rho_p BL}{36\mu q}$$

The viscosity of the air in lb/(ft · s) is

$$\mu \, (446°F) = 1.75 \times 10^{-5} \, \text{lb/(ft · s)}$$

The particle density in lb/ft^3 is

$$\rho_p = (2.65)(62.4)$$
$$= 165.4 \, \text{lb/ft}^3$$

To calculate the collection efficiency of the system at the operating conditions, the standard volumetric flowrate of contaminated air of 70.6 scfs is converted to the actual volumetric flow using Charles' law:

$$q = q_s \left(\frac{T_a}{T_s}\right) = (70.6) \left(\frac{446 + 460}{32 + 460}\right)$$
$$= 130 \, \text{acfs}$$

The collection efficiency in terms of d_p, with d_p in microns is given below. Note: To convert d_p from ft to µm, d_p is divided by (304,800) µm/ft:

$$E = \frac{v_t BL}{q} = \frac{g \rho_p BL d_p^2}{36\mu q}$$
$$= \frac{(32.2)(165.4)(10.8)(15) \, d_p^2}{(36)(1.75 \times 10^{-5})(130)(304,800)^2}$$
$$= 1.14 \times 10^{-4} \, d_p^2$$

where d_p is in microns.

Thus, for a particle diameter of 10 μm,

$$E = 1.14 \times 10^{-4} \, d_p^2$$
$$= (1.14 \times 10^{-4})(10)^2$$
$$= 1.1 \times 10^{-2}$$
$$= 1.1\%$$

The following table provides the collection efficiency for each particle size:

d_p (μm)	d_p^2 (μm²)	E (%)
93.8	8800	100.0
90	8100	92.0
80	6400	73.0
60	3600	41.0
40	1600	18.2
20	400	4.6
10	100	1.1

The size efficiency curve for the settling chamber is shown in Figure 65.

The collection efficiency of each particle size may be read from the size–efficiency curve. The product, $w_i E_i$, is then calculated for each size. The overall efficiency is equal to $\Sigma w_i E_i$. These calculations are provided in the following table.

Average d_p (μm)	Weight Fraction w_i	E_i (%)	$w_i E_i$ (%)
10	0.027	1.1	0.030
25	0.069	7.1	0.490
35	0.094	14.0	1.316
45	0.105	23.0	2.415
55	0.105	34.0	3.570
65	0.095	48.0	4.560
75	0.070	64.0	4.480
87	0.095	83.0	7.885
+94	0.340	100.0	34.000
Total	1.000		58.7

The overall collection efficiency for the settling chamber, E, is 58.7%.

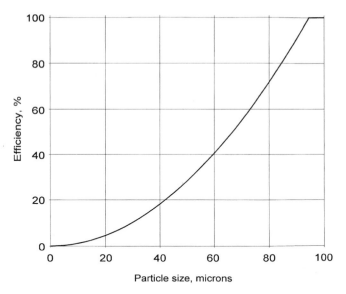

Figure 65. Size efficiency curve for settling chamber.

MCC.4   COMPLIANCE CALCULATION FOR A GRAVITY SETTLER

A gravity settler is 15 ft wide by 15 ft high by 40 ft long. In order to meet required ambient air quality standards, this unit must remove 90% of the fly ash particles entering the unit. Planned expansion will increase the flowrate to 4000 acfm with a dust loading of 30 gr/ft^3. The specific gravity of fly ash is 2.31 and the process gas stream is air at 20°C and 1.0 atm. The inlet size distribution of the fly ash is given below:

Size range (μm)	Mass Percent
0.0–10.0	1.0
10–20	1.0
20–30	3.0
30–40	15.0
40–60	20.0
60–80	25.0
80–100	20.0
100–150	15.0

Will the unit meet the specification?

The following data are provided:

$B = 15$ ft
$H = 15$ ft
$L = 40$ ft
$q = 4000$ acfm
Loading $= 30$ gr/ft^3
Specific gravity $= 2.31$; $\rho_p = 144$ lb/ft^3
$\mu = 1.23 \times 10^{-5}$ lb/(ft · s)

Solution

The throughput velocity, u, is

$$u = q/BH = 4000/[(15)(15)] = 17.778 \text{ ft/min} = 0.296 \text{ ft/s}$$

The fractional efficiency equation may be written as

$$E = \left(\frac{g d_p^2 \rho_p}{18\mu}\right)\left(\frac{LB}{q}\right)$$

$$= \left(\frac{(32.2)(144)}{(18)(1.23 \times 10^{-5})}\right)\left(\frac{(40)(15)}{(4000/60)}\right) d_p^2$$

$$= 1.88 \times 10^8 \, d_p^2 \qquad d_p \text{ in ft}$$

$$= 0.00202 \, d_p^2 \qquad d_p \text{ in } \mu m$$

The following table may now be generated:

Size Range (μm)	Average Particle Size (μm)	Mass Fraction w_i	E_i (%)	$w_i E_i$ (%)
0.0–10.0	5	0.01	5.1	0.051
10–20	15	0.01	45.45	0.455
20–30	25	0.03	100	3.0
30–40	35	0.15	100	15
40–60	50	0.20	100	20
60–80	70	0.25	100	25
80–100	90	0.20	100	20
100–150	125	0.15	100	15
				$\sum = 98.51$

The overall efficiency is 98.51%. As expected, the efficiency is high because of the coarse dust and the size of the unit. The reader should also check to ensure that

Stokes' law does in fact apply to those size ranges where the efficiency is less than 100%.

MCC.5   CHECK ON EFFICIENCY OF A GRAVITY SETTLER

A salesman from Bogus, Inc. suggests a gravity settler for a charcoal dust-contaminated air stream that you must preclean. Your supervisor has provided the particle size distribution shown below. The inlet loading is $20.00 \, \text{gr}/\text{ft}^3$ and the required outlet loading is $5.00 \, \text{gr}/\text{ft}^3$. Will the settler that the salesman has suggested do the job?

Size Range (µm)	Weight Percent
0–10	5
10–20	11
20–40	10
40–60	9
60–90	22
90–125	23
125–150	10
150+	10

Use the critical diameter ($d_p^* = 80 \, \mu\text{m}$) to calculate the size–efficiency data from the equation

$$E = kd_p^2$$

Solution

First calculate k:

$$k = \frac{E}{d_p^2} = \frac{100}{(80)^2}$$

$$= 0.01563$$

Thus, for $d_p = 5\mu\text{m}$ (average diameter for the first size range),

$$E = 0.01563 \, (5)^2$$

$$= 0.39\%$$

The following table may be generated using the above approach:

Size Range (µm)	Average Particle Size (µm)	Weight Percent	E_i (%)	$w_i E_i$ (%)
0–10	5	5	0.39	—
10–20	15	11	3.5	0.4
20–40	30	10	14	1.4
40–60	50	9	39	3.5
60–90	75	22	88	19.4
90–125	107.5	23	100	23
125–150	137.5	10	100	10
150+	130+	10	100	10
				$\sum = 67.7$

Therefore, $E = 67.7\%$.

The required efficiency, E_{reg}, is

$$E_{reg} = (I - O)/I$$

$$= (20 - 5)/20 = 75\%$$

Since $67.7 < 75\%$, the gravity settler will *not* do the job.

MCC.6 CYCLONES IN SERIES

As a graduate student you have been assigned the task of studying certain process factors in an operation that employs three cyclones in series to treat catalyst-laden gas at 25°C and 1 atm. The inlet loading to the cyclone series is 8.24 gr/ft^3 and the volumetric flowrate is 1,000,000 acfm. The efficiency of the cyclones are 93, 84, and 73%, respectively. Calculate the following:

1. Daily mass of catalyst collected (lb/day)
2. Daily mass of catalyst discharged to the atmosphere
3. Whether or not it would be economical to add an additional cyclone (efficiency = 52%) costing an additional $300,000 per year (The cost of the catalyst is $0.75 per pound.)
4. Outlet loading from the proposed fourth cyclone

Solution

The mass entering, m_i, is

$$m_i = \left(\frac{10^6 \, \text{ft}^3}{\text{min}}\right) \left(\frac{60 \, \text{min}}{\text{h}}\right) \left(\frac{24 \, \text{h}}{\text{day}}\right) \left(\frac{8.24 \, \text{gr}}{\text{ft}^3}\right) \left(\frac{1 \, \text{lb}}{7000 \, \text{gr}}\right)$$

$$= 1,695,086 \, \text{lb/day}$$

The mass collected, m_c, is

$$m_c = 1,695,086 \, (0.93) + 0.84 \, [1,695,086 \, (1 - 0.93)]$$
$$+ \, 0.73 \, [1,695,086 \, (1 - 0.93)(1 - 0.84)]$$
$$= 1,689,960 \, \text{lb/day}$$

Thus the mass discharge, m_d, is

$$m_d = 1,689,960 - 1,695,086 = 5,126 \, \text{lb/day}$$

With a fourth cyclone, the additional mass collected, m_4, is

$$m_4 = 5126 \, (0.52) = 2666 \, \text{lb/day}$$

and $5126 - 2666 = 2460 \, \text{lb/day}$ is discharged.
    The savings S is

$$S = (2666 \, \text{lb/day})(\$0.75/\text{lb})(3000 \, \text{day/yr})$$
$$= \$600,000/\text{yr}$$

Since the cyclone costs $300,000 annually, purchase it.
    The outlet loading (OL) is

$$\text{OL} = (2460 \, \text{lb/day})(1 \, \text{day/24 h})(1 \, \text{h/60 min})(1 \, \text{min}/10^6 \, \text{ft}^3)(7000 \, \text{gr/lb})$$
$$= 0.012 \, \text{gr/ft}^3$$

MCC.7 PARTICLE SIZE DEPOSITION FROM A MALFUNCTIONING CYCLONE

A cyclone on a cement plant suddenly malfunctions. By the time the plant shuts down, some dust has accumulated on parked cars and other buildings in the plant complex. The nearest affected area is 700 ft from the cyclone location, and the furthest affected area measurable on plant grounds is 2500 ft from the cyclone. What is the particle size range of the dust that has landed on plant grounds?
    On this day, the cyclone was discharging into a 6-mph wind. The specific gravity of the cement is 1.96. The cyclone is located 175 ft above the ground. Neglect effects of turbulence.

Solution

A diagram representing the system is provided in Figure 66 (S = smaller, L = larger). For air at ambient conditions,

$$\rho = 0.0741 \, \text{lb/ft}^3$$
$$\mu = 1.23 \times 10^{-5} \, \text{lb/(ft} \cdot \text{s)}$$

$$\text{Wind speed} = (6)(5289) = 31,680 \, \text{ft/h}$$

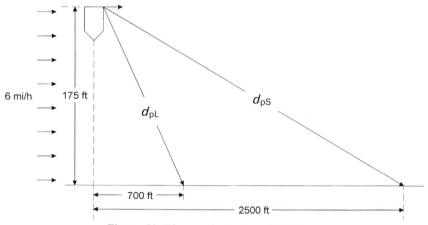

Figure 66. Diagram for Problem MCC.7.

Particle traveling times are given by

$$s = ut$$

where s = horizontal distance traveled
u = horizontal velocity
t = travel time

For the smaller particle,

$$t_S = 2500/31,680 = 0.07891 \text{ h}$$

and for the larger particle,

$$t_L = 700/31,680 = 0.02210 \text{ h}$$

Settling velocities may now be calculated from

$$v = H/t$$

where v = vertical velocity
H = vertical distance traveled

$$v_S = 175/[(0.07891)(3600)] = 0.6161 \text{ ft/s}$$
$$v_L = 175/[(0.02210)(3600)] = 2.200 \text{ ft/s}$$

To calculate d_p, assume Stokes' law to apply. For the smaller particle size,

$$d_{pS} = \left(\frac{18\mu v}{g\rho_p}\right)^{0.5}$$

$$= \left(\frac{(18)(1.23 \times 10^{-5})(0.6161)}{(32.2)(1.96)(62.4)}\right)^{0.5}$$

$$= 1.86 \times 10^{-4} \text{ ft}$$

Checking the value of K,

$$K = d_p\left(\frac{g(\rho_p - \rho)\rho}{\mu^2}\right)^{1/3}$$

$$= (1.86 \times 10^{-4})\left(\frac{(32.2)\,[(1.96)(62.4) - 0.0741](0.0741)(62.4)}{(1.23 \times 10^{-5})^2}\right)^{1/3}$$

$$= 2.32 < 3.3$$

Therefore, for the smaller particle size, Stokes' law is valid and $d_{pS} = 1.86 \times 10^{-4}$ ft. For the larger particle size,

$$d_{pL} = \left(\frac{(18)(1.23 \times 10^{-5})(2.200)}{(32.2)(1.96)(62.4)}\right)^{0.5}$$

$$= 3.52 \times 10^{-4} \text{ ft}$$

and

$$K = (3.52 \times 10^{-4})\left(\frac{(32.2)[(1.96)(62.4) - 0.0741](0.0741)}{(1.23 \times 10^{-5})^2}\right)^{1/3}$$

$$= 4.39$$

Since $3.3 < 4.39$, Stokes' law is invalid for the larger particles. Assuming the intermediate range applies,

$$d_{pL}^{1.14} = \frac{v\mu^{0.43}\rho^{0.29}}{(0.153)\,g^{0.71}\rho_p^{0.71}} = \frac{(2.200)(1.23 \times 10^{-5})^{0.43}(0.0741)^{0.29}}{(0.153)(32.2)^{0.71}[(1.96)(62.4)]^{0.71}}$$

$$= 1.465 \times 10^{-4}$$

$$d_{pL} = 4.33 \times 10^{-4} \text{ ft}$$

Checking on K

$$K = (4.33 \times 10^{-4}) \left(\frac{(32.2)[(1.96)(62.4) - 0.0741](0.0741)}{(1.23 \times 10^{-5})^2} \right)^{1/3}$$

$$= 5.39$$

Since $3.3 < 5.39 < 43.6$, the intermediate law is valid for the larger particle size. Therefore, the particle size range is

$$1.86 \times 10^{-4} \text{ ft} \leqslant d_p \leqslant 4.33 \times 10^{-4} \text{ ft}$$

or

$$5.67 \, \mu\text{m} \leqslant d_p \leqslant 132 \, \mu\text{m}$$

MCC.8   CUT DIAMETER AND OVERALL COLLECTION EFFICIENCY

An engineer was requested to determine the cut size diameter and overall collection efficiency of a cyclone given the particle size distribution of a dust from a cement kiln. Particle size distribution and other pertinent data are provided below.

Average Particle Size in Range, d_p, μm	Weight Percent
1	3
5	20
10	15
20	20
30	16
40	10
50	6
60	3
>60	7

Gas viscosity $= 0.02$ cP
Specific gravity of the particle $= 2.9$
Inlet gas velocity to cyclone $= 50$ ft/s
Effective number of turns within cyclone $= 5$
Cyclone diameter $= 10$ ft
Cyclone inlet width $= 2.5$ ft

Solution

The performance of a cyclone is often specified in terms of a cut size, d_{pc}, which is the size of the particle collected with 50% efficiency. The cut size depends on the gas and particle properties, the cyclone size, and the operating conditions. It may be calculated from

$$d_{pc} = \left(\frac{9\mu B_c}{2\pi N_t v_i \left(\rho_p - \rho \right)} \right)^{1/2}$$

where d_{pc} = cut size particle diameter (particle collected at 50% efficiency), ft
  μ = gas viscosity, lb/(ft · s)
  B_c = width of gas inlet, ft
  N_t = effective number of turns the gas stream makes in the cyclone, dimensionless
  v_i = inlet velocity, ft/s
  ρ_p = particle density, lb/ft^3
  ρ = gas density, lb/ft^3

Lapple's method provides the collection efficiency as a function of the ratio of particle diameter to cut diameter. One may use the equation

$$E = \frac{1.0}{1.0 + (d_{pc}/d_p)^2}$$

or Figure 67. (For additional details on the above equation, the reader is referred to the article by L. Theodore and V. DePaola, "Predicting Cyclone Efficiency," *J. Air Pollution Control Association*, 30, 1132–1133, 1980.)
  For the problem at hand, determine the value of $(\rho_p - \rho)$:

$$\rho_p - \rho = \rho_p$$
$$= (2.9)(62.4)$$
$$= 181 \text{ lb/ft}^3$$

Calculate the cut diameter:

$$d_{pc} = \left(\frac{9\mu B_c}{2\pi N_t v_i (\rho_p - \rho)} \right)^{1/2}$$
$$= \left(\frac{(9)(0.02)(6.72 \times 10^{-4})(2.5)}{(2\pi)(5)(50)(181)} \right)^{1/2}$$
$$= 3.26 \times 10^{-5}$$
$$= 9.94 \text{ μm}$$

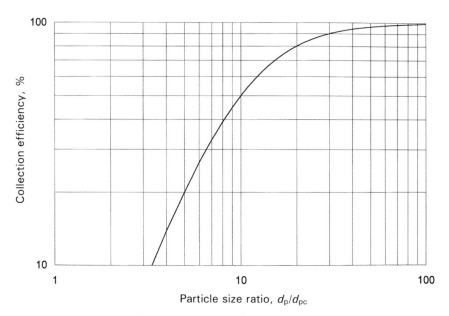

Figure 67. Collection efficiency as a function of particle size ratio.

The following table is generated using Lapple's method:

d_p (mμ)	w_i	d_p/d_{pc}	E_i (%)	$w_i E_i$ (%)
1	0.03	0.10	0	0.0
5	0.20	0.5	20	4.0
10	0.15	1.0	50	7.5
20	0.20	2.0	80	16.0
30	0.16	3.0	90	14.4
40	0.10	4.0	93	9.3
50	0.06	5.0	95	5.7
60	0.03	6.0	98	2.94
>60	0.07	—	100	7.0

Lapple's method was employed to obtain E_i. Slightly more accurate results can be obtained by employing the Theodore–DePaola equation.

Calculate the overall collection efficiency:

$$E = \sum w_i E_i = 0 + 4 + 7.5 + 16 + 14.4 + 9.3 + 5.7 + 2.94 + 7$$
$$= 66.84\% = 0.6684$$

MCC.9   PRESSURE DROP ACROSS A CYCLONE

Estimate the pressure drop across a cyclone treating a gas at ambient conditions with a velocity of 50 ft/s.

Solution

The pressure drop across a cyclone collector will generally range between 2 and 6 in. of water, and it is usually determined empirically. One method used in industrial practice is to determine the pressure drop of a geometrically similar prototype. The pressure drop is often described in units of inlet velocity heads. This inlet velocity head, in inches of water, may be expressed as follows:

$$\text{One velocity head} = 0.003 \, \rho v_i^2, \quad \text{in. } H_2O$$

where ρ is the gas density at operating conditions (lb/ft^3), and v_i is the inlet velocity (ft/s). The friction loss through cyclones encountered in practice may range from 1 to 20 inlet velocity heads, depending on the geometric proportions. For most cyclones, the friction loss is approximately 8 inlet velocity heads or, assuming an inlet velocity of 50 ft/s (typical) and a gas density of 0.075 lb/ft^3, the pressure drop across the cyclone is

$$\Delta P = 8(0.003)(0.075)(50)^2$$
$$= 4.5 \text{ in. } H_2O$$

MCC.10   CYCLONE SELECTION

A recently hired engineer has been assigned the job of selecting and specifying a cyclone unit to be used to reduce an inlet fly ash loading (with the particle size distribution given below) from 3.1 gr/ft^3 to an outlet value of 0.06. The flowrate from the coal-fired boiler is 100,000 acfm. Fractional efficiency data provided by a vendor are presented below (see Figure 68) for three different types of cyclones (multiclones).

Which type and how many cyclones are required to meet the above specifications? The optimum operating pressure drop is 3.0 in. H_2O; at this condition, the average inlet velocity may be assumed to be 60 ft/s.

Particle Diameter Range (μm)	Weight Fraction (w_i)
5–35	0.05
35–50	0.05
50–70	0.10
70–110	0.20
110–150	0.20
150–200	0.20
200–400	0.10
400–700	0.10

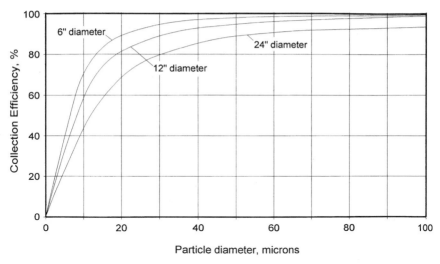

Figure 68. Fractional efficiency data.

Multiple-cyclone collectors (multiclones) are high-efficiency devices that consist of a number of small-diameter cyclones operating in parallel with a common gas inlet and outlet. The flow pattern differs from a conventional cyclone in that instead of bringing the gas in at the side to initiate the swirling action, the gas is brought in at the top of the collecting tube, and swirling action is then imparted by a stationary vane positioned in the path of the incoming gas. The diameters of the collecting tubes usually range from 6 to 24 in. with pressure drops in the 2- to 6-in. range. Properly designed units can be constructed and operated with a collection efficiency as high as 90% for particulates in the 5- to 10-μm range. The most serious problems encountered with these systems involve plugging and flow equalization.

Since the gas flow to a multiclone is axial (usually from the top), the cross-sectional area available for flow inlet conditions is given by the annular area between the outlet tubes and cyclone body. The outlet tube diameter is usually one-half the body diameter.

Solution

Calculate the required collection efficiency, E_R:

$$E_R = [(3.1 - 0.06)/3.1]\,(100)$$
$$= 98\%$$
$$= 0.98$$

Calculate the average particle size associated with each size range:

Particle Diameter Range (μm)	Average Particle Diameter (μm)
5–35	20
35–50	42.5
50–70	60
70–110	90
110–150	130
150–200	175
200–400	300
400–700	550

The following table for the 6-in. tubes provides the overall efficiency, E_6:

Average Particle Diameter (μm)	Weight Fraction w_i	Efficiency for 6-in. Tubes (%)	$E_i w_i$ for 6-in. Tubes (%)
20	0.05	89	4.45
42.5	0.05	97	4.85
60	0.10	98.5	9.85
90	0.20	99	19.8
130	0.20	100	20
175	0.20	100	20
300	0.10	100	10
550	0.10	100	10
			$E_6 = 98.95$

Since $E_6 > E_R$, the 6-in. tubes will do the job.
The following table is generated for the 12-in. tubes:

Average Particle Diameter (μm)	Weight Fraction w_i	Efficiency for 12-in. Tubes (%)	$E_i w_i$ for 12-in. Tubes (%)
20	0.05	82	4.1
42.5	0.05	93.5	4.67
60	0.10	96	9.6
90	0.20	98	19.6
130	0.20	100	20
175	0.20	100	20
300	0.10	100	10
550	0.10	100	10
			$E_{12} = 97.97$

Since the overall efficiency, $E_{12} < E_R$, the 12-in. tubes will not do the job. Thus, it will be necessary to use the 6-in. tubes for a conservative design.

Since the outlet tube diameter is one-half the body diameter, the inlet cross-sectional area (for axial flow) for each 6-in. (0.5-ft) tube will be

$$A = 0.785 \, (0.5^2 - 0.25^2)$$

$$= 0.147 \, \text{ft}^2$$

Since the velocity in each tube is 60 ft/s, the number of tubes, n, is given by

$$(60)(60)(0.147)(n) = 100,000$$

Solving for n,

$$n = 190 \, \text{tubes}$$

required in this multiple-cyclone unit. A 15×15, 14×14, or 12×16 design is recommended.

15 Electrostatic Precipitators (ESP)

ESP.1 DESIGN MODELS

An electrostatic precipitator (ESP) is being used to clean fly ash from a gas. The precipitator contains 30 ducts, with plates 12 ft high and 12 ft long. The spacing between the plates is 8 in. The gas is evenly distributed through all of the ducts. The following information can be used:

Gas volumetric flowrate $= 40,000\,\text{acfm}$

Particle drift velocity $= 0.40\,\text{ft/s}$

Use the Deutsch–Anderson equation to calculate the efficiency of the precipitator. Also, use the modified Deutsch equation with a range of exponents varying from 0.4 to 0.7 (in increments of 0.05) to calculate the efficiency of the electrostatic precipitator.

Solution

The process of electrostatic precipitation consists of corona formation around a high-tension wire, with particle charging by ionized gas molecules formed in the localized region of electrical breakdown surrounding the high-tension wire. This is followed by migration of the charged particles to the collecting electrodes. Finally, the particles collected from the collecting electrode are removed by rapping into hoppers. The approach taken by industry to size ESPs for various applications makes use of the Deutsch–Anderson equation referred to in the problem statement:

$$E = 1 - e^{-(wA/q)}$$

With the term w representing the effective migration or drift velocity. (Note: wA/q is often denoted as ϕ.) The numerical value of w is selected on the basis of experience with a particular dust, a particular set of operating conditions, and a particular design. Precipitator manufacturers usually have a specific experience file from which the precipitation rate parameter can be selected for various applications and conditions. Average values of precipitation rate parameters for various applications, and the range of values that might be expected, are presented below.

306

Application	Precipitation Rate or Drift Velocity (ft/s)
Utility fly ash	0.13–0.67
Pulp and paper mills	0.21–0.31
Sulfuric acid mist	0.19–0.25
Cement (wet process)	0.33–0.37
Cement (dry process)	0.19–0.23
Gypsum	0.52–0.64
Smelter	0.06
Open-hearth furnace	0.16–0.19
Blast furnace	0.20–0.46
Hot phosphorous	0.09
Flash roaster	0.25
Multiple hearth roaster	0.26
Catalyst dust	0.25
Cupola	0.10–0.12

Since the desired collection efficiency and gas flowrate are usually specified, the required collecting area can be determined from the Deutsch–Anderson equation once an appropriate precipitation rate parameter has been chosen.

An attempt to account for the sensitivity of w on process variables, especially for small particle size distributions, appeared in 1957 and was later revised by Allander, Matts, and Ohnfeldt, who derived the expression

$$E = 1 - e^{-(wA/q)^m}$$

The second exponent (in this, the so-called *modified Deutsch equation*) provides a more accurate prediction of performance at high-efficiency levels but can become too pessimistic in certain situations. Typical values of m range between 0.4 and 0.7, with 0.5 as the norm. This and the previous equation are employed in the solution that follows.

The collection surface area per duct, A, is

$$A = (12)(12)(2)$$
$$= 288 \, \text{ft}^2$$

The volumetric flowrate through each duct in acfs is

$$q = 40{,}000/[(30)(60)]$$
$$= 22.22 \, \text{acfs}$$

The collection efficiency using the Deutsch–Anderson model can now be calculated.

$$E = 1 - e^{-(wA/q)} = 1 - e^{-((288)(0.4)/22.22)}$$
$$= 0.9944 = 99.44\%$$

Using the modified Deutsch–Anderson (DA) equation to obtain an expression for the efficiency in terms of the exponent m leads to

$$E = 1 - e^{-(wA/q)^m} = 1 - e^{-((288)(0.4)/22.22)^m} = 1 - e^{-5.184^m}$$

The following table provides E for various values of m:

m	$E\,(\%)$
0.40	85.51
0.45	87.72
0.50	89.74
0.55	91.56
0.60	93.17
0.65	94.58
0.70	95.78

While the DA equation predicts an efficiency of 99.44%, it can be seen that the efficiency is probably somewhat lower than that. At a value for the exponent of 0.5, it appears that the ESP operates at 89.7% efficiency. If expressed in terms of penetration, the DA equation gives a value of 0.0056 while the modified DA gives a value of 0.1026. This means that 18.3 times more fly ash is passing through (not collected) the precipitator than that predicted by the DA equation. Because of this, it can be seen that the design of an ESP can be somewhat tricky.

ESP.2   MODIFIED DEUTSCH–ANDERSON EQUATION

A horizontal parallel-plate ESP consists of a single duct 24 ft high and 20 ft deep with an 11-in. plate-to-plate spacing. A collection efficiency and 88.2% is obtained with a flowrate of 4200 acfm. The inlet loading is 2.82 gr/ft^3. Calculate the following using a modified form of the Deutsch–Anderson equation, with the exponent $m = 0.5$ (i.e., exponent on ϕ):

1. The outlet loading.
2. The drift velocity for this system.
3. A revised collection efficiency if the flowrate is increased to 5400 acfm.
4. A revised collection efficiency if the plate spacing is decreased to 9 in.

Solution

The outlet loading (OL) is

$$OL = 2.82\,(1 - 0.882) = 0.333\,\text{gr/ft}^3$$

The drift velocity, w, for this system is found from the modified form of the Deutsch–Anderson equation:

$$E = 1 - e^{-\phi^m}$$

$$0.882 = 1 - e^{-\phi^{0.5}}$$

$$\phi = 4.57$$

$$w = \frac{\phi q}{A} = \frac{(4.57)(4200)}{(24)(20)(2)(60)}$$

$$= 0.333 \text{ ft/s}$$

If the flowrate is increased to 5400 acfm,

$$\phi = \frac{wA}{q} = \frac{(0.333)(20)(24)(2)}{(5400/60)}$$

$$= 3.55$$

$$E = 1 - e^{-(3.55)^{0.5}}$$

$$= 0.848$$

Since q, w, and A are all constant, the modified Deutsch–Anderson equation predicts that the efficiency does not change if the plate spacing is decreased to 9 in.

ESP.3   FRACTIONAL EFFICIENCY CURVES

Fractional efficiency curves describing the performance of a specific model of an electrostatic precipitator have been compiled by a vendor. Although you do not possess these curves, you are told that the cut diameter (50% efficiency) for a precipitator with a 10-in. plate spacing is 0.9 μm. The vendor claims that this particular model will perform with an efficiency of 98% under your operating conditions. You are asked to verify this claim and to make certain that the effluent loading does not exceed 0.2 gr/ft³.

The inlet loading is 14 gr/ft³ and the aerosol has the following particle size distribution:

Weight Range (%)	Average Particle Size (μm)
0–20	3.5
20–40	8.0
40–60	13.0
60–80	19.0
80–100	45.0

Assume a Deutsch–Anderson equation of the form

$$E = 1 - e^{-kd_p}$$

to apply.

Solution

Using the Deutsch–Anderson equation,

$$E = 1 - e^{-kd_p}$$

$$0.5 = 1 - e^{-k(0.9)}$$

$$k = 0.7702$$

The following table can now be generated:

Weight Fraction w_i	Average Particle Diameter $\langle d_i \rangle$ (µm)	E_i	$W_i E_i$
0.2	3.5	0.93250	0.18650
0.2	8	0.99789	0.19958
0.2	13	0.99996	0.19999
0.2	19	0.99999	0.20000
0.2	45	0.99999	0.20000
			$\sum = 0.98607$

Thus, the overall efficiency, E_o, is 98.607%. The outlet loading (OL) is

$$OL = (14)(1 - 0.98607) = 0.195 \, gr/ft^3 < 0.2 \, gr/ft^3$$

Therefore, the OL standard is met.

ESP.4   THREE FIELDS IN SERIES

An electrostatic precipitator is to be used to treat 100,000 acfm of a gas stream containing particulates from a hazardous waste incinerator. The proposed precipitator consists of three bus sections (fields) arranged in series, each with the same collection surface. The inlet loading has been measured as 40 gr/ft³ and a maximum outlet loading of 0.18 gr/ft³ is allowed by local Environmental Protection Agency (EPA) regulations. The drift velocity for the particulates has been experimentally determined in a similar incinerator installation with the following results:

First section (inlet): 0.37 ft/s
Second section (middle): 0.35 ft/s
Third section (outlet): 0.33 ft/s

1. Calculate the total collecting surface required based on the *average* drift velocity and the required total efficiency.
2. Find the total mass flowrate (lb/min) of particulates captured by each section using the above drift velocities.

Solution

Calculate the required total collection efficiency based on the given inlet and outlet loading.

$$E = 1 - \frac{\text{Outlet loading}}{\text{Inlet loading}}$$

$$= 1 - \frac{0.18}{40}$$

$$= 0.9955 = 99.55\%$$

Calculate the average drift velocity, w:

$$w = (0.37 + 0.35 + 0.33)/3$$

$$= 0.35 \text{ ft/s}$$

Calculate the total surface area required using the Deutsch–Anderson equation.

$$A = -\frac{\ln(1 - E)}{w/q}$$

$$= -\frac{\ln(1 - 0.9955)}{0.35/1666.7}$$

$$= 25{,}732 \text{ ft}^2$$

Calculate the collection efficiency of each section. Assume that each section has the same surface area but employs individual section drift velocities:

$$E_1 = 1 - e^{-(Aw_1/3q)} = 1 - e^{-((25,732)(0.37)/[(3)(1666.7)])}$$

$$= 0.851$$

$$E_2 = 1 - e^{-(Aw_2/3q)} = 1 - e^{-((25,732)(0.35)/[(3)(1666.7)])}$$

$$= 0.835$$

$$E_3 = 1 - e^{-(Aw_3/3q)} = 1 - e^{-((25,732)(0.33)/[(3)(1666.7)])}$$

$$= 0.817$$

Calculate the mass flowrate of particulates captured by each section using the collection efficiencies calculated above:

$$W_1 = (E_1)(\text{Inlet loading})(q)$$
$$= 3.404 \times 10^6 \text{ gr/min}$$
$$= 486.3 \text{ lb/min}$$

$$W_2 = (1 - E_1)(E_2)(\text{Inlet loading})(q)$$
$$= 4.977 \times 10^5 \text{ gr/min}$$
$$= 71.1 \text{ lb/min}$$

$$W_3 = (1 - E_1)(1 - E_2)(E_3)(\text{Inlet loading})(q)$$
$$= 8.034 \times 10^4 \text{ gr/min}$$
$$= 11.48 \text{ lb/min}$$

ESP.5   FOUR CHANNELS IN PARALLEL

A single-stage duct-type electrostatic precipitator contains five plates that are 10 ft high, 20 ft long, and spaced 9 in. apart. Air contaminated with gypsum dust enters the unit with an inlet loading of 53 gr/ft^3 and a velocity through the unit of 5 ft/s. The solids bulk density is 47 lb/ft^3.

1. Estimate the particle drift velocity, w, if the collection efficiency is 99%.
2. What is the outlet loading?
3. How many cubic feet of dust are collected per hour?
4. Determine the rapping frequency (intervals) minutes if the maximum allowable dust thickness on the plates is $\frac{1}{8}$ in. (Assume this layer is uniform over the entire plate.)
5. Due to an anticipated increase in plant capacity, a larger volumetric flowrate would result in a gas velocity of 7.5 ft/s. Determine the new efficiency.

Solution

Collecting area, A, and gas flow rate, q, are given by

$$A = (8 \text{ surfaces})(10 \text{ ft})(20 \text{ ft}) = 1600 \text{ ft}^2$$
$$q = (4 \text{ channels})(5 \text{ ft/s})(9/12 \text{ ft})(10 \text{ ft}) = 150 \text{ ft}^3/\text{s}$$

Using the Deutsch–Anderson equation,

$$w = -\frac{\ln(1 - E)}{A/q}$$
$$= -\frac{\ln(1 - 0.99)}{1600/150}$$
$$= 0.4317 \text{ ft/s}$$

The outlet loading (OL) is

$$OL = 53(1 - E) = 53(1 - 0.99) = 0.53 \, gr/ft^3$$

The volumetric flowrate of particulates captured per hour, V_p, is

$$V_p = \frac{(53.0 - 0.53) \, gr/ft^3}{(7000 \, gr/lb)(47 \, lb/ft^3)} \, (150 \, ft^3/s)(3600 \, s/h)$$

$$= 86.12 \, ft^3/h$$

The rapping cycle (RC) time is

$$RC = \frac{[(1/8)/12] \, ft \, (1600 \, ft^2)}{(86.12 \, ft^3/h)(1 \, h/60 \, min)}$$

$$= 11.61 \, min$$

The revised efficiency (for a 50% increase in flow) becomes

$$E = 1 - e^{-(Aw/q)}$$

$$= 1 - e^{-((1600)(0.43173)/[(150)(1.5)])}$$

$$= 0.954 = 95.4\%$$

ESP.6   EFFECT OF INLET DISTRIBUTION

You have been requested to calculate the collection efficiency of an electrostatic precipitator containing three ducts with plates of a given size, assuming a uniform distribution of particles. Also determine the collection efficiency if one duct is fed 50% of the gas and the other passages 25% each. Operating and design data include:

Volumetric flowrate of contaminated gas $= 4000 \, acfm$
Operating temperature and pressure $= 20°C$ and 1 atm, respectively
Drift velocity $= 0.40 \, ft/s$
Size of the plate $= 12 \, ft$ long and $12 \, ft$ high
Plate-to-plate spacing $= 8 \, in.$

Solution

Considering both sides of the plate,

$$A = (2)(12 \, ft)(12 \, ft)$$

$$= 288 \, ft^2$$

Remembering the volumetric flowrate through a passage is one third of the total volumetric flowrate,

$$q = \frac{4000}{(3)(60)}$$

$$= 22.22 \, \text{acfs}$$

Calculate the collection efficiency using the Deutsch–Anderson equation:

$$E = 1 - e^{-(wA/q)}$$

$$= 1 - e^{-((288)(0.4)/22.22)}$$

$$= 0.9944 = 99.44\%$$

This efficiency calculation assumes the gas is uniformly distributed at the inlet of the precipitator. A revised efficiency can be calculated if the flow is distributed as specified in the problem statement. First, calculate q in acfs through the middle section:

$$q = \frac{4000}{(2)(60)}$$

$$= 33.33 \, \text{acfs}$$

Calculate the collection efficiency, remembering the collection surface area per duct remains the same:

$$E_1 = 1 - e^{-((288)(0.4)/33.33)}$$

$$= 0.9684 = 96.84\%$$

Calculate q in acfs through an outer section:

$$q = \frac{4000}{(4)(60)}$$

$$= 16.67 \, \text{acfs}$$

The collection efficiency in the outside section is

$$E_2 = 1 - e^{-((288)(0.4)/16.67)}$$

$$= 0.9990 = 99.90\%$$

Calculate the new overall collection efficiency:

$$E = (0.5)(E_1) + (2)(0.25)(E_2)$$
$$= 98.37\%$$

Note that the penetration $(100 - E)$ has increased by a factor of 3. The reader, as an optional exercise should outline the calculational procedure to follow if the particle size distribution varies with each inlet duct.

ESP.7 EFFECT OF PARTICLE SIZE DISTRIBUTION

The following data are available for the proposed design of an ESP to operate at an efficiency of 97.5%:

Air volumetric flow $= 100,000$ acfm
Uniform flow distribution through 10 ducts
Duct height $= 30$ ft with 12 in. plate-to-plate spacing
Plate length $= 36$ ft
Inlet loading $= 14$ gr/ft^3
Outlet loading $= 0.35$ gr/ft^3:

In addition, the following particle size distribution and drift velocity data have been provided in terms of the average particle size (d_p) of weight fraction (x_i) in a given size range and the corresponding drift velocity (w).

d_p (µm)	w (ft/s)	Mass Fraction x_i
0.1	0.27	0.01
0.25	0.15	0.01
0.5	0.12	0.01
1.0	0.11	0.01
1.5	0.15	0.16
2.0	0.20	0.16
2.5	0.26	0.16
5.0	0.50	0.16
10.0	0.60	0.16
25.0	0.70	0.16

Determine if the proposed design will meet the desired efficiency. Also, prepare a graph of particle size vs. efficiency for the system and comment on the results.

Solution

Check if the average throughput velocity is in an acceptable range. The average velocity should be 2–8 ft/s.

$$v = (100,000)/[(10)(30)(60)]$$

$$= 5.56 \, \text{ft/s}$$

This is in the acceptable range.

The volumetric flowrate through each (one) duct is

$$q = (100,000)/10$$

$$= 10,000 \, \text{acfm}$$

$$= 166.67 \, \text{acfs}$$

Calculate the plate area in square feet for each duct. Note that both plates contribute to the collection area:

$$A = 2 \, (30)(36)$$

$$= 2160 \, \text{ft}^2$$

If the Deutsch–Anderson equation applies,

$$E = 1 - e^{-(wA/q)} = 1 - e^{-(w(2160)(60)/10,000)} = 1 - e^{-12.96 \, w}$$

Calculate the efficiency for the 0.1 μm particle size:

$$E = 1 - e^{-12.96 \, (0.27)}$$

$$= 0.9698 = 96.98\%$$

Calculate the collection efficiencies for all of the other particle size ranges. Results are presented in the following table:

d_p (μm)	E
0.1	0.9698
0.25	0.8569
0.5	0.7889
1.0	0.7596
1.5	0.8569
2.0	0.9251
2.5	0.9656
5.0	0.9985
10.0	0.9996
25.0	0.9999

Calculate the overall efficiency E of the unit:

d_p (µm)	x_i	E_i	x_iE_i
0.1	0.01	0.9698	0.009698
0.25	0.01	0.8569	0.008569
0.5	0.01	0.7889	0.007889
1.0	0.01	0.7596	0.007596
1.5	0.16	0.8569	0.137104
2.0	0.16	0.9251	0.148016
2.5	0.16	0.9656	0.154496
5.0	0.16	0.9985	0.159760
10.0	0.16	0.9996	0.159936
25.0	0.16	0.9999	0.159984
		$E = \sum x_iE_i = 0.953048$	

Thus, the efficiency is 95.30%

Since the desired efficiency is 97.5%, the proposed design is insufficient.

As is typical with particulate control, collection of large particles is highly efficient. A decline in efficiency is seen as particle size decreases. However, when the particles become very small, diffusion effects occur that actually raise the collection efficiency for these particles. See the graph in Figure 69.

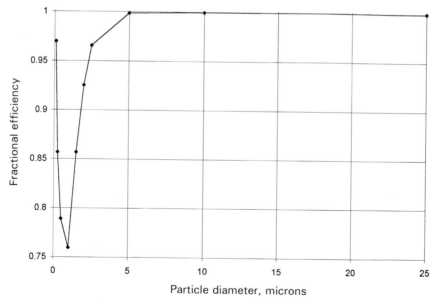

Figure 69. Effect of very small particles on collection efficiency.

ESP.8   BUS SECTION FAILURE

A precipitator consists of two bus sections, each with five plates (four passages) in a field (see Figure 70). The corona wires between any two plates are independently controlled so that the remainder of the unit can be operated in the event of a wire failure. The following operating conditions exist:

Gas flowrate: 10,000 acfm
Plate dimensions: 10 ft × 15 ft; four rows per field
Drift velocity: 19.0 ft/min; section 1
                      16.3 ft/min; section 2

1. Determine the normal operating efficiency.
2. During operation, a wire breaks in section 1. As a result, all of the wires in that row are shorted and ineffective, but the others function normally. Calculate the collection efficiency under these conditions. Assume the gas stream leaving section 1 is uniformly redistributed on entering section 2, i.e., each of the four rows is fed the same volume of gas.
3. Redo part 2 assuming the flow in (through) each of the four rows (or passages) acts in a "railroad" manner, i.e., there is no redistribution after section 1.
4. Calculate a revised efficiency if a second wire fails in a different row. Assume "railroad" flow again.

Solution

The need for series sectionalization in a precipitator arises mainly because power input needs differ at various locations in a precipitator. In the inlet sections of a precipitator, concentrations of particulate matter will be relatively heavy. This requires a great deal of power input to generate the corona discharge required for optimal particle charging: heavy concentrations of dust particles tend to suppress corona current. On the other hand, in the downstream sections of a precipitator, dust

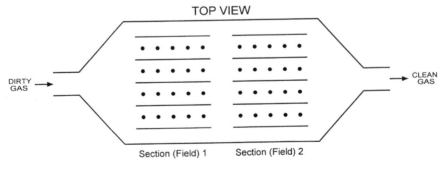

Figure 70. ESP with two bus sections and five plates.

concentrations will be lighter. As a consequence, corona current will flow more freely and particle charging will tend to be limited by sparking rather than current suppression. Hence, excessive sparking is more likely to occur at lower voltage in downstream sections; if the precipitator has only a single power set, then this sparking, under spark rate-limited control, will limit power input to the entire precipitator, including the inlet sections. This will result in insufficient power being supplied to the discharge electrodes in the inlet sections, with a consequential fall in precipitator collection efficiency in the inlet sections of the precipitator.

A remedy for this situation is to divide the precipitator into a series of independently energized electric bus sections. Each bus section has its own transformer rectifier, voltage stabilization controls, and high-voltage conductors that energize the discharge electrodes within that section. This would allow greater power input, and increased particle charging of dust and particulate precipitation than in the previously described underpowered inlet sections.

For part 1, write the equation describing the overall or total efficiency, E_T, in terms of the individual section efficiencies, E_1 and E_2:

$$E_T = 1 - (1 - E_1)(1 - E_2)$$

The equation describing the total penetration, P_T, in terms of the individual section penetrations, P_1 and P_2, is

$$P_T = P_1 P_2$$

Calculate the efficiency of section 1, E_1:

$$E_1 = 1 - e^{-(Aw/q)}$$
$$= 1 - e^{-((15)(10)(8)(19.0)/10,000)} \qquad w_1 = 19.0 \, \text{ft/min}$$
$$= 0.89772 = 89.772\%$$

Calculate the efficiency of section 2, E_2:

$$E_2 = 0.85858 \qquad w_2 = 16.3 \, \text{ft/min}$$
$$= 85.858\%$$

The total efficiency E_T is therefore

$$E_T = 1 - (1 - 0.89772)(1 - 0.85858)$$
$$= 0.98554$$
$$= 98.554\%$$

Calculate a revised total efficiency for part 2:

$$E_T = 1 - [1 - 3/4E_1][1 - E_2]$$
$$= 1 - [1 - (0.75)(0.89772)][1 - 0.85858]$$
$$= 0.95380 = 95.380\%$$

Calculate a revised total efficiency for part 3:

$$E_T = (1.0)[(0.75)(0.98554) + (0.25)(0.85858)]$$
$$= (1.0)[0.73916 + 0.21464]$$
$$= 0.95380 = 95.380\%$$

The calculation for part 4 is affected by where the second wire failure occurs. Determine a revised efficiency if the wire failure is located in section 1:

$$E_T = (1.0)[(0.5)(0.98554) + (0.5)(0.85858)]$$
$$= 0.49277 + 0.42929$$
$$= 0.92206 = 92.206\%$$

Determine a revised efficiency if the wire failure in part 4 occurs in a *different* row in section 2.

$$E_T = (1.0)[(0.5)(0.98554) + (0.25)(0.85858) + (0.25)(0.89772)]$$
$$= 0.49277 + 0.43895$$
$$= 0.93172 = 93.172\%$$

Determine a revised efficiency if the wire failure in part 4 occurs in a *same* row in section 2:

$$E_T = (1.0)[(0.75)(0.98544) + (0.25)(0.0)]$$
$$= 0.73908 + 0.0$$
$$= 0.73908 = 73.908\%$$

The reader should note the effect on efficiency of bus section failure and the location of the failure.

Parallel sectionalization provides the means for coping with different power input needs due to uneven dust and gas distributions that usually occur across the inlet face of a precipitator. Nevertheless, the gains in collection efficiency from parallel sectionalization are smaller than series sectionalization.

Bus section failure is one of the more important design operating and maintenance variables for ESPs. Detailed calculational procedures for estimating this effect are available in the literature. Two studies addressing this issue are:

Theodore, L., and Reynolds, J., "The Effect of Bus Section Failure on Electrostatic Precipitator Performance," *JAPCA*, **33**: 1202–1205, 1983.

Theodore, L., Reynolds, J., Taylor, F., Filippi, A., and Errico, S., "Electrostatic Precipitator Bus Section Failure: Operation and Maintenance," *Proceedings of the Fifth USEPA Symposium on the Transfer and Utilization of Particulate Control Technology*, Kansas City, 1984.

ESP.9 RESISTIVITY

Comment on the resistivity "problem" with ESPs.

Solution

Resistivity is a term used to describe the resistance of a medium to the flow of an electrical current. By definition, the resistivity is the electrical resistance of a dust sample $1\,cm^2$ in cross-sectional area and 1 cm thick. For ease of precipitation capture, dust resistivity values can be classified roughly into three groups:

1. Below $5.0 \times 10^3\,\Omega \cdot cm$ ($\Omega = ohm$).
2. Between 5.0×10^3 and $2.0 \times 10^{10}\,\Omega \cdot cm$.
3. Above $2.0 \times 10^{10}\,\Omega \cdot cm$ (this value is frequently referred to as the *critical resistivity*).

Particulates in group 1 are difficult to collect. They are easily charged and precipitated; upon contacting the collection electrode, however, they lose their discharge electrode polarity and acquire the polarity of the collection electrode. The particulates are then repelled into the gas stream to either escape from the precipitator or become recharged by the corona field. Examples are unburned carbon in fly ash and carbon black. If the conductive particles are coarse, they may be removed upstream from the precipitator with another collection device, e.g., a cyclone. Baffles are often designed on the collection walls to limit this precipitation-repulsion phenomenon. Particulates with resistivities in group 3 cause back-ionization or back-corona, which is a localized discharge at the collection electrode due to the surface being coated by a layer of nonconductive material. A weak back-corona will merely lower the sparkover voltage, but a strong back-corona produces a positive ion discharge at the electrode. Back-corona phenomena becomes severe with a bulk particle layer resistivity greater than $10^{12}\,\Omega \cdot cm$. Particulates with resistivities in group 2 have been shown by experiment and experience to be the most acceptable for electrostatic precipitation. The particulates do not rapidly lose their charge on contact with the collection electrode or cause back-corona. Back-corona phenomena can be decreased by treatment of the gas stream, such as altering the temperature, moisture content, or chemical composition.

Particle resistivity decreases as gas temperature increases due to enhanced volume conductivity. Resistivity may also decrease as gas temperature decreases if surface conditioning agents such as moisture or acid gases are present in the gas stream. Adsorption of these on the particle surface is favored at lower temperatures and provides a conductive path on the particle surface.

ESP.10   DESIGN PROCEDURE

Provide a design procedure for electrostatic precipitators.

Solution

No critically reviewed design procedure exists for ESPs. However, one suggested "general" design procedure (L. Theodore, personal notes) is provided below.

1. Determine or obtain a complete description of the process, including the volumetric flowrate, inlet loading, particle size distribution, maximum allowable discharge, and process conditions.
2. Calculate or set the overall collection efficiency.
3. Select a migration velocity (based on experience).
4. Calculate the ESP size (capture area).
5. Select the field height (experience).
6. Select the plate spacing (experience).
7. Select a gas throughput velocity (experience).
8. Calculate the number of gas passages in parallel.
9. Select (decide) on bus sections, fields, energizing sets, specific current, capacity of energizing set for each bus section, etc.
10. Design and select hoppers, rappers, etc.
11. Perform a capital cost analysis, including materials, erection, and startup costs.
12. Perform an operating cost analysis, including power, maintenance, inspection, capital and replacement, interest on capital, dust disposal, etc.
13. Conduct a perturbation study to optimize economics.

16 Baghouse (BAG)

BAG.1 DESCRIPTION OF FILTRATION

Describe the filtration process.

Solution

The basic filtration process may be conducted with many different types of fabric filters in which the physical arrangement of hardware and the method of removing collected material from the filter media will vary. The essential differences may be related, in general, to

1. Type of fabric
2. Cleaning mechanism
3. Equipment geometry
4. Mode of operation

Baghouse collectors are available for either intermittent or continuous operation. Intermittent operation is employed where the operational schedule of the dust-generating source permits halting the gas cleaning function at periodic intervals (regularly defined by time or by pressure differential) for removal of collected material from the filter media (cleaning). Collectors of this type are primarily utilized for the control of small-volume operations such as grinding, polishing, etc., and for aerosols of a very coarse nature. For most air pollution control installations and major dust control problems, however, it is desirable to use collectors that allow for continuous operation. This is accomplished by arranging several filter areas in a parallel-flow system and cleaning one area at a time according to some preset mode of operation.

BAG.2 CLEANING METHODS

Discuss the various cleaning methods employed in baghouses.

Solution

Baghouses may be characterized and identified according to the method used to remove collected material from the bags. Particle removal can be accomplished in a variety of ways, including shaking the bags, reversing the flow of gas through the bags, or rapidly expanding the bags by a pulse of compressed air. In general, the various types of bag cleaning methods can be divided into those involving fabric flexing and those involving a reverse flow of clean air.

Cleaning by mechanical shaking is accomplished by isolating one of several bag compartments from the air flow and vigorously shaking the bags for about a minute to dislodge the dust. For simplicity of operation, the bags are usually attached to a motor-driven oscillating carriage. Because of tensile stresses produced by this approach, strong fabric material must be used.

Reverse-flow baghouses employ an auxiliary fan that forces air through the bags in a direction opposite to that of filtration. This procedure collapses the bag and fractures the dust cake. The reverse-flow rate, which is ordinarily about the same as the face velocity during filtering, deflates the bag and helps to dislodge the dust cake from the fabric surface. It is common practice to combine shaking and reverse-flow cleaning in the same unit.

Reverse-pulse cleaning is the newest and perhaps the most efficient method to clean bags, given the proper working environment. A short pulse (about 0.1 s) of compressed air (approximately 100 psia) is injected into each bag through a venturi, causing the bag to expand while creating intense dust separating forces.

BAG.3   OVERALL COLLECTION EFFICIENCY

Calculate the overall efficiency of N compartments in a baghouse operated in parallel, if the volumetric flowrates and inlet concentrations to each compartment are $q_1, q_2, \ldots, q_N$ and $c_1, c_2, \ldots, c_N$, respectively, and the corresponding efficiencies are $E_1, E_2, \ldots, E_N$. Also, express the result in terms of the q's, the E's, and the c's.

Calculate the overall efficiency of a baghouse consisting of three compartments treating 9000 acfm of gas with an inlet loading of $4\,\mathrm{gr/ft}^3$. The first and third compartments operate at a fractional efficiency of 0.995 while the second compartment operates at a fractional efficiency of 0.990. What is the overall efficiency of the baghouse if the flow and inlet concentration are evenly distributed? Also calculate the efficiency if the following flow distribution exists:

Compartment	q (acfm)	c (gr/ft^3)	E
1	2500	3.8	0.995
2	4000	4.25	0.990
3	2500	3.8	0.995

Solution

Write the equation for the outlet concentration, c_{1o}, from module 1 in terms of c_1 and E_1:

$$E_1 = 1 - (c_{1o}/c_1)$$
$$c_{1o} = c_1(1 - E_1)$$

The equation for the inlet mass flowrate to module, $\dot{m}_1$, is

$$\dot{m}_1 = c_1 q_1$$

The equation for the outlet mass flowrate, $\dot{m}_{1o}$, from module 1 is

$$\dot{m}_{1o} = c_1(1 - E_1)q_1$$

For module i,

$$E_i = 1 - (c_{io}/c_i)$$
$$c_{io} = c_i(1 - E_i)$$
$$\dot{m}_i = c_i q_i$$
$$\dot{m}_{io} = c_i(1 - E_i)q_i$$

The equation for the overall efficiency, E, for modules $1, 2, \ldots, N$ is then

$$E = 1 - \frac{\sum \dot{m}_{io}}{\sum \dot{m}_i}$$
$$= 1 - \frac{c_1(1 - E_1)q_1 + c_2(1 - E_2)q_2 + \cdots + c_N(1 - E_N)q_N}{c_1 q_1 + c_2 q_2 + \cdots + c_N q_N}$$
$$= 1 - \frac{\sum c_i(1 - E_i)q_i}{\sum c_i q_i}$$

The companion equation for the penetration P is

$$P = \frac{c_1 P_1 q_1 + c_2 P_2 q_2 + \cdots + c_N P_{Nq_N}}{c_1 q_1 + c_2 q_2 + \cdots + c_N q_N}$$
$$= \frac{\sum c_i P_i q_i}{\sum c_i q_i}$$

If the inlet concentrations c to each module are equal, i.e.,

$$c_1 = c_2 = \cdots = c_N = c$$

the c terms can be factored out from the above equation for efficiency to yield

$$E = 1 - \frac{(1 - E_1)q_1 + (1 - E_2)q_2 + \cdots + (1 - E_N)q_N}{q_1 + q_2 + \cdots + q_N}$$

Note that the total volumetric flowrate q is given by

$$q = q_1 + q_2 + \cdots + q_N$$

Therefore,

$$E = 1 - \frac{(1 - E_1)q_1 + (1 - E_2)q_2 + \cdots + (1 - E_N)q_N}{q}$$

$$= 1 - \frac{\sum(1 - E_i)q_i}{q}$$

Equivalently,

$$P = 1 - \frac{\sum q_i P_i}{q}$$

Using the data provided in the problem statement, calculate the efficiency for the situation where the flow is equally distributed with the same inlet loading. Since $q_i = 3000$ acfm for all modules,

$$E = 1 - \frac{(2)(3000)(1 - 0.995) + (1)(3000)(1 - 0.99)}{9000}$$

$$= 0.9933 = 99.33\%$$

If neither the flow nor concentration are uniformly distributed, the general equation for the efficiency in a compartmentalized baghouse is used.

$$E = 1 - \frac{\sum c_i(1 - E_i)q_i}{\sum c_i q_i}$$

Substituting (see data),

$$E = 1 - \frac{(2)(3.8)(2500)(1 - 0.995) + (1)(4.25)(4000)(1 - 0.99)}{36,000}$$

$$= 0.9926 = 99.26\%$$

It can be seen that poor flow distribution causes the efficiency to decrease. With much larger systems, this effect can end up being significant and may cause the baghouse to be out of compliance.

BAG.4   BAGHOUSE COLLECTION

The dimensions of a bag in a filter unit are 8 in. in diameter and 15 ft long. Calculate the filtering area of the bag. If the filtering unit consists of 40 such bags and is to treat 480,000 ft^3/h of gas from an open-hearth furnace, calculate the "effective" filtration velocity in feet per minute and acfm per square foot of filter area. Also calculate the mass of particles collected daily if the inlet loading is 3.1 gr/ft^3 and the unit operates at 99.99+% collection efficiency.

Solution

Assume the bag to be cylindrical in shape with diameter D and height h. The total area of the bag is

$$A = A_{\text{curved surface}} + A_{\text{flat top}}$$
$$= \pi D h + \pi D^2/4$$
$$= \pi(\tfrac{8}{12})(15) + \pi(\tfrac{8}{12})^2/4$$
$$= 31.77 \text{ ft}^2$$

The total area for 40 bags is

$$A = (40)(31.77) = 1271 \text{ ft}^2$$

The filter velocity is then

$$v = \frac{q_G}{A} = \frac{(480,000/60)}{1271}$$
$$= 6.30 \text{ ft/min}$$

The calculation for acfm per square foot of filter area is the same. Assuming 100% collection efficiency, the mass collected daily is

$$\text{Mass collected} = q_G c_i = (480,000)(24)(3.1)/7000$$
$$= 5102 \text{ lb/day}$$

Note that 7000 gr = 1 lb.

BAG.5   ADVANTAGES AND DISADVANTAGES OF BAGHOUSES

Discuss the advantages and disadvantages associated with a baghouse.

Solution

Some of the advantages and disadvantages associated with employing a baghouse for particulate control are listed below.
    Advantages include:

1. Moderate capital cost
2. Moderate operating cost
3. Extremely high collection efficiencies
4. Dry collection
5. No resistivity problems

Disadvantages include:

1. Space requirements
2. Bag failure (to be discussed later)
3. Explosion hazards
4. Temperature dependence
5. Bag replacement

BAG.6   NUMBER OF BAGS, PRESSURE DROP, AND CLEANING FREQUENCY

A calcium hydroxide plant is required to treat the exhaust "fume" generated from the plant. The ash generated from the system is collected at the bottom of the baghouse while the exhaust gas flow of 350,000 acfm enters the baghouse with a loading of 6.0 gr/ft³. The air-to-cloth ratio is 8.0 and the operating particulate collection efficiency is 99.3%. The maximum allowable pressure drop is 10 in. H_2O. The contractor's empirical equation for the pressure drop is given by

$$\Delta P = 0.3v + 4.0cv^2t$$

where ΔP = pressure drop in inches of water
    v = filtration velocity in ft/min
    c = dust concentration in lb/ft³ of gas
    t = time in minutes since bags were cleaned

1. How many cylindrical bags, 12 in. in diameter and 30 ft high will be needed?
2. The system is designed to begin cleaning when the pressure drop reaches 10.0 in. H_2O, its maximum allowable value. How frequently should the bags be cleaned?

As particles are collected in the baghouse, the pressure drop across the fabric filtering media increases. Due in part to fan limitations, the filter must be cleaned at predetermined intervals. Dust is removed from the fabric by gravity and/or mechanical means. The fabric filters or bags are usually tubular or flat. As described earlier, the structure in which the bags hang is frequently referred to as a *baghouse*. The number of bags in a baghouse may vary from a few to several thousand. Quite often when great numbers of bags are involved, the baghouse is compartmentalized so that one compartment may be cleaned while others are still in service.

There are several equations provided in the literature to describe pressure drop in a fabric filter system. The form most often used is

$$\Delta P = \Delta P_{\text{fabric}} + \Delta P_{\text{cake}} = K_1 v + K_2 c^2 t$$

where ΔP = total pressure drop across both fabric and cake (in. H_2O)
v = superficial velocity through the bag/cake (ft/min)
c = inlet particulate loading in gas (lb/ft^3)
t = elapsed time in filtering cycle (min)
K_1 = resistance coefficient for the bag (fabric) (in. H_2O/[ft/min])
K_2 = resistance coefficient for the deposited dust (cake) (in. H_2O/[(lb/ft^3)$^2 \cdot$ min])

The size of a baghouse is primarily determined by the area of filter cloth required to filter the gases. The choice of a filtration velocity (or its equivalent, the air-to-cloth ratio (ACR or A/C) in actual cubic feet per minute of gas filtered per square foot of filter area) must take certain factors into consideration. Although the higher velocities are usually associated with the greater pressure drops, they also reduce the filter area required. Practical experience has led to the use of a series of ACRs for various materials collected and types of equipment. Ratios in current use range from < 1:1 to > 15:1. The choice depends on cleaning method, fabric, and characteristics of the particles.

For the same cleaning efficiency, felted fabrics in pulse-jet baghouses are often capable of higher air-to-cloth ratios than woven fabrics in reverse-air baghouses, thereby requiring less filter cloth area and, consequently, less space for a given air or gas volume. Woven fabrics in reverse-air baghouses usually have ACRs of 1:1 to 5:1; felted fabrics in pulse-jet baghouses usually have ratios of 3:1 to 15:1, or ratios several times those of woven fabrics. This is balanced, though, by the higher cost of the felt fabrics.

Solution

Calculate the total required surface area A of the bags if the air-to-cloth ratio is 8.0:

$$A = \text{Volumetric gas flowrate/Filtration velocity}$$
$$= 350{,}000/8$$
$$= 43{,}750 \text{ ft}^2$$

Calculate the surface area of each bag, a, and the number of the bags required, N:

$$a = \pi DL + \pi D^2/4$$
$$= (\pi)(12/12)(30) + (\pi)(12/12)^2/4$$
$$= 95 \text{ ft}^2$$

$$N = A/a$$
$$= 43{,}750/50$$
$$= 461 \text{ bags}$$

Solve the pressure drop equation explicitly for the time:

$$\Delta P = 0.3v + 4cv^2t$$
$$t = (\Delta P - 0.3v)/4cv^2$$

The concentration c is given by

$$c = 6.0/7000$$
$$= 8.57 \times 10^{-4} \text{ lb/ft}^3$$

Solving for the time yields

$$t = \frac{10 - 0.3(8)}{(4)(8.57 \times 10^{-4})(8)^2}$$
$$= 34.6 \text{ min}$$

BAG.7   BAG FAILURE

A baghouse has been used to clean a particulate gas stream for nearly 30 years. There are 600 8-in. diameter bags in the unit; 50,000 acfm of dirty gas at 250°F enters the baghouse with a loading of $5.0\,\text{gr/ft}^3$. The outlet loading is $0.03\,\text{gr/ft}^3$. Local Environmental Protection Agency (EPA) regulations state that the outlet loading should not exceed $0.4\,\text{gr/ft}^3$. If the system operates at a pressure drop of 6 in. of water, how many bags can fail before the unit is out of compliance? The Theodore–Reynolds equation applies and all the contaminated gas emitted through the broken bags may be assumed the same as that passing through the tube sheet thimble.

The effect of bag failure on baghouse efficiency can be described by the following equations:

$$P_t^* = P_t + P_{tc}$$

$$P_{tc} = \frac{0.528(\Delta P)^{0.5}}{\phi}$$

$$\phi = \frac{q}{LD^2(T + 460)^{0.5}}$$

where P_t^* = penetration after bag failure

P_t = penetration before bag failure

P_{tc} = penetration correction term; contribution of broken bags to P_t^*

ΔP = pressure drop, in. H_2O

ϕ = dimensional parameter

q = volumetric flowrate of contaminated gas, acfm

L = number of broken bags

D = bag diameter, in.

T = temperature, °F

For a detailed development of the above equation, refer to "Effect of Bag Failure on Baghouse Outlet Loading," Theodore and Reynolds, *JAPCA*, August 1979, 870–872.

Solution

Calculate the efficiency E and penetration P_t before the bag failure(s):

$$E = (\text{Inlet loading} - \text{Outlet loading})/(\text{Inlet loading})$$

$$= (5.0 - 0.03)/(5.0)$$

$$= 0.9940 = 99.40\%$$

$$P_t = 1 - 0.9940$$

$$= 0.0060 = 0.60\%$$

The efficiency and penetration, P_t^*, based on regulatory conditions are

$$E = (5.0 - 0.4)/5.0$$

$$= 0.9200 = 92.00\%$$

$$P_t^* = 1 - 0.9200$$

$$= 0.0800 = 8.00\%$$

The penetration term, P_{tc}, associated with the failed bags is then

$$P_{tc} = 0.0800 - 0.0060$$

$$= 0.0740$$

Write the equation(s) for P_{tc} in terms of the failed number of bags, L. Since

$$P_{tc} = \frac{0.528(\Delta P)^{0.5}}{\phi}$$

and

$$\phi = \frac{q}{LD^2(T + 460)^{0.5}}$$

then,

$$L = \frac{qP_{tc}}{(0.582)\Delta P^{0.5}D^2(T + 460)^{0.5}}$$

The number of bag failures that the system can tolerate and still remain in compliance is now calculated:

$$L = \frac{(50,000)(0.074)}{(0.582)(6)^{0.5}(8)^2(250 + 460)^{0.5}}$$

$$= 1.52$$

Thus, if two bags fail, the baghouse is out of compliance.

It is important to note that each bag in a set may have a different life as a result of fabric quality, bag manufacturing tolerances, location in the collector, and variation in the bag cleaning mechanism. Any one or a combination of these factors can cause bags to fail. This means that a baghouse will experience a series of intermittent bag failures until the failure rate requires total bag replacement. Typically, a few bags will fail initially or after a short period of operation due to installation damage or manufacturing defects. The failure rate should then remain very low until the operating life of the bags is approached, unless a unique failure mode is present within the system. The failure then increases, normally at a near exponential rate. Industry often describes this type of failure rate behavior as a "bathtub" curve. The reader is referred to the Theodore tutorial titled *Accident and Emergency Management*, ETS International, Roanoke, VA, 1998 for detailed information on system accidents/failures.

The proper time to replace a broken bag depends on the type of collector and the resultant effect on outlet emissions. In "inside bag collection" types of collectors, it is very important that dust leaks be stopped as quickly as possible to prevent adjacent bags from being abraded by jet streams of dust emitted from the broken bag. This is called the "domino effect" of bag failure. "Outside bag collection" systems do not have this problem, and the speed of repair is determined by whether the outlet opacity has exceeded its limits. Often, it will take several broken bags to create an opacity problem, and a convenient maintenance schedule can be employed instead of emergency maintenance.

In either type of collector, the location of the broken bag or bags has to be determined and corrective action taken. In a noncompartmentalized unit, this requires system shutdown and visual inspection. In inside collectors, bags often fail close to the bottom, near the tube sheet. Accumulation of dust on the tube sheets, the holes themselves, or unusual dust patterns on the outside of the bags often occurs. Other probable bag failure locations in reverse-air bags are near anticollapse rings or below the top cuff. In shaker bags, one should inspect the area below the top attachment. Improper tensioning can also cause early failure.

In outside collectors, which are normally top-access systems, inspection of the bag itself is difficult; however, location of the broken bag or bags can normally be found by looking for dust accumulation on top of the tube sheet, on the underside of the top-access door, or on a blow pipe.

BAG.8   FREQUENCY OF BAG FAILURE

As a recently assigned plant engineer, you are asked to troubleshoot the plant's baghouse. The baghouse is used to collect the dust created in the manufacture of an extremely expensive drug. The dust is collected and recycled into the main process. Over the past 6 months (since the baghouse was installed) the amount of dust collected has dropped off significantly without any change in the inlet loading. Since the baghouse is operated on a round-the-clock basis, i.e., 24 h per day, 7 days per week, the bags (for this unit) cannot be inspected to find the problem. The following data have been collected:

Flowrate $= 60,000$ acfm ($60°F$)

Dust loading $= 6.00$ gr/ft^3

Number of bags $= 500$

Diameter of bags $= 5.0$ in.

Pressure drop $= 9.1$ in. H_2O

Months of Operation	Amount Collected (lb/h)
New	3054.9
1	2900.4
2	2808
3	2653.8
4	2530.2
5	2376
6	1789.8

Having recently attended a class on the effects of bag failures given by the foremost authority in this field, you are asked to determine if the loss has been caused by broken bags, and, if so, how many have broken every month.

Solution

First calculate the inlet loading (IL):

$$IL = (60{,}000)(6.0)(60)/(7000)$$
$$= 3085.7 \text{ lb/h}$$

The calculations of the initial efficiency E and penetration P follow:

$$E = 3054.9/3085.7 = 0.99 = 99\%$$
$$P - 1 - 0.99 = 0.01 = 1.0\%$$

After the first month,

$$E^* = 2900/3085.7 = 0.94 = 94\%$$
$$P^* = 1 - 0.94 = 0.06 = 6\%$$

(* indicates that time has passed since baghouse installation).

The amount of increase in the penetration P_c is

$$P_c = 0.06 - 0.01 = 0.05$$

Using the equation (See Problem BAG.7),

$$\phi = 35.1 = \frac{q}{Ld^2(t+460)^{1/2}}$$

$$L = \frac{q}{\phi D^2(t+460)^{1/2}} = \frac{60{,}000}{(35.1)(5)^2(520)^{1/2}}$$

$$= 2.998 \approx 3$$

Similarly,

Month	P^* (%)	P_c (%)	ϕ	L	ΔL
1	6	5	35.1	3	3
2	9	8	21	5	2
3	14	13	15	7	2
4	18	17	10.5	10	3
5	23	22	7.5	14	4
6	42	41	4.2	25	11

The baghouse problem appears to be bag failure. Note that the efficiency drops below 90% after just 2 months.

The reader should consider whether the bag failure distribution with time is reasonable.

BAG.9 COLLECTION EFFICIENCY MODEL

Consider the situation where 50,000 acfm of gas with a dust loading of 5.0 grain/ft^3 flows through a baghouse with an average filtration velocity of 10 ft/min. The pressure drop is given by

$$\Delta P = 0.20v + 5.0\ c_i v^2 t$$

where ΔP = pressure drop, in. of H_2O
   v = filtration velocity, ft/min
   c_i = dust concentration, lb/ft^3 of gas
   t = time after bags were cleaned, min

The fan can maintain the volumetric flowrate up to a pressure drop of 5.0 in. of water. Show that the baghouse can be operated for 8.40 min between cleanings.

In an attempt to determine the efficiency of this unit at "terminal" conditions, both the fabric and deposited cake (individually) were subjected to laboratory experimentation. The following data were recorded:

Fabric alone: thickness = 0.1 in.

Concentration at $1 = 1.0$ g/cm^3
   at $2 = 0.1245$ g/cm^3

Cake alone: time $= 3.4043$ min (end of cleaning cycle)

Concentration at $1 = 1.0 \, \text{g/cm}^3$
at $2 = 0.0778 \, \text{g/cm}^3$

Using the Theodore and Reynolds collection efficiency model, determine the overall efficiency at the start and end of a cleaning cycle.

The number of variables necessary to design a fabric filter is very large. Since fundamentals cannot treat all of these factors in the design and/or prediction of performance of a filter, this determination is basically left up to the experience and judgment of the design engineer. In addition, there is no one formula that can determine whether or not a fabric filter application is feasible. A qualitative description of the filtration process is possible, although quantitatively the theories are far less successful. Theory, coupled with some experimental data, can help predict the performance and design of the unit.

As discussed previously, the state of the art of engineering process design for baghouses is the selection of filter medium, superficial velocity, and cleaning method that will yield the best economic compromise. Industry relies on certain simple guidelines and calculations, which are usually considered proprietary information, to achieve this. Despite the progress in developing pure filtration theory, and in view of the complexity of the phenomena, the most common methods of correlation are based on predicting a form of a final equation that can be verified by experiment. For example, an equation that can be used for determining the collection efficiency of a baghouse is (*Introduction to Hazardous Waste Incineration* by Santoleri, Reynolds, and Theodore, Wiley-Interscience, 2000):

$$E = 1 - e^{-(\psi L + \phi t)}$$

where $\psi =$ constant (determined by experiment) based on the fabric (ft^{-1}, or units consistent with L)

$\phi =$ constant (determined by experiment) based on the cake (s^{-1}, or units consistent with t)

$t =$ time of operation to develop the cake thickness (s)
$L =$ fabric thickness (ft)
$E =$ collection efficiency (dimensionless)

Solution

Solve the equation for the pressure drop explicitly for t:

$$t = \frac{\Delta P - 0.2v}{5c_i v^2} = \frac{5.0 - 0.2(10)}{5(5/7000)(10)^2}$$

$$= 8.4 \, \text{min}$$

Evaluate the parameter ψ with units of ft^{-1} and the parameter ϕ with units of min^{-1}:

$$\ln(c_o/c_i) = -\psi L$$
$$\ln(0.1245/1.0) = -\psi(0.1/12)$$
$$\psi = 250 \, ft^{-1}$$
$$\ln(c_o/c_i) = -\phi t$$
$$\ln(0.0778/1.0) = -\phi(8.4)$$
$$\phi = 0.304 \, min^{-1}$$

Calculate the efficiency at the end of the filtering (cleaning) cycle.

$$E = 1 - e^{-\psi L}e^{-\phi t}$$
$$= 1 - e^{-(250)(0.1/12)}e^{-(0.304)(8.4)}$$
$$= 0.9903$$
$$= 99.03\% \text{ (at end of cleaning cycle)}$$

Also calculate the efficiency at the start of the filtering (cleaning) cycle:

$$E = 1 - (0.1245)(1) \qquad t = 0$$
$$= 0.8755$$
$$= 87.55\% \text{ (at start of cycle)}$$

The reader should note that ϕ and ψ are modified inertial impaction numbers based on the cake and fabric, respectively. Unfortunately, this key equation has been totally ignored by responsible EPA individuals in this field. Taxpayers' dollars continue (for over a dozen years) to be provided to contractors whose research efforts have produced little, if any, usable results in developing quantitative equations to describe collection efficiency.

Using the above model, one can show that the exit concentration (w_e) for the combined resistance system (the fiber and the cake) is

$$w_e = w_i e^{-(\psi L + \phi 0)}$$

where w_e = exit concentration; units consistent with w_i
w_i = inlet concentration

BAG.10   FILTER BAG FABRIC SELECTION

It is proposed to install a pulse-jet fabric filter system to clean an airstream containing particulate pollutants. You are asked to select the most appropriate filter bag fabric considering performance and cost. Pertinent design and operating data, as well as fabric information, are given below.

> Volumetric flowrate of polluted airstream $= 10{,}000$ scfm ($60°$F, 1 atm)
> Operating temperature $= 250°$F
> Concentration of pollutants $= 4.00$ gr/ft^3
> Average ACR $= 2.5$ cfm/ft^2 cloth
> Collection efficiency requirement $= 99\%$

Filter Bag	A	B	C	D
Tensile strength	Excellent	Above average	Fair	Excellent
Recommended maximum temperature ($°$F)	260	275	260	220
Resistance factor	0.9	1.0	0.5	0.9
Cost per bag, ($)	26	38	10	20
Standard size	8 in. $\times$ 16 ft	10 in. $\times$ 16 ft	1 ft $\times$ 16 ft	1 ft $\times$ 20 ft

Note: No bag has an advantage from the standpoint of durability under the operating conditions for which the bag was designed.

Solution

A wide variety of woven and felted fabrics are used in fabric filters. Clean felted fabrics are more efficient dust collectors than woven fabrics, but woven materials are capable of giving equal filtration efficiency after a dust layer accumulates on the surface. When a new woven fabric is placed in service, visible penetration of dust may occur until buildup of the cake or dust layer. This normally takes from a few hours to a few days for industrial applications, depending on dust loadings and the nature of the particles.

When using woven fabrics, care must be exercised to prevent overcleaning so as not to completely dislodge the filter cake; otherwise, efficiency will drop. Over-cleaning of felted fabrics is generally impossible because they always retain substantial dust deposits within the fabric. Felted fabrics require more thorough cleaning methods than woven materials. If felted fabrics are used, filter cleaning is limited to the reverse-pulse method. When woven fabrics are employed, any cleaning technique may be used. Woven fabrics are available in a greater range of temperature and corrosion-resistant materials than felts and, therefore, cover a wider range of applications.

Bag D is eliminated since its recommended maximum temperature (220°F) is below the operating temperature of 250°F. Bag C is also eliminated since a pulse-jet fabric filter system requires the tensile strength of the bag to be at least above average.

Consider the economics for the two remaining choices. The cost per bag is $26.00 for A and $38.00 for B. The gas flowrate and filtration velocity are

$$q_a = 10,000 \left(\frac{250 + 460}{60 + 460} \right)$$

$$= 13,654 \, \text{acfm}$$

$$v_f = 2.5 \, \text{cfm/ft}^2 \, \text{cloth}$$

$$= 2.5 \, \text{ft/min}$$

The filtering (bag) area is then

$$A_c = q/v_f$$

$$= 13,654/2.5$$

$$= 5,462 \, \text{ft}^2$$

For bag A, the area and number, N, of bags are

$$A = \pi D h$$

$$= \pi(\tfrac{8}{12})(16)$$

$$= 33.5 \, \text{ft}^2$$

$$N = A_c/A$$

$$= 5,462/33.5$$

$$= 163$$

For bag B:

$$A = \pi(\tfrac{10}{12})(16)$$

$$= 41.9 \, \text{ft}^2$$

$$N = 5462/41.9$$

$$= 130$$

The total cost (TC) for each bag is as follows:
For bag A:

$$TC = N \text{ (cost per bag)}$$
$$= (163)(26.00)$$
$$= \$4{,}238$$

For bag B:

$$TC = (130)(38.00)$$
$$= \$4{,}940$$

Since the total cost for bag A is less than bag B, select bag A.

17 Venturi Scrubbers (VEN)

VEN.1 POWER REQUIREMENTS FOR A VENTURI SCRUBBER

Calculate the power requirement of a venturi scrubber treating 380,000 acfm of gas and operating at a pressure drop of 60 in. H_2O.

Solution

The gas horsepower may be calculated from

$$HP = \frac{q\,\Delta P}{6356} \qquad q = \text{acfm} \qquad \Delta P = \text{in. } H_2O$$

$$= \frac{(380,000)(60)}{6356}$$

$$= 3587 \text{ hp}$$

Alternately, the following equation may be used:

$$P_G = 0.157\,\Delta P'$$

$$= 0.157\,(60)$$

$$= 9.42 \text{ hp}/1000 \text{ acfm}$$

Thus, the power would be

$$(9.42 \text{ hp}/1000 \text{ acfm})(380,000 \text{ acfm}) = 3580 \text{ hp}$$

To determine the brake horsepower, the gas horsepower must be divided by the fan efficiency. Assuming a fan efficiency of 60%, the operating brake horsepower would be

$$\text{Brake hp} = 3580/0.60$$

$$= 5970 \text{ hp}$$

VEN.2   COLLECTION EFFICIENCY

A venturi scrubber is employed to reduce the discharge of fly ash to the atmosphere. The unit is presently treating 215,000 acfm of gas, with a concentration of 4.25 gr/ft^3, and operating at a pressure drop of 32 in. H_2O. Experimental studies have yielded the following particle size collection efficiency data:

Particle Diameter (μm)	Weight Fraction w_i	Collection Efficiency (%)
5	0.00	30
10	0.00	42
20	0.02	86
30	0.05	93
50	0.08	97
75	0.10	98.7
100	0.75	99.9+

Estimate the overall collection efficiency of the unit.

Solution

The overall efficiency of the unit can be calculated from

$$E_T = \sum w_i E_i$$

The solution is presented in tabular form below:

Particle Diameter (μm)	Weight Fraction W_i	Collection Efficiency E_i (%)	$w_i E_i$ (%)
5	0.00	30	0.00
10	0.00	42	0.00
20	0.02	86	1.72
30	0.05	93	4.65
50	0.08	97	7.76
75	0.10	98.7	9.87
100	0.75	99.9+	75.00
			$E_T = 99.00$

VEN.3   DISCHARGE FROM A VENTURI SCRUBBER

With reference to Problem VEN.2, calculate the daily mass (in tons) of fly ash collected by the scrubbing liquid and discharged to the atmosphere. Also obtain the particle size distribution of the fly ash collected and discharged to the atmosphere. Comment on the results.

Solution

The total mass entering is

$$\text{Mass}_{\text{in}} = \left(\frac{4.25 \text{ gr}}{\text{ft}^3}\right)\left(\frac{215{,}000 \text{ ft}^3}{\text{min}}\right)\left(\frac{60 \text{ min}}{\text{h}}\right)\left(\frac{24 \text{ h}}{\text{day}}\right)\left(\frac{1 \text{ lb}}{7000 \text{ gr}}\right)\left(\frac{1 \text{ ton}}{2000 \text{ lb}}\right)$$

$$= 93.968 \approx 94 \text{ tons/day}$$

$$\text{Mass}_{\text{collected}} = (0.99)(94) = 93 \text{ tons/day}$$

$$\text{Mass}_{\text{discharged}} = 94 - 93 = 1 \text{ ton/day}$$

With regard to the particle size distribution calculations, the results are presented in tabular form.

Particle Diameter (μm)	Weight Fraction W_i	E (%)	Mass Entering (tons/day)	Mass Collected (tons/day)	Mass Discharged (tons/day)
5	0.00	30	0.00	0.00	0.00
10	0.00	42	0.00	0.00	0.00
20	0.02	86	1.88	1.62	0.26
30	0.05	93	4.70	4.37	0.33
50	0.08	97	7.52	7.30	0.22
75	0.10	98.7	9.40	9.28	0.12
100	0.75	99.9	70.50	70.43	0.07
Σ			94.00	93.00	1.00

The weight fractions, w_i, for the collected and discharged streams are given below.

Particle Diameter (μm)	W_i Collected Mass	W_i Discharged Mass
5	0.000	0.00
10	0.000	0.00
20	0.017	0.26
30	0.047	0.33
50	0.078	0.22
75	0.100	0.12
100	0.757	0.07
Σ	0.999	1.00

VEN.4   LIQUID DROPLET SIZE

The following data were collected using a bench-scale venturi scrubber:

Gas rate $= 1.56\,\text{ft}^3/\text{s}$
Liquid rate $= 0.078\,\text{gal}/\text{min}$
Throat area $= 1.04\,\text{in.}^2$

Estimate the average liquid droplet size in the scrubber. Repeat the calculation using a simplified equation.

Solution

The size of the droplets generated in scrubber units affects both the collection efficiency and pressure drop, i.e., small droplet sizes requiring high-pressure atomization give greater collection efficiencies. Various correlations are available in the literature to estimate the mean liquid drop diameter from different types of atomizers under different operating conditions. These correlations are applicable to fluids within a certain range of operating conditions and properties such as the volume ratio of liquid to gas, the relative velocity of gas to liquid, the type of nozzle, the surface tension of the liquid, etc. In using one of these correlations to estimate droplet diameter, it is important to select a correlation that takes these factors into consideration.

The empirical relationship of Nukiyama and Tanasawa (NT) is probably the best known and the most widely used to predict the average droplet size in pneumatic (gas-atomized) sprays. In this type of spray the stream of liquid is broken up or atomized by contact with a high-velocity gas stream. The original NT relationship is given by

$$d_0 = \left(\frac{1920}{v_r}\right)\left(\frac{\sigma}{\rho'_L}\right)^{1/2} + (5.97)\left(\frac{\mu'_L}{(\sigma\rho'_L)^{1/2}}\right)^{0.45}\left(1000\,\frac{L'}{G'}\right)^{1.5}$$

where d_0 = average surface volume mean droplet diameter, μm
v_r = relative velocity of gas to liquid, ft/s
σ = liquid surface tension, dyn/cm
ρ'_L = liquid density, g/cm^3
μ'_L = liquid viscosity, P
L'/G' = ratio of liquid-to-gas volumetric flowrates at the venturi throat

For water scrubbing systems, one may use

$$\sigma = 72\,\text{dyn}/\text{cm}$$

$$\rho'_L = 1.0\,\text{g}/\text{cm}^3$$

$$\mu'_L = 0.00982\,\text{Pa}$$

This equation reduces to the following expression for standard air and water in a venturi scrubber:

$$d_0 = (16{,}400/v) + 1.45\ R^{1.5}$$

where $v =$ gas velocity at venturi throat, ft/s
$\quad R =$ ratio of liquid-to-gas flowrates, gal/1000 actual ft^3

From the given data,

$$G' = 1.56\ \text{ft}^3/\text{s}$$

$$L' = (0.078\ \text{gal/min})(1\ \text{ft}^3/7.48\ \text{gal})(1\ \text{min}/60\ \text{s})$$
$$= 1.74 \times 10^{-4}\ \text{ft}^3/\text{s}$$

$$A = 1.04\ \text{in.}^2 = 7.22 \times 10^{-3}\ \text{ft}^2$$

The gas velocity is

$$v_G = G'/A = 1.56/7.22 \times 10^{-3}$$
$$= 216\ \text{ft/s}$$

The liquid velocity is

$$v_L = L'/A = 1.74 \times 10^{-4}/7.22 \times 10^{-3}$$
$$= 0.0241\ \text{ft/s}$$

The relative velocity is

$$v_r = v_G - v_L \approx v_G = 216\ \text{ft/s}$$

Using the Nukiyama and Tanasawa correlation for droplet diameter,

$$d_0 = \left(\frac{1920}{v_r}\right)\left(\frac{\sigma}{\rho_L'}\right)^{1/2} + (5.97)\left(\frac{\mu_L'}{(\sigma\rho_L')^{1/2}}\right)^{0.45}\left(1000\ \frac{L'}{G'}\right)^{1.5}$$

$$= \left(\frac{1920}{216}\right)\left(\frac{72}{1'}\right)^{1/2} + (5.97)\left(\frac{0.00982}{[(72)(1)]^{1/2}}\right)^{0.45}\left((1000)\ \frac{1.74 \times 10^{-4}}{1.56}\right)^{1.5}$$

$$= 75.4\ \mu m$$

The ratio of liquid-to-gas flow rates is

$$R = \frac{(0.078 \text{ gal/min})(1 \text{ min}/60 \text{ s})(1000)}{(1.56 \text{ ft}^3/\text{s})}$$

$$= 0.833 \text{ gal}/1000 \text{ acf}$$

Using the simplified equation gives

$$d_0 = (16{,}400/v) + 1.45R^{1.5}$$

$$= (16{,}400/216) + 1.45 \, (0.833)^{1.5}$$

$$= 77.03 \text{ μm}$$

VEN.5   PRESSURE DROP

Using the data provided in Problem VEN.4, estimate the pressure drop across the bench-scale unit. Use both the Theodore and Calvert equations.

Solution

The pressure drop for gas flowing through a venturi scrubber can be estimated from knowledge of liquid acceleration and frictional effects along the wall of the equipment. Frictional losses depend largely on the scrubber geometry and usually are determined experimentally. The effect of liquid acceleration is, however, predictable. An equation (developed by Calvert) for estimating pressure drop through venturi scrubbers (given as a function of throat gas velocity and liquid-to-gas ratio) assuming that all the energy is used to accelerate the liquid droplets to the throat velocity of the gas is

$$\Delta P = 5 \times 10^{-5} \, v^2 R$$

where ΔP is the pressure drop, in. H_2O, v is the gas velocity, ft/s, and R is the liquid-to-gas ratio, gal/1000 acf. Another somewhat simpler equation that applies over a fairly wide range of values for R is given below (L. Theodore, personal notes):

$$\Delta P' = 0.8 + 0.12R$$

where $\Delta P'$ is a dimensionless pressure drop equal to the pressure drop divided by the density and velocity head $(v^2/2g_c)$.

Using the Calvert equation to estimate the pressure drop,

$$\Delta P = (5 \times 10^{-5})(216)^2(0.833)$$
$$= 1.943 \text{ in. } H_2O$$

Using Theodore's equation,

$$\Delta P' = 0.8 + (0.12)(0.833)$$
$$= 0.90$$

Thus,

$$\Delta P = (0.90)(v^2/2g_c)\rho$$
$$= (0.9)\left(\frac{216^2}{(2)(32.2)}\right)(0.0775)$$
$$= 50.5 \text{ psf}$$

Since 1 psf = 0.1922 in. H_2O,

$$\Delta P = 9.71 \text{ in. } H_2O$$

The Calvert equation significantly underpredicts the pressure drop at low values of R. Note that this equation fails when R is zero.

VEN.6   THROAT AREA

A consulting firm has been requested to calculate the throat area of a venturi scrubber to operate at a specified collection efficiency.

To achieve high collection efficiency of particulates by impaction, a small droplet diameter and high relative velocity between the particle and droplet are required. In a venturi scrubber this is often accomplished by introducing the scrubbing liquid at right angles to a high-velocity gas flow in the venturi throat (vena contracta). Very small water droplets are formed, and high relative velocities are maintained until the droplets are accelerated to their terminal velocity. Gas velocities through the venturi throat typically range from 12,000 to 24,000 ft/min. The velocity of the gases alone causes the atomization of the liquid.

Perhaps the most popular and widely used venturi scrubber collection efficiency equation is that originally suggested by Johnstone:

$$E = 1 - e^{-kR\psi^{1/2}}$$

where $E =$ fractional collection efficiency

$k =$ correlation coefficient whose value depends on the system geometry and operating conditions, typically 0.1–0.2, 1000 acf/gal

$R = q_L/q_G =$ liquid-to-gas ratio, gal/1000 acf

$\psi = C\rho_p v d_p^2/18 d_o \mu$, the inertial impaction parameter

$\rho_p =$ particle density, lb/ft^3

$v =$ gas velocity at venturi throat, ft/s

$d_p =$ particle diameter, ft

$d_o =$ droplet diameter, ft

$\mu =$ gas viscosity, (lb/ft · s)

$C =$ Cunningham correction factor

Note: Some engineers define ψ as

$$\psi = \frac{C\rho_p v d_p^2}{9 d_o \mu}$$

This change is reflected in the correlation coefficient, k.

Pertinent data are given below.

Volumetric flowrate of process gas stream $= 11{,}040$ acfm (at 68°F)

Density of dust $= 187$ lb/ft^3

Liquid-to-gas ratio $= 2$ gal/1000 ft^3

Average particle size $= 3.2 \,\mu$m (1.05×10^{-5} ft)

Water droplet size $= 48 \,\mu$m (1.575×10^{-4} ft)

Johnstone scrubber coefficient, $k = 0.14$

Required collection efficiency $= 98\%$

Viscosity of gas $= 1.23 \times 10^{-5}$ lb/(ft · s)

Cunningham correction factor $= 1.0$

Solution

Calculate the inertial impaction parameter, ψ, from Johnstone's equation:

$$E = 1 - e^{-kR\psi^{1/2}}$$

$$0.98 = 1 - e^{-(0.14)(2)\psi^{1/2}}$$

Solving for ψ,

$$\psi = 195.2$$

From the calculated value of ψ above, back calculate the gas velocity at the venturi throat, v:

$$\psi = \frac{\rho_p v d_p^2}{18 d_0 \mu}$$

$$v = \frac{18 \psi d_0 \mu}{\rho_p d_p^2} = \frac{(18)(195.2)(1.575 \times 10^{-4})(1.23 \times 10^{-5})}{(187)(1.05 \times 10^{-5})^2}$$

$$= 330.2 \, \text{ft/s}$$

Calculate the throat area, S, using gas velocity at the venturi throat, v:

$$S = q/v = (11{,}040)/[(60)(330.2)]$$

$$= 0.557 \, \text{ft}^2$$

VEN.7   THREE VENTURI SCRUBBERS IN SERIES

Three identical venturi scrubbers are connected in series. If each operates at the same efficiency and liquid-to-gas ratio, q_L/q_G, calculate the liquid-to-gas ratio, assuming the Johnstone equation to apply. Data are provided below.

E_o (overall) $= 99\%$
Inlet loading $= 200 \, \text{gr/ft}^3$
Johnstone scrubber coefficient, $k = 0.14$
Inertial impaction parameter, $\psi = 105$

Solution

First calculate the outlet loading (OL) from the last unit:

$$\text{OL} = \text{IL} \, (1 - E_o)$$

$$= 200 \, (1 - 0.99)$$

$$= 2.0 \, \text{gr/ft}^3$$

Express the individual and overall efficiencies in terms of the penetration P:

$$P_0 = 1 - E_o = 1 - 0.99 = 0.01$$

$$P_1 = 1 - E_1$$

$$P_2 = 1 - E_2$$

$$P_3 = 1 - E_3$$

Calculate the individual efficiency for each venturi scrubber, noting that the efficiencies (or penetrations) are equal:

$$P_0 = P_1 P_2 P_3 = P^3$$

$$P^3 = 0.01$$

$$P = 0.215$$

$$E = 1 - P$$

$$= 1 - 0.215$$

$$= 0.785 = 78.5\%$$

Using the Johnstone equation, solve for the liquid-to-gas ratio, q_L/q_G:

$$\ln(1 - E) = -k\left(\frac{q_L}{q_G}\right)\phi^{0.5}$$

$$\left(\frac{q_L}{q_G}\right) = -\frac{\ln(1 - E)}{k\phi^{0.5}} = -\frac{\ln(1 - 0.785)}{(0.14)(105)^{0.5}}$$

$$= 1.07 \text{ gal}/1000 \text{ acf}$$

$$= 1.07 \text{ gpm}/1000 \text{ acfm}$$

VEN.8 COMPLIANCE CALCULATIONS ON A SPRAY TOWER

Contact power theory is an empirical approach relating particulate collection efficiency and pressure drop in wet scrubber systems. The concept is an outgrowth of the observation that particulate collection efficiency in spray-type scrubbers is mainly determined by pressure drop for the gas plus any power expended in atomizing the liquid. Contact power theory assumes that the particulate collection efficiency in a scrubber is solely a function of the total power loss for the unit. The total power loss, $\mathscr{P}_T$, is assumed to be composed of two parts: the power loss of the gas passing through the scrubber, $\mathscr{P}_G$, and the power loss of the spray liquid during atomization, $\mathscr{P}_L$. The gas term can be estimated by

$$\mathscr{P}_G = 0.157 \, \Delta P$$

where $\mathscr{P}_G$ is the contacting power based on gas stream energy input in hp/1000 acfm and ΔP is the pressure drop across the scrubber in inches of water. In addition,

$$\mathscr{P}_L = 0.583 \, P_L(q_L/q_G)$$

where $\mathscr{P}_L$ is the contacting power based on liquid stream energy input in hp/1000 acfm, P_L is the liquid inlet pressure in psi, q_L is the liquid feed rate in gal/min, and q_G is the gas flowrate in ft^3/min. Then

$$\mathscr{P}_T = \mathscr{P}_G + \mathscr{P}_L$$

To correlate contacting power with scrubber collecting efficiency, the latter is best expressed as the number of transfer units. The number of transfer units is defined by analogy to mass transfer and given by

$$N_t = \ln\left(\frac{1.0}{1 - E}\right)$$

where N_t is the number of transfer units, dimensionless, and E is the fractional collection efficiency, dimensionless. The relationship between the number of transfer units and collection efficiency is by no means unique. The number of transfer units for a given value of contacting power (hp/1000 acfm) or vice versa varies over nearly an order of magnitude. For example, at 2.5 transfer units ($E = 0.918$), the contacting power ranges from approximately 0.8 to 10.0 hp/1000 acfm, depending on the scrubber and the particulate.

For a given scrubber and particulate properties, there will usually be a very distinct relationship between the number of transfer units and the contacting power. The number of transfer units for a series of scrubbers and particulates is plotted against total power consumption; a linear relation, independent of the type of scrubber, is obtained on a log-log plot. The relationship could be expressed by

$$N_t = \alpha \mathscr{P}_T^\beta$$

where α and β are the parameters for the type of particulates being collected and the scrubber unit.

A vendor proposes to use a spray tower on a lime kiln operation to reduce the discharge of solids to the atmosphere. The inlet loading is to be reduced to meet state regulations. The vendor's design calls for a certain water pressure drop and gas pressure drop across the tower. You are requested to determine whether this spray tower will meet state regulations. If the spray tower does not meet state regulations, propose a set of operating conditions that will meet the regulations. The state regulations require a maximum outlet loading of 0.05 gr/ft³. Assume that contact power theory applies. Operating and design data are provided:

Gas flowrate = 10,000 acfm
Water rate = 50 gal/min
Inlet loading = 5.0 gr/ft³
Maximum gas pressure drop across the unit = 15 in. H_2O
Maximum water pressure drop across the unit = 100 psi

The vendor's design and operating data are also available:

$\alpha = 1.47$

$\beta = 1.05$

Water pressure drop = 80 psi

Gas pressure drop across the tower = 5.0 in. H_2O

Solution

Calculate the contacting power based on the gas stream energy input, $\mathscr{P}_G$, in hp/1000 acfm:

$$\mathscr{P}_G = (0.157)\,\Delta P$$
$$= (0.157)(5.0)$$
$$= 0.785\,\text{hp}/1000\,\text{acfm}$$

Calculate the contacting power based on the liquid stream energy input, $\mathscr{P}_L$, in hp/1000 acfm:

$$\mathscr{P}_L = 0.583 P_L (q_L/q_G)$$
$$= (0.583)(80)(50/10{,}000)$$
$$= 0.233\,\text{hp}/1000\,\text{acfm}$$

The total power loss, $\mathscr{P}_T$, in hp/1000 acfm is then

$$\mathscr{P}_T = \mathscr{P}_G + \mathscr{P}_L$$
$$= 0.785 + 0.233$$
$$= 1.018\,\text{hp}/1000\,\text{acfm}$$

The number of transfer units, N_t, is

$$N_t = \alpha \mathscr{P}_T^{\beta}$$
$$= (1.47)(1.018)^{1.05}$$
$$= 1.50$$

The collection efficiency can be calculated based on the design data given by the vendor:

$$N_t = \ln\left(\frac{1.0}{1 - E}\right)$$

or

$$E = 1 - e^{-N_t} = 1 - e^{-1.50}$$
$$= 77.7\%$$

The collection efficiency required by state regulations, E_s, is

$$E_s = \frac{\text{Inlet loading} - \text{Outlet loading}}{\text{Inlet loading}} (100)$$

$$= \frac{5.0 - 0.05}{5.0} (100)$$

$$= 99.0\%$$

Since $E_s > E$, the spray tower does not meet the regulations.

One may now propose a set of operating conditions that will meet the regulations:

$$N_t = \ln\left(\frac{1.0}{1-E}\right) = \ln\left(\frac{1.0}{1-0.99}\right)$$

$$= 4.605$$

The total power loss, $\mathscr{P}_T$, in hp/1000 acfm is

$$N_t = \alpha \, \mathscr{P}_T^{\beta}$$

$$4.605 = (1.47)(\mathscr{P}_T)^{1.05}$$

Solving for $\mathscr{P}_T$

$$\mathscr{P}_T = 2.96 \, \text{hp}/1000 \, \text{acfm}$$

Calculate the contacting power based on the gas stream energy input, $\mathscr{P}_G$, using a ΔP of 15 in. H_2O:

$$\mathscr{P}_G = 0.157 \, \Delta P$$

$$= (0.157)(15)$$

$$= 2.355 \, \text{hp}/1000 \, \text{acfm}$$

The liquid stream energy input, $\mathscr{P}_L$, is then

$$\mathscr{P}_L = \mathscr{P}_T - \mathscr{P}_G$$

$$= 2.96 - 2.355$$

$$= 0.605 \, \text{hp}/1000 \, \text{acfm}$$

Calculate q_L/q_G, in gal/acf, using P_L in psi:

$$q_L/q_G = \mathscr{P}_L/[(0.583)(P_L)]$$
$$= (0.605)/[(0.583)(100)]$$
$$= 0.0104$$

The new water flow rate, q'_L in gal/min, is therefore

$$q'_L = (q_L/q_G)(10{,}000 \, \text{acfm})$$
$$= (0.0104)(10{,}000 \, \text{acfm})$$
$$= 104 \, \text{gal/min}$$

The new set of operating conditions that will meet the regulations are

$$\Delta P = 15 \, \text{in. H}_2\text{O}$$
$$P_L = 100 \, \text{psi}$$
$$q'_L = 104 \, \text{gal/min}$$
$$\mathscr{P}_T = 2.96 \, \text{hp}/1000 \, \text{acfm}$$

Unlike the Johnstone equation approach, this method requires specifying two coefficients. The validity and accuracy of the coefficients available from the literature for the contact power theory equations have been questioned. Some numerical values of α and β for specific particulates and scrubber devices are provided below.

Aerosol	Scrubber Type	α	β
Raw gas (lime dust and soda fume)	Venturi and cyclonic spray	1.47	1.05
Prewashed gas (soda fume)	Venturi, pipe line, and cyclonic spray	0.915	1.05
Talc dust	Venturi	2.97	0.362
Black liquor recovery furnace fume	Venturi and cyclonic spray	1.75	0.620
Phosphoric acid mist	Venturi	1.33	0.647
Foundry cupola dust	Venturi	1.35	0.621
Open-hearth steel furnace fume	Venturi	1.26	0.569
Talc dust	Cyclone	1.16	0.655
Ferrosilicon furnace fume	Venturi and cyclonic spray	0.870	0.459
Odorous mist	Venturi	0.363	1.41

VEN.9   CALCULATIONS ON A VENTURI SCRUBBER

A venturi scrubber is being designed to remove particulates from a gas stream. The maximum gas flowrate of 30,000 acfm has a loading of $4.8\,gr/ft^3$. The average particle size is $1.2\,\mu m$ and the particle density is $200\,lb/ft^3$. Neglect the Cunningham correction factor. The Johnstone coefficient, k, for this system is 0.15. The proposed water flowrate is $180\,gal/min$ and the gas velocity is $250\,ft/s$.

1. What is the efficiency of the proposed system?
2. What would the efficiency be if the gas velocity were increased to $300\,ft/s$?
3. Determine the pressure drop for both gas velocities. Assume Calvert's equation to apply.
4. Determine the daily mass of dust collected and discharged for each gas velocity.
5. What is the discharge loading in each case?

Solution

1. The ratio of liquid-to-gas flowrates is given by

$$R = (180)(1000)/(30,000)$$
$$= 6.0\,gal/1000\,acf = 6.0\,gpm/acfm$$

and

$$v_G = 250\,ft/s$$
$$d_p = 1.2\,\mu m = 3.937 \times 10^{-6}\,ft$$
$$\rho_p = 200\,lb/ft^3$$
$$\mu = 1.23 \times 10^{-5}\,lb/(ft \cdot s)$$

Assume the Nukiyama–Tanasawa (NT) equation to apply:

$$d_0 = (16,400/v) + 1.45\,R^{1.5}$$
$$= (16,400/250) + 1.45\,(6.0)^{1.5}$$
$$= 86.91\,\mu m$$

$$N_I = \frac{d_p^2 \rho_p v}{9\mu d_0}$$

$$= \frac{(3.937 \times 10^{-6})^2(200)(250)}{(9)(1.23 \times 10^{-5})(2.85 \times 10^{-4})}$$

$$= 24.56$$

$$E = 1 - e^{-kR\sqrt{N_I}}$$

$$= 1 - e^{-(0.15)(6)\sqrt{24.56}}$$

$$= 0.9884 = 98.84\%$$

2. If v were increased to $300\,\text{ft/s}$,

$$d_0 = (16{,}400/300) + 1.45(6.0)^{1.5}$$
$$= 75.98\,\mu\text{m}$$

$$N_1 = \frac{(3.937 \times 10^{-6})^2(200)(300)}{(9)(1.23 \times 10^{-5})(2.85 \times 10^{-4})}$$
$$= 29.48$$

$$E = 1 - e^{-(0.15)(6)\sqrt{29.48}}$$
$$= 99.24\%$$

3. The pressure drops are given by (see Problem VEN.5)

$$\Delta P_a = (5 \times 10^{-5})(250)^2(6) = 18.75\,\text{in. H}_2\text{O}$$
$$\Delta P_b = (5 \times 10^{-5})(300)^2(6) = 27\,\text{in. H}_2\text{O}$$

4. The total mean loading, TML, is

$$\text{TML} = (4.8\,\text{gr/ft}^3)(30{,}000)(60)(24)/7000$$
$$= 29{,}600\,\text{lb/day} = 14.81\,\text{tons/day}$$

For $v = 250\,\text{ft/s}$,

$$\text{Dust collected} = (0.9884)(29{,}600)$$
$$= 29{,}300\,\text{lb/day}$$

$$\text{Dust discharged} = 344\,\text{lb/day}$$

For $v = 300\,\text{ft/s}$,

$$\text{Dust collected} = (0.9924)(29{,}600)$$
$$= 29{,}400\,\text{lb/day}$$

$$\text{Dust discharged} = 225\,\text{lb/day}$$

5. The discharge loading (DL) for $v = 250 \, \text{ft/s}$ is

$$DL = (4.8)(1 - E) = (4.8)(1 - 0.9884)$$
$$= 0.056 \, \text{gr/ft}^3$$

and for $v = 300 \, \text{ft/s}$ is

$$DL = (4.8)(1 - E) = (4.8)(1 - 0.9924)$$
$$= 0.036 \, \text{gr/ft}^3$$

VEN.10   OPEN-HEARTH FURNACE APPLICATION

The installation of a venturi scrubber is proposed to reduce the discharge of particulates from an open-hearth steel furnace operation. Preliminary design information suggests water and gas pressure drops across the rubber of 5.0 psia and 36.0 in. of H_2O, respectively. A liquid-to-gas ratio of 6.0 gpm/1000 acfm is usually employed with this industry. Estimate the collection efficiency of the proposed venturi scrubber. Assume contact power theory to apply with α and β given by 1.26 and 0.57, respectively. Recalculate the collection efficiency if the power requirement on the liquid side is neglected.

Solution

Due to the low water pressure drop, it can be assumed that

$$\mathscr{P}_G \ggg \mathscr{P}_L \qquad \mathscr{P}_T \approx \mathscr{P}_G$$

with

$$\mathscr{P}_G = 0.157 \, (\Delta P)$$

Solving for $\mathscr{P}_G$ gives

$$\mathscr{P}_G = (0.157)(36)$$
$$= 5.65 \, \text{hp}/1000 \, \text{acfm}$$

The number of transfer units is calculated from

$$N_t = \alpha \mathscr{P}_T^{\beta}$$
$$= (1.26)(5.65)^{0.57}$$
$$= 3.38$$

The collection efficiency can now be calculated:

$$N_t = 3.38 = \ln\left(\frac{1}{1-E}\right)$$

$$E = 0.966 = 96.6\%$$

Since the power requirement on the liquid side is neglected, the efficiency remains the same.

18 Hybrid Systems (HYB)

HYB.1 DESCRIPTION OF HYBRID SYSTEMS

Briefly describe hybrid systems.

Solution

Hybrid systems are defined as those types of control devices that involve combinations of control mechanisms, for example, fabric filtration combined with electrostatic precipitation. Unfortunately, the term *hybrid system* has come to mean different things to different people. The two most prevalent definitions employed today for hybrid systems are:

1. Two or more pieces of different air pollution control equipment connected in series, e.g., a baghouse followed by an absorber.
2. An air pollution control system that utilizes two or more collection mechanisms simultaneously to enhance pollution capture, e.g., an ionizing wet scrubber (IWS), which will be discussed shortly.

HYB.2 DRY SO$_2$ SCRUBBER

Estimate the water requirement of a spray dryer (dry SO$_2$ scrubber) at a coal-fired incineration facility that treats 150,000 lb/h of a flue gas at 2180°F. Assume an approach temperature to the adiabatic saturation temperature (AST) of 40°F. The AST can be assumed to be 180°F.

Solution

A spray dryer flue gas desulfurization (FGD) operation consists of four major steps:

1. Absorbent preparation
2. Absorption and drying
3. Solids collection
4. Solids disposal

Flue gas exiting the process (usually the combustion air preheater) comes in contact with an alkaline–water solution in a spray dryer. The flue gas passes through a contact chamber, and the solution or slurry is sprayed into the chamber with a rotary or nozzle atomizer. The heat of the flue gas evaporates the water in the atomized droplets while the droplets absorb SO_2 from the flue gas. The SO_2 reacts with the alkaline reagent to form solid-phase sulfite and sulfate salts. Most of the solids (and any fly ash present) are carried out of the dryer in the exiting flue gas. The rest fall to a hopper at the bottom of the dryer. With spray drying, in contrast to wet FGD, the flue gas is not saturated with moisture after the absorption step. However, the gas approaches within 20–50°F (11–28°C) of the adiabatic saturation temperature (AST). The water requirement to "cool" the gases to a temperature approaching the AST is an important design and operational requirement.

Note: Normally, cooling by liquid quenching is essentially accomplished by introducing a liquid (usually water) directly to the hot gases. When the water evaporates, the heat of vaporizing the water is obtained at the expense of the hot combustion gas, resulting in a reduction in the gas temperature. The temperature of the combustion gases discharged from the unit is at the adiabatic saturation temperature of the combustion gas if the operation is adiabatic and the gas leaves the unit saturated with water vapor. (A saturated gas contains the maximum water vapor possible at the temperature; any increase in water content will result in condensation.) Simple calculational and graphical procedures are available for estimating the adiabatic saturation temperature of a gas (see *Introduction to Hazardous Waste Incineration*, Santoleri, Reynolds, and Theodore, Wiley-Interscience, 2000).

For hot combustion gases being cooled approximately 2000°F, the quench water requirement may be estimated by (personal notes, L. Theodore)

$$w_{water} = \tfrac{1}{2} w_{flue}$$

where w_{water}, w_{flue} are the water and flue gas flow rates, respectively, in consistent units. The AST of most combustion gases is approximately 175°F.

Calculate the discharge temperature T from the spray dryer section of the dry scrubber:

$$T = 40 + AST$$
$$= 40 + 180 = 220°F$$

The dry scrubber water requirement rate is approximately

$$w_{water} = \tfrac{1}{2} w_{flue}$$
$$= (0.5)(150,000)$$
$$= 75,000 \text{ lb/h}$$

HYB.3   LIME REQUIREMENT FOR A SPRAY DRYER

Combustion of a hazardous waste produces 15,000 acfm of flue gas at 700°F and 1 atm. You have been asked to calculate lime requirements for this process. HCl and SO_2 concentrations are 10,000 and 250 ppm, respectively. HCl must be controlled to 99% collection efficiency or 4 lb/h. SO_2 emissions are to be controlled at 70% collection efficiency. A spray dryer is used to control the HCl and SO_2 emissions. $Ca(OH)_2$ is the sorbent that will react with HCl and SO_2 to form $CaCl_2$ and $CaSO_4$, respectively. Assume that it is necessary to provide 10% excess lime feed for the required HCl removal and 30% excess lime feed for total SO_2 removal.

What is the required feed rate of $Ca(OH)_2$? What is the total mass production rate of solids from the spray dryer? Assume that the excess solids in the spray dryer are $Ca(OH)_2$. Also size the spray dryer if the residence time (based on actual inlet conditions) is 10 s and the length-to-diameter ratio of the unit is 1.75 (neglecting hopper volume).

Solution

Determine the mass flowrates of HCl and SO_2 in the flue gas:

$$PV = nRT$$

$$Pq = \dot{n}RT$$

$$\dot{m} = \dot{n}(MW) = Pq(MW)/RT$$

$$\dot{m}_{HCl} = \frac{(1\,\text{atm})(0.01)(15,000\,\text{acfm})(36.5\,\text{lb/lbmol})}{[0.7302\,\text{atm} \cdot \text{ft}^3/(\text{lbmol} \cdot {}^\circ R)]\,(460 + 700\,^\circ R)(1\,\text{h}/60\,\text{min})}$$

$$= 387.8\,\text{lb HCl}/\text{h}$$

$$\dot{m}_{SO_2} = \frac{(1\,\text{atm})(0.00025)(15,000\,\text{acfm})(64\,\text{lb/lbmol})}{[0.7302\,\text{atm} \cdot \text{ft}^3/(\text{lbmol} \cdot {}^\circ R)]\,(460 + 700\,^\circ R)(1\,\text{h}/60\,\text{min})}$$

$$= 17.0\,\text{lb SO}_2/\text{h}$$

Determine which appropriate regulation applies for HCl control; $(0.01)(387.8\,\text{lb/h}) = 3.88\,\text{lb/h}$ which is less than 4 lb/h. Therefore, the 4 lb/h rule applies.

Write the two balanced chemical reaction equations (one for HCl and one for SO_2):

$$1.3Ca(OH)_2 + SO_2 + \frac{1}{2}O_2 \rightarrow CaSO_4 + H_2O + 0.3Ca(OH)_2$$

$$1.1Ca(OH)_2 + 2HCl \rightarrow CaCl_2 + 2H_2O + 0.1Ca(OH)_2$$

Determine the molar amount of $Ca(OH)_2$ needed for neutralization. SO_2 removal requires 1.3 mol lime/mol SO_2, while HCl removal requires 1.1/2 or 0.55 mol lime/mol HCl.

Determine the feed rate of $Ca(OH)_2$ required in lbmol/h:

$$\text{Lime feed rate} = \left(\frac{17 \text{ lb } SO_2/h}{64 \text{ lb } SO_2/\text{lbmol}}\right)\left(\frac{1.3 \text{ lbmol lime}}{\text{lbmol } SO_2}\right)$$

$$+ \left(\frac{(387.3 - 4.0) \text{ lb HCl}/ h}{36.45 \text{ lb HCl}/\text{lbmol}}\right)\left(\frac{0.55 \text{ lbmol lime}}{\text{lbmol HCl}}\right)$$

$$= 6.13 \text{ lbmol lime}/ h$$

Calculate the $Ca(OH)_2$ feed rate in lb/h:

$$\text{Lime feed rate} = (6.13)(74)$$
$$= 453 \text{ lb}/ h$$

Determine the production rate of $CaSO_4$ solids in lb/h. There is one mol of $CaSO_4$ produced per mol of SO_2 reacted. Since 70% of the SO_2 reacts,

$$\text{Production rate of } CaSO_4 = (17)(0.7)/64 = 0.186 \text{ lbmol}/ h$$
$$= (0.186)(136) = 25.3 \text{ lb/h}$$

Determine the $CaCl_2$ produced in lb/h. One mole of $CaCl_2$ is produced per 2 mol of HCl reacted; 4 lb HCl are not reacted:

$$\text{Production rate} = (387.3 - 4.0)(0.5)/36.45 = 5.258 \text{ lbmol/h}$$
$$= (5.258)(110.9) = 583.1 \text{ lb/h}$$

Determine the unreacted $Ca(OH)_2$ remaining in lb/h. One mol of $Ca(OH)_2$ is required to react in the production of either 1 mol of $CaSO_4$ or 1 mol of $CaCl_2$. Using previous results,

$$\text{Ca }(OH)_2 \text{ unreacted} = 6.12 - 5.258 - 0.186 = 0.676 \text{ lbmol/h}$$
$$= (0.676)(74) = 50.0 \text{ lb/h}$$

Calculate the total solids produced in lb/h:

$$\text{Total solids} = CaCl_2 + CaSO_4 + Ca(OH)_2 \text{ (unreacted)}$$
$$= 583.1 + 25.3 + 50.0$$
$$= 658.4 \text{ lb/h}$$

Calculate the volume of the spray dryer in ft^3:

$$V = (15,000)(10)/(60)$$

$$= 2500 \, ft^3$$

Size the spray dryer. Assume a cylindrical shape with $L =$ length and $D =$ diameter.

$$V = \pi D^2 L/4$$

Noting that L has been specified to be $1.75D$,

$$V = \pi D^2 (1.75D)/4$$

$$D^3 = (2500)(4)/(\pi)(1.75)$$

$$= 1818.9 \, ft^3$$

$$D = 12.2 \, ft$$

$$L = (1.75)(12.2)$$

$$= 21.4 \, ft$$

HYB.4   SPRAY DRYER VS. WET SCRUBBER

List the advantages the spray dryer has over traditional wet scrubbers.

Solution

Among the inherent advantages that the spray dryer enjoys over wet scrubbers are

1. Lower capital costs
2. Lower draft losses
3. Reduced auxiliary power
4. Reduced water consumption, with liquid-to-gas (L/G) ratios significantly lower than those of wet scrubbers
5. Continuous, two-stage operation

HYB.5   WET ELECTROSTATIC PRECIPITATOR (ESP) DESIGN

An exhaust stream from an industrial operation with a particulate loading of $5.0 \, gr/ft^3$ and a flowrate of 660 acfm is to be treated with a wet tubular ESP. Based on the data provided below, design the unit.

Outlet loading (required) $= 0.08\,\mathrm{gr/ft^3}$
Migration (drift) velocity $= 0.52\,\mathrm{ft/s}$
Cylindrical tube diameter $= 6\,\mathrm{in.}$
Suggested average throughput velocity $= 3\,\mathrm{ft/s}$
Water requirement $= 10\,\mathrm{gal/(min \cdot tube)}$

Assume the Deutsch–Anderson equation to apply.

Tubular precipitators, generally used for collecting mists or fogs, consist of cylindrical collection electrodes with discharge electrodes located in the center of the cylinders. Dirty gas flows into the cylinder where precipitation occurs. The negatively charged particles migrate to and are collected on grounded collecting tubes. The collected dust or liquid is removed by washing the tubes with water sprays located directly above the tubes. (These precipitators have also been referred to as *water-walled ESPs*.) Tube diameters typically vary from 0.5 to 1 ft (0.15 to 0.31 m), with length usually ranging from 6 to 15 ft (1.85 to 4.6 m).

Solution

The required collection efficiency E is

$$E = (5.0 - 0.08)/5.0$$
$$= 0.984 = 98.4\%$$

Apply the DA equation and calculate the required collection area:

$$E = 1 - e^{-(Aw/q)}$$

$$A = -\left(\frac{q}{w}\right)\ln(1 - E)$$

$$= -\left(\frac{660}{(0.52)(60)}\right)\ln(1 - 0.984)$$

$$= 87.5\,\mathrm{ft^2}$$

The cross-sectional area of each tube, A_C, is

$$A_C = (\pi)(D)^2/4$$
$$= (\pi)(6/12)^2/4$$
$$= 0.196\,\mathrm{ft^2}$$

Calculate the volumetric flowrate of exhaust gas, q_1, passing through one tube:

$$q_1 = vA$$
$$= (3)(0.196) = 0.588 \, \text{acfs}$$
$$= 35.3 \, \text{acfm}$$

Determine the required number of tubes, n:

$$n = q/q_1$$
$$= 660/35.3$$
$$= 18.7 = 19 \, \text{tubes}$$

Calculate the length of each tube:

$$n\pi DL = 87.5 \, \text{ft}^2$$
$$L = 87.5/[(\pi)(0.5)(19)]$$
$$= 2.93 \approx 3 \, \text{ft}$$

Calculate the water requirement, W, in gal/day:

$$W = (10)(19)(60)(24)$$
$$= 273,600 \, \text{gal/day}$$

The design appears adequate although the tube length is a bit short. This can be compensated for by operating with a higher throughput velocity. This in turn will correspondingly decrease the required number of tubes.

HYB.6   ADVANTAGES AND DISTADVANTAGES OF WET ELECTROSTATIC PRECIPITATORS

List the advantages and disadvantages of wet electrostatic precipitator (WEP) usage.

Solution

Some of the advantages of a WEP include:

1. Simultaneous gas adsorption and dust removal.
2. Low energy consumption.
3. No dust resistivity problems.
4. Efficient removal of fine particles.

Disadvantages of the WEP are the following:

1. Low gas absorption efficiency.
2. Sensitivity to changes in flowrate.
3. Dust collection is wet.

HYB.7   IONIZING WET SCRUBBER CALCULATIONS

A three-stage Ceilcote ionizing wet scrubber (IWS) is currently treating an 8.0-ft/s discharge stream from a hospital waste incinerator. The inlet (from the incinerator) and outlet particulate loadings to/from the IWS are 1.23 and 0.017 gr/dscf, respectively, and it operates at a pressure drop of 4.75 in. H_2O. Ceilcote provided the following design data on this system:

> Pressure drop = 1.55 in. H_2O/stage
> Number of stages = 3
> Collection efficiency of each stage = 74%
> Average bulk throughput velocity = 8.0 ft/s
> Length of packing per stage = 5 ft

Compare the operating conditions of pressure drop and collection efficiency with the design specifications provided by the vendor.

Solution

Note: Much of the following writeup has been drawn (with permission) from the Ceilcote literature (Ceilcote Air Pollution Control, *Strangeville*, OH).

The term *ionizing wet scrubber* was first used by the Ceilcote Co., located in Berea, Ohio, and has found wide application in the air pollution control field. This system is a proven means for the removal of pollutants from industrial process gas streams. The IWS combines the established principles of electrostatic particle charging, image force attraction, inertial impaction, and gas absorption to collect submicron solid particles, liquid particles, and noxious and malodorous gases simultaneously. The IWS system requires little energy and its collection efficiency is high for both submicron and micron size particles.

The ionizing wet scrubber utilizes high-voltage ionization to electrostatically charge particulate matter in the gas stream before the particles enter a Tellerette packed scrubber section where they are removed by attraction of the charged particles to neutral surfaces. Larger particles equal to or greater than 3–5 μm are collected through inertial impaction. As small particles flow through the scrubber, they pass close to the surfaces of the Tellerettes and scrubbing liquid droplets. The electrostatic charges on the particles cause them to be attracted to these neutral surfaces by image force attraction. All particles are eventually washed out of the scrubber with the exit liquor. Noxious and malodorous gases are also absorbed and reacted in the same scrubbing liquor.

The IWS system utilizes Tellerette packing as its collection surface to achieve particulate removal. Scrubbing liquid droplets also act as collection surfaces. Particles of any size or composition are collected by the IWS. Fine particles (0.05–2 μm) are collected with high efficiency as well as coarse particles (2 μm and larger), regardless of their composition (organic or inorganic with either high or low resistivity).

Particle collection efficiency over long-term service remains consistently high. Interestingly, the percent of particulate removed varies little with load and particle size distribution over a wide range. As particulate load increases, the percent removed remains nearly constant. In addition, the collection efficiency for fine particles is nearly as great as for coarse particles. The IWS system also simulta-neously absorbs gases. Noxious gases are removed through physical absorption and/or adsorption that is accompanied by chemical reaction. Pressure drop through a single-stage IWS is only 0.5–1.5 in. of H_2O. Energy for particle charging is low— approximately 0.2–0.4 kVA per 1000 cfm. The shell and most internal parts of the IWS are commonly fabricated of Duracor (fiberglass-reinforced plastic) and thermo-plastic materials. This predominance of plastic construction assures corrosion-free operation in the presence of acid gases such as HCl, HF, Cl_2, NH_3, SO_2, and SO_3. For noncorrosive applications, metallic construction is also available. Factory-assembled modules, available in standard capacities from 900 to 54,000 acfm, can also be grouped together to handle virtually any gas volume.

With reference to the problem statement, calculate the overall pressure drop, ΔP, across the unit based on vendor data:

$$\Delta P = (3)(1.55)$$
$$= 4.65 \text{ in. } H_2O$$

Compare the above result with the actual pressure drop:

$$DIFF = 4.75 - 4.65 = 0.10 \text{ in. } H_2O$$

$$\%DIFF = (1.10/4.75)100$$
$$= 2.1\% = 0.021$$

The operating particulate collection efficiency E of the unit is

$$E = (1.23 - 0.017)/1.23$$
$$= 0.9862 = 98.62\%$$

Calculate the design collection efficiency based on design data:

$$P = (1 - 0.74)^3$$
$$= (0.26)^3$$
$$= 0.0176$$

$$E = 1 - P$$
$$= 1 - 0.0176$$
$$= 0.9824 = 98.24\%$$

The unit appears to be operating at an efficiency slightly above that indicated by the vendor.

HYB.8   TWO-STAGE IWS SYSTEM

Ceilcote has submitted the design for a two-stage IWS system for particulate control of a proposed hazardous waste incineration (HWI) facility that is early in the permit review process. The following data (estimated) have been provided for the facility:

Inlet loading $= 1.17\,\text{gr/dscf}$ corrected to 7% oxygen
Outlet loading $= 0.015\,\text{gr/dscf}$ corrected to 7% oxygen
Flue gas flowrate $= 20,000\,\text{acfm}$

Note: The federal particulate regulation (at the time of submission of the design) for HWI facilities is 0.08 gr/dscf corrected to 7% oxygen (or corrected to 50% excess air). State and/or local regulations can be lower; most state regulations are 0.03, but there are some states requiring that a 0.015 level be met. The reader is referred to *Introduction to Hazardous Waste Incineration* by Santoleri, Reynolds and Theodore (Wiley-Interscience, 2000) for additional details on these regulations and HWIs in general.
    The following performance data are available on IWS systems from Ceilcote:

Average Throughput Velocity (ft/s)	Collection Efficiency (%)
8.0	75
7.0	80
6.0	85
5.0	90
4.0	94

The following equipment data are also provided:

Unit	Frontal (face) Area (ft^2)
IWS-400	40
IWS-500	50
IWS-600	60
IWS-700	70
IWS-800	80
IWS-900	90
IWS-1000	100

Design (size) the IWS unit based on the information given above.

Solution

First, calculate a velocity that will provide a collection efficiency that will meet regulation specifications. Use the velocity to then determine the frontal area requirement, which will in turn effectively size the unit.

The required overall particulate collection efficiency E_0 is

$$E_0 = (1.17 - 0.015)/1.17$$
$$= 0.9872 = 98.72\%$$

For a two-stage unit, assuming equal penetrations,

$$1 - E_0 = P_1 P_2 = P^2$$

Thus,

$$P^2 = 1 - 0.9872 = 0.0128$$
$$P = 0.1131$$

and, the single-stage efficiency is

$$E = 1 - P$$
$$= 1 - 0.1131$$
$$= 0.8869 = 88.69\%$$

Using the vendor's velocity–efficiency data and linearly interpolating yields the following velocity:

$$v = 5.262 \text{ ft/s}$$

The face area requirement of the unit, A, in ft^2 is

$$A = 20{,}000/[(5.262)(60)]$$
$$= 63.3 \text{ ft}^2$$

Since the unit must meet or exceed the regulatory requirement, select the IWS-700.
  Because the size of the unit selected exceeds the required collection efficiency, calculate a revised efficiency and the corresponding throughput velocity:

$$v = 20{,}000/[(70)(60)]$$
$$= 4.76 \text{ ft/s}$$

The corresponding single-stage efficiency (linearly interpolating) is

$$E = 90 + (4.76 - 5.0)(94 - 90)/(4.0 - 5.0)$$
$$= 90.96\%$$

The overall efficiency then becomes

$$E_0 = 1 - (1 - 0.9096)^2$$
$$= 0.9918 = 99.18\%$$

19 Combustion (CMB)

CMB.1 DESCRIPTION OF COMBUSTION EQUIPMENT

Briefly describe some of the terms used to describe combustion equipment.

Solution

Afterburning is the most common term used to describe the combustion process employed to control gaseous emissions. The term *afterburner* is appropriate only to describe a thermal oxidizer used to control gases coming from a process where combustion was not complete. Incinerators are used to combust solid, liquid, and gaseous materials; when used in this handbook, the term *incinerator* will refer to combustion of waste streams. Other terms used to describe combustion equipment include *oxidizer, reactor, chemical reactor, combustor,* etc.

CMB.2 THE THREE *T*'S

Describe the combustion process in terms of the "three *T*'s."

Solution

To achieve complete combustion once the air (oxygen), waste (pollutants), and fuel have been brought into contact, the following conditions must be provided: a temperature high enough to ignite the waste/fuel mixture, turbulent mixing of the air and waste/fuel, and sufficient residence time for the reaction to occur. These three conditions are referred to as the "three *T*'s of combustion." Time, temperature, and turbulence govern the speed and completeness of reaction. They are not usually independent variables since changing one can affect the other two.

The rate at which a combustible compound is oxidized is greatly affected by temperature. The higher the temperature, the faster the oxidation reaction will proceed. The chemical reactions involved in the combination of a fuel and oxygen can occur even at room temperature, but very slowly. For this reason, a pile of oily rags can be a fire hazard. Small amounts of heat are liberated by the slow oxidation of the oils. This, in turn, raises the temperature of the rags and increases the oxidation rate, liberating more heat. Eventually, a full-fledged fire can break out. For combustion processes, ignition is accomplished by adding heat to speed up the oxidation process. Heat is needed to combust any mixture of air and fuel until the ignition temperature of the mixture is reached. By gradually heating a mixture of fuel and air, the rate of reaction and energy released will gradually increase until the reaction no longer depends on the outside heat source. More heat is being generated than is lost to the surroundings. The ignition temperature must be reached or exceeded to ensure complete combustion. To maintain combustion of a waste, the amount of energy released by the combusted waste must be sufficient to heat the incoming waste (and air) up to its ignition temperature; otherwise, a fuel must be added. The ignition temperature of various fuels and compounds can be found in combustion handbooks. These temperatures are dependent on combustion conditions and therefore should be used only as a guide. Most incinerators operate at a higher temperature than the ignition temperature, which is a minimum. Thermal destruction of most organic compounds occurs between 590 and 650°C (1100 and 1200°F). However, most hazardous waste incinerators are operated at 1800–2200°F to ensure near complete conversion of the waste.

Generally, ignition depends on:

1. Concentration of combustibles in the waste stream
2. Inlet temperature of the waste stream
3. Rate of heat loss from the combustion chamber
4. Residence time and flow pattern of the waste stream
5. Combustion chamber geometry and materials of construction

Time and temperature affect combustion in much the same manner as temperature and pressure affect the volume of a gas. When one variable is increased, the other may be decreased with the same end result. With a higher temperature, a shorter residence time can achieve the same degree of oxidation. The reverse is also true; a higher residence time allows the use of a lower temperature. In describing incinerator operation, these two terms are always mentioned together. The choice between higher temperature or longer residence time is based on economic considerations. Increasing residence time involves using a larger combustion chamber resulting in a higher capital cost. Raising the operating temperature increases fuel usage, which also adds to the operating costs. Fuel costs are the major operating expense for most incinerators. Within certain limits, lowering the temperature and adding volume to increase residence time can be a cost-effective alternative method of operation. The residence time of gases in the combustion chamber may be calculated from

$$t = V/q$$

where t is the residence time (s), V is the chamber volume (ft^3), and q is the gas volumetric flow rate at combustion conditions within the unit (ft^3/s). Adjustments to the flowrate must include outside air added for combustion.

Proper mixing is important in combustion processes for two reasons. First, for complete combustion to occur every mass of waste and fuel must come in contact with air (oxygen). If not, unreacted waste and fuel will be exhausted from the stack. Second, not all of the fuel or waste stream is able to be in direct contact with the burner flame. In most incinerators, a portion of the waste stream may bypass the flame and be mixed at some point downstream of the burner with the hot products of combustion. If the two streams are not completely mixed, a portion of the waste stream will not react at the required temperature and incomplete combustion will occur. A number of methods are available to improve mixing the air and waste (combustion) streams. Some of these include the use of refractory baffles, swirl-fired burners, and baffle plates. The problem of obtaining complete mixing is not easily solved. Unless properly designed, many of these mixing devices may create "dead spots" and reduce operating temperatures. Merely inserting obstructions to increase turbulence is not necessarily the answer. The process of mixing flame and waste stream to obtain a uniform temperature for decomposition of wastes is the most difficult part in the design of the incinerator.

In addition, oxygen is necessary for combustion to occur. To achieve complete combustion of a compound, a sufficient supply of oxygen must be present to convert all of the carbon to CO_2. This quantity of oxygen is referred to as the *stoichiometric* or *theoretical* amount. The stoichiometric amount of oxygen is determined from a balanced chemical equation summarizing the oxidation reactions. If an insufficient amount of oxygen is supplied, the mixture is referred to as *rich*. In a rich mixture, there is not enough oxygen to combine with all the fuel and waste so that incomplete combustion occurs. If more than the stoichiometric amount of oxygen is supplied, the mixture is referred to as *lean*. The excess oxygen plays no part in the oxidation reaction and passes through the incinerator. Oxygen for the combustion process is supplied by using air. Since air is essentially 79% nitrogen and 21% oxygen (by volume), a larger volume of air is required than if pure oxygen were used.

CMB.3 TYPES OF COMBUSTION UNITS FOR GASEOUS POLLUTANT CONTROL

Describe the three types of combustion equipment employed by industry for gaseous pollutant control.

Solution

Equipment used to control waste gases by combustion can de divided into three general categories: direct combustion or flaring, thermal oxidation, and catalytic oxidation. A direct combustor or flare is a device in which air and all the combustible waste gases react at the burner. Complete combustion must occur almost instantaneously since there is no residence chamber. Therefore, the flame temperature is the most important variable in flaring waste gases. In contrast, in thermal oxidation, the combustible waste gases pass over or around a burner flame into a residence chamber where oxidation of the waste gases is completed. Catalytic oxidation is very similar to thermal oxidation. The main difference is that after passing through the flame area, the gases pass over a catalyst bed that promotes oxidation at a lower temperature than does thermal oxidation. Details on these three control devices is given below.

Afterburners can be used over a fairly wide range of organic vapor concentrations. The concentration of the organics in air must be substantially below the lower flammable level (lower explosive limit). As a rule, a factor of 4 is employed for safety precautions. Reactions are conducted at elevated temperatures to ensure high chemical reaction rates for the organics. To achieve this temperature, it is necessary to preheat the feed stream with auxiliary energy. Along with the contaminant-laden gas stream, air and fuel are continuously delivered to the reactor where the fuel is combusted with air in the firing unit (burner). The burner may utilize the air in the process waste stream as the combustion air for auxiliary fuel, or it may use a separate source of outside air for this purpose. The products of combustion and the unreacted feed stream are intensely mixed and enter the reaction zone of the unit. The pollutants in the process gas stream are then reacted at the elevated temperature. The unit requires operating temperatures in the 1200–2000°F range for combustion of most pollutants. A residence time of 0.2–2.0 s is recommended in the literature, but this factor is primarily dictated by kinetic considerations. A length-to-diameter ratio of 2.0–3.0 is usually employed. The end products are continuously discharged at the outlet of the reactor. The average gas velocity can range from as low as 10 ft/s to as high as 50 ft/s. (The velocity increases from inlet to outlet due to the increase in the number of moles of gas and the increase in volume due to the higher temperature.) These high velocities are required to prevent settling of particulates (if present) and to minimize the dangers of flashback and fire hazards.

The fuel is usually natural gas. The energy liberated by reaction may be directly recovered in the process or indirectly recovered by suitable external heat exchange. This should be included in a design analysis since energy is the only commodity of value that is usually derived from the combustion process.

Catalytic reactors are an alternative to thermal reactors. If a solid catalyst is added to the reactor, the reaction is said to be heterogeneous. For simple reactions, the effect of the presence of a catalyst is to:

1. Increase the rate of reaction
2. Permit the reaction to occur at a lower temperature
3. Permit the reaction to occur at a more favorable pressure
4. Reduce the reactor volume
5. Increase the yield of a reactant(s) relative to the other components

In a typical catalytic reactor, the gas stream is delivered to the reactor continuously by a fan at a velocity in the 10- to 30-ft/s range but at a lower temperature—usually 650–800°F—than a thermal unit. A length-to-diameter ratio less than 0.5 is usually employed. The gases, which may or may not be preheated, pass through the catalyst bed where the reaction occurs. The combustion products, which are again made up of water vapor, carbon dioxide, inerts, and unreacted vapors are continuously discharged from the outlet at a higher temperature. Energy savings can again be achieved with heat recovery from the exit stream.

CMB.4   COMBUSTION OF BUTANOL

The offensive odor of butanol can be removed from stack gases by its complete combustion to carbon dioxide and water. It is of interest that the incomplete combustion of butanol actually results in a more serious odor pollution problem than the original one. Write the equations showing the two intermediate malodorous products formed if butanol undergoes incomplete combustion.

Solution

The malodorous products are butyraldehyde (C_4H_8O) and butyric acid (C_3H_7COOH), which can be formed sequentially as follows:

$$C_4H_9OH + \tfrac{1}{2}O_2 \rightarrow C_4H_8O + H_2O$$
$$C_4H_8O + \tfrac{1}{2}O_2 \rightarrow C_3H_7COOH$$

or the acid can be formed directly as follows:

$$C_4H_9OH + O_2 \rightarrow C_3H_7COOH + H_2O$$

For complete combustion:

$$C_4H_9OH + 6O_2 \rightarrow 4CO_2 + 5H_2O$$

CMB.5   GROSS HEATING VALUE

The composition and combustion properties of a gas mixture are given below:

Component	Mole Fraction x_i	Gross Heating Value (Btu/scf)
N_2	0.0515	0
CH_4	0.8111	1013
C_2H_6	0.0967	1792
C_3H_8	0.0351	2590
C_4H_{10}	0.0056	3370
Σ	1.0000	

Determine the gross heating value (HV_G) of this gas.

Solution

The gross heating value (HV_G) represents the enthalpy change or heat released when a gas is stoichiometrically combusted at 60°F, with the final (flue) products at 60°F and any water present in the liquid state. Stoichiometric combustion requires that no oxygen be present in the flue gas following combustion of the hydrocarbons.

Using values from the table, the gross heating value of the gas mixture, HV_G in Btu/scf is

$$HV_G = \Sigma x_i HV_{Gi}$$
$$= (0.0515)(0) + (0.8111)(1013) + (0.0967)(1792) + (0.0351)(2590)$$
$$+ (0.0056)(3370)$$
$$= 1105\,\text{Btu/scf}$$

where x_i = mole fraction of the ith component

The gas described in this example is typical of natural gas.

The *net heating value* (HV_N) is similar to HV_G except the water is in the vapor state. The net heating value is also known as the *lower heating value* and the gross heating value is known as the *higher heating value*.

CMB.6   THERMAL AFTERBURNER DESIGN

Provide a design procedure for thermal afterburners.

Solution

There are two key calculations associated with combustion devices. These include determining:

1. The fuel requirements
2. The physical dimensions of the unit

Both these calculations are interrelated. The general procedure to follow, with pertinent equations, is given below. It is assumed that the process gas stream flowrate, inlet temperature, and the combustion temperature are known. The required residence time is also specified. Primary (outside the process) air is employed for combustion.

Note: This design procedure was originally developed by Dr. Louis Theodore in 1985 and later published in 1988. These materials recently appeared in the 1992 and 2000 Air and Waste Management Association texts published by Van Nostrand Reinhold and John Wiley, respectively, titled *Air Pollution Engineering Manual.*

This was done without properly acknowledging the author, Dr. Louis Theodore, and without permission from the original publisher.

1. Calculate the heat load required to raise the process gas stream from its inlet temperature to the operating temperature of the combustion device:

$$\dot{Q} = \Delta \dot{H}$$

2. Correct the heat load term for any radiant losses (RL):

$$\dot{Q} = (1 + \text{RL})(\Delta \dot{H}) \qquad \text{RL} = \text{fractional basis}$$

3. Assuming natural gas of known heating value, HV_G, is the fuel, calculate the available heat at the operating temperature. For engineering purposes, one may use a short-cut method that bypasses a detailed calculation.

$$HA_T = (HV_G)(HA_T/HV_G)_{\text{ref}}$$

The subscript "ref" refers to a reference fuel. For natural gas with a reference HV_G of 1059 Btu/scf, the available heat (assuming stoichiometric air) is given by (L. Theodore, personal notes):

$$(HA_T)_{\text{ref}} = -0.237T + 981 \qquad T = {}^\circ F$$

The *available heat* is defined as the quantity of heat released within a combustion chamber minus (1) the sensible heat carried away by the dry flue gases, and (2) the latent heat and sensible heat carried away in water vapor contained in the flue gases. Thus, the available heat represents the net quantity of heat remaining for useful heating.

4. Calculate the flowrate of natural gas required, q_{NG}:

$$q_{NG} = \dot{Q}/HA \qquad \text{consistent units}$$

5. Determine the volumetric flowrates of both the process gas stream, q_p, and the flue products of combustion of the natural gas, q_c, at the operating temperature:

$$q_T = q_p + q_c$$

A good estimate for q_c is

$$q_c = (11.5)q_{NG}$$

6. The cross-sectional area of the combustion device is given by

$$S = q_T/v$$

where v is the throughput velocity.

7. The residence time of gases in the combustion chamber may be calculated from

$$t = V/q_T$$

where t is the residence time (s), V is the chamber volume (ft^3), and q_T is the gas volumetric flowrate at combustion conditions in the chamber (acfs).

Adjustments to both the gas volumetric flowrate and fuel rate must be performed if secondary (from process gas stream) air rather than primary (outside) air is added for combustion.

CMB.7   CALCULATIONS ON NATURAL GAS COMBUSTION

As an air pollution control engineer you have been requested to evaluate the gross heating value of a natural gas of a given composition. You are also to determine the available heat of the natural gas at a given temperature, the rate of auxiliary fuel (natural gas) required to heat a known amount of contaminated air to a given temperature, the dimensions of an afterburner treating the contaminated airstream, and the residence time. Operating and pertinent design data are provided below.

Natural gas composition (mole or volume fraction):

Component	Mole Fraction
N_2	0.0515
CH_4	0.8111
C_2H_6	0.0967
C_3H_8	0.0351
C_4H_{10}	0.0056
Σ	1.0000

Gas velocity $= 20$ ft/s
Length-to-diameter ratio of the afterburner $= 2.0$
Temperature of dry natural gas $= 60°$F
Volumetric flowrate of contaminated air $= 5000$ scfm ($60°$F, 1 atm)

Enthalpy data are provided in the table below.

Enthalpy of Combustion Gases (Btu/lbmol)

T (°F)	N_2	Air (MW = 28.97)	CO_2	H_2O
32	0	0	0	0
60	194.9	194.6	243.1	224.2
77	312.2	312.7	392.2	360.5
100	473.3	472.7	597.9	545.3
200	1,170	1,170	1,527	1,353
300	1,868	1,870	2,509	2,171
400	2,570	2,576	3,537	3,001
500	3,277	3,289	4,607	3,842
600	3,991	4,010	5,714	4,700
700	4,713	4,740	6,855	5,572
800	5,443	5,479	8,026	6,460
900	6,182	6,227	9,224	7,364
1000	6,929	6,984	10,447	8,284
1200	8,452	8,524	12,960	10,176
1500	10,799	10,895	16,860	13,140
2000	14,840	14,970	23,630	18,380
2500	19,020	19,170	30,620	23,950
3000	23,280	23,460	37,750	29,780

Source: Kobe, Kenneth, A, and Long, Ernest G., "Thermochemistry for the Petroleum Industry," *Petroleum Refiner*, **28**, (11), November, 1949, p. 129, Table 9.

It is required to heat the contaminated air from 200°F to 1200°F.

Solution

The gross heating value of this natural gas was determined in Problem CMB.5.

$$HV_G = \Sigma x_i HV_{Gi}$$

$$= 0.0515)(0) + (0.8111)(1013) + (0.0967)(1792) + (0.0351)(2590)$$

$$+ (0.0056)(3370)$$

$$= 1105 \text{ Btu/scf natural gas}$$

The balanced chemical combustion equations for each of the four components of the natural gas using 1 scf of natural gas as a basis is

$$0.8111CH_4 + 1.6222O_2 \rightarrow 0.8111CO_2 + 1.6222H_2O$$

$$0.0967C_2H_6 + 0.3385O_2 \rightarrow 0.1934CO_2 + 0.2901H_2O$$

$$0.0351C_3H_8 + 0.1755O_2 \rightarrow 0.1053CO_2 + 0.1404H_2O$$

$$0.0056C_4H_{10} + 0.0364O_2 \rightarrow 0.0224CO_2 + 0.0280H_2O$$

The number of standard cubic feet for each of the following components of combustion may now be determined.

For O_2: $1.6222 + 0.3385 + 0.1755 + 0.0364 = 2.172$ scf/scf natural gas
For CO_2: $0.8111 + 0.1934 + 0.1053 + 0.0224 = 1.132$ scf/scf natural gas
For H_2O: $1.6222 + 0.2901 + 0.1404 + 0.0280 = 2.081$ scf/scf natural gas
For N_2: $0.0515 + (79/21)(2.172) = 8.222$ scf/scf natural gas

The total cubic feet of combustion products per scf of natural gas burned are the sum of scf CO_2/scf of natural gas, scf H_2O/scf of natural gas, and scf N_2/scf of natural gas:

Total cubic feet of combustion products $= 1.132 + 2.081 + 8.222$

$$= 11.435 \text{ scf of products/scf of natural gas}$$

From the table in the problem statement, the following values of enthalpies at 60°F and 1200°F are obtained.
   For CO_2:

$$\Delta H_{CO_2} = H \text{ at } 1200°F - H \text{ at } 60°F$$

$$= 12,960 - 243.1$$

$$= 12,720 \text{ Btu/lbmol}$$

Since there are 379 scf per lbmol of any ideal gas and 1.132 scf CO_2/scf natural gas,

$$\Delta H_{CO_2} = (12,720)(1.132)/(379)$$

$$= 38.0 \text{ Btu/scf of natural gas}$$

For N_2:

$$\Delta H_{N_2} = (8452 - 194.9)(8.222)/(379)$$

$$= 179.1 \text{ Btu/scf of natural gas}$$

For H_2O (g):

$$\Delta H_{H_2O} = (10,176 - 224.2)(2.081)/(379)$$

$$= 54.6 \text{ Btu/scf of natural gas}$$

Determine the amount of heat required to take the products of combustion from 60 to 1200°F ($\Sigma \Delta H$).

Since the water present in the combustion product is vapor,

$$\Delta H_{\lambda} = \frac{(1060 \text{ Btu/lb})(18 \text{ lb/lbmol})(2.081 \text{ scf } H_2O/\text{scf natural gas})}{379 \text{ scf/lbmol natural gas}}$$

$$= 104.8 \text{ Btu/scf of natural gas}$$

Therefore,

$$\Sigma \Delta H = \Delta H_{CO_2} + \Delta H_{N_2} + \Delta H_{H_2O} + \Delta H_{\lambda}$$

$$= 38.0 + 179.1 + 54.6 + 104.8$$

$$= 376.5 \text{ Btu/scf of natural gas}$$

The available heat (HA) of natural gas at 1200°F in Btu/scf natural gas is

$$HA = HV_G - \Sigma \Delta H$$

$$= 1105 - 376.5$$

$$= 728.5 \text{ Btu/scf of natural gas}$$

The enthalpy change of air going from 200 to 1200°F is

$$\Delta H_{\text{air}} = H \text{ at } 1200°F - H \text{ at } 200°F$$

$$= 8524 - 1170$$

$$= 7354 \text{ Btu/lbmol}$$

The heat rate, $\dot{Q}$, required to heat 5000 scfm of air from 200 to 1200°F in Btu/min is

$$\dot{Q} = (5000 \text{ scfm})(7354 \text{ Btu/lbmol})/(379 \text{ scf/lbmol})$$

$$= 97,018 \text{ Btu/min}$$

The fuel requirement is therefore

$$q_{NG} = \dot{Q}/HA$$

$$= 97,018/728.5$$

$$= 133.2 \text{ scfm}$$

To size the unit, first calculate the volumetric flowrate of the products of combustion (CP) in scfm:

$$q_{CP} = 11.435 \; \frac{\text{scf products}}{\text{scf natural gas}} \; (133.2 \text{ scfm natural gas})$$

$$= 1523 \text{ scfm}$$

Also calculate the total flue gas flowrate, q_T, in scfm:

$$q_T = 5000 + 1523$$

$$= 6523 \text{ scfm}$$

Determine the total flue gas flowrate at 1200°F in acfm using the ideal gas law:

$$q_T = (6523)(1200 + 460)/(60 + 460)$$

$$= 20{,}823 \text{ acfm}$$

$$= 347 \text{ acfs}$$

The diameter D and length L of the afterburner in feet are

$$D = (4q_T/v\pi)^{1/2}$$

$$= [(4)(347)/(20)(\pi)]^{1/2}$$

$$= 4.7 \text{ ft}$$

$$L = (2.0)(4.7)$$

$$= 9.4 \text{ ft}$$

Therefore, the residence time t for the gases in the afterburner in seconds is

$$t = L/v$$

$$= (9.4)/(20)$$

$$= 0.47 \text{ s}$$

CMB.8   AVAILABLE HEAT, FUEL FLOWRATE, AND RESIDENCE TIME

A process gas stream (assume air) of 1.75×10^6 acfm at 32°F and 1 atm is to be incinerated at 1900°F in a process boiler that is 30 ft × 30 ft × 40 ft. Natural gas (assume reference) is to be employed with stoichiometric (primary) air. The stoichiometric combustion of natural gas produces 11.4 mol of flue gas per mole of natural gas.

1. Find the available heat in Btu/min.
2. Calculate the required natural gas flowrate in scfm (60°F, 1 atm).
3. Does the unit meet the required residence time of 0.7 s?

Solution

The molar flowrate and enthalpy change are first calculated. Note that one lbmol of an ideal gas occupies 359 ft^3 at 32°F and 1 atm.

$$\dot{n} = (1.75 \times 10^6 \text{ ft}^3/\text{min})/(359 \text{ ft}^3/\text{lbmol}) = 4.875 \times 10^3 \text{ lbmol/min}$$

$$\Delta\dot{H} = (4.87 \times 10^3)(14{,}115 - 0.0)$$

$$= 6.88 \times 10^7 \text{ Btu/min}$$

Using Theodore's equation presented in Problem CMB.6,

$$HA_{1900°F} = -0.237(1900) + 981$$

$$= 530.7 \text{ Btu/scf } (60°F)$$

The natural gas flowrate is then

$$q_{NG} = \Delta\dot{H}/HA_{1900°F}$$

$$= 6.88 \times 10^7/530.7$$

$$= 1.30 \times 10^5 \text{ scfm}$$

The total volumetric flowrate of the flue gas following may now be calculated:

$$q_{1900°F} = (1.75 \times 10^6)\left(\frac{1900 + 460}{32 + 460}\right) + (1.30 \times 10^5)(11.4)\left(\frac{1900 + 460}{60 + 460}\right)$$

$$= 1.512 \times 10^7 \text{ acfm} = 2.52 \times 10^5 \text{ acfs}$$

The volume of the system is

$$V = (30)(30)(40)$$

$$= 36{,}000 \text{ ft}^3$$

Thus, the residence time is

$$t = 36,000/2.52 \times 10^5$$
$$= 0.142 \text{ s}$$

The unit does *not* meet the required residence time of 0.7 s.

CMB.9   PLAN REVIEW OF A DIRECT FLAME AFTERBURNER

A regulatory agency engineer must review plans for a permit to construct a direct flame afterburner (Figure 71) serving a lithographer. Review is for the purpose of judging whether the proposed system, when operating as it is designed to operate, will meet emission standards. The permit application provides operating and design data. Agency experience has established design criteria that, if met in an operating system, typically ensure compliance with standards. Operating data from the permit application are provided below.

Application = lithography

Effluent exhaust volumetric flowrate = 7000 scfm (60°F, 1 atm)

Exhaust temperature = 300°F

Hydrocarbons in effluent air to afterburner (assume hydrocarbons to be toluene) = 30 lb/h

Afterburner entry temperature of effluent = 738°F

Afterburner heat loss = 10% in excess of calculated heat load

Afterburner dimensions = 4.2 ft in diameter, 14 ft in length

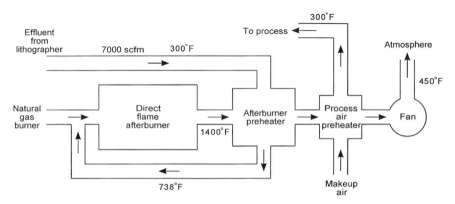

Figure 71. Direct flame afterburner system.

Agency Design Criteria

Afterburner temperature $= 1300–1500°F$

Residence time $= 0.3–0.5$ s

Afterburner velocity $= 20–40$ ft/s

Standard Data

Gross heating value of natural gas $= 1059$ Btu/scf of natural gas

Combustion products per cubic foot of natural gas burned $= 11.5$ scf/scf natural gas

Available heat of natural gas at $1400°F = 600$ Btu/scf of natural gas

Molecular weight of toluene $= 92$

Average heat capacity of effluent gases at $738°F$ (above $0°F) = 7.12$ Btu/ (lbmol · °F)

Average heat capacity of effluent gases at $1400°F$ (above $0°F) = 7.38$ Btu/ (lbmol · °F)

Volume of air required to combust natural gas $= 10.33$ scf air/scf natural gas

Solution

The design temperature is already within agency criteria. To determine the fuel requirement for the afterburner, first calculate the total heat load (heating rate), $\dot{Q}$, required to raise 7000 scfm of the effluent stream from 738 to 1400°F in Btu/min:

$$\dot{n} = (7000 \text{ scfm})/(379 \text{ scf/lbmol})$$
$$= 18.47 \text{ lbmol/min}$$

$$\dot{Q} = \dot{n}[C_{P2}(T_2 - T_b) - C_{P1}(T_1 - T_b)]$$
$$= 18.47[(7.38)(1400 - 0) - (7.12)(738 - 0)]$$
$$= 93,780 \text{ Btu/min}$$

Calculate the actual heat load required, accounting for a 10% heat loss, in Btu/min:

$$\text{Actual heat load} = (1.1)\dot{Q}$$
$$= (1.1)(93,780)$$
$$= 103,200 \text{ Btu/min}$$

Calculate the rate of natural gas required to supply the actual heat required to heat 7000 scfm of the effluent from 738 to 1400°F in scfm:

$$q_{NG} = \dot{Q}/HA$$
$$= (103,200)/(600)$$
$$= 172.0 \text{ scfm}$$

Determine the total volumetric flowrate through the afterburner, q_T, by first calculating the volumetric flowrate of the combustion products of the natural gas, q_1, in scfm:

$$q_1 = q_{NG}\left(11.5 \; \frac{\text{scf combustion products}}{\text{scf natural gas}}\right)$$
$$= (172.0)(11.5)$$
$$= 1978 \text{ scfm}$$

Also note that the volumetric flowrate of the effluent is 7000 scfm.

The volumetric flowrate of air required to combust the natural gas required, q_2, in scfm is

$$q_2 = q_{NG}\left(10.33 \; \frac{\text{scf air}}{\text{scf natural gas}}\right)$$
$$= (172.0)(10.33)$$
$$= 1776 \text{ scfm}$$

Calculate the total volumetric flowrate through the afterburner, q_T, in scfm. Since primary air is employed in the combustion of the natural gas, q_2 is not subtracted from q_T. Thus, the q_2 calculation is not required in this solution:

$$q_T = 7000 + q_1$$
$$= 7000 + 1978$$
$$= 8978 \text{ scfm}$$
$$= (8978)(1400 + 460)/(60 + 460)$$
$$= 32,110 \text{ acfm}$$

To determine if the afterburner velocity meets the agency criteria, first calculate the cross-sectional area of the afterburner, S, in ft^2.

$$S = \pi D^2 / 4$$
$$= (\pi)(4.2)^2 / 4$$
$$= 13.85 \, ft^2$$

The afterburner velocity, v, in ft/s is then

$$v = q_T / S$$
$$= (32{,}110)/(13.85)$$
$$= 2318 \, ft/min$$
$$= 38.6 \, ft/s$$

Thus, the afterburner velocity is within the agency criterion.
    The residence time t is then

$$t = L/v$$
$$= 14/38.6$$
$$= 0.363 \, s$$

This also meets the agency criterion.

CMB.10   THERMAL VS. CATALYTIC COMBUSTION

Provide a comparative analysis—in terms of advantages and disadvantages—between thermal and catalytic combustion devices.

Solution

Some of the advantages of catalytic combustion reactors over thermal reactors are:

1. Lower fuel requirements
2. Lower operating temperatures
3. Little or no insulation requirements
4. Reduced fire hazards
5. Reduced flashback problems

The disadvantages include:

1. Higher initial cost
2. Catalyst poisoning
3. (Large) particles must first be removed
4. Some liquid droplets must first be removed
5. Catalyst regeneration problems

CMB.11   CATALYSIS IN AIR POLLUTION CONTROL

Discuss the subject of catalysis as it applies to air pollution control equipment.

Solution

A *catalyst* is a substance that speeds a chemical reaction without undergoing a change itself. In catalytic incineration, a waste gas is passed through a layer of catalyst, referred to as *the catalyst bed*. The catalyst causes the oxidation reaction to proceed at a faster rate and at lower temperatures than in thermal oxidation. A catalytic oxidizer (catalytic incinerator) operating in a 370–480°C (700–900°F) range can achieve the same efficiency as a thermal incinerator operating at between 700 and 820°C (1300 and 1500°F). This can result in a 40–60% fuel savings, which substantially reduces operating costs. It should be noted that catalytic oxidation produces the same end products (usually CO_2 and H_2O) and gives off the same heat of combustion as does thermal incineration.

The most effective and commonly used catalysts for oxidation reactions come from the noble metals group. Platinum, either alone or in combination with other noble metals, is by far the most commonly used. Platinum is desirable because it gives a high oxidation activity level at low temperatures, is stable at high temperatures, and is chemically inert. Palladium is another noble metal that exhibits these properties and is sometimes used in catalytic incinerators.

Since catalytic oxidation is a surface reaction, an inexpensive support material is normally coated with the noble metal. The support material can be made of ceramic, such as alumina, silica–alumina, or of a metal, such as nickel–chromium. The support material is arranged in a matrix shape to provide high surface area, low pressure drop, uniform flow of the waste gas through the catalyst bed, and a structurally stable surface. Structures that provide these characteristics are pellets, honeycomb matrices, or mesh matrices.

The main problem in catalytic incineration is deactivation—the reduction in the effectiveness of the catalyst. Certain contaminants, if present in the waste gas stream, cause a loss of catalyst activity.

Particulate matter in the waste gas stream will coat the surface of the catalyst, reducing its effectiveness. Certain metals (such as phosphorous, bismuth, arsenic, antimony, mercury, zinc, lead, and tin) can chemically combine with the catalyst, thereby deactivating the catalyst. Deactivation of the catalyst in this manner is referred to as *catalyst poisoning*. Sulfur and halogen compounds can also cause a reduction in catalyst effectiveness. However, catalysts absorbing these compounds can be regenerated. Finally, all catalysts deteriorate with normal use. A catalyst bed normally lasts from 3 to 5 years before it must be replaced.

CMB.12 PLAN REVIEW OF A CATALYTIC AFTERBURNER

Plans have been submitted for a catalytic afterburner. The installed afterburner is to incinerate a 3000-acfm contaminated gas stream discharged from a direct-fired paint baking oven at 350°F. The following summarizes the data taken from the plans:

Data Sheet

Exhaust flowrate from oven: 3000 acfm

Exhaust gas temperature from oven: 350°F

Solvent emission to afterburner: 0.3 lb/min

Final temperature in afterburner: 1000°F

Gross heating value of natural gas: 1100 Btu/scf

Total heat requirement: 26884 Btu/min

Natural gas requirement: 35.0 scfm

Furnace volume: 46.0 ft^3

Exhaust flow rate from afterburner at 1000°F: 6350 cfm

Gas velocity through catalytic bed: 8.6 ft/s

Number of type A 19 × 24 × 3.75 in. catalyst elements: 4

The following additional information and rules of thumb may be required to review the plans for the catalytic afterburner:

1. Heat will be recovered from the afterburner effluent, but that process will not be considered in this problem.
2. Catalytic afterburner operating temperatures of approximately 950°F have been found sufficient to control emissions from most process ovens.
3. Preheat burners are usually designed to increase the temperature of the contaminated gases to the required catalyst discharge gas temperature without regard to the heating value of the contaminants (especially if considerable concentration variation occurs).
4. A 10% heat loss is usually a reasonable estimate for an afterburner. This may be accounted for by dividing the calculated heat load by 0.9.
5. The properties of the contaminated effluent may usually be considered identical to those of air.

6. The natural gas is combusted using near stoichiometric (0% excess) external air.

7. The catalyst manufacturer's literature suggests a superficial gas velocity through the face surface of the catalyst element (in this case 19×24 in.) of $10\,\text{ft/s}$.

Three key questions are to be considered:

1. Is the operating temperature adequate for efficiency control?
2. Is the fuel requirement adequate to maintain the operating temperature?
3. Is the catalyst section properly sized?

Solution

The plans indicate a combustion temperature of $1000°F$; this is acceptable when compared to a $950°F$ rule-of-thumb temperature.

To determine if the fuel requirement is adequate to maintain the operating temperature, first calculate the lbmol/min of gas to be heated from 350 to $1000°F$.

$$(3000\,\text{acfm})[(460 + 60)/(460 + 350)] = 1926\,\text{scfm}$$

$$1926/379 = 5.08\,\text{lbmol/min}$$

Also determine the heat requirement in Btu/min to raise the gas stream (air) temperature from 350 to $1000°F$. See table in Problem CMB.7.

H at $350°F = 2222\,\text{Btu/lbmol}$
H at $1000°F = 6984\,\text{Btu/lbmol}$

$$\dot{Q} = \dot{n}\,\Delta H$$
$$= (5.08)(6984 - 2222)$$
$$= 24{,}196\,\text{Btu/min}$$

The total heat requirement, $\dot{Q}_T$, in Btu/min is then

$$\dot{Q}_T = \dot{Q}/0.9$$
$$= 24{,}196/0.9$$
$$= 26{,}884\,\text{Btu/min}$$

The available heat of the natural gas in Btu/scf at 1000°F may be calculated from the following equation:

$$HA_{1000}/HV_G = (HA_{1000}/HV_G)_{ref\ fuel}$$

$$HA_{1000} = 1100(745/1059)$$

$$= 774\ Btu/scf\ natural\ gas$$

The natural gas (NG) requirement in scfm may be calculated from the last two results:

$$Natural\ gas\ requirement = \frac{Total\ heat\ requirement}{Available\ heat\ in\ fuel}$$

$$q_{NG} = \dot{Q}_T/HA$$

$$= 26,884/774$$

$$= 34.7\ scfm$$

Thus, the natural gas requirement agrees with the given value.

To determine if the catalyst section is sized properly, first calculate the volume of flue products at 1000°F from the combustion of the natural gas:

$$q_c = (11.45)(35)[(460 + 1000)/(460 + 60)]$$

$$= 1125\ acfm$$

Also calculate the volume of contaminated gases at 1000°F:

$$q = (3000)[(460 + 1000)/(460 + 350)]$$

$$= 5407\ acfm$$

The total volumetric gas rate at 1000°F is then

$$q_T = q_c + q$$

$$= 1125 + 5407$$

$$= 6532\ acfm$$

Calculate the number of 19 × 24 × 3.75 in. catalyst elements, N, required:

$$N = (6532)(144)/[(19)(24)(600)]$$

$$= 3.44$$

The catalyst section is sized properly. It is seen that four elements have been specified; this is a conservative design allowing a slightly slower gas flow through the elements. With four elements, the superficial velocity is reduced to 8.6 ft/s. This too is in agreement with the design specification.

20 Absorption (ABS)

ABS.1   DESCRIPTION OF ABSORPTION

Briefly describe the absorption process.

Solution

Gas absorption, as applied to the control of air pollution, is concerned with the removal of one or more pollutants from a contaminated gas stream by treatment with a liquid. The necessary condition is the solubility of these pollutants in the absorbing liquid. The rate of transfer of the soluble constituents from the gas to the liquid phase is determined by diffusional processes occurring on each side of the gas–liquid interface.

Consider, for example, the process taking place when a mixture of air and sulfur dioxide is brought into contact with water. The SO_2 is soluble in water, and those molecules that come into contact with the water surface dissolve fairly rapidly. However, the SO_2 molecules are initially dispersed throughout the gas phase, and they can only reach the water surface by diffusing through the air, which is substantially insoluble in the water. When the SO_2 at the water surface has dissolved, it is distributed throughout the water phase by a second diffusional process. Consequently, the rate of absorption is determined by the rates of diffusion in both the gas and liquid phases.

Equilibrium is another extremely important factor to be considered in controlling the operation of absorption systems. The rate at which the pollutant will diffuse into an absorbent liquid will depend on the departure from equilibrium that is maintained. The rate at which equilibrium is established is then essentially dependent on the rate of diffusion of the pollutant through the nonabsorbed gas and through the absorbing liquid. Equilibrium concepts and relationships are considered in a later problem.

The rate at which the pollutant mass is transferred from one phase to another depends also on a so-called mass transfer, or rate, coefficient, which equates the quantity of mass being transferred with the driving force. As can be expected, this transfer process would cease upon the attainment of equilibrium.

Gas absorption can be viewed as a mass transfer, or diffusional operation, characterized by a transfer of one substance through another, usually on a molecular scale. The mass transfer process may be considered the result of a concentration difference driving force, the diffusing substance moving from a place of relatively high to one of relatively low concentration. The rate at which this mass is transferred depends to a great extent on the diffusional characteristics of both the diffusing substance and the medium.

The principal types of gas absorption equipment may be classified as follows:

1. Packed columns (continuous operation)
2. Plate columns (staged operation)
3. Miscellaneous

Of the three categories, the packed column is by far the most commonly used for the absorption of gaseous pollutants. It might also be mentioned at this time that the exhaust (cleaned gas) from an absorption air pollution control system is usually released to the atmosphere through a stack. To prevent condensation in and around the stack, the temperature of this exhaust gas should be above its dew point. A general rule of thumb is to ensure that the exhaust gas stream temperature is approximately 50°F above its dew point.

ABS.2 SOLVENT SELECTION

List 10 factors that should be considered when choosing a solvent for a gas absorption column that is to be used as an emission control device.

Solution

Solvent selection for use in an absorption column for gaseous pollutant removal should be based upon the following criteria:

1. *Solubility of the gas in the solvent:* High solubility is desirable as it reduces the amount of solvent needed. Generally, a polar gas will dissolve best in a polar solvent and a nonpolar gas will dissolve best in a nonpolar solvent.
2. *Vapor pressure:* A solvent with a low vapor pressure is preferred to minimize loss of solvent.
3. *Corrosivity:* Corrosive solvents may damage the equipment. A solvent with low corrosivity will extend equipment life.
4. *Cost:* In general, the less expensive, the better. However, an inexpensive solvent is not always the best choice if it is too costly to dispose of and/or recycle after it has been used.
5. *Viscosity:* Solvents with low viscosity offer benefits such as better adsorption rates, better heat transfer properties, lower pressure drops, lower pumping costs, and improved flooding characteristics in absorption towers.
6. *Reactivity:* Solvents that react with the contaminant gas to produce an unreactive product are desirable since the scrubbing solution can be recircu-

lated while maintaining high removal efficiencies. Reactive solvents should produce few unwanted side reactions with the gases that are to be absorbed.

7. *Low freezing point:* Resistance to freezing lessens the chance of solid formation and clogging of the column. See 2 on boiling point (vapor pressure).

8. *Availability:* If the solvent is "exotic," it generally has a higher cost and may not be readily available for long-term continuous use. Water is often the natural choice based on this criteria.

9. *Flammability:* Lower flammability or nonflammable solvents decrease safety problems.

10. *Toxicity:* A solvent with low toxicity is desirable.

ABS.3   OUTLET CONCENTRATION FROM A SPRAY TOWER

A waste incinerator emits 300 ppm HCl with peak values of 600 ppm. The air flow is a constant 5000 acfm at 75°F and 1 atm. Only sketchy information was submitted with the scrubber permit application to the state for a spray tower. You are requested to determine if the spray unit is satisfactory.

Data

Emission limit $= 30$ ppm HCl
Maximum gas velocity allowed through the tower $= 3$ ft/s
Number of sprays $= 6$
Diameter of the tower $= 10$ ft

The number of transfer units, N_{OG}, to meet the regulations, assuming the worst scenario, is

$$N_{OG} = \ln(y_1/y_2)$$

where $y_1 =$ inlet mole fraction
$y_2 =$ outlet mole fraction

As a rule of thumb in a spray tower, the N_{OG} of the first or top spray is 0.7. Each lower spray will have only about 60% of the N_{OG} of the spray above it. The absorption that occurs in the inlet duct adds no height but has an N_{OG} of 0.5.

Solution

First, check the gas velocity:

$$v = \frac{q}{A} = \frac{q}{\pi D^2/4}$$

$$= \frac{5000/60}{\pi (10)^2/4}$$

$$= 1.06 \text{ ft/s} < \text{maximum of 3 ft/s}$$

The number of transfer units required is

$$N_{OG} = \ln(600/30)$$
$$= 3.00$$

For a tower with five spray sections,

Spray Section	N_{OG}
Top	0.70
Second	0.42
Third	0.25
Fourth	0.15
Fifth	0.09
Sixth (inlet)	0.50
	$\Sigma = 2.11$

Therefore,

$$2.11 = \ln(600/y_2)$$
$$y_2 = 72.45 \text{ ppm} > 30 \text{ ppm HCl}$$

Therefore, the outlet concentration does not meet the specification!

ABS.4   ABSORPTION PRINCIPLES

You are given experimental data for an absorption system to be used for scrubbing ammonia (NH_3) from air with water. The water rate is 300 lb/min and the gas rate is 250 lb/min at 72°F. The applicable equilibrium data for the ammonia–air system is shown in the following table (R. H. Perry and D. W. Green, Ed., *Perry's Chemical Engineers' Handbook*, 7th Edition, McGraw Hill, New York, 1996.)

Equilibrium Partial Pressure (mm Hg)	NH_3 Concentration (lb NH_3/100 lbH_2O)
3.4	0.5
7.4	1.0
9.1	1.2
12.0	1.6
15.3	2.0
19.4	2.5
23.5	3.0

The air to be scrubbed has 1.5% (weight basis) NH_3 at 72°F and 1 atm pressure and is to be vented with 95% of the ammonia removed. The inlet scrubber water is ammonia free.

1. Plot the equilibrium data in mole fraction units.
2. Perform the material balance and plot the operating line on the equilibrium plot.

Solution

In gas absorption operations the equilibrium of interest is that between a relatively nonvolatile absorbing liquid (solvent) and a solute gas (usually the pollutant). As described earlier, the solute is ordinarily removed from a relatively large amount of a carrier gas that does not dissolve in the absorbing liquid. Temperature, pressure, and the concentration of solute in one phase are independently variable. The equilibrium relationship of importance is a plot (or data) of x, the mole fraction of solute in the liquid, against y^*, the mole fraction in the vapor in equilibrium with x. For cases that follow Henry's law, Henry's law constant m, can be defined by the equation

$$y^* = mx$$

The usual operating data to be determined or estimated for isothermal systems are the liquid rate(s) and the terminal concentrations or mole fractions. An operating line that describes operating conditions in the column is obtained by a mass balance around the column (as shown in Figure 72).

$$\text{Total moles in} = \text{Total moles out}$$

$$G_{m1} + L_{m2} = G_{m2} + L_{m1}$$

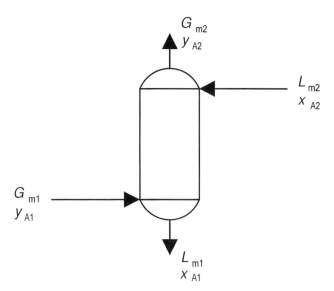

Figure 72. Material balance for the absorption of component A in an absorption column

For component A, the mass (or mole) balance becomes

$$G_{m1} y_{A1} + L_{m2} x_{A2} = G_{m2} y_{A2} + L_{m1} x_{A1}$$

Assuming $G_{m1} = G_{m2}$ and $L_{m1} = L_{m2}$ (reasonable for most air pollution control applications where contaminant concentrations are usually extremely small), then

$$G_m y_{A1} + L_m x_{A2} = G_m y_{A2} + L_m x_{A1}$$

and rearranging

$$L_m/G_m = (y_{A1} - y_{A2})/(x_{A1} - x_{A2})$$

This is the equation of a straight line known as the *operating line*. On x, y coordinates, it has a slope of L_m/G_m and passes through the points (x_{A1}, y_{A1}) and (x_{A2}, y_{A2}) as indicated in Figure 73.

In the design of most absorption columns, the quantity of gas to be treated G_m, the terminal concentrations y_{A1} and y_{A2}, and the composition of the entering liquid x_{A2} are ordinarily fixed by process requirements; however, the quantity of liquid solvent to be used is subject to some choice. Should this quantity already be specified, the operating line in the figure is fixed. If the quantity of solvent is unknown, the operating line is consequently unknown. This can be obtained through setting the minimum liquid-to-gas ratio.

With reference to Figure 73, the operating line must pass through point A and must terminate at the ordinate y_{A1}. If such a quantity of liquid is used to determine operating line AB, the existing liquid will have the composition x_{A1}. If less liquid is

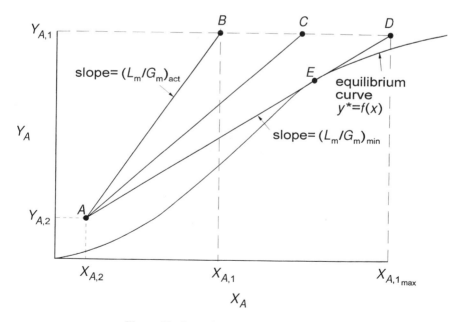

Figure 73. Operating and equilibrium lines.

used, the exit liquid composition will clearly be greater, as at point C, but since the driving forces (displacement of the operating line from the equilibrium line) for mass transfer are less, the absorption is more difficult. The time of contact between gas and liquid must then be greater, and the absorber must be correspondingly taller.

The minimum liquid used corresponds to the operating line AD, which has the greatest slope for any line touching the equilibrium curve and is tangent to the curve at E. At point E, the diffusional driving force is zero, the required contact time for the concentration change desired is infinite, and an infinitely tall column results. This then represents the limiting liquid-to-gas ratio.

The importance of the minimum liquid-to-gas ratio lies in the fact that column operation is frequently specified as some factor of the minimum liquid–gas ratio. For example, a typical situation frequently encountered is that the slope of the actual operating line, $(L_m/G_m)_{act}$, is 1.5 times the minimum, $(L_m/G_m)_{min}$.

Employing the data provided in the problem statement, convert the equilibrium partial pressure data and the liquid concentration data to mole fractions:

Gas Mole Fraction y	Liquid Mole Fraction x
0.00447	0.0053
0.00973	0.0106
0.0120	0.0127
0.0158	0.0169
0.0201	0.0212
0.0255	0.0265
0.0309	0.0318

Plotting the mole fraction values on the graph in Figure 74 results in a straight line. The slope of the equilibrium line is approximately 1.0.

Convert the liquid and gas rates to lbmol/min:

$$L = 300/18$$

$$= 16.67 \, \text{lbmol/min}$$

$$G = 250/29$$

$$= 8.62 \, \text{lbmol/min}$$

Determine the inlet and outlet mole fractions for the gas, y_1 and y_2, respectively:

$$y_1 = \frac{1.5/17}{1.5/17 + 98.5/29}$$

$$= 0.0253$$

$$y_2 = \frac{(0.05)(1.5/17)}{(0.05)(1.5/17) + (98.5/29)}$$

$$= 0.0013$$

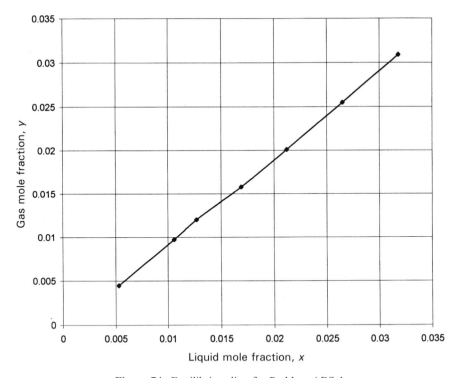

Figure 74. Equilibrium line for Problem ABS.4.

The inlet liquid mole fraction, x_2, is given as 0, and the describing equation for x_1, the outlet liquid mole fraction, is

$$x_1 = (G/L)(y_1 - y_2) + x_2$$

$$= (8.62/16.67)(0.0253 - 0.0013) + 0$$

$$= 0.0124$$

One may now use the inlet and outlet mole fractions to plot the operating line on the graph in Figure 75.

The slope of the operating line is

$$\text{Slope} = \frac{0.0253 - 0.0013}{0.0124 - 0.0}$$

$$= 1.936$$

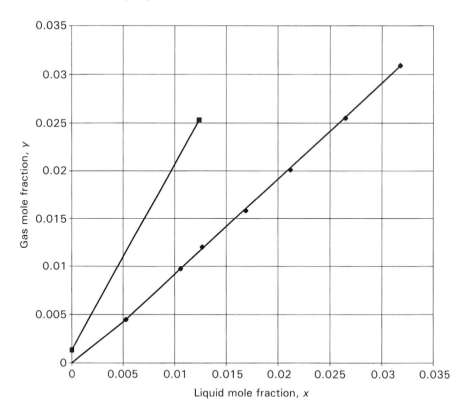

Figure 75. Operating line for Problem ABS.4.

ABS.5 PACKING HEIGHT

Pollution Unlimited, an Aldo Leone Corporation, has submitted design plans to Theodore Consultants for a packed ammonia scrubber on an airstream containing NH_3. The operating and design data provided by Pollution Unlimited, Inc. are given below. Theodore Consultants remember reviewing plans for a nearly identical scrubber for Pollution Unlimited, Inc. in 1988. After consulting old files, the consultants find all the conditions were identical except for the gas flowrate. What recommendation should be made?

Tower diameter $= 3.57$ ft
Packed height of column $= 8$ ft
Gas and liquid temperature $= 75°$F inlet
Operating pressure $= 1.0$ atm
Ammonia-free liquid flowrate (mass flux or mass velocity) $= 1000$ lb/(ft$^2 \cdot$ h)
Gas flowrate $= 1575$ acfm
Gas flowrate in 1988 plan $= 1121$ acfm

Inlet NH_3 gas concentration $= 2.0$ mol %

Air density $= 0.0743$ lb/ft^3

Molecular weight of air $= 29$

Henry's law constant, $m = 0.972$

Molecular weight of water $= 18$

Figure 76 (packing "A" is used)

Colburn chart

Emission regulation $= 0.1\%$ NH_3 (by mole or volume)

Packing height $= H_{OG}N_{OG}$

Solution

Calculate the cross-sectional area of the tower, S, in ft^2:

$$S = \pi D^2 / 4$$

$$= (\pi)(3.57)^2 / (4)$$

$$= 10.0 \, \text{ft}^2$$

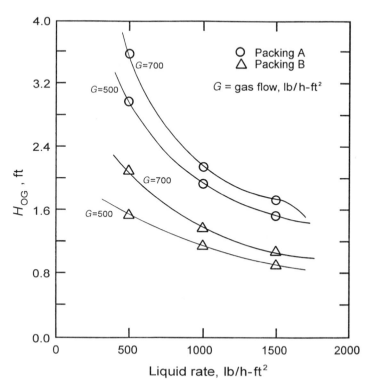

Figure 76. H_{OG} vs. liquid rate for ammonia–water absorption system.

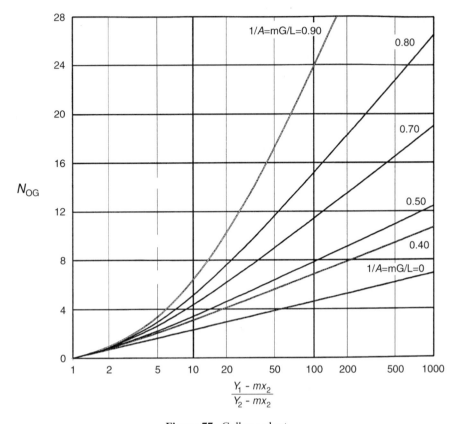

Figure 77. Colburn chart.

Calculate the gas molar flux (molar flowrate per unit cross section) and liquid molar flux in lbmol/(ft² · h):

$$G_m = q_\rho / S(MW)_G$$

$$= (1575)(0.0743)/[(10.0)(29)]$$

$$= 0.404 \text{ lbmol}/(\text{ft}^2 \cdot \text{min})$$

$$= 24.2 \text{ lbmol}/(\text{ft}^2 \cdot \text{h})$$

$$L_m = L/(MW)_L$$

$$= (1000)/(18)$$

$$= 55.6 \text{ lbmol}/(\text{ft}^2 \cdot \text{h})$$

The value of mG_m/L_m is therefore

$$mG_m/L_m = (0.972)(24.2/55.6)$$
$$= 0.423$$

The absorption factor, A, is defined as

$$A = L_m/(mG_m) = 1/0.423 = 2.364$$

The value of $(y_1 - mx_2)/(y_2 - mx_2)$ or X is

$$\frac{y_1 - mx_2}{y_2 - mx_2} = \frac{0.02 - (0.972)(0)}{0.001 - (0.972)(0)}$$
$$= 20.0$$

N_{OG} is calculated from Colburn's equation (or read from Colburn's chart, Figure 77),

$$N_{OG} = \frac{\ln((y_1 - mx_2)/(y_2 - mx_2) \,(1 - \{1/A\}) + \{1/A\})}{1 - \{1/A\}}$$
$$= \frac{\ln(20.0)(1 - \{1/2.364\}) + \{1/2.364\}}{1 - \{1/2.364\}}$$
$$= 4.30$$

To calculate the height of an overall gas transfer unit, H_{OG}, first calculate the gas mass velocity, G, in lb/ft$^2 \cdot$ h.

$$G = q_\rho/S$$
$$= (1575)(0.0743)/10.0$$
$$= 11.7\,\text{lb}/(\text{ft}^2 \cdot \text{min})$$
$$= 702\,\text{lb}/(\text{ft}^2 \cdot \text{h})$$

From the figure,

$$H_{OG} = 2.2\,\text{ft}$$

The required packed column height, Z, in feet is

$$Z = N_{OG}H_{OG}$$
$$= (4.3)(2.2)$$
$$= 9.46\,\text{ft}$$

The application should be rejected.

ABS.6   TOWER HEIGHT AND DIAMETER

A packed column is used to absorb a toxic pollutant from a gas stream. From the data given below, calculate the height of packing and column diameter. The unit operates at 50% of the flooding gas mass velocity, the actual liquid flowrate is 40% more than the minimum, and 95% of the pollutant is to be collected. Employ the generalized correlation provided in Figure 78 to estimate the column diameter.

Gas mass flowrate $= 3500\,\text{lb/h}$

Pollutant concentration in inlet gas stream $= 1.1\,\text{mol}\,\%$

Scrubbing liquid $=$ pure water

Packing type $= 1$ in. Raschig rings; packing factor, $F = 160$

H_{OG} of the column $= 2.5\,\text{ft}$

Henry's law constant, $m = 0.98$

Density of gas (air) $= 0.075\,\text{lb/ft}^3$

Density of water $= 62.4\,\text{lb/ft}^3$

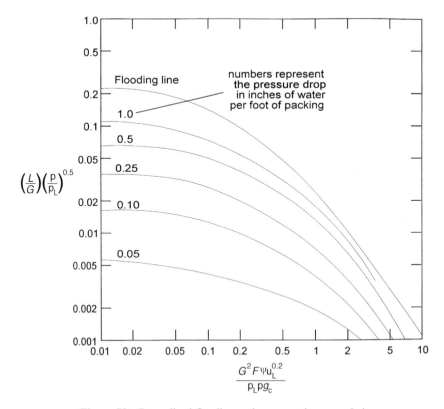

Figure 78. Generalized flooding and pressure drop correlation.

Viscosity of water $= 1.8\,\text{cP}$

The ordinate and abscissa of the graph in Figure 78 are dimensionless numbers

$$\left(\frac{L}{G}\right)\left(\frac{\rho}{\rho_L}\right)^{0.5}$$

$$\frac{G^2 F \psi \mu_L^{0.2}}{\rho_L \rho g_c}$$

where:

G = mass flux (mass flowrate per unit cross-sectional area) of gas stream
L = mass flux of liquid stream
F = packing factor
ψ = ratio of the specific gravity of the scrubbing liquid to that of water
μ_L = viscosity of liquid phase
ρ_L = density of liquid phase
ρ = density of gas phase
g_c = Newton's law proportionality factor

Solution

To calculate the number of overall gas transfer units, N_{OG}, first calculate the equilibrium outlet concentration, x_1^*, at $y_1 = 0.011$:

$$x_1^* = y_1/m$$

$$= 0.011/0.98$$

$$= 0.0112$$

Determine y_2 for 95% removal:

$$y_2 = \frac{(0.05)\,y_1}{(1 - y_1) + (0.05)\,y_1}$$

$$= \frac{(0.05)(0.011)}{(1 - 0.011) + (0.05)(0.011)}$$

$$= 5.56 \times 10^{-4}$$

The minimum ratio of molar liquid flowrate to molar gas flowrate, $(L_m/G_m)_{min}$, is determined by a material balance:

$$(L_m/G_m)_{min} = (y_1 - y_2)/(x_1^* - x_2)$$

$$= (0.011 - 5.56 \times 10^{-4})/(0.0112 - 0)$$

$$= 0.933$$

The actual ratio of molar liquid flow rate to molar gas flowrate L_m/G_m is

$$L_m/G_m = (1.40)(L_m/G_m)_{min}$$
$$= (1.40)(0.933)$$
$$= 1.306$$

In addition,

$$(mG_m)/L_m = (0.98)/(1.306) = 0.7504$$

The absorption factor, A, is defined as

$$A = L_m/(mG_m) = 1/0.7504 = 1.333$$

The value of $(y_1 - mx_2)/(y_2 - mx_2)$ is

$$\frac{y_1 - mx_2}{y_2 - mx_2} = \frac{0.011 - (0.98)(0)}{5.56 \times 10^{-4} - (0.98)(0)} = 19.78$$

Then N_{OG} is calculated from Colburn's equation (or read from Colburn's chart—Figure 77),

$$N_{OG} = \frac{\ln((y_1 - mx_2)/(y_2 - mx_2)\,(1 - \{1/A\}) + \{1/A\})}{1 - \{1/A\}}$$
$$= \frac{\ln[(19.78)(1 - \{1/1.333\}) + \{1/1.333\}]}{1 - \{1/1.333\}}$$
$$= 6.96$$

The height of packing, Z, is then

$$Z = N_{OG}H_{OG}$$
$$= (6.96)(2.5)$$
$$= 17.4 \text{ ft}$$

To determine the diameter of the packed column, the ordinate of Figure 78, is first calculated:

$$\left(\frac{L}{G}\right)\left(\frac{\rho}{\rho_L}\right)^{0.5} = \left(\frac{L_m}{G_m}\right)\left(\frac{18}{29}\right)\left(\frac{\rho}{\rho_L}\right)^{0.5} = (1.306)\left(\frac{18}{29}\right)\left(\frac{0.075}{62.4}\right)^{0.5}$$
$$= 0.0281$$

The value of the abcissa at the flooding line is determined from Figure 78:

$$\frac{G^2 F \psi \mu_L^{0.2}}{\rho \rho_L g_c} = 0.21$$

The flooding gas mass velocity, G_f, in $\mathrm{lb}/(\mathrm{ft}^2 \cdot \mathrm{s})$, is

$$G_f = \left(\frac{0.21 \, \rho_L \rho g_c}{F \psi \mu_L^{0.2}}\right)^{1/2} = \left(\frac{(0.21)(62.4)(0.075)(32.2)}{(160)(1)(1.8)^{0.2}}\right)^{1/2}$$

$$= 0.419 \, \mathrm{lb}/(\mathrm{ft}^2 \cdot \mathrm{s})$$

The actual gas mass velocity, G_{act}, in $\mathrm{lb}/(\mathrm{ft}^2 \cdot \mathrm{s})$, is

$$G_{act} = (0.5)(0.419)$$

$$= 0.2095 \, \mathrm{lb}/(\mathrm{ft}^2 \cdot \mathrm{s})$$

$$= 754 \, \mathrm{lb}/(\mathrm{ft}^2 \cdot \mathrm{h})$$

Calculate the diameter of the column in feet:

$$D = [(4m)/(G_{act}\pi)]^{1/2}$$

$$= [(4)(3500)/(754)(\pi)]^{1/2}$$

$$= 2.43 \, \mathrm{ft}$$

The column height (packing) and diameter are 17.4 and 2.43 ft, respectively.

ABS.7 PACKED TOWER ABSORBER DESIGN WITH NO DATA

A 1600-acfm gas stream is to be treated in a packed tower containing ceramic packing. The gas stream contains 100 ppm of a toxic pollutant that is to be reduced to 1 ppm. Estimate the tower's cross-sectional area, diameter, height, pressure drop, and packing size. Use the procedure outlined in Problem MTO. 4 in Chapter 8.

Solution

Key information from Problem MTO.4 is provided below. For ceramic packing:

Packing Height, Z (ft)

Removal Efficiency (%)	Ceramic Packing size (in.)				
	1.0	1.5	2	3	3.5
63.2	2.0	2.5	3.0	4.5	5.5
77.7	3.0	3.7	4.5	6.75	8.25
86.5	4.0	5.0	6.0	9.0	11.0
90	4.6	5.75	6.9	10.4	12.7
95	6.0	7.5	9.0	13.5	16.5
98	7.8	9.8	11.7	17.6	21.5
99	9.2	11.5	13.8	20.7	25.3
99.5	10.6	13.25	15.9	23.8	29.1
99.9	13.8	17.25	20.7	31.1	38.0
99.99	18.4	23.0	27.6	41.4	50.7

The equation for the cross-sectional area of the tower, S, in terms of the gas volumetric flowrate, q, in acfs is

$$S(\text{ft}^2) = q(\text{acfs})/4$$

An equation to estimate the tower packing pressure drop, ΔP, in terms of Z is

$$\Delta P \text{ (in. H}_2\text{O)} = (0.2) Z \qquad Z = \text{ft}$$

The following packing size(s) is (are) recommended:

For $D \approx 3$ ft, use 1-in. packing
For $D < 3$ ft, use < 1-in. packing
For $D > 3$ ft, use > 1-in. packing

As a rule, recommended packing size increases with lower diameter.
    For the problem at hand,

$$S = 1600/4 = 400 \text{ ft}^2$$

The diameter D is

$$D = (4S/\pi)^{0.5}$$
$$= [(4)(400)/\pi]^{0.5}$$
$$= 22.6 \text{ ft}$$

For a tower this large, the 3.5-in. packing should be used.

The removal efficiency (RE) is

$$RE = (100 - 1)/100 = 0.99 = 99\%$$

For 99% RE and a packing size of 3.5 in., the required height is 25.3 ft. The pressure drop is

$$\Delta P = (0.2)(25.3) = 5.06 \text{ in. } H_2O$$

ABS.8   TWO ABSORBERS TO REPLACE ONE

The calculations for an absorber indicate that it would be excessively tall, and the five schemes in the diagrams in Figure 79 are being considered as a means of using two shorter absorbers. Make freehand sketches of operating lines, one for each scheme showing the relation between operating lines for the two absorbers and the equilibrium curve. Mark the concentrations on the figure for each diagram. No calculations are required. Assume dilute solutions.

Solution

Operating lines for each scheme are provided in Figure 79.

1.

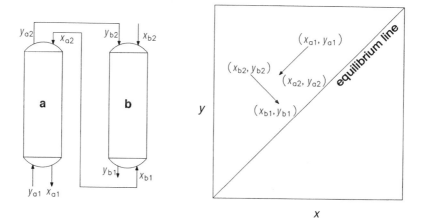

2.

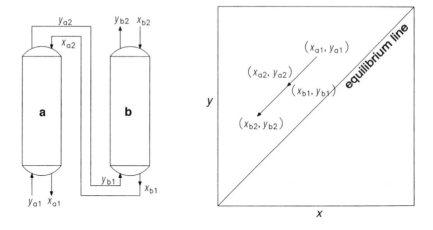

3.

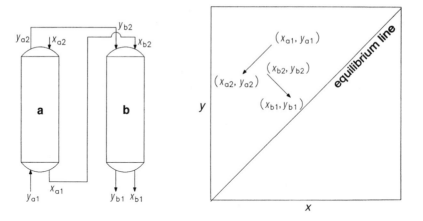

Figure 79. Answers to Problem ABS.8.

4.

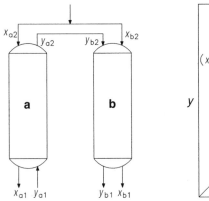 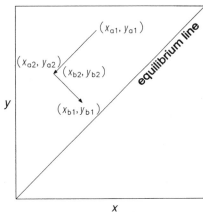

5.

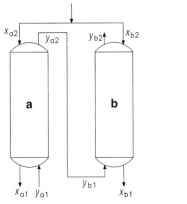 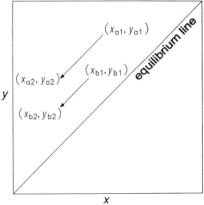

Figure 79. (*continued*) Answers to Problem ABS.8.

ABS.9   DESIGN PROCEDURE FOR AN ABSORPTION COLUMN

Provide a general design procedure for an absorption column.

Solution

The general design procedure consists of a number of steps that have to be taken into consideration. These include:

1. Solvent selection
2. Equilibrium data evaluation
3. Estimation of operating data (usually consisting of a mass and energy balance, where the energy balance decides whether the absorption process can be considered as isothermal or adiabatic)
4. Column selection (should the column selection not be obvious or specified, calculations must be carried out for the different types of columns and the final selection based on economic considerations)
5. Calculation of column diameter (for packed columns, this is usually based on flooding conditions, and for plate columns on the optimum gas velocity or the liquid-handling capacity of the plate)
6. Estimation of column height or the number of plates (for packed columns, the column height is obtained by multiplying the number of transfer units, obtained from a knowledge of equilibrium and operating data, by the height of a transfer unit; for plate columns, the number of theoretical plates is determined from the plot of equilibrium and operating lines. This number is then divided by the estimated overall plate efficiency to give the number of actual plates, which in turn allows the column height to be estimated from the plate spacing)
7. Determination of pressure drop through the column (for packed columns, correlations dependent on packing type, column operating data, and physical properties of the constituents involved are available to estimate the pressure drop through the packing; for plate columns, the pressure drop per plate is obtained and multiplied by the number of plates)

ABS.10   PACKED VS. PLATE COLUMN

Briefly describe the differences between a packed column and a plate column.

Solution

Of the various types of gas absorption devices mentioned, packed columns and plate columns are the most commonly used in industrial practice. Although packed columns are used more often in air pollution control, both have their special area of usefulness, and the relative advantages and disadvantages of each are worth considering. In general:

1. The pressure drop of the gas passing through the packed column is smaller.
2. The plate column can treat an arbitrarily low liquid feed and permits a higher gas feed than the packed column. It can also be designed to handle liquid rates that would ordinarily flood the packed column.
3. If the liquid deposits a sediment, the plate column is more advisable. By fitting the column with manholes, the plate column can be cleaned of

accumulated sediment that would clog many packing materials and warrant necessary costly removal and refilling of the column. Packed columns are also susceptible to plugging if the gas contains particulate contaminants.

4. In mass transfer processes accompanied by considerable heat effects, cooling or heating the liquid is accomplished more easily in the plate column. A system of pipes immersed in the liquid can be placed on the plates between the caps, and heat can be removed or supplied through the pipe wall directly to the area in which the process is taking place. The solution of the same problem for a packed column leads to the division of this process into a number of sections, the cooling or heating of the liquid taking place between these sections.

5. The total weight of the plate column is usually less than for the packed column designed for the same capacity.

6. A well-installed plate column avoids serious channeling difficulties ensuring good, continuous contact between the gas and liquid throughout the column.

7. In highly corrosive atmospheres, the packed column is simpler and cheaper to construct.

8. The liquid holdup in the packed column is considerably less than in the plate column.

9. Temperature changes are apt to do more damage to a packed column than to a plate column.

10. Plate columns are advantageous for absorption processes with an accompanying chemical reaction (particularly when it is not very rapid). The process is favored by a long residence time of the liquid in the column and by easier control of the reaction.

11. Packed columns are preferred for liquids with high foaming tendencies.

12. The relative merits of the plate column and packed column for a specified purpose are properly determined only by comparison of the actual cost figures resulting from a detailed design analysis for each type. Most conditions being equal, packed columns in the smaller sizes (diameters up to approximately 2–3 ft) are on the average less expensive. In the larger sizes, plate columns tend to be the more economical.

21 Adsorption (ADS)

ADS.1 DESCRIPTION OF ADSORPTION

Describe the adsorption process.

Solution

Adsorption is a mass transfer process in which gas molecules are removed from an airstream because they adhere to the surface of a solid. In an adsorption system, the contaminated airstream is passed through a layer of solid particles referred to as the *adsorbent bed*. As the contaminated airstream passes through the adsorbent bed, the pollutant molecules adsorb or "stick" to the surface of the solid adsorbent particles. Eventually the adsorbent bed becomes "filled" or saturated with the pollutant. The adsorbent bed must then be disposed of and replaced, or the pollutant gases/vapors must be desorbed before the adsorbent bed can be reused.

The process of adsorption is analogous to using a sponge to mop up water. Just as a sponge soaks up water, a porous solid (the adsorbent) is capable of capturing gaseous pollutant molecules. The airstream carrying the pollutants must first be brought into contact with the adsorbent. The pollutant molecules then diffuse into the pores of the adsorbent (internal surface) where they are adsorbed. The majority of the gas molecules are adsorbed on internal pore surfaces.

The relation between the amount of substance adsorbed by an adsorbent and the equilibrium gas (or vapor) partial pressure or concentration at constant temperature is called the *adsorption isotherm*. The adsorption isotherm is the most important and by far the most often used of the various equilibria data that are available.

ADS.2 TYPES OF ADSORBENTS

Briefly describe the four major adsorbents employed in air pollution control.

Solution

Four important adsorbents widely used industrially will be considered briefly, namely, activated carbon, activated alumina, silica gel, and molecular sieves. The first three of these are amorphous adsorbents with a nonuniform internal structure.

Molecular sieves, however, are crystalline and have, therefore, an internal structure of regularly spaced cavities with interconnecting pores of definite size. Details of the properties peculiar to the various materials are best obtained directly from the manufacturer. The following is a brief description of these principal adsorbents.

Activated Carbon   Charcoal, the raw material employed in producing activated carbon, is obtained by the carbonization of wood. Various raw materials have been used in the preparation of adsorbent chars, resulting in the development of active carbon, a much more adsorbent form of charcoal. Industrial manufacture of activated carbon is today largely based on nut shells or coal, which are subjected to heat treatment in the absence of air followed by steam activation at high temperatures. Other substances of a carbonaceous nature also used in the manufacture of active carbons include wood, coconut shells, peat, and fruit pits. Zinc chloride, magnesium chloride, calcium chloride, and phosphoric acid have also been used in place of steam as activating agents. Some approximate properties of typical granular adsorbent carbons include:

Bulk density: 22–34 lb/ft^3
Heat capacity: 0.27–0.36 Btu/(lb · °F)
Pore volume: 0.56–1.20 cm^3/g
Surface area: 600–1,600 m^2/g
Average pore diameter: 15–25 Å
Regeneration temperature (steaming): 100–140°C
Maximum allowable temperature: 150°C

Gas (or vapor) adsorbent carbons find primary application in solvent recovery (hydrocarbon vapor emissions), odor elimination, and gas purification.

Activated Alumina   Activated alumina (hydrated aluminum oxide) is produced by special heat treatment of precipitated or native aluminas or bauxite. It is available in either granule or pellet form with the following typical properties:

Density in bulk: granules: 38–42 lb/ft^3
                   pellets: 54–58 lb/ft^3
Specific heat: 0.21–0.25 Btu/(lb · °F)
Pore volume: 0.29–0.37 cm^3/g
Surface area: 210–360 m^2/g
Average pore diameter: 18–48 Å
Regeneration temperature: 200–250°C
Stable up to: 500°C

Activated alumina is mainly used for the drying of gases, and it is particularly useful for the drying of gases under pressure.

Silica Gel The manufacture of silica gel consists of the neutralization of sodium silicate by mixing with dilute mineral acid, washing the gel formed to remove salts produced during the neutralization reaction, followed by drying, roasting, and grading processes. The name "gel" arises from the jellylike form of the material during one stage of its production. It is generally used in granular form, although bead forms are available. The material has the following typical physical properties:

Bulk density: 44–46 lb/ft^3
Heat capacity: 0.22–0.26 Btu/(lb · °F)
Pore volume: 0.37 cm^3/g
Surface area: 750 m^2/g
Average pore diameter: 22 Å
Regeneration temperature: 120–250°C
Stable up to: 400°C

Silica gel also finds primary use in gas drying, although it also finds application in gas desulfurization and purification.

Molecular Sieves Unlike the amorphous adsorbents (activated carbon, activated alumina, and silica gel), molecular sieves are crystalline, being essentially dehydrated zeolites, i.e., aluminosilicates in which atoms are arranged in a definite pattern. The complex structural units of molecular sieves have cavities at their centers to which access is by pores or windows. For certain types of crystalline zeolites, these pores are precisely uniform in diameter. Due to the crystalline porous structure and precise uniformity of the small pores, adsorption phenomena only takes place with molecules that are of small enough size and of suitable shape to enter the cavities through the pores. The fundamental building block is a tetrahedron of four oxygen anions surrounding a smaller silicon or aluminum cation. The sodium ions or other cations serve to make up the positive charge deficit in the alumina tetrahedra. Each of the four oxygen anions is shared, in turn, with another silica or alumina tetrahedron to extend the crystal lattice in three dimensions. The resulting crystal is unusual in that it is honeycombed with relatively large cavities, each cavity connected with six adjacent ones through apertures or pores.

The very strong adsorptive forces in molecular sieves are due primarily to the cations that are exposed in the crystal lattice. These cations act as sites of strong localized positive charge that electrostatically attract the negative end of polar molecules. The greater the dipole moment of the molecule, the more strongly it will be attracted and adsorbed. Polar molecules are generally those that contain O (oxygen), S (sulfur), Cl (chlorine), or N (nitrogen) atoms and are asymmetrical. For example, molecular sieves will adsorb carbon monoxide in preference to argon. Under the influence of the localized, strong positive charge on the cations, molecules can have dipoles induced in them. The polarized molecules are then strongly adsorbed due to the electrostatic attraction of the cations. The more unsaturated the molecule, the more polarizable it is and the more strongly it is adsorbed. Thus, molecular sieves will effectively remove acetylene from olefins, and ethylene or propylene from saturated hydrocarbons.

The sieves have the following typical properties:

	Anhydrous Sodium Alumino silicate	Anhydrous Calcium Alumino silicate	Anhydrous Alumino silicate
Type	4A	5A	13X
Density in bulk (lb/ft^3)	44	44	38
Specific heat [Btu/(lb · °F)]	0.19	0.19	
Effective diameter of pores (Å)	4	5	13
Regeneration temperature	200–300°C	200–300°C	200–300°C
Stable up to (short period)	600°C	600°C	600°C

ADS.3   ACTIVATED CARBON

List and briefly discuss several design or process factors that affect activated carbon effectiveness in controlling gaseous pollutants.

Solution

Seven factors that affect activated carbon effectiveness are as follows:

1. *Temperature:* Higher temperatures will result in less adsorption of the gas.
2. *Pressure:* The higher the gas pressure, the more adsorbate will be adsorbed.
3. *Molecular weight of the gas:* The higher the molecular weight of the gas, the greater the retention of the gas by the adsorbent.
4. *Presence of other gases:* If gases are present that can be adsorbed but for which adsorption is not desired, then the capacity of the adsorbent to remove the desired gases will be decreased. Water and carbon dioxide are often problems in this regard.
5. *Decomposition and polymerization:* Activated carbon at elevated temperatures may act as a catalyst for the polymerization and/or decomposition of some organic compounds. The products of these polymerizations and/or decomposition reactions could interfere with the adsorption of the desired gases and the regeneration of the adsorbent.
6. *The size and surface area of the adsorbent particles:* Smaller particles have larger service areas per unit mass; this results in a higher equilibrium capacity, but also in a higher pressure drop (flowthrough bed).

7. *Velocity of the gas:* In general, the lower the gas velocity, the higher the collection efficiency and the lower the pressure drop.

ADS.4   CARBON DIOXIDE ADSORPTION

The carbon dioxide adsorption data on Columbia (Columbia is a registered trademark of Union Carbide Corporation) activated carbon are presented below at a temperature of 50°C. Determine the constants of both the Freundlich equation and the Langmuir equation.

Equilibrium Capacity (cm^3/g)	Partial Pressure CO_2 (atm)
30	1
51	2
67	3
81	4
93	5
104	6

Solution

As described earlier, the relation between the amount of substance adsorbed by an adsorbent and the equilibrium pressure or concentration at constant temperature is called the *adsorption isotherm*. The adsorption isotherm is the most important and by far the most often used of the various equilibria data that can be measured. To represent the variation of the amount of adsorption per unit area or unit mass with pressure, Freundlich proposed the equation

$$Y = kp^{(1/n)}$$

where Y = weight or volume of gas (or vapor) adsorbed per unit area or unit mass of adsorbent
p = equilibrium partial pressure (vapor pressure) of adsorbed gas
k, n = empirical constants dependent on the nature of the solid and adsorbate, and on the temperature

The above equation may be written as follows. Taking logarithms of both sides,

$$\log Y = \log k + (1/n) \log p$$

If $\log Y$ is now plotted against $\log p$, a straight line should result with a slope equal to $1/n$ and a $\log Y$ intercept equal to $\log k$. Although the requirements of the equation are often met satisfactorily at low pressures, at higher pressures experimental points tend to deviate from a straight line, indicating that this equation does not have general applicability in reproducing adsorption of gases (or vapors) by solids.

A much better equation for type I isotherms (see Figure 80) was deduced by Langmuir from theoretical considerations:

$$Y = \frac{ap}{1 + bp}$$

The preceding equation is the Langmuir adsorption isotherm. The constants a and b are characteristic of the system under consideration and are evaluated from experimental data. Their magnitude also depends on the temperature. At any one temperature the validity of the Langmuir adsorption equation can be verified most conveniently by first dividing both sides of this equation by p and then taking reciprocals. The result is

$$\frac{p}{Y} = \frac{1}{a} + \frac{b}{a}p$$

Since a and b are constants, a plot of p/Y vs. p should yield a straight line with slope equal to b/a and an ordinate intercept equal to $1/a$.

For the Freundlich equation, the following table can be generated using the data in the problem statement:

Equilibrium Capacity (cm^3/g)	Partial Pressure CO_2 (atm)	$\log Y$	$\log P$
30	1	1.477	0.000
51	2	1.708	0.301
67	3	1.826	0.477
81	4	1.909	0.602
93	5	1.969	0.699
104	6	2.017	0.778

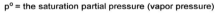

p^o = the saturation partial pressure (vapor pressure)

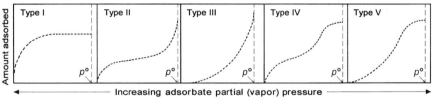

Figure 80. Types of adsorption isotherms.

A plot of $\log Y$ vs. $\log p$ yields the equation (see Figure 81)

$$Y = 30p^{0.7}$$

For the Langmuir equation,

$$\frac{p}{Y} = \frac{1}{a} + \frac{b}{a}p$$

p/Y	p
0.033	1
0.039	2
0.045	3
0.049	4
0.054	5
0.058	6

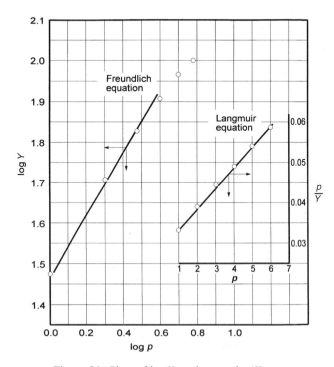

Figure 81. Plots of $\log Y$ vs. $\log p$ and p/Y vs. p.

A plot of p/Y vs. p yields the equation (see Figure 81)

$$\frac{p}{Y} = \frac{3.57p}{(1 + 0.186)p}$$

The Langmuir equation appears to fit the data better. Regressing the data for both equations and generating regression coefficients are left as exercises for the reader.

ADS.5   BREAKTHROUGH AND WORKING CAPACITIES

Calculate both the breakthrough capacity and the working capacity of an adsorption bed given the saturation (equilibrium) capacity (SAT), mass transfer zone (MTZ), and HEEL data provided below:

Depth of adsorption bed, $Z = 3$ ft
SAT $= 39\% = 0.39$; lb solvent/lb adsorbent
MTZ $= 4$ in.
HEEL $= 2.5\% = 0.025$; lb solvent/lb adsorbent

Solution

The following is offered before proceeding to the solution to this problem. Fixed-bed adsorbers are the usual choice for the control of gaseous pollutants when adsorption is the desired method of control. Consider a binary solution containing a strongly adsorbed solute (gaseous pollutant) at concentration C_0 (see Figure 82). The gas stream containing the pollutant is to be passed continuously down through a relatively deep bed of adsorbent that is initially free of adsorbate. The top layer of adsorbent, in contact with the contaminated gas entering, at first adsorbs the pollutant rapidly and effectively, and what little pollutant is left in the gas is substantially all removed by the layers of adsorbent in the lower part of the bed. At this point in time the effluent from the bottom of the bed is practically pollutant-free at C_1. The top layer of the bed is practically saturated, and the bulk of the adsorption takes place over a relatively narrow adsorption zone (defined as the mass transfer zone, MTZ) in which there is a rapid change in concentration. At some later time roughly half of the bed is saturated with the pollutant, but the effluent concentration C_2 is still substantially zero. Finally at C_3, the lower portion of the adsorption zone has reached the bottom of the bed, and the concentration of pollutant in the effluent has suddenly risen to an appreciable value for the first time. The system is said to have reached the "breakpoint." The pollutant concentration in the effluent gas stream now rises rapidly as the adsorption zone passes through the bottom of the bed and C_4 has just about reached the initial value C_0. At this point the bed is almost fully saturated with pollutant. The portion of the curve between C_3 and C_4 is termed the "breakthrough" curve.

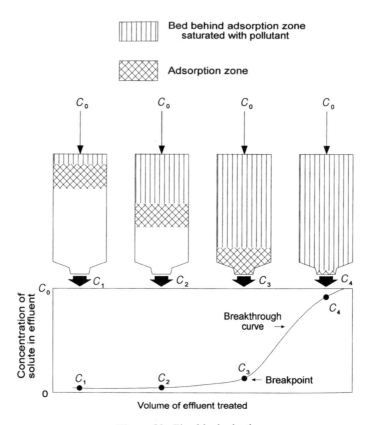

Figure 82. Fixed-bed adsorber.

The usual procedure in practice is to work with a term defined as the *working charge* (or *working capacity*). It provides a numerical value for the actual adsorbing capacity of the bed of height Z under operating conditions. If experimental data are available, the working charge (WC) may be estimated from

$$\text{WC} = \text{SAT}\left(\frac{Z - \text{MTZ}}{Z}\right) + (0.5)\text{SAT}\left(\frac{\text{MTZ}}{Z}\right) - \text{HEEL}$$

where SAT is the equilibrium capacity and HEEL is the residual adsorbate present in the bed following regeneration. (The breakthrough capacity is given in the above equation without the inclusion of the HEEL term.)

Regarding the problem, first calculate the breakthrough capacity (BC):

$$\text{BC} = \frac{(0.5)(0.39)(4) + (0.39)(36 - 4)}{36}$$

$$= 0.368 = 36.8\%$$

Calculate the working capacity (WC):

$$WC = 36.8 - 2.5$$
$$= 34.3\% = 0.343$$

Since much of the above data is rarely available, or just simply ignored, the working charge may be taken to be some fraction, f, of the saturated (equilibrium) capacity of the adsorbent, i.e.,

$$WC = (f)(SAT) \quad 0 \leqslant f \leqslant 1.0$$

Note that the notation, CAP, is often employed in place of SAT (see next problem).

ADS.6   CALCULATIONS ON AN ADSORPTION CANISTER

Small volatile organic compound (VOC) emission sources often use activated carbon, which is available in canisters or drums. An example of this is a modified form of Carbtrol model G-1, which is suitable for low air flow rates. The drum is not regenerated on site; it is returned to the manufacturer and a new drum is delivered.

A small pilot-scale reactor uses this modified model G-1 adsorber to capture methylene chloride emissions in a 50-acfm nitrogen purge source. The following operating and design data are provided:

Volumetric flowrate of nitrogen purge $= 50\,acfm$
Molecular weight of methylene chloride $= 85$
Operating temperature $= 70°F$
Operating pressure $= 1.0\,atm$
Saturation capacity $= 30\,lb\ CH_2Cl_2/100\,lb\cdot°C$
Methylene chloride concentration $= 500\,ppm$
Weight of carbon in drum $= 200\,lb$
Height of adsorbent in drum $= 24\,in.$
Adsorption time $= 6\,h$
Mass transfer zone (MTZ) $= 2\,in.$

Based on the above data and information, estimate the number of purge stream batches that this G-1 model adsorber canister can treat to breakthrough.

Solution

Calculate the working charge (WC) of the carbon drum (canister). Note that the HEEL is zero since this is a batch generation:

$$WC = CAP\left(\frac{Z - MTZ}{Z}\right) + (0.5)CAP\left(\frac{MTZ}{Z}\right)$$

$$= (0.30)\left(\frac{24 - 2}{24}\right) + (0.5)(0.30)\left(\frac{2}{24}\right)$$

$$= 0.2875 \text{ lb MeCl}_2/\text{lb carbon}$$

The drum capacity (DC) in pounds of methylene chloride is

$$DC = (WC)(\text{carbon weight})$$

$$= (0.2875)(200)$$

$$= 57.5 \text{ lb MeCl}_2$$

Determine the mole fraction of methylene chloride (MEC) in the purge stream:

$$y_{MEC} = 500/10^6$$

$$= 0.0005$$

The volumetric flowrate of methylene chloride (MEC) may now be calculated:

$$q_{MEC} = y_{MEC}q$$

$$= (0.0005)(50 \text{ acfm})(60)$$

$$= 1.5 \text{ acfh}$$

The density of the methylene chloride vapor in lb/ft^3 is

$$\rho_{MEC} = \frac{P\,(MW)}{RT}$$

$$= \frac{(1)(85)}{(0.73)(70 + 460)}$$

$$= 0.220 \text{ lb/ft}^3$$

The weight W of methylene chloride emitted per batch is then

$$W = (q_{MEC})(\rho_{MEC})(6\,h/batch)$$
$$= (1.5)(0.220)(6)$$
$$= 1.98\,lb/batch$$

The maximum number of batches per canister (MBC) is

$$MBC = DC/W$$
$$= 57.5/1.98$$
$$= 29$$

ADS.7 MULTICOMPONENT ADSORPTION

Provide an equation to estimate the saturation capacity, SAT, for a multicomponent mixture.

Solution

For multicomponent adsorption the saturation capacity may be calculated from the following (personal notes, L. Theodore):

$$SAT = \frac{1.0}{\displaystyle\sum_{i=1}^{n}\left(\frac{w_i}{SAT_i}\right)}$$

where n = number of components

w_i = mass fraction of i in n components (not including carrier gas)

SAT_i = equilibrium capacity of component i

For a two-component (A, B) system, the above equation reduces to:

$$SAT = \frac{(SAT)_A(SAT)_B}{w_A\,(SAT)_B + w_B\,(SAT)_A}$$

ADS.8 PERFORMANCE OF A TWO-BED CARBON ADSORPTION SYSTEM

A two-bed carbon adsorption system is being used to control odors emitting from a drum filling operation. The material being drummed is a high-purity grade of pyridine (C_5H_5N) that has a human detection level of 100 ppm. It has been reported that an odor can be detected from outside the drumming area when the equipment is in service, i.e., when drums are being filled. Discussions with operating personnel have indicated that the adsorption system is the source of the odor. You are requested to determine if the adsorption equipment/emission is the source of the odor or if the equipment is capable of containing/controlling the pyridine emission. Design and actual operating data are provided below.

1. The adsorption units are twin horizontal units with face dimensions for flow of 5 by 12 ft. Each unit contains new 4×6 mesh activated carbon B that was installed one month ago. The measured bed height is 12 in.

2. The carbon manufacturer maintains that the breakthrough capacity of the carbon is 0.49 lb pyridine/lb carbon and that the carbon has a bulk density of $25 \, lb/ft^3$.

3. Laboratory tests performed by plant personnel indicate that the carbon contains a HEEL of approximately 0.03 lb pyridine/lb carbon when regenerated with 4.0 lb of steam/lb pyridine at 10 psig.

4. The ventilation blower for the drum filling station has a flow of 5000 acfm at 25°C and 14.7 psia and contains a pyridine concentration of 2000 ppm (plant hygienist data).

5. The drum filling operation operates on a 24-h/day basis and the adsorption units are operated on an 8-h adsorption, 5-h regeneration cycle, with 3 h for cooling and standby. The steam used during the 5-h regeneration cycle was determined to be 2725 lb (mass flow meter).

6. The adsorption unit was designed based on a pressure drop through the bed following the relationship:

$$\Delta P = 0.37Z(v/100)^{1.56}$$

where Z is the bed depth in inches, v is the velocity in ft/min, and the pressure drop is in inches of water. The measured operational pressure drop is 3.3 in. of water.

7. The fractional fan efficiency, E_f, is 0.58.

To evaluate the adsorber's performance please determine the following:

1. The mass of pyridine to be captured in the adsorption period
2. The working capacity of the carbon B
3. The mass and volume of carbon that should be used in each unit
4. The required bed height
5. The design pressure drop through the bed using the required bed height for full capture
6. The horsepower requirement for this process

7. The required steam to regenerate the bed to the HEEL level of 0.03 lb
   pyridine/lb carbon

Solution

Calculate the mole fraction of pyridine (P) in the gas stream:

$$y_P = 2000/10^6$$
$$= 0.0020$$

Calculate the volumetric flowrate of P in acfm:

$$q_P = y_P q$$
$$= (0.0020)(5000)$$
$$= 10.0 \, \text{acfm}$$

Determine the density of the P vapor at the operating conditions:

$$\rho = \frac{P\,(\text{MW})}{RT} = \frac{(14.7)(79)}{(10.73)(537)}$$
$$= 0.2015 \, \text{lb/ft}^3$$

The mass of P collected during the adsorption period is then

$$m_P = (10)(0.2015)(8)(60)$$
$$= 967.2 \, \text{lb}$$

Estimate the working capacity (WC) of carbon B for this system:

$$\text{WC} = \text{BC} - \text{HEEL}$$
$$= 0.49 - 0.03$$
$$= 0.46 \, \text{lb P/lb carbon } B$$

Calculate the mass of carbon that should be used for each unit:

$$m_{\text{AC}} = m_P/\text{WC}$$
$$= 967.2/0.46$$
$$= 2103 \, \text{lb carbon } B$$

The volume of activated carbon, V_{AC}, is

$$V_{AC} = m_{AC}/\rho_B$$
$$= 2103/25$$
$$= 84.1 \text{ ft}^3$$

The cross-sectional area A of the carbon that is presently available for flow is

$$A = (5)(12) = 60 \text{ ft}^2$$

Since the bed height is 12 in. or 1.0 ft, the critical volume currently employed is 60 ft^3. Because this is below the required 84 ft^3, the odor problem is present; the equipment is not capable of controlling the emission.

The "required" height of the adsorbent in the unit is

$$Z = V_{AC}/A$$
$$= 84.1/60$$
$$= 1.40 \text{ ft}$$
$$= 16.8 \text{ in.}$$
$$\approx 17.0 \text{ in.}$$

Estimate the pressure drop across the adsorbent in inches of H_2O.

$$\Delta P = 0.37Z\left(\frac{v}{100}\right)^{1.56} = 0.37Z\left(\frac{q/A}{100}\right)^{1.56}$$
$$= (0.37)(17)\left(\frac{5000/60}{100}\right)^{1.56}$$
$$= 4.73 \text{ in. } H_2O$$

The total pressure drop across the bed in lb$_f$/ft^2 is then

$$\Delta P_{total} = (4.73)(5.2)$$
$$= 24.6 \text{ lb}_f/\text{ft}^2$$

while the HP requirement is

$$HP = \frac{(24.6)(5000)}{(60)(550)(0.58)}$$

$$= 6.43 \, HP$$

Finally, the steam requirement for regeneration is

$$m_{steam} = (4.0)(967.2)$$

$$= 3869 \, lb \text{ steam during regeneration}$$

The actual operating steam rate (2725 lb for 5 hours) is below the required value.

ADS.9   MOLECULAR SIEVE REGENERATION

Use the graph in Figure 83 to solve the following problem.

1. Estimate the pounds of CO_2 that can be adsorbed by 100 lb of Davison 4A Molecular Sieve from a discharge gas mixture at 77°F and 40 psia containing 10,000 ppmv (ppm by volume) CO_2.

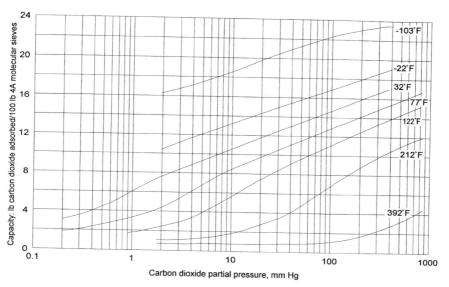

Figure 83. Vapor–solid equilibrium isotherms.

2. What percentage of this adsorbed vapor would be recovered by passing superheated steam at a temperature of 392°F through the adsorbent until the partial pressure of the CO_2 in the stream leaving is reduced to 1.0 mm Hg?
3. What is the residual CO_2 partial pressure in a gas mixture at 77°F in contact with the freshly stripped sieve in part 2?

Solution

Calculate the mole fraction of CO_2 in the discharge gas mixture:

$$y_{CO_2} = ppm/10^6$$
$$= 10,000/10^6$$
$$= 0.01$$

Also determine the partial pressure of CO_2 in psia and mm Hg:

$$p_{CO_2} = y_{CO_2} P$$
$$= (0.01)(40)$$
$$= 0.4 \, psia$$
$$= (0.4)(760/14.7)$$
$$= 20.7 \, mm \, Hg$$

Estimate the adsorbent capacity, SAT, at 77°F:

$$SAT = 9.8 \, lb \, CO_2/100 \, lb \, sieve \, (from \, Figure \, 83)$$

Also estimate the adsorbent capacity at 392°F and 1.0 mm Hg. Note that this represents the HEEL:

$$HEEL = 0.8 \, lb \, CO_2/100 \, lb \, sieve$$

The amount of CO_2 recovered is therefore

$$CO_2 \, recovered = 9.8 - 0.8$$
$$= 9.0 \, lb \, CO_2/100 \, lb \, sieve$$

while the percent recovery is

$$\% \, recovery = (9.0/9.8)100$$
$$= 91.8\%$$

Note that this represents the percent recovery relative to the HEEL.

Estimate the partial pressure of CO_2 in mm Hg in equilibrium at 77°F with sieve containing 0.8 lb CO_2/100 lb sieve (the HEEL):

$$p_{CO_2} \approx 0.05 \text{ mm Hg}$$

The equilibrium CO_2 concentration may be converted to ppm:

$$\text{ppm} = (p_{CO_2}/P)10^6$$
$$= (0.05)(14.7/760)(10^6)/40$$
$$= 24.1$$

The reader is left the exercise of calculating the percent recovery based on inlet and outlet concentrations (ppm).

ADS.10   DESIGN OF A FIXED-BED ADSORBER

Provide a design procedure for a fixed-bed adsorber.

Solution

A rather simplified overall design procedure for a system adsorbing an organic that consists of two horizontal units (one on/one off) that are regenerated with steam is provided below.

Note: This design procedure was originally developed by Dr. Louis Theodore in 1985 and later published in 1988. These materials recently appeared in the 1992 Air and Waste Management Association text published by Van Nostrand Reinhold titled *Air Pollution Engineering Manual*. This was done without properly acknowledging the author, Dr. Louis Theodore, and without permission from the original publisher.

1. Select adsorbent type and size.
2. Select cycle time; estimate regeneration time; set adsorption time equal to regeneration time; set cycle time equal to twice the regeneration time; generally, try to minimize regeneration time.
3. Set velocity; v is usually 80 ft/min but can increase to 100 ft/min.
4. Set the steam/solvent ratio.
5. Calculate (or obtain) working capacity (WC) for above.

6. Calculate the amount of solvent adsorbed (M_s) during one-half the cycle time (t_{ads}).

$$m_s = qc_i t_{ads}$$

where c_i = inlet solvent concentration, mass/volume
q = volume rate of flow, volume/time

7. Calculate the adsorbent required, m_{AC}:

$$m_{AC} = m_s/WC$$

8. Calculate the adsorbent volume requirement, V_{AC}:

$$V_{AC} = m_{AC}/\rho_B$$

9. Calculate the face area of the bed, A_{AC}:

$$A_{AC} = q/v$$

10. Calculate the bed height, Z.

$$Z = V_{AC}A_{AC}$$

11. Estimate the pressure drop from the graph in Figure 84 or a suitable equation.
12. Set the L/D (length-to-diameter) ratio. Calculate L and D, noting that

$$A = LD$$

Constraints: $L < 30$ ft, $D < 10$ ft; L/D of 3 to 4 acceptable if $v < 30$ ft/min
13. Design (structurally) to handle if filled with water.
14. Consider designing vertically if $q < 2500$ actual cubic feet per minute (acfm). Consider designing horizontally if $q > 7500$ acfm.

ADS.11   SIZING A CARBON ADSORBER

A printing company must reduce and recover the amount of toluene it emits from its Rotograve printing operation. The company submits some preliminary information on installing a carbon adsorption system. You, the primary consultant, are given the following information:

Air flow = 20,000 acfm (77°F, 1 atm)

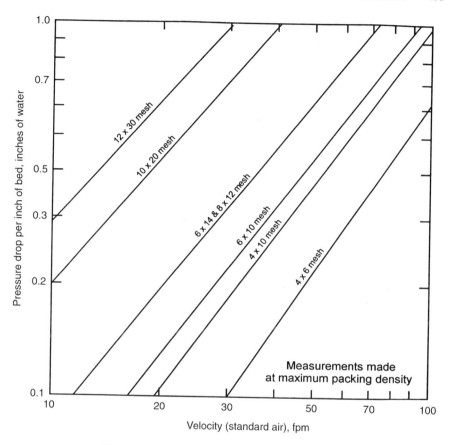

Figure 84. Activated carbon pressure drop curves.

Adsorption capacity for toluene is 0.175 lb toluene/lb activated carbon

Operation is at 10% of LEL (lower explosivity limit) for toluene in the exit air from printer

LEL for toluene = 1.2%

Toluene molecular weight = 92.1

Carbon bulk density (4 × 6 mesh) = 30 lb/ft³

Working charge is 60% of saturation capacity

Regeneration is just under one hour; assume 1.0 h

Maximum velocity through adsorber is 100 fpm

Determine the minimum size of adsorber you would recommend for a 1 × 1 system. Calculations should include the pertinent dimensions of the adsorber, the amount of carbon, the depth of the bed and an estimate of the pressure drop. Also calculate the fan horsepower if the blower/motor efficiency is 58%.

Solution

Initially, base the calculations on 1-h regeneration time so that 1 h of adsorption is available. Key calculations and results are provided below for the toluene (TOL) and activated carbon (AC).

$$V_{TOL} = (20,000)(0.10)(0.012)$$
$$= 24\,\text{acfm}$$

$$\dot{m}_{TOL} = \frac{(24)(492/537)(92.1)(60)}{359}$$
$$= 338\,\text{lb/h}$$

$$p_{TOL} = (24/20,000)(14.7)$$
$$= (0.0012)(14.7)$$
$$= 0.01764\,\text{psia}$$

$$SAT = 17.5\% = 0.175\,\text{lb}_{TOL}/\text{lb}_{AC}$$

$$WC = (0.175)(0.60)$$
$$= 0.105\,\text{lb}_{TOL}/\text{lb}_{AC}$$
$$= 10.5\,\text{lb}_{TOL}/100\,\text{lb}_{AC}$$

$$m_{AC} = (338/0.105)(1.0)$$
$$= 3220\,\text{lb}_{AC}\ \text{for one bed}$$
$$= 6440\,\text{lb}_{AC}\ \text{for both beds}$$

$$V_{AC} = 3220/30 = 107\,\text{ft}^3$$

$$A_{AC} = 20,000/100 = 200\,\text{ft}^2$$

$$H = 107/200$$
$$= 0.535\,\text{ft} = 6.4\,\text{in.}$$

Suggest a horizontal 10-ft diameter by 20-ft long design. Since $\Delta P = 0.625$ in. H_2O/in. bed,

$$\Delta P_{TOTAL} = (0.625)(6.4)$$

$$= 4.0 \text{ in. } H_2O$$

$$HP = \frac{(20,000)(4.0)(5.2)}{(0.58)(33,000)}$$

$$= 22 \text{ HP}$$

Note: This represents a marginal design since H is slightly higher than 0.5 ft.

ADS.12   BREAKTHROUGH TIME CALCULATION

A degreaser ventilation stream contaminated with trichloroethylene (TCE) is treated through the use of a horizontal carbon bed adsorber. The adsorber is normally designed to operate at a gas flow of 8000 scfm (60°F, 1 atm), and the concentration of TCE at the adsorber inlet is 1500 ppmv. The efficiency of the adsorber is 99% under normal design conditions. Design parameters are as follows:

Actual conditions: 25 psia, 90°F
SAT $= 35\%$
$Z =$ depth of bed $= 2.5$ ft
$L =$ length of adsorber $= 25$ ft
$D =$ diameter of adsorber $= 8$ ft
MTZ $= 5$ in.
HEEL $= 2.0\%$
Bulk density of carbon bed $= 35$ lb/ft^3

1. Determine the time before breakthrough occurs.
2. Recalculate the time before breakthrough occurs, based on the following transient condition. The adsorber system is on line for one hour at the above normal design conditions when the inlet concentration of TCE rises to an average value of 2500 ppmv due to a malfunction in the degreaser process; the efficiency also drops to 97.5% during this time. Assume the SAT remains the same.

Solution

Key calculations for part 1 are first provided:

$$q = 8000 \left(\frac{14.7}{25} \right) \left(\frac{90 + 460}{60 + 460} \right)$$

$$= 4975 \, \text{acfm}$$

$$q_{\text{TCE}} = (1500 \times 10^{-6})(4975)$$

$$= 7.46 \, \text{acfm}$$

$$\dot{m}_{\text{TCE}} = \frac{P(\text{MW}) q_{\text{TCE}}}{RT} = \frac{(25)(131.5)(7.46)}{(10.73)(90.460)}$$

$$= 4.16 \, \text{lb/min}$$

$$\text{BC} = \frac{0.5(C_s)(\text{MTZ}) + (C_s)(Z - \text{MTZ})}{Z}$$

$$= \frac{0.5 \, (0.35)(5/12) + (0.35)[2.5 - (5/12)]}{2.5}$$

$$= 0.32$$

$$\text{WC} = \text{BC} - \text{HEEL} \pm \text{SF (safety factor)}$$

$$= 0.32 - 0.02 - 0$$

$$= 0.30 = 30\%$$

$$V_C = (25)(8)(2.5)$$

$$= 500 \, \text{ft}^3$$

$$m_C = (500)(35)$$

$$= 17,500 \, \text{lb}$$

$$t \, (\text{to saturation}) = \frac{(\text{WC})(m_c)}{\dot{m}E} = \frac{(0.30)(17,500)}{(4.16)(0.99)}$$

$$= 1275 \, \text{min} \approx 21 \, \text{h}$$

For transient conditions (part 2),

$$m_{TCE} \text{ (in carbon)} = (0.30)(500)(35)$$
$$= 5250 \text{ lb}_{TCE}, \text{ maximum}$$

$$m_{TCE} \text{ (flow, first hour)} = (4.16)(60)(0.99)$$
$$= 247.1 \text{ lb}_{TCE} \text{ captured}$$

The remaining capacity of the bed after the first hour is now

$$m_{TCE} \text{ (after first hour)} = 5250 - 247.1$$
$$= 5003 \text{ lb}_{TCE}$$

$$t \text{ (to transient saturation)} = \frac{5003}{\dot{m}_{TCE,transient} (0.975)}$$

$$\dot{m}_{TCE,transient} = (4.16)(2500/1500)$$
$$= 6.93 \text{ lb/min}$$

$$t = \frac{5003}{(6.93)(0.975)}$$
$$= 740 \text{ min} = 12.34 \text{ h}$$

The time to breakthrough, following the transient period, is

$$t_B = 60 + 740.1$$
$$= 801.1 \text{ min} = 13.34 \text{ h}$$

Thus, the time to breakthrough has been reduced from 21 to 13.3 h.

PART IV
Solid Waste

Christopher Ruocco

22 Regulations (REG)

REG.1   SOLID WASTE DISPOSAL ACT

Describe the objectives of the Solid Waste Disposal Act of 1965.

Solution

The objectives of the Solid Waste Disposal Act of 1965 were:

- Promote proper solid waste management and recycling practices.
- Provide technical and financial assistance to municipalities to plan and implement solid waste management programs.
- Create a national research and development program for improved solid waste management practices and technology.
- Create guidelines for proper handling and disposal of solid waste.
- Provide technical grants for the design and operation of solid waste management facilities.

REG.2   HAZARDOUS AND SOLID WASTE ACT

1. In passing the Hazardous and Solid Waste Act (HSWA) amendments of 1984, Congress required hazardous waste generators to certify that they had a program in place to reduce the volume and/or toxicity of the waste they generated or managed. Explain why Congress did not include a corresponding enforcement section in the law.
2. Explain the difference between a "hazardous substance" and an "extremely hazardous substance."

Solution

1. There were two major reasons for a lack of enforcement/inspection of industrial waste volume/toxicity reduction program certification. The first was that Congress was convinced that the U.S. Environmental Protection Agency (EPA) had neither enough personnel nor personnel with adequate

expertise to evaluate industrial process systems. The second was that Congress was not convinced that this enforcement could be carried out without placing U.S. industry at a competitive disadvantage in world markets.

To resolve these problems, Congress did include an extensive research program in HSWA to augment the certification requirement, but left the responsibility of truthfulness with the industrial producers of hazardous waste. There does remain the basic requirement to truthfully report program activities by hazardous waste generators.

2. From a regulatory perspective, a *hazardous substance* is a material that may cause damage to public health and/or the environment when released to the surroundings. An *extremely hazardous substance* identifies a material that can cause irreversible damage to public health and/or the environment after only a single exposure.

REG.3 PERFORMANCE STANDARDS FOR HAZARDOUS WASTE INCINERATORS

The basic thrust of the 1984 incineration regulations requires that all hazardous waste incinerators meet certain performance standards. Describe the three key standards in technology.

Solution

1. *Principal Organic Hazardous Constituents (POHC).* The DRE for a given POHC is defined as the mass percentage of the POHC removed from the waste. The POHC performance standard requires that the DRE for each POHC *designated* in the permit be 99.99% or higher.

$$DRE \geqslant 99.99\% \qquad \text{percentage basis}$$
$$\geqslant 0.9999 \qquad \text{fractional basis}$$

The DRE performance standard implicitly requires a sampling and analysis to measure the amounts of the *designated* POHC(s) in both the waste stream and the stack effluent gas during a trial burn.

2. *Hydrochloric Acid.* The limitation for HCl emissions from the stack of an incinerator is specified in two parts; if the first part is not met, the second is required. These specifications are

   a. for the gas leaving the incinerator,

$$\dot{m}_{HCl} \leqslant 4\,\text{lb/h}\,(1.8\,\text{kg/h})$$

or

b. for the scrubber,

$$E_{HCl} \geqslant 99\%$$

A scrubber efficiency of 99% allows a maximum of 1% of the HCl entering the scrubber to be emitted to the atmosphere. The latest regulations covered in the *Federal Register* (April 27, 1990) require HCl emissions to be based on a risk assessment, the same as metals. Most often, this may be well above the 99% requirement. This sampling standard implicitly requires, in some cases, sampling and analysis to measure the HCl in the stack gas (see Problem STC.5 in Chapter 5).

3. *Particulates*. Stack emissions of particulate matter are limited to a concentration (c) of 0.08 grains per dry standard cubic foot (gr/dscf) for the stack gas corrected to 7% O_2 dry volume (or approximately 50% excess air):

$$c_{part} = 0.08 \, gr/dscf$$

The term *excess air* is discussed in Problem STC.3 in Chapter 5; it is a measure of the amount of air supplied to the incinerator over and above that needed for complete combustion of the waste and/or auxiliary fuel. (Although essentially the same, note that some of the calculations for this and later problems are based on 50% excess air and not 7% O_2 since the principles can be more easily demonstrated.) If $>$ or $< 50\%$ excess air (or 7% O_2 dry volume) is being used, the measured particulate concentration in grains per dry standard cubic foot must first be adjusted before comparison with the 0.08 gr/dscf standard. This adjustment is made by calculating what the concentration would be if 50% excess air was used. In this way, a decrease of particulate concentration due solely to increasing air flow in the stack is not rewarded, and an increase of particulate concentration due solely to reduction of air flow in the stack is not penalized. (See Problem REG. 8 of this chapter and Problem HWI 2 of Chapter 27.)

Compliance with these performance standards is documented by a *trial burn* of the facility's waste streams. (The reader should note that new standards were proposed in April of 1999. No additional information was available at the time of the preparation of this book).

REG.4   PERMIT PROCESS FOR INCINERATORS

Briefly describe the major strong points of the permit process for incinerator systems under RCRA. Include the perspective of the public, the regulatory agency, and the permittee.

Solution

The permit process is designed to make maximum information available to the public through the public hearing process. It ensures preparation and delivery of this information in an organized and descriptive manner. The permit process assures that all information used to choose a particular hazardous waste incineration system is public and available.

The permit process is designed to present to the regulators pertinent information that will allow them to make informed decisions in a reasonably consistent manner that, with proper coordination, utilizes a minimum of resources.

The application process requires basic information to demonstrate that the design and operation of the incinerator will result (see previous problem) in the destruction of principal organic hazardous constituents (POHCs) by 99.99%; a hydrogen chloride scrubbing efficiency of 99% or, as described in the previous problem, $<4\,\text{lb/h}$ maximum HCl emissions; and particulate emissions $<0.08\,\text{gr/dscf}$ corrected to 50% excess air (EA). These may be written in equation form as:

$$E_{POHC} = 99.99\%$$

$$E_{HCl} = 99\% \quad \text{or} \quad \dot{m}_{HCl} < 4\,\text{lb/h} \quad \text{(see previous problem)}$$

$$c_{particulate} < 0.08\,\text{gr/dscf (see previous problem)}$$

The permit, in general, restricts the area of technical information that the permittee must develop, and most importantly, it serves as a shield of protection for the permittee against legal action so long as the permittee maintains compliance. This shield protects against arbitrary regulatory actions and unjustifiable public suits. It is important to note that it does not shield against actions that lead to endangerment of public health or the environment.

REG.5   LIABILITY

The term *liability* is closely tied to the system of values and ethics that have developed in this country. In concise language define and give an example of the terms *liability, strict liability,* and *joint and several liability.* Explain how the interpretation of liability affects waste incineration permitting, application, and design.

Solution

Liability implies responsibility for an action. An individual may be held liable for a result if, in the mind of the normal, prudent person, the individual failed to exercise due caution. Examples include driving too fast, losing control of a car, and causing damage to property or persons.

Strict liability implies responsibility without regard to prudence or care, i.e., without regard to negligence. Such standards are imposed for a variety of activities, such as handling dynamite, statutory rape, or hazardous waste management. These standards require that proper caution be exercised at all times. Defenses available, if harm results, are limited. These standards are the basis for training requirements imposed upon the permittee who is an owner/operator of hazardous waste treatment, storage, or disposal facilities.

Joint and several liability is an assignment of responsibility when two or more persons fail to exercise the proper care and a division of harm is not possible. If two hunters fire their weapons and a person is killed, and there is no way to determine which projectile caused the harm, both hunters may be each held liable for the harm to the aggrieved party. In the case of joint and several strict liability, each party who managed a waste may be responsible for mitigating damages caused by the waste. For example, the generator, the transporter, the storage facility, and the incinerator operator may each individually or collectively be responsible for damages caused by mismanagement of a waste.

These provisions provide a tremendous impetus to hazardous waste generators to dispose of their waste on site under carefully controlled conditions. This concept of liability also burdens the generator with the threat of future costs as a result of someone else's improper actions. These values follow directly from our system of government, which was created to assure that individual citizens do not suffer loss of property and freedoms (health) by the actions of others. The need for a careful choice of a contractor to carry out waste management and disposal responsibilities is also highlighted by these provisions.

REG.6   COLLECTION EFFICIENCY CALCULATIONS

The hazardous waste flow rate into a treatment device is 100 lb/h. Calculate the waste rate leaving the unit to achieve a collection efficiency of

1. 95%
2. 99%
3. 99.9%
4. 99.99%
5. 99.9999%

Solution

The definition of collection efficiency, E, in terms of mass flowrate in, $\dot{m}_{in}$ and mass flowrate out, $\dot{m}_{out}$, is

$$E = \left(\frac{\dot{m}_{in} - \dot{m}_{out}}{\dot{m}_{in}} \right)(100)$$

The above equation may be rewritten for $\dot{m}_{out}$:

$$\dot{m}_{out} = \dot{m}_{in}(1 - E/100)$$

1. The mass flowrate out, $\dot{m}_{out}$, for an E of 95% is

$$\dot{m}_{out} = 100(1 - 95/100)$$
$$= 5 \, lb/h$$

2. The mass flowrate out, $\dot{m}_{out}$, for an E of 99% is

$$\dot{m}_{out} = 100(1 - 99/100)$$
$$= 1 \, lb/h$$

3. The mass flowrate out, $\dot{m}_{out}$, for an E of 99.9% is

$$\dot{m}_{out} = 100(1 - 99.9/100)$$
$$= 0.1 \, lb/h$$

4. The mass flowrate out, $\dot{m}_{out}$, for an E of 99.99% is

$$\dot{m}_{out} = 100(1 - 99.99/100)$$
$$= 0.01 \, lb/h$$

5. The mass flowrate out, $\dot{m}_{out}$, for an E of 99.9999% is

$$\dot{m}_{out} = 100(1 - 99.9999/100)$$
$$= 0.0001 \, lb/h$$

REG.7   CORRECTION TO 7% OXYGEN

The carbon monoxide (CO) concentration in the stack from a solid hazardous waste treatment facility is measured at 20 ppmv at a temperature of 175°F. The oxygen concentration in the stack is measured to be 12% by volume on a wet basis. The water content of the stack gas is 10 mol %.

1. What is the corrected CO content in the stack gas?
2. Similarly, the particulate concentration in the stack is measured at 20 mg/dscf at a stack oxygen concentration of 12% on a dry basis at 70°F. What is the corrected particulate concentration?

3. Federal regulations (in SI units) require that corrected particulate concentrations do not exceed 180 mg/dscm corrected to 7% O_2. Does this stack meet
   the regulations?

Many current waste regulations require that emissions be corrected to 7% oxygen
in the stack on a dry basis. The correction formula provided in the regulations (see
40 CFR Part 264.343) is as follows:

$$P_C = P_M[14/(21 - Y)]$$

where P_C = corrected concentration
  P_M = measured concentration
  Y = O_2 concentration in stack, dry basis

Solution

Determine the percent by volume of CO and N_2 in the stack:

  O_2 concentration = 12%
  H_2O content = 10%

Thus, CO_2 and N_2 content is $(100 - 22)\% = 78\%$.
  Calculate the O_2 concentration in the stack in mole fraction units.
  Basis 100 mol:

$$Y = O_2 \text{ concentration in the stack} = 12/90$$

$$= 0.1333 = 13.33\%$$

Calculate the corrected CO content in the stack gas in ppm using the equation given
above.
  For CO:

$$P_C = P_M[14/(21 - Y)]$$

$$= 20[14/(21 - 13.33)]$$

$$= 36.5 \text{ ppm CO (corrected to 7\% } O_2)$$

Calculate the corrected particulate concentration in the stack in mg/dscf.
  For particulates:

$$P_C = 20[14/(21 - 12)]$$

$$= 31.1 \text{ ppm CO (corrected to 7\% } O_2)$$

Determine if the stack meets regulations.

Convert mg/dscf to mg/dscm:

$$(31.1 \, \text{mg/dscf})(35.3 \, \text{ft}^3/\text{m}^3) = 1098 \, \text{mg/dscm}$$

Therefore, this stack does *not* meet regulations.

REG.8   PARTICULATE LOADING

A hazardous waste incinerator is burning an aqueous slurry of soot (carbon) with the production of a small amount of fly ash. The waste is 70% water by mass and is burned with 0% excess air (EA). The flue gas contains 0.25 grains (gr) of particulates in each 7 ft^3 (actual) at 620°F. For regulation purposes, calculate the particulate concentration in the flue gas in gr/acf, in gr/scf, and in gr/dscf.

Solution

Calculate the particulate concentration in the flue gas per actual cubic foot:

$$0.25 \, \text{gr}/7.0 \, \text{acf} = 0.0357 \, \text{gr/acf}$$

Convert actual cubic feet to standard cubic feet:

$$60°F = 520°R \qquad 620°F = 1080°R$$

$$7.0 \, \text{acf}(520/1080) = 3.37 \, \text{scf}$$

Calculate particulate concentration in standard cubic feet:

$$0.25 \, \text{gr}/3.37 \, \text{scf} = 0.074 \, \text{gr/scf}$$

To determine the volume fraction of water in the flue gas, the reaction equation for the combustion of the waste is

$$C + H_2O + O_2 \rightarrow CO_2 + H_2O$$

The number of moles of each component becomes

C: $30 \, \text{lb}/(12 \, \text{lb/lbmol}) = 2.5 \, \text{lbmol}$
H_2O: $70 \, \text{lb}/(18 \, \text{lb/lbmol}) = 3.89 \, \text{lbmol}$

Calculate the moles of CO_2 and H_2O in the flue gas:

$$CO_2 = 2.5 \, \text{lbmol}$$

$$H_2O = 3.89 \, \text{lbmol}$$

The stoichiometric (0% excess air) air requirement for this solid waste can now be determined:

O_2:   2.5 lbmol
N_2: $(79/21)(2.5\,\text{lbmol}) = 9.4\,\text{lbmol}$
Air: $9.4 + 2.5 = 11.9$

The oxygen and nitrogen in the flue gas are therefore

$$O_2 = 0.0\,\text{lbmol}$$

$$N_2 = 9.4\,\text{lbmol}$$

Dividing the moles of H_2O by the total moles yields the mole fraction of H_2O in the flue gas:

$$H_2O \text{ fraction} = \text{lbmol } H_2O/(CO_2 + H_2O + N_2)\,\text{lbmol}$$

$$= 3.89/(2.5 + 3.89 + 9.4)$$

$$= 0.25 \text{ (by mole or volume)}$$

The molar quantity of dry gas is obtained by subtracting the moles of H_2O from the total moles to yield

$$\text{Moles dry gas} = (2.5 + 3.89 + 9.4) - 3.8 = 12.0\,\text{lbmol}$$

The dry volume in dry standard cubic feet (dscf) is obtained by subtracting the volume of H_2O from the total volume to yield

$$\text{Dry volume} = 3.37\,(1 - 0.25) = 2.53\,\text{dscf}$$

The particulate concentration on a dscf basis becomes

$$0.25\,\text{gr}/2.53\,\text{dscf} = 0.099\,\text{gr/dscf}$$

REG.9   PARTICULATE AND HCl EMISSIONS

A hazardous waste incinerator is burning a waste mixture containing solids with 50% excess air at 2100°F with a residence time of 2.5 s. The stack gas flowrate was determined to be 14,280 dscfm. The composition of contaminants in the stack gas is given below:

Compound	Inlet (lb/h)	Outlet (lb/h)
Toluene	860	0.20
Chlorobenzene	450	0.02
Dichlorobenzene	300	0.03
HCl		4.2
Particulates		9.65
Pb		0.05 lb/100 lb (TOC)

TOC = total organic compounds

1. Calculate the DREs for toluene, chlorobenzene, dichlorobenzene, and HCl.
2. Calculate the discharge concentrations of Pb and particulates.
3. Is the unit in compliance with present federal regulations? Assume chlorobenzene and dichlorobenzene are the POHCs (principal organics) for this incinerator.

Solution

1. DREs for each compound are calculated as follows:
Toluene:

$$100\left(\frac{860 \text{ lb in/h} - 0.20 \text{ lb out/h}}{860 \text{ lb in/h}}\right) = 99.98\%$$

Chlorobenzene:

$$100\left(\frac{450 \text{ lb in/h} - 0.02 \text{ lb out/h}}{450 \text{ lb in/h}}\right) = 99.996\%$$

Dichlorobenzene:

$$100\left(\frac{300 \text{ lb in/h} - 0.03 \text{ lb out/h}}{300 \text{ lb in/h}}\right) = 99.99\%$$

Molar HCl production is calculated based on all Cl coming into the incinerator, and assuming that all Cl is converted to HCl upon combustion.

For chlorobenzene, the amount of HCl produced is calculated assuming:

$$C_6H_5Cl + 7O_2 \rightarrow 6CO_2 + HCl + 2H_2O$$

HCl produced from chlorobenzene

$$= \frac{450 \text{ lb/h}}{6(12 \text{ lb C/lbmol}) + 5(1 \text{ lb H/lbmol}) + 1(35.45 \text{ lb Cl/lbmol})}$$

$$= 4.00 \text{ lbmol HCl/h}$$

For dichlorobenzene, the amount of HCl produced is calculated assuming:

$$C_6H_4Cl_2 + 6.5O_2 \rightarrow 6CO_2 + 2HCl + H_2O$$

HCl produced from dichlorobenzene

$$= \frac{300 \text{ lb/h}}{6(12 \text{ lb C/lbmol}) + 4(1 \text{ lb H/lbmol}) + 2(35.45 \text{ lb Cl/lbmol})}$$

$$= 2.04 \text{ lbmol HCl/h}$$

Therefore,

Total amount of HCl produced

$$= (4.00 + 2.04) \text{ lbmol } (36.45 \text{ lb/lbmol}) = 220.2 \text{ lb HCl/h}$$

The DRE for HCl is

$$100 \left(\frac{220.2 \text{ lb HCl in/h} - 4.22 \text{ lb HCl out/h}}{220.2 \text{ lb HCl in/h}} \right) = 98.09\%$$

2. The amount of Pb in the waste mixture is based on the TOC content of the waste as given by:
    TOC for toluene (C_7H_8):

$$\% \text{ TOC in toluene} = \frac{\text{Mass of C in toluene}}{\text{Total mass of toluene}} = \frac{7(12)}{7(12) + 8(1)}$$

$$= 0.913$$

$$\text{Toluene TOC emission rate} = (0.913)(860 \text{ lb toluene/h})$$
$$= 785 \text{ lb TOC/h}$$

TOC for chlorobenzene (C_6H_5Cl, CB):

$$\% \text{ TOC in CB} = \frac{\text{Mass of C in CB}}{\text{Total mass of CB}} = \frac{6(12)}{6(12) + 5(1) + 1(35.45)}$$

$$= 0.64$$

$$\text{Chlorobenzene TOC emission rate} = (0.64)\,450\,\text{lb CB/h}$$
$$= 288\,\text{lb TOC/h}$$

TOC for dichlorobenzene ($C_6H_4Cl_2$, DCB):

$$\% \text{ TOC in DCB} = \frac{\text{Mass of C in DCB}}{\text{Total mass of DCB}} = \frac{6(12)}{6(12) + 4(1) + 2(35.45)}$$

$$= 0.49$$

$$\text{Dichlorobenzene TOC emission rate} = (0.49)\,300\,\text{lb DCB/h}$$
$$= 147\,\text{lb TOC/h}$$

The amount of Pb in the flue gas is:

$$\text{Total emission of TOC} = (785 + 288 + 147\,\text{lb TOC/h})$$
$$= 1220\,\text{lb TOC/h}$$

$$\text{Total mass of Pb} = (1220\,\text{lb TOC/h})(0.005\,\text{lb PB/lb TOC})$$
$$= 6.1\,\text{lb Pb/h}$$

Concentration of Pb in the incinerator flue gas is

$$\frac{(6.1\,\text{lb Pb/h})(1\,\text{h}/60\,\text{min})}{14{,}280\,\text{dscfm}}$$

$$= 7.12 \times 10^{-6}\,\text{lb/dscf}$$

$$= 7.12 \times 10^{-6}\,\text{lb/dscf}\left(\frac{1.6033 \times 10^{10}\,\mu\text{g/m}^3}{\text{lb/dscf}}\right)$$

$$= 114{,}147\,\mu\text{g/m}^3$$

The particulate concentration in the flue gas is calculated as follows:

$$\text{Total emission of particulates} = 9.65\,\text{lb/h}\,(7000\,\text{gr/lb})$$
$$= 67{,}550\,\text{gr/h}$$

The concentration of particulates in the incinerator flue gas is

$$\frac{67,550\,\text{gr/h}}{14,280\,\text{dscfm}\,(60\,\text{min/h})} = 0.079\,\text{gr/dscf}$$

3. The unit is in compliance with present federal regulations for POHCs and particulates but is *NOT* in compliance for HCl emissions.

REG.10   COMPLIANCE STACK TEST

A compliance stack test on a facility yields the results below. Determine whether the incinerator meets the state particulate standard of 0.05 gr/dscf. Estimate the amount of particulate matter escaping the stack, and indicate the molecular weight of the stack gas. Use standard conditions of 70°F and 1 atm pressure.

Volume sampled	35 dscf
Diameter of stack	2 ft
Pressure of stack gas	29.6 in. Hg
Stack gas temperature	140°F
Mass of particulate collected	0.16 g
% moisture in stack gas	7% (by volume)
% O_2 in stack gas (dry)	7% (by volume)
% CO_2 in stack gas (dry)	14% (by volume)
% N_2 in stack gas (dry)	79% (by volume)
Pitot tube factor (k)	0.85

Pitot tube measurements made at eight points across the diameter of the stack provided values of 0.3, 0.35, 0.4, 0.5, 0.5, 0.4, 0.3, and 0.3 in. of H_2O.

Use the following equations for S-type pitot tube velocity, v (m/s), measurements:

$$v = k\sqrt{2gH}$$

$$= k\sqrt{2g\,\frac{\rho_1}{\rho_a}\,(0.0254)h}$$

where g = gravitational acceleration 9.81 m/s^2
H = fluid velocity head, in. H_2O
ρ_1 = density of manometer fluid, 1000 kg/m^3
ρ_a = density of flue gas, 1.084 kg/m^3
h = mean pitot tube reading, in. H_2O

Solution

The particulate concentration in the stack is

$$\text{Particulate concentration} = \frac{0.16 \text{ g collected}}{35 \text{ dscf sampled}} \left(\frac{15.43 \text{ gr}}{g} \right)$$

$$= 0.0706 \text{ gr/dscf}$$

Since this does exceed the particulate standard of 0.05 gr/dscf, *the facility is not in compliance.*

The actual particulate emission rate is the product of the stack flowrate and the stack flue gas particulate concentration. The stack flowrate is calculated from the velocity measurements provided in the problem statement using the second velocity equation given:

$$v = 0.85 \sqrt{ 2\left(\frac{9.81 \text{ m}}{s^2}\right)\left(\frac{1000 \text{ kg/m}^3}{1.084 \text{ kg/m}^3}\right)0.0254h}$$

$$= 0.85(21.4)\sqrt{h}$$

$$= 0.85(21.4)(0.6142)$$

$$= 11.2 \text{ m/s} = 36.75 \text{ fps}$$

Stack flowrate $= v$ (cross-sectional area)

$$v = 36.75 \text{ fps}\left(\frac{\pi}{4}\right)(2 \text{ ft})^2$$

$$= 115.45 \text{ acfs} = 6.924 \text{ acfm}$$

Dry volumetric flowrate $= (1 - 0.07) \times 6924 \text{ acfm} = 6439 \text{ dacfm}$

Correct to standard conditions of 70°F and 1 atm pressure:

Standard volumetric flowrate

$$= 6439 \text{ dacfm}\left(\frac{530°\text{R}}{600°\text{R}}\right)\left(\frac{29.6 \text{ psi}}{29.9 \text{ psi}}\right)$$

$$= 5631 \text{ dscfm}$$

Particulate emission rate $= 0.0706 \text{ gr/dscf}(5631 \text{ dscfm})$

$$= 398 \text{ gr/min}$$

$$= 398 \text{ gr/min}\left(\frac{1 \text{ lb}}{7000 \text{ gr}}\right) = 0.0569 \text{ lb/min}$$

$$= (0.0569 \text{ lb/min})(1440 \text{ min/day}) = 81.9 \text{ lb/day}$$

The molecular weight of flue gas is based on the mole fraction of the flue gas components. The flue gas is 7% water and 93% other components by volume. On a dry basis, the flue gas molecular weight is

$$\text{MW} = 0.07 \, O_2 \left(\frac{32 \, \text{lb}}{\text{lbmol}} \right) + 0.14 \, CO_2 \left(\frac{44 \, \text{lb}}{\text{lbmol}} \right) + 0.79 \, N_2 \left(\frac{28 \, \text{lb}}{\text{lbmol}} \right)$$

$$= 30.52 \, \text{lb/lbmol}$$

The average molecular weight of the stack gas on an actual (wet) basis is then

$$\text{Average MW} = 0.07 \, \text{water} \left(\frac{18 \, \text{lb}}{\text{lbmol}} \right) + 0.93 \, \text{other components} \left(\frac{30.52 \, \text{lb}}{\text{lbmol}} \right)$$

$$= 29.64 \, \text{lb/lbmol}$$

23 Characteristics (CHR)

CHR.1 ELEMENTS OF SOLID WASTE DISPOSAL

1. Identify and describe the six elements of solid waste disposal.
2. What are the four elements included in the hierarchy for the planning of waste management systems? How can each element affect the effectiveness of solid waste management systems?

Solution

1. The six elements of solid waste disposal are:
   a. *Waste generation:* Activities that result in the creation of solid waste.
   b. *Waste handling and separation*: Handling of solid waste prior to containerization.
   c. *Collection:* The gathering and transport of solid waste from point sources.
   d. *Separation, processing and transformation:* Preparation and treatment of solid waste.
   e. *Transfer and transport:* The transport of solid waste to final disposal sites.
   f. *Disposal:* The final fate of solid wastes.
2. The four elements included in the hierarchy for the planning of waste management systems are:
   a. *Source reduction:* Involves reducing the amount and toxicity of wastes generated at a point source. This element is the most effective means of reducing the quantity of solid waste.
   b. *Recycling:* Reduces demand on natural resources and solid waste handling systems.
   c. *Waste transformation:* Physical, chemical, and/or biological treatment that is designed to improve the efficiency of a solid waste management operation.
   d. *Ultimate disposal:* Identified as the final means of dealing with solid wastes. Several methods of disposal can be implemented to deal with various types of solid wastes.

The above hierarchy is based on pollution prevention principles (see part VI, Pollution Prevention problems, additional details).

CHR.2   SOLID WASTE MANAGEMENT PLAN

1. What factors should be considered in the design of a solid waste management plan?
2. Describe the challenges that presently face the implementation of proper solids waste management programs.

Solution

1. First, the factors that need to be considered in the design of a solid waste management plan are processing strategy and technology, plan flexibility, and plan monitoring. Several programs and treatment technologies are available for solid waste disposal. Second, a management plan must be designed with the ability to cope with future changes in the types and amounts of solid waste. Finally, solid waste management plans require constant monitoring and evaluation for optimal performance.
2. One challenge facing proper solid waste management implementation is to change established public habits that promote consumption. Another challenge is the promotion of waste reduction and recycling practices at point sources. A third challenge involves making landfills safer to the surrounding public and environment. A fourth challenge is the development of new technologies that reduce demand for natural resources and reduce solid waste generation.

CHR.3   EFFECTS OF LAND POLLUTION ON ANIMALS

How can land pollution affect the health of plants and animals?

Solution

The effects of land pollution on animals are kidney and liver damage, brain and nerve damage, acid burns, cancer, and genetic disorders. Pollutants can infect local water supplies and act as fuels for potential fires. Land pollutants affect plants by hindering growth.

CHR.4   MOST COMMON AIR POLLUTANTS

What are the three most common air pollutants? How does each pollutant affect human health?

Solution

1. *Sulfur dioxide:* Irritates the human respiratory system and promotes respiratory diseases.
2. *Carbon monoxide:* Reduces the capacity of blood to carry oxygen to cells.
3. *Nitrogen oxides* (i.e., NO, NO_2): Attacks lung tissue.

CHR.5   EFFECT OF TEMPERATURE ON WASTE STORAGE

The physical state of hazardous material in a storage or transport container is an important factor in considering the fate or transport into the environment during an accidental spill or discharge situation. Define the conditions of container temperature (T_c) and ambient temperature (T_a) that will maintain the status of a chemical with the following physical characteristics. Note that MP and BP are the melting point and boiling point of the chemical, respectively.

Problem	Physical State of Material	MP/BP	Container Conditions
Example	Cold or refrigerated solid	$MP > T_a$	$T_c < MP$ and T_a
a	Solid		
b	Warm hot liquid		
c	Cold liquid		
d	Liquid		
e	Hot liquid		
f	Hot or warm compressed gas or vapor over hot liquid		
g	Compressed liquefied gas		
h	Hot or warm compressed gas or compressed liquefied gas		

Solution

The following is a summary of the solution to this problem.

Problem	Physical State of Material	MP/BP	Container Conditions
Example	Cold or refrigerated solid	$MP > T_a$	$T_c < MP$ and T_a
a	Solid	$MP > T_a$	T_c near T_a
b	Warm hot liquid	$BP > T_a$	$T_c > MP$ and $T_a < BP$
c	Cold liquid	$MP < T_a$	$T_a > MP < T_a$ and BP
d	Liquid	$BP > T_a$	T_c near T_a
e	Hot liquid	$BP > T_a$	$BP > T_c > T_a$
f	Hot or warm compressed gas or vapor over hot liquid	$BP > T_a$	$T_c > BP$ and T_a
g	Compressed liquefied gas	$BP < T_a$	T_c near T_a
h	Hot or warm compressed gas or compressed liquefied gas	$BP < T_a$	$T_c > BP$ and T_a

CHR.6   DECOMPOSITION OF ALKYL DICHLOROBENZENES

The general formula for the alkyl dichlorobenzenes is

$$C_nH_{2n-8}Cl_2 \qquad \text{where } n > 6$$

Write a balanced, general chemical equation for the decomposition in the presence of oxygen of alkyl dichlorobenzenes.

Solution

The general balanced equation for the complete combustion of the alkyl dichlorobenzenes is as follows:

$$C_nH_{2n-8}Cl_2 + (1.5n - 2.5)O_2 \rightarrow nCO_2 + 2HCl + (n - 5)H_2O$$

where $n > 6$.

CHR.7   HEAT OF COMBUSTION CALCULATION

Compare the heat of combustion of 1 mol of benzene (C_6H_6) with the combined heats of combustion of 6 mol of solid carbon and 3 mol of H_2.

Solution

The following data are provided by J. Santoleri, J. Reynolds, and L. Theodore (*Introduction to Hazardous Waste Incineration*, 2nd ed., Wiley-Interscience, New York, 2000):

$$\Delta H^{\circ}_{c[C_6H_6(g)]} = -789,080 \text{ cal/gmol}$$

$$\Delta H^{\circ}_{c[C]} = -94,052 \text{ cal/gmol}$$

$$\Delta H^{\circ}_{c[H_2]} = -68,317 \text{ cal/gmol}$$

The heat of combustion of 6 mol of carbon and three moles of hydrogen molecules is

$$\Delta H_c = 6(-94,052 \text{ cal/gmol}) + 3(-68,317 \text{ cal/gmol}) = -769,263 \text{ cal/gmol}$$

The error involved in such a calculation is

$$\% \text{ Error} = \frac{-798,080 - (-769,263)}{-789,080} = 0.025 = 2.5\%$$

CHR.8 HEATING VALUES ESTIMATED BY DULONG'S EQUATION

Calculate the net heating value (NHV) of methane, chloroform, benzene$_{(g)}$, chlorobenzene, and hydrogen sulfide. This assumes that the water product is in the vapor state. Compare these values with those calculated using Dulong's equation. Calculate the relative percent difference between the "true" NHVs as determined by thermodynamic calculations and the "estimated" values calculated using Dulong's equation. Dulong's equation can be written as follows:

$$NHV \approx 14{,}000\,m_C + 45{,}000(m_H - \tfrac{1}{8}m_O) - 760(m_{Cl}) + 4500(m_S)$$

where m_i is the mass fraction of component i.

Solution

The first step in the solution of this problem is to write the balanced oxidation reaction equation for each compound, and from these balanced equations, calculate the standard heat of combustion as follows:

$$CH_4: \quad CH_4 + 2O_2 \rightarrow CO_2 + 2H_2O_{(g)}$$

$$CHCl_3: \quad CHCl_3 + \tfrac{7}{6}O_2 \rightarrow CO_2 + \tfrac{1}{3}H_2O + \tfrac{1}{3}HCl + \tfrac{4}{3}Cl_2$$

$$C_6H_6: \quad C_6H_6 + \tfrac{15}{2}O_2 \rightarrow 6CO_2 + 3H_2O_{(g)}$$

$$C_6H_5Cl: \quad C_6H_5Cl + 7O_2 \rightarrow 6CO_2 + 2H_2O_{(g)} + HCl$$

$$H_2S: \quad H_2S + \tfrac{3}{2}O_2 \rightarrow SO_2 + H_2O_{(g)}$$

For the methane reaction, the heat of combustion is calculated from heats of formation, as follows:

$$\Delta H_c^\circ = (-94{,}052 \text{ cal/gmol } CO_2) + 2(-57{,}798 \text{ cal/gmol } H_2O_{(g)})$$

$$- (-17{,}889 \text{ cal/gmol } CH_4)$$

$$= -191{,}759 \text{ cal/gmol}$$

Heat of combustion values for the balance of these compounds are calculated in a similar manner and are found in R. H. Perry and D. W. Green, Ed., *Perry's Chemical Engineers' Handbook*, 7th Edition, McGraw Hill, New York, 1996. The results of these calculations are summarized in the table below using the conversion 1.8 cal/g = Btu/lb.

Compound	NHV (gal/gmol)	MW (g/gmol)	NHV (cal/g)	NHV (Btu/lb)
CH_4	191,759	16	11,985	21,593
$CHCl_3$	96,472	119.5	807	1,453
C_6H_6	717,886	78	9,204	16,583
C_6H_5Cl	714,361	112.5	6,350	11,441
H_2S	123,943	34	3,645	6,567

The following results from Dulong's equation can be summarized as:

Compound	Mass %C	Mass %H	Mass %O	Mass %Cl	Mass %S	NHV (Btu/lb)
CH_4	0.75	0.25	0	0	0	21,750
$CHCl_3$	0.10	0.0084	0	0.891	0	1,101
C_6H_6	0.92	0.08	0	0	0	16,480
C_6H_5Cl	0.64	0.04	0	0.32	0	10,520
H_2S	0	0.06	0	0	0.94	6,930

Based on these calculations, the difference between the thermodynamically based NHV values and those estimated using Dulong's equation can be summarized as:

Compound	NHV_{thermo} (Btu/lb)	NHV_{Dulong} (Btu/lb)	% Difference
CH_4	21,593	21,750	0.73
$CHCl_3$	1,453	1,101	6.7
C_6H_6	16,5883	16,480	0.62
C_6H_5Cl	11,441	10,520	8.05
H_2S	6,567	6,930	5.53

CHR.9   VOST AND SEMI-VOST SAMPLING METHODS

Principal organic hazardous constituents (POHCs) are monitored using the volatile organic sampling train (VOST) and semivolatile organic sampling train (Semi-VOST). The VOST method is intended for POHCs with boiling points from 30 to 100°C, while the Semi-VOST is designed for POHCs with boiling points > 100°C.

1. For the following POHCs determine whether one would expect them to be detected by the VOST or Semi-VOST method:

Carbon tetrachloride	Heptachlor epoxide
Hexachlorobenzene	Chlordane
Decachlorobiphenyl	Pentachlorophenol
1,1,1-Trichloroethane	Kepone
Hexachlorobutadiene	Tetrachloroethylene

2. Will the distribution of compounds between methods be absolute, i.e., all or nothing of a given chemical in one method or another? Why or why not?

3. These protocols allow for an accuracy of ±50% in final POHC concentrations. If the true value for a given POHC was just small enough to produce a destruction and removal efficiency (DRE) of 99.99%, what is the tolerance of the measured DRE that could arise for a single determination based upon the

allowable analytical tolerance? Assume a fixed mass flow into the incinerator of 1.0 kg/h.

Solution

1. Examining the boiling point (BP) data for the compounds listed and comparing these values to the criteria of $30°C < BP < 100°C$ for VOST and $BP > 100°C$ for Semi-VOST, the following distribution can be given:

VOST	Semi-VOST
Carbon tetrachloride	Hexachlorobenzene
1,1,1-Trichloroethane	Heptachlor epoxide
Tetrachloroethylene	Chlordane
	Decachlorobiphenyl
	Pentachlorophenol
	Kepone
	Hexachlorobutadiene

2. The distribution will not be mutually exclusive. Some of the less volatile compounds from the VOST procedure will be collected in the Semi-VOST system and vice versa. Compounds will have substantial vapor pressures below their boiling points so the dividing line between analytical methods is not distinct.

3. If the true value of the DRE is 0.9999, then the true value of the mass flow out of the incinerator is

$$w_{out} = (0.0001)(1.0\,kg/h) = 0.1\,g/h$$

If the analytical tolerance is the true value ±50%, then the range of the mass flow out of the incinerator will be 0.05 g/h to 0.15 g/h. Because the DRE is defined as

$$DRE = \left(1 - \frac{w_{out}}{w_{in}}\right)100$$

then the allowable measured range of the true DRE will be 99.985–99.995% for a true DRE of 99.99%.

CHR.10   RANKING OF POHCS

1. The Environmental Protection Agency (EPA) proposes that the difficulty of incinerating individual POHCs can be predicted based upon the heat of combustion value for each POHC. Using this rationale, rank the 10 POHCs from Problem CHR.9 in order of most to least difficult to incinerate.

2. Comment on the suitability of using a thermodynamic parameter, such as heat of combustion, for predicting incinerability under reaction conditions with reaction times of only ≈ 2 s. Propose an alternative concept that might be more suitable if data were available?

Solution

1. The difficulty of incinerating individual POHCs from Problem CHR.9 in order of most to least difficult to incinerate are:

		kcal/g
Supposedly most	Carbon tetrachloride	0.24
difficult to incinerate	Tetrachloroethylene	1.19
	Hexachlorobenzene	1.79
	1,1,1-Trichloroethane	1.99
	Pentachlorophenol	2.09
	Hexachlorobutadiene	2.12
	Kepone	2.15
Supposedly least	Decachlorobiphenyl	2.31
difficult to incinerate	Heptachlor epoxide	2.70
	Chlordane	2.71

2. A thermodynamic property may be appropriate for predicting equilibrium conditions, but the rapid reactions in an incinerator will be governed by kinetic considerations as well. Consequently, a measure of incinerability based on reaction energy and destruction kinetics would be a more relevant indicator of the difficulty of combustion for a given chemical.

24 Nuclear/Radioactive Waste (NUC)

NUC.1 CLASSIFICATION OF RADIOACTIVE WASTES

Provide answers to the following questions:

1. What is a radioactive waste?
2. What are the different categories of radioactive wastes?
3. Identify at least three sources of radioactive wastes.
4. Describe methods by which the volume of radioactive waste may be minimized.

Solution

1. Radioactive wastes are waste materials that consist of unstable isotopes. Over time, the materials decay to a more stable form (or element) emitting potentially harmful energy in the process.

2. Radioactive wastes are present in several forms. These forms are identified as follows:
   - High-level waste (HLW)
   - Low-level waste (LLW)
   - Transuranic waste (TRU)
   - Uranium mine and mill tailings
   - Mixed wastes
   - Naturally occurring radioactive wastes

3. Sources of radioactive wastes include:
   - Nuclear power
   - Government waste (nuclear defense)
   - Medical radiotherapy/hospitals
   - Mining waste (particularly phosphate mining)
   - Normally occurring radioactive materials
   - Industrial waste

4. One method of radioactive waste minimization is to compact the waste into a smaller, more densely packed volume. A second method is by incineration, specifically of exposed organic materials. A third method of waste volume reduction is by either dewatering (filtration) and/or evaporation for water removal and recovery.

NUC.2   NUCLEAR ACCIDENTS

Three historical events have been recognized as significant radiological accidents. These events include the meltdown at Chernobyl, the partial meltdown at Three Mile Island, and an incident of radioactive waste mishandling in Brazil. Elaborate on the events that occurred in Brazil. What were the impacts on the public health of the local community?

Solution

A stainless steel container holding a small amount of radioactive cesium had been abandoned by a medical clinic in Brazil and ended up in a junkyard. Local inhabitants, in an attempt to salvage junk, discovered the container and forced it open to inspect the contents. Discovering the luminescent cesium, the locals used the materials as costume glitter, exposing their skin, face, and clothing to direct contact with the radioactive material. People exposed directly to the cesium suffered from severe illness and death. The release of approximately 1 g of the radioactive cesium generated over 40 tons of additional radioactive waste from exposed homes and clothing.

NUC.3   MECHANISMS OF RADIOACTIVE TRANSFORMATIONS

Identify the three mechanisms of radioactive transformation of unstable elements.

Solution

Radioactive transformations are accomplished by several different mechanisms, most importantly alpha particle, beta particle, and gamma ray emissions. Each of these mechanisms are spontaneous nuclear transformations. The result of these transformations is the formation of different stable elements. The kind of transformation that will take place for any given radioactive element is a function of the type of nuclear instability as well as the mass/energy relationship. The nuclear instability is dependent on the ratio of neutrons to protons; a different type of decay will occur to allow for a more stable daughter product. The mass/energy relationship states that for any radioactive transformations the laws of conservation of mass and the conservation of energy must be followed.

An alpha particle is an energetic helium nucleus. The alpha particle is released from a radioactive element with a neutron-to-proton ratio that is too low. The helium nucleus consists of two protons and two neutrons. The alpha particle differs from a helium atom in that it is emitted without any electrons. The resulting daughter product from this type of transformation has an atomic number that is two less than its parent and an atomic mass number that is four less. Below is an example of alpha decay using polonium (Po); polonium has an atomic mass number (protons and neutrons) and atomic number of 210 and 84, respectively.

$$_{84}Po^{210} \rightarrow \ _2He^4 + \ _{82}Pb^{206}$$

The terms He and Pb represent helium and lead, respectively. This is a useful example because the lead daughter product is stable and will not decay further. The neutron-to-proton ratio changed from 1.50 to 1.51, just enough to result in a stable element. Alpha particles are known as having a high LET or linear energy transfer. The alphas will only travel a short distance while releasing energy. A piece of paper or the top layer of skin will stop an alpha particle. So, alpha particles are not external hazards but can be extremely hazardous if inhaled or ingested.

Beta particle emission occurs when an ordinary electron is ejected from the nucleus of an atom. The electron (*e*), appears when a neutron (*n*) is transformed into a proton within the nucleus:

$$_0n^1 \rightarrow \ _1H^1 + \ _{-1}e^0$$

Note that the proton is shown as a hydrogen (H) nucleus. This transformation must conserve the overall charge of each of the resulting particles. Contrary to alpha emission, beta emission occurs in elements that contain a surplus of neutrons. The daughter product of a beta emitter remains at the same atomic mass number but is one atomic number higher than its parent. Many elements that decay by beta emission also release a gamma ray at the same instant. These elements are known as *beta–gamma emitters*. Strong beta radiation is an external hazard because of its ability to penetrate body tissue.

Similar to beta decay is positron emission, where the parent emits a positively charged electron. Positron emission is commonly called *beta-positive decay*. This decay scheme occurs when the neutron-to-proton ratio is too low and alpha emission is not energetically possible. The positively charged electron, or positron, will travel at high speeds until it interacts with an electron. Upon contact, each of the particles will disappear and two gamma rays will result. When two gamma rays are formed in this manner, it is called *annihilation radiation*.

Unlike alpha and beta radiation, gamma radiation is an electromagnetic wave with a specified range of wavelengths. Gamma rays cannot be completely shielded against but can only be reduced in intensity with increased shielding. Gamma rays typically interact with matter through the photoelectric effect, Compton scattering, pair production, or direct interactions with the nucleus.

NUC.4   FIELD SAMPLING AND MEASUREMENT OF RADIOACTIVE WASTES

Describe the varying individual investigations that can be included in a field sampling and measurements program involving nuclear and radioactive wastes.

Solution

1. Characterization of radiologically contaminated wastes stored or disposed of onsite above or below ground. These may be found in tanks, drums, lagoons, impoundments, piles, pits, or a variety of wood or cardboard containers. The wastes may also be dispersed on the surface or mixed with other media (i.e., soil, water).

2. Hydrogeologic investigations to assess the horizontal and vertical distribution of radiological constituents in the underlying groundwater. This information will, in turn, permit the evaluation of the short- and long-term potential for contaminant dispersion in the groundwater both on- and offsite, and also provide a basis for evaluating the suitability of a site for long-term waste containment and isolation from the environment.

3. Surface and subsurface soils investigations to assess the location and extent of contamination from each significant constituent. In many cases, the prior mixing of the wastes with the soil makes the soil characterization a critical aspect of defining the source strength and extent.

4. Surface water investigation to evaluate the extent of contribution from the source to contamination of local surface water bodies. The sampling program should attempt to distinguish between contributions from runoff, deposition, or cross contamination from groundwater sources. In addition, both in the case of the surface and groundwater investigations, samples should be taken from up-gradient and down-gradient of the contaminant source to isolate the source contribution. Surface water investigations will also include collection and analysis of sediment samples since the sediment frequently acts as a preferential concentrator of contaminants.

5. Air investigations to determine the tendency of airborne particulates and gases to be released into the atmosphere, the on- and offsite migration and deposition patterns, and particularly the concentrations at significant locations such as site boundaries and local residences. In conjunction with the contaminant measurements program, it is also essential that local wind patterns (directions and stability classes) be determined to provide input to the airborne dispersion modeling program. Where wastes contain constituents of the uranium or thorium decay chains, it is necessary that the airborne releases of radon or thoron gas, respectively, be measured in terms of fluxes from the contaminated surface and concentrations at significant receptor points. The values establish the starting point for cleanup activities that will reduce the fluxes to essentially background levels.

6. Local flora and fauna analysis to permit determination as to whether the contaminants have entered the food chain and to assess the tendency of various species to concentrate or eliminate individual contaminants. In some cases, it is necessary to supplement the field investigations with controlled bench- or pilot-scale studies. These studies may be performed to simulate a mobilization or dispersion mechanism, or the complex chemical interactions between the waste form, surrounding matrix, or soil pathways, and/or the effectiveness of certain technologies in preventing migration or providing the required level of isolation. These pilot studies are often defined as feedback and obtained from the assessment of remedial alternatives.

NUC.5   REQUIRED NHV FOR THE INCINERATION OF A RADIOACTIVE WASTE

A radioactive mixture is to be burned in an incinerator at an operating temperature of $1900°F$. Calculate the minimum net heating value (NHV) of the mixture in Btu/lb if 0, 20, 40, 60, 80, and 100% excess air is employed. Use the following equation to perform the calculations:

$$NHV = \frac{(0.3)(T-60)}{[1-(1+EA)(7.5 \times 10^{-4})(0.3)(T-60)]}$$

Solution

For this problem,

$$T = 1900°F$$

For 0% excess air:

$$NHV = \frac{(0.3)(1900-60)}{[1-(1)(7.5 \times 10^{-4})(0.3)(1900-60)]}$$

$$= 942 \, Btu/lb$$

Similarly,

$$NHV \, (20\% \text{ excess air}) = 1097 \, Btu/lb$$

$$NHV \, (40\% \text{ excess air}) = 1313 \, Btu/lb$$

$$NHV \, (60\% \text{ excess air}) = 1635 \, Btu/lb$$

$$NHV \, (80\% \text{ excess air}) = 2166 \, Btu/lb$$

$$NHV \, (100\% \text{ excess air}) = 3209 \, Btu/lb$$

NUC.6 POTENTIAL LIABILITY

Ruocco Chemical Company transports slurry containing a solid nuclear waste slurry to a disposal site. On average, Ruocco's hauling trucks carry 4 tons of waste per trip for a total of 32,000 tons per year. In the event of a truck overturn, it can be assumed that 2 tons of the waste is spilled. US Department of Transportation (DOT) statistics indicate that 1 out of 4000 waste hauling trucks overturn during an average trip. Studies also indicate that cleanups resulting from transportation spills cost at a minimum as much as $10,000 per ton. Calculate the minimum total potential liability of producing and "disposing" of the waste in this manner. Express the answer on both an annual and per ton basis.

Solution

The potential liability (PL) is solved by the following equation:

$$PL = \left(\frac{\$10,000}{ton\ spilled}\right)\left(\frac{x\ tons\ spilled}{overturn}\right)\left(\frac{u\ tons\ of\ waste}{yr}\right)\left(\frac{1\ trip}{z\ tons}\right)\left(\frac{spill}{4000\ trips}\right)$$

where $x = 2$ tons spilled/overturn
$u = 32,000$ tons of waste/yr
$z = 4$ tons

$$PL = \left(\frac{\$10,000}{ton\ spilled}\right)\left(\frac{2\ tons\ spilled}{overturn}\right)\left(\frac{32,000\ tons\ of\ waste}{yr}\right)\left(\frac{1\ trip}{4\ tons}\right)\left(\frac{spill}{4000\ trips}\right)$$

$$= \$40,000\ per\ year$$

On a ton basis, the following can be obtained:

$$ton\ PL = \left(\frac{\$40,000}{yr}\right)\left(\frac{yr}{32,000\ tons\ of\ waste}\right)$$

$$= \$1.25\ per\ ton$$

NUC.7 POWER REQUIREMENT CALCULATION

Estimate the power required for a conveyor belt system transporting 2.5 tons per hour of radioactive contaminated solids from a location in a utility to a storage bin. The process unit end of the conveyor is at ground level while the top of the storage bin is 25 ft high. The total length of the belt system is 75 ft. Typical power requirements for flat conveyors of this capacity may be assumed to be 2.13 hp per 100 ft. Assume a safety factor of 10%.

Solution

Calculate the power required to lift the solids:

$$\text{Power} = (\text{Mass flow})(\text{Height})(g/g_c)$$
$$= (2.5 \text{ tons/h})(2000 \text{ lb/ton})(25 \text{ ft})(1 \text{ lb}_f/\text{lb})$$
$$= 125,000 \text{ ft} \cdot \text{lb}_f/\text{h}$$
$$= \left(\frac{125,000 \text{ ft} \cdot \text{lb}_f}{h}\right)\left(\frac{1 \text{ h}}{3600 \text{ s}}\right)\left(\frac{\text{hp}}{500 \text{ ft} \cdot \text{lb}_f/\text{s}}\right)$$
$$= 0.063 \text{ hp}$$

Calculate the power required to operate the conveyor if it were level:

$$\text{Power} = (2.13/100 \text{ ft})(75 \text{ ft})$$
$$= 1.60 \text{ hp}$$

The total power is

$$\text{Total power} = (0.063 + 1.60) \text{ hp} = 1.663 \text{ hp}$$

The total power, employing a 10% safety factors therefore

$$\text{Total power} = 1.663 \text{ hp} (1.10) = 1.83 \text{ hp}$$

Most nuclear wastes are in liquid or solid form. If the waste is solid, it may be transported by various methods: in drums using forklifts, on conveyor belts, suspended in gases in pneumatic conveyors, suspended in liquids as slurries, in screw conveyors, in bucket elevators, etc.

NUC.8   RADIOACTIVE DECAY

Exponential decay can be described using either a reaction rate coefficient (k) or a half-life (τ). The equation that relates these two parameters is as follows:

$$\tau = \frac{0.693}{k}$$

Exponential decay can be expressed by the following equation:

$$N = N_o e^{-kt}$$

where k = reaction rate coefficient (time^{-1})
N_0 = initial amount
N = amount at time t

Determine how much of a 100-g sample of Po-210 is left after 5.52 days using:

1. The reaction rate coefficient
2. The half-life
3. Calculate the percent error between the two methods.

The half-life for Po-210 is 1.38 days.

Solution

1. Determine the reaction rate coefficient:

$$k = \frac{0.693}{\tau} = \frac{0.693}{1.38 \text{ day}}$$
$$= 0.502/\text{ day}$$

The amount of substance left after 5.52 days is as follows:

$$N = N_0 e^{-Kt} = (100)e^{-(0.502)(5.52)}$$
$$= 6.26 \text{ g}$$

2. The first step is to determine how many half-lives the 100-g sample has undergone in the given time period:

$$\text{Number of half-lives} = (5.52 \text{ days})/(1.38 \text{ days}) = 4.0$$

Therefore, in a 5.52-year period, the 100-g sample has undergone four half-lives.

The amount of the substance left after one half-life is calculated as follows:

$$(0.5)^1 (100 \text{ g}) = 50 \text{ g}$$

Therefore, the amount of substance left after four half-lives is

$$(0.5)^4 (100 \text{ g}) = 6.25 \text{ g}$$

3. Since the two equations are, in principle, identical to each other, there is no difference between the two values. The small difference between the two results arises because of roundoff errors.

NUC.9   PROBABILITY OF EXCESSIVE Pb CONCENTRATION IN TRANSPORT DRUMS

Tests indicate that radioactive sludge waste arriving in 55-gal drums to a storage facility has a mean lead content of 15 ppm with a standard deviation of 12 ppm. The drums are unloaded into a 300-gal receiving tank. The storage facility is required to keep the lead concentration at or below 20 ppm in order to meet the required standard. Assume that the lead content from one drum to the next are not correlated, and that the tank is nearly full. What is the probability that the lead content in any drum exceeds 45 ppm?

Solution

Refer to Chapter 52 for normal distribution information and equations.

For a standard deviation of 12 ppm and a mean of 15 ppm, a lead concentration of 45 ppm represents $(45-15)/12$ or 2.5 standard deviations above (displaced from) the mean.

In Figure 85, the area in the "right-hand tail" (above z_0) of the normal distribution curve represents the probability that the variable (an event) is z_0 standard deviations above the mean. Applied to this problem, the area in the right-hand tail (above a z_0 of 2.5) is the probability that the lead content in any drum exceeds 45 ppm. From the table, the area in the right-hand tail corresponding to a z_0 of 2.5, is 0.006 or 6% of the total area under the curve. The probability that lead content in a drum exceeds 45 ppm is therefore 0.6%.

z_0	Next Decimal Place of z_0									
	0	1	2	3	4	5	6	7	8	9
0.0	0.500	0.496	0.492	0.488	0.484	0.480	0.476	0.472	0.468	0.464
0.1	0.460	0.456	0.452	0.448	0.444	0.440	0.436	0.433	0.429	0.425
0.2	0.421	0.417	0.413	0.409	0.405	0.401	0.397	0.394	0.390	0.386
0.3	0.382	0.378	0.374	0.371	0.367	0.363	0.359	0.356	0.352	0.348
0.4	0.345	0.341	0.337	0.334	0.330	0.326	0.323	0.319	0.316	0.312
0.5	0.309	0.305	0.302	0.298	0.295	0.291	0.288	0.284	0.281	0.278
0.6	0.274	0.271	0.268	0.264	0.261	0.258	0.255	0.251	0.248	0.245
0.7	0.242	0.239	0.236	0.233	0.230	0.227	0.224	0.221	0.218	0.215
0.8	0.212	0.209	0.206	0.203	0.200	0.198	0.195	0.192	0.189	0.187
0.9	0.184	0.181	0.179	0.176	0.174	0.171	0.189	0.166	0.164	0.161
1.0	0.159	0.156	0.154	0.152	0.149	0.147	0.145	0.142	0.140	0.138
1.1	0.136	0.133	0.131	0.129	0.127	0.125	0.123	0.121	0.119	0.117
1.2	0.115	0.113	0.111	0.109	0.107	0.106	0.104	0.102	0.100	0.099
1.3	0.097	0.095	0.093	0.092	0.090	0.089	0.087	0.085	0.084	0.082
1.4	0.081	0.079	0.078	0.076	0.075	0.074	0.072	0.071	0.069	0.068
1.5	0.067	0.066	0.064	0.063	0.062	0.061	0.059	0.058	0.057	0.056
1.6	0.055	0.054	0.053	0.052	0.051	0.049	0.048	0.047	0.046	0.046

	0	1	2	3	4	5	6	7	8	9
1.7	0.045	0.044	0.043	0.042	0.041	0.040	0.039	0.038	0.038	0.037
1.8	0.036	0.035	0.034	0.034	0.033	0.032	0.031	0.031	0.030	0.029
1.9	0.029	0.028	0.027	0.027	0.026	0.026	0.025	0.024	0.024	0.023
2.0	0.023	0.022	0.022	0.021	0.021	0.020	0.020	0.019	0.019	0.018
2.1	0.018	0.017	0.017	0.017	0.016	0.016	0.015	0.015	0.015	0.014
2.2	0.014	0.014	0.013	0.013	0.013	0.012	0.012	0.012	0.011	0.011
2.3	0.011	0.010	0.010	0.010	0.010	0.009	0.009	0.009	0.009	0.008
2.4	0.008	0.008	0.008	0.008	0.007	0.007	0.007	0.007	0.007	0.006
2.5	0.006	0.006	0.006	0.006	0.006	0.005	0.005	0.005	0.005	0.005
2.6	0.005	0.005	0.004	0.004	0.004	0.004	0.004	0.004	0.004	0.004
2.7	0.003	0.003	0.003	0.003	0.003	0.003	0.003	0.003	0.003	0.003
2.8	0.003	0.002	0.002	0.002	0.002	0.002	0.002	0.002	0.002	0.002
2.9	0.002	0.002	0.002	0.002	0.002	0.002	0.002	0.001	0.001	0.001

z_0	Detail of Tail ($._2135$, for example, means 0.00135)									
2.	$0._1228$	$0._1179$	$0._1139$	$0._1107$	$0._2820$	$0._2621$	$0._2466$	$0._2347$	$0._2256$	$0._2187$
3.	$0._2135$	$0._3968$	$0._3687$	$0._3483$	$0._3337$	$0._3233$	$0._3159$	$0._3108$	$0._4723$	$0._4481$
4.	$0._4317$	$0._4207$	$0._4133$	$0._5854$	$0._5541$	$0._5340$	$0._5211$	$0._5130$	$0._6793$	$0._6479$
5.	$0._6287$	$0._6170$	$0._7996$	$0._7579$	$0._7333$	$0._7190$	$0._7107$	$0._8599$	$0._8332$	$0._8182$
0	1	2	3	4	5	6	7	8	9	

[a] From R. J. Woonacott and T. H. Woonacott, *Introductory Statistics*, 4th ed., Wiley, New York, 1985.

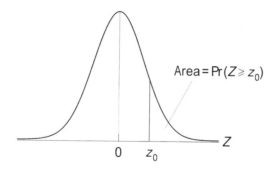

Figure 85. Standard normal, cumulative probability in right-hand tail (for negative values of z, areas are found by symmetry)[a]

NUC.10   PROBABILITY OF EXCESSIVE Pb CONCENTRATION IN RECEIVING TANK

With reference to Problem NUC.9, (1) what is the probability that the lead content in the receiving tank exceeds 20 ppm? (2) State the limitations of the analysis, and how these limitations could be accommodated.

Solution

1. To determine the standard deviation of the mean lead concentration in the receiving tank, a basic theorem in probability called the *central limit theorem* must be employed. If σ represents the standard deviation of the lead concentrations in the drums, then the standard deviation of the mean lead concentration in the receiving tank is given by $\sigma/\sqrt{n}$ where n is the number of drums:

$$n = 300\,\text{gal}/(55\,\text{gal}/\text{drum}) = 5.45\,\text{drums}$$

The standard deviation of the mean is then

$$\frac{\sigma}{\sqrt{n}} = \frac{12\,\text{ppm}}{\sqrt{5.45}} = 5.1\,\text{ppm}$$

For a standard deviation of 5.1 ppm and a mean of 15 ppm, a lead concentration of 20 ppm is $(20-15)/5.1$ or 0.98 standard deviations above the mean.

Again from the normal distribution table, for $z_0 = 0.98$, the probability that the mean lead content in the receiving tank exceeds 20 ppm is 0.164 or 16.4%.

2. The limitations in the analysis and possible accommodations are:

   a. The critical assumption made is that the lead content of the drums and waste shipments are not correlated in time. In other words, it has been assumed that it is unlikely that the facility will receive consecutive shipments of waste with a lead content higher than the mean. With multiple drums from a given source, it is highly likely, however, that the drum contents will be correlated.

   b. The distribution of the lead content in the problem was assumed to be normal. Concentrations often have skewed distributions, e.g., log-normal.

   c. With additional information, e.g., autocorrelation and skewness of the concentration distribution, much more complex and realistic problems can be solved using queuing or inventory techniques. Oversizing the tank would provide a margin of safety.

25 Superfund (SUP)

SUP.1 COMPREHENSIVE ENVIRONMENTAL RESPONSE, COMPENSATION AND LIABILITY ACT

Answer the following questions.

1. In 1980, the Comprehensive Environmental Response, Compensation and Liability Act (CERCLA) created the Superfund. What were the objectives of the Superfund?
2. Identify and describe the three concepts that make CERCLA a unique law?
3. The Environmental Protection Agency (EPA) has four options to achieve compliance with the objectives under CERCLA. What are these options?

Solution

1. In 1980, CERCLA put aside $1.6 billion in a trust fund that was to be used to fund cleanup efforts for the nation's highest risk hazardous waste sites. CERCLA established the means by which the highest risk sites were identified and ranked on a national priorities list (NPL) for cleanup.
2. The three concepts that make CERCLA a unique law are:
   a. *Ex post facto:* According to ex post facto, any party can be liable for past waste disposal techniques even though they may have been once legal but are now illegal and have led to contamination of a specific area.
   b. *Landowner liability:* Any landowner who does not perform an "all appropriate inquiry" may be found liable for hazardous waste contamination despite whether they committed the act or not.
   c. *Joint and several liability:* Liability may be shared among several potentially responsible parties (PRP) who have occupied the property in the past.
3. The four options the EPA has to achieve compliance with under CERCLA are:
   a. *Voluntary compliance:* Responsible party accepts liability for cleanup of a hazardous waste site.
   b. *Enforcement agreement:* Either a judicial consent decree or administrative order that identifies a party's responsibilities with regards to a site cleanup.

   c. *Unilateral administrative order:* A decree that forces a PRP to take responsibility for a site cleanup.

   d. *Lawsuit:* EPA can bring charges against a PRP who then faces punitive damages up to three times the cost of the cleanup.

SUP.2   NATURAL PRIORITIES LIST

Answer the following questions:

1. What three criteria are used to place a hazardous waste site on the natural priorities list (NPL)?
2. Identify the steps for cleanup of a hazardous waste site once it has been put on the national priorities list.
3. Describe the steps the public may take to assist the EPA in finding a PRP or influencing the handling of an NPL cleanup.

Solution

1. The three criteria used to place a hazardous waste site on the NPL are:
   a. The site receives an advisory from the Agency of Toxic Substances and Disease Registry that recommends that local residents relocate away from the site.
   b. The site receives a score of 28.5 or higher on the hazard ranking system.
   c. The site is selected as a state's top cleanup priority.
2. Note that a total of 10 steps have been identified as steps in the clean-up of a hazardous waste site. The first 5 deal with the steps leading up to placement on the NPL. The second 5 refer to actions taken following placement on the NPL. These 5 steps are listed as follows:
   a. Remedial investigation and feasibility study (RI/FS)
   b. Remedy selection/record of decision
   c. Remedial design
   d. Remedial action
   e. Project close out
3. The steps the public may take to assist the EPA in finding a PRP or influencing the handling of an NPL cleanup are:
   a. Report illegal hazardous waste dumping.
   b. Petition the EPA to perform a preliminary assessment of a potentially affected area.
   c. Collaborate cleanup discussions.
   d. Solicit information from the EPA on programs or a specific site.

e. Take part in EPA-sponsored community involvement programs.

SUP.3   SUPERFUND LEGISLATION

Answer the following questions:

1. Describe various response actions and remedial technologies.
2. List the principal federal agencies involved in feasibility studies.
3. List the principal federal legislative acts and policies applicable to uncontrolled hazardous waste sites.

Solution

1. Various response actions and remedial technologies are:

No action	*Control of gas migration*
	Capping: containment; collection
Site access control	*Removal of materials*
Fencing; land-use limits	Excavation; grading; capping
Surface water control	*Off site treatment*
Capping; grading; revegetation; collection systems	Physical, biological, and chemical
Groundwater or leachate control	*Water or sewer line decontamination*
Capping; containment; pumping	In situ cleaning; replacement; alternative supply
Onsite treatment	
Incineration; physical, biological, and chemical treatment	

2. The principal federal agencies involved in feasibility studies are:
   - Environmental Protection Agency
   - Federal Emergency Management Agency
   - Department of Health & Human Services
   - Army Corps of Engineers
   - Department of the Interior
     Geological Survey
     Bureau of Land Management
     Fish & Wildlife Services
   - Department of Transportation
     Coast Guard

3. The principal federal legislative acts and policies applicable to uncontrolled hazardous waste sites are:

- Comprehensive Environmental Response, Compensation, and Liability Act (CERCLA) PL 96-510, 42 U.S.C. 9601
- Superfund Amendments and Reauthorization Act of 1986 (SARA) PL 99-499
- Resource Conservation and Recovery Act (RCRA) PL 94-580, as amended, 41 U.S.C. 6901
- Safe Drinking Water Act (SDWA) PL 93-523 as amended, 42 U.S.C. 300f *et seq.*
- Clean Water Act (CWA) PL 92-500, as amended, 33 U.S.C. 1251 *et seq.*
- Ground Water Protection Agency (GWPS) U.S. EPA Report WH 550, November 1980
- Clean Air Act (CAA) PL 90-148 as amended, 42 U.S.C. 7401 *et seq.*

SUP.4   HEAT RELEASE RATE CALCULATION

A rotary kiln at a Superfund site is designed to treat contaminated soil with a nominal heat release rate (HRR) of 15,000 Btu/(h · ft^3), an inside diameter of 8 ft, and a length of 30 ft. What is the design heat rate ($\dot{Q}$) in Btu/h?

Solution

$$\text{Kiln volume, } V = \frac{30\pi(8)^2}{4} = 1508 \text{ ft}^3$$

The design heat rate is given by

$$\dot{Q} = (V)(\text{HRR})$$

$$= (1508 \text{ ft}^3)\left(\frac{15,000 \text{ Btu}}{\text{h} \cdot \text{ft}^3}\right) = 22,620,000 \text{ Btu/h}$$

SUP.5   BATCH FEED TO A KILN

Normally a waste having a heating value of 750 Btu/lb is burned in a kiln. It is fed continuously through an auger. Occasionally a solid waste consisting of contaminated polyethylene pellets is "batch" fed to the kiln in 30-gal fiber containers. The pellets have a bulk density of 50 lb/ft^3 and a heating value (HV) of 18,350 Btu/lb. A single container is consumed in 6.5 min. Assume the kiln described in Problem SUP.4 is employed in this operation.

1. Will the kiln operate within its design parameters?
2. Considering your answer to part 1, would you expect any combustion problems?
3. How would you correct the problem if you were the operator?

Solution

1. To determine if the kiln operates within its design parameters, the actual rate should be compared to the design value. The mass m in each container is

$$m = (30 \text{ gal})\left(\frac{8.35 \text{ lb}}{\text{gal}}\right)\left(\frac{50}{62.4}\right) = 201 \text{ lb}$$

The actual heat release, Q, is then

$$Q = m(\text{HV})$$

$$= (201 \text{ lb})\left(\frac{18{,}350 \text{ Btu}}{\text{lb}}\right)$$

$$= 3{,}688{,}000 \text{ Btu heat release}$$

But this will occur in 6.5 min; therefore the equivalent hourly heat rate, $\dot{Q}$, is

$$\dot{Q} = \frac{60 \text{ min/h}}{6.5 \text{ min}} (3{,}688{,}000 \text{ Btu}) = 34 \text{ million Btu/h}$$

This is about 1.5 times the design heat rate of 22.6 million Btu/h so the answer to part 1 is No.
2. There will not be enough air in the system to burn the polyethylene waste at that rate so the kiln will produce a dense black smoke.
3. To correct the situation, package the polyethylene in smaller batches. The batch size will have to be smaller by at least the same factor (1.5). Therefore,

$$m = 201/1.5 = 134 \text{ lb/batch}$$

$$Q = (134)(18{,}350) = 2{,}459{,}000 \text{ Btu heat release}$$

The heat rate is then

$$\dot{Q} = (60/6.5)(2{,}458{,}900) = 33{,}697{,}538 \text{ Btu/h}$$

This is almost the same as the design value. Good practice would suggest that a safety margin of at least 10% be used so about 120 lb/batch is a good size.

Also, the containers must be fed at a rate not greater than one every 6.5 min, e.g., 9 per hour.

SUP.6 SIZING OF A ROTARY KILN

A rotary kiln incinerator is operating with an average energy release rate (HRR) of 28,000 Btu/(h · ft³) of furnace volume. During operation 4500 lb/h of a solid waste from a Superfund site with an approximate heating value of 8000 Btu/lb is to be combusted. Assume the L/D ratio of the rotary kiln to be 3.5.

1. Calculate the furnace volume required.
2. What are the dimensions of the kiln?

Solution

1. Calculate the heat released by the waste:

$$\dot{Q} = (4500)(8000) = 3.60 \times 10^7 \, \text{Btu/h}$$

The volume required is

$$V = \frac{\dot{Q}}{\text{HRR}}$$

$$= \frac{3.60 \times 10^7}{28,000} = 1286 \, \text{ft}^3$$

2. The dimensions of the kiln for an L/D ration of 3.5 are obtained as follows:

$$V = \left(\frac{\pi}{4}\right) D^2 L$$

$$= \left(\frac{\pi}{4}\right) D^2 (3.5D) = 2.75 \, D^3$$

$$D = \left(\frac{V}{2.75}\right)^{1/3}$$

$$= \left(\frac{1286}{2.75}\right)^{1/3}$$

$$= 7.76 \, \text{ft}$$

$$L = 3.5D = 3.5(7.76) = 27.2 \, \text{ft}$$

SUP.7   CALCULATION OF RECEIVING TANK SIZE

Tests indicate that a sludge waste arriving in 55-gal drums to a treatment facility have a mean lead content of 11 ppm with a standard deviation of 10 ppm. The drums are unloaded into a 250-gal receiving tank. The facility is required to keep the lead concentration entering the incinerator at or below 15 ppm in order to meet the required particulate emission levels. Assume that the lead contents from one drum to the next are not correlated and that the tank is nearly full. What size should the receiving tank be to ensure, with 98% confidence, that the facility treats a waste with a mean lead concentration below 15 ppm?

Solution

For this condition, the probability that the Pb concentration in the receiving tank exceeds 15 ppm is

$$1.0 - 0.98 = 0.02 \quad \text{or} \quad 2.0\%$$

The value of z_0 from the standard normal table (see Problem NUC.9 in Chapter 24) is then 2.05, which corresponds to the number of standard deviations above the mean tank concentration. According to the *central limit theorem*, the standard deviation of the mean is given by

$$\frac{\sigma}{\sqrt{n}} = \frac{10\,\text{ppm}}{\sqrt{n}}$$

where n is the number of drums. Therefore, for a 2% probability that the mean concentration in the tank exceeds 15 ppm, the number of drums can be found by solving

$$2.05 = \frac{15 - 11}{(10/\sqrt{n})}$$

or

$$n = 26.3$$

The tank volume V must then be

$$V = 26.3\,\text{drums} \,(55\,\text{gal}/\text{drum})$$
$$= 1446\,\text{gal}$$

SUP.8 FEASIBILITY OF LANDFILLING

The total amount of contaminated soil at the RAT (Ruocco and Theodore) company site is approximately 80,000 tons. Evaluate the cost of landfilling versus stabilization for the management of the hazardous waste at this site.

Each truck can carry 25,000 lb to the nearest suitable landfill site at Theodore Estates, at a distance of 750 miles. The trucking cost per mile is $2.50. The total stabilization cost is $62 per ton. Identify advantages and disadvantages of landfilling for this site.

Solution

Stabilization cost estimate:

$$80,000 \text{ tons} \times \$62/\text{ton} = \$4,960,000$$

Trucking cost estimate:

$$80,000 \text{ tons} \left(\frac{2000 \text{ lb}}{\text{ton}} \right) \left(\frac{\text{truck}}{25,000 \text{ lb}} \right) = 6400 \text{ trucks}$$

$$6400 \text{ trucks} \left(\frac{750 \text{ miles}}{\text{truck}} \right) \left(\frac{\$2.50}{\text{mile}} \right) = \$12,000,000$$

Because trucking costs alone are $12 million, stabilization is certainly more cost effective.

Advantages of landfilling: Landfilling permits the actual reclamation of the contaminated land at the Ruocco and Theodore company site and eliminates one site from the national inventory of contaminated sites.

Disadvantages of landfilling: Valuable landfill space is being used by soils amenable to other treatment, precluding the use of that space by wastes better suited for landfill.

SUP.9 CHOICE OF AN INCINERATION SYSTEM

A Superfund site has been studied and is ready for remedial action. The site contains approximately 40,000 cubic yards of soil contaminated by wood treating operations. The plant is in flat terrain in a rural location. The contaminants are from coal tar compounds and metals used to treat wood. The following data have been summarized from the remedial investigation/feasibility study (RI/FS).

Item	Value
Density of soil	$1.5 \, \text{ton/yd}^3$
Moisture content	10%
Heating value	1,000 Btu/lb
Ash content	85%
Soil makeup	
Clay content	80%
Sand and gravel	20%
Major contaminants	
Creosote PAH compounds	30,000 ppm
Pentachlorophenol (PCP)	100 ppm
Chromium	1 ppm
Arsenic	0.5 ppm
Ash fusion temperature	2,100°F

The project is bound by RCRA regulations, not TSCA (Toxic Substances Control Act), as the PCP content is low. The schedule calls for the production burn period to be less than 8 months. The required ash quality goal is < 5 ppm, total PAH (polyaromatic hydrocarbons) compounds.

Based on the above information, choose an appropriate incineration system. Provide the following information:

1. Size the system in terms of throughput in tons per hour.
2. Specify type of incinerator (e.g., rotary kiln, fluid bed, infrared, indirect fired desorber).
3. Type of pollution control (e.g., venturi scrubber, baghouse).
4. List primary RCRA required stack tests.

Solution

1. System size is based on 8 months of production burn time. This is 5840 operating hours. Dividing the tonnage (1.5 × 40,000 yards) by this value results in a feed rate of 10 tons per hour (tph). Applying a 65% capacity utilization (to account for forced and planned outages) for this project duration results in a capacity of 16 tph.
2. The predominant type of incinerator used for this type of project is the rotary kiln. A co-current (cools faster at inlet) type is preferred for feedstocks with high heating values. A fluid bed could also be used due to the high boiling points of the PAH compounds, the stringent ash quality goal, and relatively high PAH concentrations; a low-temperature desorber would not be a good choice.

3. The dust loading will be high as the feedstock is clay (which is very fine) plus metals. Therefore, the best choice for pollution control would be a baghouse to remove the particulate. A venturi scrubber would be a poor choice as it has difficulty in removing fine particulates and the scrubber water would have high suspended solids, requiring a high blowdown rate.

    Is an acid gas absorber needed? The pentachlorophenol feed rate at 16 tph of soil is 3.2 lb/h. It is not all chlorine and will produce less than 4 lb/h of HCl; hence, no acid gas absorber is required. As long as a nonchlorinated principal organic hazardous constituents (POHC) (e.g., naphthalene) is used in the destruction and removal efficiency (DRE) tests, HCl should not be a problem. Had HCl been more than the allowable limit, it could be taken care of with a wet acid gas absorber placed downstream of the baghouse or a spray dryer upstream of the baghouse (using a lime slurry as the reagent) or dry reagent injection into the baghouse if only low levels of acid gas were to be removed.

4. Primary RCRA regulations call for a DRE of at least 99.99% of all POHCs. The regulations also require 99% removal of HCl from the incinerator flue gas or a maximum of 4 lb/h (1.8 kg/h). Federal regulations require a maximum particulate emission of 0.08 gr/dscf corrected to 7% oxygen in the flue gas. In addition, many states require BACT (best available control technology). Metals emissions are dealt with in the next problem.

SUP.10   DESIGN OF AN INCINERATION SYSTEM

With reference to Problem SUP.9, answer the following:

1. Specify the stack height required to pass Tier 1 limits for metals.
2. Specify the type of feed preparation system.
3. Specify the type of ash handling system.
4. Specify the appropriate temperature for primary and secondary combustion chambers.
5. Estimate the stack gas volume in scfm.

Solution

1. Metals regulations are contained in the EPA "Guidance on Metals and Hydrogen Chloride Controls for Hazardous Waste Incinerators" issued in August, 1989. Chromium has the lowest emission limit of any metal on the list. Since its concentration is more than that of arsenic, the chromium limits will prevail. To solve for the effective stack height, first calculate the chrome feed rate: $32,000 \text{ lb/h} \times 1 \text{ ppm}/1,000,000 = 0.032 \text{ lb/h}$. Checking the Tier 1 tables for rural locations shows a limit of 0.033 lb/h for a 70-m effective stack height. If effective stack height is 70 m or more, no Tier 2 stack testing will be

required. The actual stack could be shortened by subtracting the plume rise, found by calculating the stack outlet temperature and velocity and using the appropriate table in the metals guidance document.

2. While soil data are limited, the large amount of clay suggests that the material could be sticky when wet and will, when heated, be fine and dusty. Feed handling would consist of oversized objects, a soil storage building to keep moisture to a minimum and conveyor belts to bring the material to the incinerator. A weigh belt or other device would monitor and record soil feed rates. The final feed device to the kiln could be dual augers, or if the material was sticky, a feed chute or slinger belt. The feed could also be blended with a bulking reagent to decrease moisture content. The mixing would take place in a pug mill.

3. The ash will be hot and fine. A rotary cooler would be the best choice to promote cooling and consolidation. It will be better than the alternative, a wet ash quench system which might have a problem in quenching fine clays and getting them out of the quench via the usual method, a drag chain conveyor.

4. The primary chamber temperatures must be held down by excess air or water sprays to prevent the solids temperature from approaching the ash fusion temperature. A minimum temperature to ensure clean ash passing the ash quality goal is 1200°F. A better operating temperature would be 1400°F for the ash outlet temperature, and assuming a 100°F difference between the ash and gas, a 1500°F outlet gas temperature. These temperatures are a good starting point for start-up and tests and could be raised during the production burn as long as slagging of the ash did not occur. A secondary combustion tempera-ture of 1700–1800°F is appropriate for > 99.99% DRE.

5. While accurate estimation of flue gas volumes requires heat and material balance and iterative calculations, a few rules of thumb can be used to produce a rough estimate. The heat required to raise the soil temperature to 1400°F is equal to the weight of soil (moisture free, $32,000 \times 0.9$) times the temperature difference ($1400 - 60$) times the specific heat [approximately $0.30\,Btu/(lb°F)$], or 12 MBtu/h (1 MBtu $= 10^6$ Btu). To vaporize the 3200 lb/h water, 3 MBtu/h is required. Available heat at a kiln exhaust gas temperature of 1500°F at 50% excess air is 45%, so a total of 33 MBtu/h is required. Adding in radiation loss brings this to 35 MBtu/h. The secondary combustion chamber takes approxi-mately an equal amount of heat, or 35 Btu/h. Adding these two produces total heat input of 70 MBtu/h.

The bulk of the stack gas volume is due to nitrogen in the combustion air. To find the combustion air requirement, use the rule of thumb that one scfm of air is equal to one Btu/h divided by 6000. This is equivalent to 168 scfm/MBtu. Doubling that for 100% excess air yields 23,000 scfm of combustion air at 70 MBtu/h fuel input. The fuel weighs about 50 lb/MBtu, and this will add 8% to the stack gas at 0% excess air, or 4% at 100% excess air, yielding a total figure of about 25,000 scfm of stack gas volume.

While the stack gas is not air, the density is close to that of air. Hence, the scfm estimated above includes the water vapor added in any cooling or quenching process.

26 Municipal Waste (MUN)

MUN.1 MUNICIPAL LANDFILLS

Provide answers to the following questions:

1. What are some reasons given for the opposition to the construction of new municipal landfills?
2. Identify several restrictions with respect to location of a new municipal landfill.
3. Identify several restrictions with respect to operations of a new municipal landfill.
4. What is the most significant resistance to the development of a recycling program?

Solution

1. Public opposition to new landfills as displayed by the "not in my backyard" attitude can be attributed to past problems with older landfills. Poor designs and locations have been responsible for the contamination of nearby ground-water sources. Ground contaminations by older landfills require extensive cleanup efforts at large expense.
2. Several restrictions with respect to location of a new municipal landfill are:
   a. Landfills must be constructed away from water sources and wetlands.
   b. They must be resistant to damage by flooding and should be located away from areas prone to earthquakes, landslides, mudslides, and sinkholes.
   c. Municipal landfills are not to be located in areas where the added bird population can interfere with local air traffic.
3. Several restrictions with respect to operations of a new municipal landfill are:
   a. Municipal landfills must be operated to minimize the amount of hazardous waste that can be deposited.
   b. Landfills must be covered daily with a fresh layer of soil to control local pest and animal populations.
   c. Landfills must monitor methane emissions and must have a plan to deal with excess emissions.

    d. No discharge, accidental or intentional, is allowed from a landfill to a local water body.

    e. Landfills must restrict any unauthorized dumping and prohibit burning of waste.

4. The most significant resistance to development of a recycling program is operating costs associated with collection, processing, and disposal of residual matter. Other operating costs include the cost for new equipment, labor, and utilities. Due to these costs, materials such as plastics, cardboard, and paper use are often more cost effective as virgin material rather than as recycled products.

MUN.2   MUNICIPAL WASTE MANAGEMENT

Answer the following two-part question:

1. Federal, state, and local governments have adopted an integrated approach to municipal waste management. Describe the three waste management techniques that support this approach.

2. What are the roles of the federal, state, and local governments in establishing regulations for municipal solid waste disposal facilities?

Solution

1. The three waste management techniques include:
    a. Reduce the overall toxicity of solid wastes by reducing its volume.
    b. Increase recycling efforts to prevent losing renewable resources.
    c. Improve landfill design and management.

2. The role of the federal government with respect to the regulation of municipal landfills is to set minimum national standards for states and local governments and to evaluate and monitor the effectiveness of state and local agencies in managing solid waste facilities. It is the responsibility of state and local governments to implement and enforce the national standards and the needs of the local community.

MUN.3   AMOUNT OF WASTE GENERATED BY A MUNICIPALITY

A municipality in the Midwest has a population of 50,000 and generates $100,000 \, \text{yd}^3$ of municipal waste annually. The waste is made up of 30% compacted waste and 70% uncompacted waste. Assume that the waste has a density of $1,000 \, \text{lb/yd}^3$

compacted, and 400 lb/yd³ uncompacted. How many pounds of waste are generated by this city each year? By each person each year?

Solution

Based on the waste densities given in the problem statement, the following generation rates are determined:

$$\text{Waste generated/yr} = (0.3)(100,000 \text{ yd}^3)\left(\frac{1,000 \text{ lb}}{\text{yd}^3}\right)$$

$$+ (0.7)(100,000 \text{ yd}^3)\left(\frac{400 \text{ lb}}{\text{yd}^3}\right)$$

$$= 30,000,000 \text{ lb} + 28,000,000 \text{ lb}$$

$$= 58,000,000 \text{ lb/yr}$$

$$\text{Per capita generation rate} = \frac{58,000,000 \text{ lb/yr}}{50,000 \text{ people}}$$

$$= 1160 \text{ lb/person} \cdot \text{yr}$$

$$= 3.2 \text{ lb/person} \cdot \text{day}$$

MUN.4 CHARACTERISTICS OF MUNICIPAL SOLID WASTE

Besides economic considerations and site specifications, what are the major characteristics of municipal solid waste that are employed for the selection of the treatment and disposal processes listed below?

1. Composting
2. Incineration
3. Sanitary landfill

Solution

The characteristics relevant to the treatment and disposal options listed in the problem statement are summarized below:

1. Composting: Biodegradable/nonbiodegradable fraction, moisture content, carbon-to-nitrogen ratio
2. Incineration: Heating value, moisture content, inorganic fraction, metal content, alkali earth metal content

3. Sanitary landfill: Biodegradable to nonbiodegradable fraction, liquid content, hazardous material content

MUN.5   CALCULATION OF AVERAGE WASTE DENSITY

An analysis of a solid waste generator has revealed that the waste is composed (by volume) of 20% supermarket waste, 15% plastic-coated paper waste, 10% polystyrene, 20% wood, 10% vegetable food waste, 10% rubber, and 10% hospital waste. What is the average density of this solid waste in lb/yd^3? Use the following as discarded waste densities (in lb/yd^3) for each of the components of the generator's waste: supermarket waste, 100; plastic-coated paper waste, 135; polystyrene, 175; wood, 300; vegetable food waste, 375; rubber-synthetics, 1,200; and hospital waste, 100.

Solution

Based on the volume percent composition and waste densities given in the problem statement, the following average discarded waste density is estimated to be

$$\text{Average density} = (0.2)(100) + (0.15)(135) + (0.15)(175)$$
$$+ (0.2)(300) + (0.1)(375) + (0.1)(1200) + (0.1)(100)$$
$$= 294 \, lb/yd^3$$

MUN.6   MOISTURE CONTENT

The design of garbage collection vehicles differs from county to county and country to country based on location, local culture, etc. Many factors influence the design of these vehicles, with moisture content being one of the most important considerations in the hauling of municipal waste. The general formula for calculating the moisture content of solid waste is as follows:

$$\text{Moisture content} = \left(\frac{A - B}{A}\right)(100\%)$$

where A = weight of sample as delivered, kg
B = weight of sample after drying, kg

Use the data provided below on waste component moisture content and the weight composition of Ruocco County, Italy, waste to determine the average moisture content of the municipal solid waste. Base the calculation on a 100-kg sample.

Typical Moisture Content of Municipal Solid Waste

Component	Moisture Content (wt %)
Food waste	70
Paper	6
Garden waste	60
Plastic	2
Textiles	10
Wood	20
Glass	2
Metals	3

Composition of Ruocco County, Italy, Solid Waste on a Weight Percent Basis

Component	Composition (wt %)
Food waste	20
Paper	22
Garden waste	18
Plastic	3
Textiles	7
Wood	25
Glass	2
Metals	3

Solution

The dry weight of each component can be determined using the following equation:

$$\text{Dry weight} = (\text{Discarded weight})(100 - \%\text{Moisture})/100$$

The discarded weight for each component can be determined using the following equation:

$$\text{Discarded weight} = (\text{Total waste weight})(\text{wt \% of component})$$

Finally the average moisture content of the waste is determined as:

$$\text{Avg. \% moisture} = [(\text{Discarded weight} - \text{Dry weight})/100 \text{ kg}](100\%)$$

Based on the %moisture content data and waste composition data provided in the problem statement, and assuming a total waste weight of 100 kg, the calculations presented below result for each waste component given in the problem statement.

Composition, Weight Percent, Component Discarded Weight, and Component Dry Weight for the Municipal Solid Waste from Ruocco County, Italy

Component	Wt %	Component Discarded Weight (kg)	Component Dry Weight (kg)
Food waste	20	20	$(1.0 - 0.7)(20) = 6.00$
Paper	22	22	$(1.0 - 0.6)(22) = 20.68$
Garden waste	18	18	$(1.0 - 0.6)(18) = 7.20$
Plastic	3	3	$(1.0 - 0.02)(3) = 2.91$
Textiles	7	7	$(1.0 - 0.1)(7) = 6.30$
Wood	25	25	$(1.0 - 0.2)(25) = 20.00$
Glass	2	2	$(1.0 - 0.02)(2) = 1.96$
Metals	3	3	$(1.0 - 0.03)(3) = 2.91$
Total	100		$67.96 = 68$
Average %moisture $= [(100-68)/100](100\%) = 32\%$			

MUN.7   DENSITY OF A COMPACTED WASTE

The Jefferson County Solid Waste Management Corporation analysis of its solid waste includes the following major components: weight, volume, and compaction factors. Determine the density of well-compacted waste as delivered to a landfill. (*Note:* waste components and compaction factors vary from place to place and may not be identical to those shown below for Jefferson County.)

Composition, Weight, Volume, and Compaction Factor of Solid Waste from Jefferson County, Mississippi

Component	Weight (kg)	Volume as Discarded (m^3)	Compaction Factor
Food waste	250	1.00	0.33
Paper	300	3.75	0.15
Garden waste	250	1.18	0.20
Plastic	50	0.80	0.10
Textiles	60	9.61	0.15
Wood	50	0.34	0.30
Glass	20	0.10	0.40
Metals	20	0.10	0.30
Total	1000		

Note: Compaction factor represents the ratio of the resultant volume to the original volume.

Solution

The compacted volume of each component in the landfill can be determined using the following equation:

$$\text{Compacted volume} = (\text{Discarded Volume})(\text{Compaction factor})$$

Based on the weight, discarded volume, and compaction factor data provided in the problem statement, the calculations presented in the table below result for each waste component. Finally, the average density of the well-compacted mixed waste delivered to the landfill can be determined as:

$$\text{Average density} = (\text{Discarded weight})/(\text{Total compacted volume})$$

Results are provided below.

Composition, Weight, Volume, Compaction Factor, and Compacted Volume of Solid Waste from Jefferson County, Mississippi

Component	Weight (kg)	Volume as Discarded (m^3)	Compaction Factor	Compacted Volume (m^3)
Food waste	250	1.00	0.33	0.33
Paper	300	3.75	0.15	0.56
Garden waste	250	1.18	0.20	0.24
Plastic	50	0.80	0.10	0.08
Textiles	60	9.61	0.15	1.44
Wood	50	0.34	0.30	0.10
Glass	20	0.10	0.40	0.004
Metals	20	0.10	0.30	0.018
Total	1000	—	—	2.77

MUN.8   REQUIRED LANDFILL AREA

Estimate the required landfill area for a community with a population of 260,000. Assume that the following conditions apply:

1. Solid waste generation $= 7.6 \, \text{lb/capita} \cdot \text{day}$
2. Compacted specific weight of solid wastes in landfill $= 830 \, \text{lb/yd}^3$
3. Average depth of compacted solid wastes $= 60 \, \text{ft}$

Solution

Determine the daily solid wastes generation rate in tons per day:

$$\text{Generation rate} = \frac{(260{,}000 \text{ people})(7.6 \text{ lb/capita} \cdot \text{day})}{2000 \text{ lb/ton}}$$

$$= 988 \text{ ton/day}$$

The required area is determined as follows:

$$\text{Volume required/day} = \frac{(988 \text{ ton/day})(2000 \text{ lb/ton})}{830 \text{ lb/yd}^3}$$

$$= 2381 \text{ yd}^3/\text{day}$$

$$\text{Area required/yr} = \frac{(2381 \text{ yd}^3/\text{day})(365 \text{ day/yr})(27 \text{ ft}^3/\text{yd}^3)}{(20 \text{ ft})(43{,}650 \text{ ft}^2/\text{acre})}$$

$$= 26.88 \text{ acre/yr}$$

$$\text{Area required/day} = \frac{(2381 \text{ yd}^3/\text{day})(27 \text{ ft}^3/\text{yd}^3)}{(20 \text{ ft})(43{,}650 \text{ ft}^2/\text{acre})}$$

$$= 0.074 \text{ acre/day}$$

The actual site requirements will be greater than the value computed because additional land is required for a buffer zone, office and service building, access roads, utility access, and so on. Typically, this allowance varies from 20 to 40%. Thus, if an allowance of 30% is employed, the daily area requirement becomes

$$\text{Area required/day} = (0.074)(1.3)$$

$$= 0.096 \text{ acre/day}$$

A more rigorous approach to the determination of the required landfill area involves consideration of the contours of the completed landfill and the effects of gas production and overburden compaction.

MUN.9   CALCULATIONS ON A MSW INCINERATOR

A small community in New York incinerates 10,000 lb/h of municipal solid waste (MSW) at a central disposal facility using 100% excess air. The waste has a heating value of 6500 Btu/lb, an ash content of 10 wt %, and a moisture content of 25 wt %.

It may be assumed that 7.5 lb of dry gas is generated for every 10,000 Btu of waste burned in the incinerator. In addition, 0.51 lb of moisture is produced for every 10,000 Btu of waste fired. Air cools the incinerator shell at a flow of 10,000 lb/h (this air is discharged to the atmosphere at 400°F). Assume the air entering the incinerator has a humidity of 0.015 lb of moisture per pound of dry air and the ash produced has a heating value of 100 Btu/lb (assume no fly ash). The radiation heat loss from the shell is 1% of the total heating value of the waste. The temperature of the products of combustion (the flue gas) must not be less than 2000°F. The following information is required:

1. Calculate the inlet and outlet flowrates in lb/h on both a dry and a wet basis.
2. Determine the amount of ash remaining after combustion.
3. Calculate the volatile heating value of the waste.
4. Determine if supplemental fuel must be added to the incinerator.
5. Comment on the ultimate disposition of the remaining solid waste (ash).

Solution

Determine the moisture feed rate in lb/h:

$$\text{Moisture feed rate} = 0.25(10,000 \text{ lb/h})$$
$$= 2500 \text{ lb/h}$$

Calculate the dry feed rate to the incinerator in lb/h:

$$\text{Dry feed rate} = (10,000 - 2500) \text{ lb/h}$$
$$= 7500 \text{ lb/h}$$

Determine the amount of ash remaining after combustion in lb/h:

$$\text{Ash rate} = (0.1)(10,000 \text{ lb/h})$$
$$= 1000 \text{ lb/h}$$

The ash may be treated and/or sent to a second landfill.
    The amount of feed that is combusted (volatile) is therefore

$$\text{Volatile} = (7500 - 1000) \text{ lb/h}$$
$$= 6500 \text{ lb/h}$$

Determine the total heating value of the waste charged to the incinerator in Btu/h:

$$\text{Heating value} = (10,000 \text{ lb/h})(6500 \text{ Btu/lb})$$
$$= 65.0 \text{ MBtu/h}$$

Calculate the heating value per lb of volatile waste in Btu/h:

$$\text{Volatile waste heating value} = \frac{65.0 \text{ MBtu/h}}{6500 \text{ lb/h}}$$
$$= 10,000 \text{ Btu/lb}$$

Calculate the amount of dry gas produced from combustion in lb/h:

$$\text{Dry gas} = (7.5 \text{ lb/10,000 Btu})(65.0 \text{ MBtu/h})$$
$$= 48,750 \text{ lb/h dry gas}$$

Determine the amount of moisture produced from combustion in lb/h:

$$\text{Combustion moisture} = (0.51 \text{ lb/10,000 Btu})(65.0 \text{ MBtu/h})$$
$$= 3315 \text{ lb/h}$$

Therefore the total amount of gas (wet basis) leaving the incinerator due to combustion in lb/h is

$$\text{Total combustion gas} = 48,750 \text{ lb/h dry gas} + 3315 \text{ lb/h wet gas}$$
$$= 52,065 \text{ lb/h}$$

Calculate the stoichiometric air requirement for the incinerator in lb/h:

$$\text{Stoichiometric air} = (52,065 - 6500) \text{ lb/h}$$
$$= 45,565 \text{ lb/h}$$

Determine the amount of excess air in lb/h:

$$\text{Excess air} = 1.0(45,565 \text{ lb/h})$$
$$= 45,565 \text{ lb/h}$$

The total air requirement in lb/h is then

$$\text{Total air} = (45{,}565 + 45{,}565)\,\text{lb/h}$$
$$= 91{,}130\,\text{lb/h}$$

Calculate the amount of moisture in the entering air supply in lb/h:

$$\text{Inlet air moisture} = (0.015\,\text{lb/lb dry air})\,(91{,}130\,\text{lb/h})$$
$$= 1367\,\text{lb/h}$$

The total outlet moisture flowrate in lb/h is obtained by summing the three contributing terms:

$$\text{Total outlet moisture rate} = (2500 + 3315 + 1367)\,\text{lb/h}$$
$$= 7182\,\text{lb/h}$$

The total dry gas flowrate exiting the system in lb/h can also be determined:

$$\text{Outlet dry gas rate} = (48{,}750 + 45{,}565)\,\text{lb/h}$$
$$= 94{,}315\,\text{lb/h}$$

Calculate the heat loss in Btu/h due to the ash discharge:

$$\text{Ash heat loss} = (1000\,\text{lb/h})(100\,\text{Btu/lb})$$
$$= 0.1\,\text{MBtu/h}$$

Calculate the heat loss in Btu/h due to the air used to cool the incinerator shell. The enthalpy of air at 400°F is 81.8 Btu/lb:

$$\text{Cooling air heat loss} = 10{,}000\,\text{lb/h}\,(81.8\,\text{Btu/lb})$$
$$= 0.82\,\text{MBtu/h}$$

Determine the heat loss in Btu/h due to radiation:

$$\text{Radiation loss} = 0.01(65.0\,\text{MBtu/h})$$
$$= 0.65\,\text{MBtu/h}$$

Calculate the amount of heat absorbed in Btu/h by the humidity of the inlet air supply:

$$\text{Humidity correction} = 960\,\text{Btu/lb}\ (1367\,\text{lb/h})$$
$$= 1.31\,\text{MBtu/h}$$

The total heat loss of the incinerator in Btu/h is then

$$\text{Total heat loss} = (0.1 + 0.82 + 0.65 - 1.31)\,\text{MBtu/h}$$
$$= 0.26\,\text{Btu/h}$$

Calculate the outlet heat content of the flue gas in Btu/h:

$$\text{Outlet heat content} = (65.0 - 0.26)\,\text{MBtu/h}$$
$$= 64.74\,\text{MBtu/h}$$

From tables, the enthalpies of dry air and moisture at 2000°F:

$$\text{Enthalpy of dry air} = 513.0\,\text{Btu/lb}$$
$$\text{Enthalpy of moisture} = 2060.0\,\text{Btu/lb}$$

Calculate the outlet heat content of the flue gas at 2000°F:

$$\text{Outlet heat content at 2000°F} = (7182\,\text{lb/h})(2060.0\,\text{Btu/lb})$$
$$+ (94{,}315\,\text{lb/h})(513\,\text{Btu/lb})$$
$$= 63.2\,\text{MBtu/h}$$

Since the outlet heat content of 64.74 MBtu/h is greater than the outlet heat content at 2000°F, supplemental fuel is not necessary.

Municipal solid waste (MSW) is usually incinerated at a central disposal facility. Central disposal facilities are often used for energy generation. While these facilities have largely been popular in Europe, more central disposal facilities are appearing in the United States since the cost and availability of traditional energy sources are no longer reliable.

A secondary combustion chamber is necessary on municipal solid waste incinerators since a primary chamber alone will not provide the necessary time, temperature, and turbulence to destroy the organic waste components to the required destruction and removal efficiency (DRE). The primary chamber serves to volatilize the organic waste fraction while the secondary chamber heats the vaporized organics to an adequate temperature for oxidation. The oxidation process occurs in the 1800–

2200°F range, and the flue gas residence time in the secondary chamber is normally at least 2 s.

Most central disposal facilities utilize mass-burning systems; other facilities use controlled air incineration techniques. Mass-burning technology is described in more detail in the following problem (MUN.10).

MUN.10   INCOME FROM AN MSW INCINERATION SYSTEM

A mass-burn incineration system has been chosen to incinerate the municipal solid waste in a community. The incinerator will be located at a central disposal system and will provide steam to a local utility. Two rotary kiln incinerators will continuously incinerate 250 tons/day each of MSW at 1800°F with a waste heating value of 5000 Btu/lb. To generate the steam, a waste heat boiler is to be installed at the discharge of each incinerator. Each boiler generates steam at 150 psia. Feedwater is provided to each boiler at 220°F and the blowdown leaving each boiler is 4% of the feedwater entering each boiler. The exhaust gas temperature from each boiler is 400°F and the boiler radiation loss is 0.5% of the total boiler input. The kilns have slopes of 0.04 ft/ft and length-to-diameter ratios of 5:1. Steam is sold to the local utility at $2.50 per 1000 lb of steam. A tipping fee (the cost of dumping waste at the facility) of $20.00 per ton is charged. Assume a kiln heat release rate of 25,000 Btu/(h · ft^3). The products of combustion leaving each incinerator are as follows:

CO_2:   2500 lb/h
H_2O: 5000 lb/h
N_2:   12,500 lb/h
O_2:   500 lb/h
HCl: 250 lb/h
Ash: 83.5 lb/h

(Ash has a heat content of 522 Btu/lb at 1800°F and 102 Btu/lb at 400°F.)

From the steam tables, enthalpies for the steam, feedwater, and blowdown are as follows:

Steam at 150 psia = 1194 Btu/lb
Feedwater at 220°F = 188 Btu/lb
Blowdown at 150 psia and 358°F = 330.7 Btu/lb

1. Calculate the inside diameter of each kiln.
2. Determine the steam generation rate in tons/day, the amount of heat in Btu/h in the steam, and the boiler efficiency.
3. Calculate the yearly income of the facility based on the steam charge and the tipping fee.

4. Recalculate the yearly income if the steam is sold at $4.00 per 1000 lb of steam and the tipping fee is increased to $25.00 per ton. Discuss the disadvantages of increasing these costs.

Solution

Determine the heat release in Btu/h of each kiln:

$$\dot{Q} = (5000\,\text{Btu/lb})(250\,\text{ton/day})(2000\,\text{lb/ton})(\text{day/24 h})$$
$$= 104,166,667\,\text{Btu/h}$$

Calculate the volume of each kiln:

$$V = (104,166,667\,\text{Btu/h})/[25,000\,\text{Btu/(h}\cdot\text{ft}^3)]$$
$$= 4167\,\text{ft}^3$$

Calculate the inside diameter, D, of each kiln:

$$L = 5D$$

$$V = \left(\frac{\pi(D)^2}{4}\right)(L)$$

$$= \left(\frac{\pi(D)^2}{4}\right)(5D)$$

$$= \left(\tfrac{5}{4}\right)\left(\pi(D)^3\right)$$

$$D = \left(\frac{4V}{5\pi}\right)^{1/3}$$

$$= \left[\frac{4(4167\,\text{ft}^3)}{5\pi}\right]^{1/3}$$

$$= 10.2\,\text{ft}$$
$$\approx 10.5\,\text{ft}$$

Determine the length, L, of each kiln:

$$L = 5D$$
$$= 5(10.5\,\text{ft})$$
$$= 52.5\,\text{ft}$$

Determine the heat (enthalpy) content of the flue gas at the entrance and the exit of the boiler:

$$\text{Heat content} = (\text{Mass flowrate})(\text{Enthalpy})$$

Entrance at 1800°F		Exit at 400°F
CO_2	$(2500)(470.9) = 1{,}177{,}250\ \text{Btu/h}$	$(2500)(75.3) = 188{,}250\ \text{Btu/h}$
H_2O	$(5000)(1947) = 9{,}375{,}000\ \text{Btu/h}$	$(5000)(1213) = 6{,}065{,}000\ \text{Btu/h}$
N_2	$(12{,}500)(465) = 5{,}812{,}500\ \text{Btu/h}$	$(12{,}500)(85) = 1{,}062{,}500\ \text{Btu/h}$
O_2	$(500)(430.7) = 215{,}350\ \text{Btu/h}$	$(500)(76.2) = 38{,}100\ \text{Btu/h}$
HCl	$(250)(349) = 87{,}250\ \text{Btu/h}$	$(250)(64.9) = 16{,}225\ \text{Btu/h}$
Ash	$(83.5)(522) = 43{,}587\ \text{Btu/h}$	$(83.5)(102) = 8{,}517\ \text{Btu/h}$

Enthalpies appearing above in parentheses were drawn from the literature.

Calculate the total mass flow in lb/h and the total energy flow in Btu/h at both the entrance and exit of the boiler.

$$\text{Total mass flow} = 20{,}830\ \text{lb/h}$$
$$\text{Total energy flow at entrance} = 17{,}071{,}000\ \text{Btu/h}$$
$$\text{Total energy flow at exit} = 7{,}379{,}000\ \text{Btu/h}$$

Calculate the heat loss:

$$\text{Heat loss} = 0.005(17{,}071{,}000\ \text{Btu/h})$$
$$= 85{,}360\ \text{Btu/h}$$

Write an energy balance on the boiler:

Heat in = Heat out

Flue gas in + Feedwater = Flue gas out + Heat loss + Steam + Blowdown

$(17{,}071{,}000\ \text{Btu/h}) + (\text{Feedwater rate})(188\ \text{Btu/lb}) = (7{,}379{,}000\ \text{Btu/h})$

$\quad + (85{,}360\ \text{Btu/hr}) + (\text{Steam rate})(1194\ \text{Btu/lb}) + (\text{Blowdown rate})(330.7\ \text{Btu/lb})$

Write a mass balance on the boiler including only the steam, feedwater, and blowdown.

$$\text{Feedwater rate} = \text{Steam rate} + \text{Blowdown rate}$$

Solve the mass balance equation above for the steam flowrate in terms of feedwater flowrate only.

$$\text{Feedwater rate} = \text{Steam rate} + 0.04(\text{Feedwater rate})$$
$$\text{Steam rate} = 0.96(\text{Feedwater rate})$$

Substituting the above result into the energy balance equation leads to:

$$(17,071,000 \text{ Btu/h}) + (\text{Feedwater rate})(188 \text{ Btu/lb})$$
$$= (7,379,000 \text{ Btu/h}) + (85,360 \text{ Btu/h})$$
$$+ [0.96 \ (\text{Feedwater rate})(1194 \text{ Btu/lb})]$$
$$+ [0.04(\text{Feedwater rate}) \ (330.7 \text{ Btu/lb})]$$

Calculate the feedwater mass rate in lb/h:

$$9,607,000 \text{ Btu/h} = [(0.96)(\text{Feedwater rate})(1194 \text{ Btu/lb})]$$
$$+ [(0.04)(\text{Feedwater rate})(330.7 \text{ Btu/lb})]$$
$$- (\text{Feedwater rate})(188 \text{ Btu/lb})$$

$$9,607,000 \text{ Btu/h} = 971.5 \text{ Btu/lb}(\text{Feedwater rate})$$

$$\text{Feedwater rate} = 9889 \text{ lb/h}$$

Determine the total steam generation rate in tons/day from both boilers:

$$\text{Steam generation rate} = 0.96(\text{Feedwater rate})$$
$$= 0.96(9889 \text{ lb/h})(24 \text{ h/day})(\text{ton}/2000 \text{ lb})$$
$$= 114 \text{ ton/day per boiler}$$

For both boilers,

$$\text{Total steam generation rate} = 2(114 \text{ ton/day})$$
$$= 228 \text{ ton/day}$$

Calculate the amount of heat in Btu/h in the steam from each boiler:

$$\text{Steam heat amount} = (\text{Steam generation rate})(\text{Steam enthalpy})$$
$$= (9493 \text{ lb/h})(1194 \text{ Btu/lb})$$
$$= 11,330,000 \text{ Btu/h}$$

Calculate the efficiency of each boiler:

$$\text{Boiler efficiency} = [(11{,}330{,}000\ \text{Btu/h})/(17{,}071{,}000\ \text{Btu/h})](100)$$
$$= 66.4\%$$

Calculate the yearly income of the facility based on the steam charge:

$$\text{Steam income} = (\$2.50/1000\ \text{lb steam})(228\ \text{ton steam/day})$$
$$(365\ \text{day/yr})(2000\ \text{lb steam/ton steam})$$
$$= \$416{,}100/\,\text{yr}$$

Calculate the yearly income of the facility based on the tipping fee:

$$\text{Tipping fee income} = (\$20.00/\text{ton})(500\ \text{ton/day})(365\ \text{day/yr})$$
$$= \$3{,}650{,}000/\text{yr}$$

Calculate the total yearly income of the facility.

$$\text{Total yearly income} = \text{Steam income} + \text{Tipping fee income}$$
$$= \$416{,}100/\text{yr} + \$3{,}650{,}000/\text{yr}$$
$$= \$4{,}066{,}100/\text{yr}$$

Recalculate the total yearly income of the facility based on the rates given in question 4 of the problem statement and data.

$$\text{Steam income} = (\$4.00/1000\ \text{lb steam})(228\ \text{tons steam/day})$$
$$(365\ \text{day/yr})(2000\ \text{lb steam/ton steam})$$
$$= \$65{,}760/\text{yr}$$
$$\text{Tipping fee income} = (\$25.00/\text{ton})(500\ \text{tons/day})(365\ \text{days/yr})$$
$$= \$4{,}562{,}500/\text{yr}$$
$$\text{Total yearly income} = \text{Steam income} + \text{Tipping fee income}$$
$$= \$665{,}760/\text{yr} + \$4{,}562{,}500/\text{yr}$$
$$= \$5{,}228{,}260/\text{yr}$$

Low steam charges and tipping fees are necessary to encourage sales. In the United States, steam production from incineration is not considered conventional although there are successful incineration facilities in the United States that produce steam.

Low rates would attract as many customers as possible. Therefore, it would probably be disadvantageous for a facility to charge excessively high steam and tipping fees.

The two leading methods of generating energy from the incineration of municipal solid waste (MSW) are the mass-burn system and the refuse derived fuel (RDF) system. The mass-burn system incinerates unprocessed MSW to recover energy and the RDF system processes unprocessed MSW into a usable fuel prior to incineration. Both methods use either starved-air modular, stoker grates, rotary kiln, or fluidized-bed units for incineration. While the mass-burn system is currently more widely utilized, both systems may be used for large waste capacities.

Many considerations are necessary when deciding if heat recovery equipment should be included in an incineration facility. Unless a practical use for the recovered heat exists, it is usually not advisable to include heat recovery equipment at a facility since the equipment is expensive. If an incineration facility operates at a large capacity, the generation of power from heat recovery equipment is generally economical. Municipal solid waste incineration facilities are widely used to produce steam for electric power generation.

27 Hazardous Waste Incineration (HWI)

HWI.1 DESTRUCTION AND REMOVAL EFFICIENCY (DRE)

By federal law (as described in Chapter 22), hazardous waste incinerators must meet a minimum destruction and removal requirement, i.e., that the principal organic hazardous constituents (POHCs) of the waste feed be incinerated with a minimum destruction and removal efficiency (DRE) of 99.99% ("four nines"). The *destruction and removal efficiency* is the fraction of the inlet mass flowrate of a particular chemical that is destroyed and removed in the incinerator or, equivalently,

$$\text{DRE} = 1 - \frac{\text{Mass flow of chemical out}}{\text{Mass flow of chemical in}}$$

A feed stream to a rotary kiln incinerator contains 245 kg/h of solid waste. Calculate the maximum outlet flowrate of solid waste from the incinerator allowed by law.

Solution

Substituting data given in the problem statement into the DRE equation provides the means of determining an acceptable mass flowrate of chemical out of the incinerator as follows:

$$\text{DRE} = 1 - \frac{\text{Mass flow of chemical out}}{\text{Mass flow of chemical in}}$$

$$0.9999 = 1 - \frac{\text{Mass flow of chemical out}}{245 \text{ kg/h}}$$

$$\text{Mass flow of chemical out} = (1 - 0.9999)(245 \text{ kg/h})$$

$$= (0.0001)(245 \text{ kg/h})$$

$$= 0.0245 \text{ kg/h}$$

HWI.2   REGULATION OF PARTICULATE EMISSIONS

Some state regulations limit stack emissions of particulate material from a hazardous waste incinerator to an outlet loading of 0.08 gr/dscf (grains of particulate per dry standard cubic foot of stack gas), corrected to 7% oxygen (or an approximate excess air level of 50%).

1. Describe the effect on the outlet particulate loading of each of the three following actions as far as increasing or decreasing the measured value of the outlet loading. Give a reason for each answer.
   a. Removing water from the gas (*Note*: The incineration of organic material almost always produces water as a product.)
   b. Lowering the temperature of the gas from its stack temperature of 800°F to the standard temperature of 60°F
   c. Using 50% excess air for the incineration instead of 100% excess air
2. From the three answers provided to part 1, explain why the terms "dry," "standard," and "corrected to 7% oxygen" are used in the regulations.

Solution

Once again the reader may refer to Chapter 22 for additional details.

1. Removing water from the gas decreases the gas volume without affecting the mass of particulate matter. Since the particulate loading is the mass of particulate matter divided by the gas volume, the particulate loading increases.
   Lowering the temperature also decreases the gas volume without affecting the particulate mass. Therefore, the loading increases. The less excess air used for the combustion, the less oxygen and nitrogen contribute to the gas stream. The gas volume is therefore lower and the particulate loading increases.
2. "Dry" is used so that extra water cannot be added to the gas stream to artificially lower the outlet particulate loading. "Standard" is used so that the outlet particulate loading is not a function of temperature, i.e., the loading cannot be artificially lowered by raising the temperature. "Converted to 7% oxygen" is used so that the outlet particulate loading cannot be artificially lowered by using an inordinate amount of excess air.

HWI.3   INCINERATOR OUTPUT LOADING

An incinerator is burning a hazardous waste slurry using 50% excess air. The stack gas at 800°F has a flowrate of 10,000 acfm (actual cubic feet per minute) with a particulate mass flow of 20 gr/min and a water content of 10% by volume. The maximum allowed outlet loading for particulates is 0.08 gr/dscf (grains per dry standard cubic foot), corrected to 7% oxygen (or an approximate excess air level of 50%).

1. Calculate the outlet loading in gr/acf (grains per actual cubic foot).
2. Calculate the outlet loading in gr/scf (grains per standard cubic foot) noting that standard temperature is 60°F.
3. Calculate the outlet loading in gr/dscf.
4. As far as particulates are concerned, is the incinerator in compliance?

Solution

1. The outlet loading (OL) in gr/acf is given by:

$$OL_{gr/acf} = \frac{150\,gr/min}{10,000\,ft^3/min} = 0.015\,gr/acf$$

2. From the ideal gas law, the volume is proportional to the absolute temperature. Noting that $T(°R) = T(°F) + 460$, the outlet loading in gr/scf is

$$OL_{gr/scf} = (0.015\,gr/ft^3)\left(\frac{800 + 460°R}{60 + 460°R}\right) = 0.036\,gr/scf$$

3. Since the gas stream is 10% water by volume, the outlet loading in gr/dscf is given by:

$$OL_{gr/dscf} = (0.036\,gr/ft^3)\left(\frac{100\,ft^3_{wet}}{90\,ft^3_{dry}}\right) = 0.0403\,gr/dscf$$

4. Because 50% excess air (approximately 7% O_2) is used, the outlet loading is 0.0403 gr/dscf corrected to 7% O_2. This is below the allowed maximum particulate concentration of 0.08. Therefore, the incinerator is in compliance.

HWI.4   DETERMINATION OF COMPLIANCE FOR PARTICULATE EMISSIONS

A hazardous waste incinerator is burning an aqueous slurry of soot (i.e., carbon) with the production of a small amount of fly ash. The waste is 70% water by mass and is burned with 0% excess air (EA). The flue gas generated contains 0.30 gr of particulates in each 8.0 ft³ (actual) at 580°F.

If the state regulations require the particulate emissions be less than 0.08 gr/dscf corrected to 50% EA, is this incinerator in compliance or must additional particulate control measures be taken? Assume that when the flue gas passes through a waste heat boiler no water condensation occurs.

Solution

The particulate concentration in the flue gas is

$$0.30 \, \text{gr}/8.0 \, \text{acf} = 0.0375 \, \text{gr/acf}$$

Using Charles' law, one can calculate the volume at $60°F = 520°R$:

$$(8.0 \, \text{acf})(520°R/1040°R) = 4.0 \, \text{scf}$$
$$\text{Particulate concentration} = 0.30 \, \text{gr}/4.0 \, \text{scf}$$
$$= 0.075 \, \text{gr/scf}$$

The volume fraction of water in the flue gas is obtained from the mass composition of the waste and the balanced equation:

$$C + H_2O + O_2 \rightarrow CO_2 + H_2O$$

For each 100 lb of waste (30 lb C and 70 lb water), the following molar quantities are input to the incinerator

C: 30 lb/(12 lb/lbmol) = 2.5 lbmol
Water: 70 lb/(18 lb/lbmol) = 3.89 lbmol
Oxygen: 2.5 lbmol
Nitrogen: (79/21) (2.5 lbmol) = 9.40 lbmol
Fraction of water in flue gas y_{FC}, is

$$y_{FG} = \frac{3.89}{2.5 + 3.89 + 9.4} = 0.25$$

The molar quantity of dry gas, n_{DG} is

$$n_{DG} = (2.5 + 9.4) = 11.9 \, \text{lbmol}$$

The dry volume is calculated by subtracting the volume of water from the total gas volume shown as:

$$(4.0 \, \text{scf})(1 - 0.25) = 3.0 \, \text{dscf}$$

The particulate concentration on a dscf basis is then calculated as:

$$\text{Particulate concentration} = 0.30 \, \text{gr}/3.0 \, \text{dscf}$$
$$= 0.10 \, \text{gr}/\text{scf}$$

To correct to 50% EA, 50% more N_2 and O_2 are added to the flue gas:

Excess nitrogen added $= 0.5(79/21) \, (2.5 \, \text{lbmol}) = 4.7 \, \text{lbmol}$
Excess oxygen added $= 0.5(2.5) = 1.25 \, \text{lbmol}$

The total moles of dry flue gas is increased by the EA to yield a total of

$$2.5 + 9.4 + 4.7 + 1.25 = 17.85 \, \text{lbmol}$$

The total flue gas volume can then be calculated based on the ratio of total moles of flue gas at 50% and at 0% EA.

$$3.0 \, \text{dscf} \, (17.85 \, \text{lbmol}/11.89 \, \text{lbmol}) = 4.5 \, \text{dscf} \, (50\% \, \text{EA})$$

The particulate concentration on a dry weight basis, corrected to 50% EA is:

$$0.30 \, \text{gr}/4.5 \, \text{dscf} \, (50\% \, \text{EA}) = 0.067 \, \text{gr}/\text{dscf} \, (50\% \, \text{EA})$$

Since this does not exceed the state particulate standard, the incinerator is in compliance. The reader should note that the particulate standard most often employed today is in the 0.015–0.03 gr/dscf (corrected to 50% EA) range.

HWI.5   CONTROL OF HCl AND PARTICULATES

A state incinerator emissions limit requires 99% HCl control and allows 0.07 gr/dscf at 68°F, corrected to 50% EA. An incinerator is to burn 5 tons/h hazardous sludge waste containing 2% Cl, 80% C, 5% inerts, and the balance H_2O by weight. Perform the following calculations:

1. Calculate the maximum mass emission rate of equivalent HCl in lb/h that may be emitted.
2. Calculate the maximum mass emission rate of particulates in lb/h that may be emitted. What is the actual particulate emission rate if all the inerts are emitted from the stack as fly ash?
3. Determine the combustion efficiency of this incinerator if a stack test indicates that the flue gas contains 12% CO_2 and 20 ppm CO.

Solution

The incinerator receives 5 tons/h of waste with 2% Cl content. Assuming all chlorine is converted to HCl, the amount of HCl formed is given as:

$$\text{Cl in feed} = (5\,\text{ton/h})(2000\,\text{lb/ton})(0.02\,\text{lb Cl/lb waste})$$

$$= 200\,\text{lb Cl/h}$$

$$\text{HCl formed} = [(200\,\text{lb Cl/h})(36.5\,\text{lb HCl})]/(35.5\,\text{lb Cl})$$

$$= 205.6\,\text{lb HCl/h}$$

The maximum permissible mass emission rate of HCl at 99% control is

$$(205.6\,\text{lb HCl/h})(1 - 0.99) = 2.06\,\text{lb HCl/h emitted}$$

To calculate particulate emissions at 50% EA, a 1-mole C basis is used along with the following balanced equation:

$$C + 1.5O_2 + 1.5(79/21)N_2 = CO_2 + 0.5O_2 + 1.5(79/21)N_2$$

The volume of flue gas generated per lbmol of C combusted is calculated from the ideal gas law:

$$V = \frac{nRT}{P}$$

$$= \frac{[1 + 0.5 + 1.5(3.76)]\,\text{lbmol}(460 + 68)^\circ R[0.7302\,\text{atm} \cdot \text{ft}^3/(\text{lbmol} \cdot {}^\circ R)]}{1\,\text{atm}}$$

$$= 2754\,\text{dscf}$$

Based on the required incinerator emission standard of 0.07 gr/dscf, the maximum emission rate allowed at 50% EA is as follows. The volume of gas generated per hour, q, can be calculated:

$$q = \left(5\,\frac{\text{ton fuel}}{\text{h}}\right)\left(2000\,\frac{\text{lb fuel}}{\text{ton fuel}}\right)\left(0.8\,\frac{\text{lb C}}{\text{lb fuel}}\right)\left(\frac{1\,\text{lbmol C}}{12\,\text{lb C}}\right)\left(\frac{2754\,\text{dscf flue gas}}{\text{lbmol C}}\right)$$

$$= 1,836,000\,\text{dscf/h}$$

The allowable particulates discharge rate is

$$\dot{m}_{\text{allowed}} = (1,836,000\,\text{dscf/h})(0.07\,\text{gr particulates/dscf})(1\,\text{lb}/7000\,\text{gr})$$

$$= 18.36\,\text{lb particulates/h}$$

The actual particulates emission rate is

$$\dot{m}_{actual} = (5 \text{ ton fuel/h})(2000 \text{ lb/ton})(0.05 \text{ lb ash/lb fuel})$$
$$= 500 \text{ lb particulates/h}$$

This is well above the allowable limit, requiring particulate removal to meet permit requirements.

The actual combustion efficiency can be calculated based on the ratio of carbon dioxide to total effluent carbon species in the flue gas effluent, and is

$$\text{Combustion efficiency} = \left(\frac{CO_2}{CO_2 + CO}\right)$$
$$= \left(\frac{12}{12 + 0.0020}\right)(100)$$
$$= 99.98\%$$

HWI.6   THEORETICAL FLAME TEMPERATURE ESTIMATION

Estimate the theoretical flame temperature of a waste mixture containing 25% cellulose, 35% motor oil, 15% water (vapor), and 25% inerts, by mass. Assume 5% radiant heat losses. The flue gas contains 11.8% CO_2, 13% CO, and 10.4% O_2 (dry basis) by volume.

NHV of cellulose = 14,000 Btu/lb
NHV of motor oil = 25,000 Btu/lb
NHV of water = 0 Btu/lb
NHV of inerts (effective) = − 1,000 Btu/lb

Assume the average heat capacity of the flue gas is 0.325 Btu/(lb · °F). Employ the Theodore–Reynolds equation to perform this calculation:

$$T = 60 + \frac{\text{NHV}}{0.325[1 + (1 + \text{EA})(7.5 \times 10^{-4})(\text{NHV})]}$$

Solution

Determine the net heating value (NHV) for the mixture:

$$\begin{aligned}
\text{NHV} &= 0.25(14{,}000\,\text{Btu/lb}) + 0.35(25{,}000\,\text{Btu/lb}) \\
&\quad + 0.15(0.0\,\text{Btu/lb}) + 0.25(-1000\,\text{Btu/lb}) \\
&= 12{,}000\,\text{Btu/lb}
\end{aligned}$$

Determine the excess air employed. If Y is the % by volume of O_2 (dry basis), then

$$\begin{aligned}
\text{EA} &= 0.95Y/(21 - Y) \\
&= 0.95(10.4)/(21 - 10.4) \\
&= 0.932
\end{aligned}$$

Estimate the flame temperature using the Theodore–Reynolds equation.

$$\begin{aligned}
T &= 60 + \frac{\text{NHV}}{0.325[1 + (1 + \text{EA})(7.5 \times 10^{-4})(\text{NHV})]} \\
&= 60 + \frac{12{,}000}{0.325[1 + (1 + 0.932)(7.5 \times 10^{-4})(12{,}000)]} \\
&= 2068^\circ\text{F}
\end{aligned}$$

HWI.7   EFFECT OF PCB CONCENTRATION ON COMBUSTION

Polychlorinated biphenyls (PCBs) are mixed with some solids and waste oil for burning in a rotary kiln incinerator. What will happen as the percentage of PCBs is increased in the waste mixture assuming that all other variables (i.e., excess air, heat loss, feed rate) are kept constant?

Solution

As the percentage of PCB increases in the waste mixture, the following will occur:

1. Incinerator temperature will decrease because of the low heat of combustion of PCBs.
2. The efficiency of combustion will decrease because as the combustion temperature decreases the PCBs are more difficult to incinerate at a lower temperature.
3. The concentration of Cl_2 in the flue gas will increase; at lower temperatures, the $HCl \Leftrightarrow Cl_2$ equilibrium is shifted toward Cl_2.

HWI.8   RESIDENCE TIME

One incinerator at a hazardous waste facility operates at 2200°F with a combustion gas flowrate of 40,000 acfm. If the incinerator is 10 ft wide, 12 ft deep, and 22 ft high, calculate the maximum residence time in the incinerator. The combustion gas flowrate through a second incinerator at the facility, which is operating at 2000°F, is 5,000 scfm (60°F). If the incinerator requires a minimum residence time of 2 s, calculate the required volume of this incinerator.

Solution

Calculate the volume, V, of the first incinerator:

$$V = \text{(width)(length)(depth)}$$
$$= (10)(12)(22)$$
$$= 2640 \text{ ft}^3$$

Calculate the maximum residence time, θ:

$$\theta = V/q_a$$
$$= (2460/40,000)(60)$$
$$= 3.96 \text{ s}$$

Calculate the actual combustion gas flowrate for the second incinerator:

$$q_a = q_s(T_a/T_s)$$
$$= (5000)(2000 + 460)/(60 + 460)$$
$$= 23,650 \text{ acfm}$$

The volume required is therefore

$$V = q_a\theta$$
$$= (23,650)(2)/60$$
$$= 788 \text{ ft}^3$$

Note that the residence time is a function of the temperature since the volumetric flowrate is linearly related to the absolute temperature. Thus, the higher the operating temperature the shorter the residence time. In addition, the gas residence time calculated above by dividing the volume of the incinerator chamber by the combustion gas flowrate is an approximate value. This does not include flowrate variations that arise due to chemical reaction and temperature changes through the unit. Thus, a residence time distribution exists with real systems.

HWI.9 pH CONTROL

Process considerations require pH control in a 50,000-gal storage tank used for incoming waste mixtures (including liquid plus solids) at a hazardous waste incinerator. Normally, the tank is kept at neutral pH. However, operation can tolerate pH variations from 6 to 8. Waste arrives in 5000-gal shipments. Assume that the tank is completely mixed, contains 45,000 gal when the shipment arrives, the incoming acidic waste is fully dissociated, and that there is negligible buffering capacity in the tank.

1. What is the pH of the most acidic waste shipment that can be handled without neutralization?
2. What is the pH of the most acidic waste shipment that can be handled without neutralization if the storage tank volume is 100,000 gal, and contained 95,000 gal of neutral waste when the shipment arrived?

Solution

1. The pH of the most acidic waste shipment that can be handled without neutralization is calculated as follows: 5000 gal of waste with a $[H^+] = X$ is diluted by 45,000 gal at pH $= 7$ or $[H^+] = 10^{-7}$. The minimum pH of 6 that can be tolerated is equivalent to a $[H^+] = 10^{-6}$. From an ion balance:

$$[H^+] = 10^{-6} = (5000/50,000)X + (45,000/50,000)(10^{-7})$$

$$X = \left(\frac{50,000}{5000}\right)\left[10^{-6} - \frac{45,000(10^{-7})}{5000}\right]$$

$$= 0.91 \times 10^{-6}$$

$$pH = 6.04$$

2. With a tank volume of 100,000 gal, the solution is as follows:

$$10^{-6} = (5000/100,000)X + (95,000/100,000)(10^{-7})$$

$$X = \left(\frac{100,000}{5000}\right)\left[10^{-6} - \frac{95,000(10^{-7})}{100,000}\right]$$

$$= 1.81 \times 10^{-5}$$

$$pH = 4.74$$

HWI.10   DRE AND NHV CALCULATIONS

A mixture of trichloroethylene, tetrachlorethylene, dichlorofluoromethane, and phthaloyl chloride in No. 2 fuel oil (1% sulfur) is fired during a trial burn in a liquid injection incinerator. The facility is equipped with a quench tower, venturi scrubber, and packed-bed caustic scrubber.

Each of the four organic compounds makes up 5% (by weight) of the waste/fuel feed of 5000 lb/h. The excess air is 35%. The stack gas flowrate, corrected to 7% oxygen and standard conditions (1 atm, 25°), is 21,300 dscfm.

1. Calculate the destruction and removal efficiency (DRE) for each hazardous constituent using the flue gas measurements below:

C_2HCl_3	0.03 ppm
C_2Cl_4	0.02 ppm
$CHCl_2F$	0.30 ppm
$C_3H_4Cl_2O$	1.8 ppm

2. Use Dulong's equation (see CHR.8 in Chapter 23) to calculate the NHV of the mixture in Btu/lb. The NHV of No. 2 fuel oil is 18,650 Btu/lb.

Solution

1. As described earlier, the DRE for each hazardous material is related to its mass flowrate into, $\dot{m}_i$, the HWI and the mass flowrate out of, $\dot{m}_o$, the stack by the equation:

$$DRE = \left(1 - \frac{\dot{m}_o}{\dot{m}_i}\right)100$$

where, $\dot{m}_o$ is equal to the product of the mass concentration in the stack, ρ_o, and the volumetric flowrate, c_o, out the stack. Each stack gas concentration, c_o, in ppm can be shown to be converted to ρ_o by the following equation at 25°C and 1 atm pressure:

$$\rho_o = 40.9c_o(MW)$$

where ρ_o is in μg/m³ and MW is the gram molecular weight.

The inlet mass of each compound is 5% of the inlet waste feed = 0.05(5000 lb/h) = 250 lb/h.

With the above information the table below was constructed.

	$\dot{m}_i$ (lb/h)	MW	c_o (ppm)	ρ_o (g/m³)	$\dot{m}_o$ (lb/min)	DRE (%)
C_2HCl_3	250	131.5	0.03	161.35	2.147×10^{-4}	99.9999
C_2Cl_4	250	166	0.02	135.79	1.807×10^{-4}	99.9999
$CHCl_2F$	250	103	0.30	1263.8	1.682×10^{-3}	99.9993
$C_3H_4Cl_2O$	250	187	1.8	13,766.9	1.83×10^{-2}	99.9927

2. Assume F and Cl have essentially the same influence on the waste NHV. The calculation of the mass fraction of each element in each waste compounds is illustrated (for carbon in C_2HCl_3) below:

$$\text{Mass fraction C} = \left(\frac{(2)(12)}{131.5}\right) = 0.1825$$

$$\dot{m}_C = (0.1825)(250) = 45.6 \text{ lb/h}$$

The remaining values in the table are calculated in an identical manner.

	$\dot{m}_i$ (lb/h)	MW	$\dot{m}_C$	$\dot{m}_H$	$\dot{m}_{Cl}$ (&) $\dot{m}_F$	$\dot{m}_O$
C_2HCl_3	250	131.5	45.6	1.9	202.5	
C_2Cl_4	250	166	36.1		213.9	
$CHCl_2F$	250	103	29.13	2.43	218.45	
$C_3H_4Cl_2O$	250	187	48.1	5.35	94.92	21.4
Total	1000		158.9	9.68	729.77	21.4

In this table, a basis of 1000 lb/h total waste flow of the four compounds was used. The NHV can be calculated for this four-compound mixture using Dulong's equation:

$$\text{NHV} = 14,000x_C + 45,000[x_H - (x_O/8)] - 760x_{Cl} + 4500x_S$$

where $x_i = $ mass fraction of i.

The mass fractions are calculated by dividing the $\dot{m}_i$ for each element by 1000 lb/h.

$$\text{NHV} = 14,000(0.159) + 45,000[0.00968 - (0.0214/8)] - 760(0.7298) + 0$$

$$= 1987 \text{ Btu/lb}$$

The NHV of the entire waste mixture, i.e., the 4 compounds and the No. 2 fuel oil, can be calculated as follows:

$$NHV_{mixture} = 0.8(18,650 \text{ Btu/lb fuel oil})$$
$$+ 0.2(1987 \text{ Btu/lb waste compounds})$$
$$= 15,320 \text{ Btu/lb}$$

HWI.11   NEUTRALIZATION WITH SODA ASH

A hazardous waste incinerator has been burning a certain mass of dichlorobenzene ($C_6H_4Cl_2$) per hour, and the HCl produced was neutralized with solid soda ash (Na_2CO_3). If the incinerator switches to burning an equal mass of mixed tetra-chlorobiphenyls ($C_{12}H_6Cl_4$), by what factor will the consumption of soda ash be increased?

Solution

The balanced stoichiometric reaction for the oxidation of dichlorobenzene is shown below:

$$C_6H_4Cl_2 + 6.5O_2 \rightarrow 6CO_2 + H_2O + 2HCl$$

Therefore, for 1 lb of dichlorobenzene (DCB), the following mass of HCl is produced:

$$\left(\frac{1 \text{ lb}}{147 \text{ lb/lbmol DCB}}\right)\left(\frac{2 \text{ mol HCl}}{\text{mol DCB}}\right) = 0.0136 \text{ lbmol HCl produced}$$

The balanced stoichiometric reaction for the oxidation of tetrachlorobiphenyl is as shown below:

$$C_{12}H_6Cl_4 + 12.5O_2 \rightarrow 12CO_2 + H_2O + 4HCl$$

Therefore, for 1 lb of tetrachlorobiphenyl (TCB), the following mass of HCl is produced:

$$\left(\frac{1 \text{ lb}}{290 \text{ lb/lbmol TCB}}\right)\left(\frac{4 \text{ mol HCl}}{\text{mol DCB}}\right) = 0.0138 \text{ lbmol HCl produced}$$

Thus, the amount of acid produced does not change significantly ($\approx 1.5\%$), and neither will the amount of base required for neutralization.

HWI.12 DEGREE OF DISSOCIATION

Compute the degree of dissociation (extent of reaction) of bromine at 527°C and 1 atm for the reaction

$$Br_2(g) \leftrightarrow 2Br(g)$$

The standard free energy of reaction in cal/gmol Br_2 is given by

$$\Delta G_T^\circ = 53,424 - 2.6(T)\ln(T) + (0.0005)T^2 - (5.0)T \qquad T = K$$

For the dissociation of bromine at 527°C, what is the effect of increasing the pressure to 10.0 atm? Assume ideal gas behavior.

Solution

The reader is referred to Chapter 9 for chemical reaction equilibrium details and equations.

Calculate ΔG_T° at a temperature of 527°C:

$$\Delta G_T^\circ = 53,424 - 2.6(T)\ln(T) + (0.0005)T^2 - (5.0)T$$

For $T = 800\,\mathrm{K}$,

$$\Delta G_{527}^\circ = +35,840 \text{ cal/gmol}$$

Calculate K at 527°C.

$$\Delta G^\circ = -RT\ln K$$

$$\ln K = \frac{-\Delta G^\circ}{RT}$$

$$= \frac{-35,840}{(1.987)(800)}$$

$$= -22.55$$

$$K = 1.615 \times 10^{-10}$$

Write the equation for K in terms if K_y and P:

$$K = K_y P^{\Delta n} \qquad P = 1 \text{ atm} \qquad \Delta n = 2 - 1 = 1$$

$$K_y = 1.615 \times 10^{-10}$$

Express the K_y in terms of the equilibrium conversion variable z. Initially,

$$n_{Br_2,0} = 1.0 \text{ (assumed basis)}$$

$$n_{Br,0} = 0$$

At equilibrium,

$$n_{Br} = 2z$$

$$n_{Br_2} = 1 - z$$

$$n_{total} = 1 + z$$

$$y_{Br} = \frac{2z}{1+z}$$

$$y_{Br_2} = \frac{1-z}{1+z}$$

Since $P = 1$ atm,

$$K_y = \frac{y_{Br}^2}{y_{Br_2}}$$

$$1.615 \times 10^{-10} = \frac{\left(\dfrac{2z}{1+z}\right)^2}{\left(\dfrac{1-z}{1+z}\right)}$$

$$= \frac{4z^2}{1-z^2}$$

Solving the quadratic equation analytically yields

$$z = 6.354 \times 10^{-6}$$

At 10 atm,

$$K = K_y P^{\Delta n}$$

$$\Delta n = 2 - 1 = 1$$

$$K_y = K/10 = 1.615 \times 10^{-9}$$

For this condition,

$$z = 2.01 \times 10^{-6}$$

It is concluded that a negligible amount of bromine is formed at both pressures.

28 Hospital/Medical Waste (MED)

MED.1 CLASSIFICATION OF MEDICAL WASTE

Answer the following questions:

1. Describe the difference between a medical waste and an infectious waste.
2. Identify the seven classes of regulated medical wastes.
3. By what means are hospital waste disposed?
4. What steps can be taken by hospital operations to minimize occupational hazards, contamination, and infection?

Solution

1. A medical waste is any solid waste that is generated in the diagnosis, treatment, or immunization of human beings or animals. A medical waste can be infectious or noninfectious. Infectious waste is a medical waste that contains pathogenic microorganisms that can cause disease.
2. The seven classes of regulated medical wastes are:
   a. Cultures and stocks
   b. Pathological waste
   c. Blood and blood products
   d. Sharps
   e. Animal waste
   f. Isolation wastes
   g. Unused sharps
3. The predominant means of disposing of hospital waste is incineration (almost 35% of all hospital waste is incinerated). Other means of treatment or disposal include autoclaving, sanitary landfilling, chemical disinfection, thermal inactivation, ionizing radiation, gas vapor sterilization, segregation, and bagging.
4. The steps that may be taken by hospital operations to minimize occupational hazards, contamination, and infection include:
   a. Enclose medical waste at point of generation.
   b. Construct carts and equipment that have sanitary construction.

   c. Construct and operate chutes to minimize microbiological contamination.

   d. Reduce dangers of medical waste handling by using personal protective equipment.

   e. Require higher training and qualifications for personnel handling medical waste.

   f. Improve incinerator operation through training.

   g. Implement safe management of hazardous wastes.

MED.2   INFECTIOUS WASTE MANAGEMENT PLANS

Answer the following questions:

1. What factors should be taken into consideration when designing an infectious waste management plan for an individual facility?

2. Future infectious hospital waste management plans should include procedures that will deal with emergency situations. Describe three scenarios that could use emergency response procedures.

Solution

1. Infectious waste management plans are designed around the individual needs of a facility. Three main factors, which must be considered in the custom designing of an infectious waste management plan, are:

   a. Facility location

   b. Size of the facility

   c. Operation budget

   Other factors that are important include the nature of the infectious waste, waste quantity, the availability of both onsite and offsite equipment for treatment of the waste, regulatory constraints, and operating costs.

2. Three possible scenarios include:

   a. Liquid infectious waste spills

   b. Rupture of infectious waste containers

   c. Failure of infectious waste treatment equipment

MED.3   INCINERATION REQUIREMENTS

In many cases, medical waste can be converted to inorganic matter or changed in form by high-temperature processes in the presence of oxidizing agents. Such thermal processes can result in the partial or complete reduction in the degree of hazard of a material. The general class of such thermal processes when oxygen is the oxidizing agent is termed *incineration*. What considerations must be present in an incinerator to provide successful conversion of a medical waste material?

Solution

For incineration to be successful in the destruction of a medical waste material, the following characteristics must exist:

- Adequate free oxygen must always be available in the combustion zone.
- Sufficient turbulence must exist within the incinerator to ensure the constant mixing of waste and oxygen.
- Adequate combustion temperatures must be maintained. Exothermic combustion reactions must supply enough heat to raise the burning mixture to a sufficient temperature to destroy all of the organic components of the waste.
- The duration of exposure of the waste material to combustion temperatures must be long enough to ensure that even the slowest combustion reaction is completed. In other words, transport of the burning mixture through the high-temperature region must occur for a sufficient period of time to allow reactions to go to completion.

MED.4   FLY ASH CALCULATION

An incinerator burning medical waste contains 7.8% ash. The hospital produces approximately 540 lb waste every hour on a round-the-clock basis. How much ash needs to be disposed of annually, if the 25% of the ash "flies" during the burning process?

Solution

The rate of ash output from the incinerator in pounds/hour is

$$(0.078)(540 \, \text{lb/h}) = 39 \, \text{lb/h}$$

The rate of ash output per day would then be

$$(24)(39)(0.75) = 702 \, \text{lb/day}$$

The rate of ash output per yr would be

$$(365)(702) = 256,000 \, \text{lb/yr}$$

MED.5   MERCURY REMOVAL

Medical sludge containing mercury is burned in an incinerator. The mercury feed rate is 9.2 lb/h. The resulting 500°F product (40,000 lb/h of gas; $MW = 32$) is quenched with water to a temperature of 150°F. The resulting stream is filtered to remove all particulates. What happens to the mercury? Assume the process pressure is 14.7 psi and that the vapor pressure of Hg at 150°F is 0.005 psi.

Solution

For the mercury to be removed by the filter, it must condense and form particles. Therefore, the question to be answered relates to the partial pressure of mercury during removal compared to its vapor pressure at 150°F:

$$\text{Molar flowrate of Hg} = (9.2 \, \text{lb/h})/(200.6 \, \text{lb Hg/lbmol})$$

$$= 0.046 \, \text{lbmol/h}$$

$$\text{Molar flowrate of gas} = (40,000 \, \text{lb gas})/(32 \, \text{lb gas/lbmol})$$

$$= 1250 \, \text{lbmol/h}$$

$$y = \frac{\text{lbmol Hg}}{\text{lbmol Hg} + \text{lbmol gas}} = \frac{0.046 \, \text{lbmol/h}}{0.046 \, \text{lbmol/h} + 1250 \, \text{lbmol/h}} = 3.68 \times 10^{-5}$$

$$\text{Partial pressure} = y_i P$$

$$p_i = y_i \, (14.7 \, \text{psia}) = 3.68 \times 10^{-5} \, (14.7 \, \text{psia}) = 5.4 \times 10^{-4} \, \text{psia}$$

Since the partial pressure is much less than the vapor pressure, mercury will NOT condense and thus will NOT be removed by the filter.

MED.6   MERCURY EMISSIONS

An incinerator burns mercury-contaminated waste. The waste material has an ash content of 1%. The solid waste feed rate is 1000 lb/h and the gas flowrate is 20,000 dscfm. It is reported that the average mercury content in the particulates was 2.42 µg/g when the vapor concentration was 0.3 mg/dscm. For the case where incinerator emissions meet the particulate standard of 0.08 gr/dscf (0.1832 g/dscm) with a 99.5% efficient electrostatic precipitator (ESP), calculate:

1. The amount of mercury bound to the fly ash, which is captured in the ESP in grams/day.
2. The amount of mercury leaving the stack as a vapor and with the fly ash in grams/day.

Solution

1. The amount of ash leaving the stack is

$$\left(\frac{0.08\,\text{gr}}{\text{dscf}}\right)\left(\frac{1\,\text{lb}}{7000\,\text{gr}}\right)\left(\frac{20{,}000\,\text{dscf}}{1\,\text{min}}\right)\left(\frac{60\,\text{min}}{\text{h}}\right)\left(\frac{24\,\text{h}}{\text{day}}\right) = 329\,\text{lb/day}$$

The amount of ash collected in the ESP is

$$(329\,\text{lb/day})/(1 - 0.995\,\text{collected}) = 65{,}800\,\text{lb/day}$$

2. The amount of mercury leaving the stack with the fly ash is

$$(329\,\text{lb ash/day})(2.42 \times 10^{-6}\,\text{g Hg/g ash}) = 7.96 \times 10^{-4}\,\text{lb Hg/day}$$
$$= 0.361\,\text{g Hg/day}$$

The amount of mercury leaving the stack as vapor is

$$\left(\frac{0.3 \times 10^{-3}\,\text{g Hg}}{\text{dscm}}\right)\left(\frac{20{,}000\,\text{dscf}}{1\,\text{min}}\right)\left(\frac{1\,\text{m}^3}{35.3\,\text{ft}^3}\right)\left(\frac{60\,\text{min}}{\text{h}}\right)\left(\frac{24\,\text{h}}{\text{day}}\right) = 244.8\,\text{g/day}$$

Total mercury leaving the stack $= 244.8 + 0.361$
$$= 245.2\,\text{g/day}$$

MED.7   WASTE BLENDING

Two hospital wastes are received and stored in separate tanks at an incineration facility. The first, a sludge with an net heating value (NHV) of 6000 Btu/lb contains 2% Cd by weight. The second, a mercury-contaminated waste with an NHV of 8000 Btu/lb, contains 8% Cd by weight. A minimum of 1000 lb/h of each waste is to be incinerated. Because of pump limitations, no more than 5000 lb/h of each waste can be utilized.

To achieve the required destruction and removal efficiency (DRE), the facility requires a minimum of 8000 Btu/lb in the waste stream, which may be obtained by adding fuel oil with an NHV of 15,000 Btu/lb. The incinerator can operate with a heat rate between 25 and 40 million Btu/h. A graphical approach involving the plotting of the two principal variables, $\dot{m}_{\text{sludge}}$ and $\dot{m}_{\text{plate}}$ (plating waste), may aid in the following analysis.

Determine the maximum amount of mercury-contaminated waste that can be incinerated.

Solution

Let $\dot{m}_{plate}$ and $\dot{m}_{sludge}$ represent the flowrates of the two wastes to the hazardous waste incinerator in pounds/hour. The restrictions on the available blending options are shown in Figure 86 using numbered lines for the constraints.

Line 1	Limit of 5000 lb/h of plating waste	$\dot{m}_{plate} < 5000$
Line 2	Minimum of 1000 lb/h of plating waste	$\dot{m}_{plate} > 1000$
Line 3	Limit of 5000 lb/h of plating waste	$\dot{m}_{sludge} < 5000$
Line 4	Minimum of 1000 lb/h of sludge	$\dot{m}_{sludge} > 1000$
Line 5	Maximum heat rate of 40 MBtu/h	$\dot{m}_{sludge} = 3890 - \left(\dfrac{3890}{5000}\right)\dot{m}_{plate}$
Line 6	Minimum heat rate of 25 MBtu/h	$\dot{m}_{sludge} = 2431 - \left(\dfrac{2431}{3125}\right)\dot{m}_{plate}$

Lines 5 and 6 are determined by relaxing all constraints and assuming that only sludge is to be incinerated.

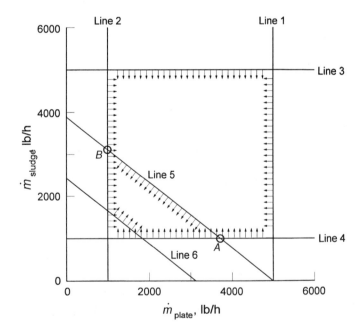

Figure 86. Diagram showing six constraints for Problem MED.7.

The permissible area in Figure 86 is the quadrilateral bounded by lines 2, 4, 5, and 6. The point farthest to the right (point A) represents the maximum rate at which plating waste can be burned. This occurs where constraints 4 and 5 intersect, i.e.,

$$1000 \text{ lb/h} = 3890 \text{ lb/h} - \left(\frac{3890 \text{ lb/h}}{5000 \text{ lb/h}}\right)\dot{m}_{\text{plate}}$$

$$\dot{m}_{\text{plate}} = 3715 \text{ lb/h}$$

MED.8   MAXIMUM AMOUNT OF SLUDGE TO BE INCINERATED

Refer to Problem MED.7. Determine the maximum amount of sludge that can be incinerated, along with the required supplemental fuel oil necessary for this waste.

Solution

In the permissible area in Figure 86, the highest point (point B) represents the maximum rate at which the sludge can be burned. This occurs where constraints 2 and 5 intersect, i.e.:

$$\dot{m}_{\text{sludge}} = 3890 \text{ lb/h} - \left(\frac{3890 \text{ lb/h}}{5000 \text{ lb/h}}\right)1000 = 3112 \text{ lb/h}$$

To determine the fuel oil requirements at this sludge loading, the NHV of the waste is first calculated as:

$$\left(\frac{3112 \text{ lb/h}}{3112 \text{ lb/h} + 1000 \text{ lb/h}}\right)\left(\frac{6000 \text{ Btu}}{\text{lb}}\right) + \left(\frac{1000 \text{ lb/h}}{3112 \text{ lb/h} + 1000 \text{ lb/h}}\right)\left(\frac{8000 \text{ Btu}}{\text{lb}}\right)$$

$$= 6490 \text{ Btu/lb}$$

Then the fraction of the fuel, F, required to bring the mixture to 8000 Btu/lb is calculated as follows:

$$(6490 \text{ Btu/lb})(1 - F) + (15{,}000 \text{ Btu/lb})F = 8000 \text{ Btu/lb}$$

$$F = 0.18$$

The supplemental fuel requirement at 40 MBtu/h is

$$0.18\ (40 \times 10^6 \text{ Btu/h})(1 \text{ lb}/15{,}000 \text{ Btu}) = 480 \text{ lb/h}$$

MED.9   MAXIMUM AMOUNT OF MERCURY-CONTAMINATED WASTE TO BE INCINERATED

Refer to Problem MED.7.

1. If no more than 150 lb/h of mercury can be incinerated, what is the maximum amount of mercury-contaminated waste that can be incinerated?
2. Discuss the impact that the mercury and other constraints impose on the operation of this incineration facility.

Solution

1. A seventh constraint is required to keep the Hg < 150 lb/h. This may be represented by

$$0.02 \, \dot{m}_{\text{sludge}} + 0.08 \, \dot{m}_{\text{plate}} < 150 \, \text{lb/h}$$

This constraint is shown as line 7 in Figure 87. The equation of this line is

$$\dot{m}_{\text{sludge}} = \frac{150}{0.02} - \left(\frac{0.08}{0.02}\right) \dot{m}_{\text{plate}}$$

Given all seven constraints, the permissible area is then the quadrilateral bounded by lines 2, 5, 6, and 7. The farthest point to the right in this area (point C) on the figure is the maximum incineration rate of the plating waste. This occurs where constraints 6 and 7 intersect, i.e.,

$$7500 \, \text{lb/h} - 4\dot{m}_{\text{plate}} = 2431 \, \text{lb/h} - \left(\frac{2431 \, \text{lb/h}}{3125 \, \text{lb/h}}\right) \dot{m}_{\text{plate}}$$

$$\dot{m}_{\text{plate}} = 1573 \, \text{lb/h}$$

2. The shaded area on the figure represents the permissible operating range for this unit given all seven constraints. The operating range does not allow much flexibility. The solution is deterministic, and variation in waste, operator error, waste analysis uncertainty, etc. are unaccounted for. Low-intensity burners may allow a decrease in the required NHV of the waste and permit additional flexibility.

This graphical technique is adequate for the two variables investigated in this problem. This problem can also be solved using optimization techniques such as linear programming, a technique that can handle numerous constraints and variables

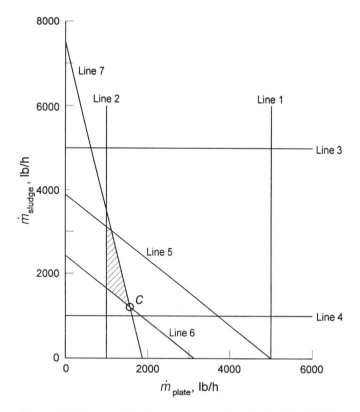

Figure 87. Diagram showing seven constraints for Problem MED.9.

and can be used, for example, to minimize supplemental fuel costs while operating within given system constraints.

MED.10   COMBUSTION TEMPERATURE

Estimate the combustion temperature of an incinerator contaminated with hospital alcohol, 50% ethyl alcohol + 50% water and solids (may treat all as water) by weight using 40% EA. To simplify the calculations, assume an average flue gas heat capacity for N_2, CO_2, and O_2 of 0.26 and for H_2O vapor of 0.5 Btu/(lb · °F).

Solution

The stoichiometry for the problem is as follows:

$$C_2H_5OH + 3O_2 \rightarrow 2CO_2 + 3H_2O$$

Assuming a 1-lb waste basis,

$$\text{Stoichiometic } O_2 = (0.5)(3 \text{ lbmol}) \left(\frac{\dfrac{32 \text{ lb}}{\text{lbmol } O_2}}{\dfrac{46 \text{ lb}}{\text{lbmol EtOH}}} \right) = 1.04 \text{ lb } O_2$$

Then, the oxygen requirement at 40% excess air $(EA) = 1.46 \text{ lb } O_2/\text{lb}$ waste. The composition of the flue gas is given as:

$$N_2: \left(\frac{1.46 \text{ lb } O_2}{\text{lb waste}} \right) \left(\frac{28 \text{ lb}/\text{lbmol } N_2}{32 \text{ lb}/\text{lbmol } O_2} \right) \left(\frac{79 \text{ lbmol } N_2}{21 \text{ lbmol } O_2} \right) = 4.806 \text{ lb } N_2/\text{lb waste}$$

$$CO_2: \left(\frac{0.5 \text{ lb EtOH}}{\text{lb waste}} \right) (2) \left(\frac{44 \text{ lb } CO_2/\text{lbmol}}{46 \text{ lb EtOH}/\text{lbmol}} \right) = 0.956 \text{ lb } CO_2/\text{lb waste}$$

$$O_2: (0.4 \, EA)(1.04) = 0.416 \text{ lb } O_2/\text{lb waste}$$

$$H_2O: \left(\frac{0.5 \text{ lb } H_2O}{\text{lb waste}} \right) + \left(\frac{0.5 \text{ lb EtOH}}{\text{lb waste}} \right) (3) \left(\frac{18 \text{ lb } H_2O/\text{lbmol}}{46 \text{ lb EtOH}/\text{lbmol}} \right)$$

$$= 1.09 \text{ lb } H_2O/\text{lb waste}, 0.5 \text{ lb of which comes in with waste.}$$

$$\text{Total } N_2 + CO_2 + O_2 \text{ in the dry flue gas} = 4.806 + 0.956 + 0.416$$

$$= 6.18 \text{ lb}/\text{lb waste}$$

NHV for ethyl alcohol $= 11.929 \text{ Btu}/\text{lb}$

The NHV for the combined water/ethyl alcohol waste is

$$NHV = \left(\frac{0.5 \text{ lb EtOH}}{\text{lb waste}} \right) \left(\frac{11,929 \text{ Btu}}{\text{lb EtOH}} \right) - \left(\frac{0.5 \text{ lb } H_2O}{\text{lb waste}} \right) \left(\frac{1 \text{ Btu}}{\text{lb } H_2O \cdot {}^\circ F} \right) (212 - 60)$$

$$- \left(\frac{0.5 \text{ lb } H_2O}{\text{lb waste}} \right) \left(\frac{970 \text{ Btu}}{\text{lb } H_2O} \right) - \left(\frac{0.5 \text{ lb } H_2O}{\text{lb waste}} \right) \left(\frac{0.5 \text{ Btu}}{\text{lb } H_2O \cdot {}^\circ F} \right) (T - 212)$$

$$= 5403.5 \text{ Btu}/\text{lb waste} - 0.25(T - 212)$$

$$= 5456.5 - 0.25(T) \text{ Btu}/\text{lb waste}$$

The reader is referred to Chapter 9 for details on the following calculation. The energy of the waste is used to raise the temperature of the combustion products, and this change in enthalpy of the combustion products is calculated as:

$$\Delta H_p = \left(\frac{6.18 \text{ lb gas}}{\text{lb waste}}\right)\left(\frac{0.26 \text{ Btu}}{\text{lb gas} \cdot {}^\circ\text{F}}\right)(T - 60)$$

$$+ \left(\frac{0.59 \text{ lb } H_2O}{\text{lb waste}}\right)\left(\frac{1 \text{ Btu}}{\text{lb gas} \cdot {}^\circ\text{F}}\right)(212 - 60)$$

$$+ \left(\frac{0.59 \text{ lb } H_2O}{\text{lb waste}}\right)\left(\frac{970 \text{ Btu}}{\text{lb gas}}\right) + \left(\frac{0.59 \text{ lb } H_2O}{\text{lb waste}}\right)\left(\frac{0.5 \text{ Btu}}{\text{lb gas} \cdot {}^\circ\text{F}}\right)(T - 212)$$

$$= 1.607 \, (T - 60) + 89.68 + 572.3 + 0.295 \, (T - 212)$$

Equating NHV to ΔH_p yields the following relationship and solution:

$$5456.5 - 0.25(T) = 1.902(T) + 503.02$$
$$T = 2302{}^\circ\text{F}$$

MED.11   CHOOSING AN INCINERATION SYSTEM

A city has decided to incinerate all hospital and infectious wastes at a central location due to new stringent air emission regulations. Two hospitals, A and B, each containing 500 beds, along with a 50-bed nursing home and an animal research center, contribute their waste to the central facility. Hospital A contributes 20 lb/bed/day of waste while hospital B contributes 15 lb/bed/day. The nursing home contributes 10 lb/bed/day of waste, and the animal research center provides 250 lb/day of waste to the incineration facility. The compositions by weight of the wastes from the contributing facilities are as follows:

Waste Type	Hospital A	Hospital B	Nursing Home	Animal Research
Trash	55	55	65	30
Plastic	30	15	20	15
Garbage	10	10	10	5
Pathological	5	20	5	50

Hospital A and the nursing home have a high percentage of plastics, and hospital B and the animal research facility have high percentages of pathological (infectious) waste.

1. What are some considerations that are necessary before an incineration system is selected?

2. Compare controlled-air incineration and rotary kiln incineration for the central facility.
3. Would heat recovery be viable for this central facility?

Solution

1. The total quantity of waste received must first be determined. Continuous or batch operation of the waste must be chosen. Biomedical waste incinerators are normally batch-operated, but since the quantity of waste received at this facility is large, continuous operation may make more sense.

Since the wastes from hospital A and the nursing home contain a high percentage of plastics, excessive acid gas emissions are possible. Therefore, acid gas emissions control should be considered.

The high percentage of pathological waste from hospital B and the animal research facility would result in lower waste heating values. Operators must be trained to account for variations in the heating value of the feed to the incinerator. Batch operation should still be considered an option due to the expected variation of the incinerator feed heating value.

The principal organic hazardous constituents (POHCs) of the waste must be destroyed to the required destruction and removal efficiency. HCl and particulate emissions must comply with state and federal regulations for hospital waste incinerators.

Since a high percentage of plastics occurs in the waste stream fed to the incinerator, a suitable method of treating the acid gases produced must be chosen. This may entail reacting the acid gases with an alkaline material. Therefore, a spray dryer should be considered to neutralize the acid gases.

2. Controlled-air incinerators have a lower capital cost since they do not normally require air pollution control equipment (unless acid gas emissions are excessive). But since this facility must comply with new stringent regulations and acid gas emissions may be excessive, a rotary kiln system may be more advantageous.

It is expected that the heating value of the waste fed to the incinerator may vary substantially. Therefore, a rotary kiln incinerator would be more appropriate since it can handle a large range of waste heating values. The residence time of a rotary kiln incinerator may be changed by adjusting the rotational speed of the kiln.

Since this facility is large compared to traditional hospital waste incinerators, the incinerator may be operated continuously. Controlled-air incinerators are normally operated in batch mode.

All factors suggest that a rotary kiln incineration unit would be more appropriate for this facility than a controlled-air incineration system.

3. The facility operates at over 9 tons per day. While this is not large compared to municipal solid waste incineration facilities, if the facility expands in the future, heat recovery may be a viable option. But heat recovery is only economically advantageous if customers exist to purchase the recovered heat (normally steam) or if there is an onsite need for the recovered heat.

Biomedical wastes are not only generated by hospitals. Animal research facilities, research centers, universities, rest homes, and veterinary clinics also generate pathological (infectious) waste. Pathological waste includes animal carcasses, contaminated laboratory wastes, hypodermic needles, contaminated food and equipment, blood products, and even dialysis unit wastes. Normally, biomedical wastes are incinerated along with other wastes generated by the facilities such as paper and plastic.

In addition to the above, an increase in plastics in hospital waste streams has occurred during the last decade. Plastics may account for as much as 30% of a hospital waste stream. Unfortunately, incinerating plastics normally increases the chlorine content of the exiting flue gas. This creates a need for air pollution control devices to remove chlorine compounds.

In the past, controlled-air incinerators have been the most popular incinerators for biomedical waste destruction. A controlled-air incinerator is a two-chamber, hearth-burning, pyrolytic unit. The primary chamber receives the waste and burns it with less than stoichiometric air. Volatiles released in the primary chamber are burned in the secondary combustion chamber. These units result in low fly ash generation and low particulate emissions. In addition, they have a low capital cost and may be batch operated. They normally do not require air pollution control equipment unless acid gas emissions are excessive.

MED.12   COMPLIANCE DETERMINATION FOR A HOSPITAL INCINERATOR

A hospital in the state of Pennsylvania is currently incinerating its waste in a modular incinerator at a temperature of 1800°F and a residence time of 2 s. Regulations for particulate emissions for hospital waste incinerators in Pennsylvania are as follows:

Capacity (lb/h)	Particulate Emissions Standard (gr/dscf)
≤ 500	0.08 (corrected to 7% O_2)
500–2000	0.03 (corrected to 7% O_2)
≥ 2000	0.015 (corrected to 7% O_2)

Hydrogen chloride (HCl) emissions must not exceed 4 lb/h for incinerators operating at capacities of 500 lb/h or less. For larger incinerators, HCl emissions must not exceed 30 ppmdv (corrected to 7% O_2).

The hospital generates 20 lb/bed · day of waste of which 10% is infectious (the remaining 90% is paper, cardboard, etc.). It also produces 0.056 lb/bed · day of Resource Conservation and Recovery Act (RCRA) hazardous wastes. Two hundred beds are present in the hospital. The RCRA hazardous wastes are:

Hazardous Waste Component	Wt %
Methyl alcohol	12.5
Polyvinyl chloride (PVC)	75
Xylene	12.5

A trial burn is conducted with the waste composition in the preceding table. The designated principal organic hazardous constituents (POHCs) for the trial burn are methyl alcohol, polyvinyl chloride, and xylene. The results of the trial burn are:

Hazardous Waste Component	Outlet mass flowrate (lb/h)
Methyl alcohol	0.0001
Polyvinyl chloride (PVC)	0.0002
Xylene	0.00001
HCl	3.2
Particulates	5.64

The total stack gas flowrate is 10,000 dscfm. Determine

1. The hazardous waste generation rate in pounds/month.
2. The regulations the incinerator must comply with (state hospital regulations or combined state and hospital and RCRA regulations).
3. The total incinerator capacity in pounds/hour.
4. Is the incinerator in compliance?

Solution

1. Calculate the hazardous waste generation rate in pounds/month:

$$\text{Hazardous waste rate} = (0.056\,\text{lb/bed} \cdot \text{day})(200\,\text{beds})$$
$$= 11.2\,\text{lb/day}$$
$$= 11.2\,\text{lb/day}\,(30\,\text{days/mo})$$
$$= 336\,\text{lb/mo}$$

2. Determine if the incinerator must comply with RCRA regulations. Since the hazardous waste generation rate of 336 lb/mo is greater than the 220 lb/mo RCRA regulation, the incinerator must comply with RCRA regulations in addition to the state regulations for hospital incinerators.

3. Calculate the amount of nonhazardous waste produced in pounds/day:

$$\text{Nonhazardous waste rate} = (20\ \text{lb/bed} \cdot \text{day})(200\ \text{beds})$$
$$= 4000\ \text{lb/day}$$

Determine the total incinerator capacity in pounds/hour:

$$\text{Total capacity} = \text{Hazardous waste rate} + \text{Nonhazardous waste rate}$$
$$= (11.2\ \text{lb/day} + 4000\ \text{lb/day})(\ \text{day/24 h})$$
$$= 167.1\ \text{lb/h}$$

Determine the inlet mass rate of methyl alcohol to the incinerator in pounds/hour:

$$\dot{m}_{\text{in}} = (0.125)(11.2\ \text{lb/day})(\text{day/24 h})$$
$$= 0.0583\ \text{lb/h}$$

Calculate the destruction efficiency for the hazardous component methyl alcohol (MeOH):

$$\text{DRE}_{\text{MeOH}} = [(0.0583\ \text{lb/h} - 0.0001\ \text{lb/h})/(0.0583\ \text{lb/h})](100)$$
$$= 99.827\%$$

Calculate the destruction efficiency for the hazardous component xylene:

$$\dot{m}_{\text{in}} = (0.125)(11.2\ \text{lb/day})(\text{day/24 h})$$
$$= 0.0583\ \text{lb/h}$$

$$\text{DRE}_{\text{xylene}} = [(0.0583\ \text{lb/h} - 0.00001\ \text{lb/h})/(0.0583\ \text{lb/h})]\ (100)$$
$$= 99.983\%$$

Calculate the destruction efficiency for the hazardous component polyvinyl chloride (PVC):

$$\dot{m}_{\text{in}} = (0.75)(11.2\ \text{lb/day})(\text{day/24 h})$$
$$= 0.350\ \text{lb/h}$$

$$\text{DRE}_{\text{PVC}} = [(0.350\ \text{lb/h} - 0.0002\ \text{lb/h})/(0.350\ \text{lb/h})]\ (100)$$
$$= 99.943\%$$

Since the total incineration capacity is less than 500 lb/h, the 4 lb/h regulation applies. The trial burn resulted in an HCl emission rate of 3.2 lb/h.

Calculate the outlet loading (OL) of the particulates in gr/dscf.

$$OL = [(5.64\,\text{lb/h})(7000)]/[(10,000\,\text{dscfm})(60)]$$

$$= 0.0658\,\text{gr/dscf}$$

4. Determine if the incinerator is in compliance.

POHCs: Since all of the POHCs had DREs < 99.99%, the incinerator is out of compliance.

HCl: The incinerator complies with the 4 lb/h limit on HCI emissions.

Particulates: The outlet loading of 0.0658 gr/dscf is less than the 0.08 gr/dscf limit for incinerators operating at a capacity less than 500 lb/h.

The incinerator is not in compliance.

Note that the pathological waste in the waste stream in this problem is the most difficult waste to destroy since its heating value is low. Hospital waste incinerators must be designed to destroy pathological and infectious waste, not paper waste alone. The contents of a hospital waste stream are normally more complex than shown in this problem. Other hazardous components may include pentane, diethyl ether, acetone, methyl cellosolve, and other laboratory wastes.

Each state has its own regulations concerning hospitals. Other regulations must also be complied with in addition to those stated in this problem. The pertinent state agencies should be contacted for a list of the detailed regulations for hospital waste incinerators.

PART V
Water Quality and Wastewater Treatment

Joseph Flesche

29 Regulations (REG)

REG.1 DEFINITIONS

Answer the following questions:

1. Identify key legislation directed at controlling wastewater discharges from municipalities. Briefly define the priorities of each legislation.
2. What is the purpose of the *Safe Drinking Water Act* (SDWA)?
3. What is non-point-source (NPS) pollution? Why is NPS pollution difficult to deal with?
4. Highlight the difference between a "direct discharger" and an "indirect discharger."

Solution

1. The *Clean Water Act* (Federal Water Pollution Act of 1972): created water quality standards for point discharges, established timelines and penalties, and regulated toxic discharges.

   The *Resource Conservation and Recovery Act* (RCRA): is concerned with the safe disposal of solid wastes.

   The *Marine Protection, Research and Sanctuaries Act*: prohibits sewage discharge by ocean barge dumping.

   The *Water Quality Act of 1987*: strengthened federal water quality regulations, regulated toxic discharges, expanded treatment to nonpoint sources, and established new stormwater restrictions.

2. The purpose of the SDWA is to establish *maximum containment level goals* (MCLGs) and *national primary drinking water regulations* (NPDWRs) that are designed to protect the public from the contamination of drinking water.

3. NPS pollution is described as pollution that enters water supplies over a large geographical area from a variety of human activities and surface runoff. NPS pollution is difficult to deal with because it is widespread and tough to isolate. NPS pollution appears often in surges that are associated with periods of rainfall and snowmelt runoff.

4. A direct discharger is an industrial plant that discharges its effluent wastewater

directly to a surrounding water source with no intermediate means of treatment. An indirect discharger is a plant that first discharges to a *publicly owned treatment works* (POTW) facility prior to release to the environment.

REG.2 PROTECTION OF THE WATER SUPPLY

Answer the following questions:

1. Describe the present-day stresses that threaten natural wetlands and their habitats.
2. Identify two methods for the potential transfer of toxic materials from water bodies to humans.
3. List examples of industries that are recognized as traditional sources of industrial wastewater discharges.
4. Describe the roles of the Environmental Protection Agency (EPA) and the U.S. Army Corps of Engineers with respect to protection of the national water supply.

Solution

1. Wetlands are susceptible to contamination from industrial and municipal sources that introduce pollutants into the habitat. Large construction projects such as dams and canals can rapidly alter the conditions within a specific wetland beyond the habitat's capacity to recover. Wetlands are often threatened by developers who utilize the space for residential housing and commercial building. Manmade wastewater and surface drainage projects have contaminated wetlands by increasing concentrations of metal-bearing leachates to harmful levels.
2. Two methods that involve the transfer of toxic materials from water bodies to humans are:

   Ingestion of chemicals found in drinking water
   Ingestion of food sources grown in aquatic environments

3. Industrial pollutants include
   Textile
   Tannery
   Laundry
   Cannery
   Dairy
   Brewery, winery, distillery

Pharmaceutical

Meat packing and rendering

Petrochemical

Beet sugar

Metal fabrication

Wood fibers

Chemical

Liquid materials

Nuclear power

Energy

4. The primary role of the U.S. Army Corps of Engineers in protecting national water sources is to issue and control permits for the discharge of dredge and/or fill material. The EPA has a number of duties, which include establishing the jurisdiction of the regulation, setting standards for the release of water pollutants, and prohibiting the discharge of potentially harmful substances. The EPA also assists the U.S. Army Corps of Engineers in reviewing applications for water discharge permits.

REG.3   CLEAN WATER ACT AND PWPS

Describe the interrelationship between the *Clean Water Act* (CWA) and *priority water pollutants* (PWPs).

Solution

The Clean Water Act addresses a large number of issues related to water pollution management. Control of industrial wastewaters is primarily the responsibility of the *National Pollutant Discharge Elimination System* (NPDES), originally established by Public Law 92-500. Any municipality or industry that discharges wastewater in the United States must obtain a discharge permit under the regulations set forth by the NPDES. Under this system, there are three classes of pollutants (conventional pollutants, priority pollutants, and nonconventional/nonpriority pollutants). Conventional pollutants are substances such as biochemical oxygen demand (BOD), suspended solids (SS), pH, oil and grease, and coliforms. Priority pollutants were so designated on a list of 129 substances originally set forth in a consent decree between the Environmental Protection Agency and several environmental organizations. This list was incorporated into the 1977 amendments and has since been reduced to 126 substances. Most of the substances on this list are organics, but it does include most of the heavy metals. These substances are generally considered to be toxic. However, the toxicity is not absolute; it primarily depends on the concentration. In recent years, pollution prevention programs have been implemented to reduce their use in industrial processes. *Regulatory Chemicals Handbook*, by Spero et al. (Marcel Dekker, New York, 2000) provides extensive details on these priority (water) pollutants. The third class of pollutants could include any pollutant not in the first two categories. Examples of substances that are presently regulated in the third category are nitrogen, phosphorus, and sodium.

REG.4   WATER QUALITY STANDARDS

Discuss the two primary criteria employed to set water quality standards.

Solution

Water quality standards are usually based on one of two primary criteria: stream standards or effluent standards. Stream standards are based upon receiving-water quality concentrations. These standards are determined from threshold values of specific pollutants. The waste being discharged into a stream must not cause the pollutant concentration to exceed the threshold value.

Effluent standards establish the concentration of pollutants that can be discharged (the maximum concentration of a pollutant, mg/L or the maximum load, lb/day) to a receiving water or upon the degree of treatment required for a given type of wastewater discharge, no matter where in the United States the waste originates. These effluent limitations are related to the characteristics of the discharger, not to that of the receiving stream.

It should be noted that the water quality to be attained is not static but is often subject to modification with a changing municipal and industrial environment. For example, as the carbonaceous organic load is removed by wastewater treatment, the detrimental effect of nitrification may also become a serious problem. These considerations may require an upgrading of the degree of treatment provided for waste discharges over time and suggest that the assessment of water quality conditions and water quality demands must be an ongoing process, i.e., it is time variant or time dependent.

REG.5   CHEMICAL FORMULAS

With reference to PWPs, provide the chemical formula for the following 10 pollutants:

Anthracene
Chrysene
Diethyl phthalate
Methylene chloride
Naphthalene

2-Nitrophenol
Pentachlorophenol
PCB-1016
Tetrachloroethylene
Vinyl chloride

Solution

The chemical formula for each pollutant (except PCB-1016) is provided below.

Anthracene: $C_{14}H_{10}$
Chrysene: $C_{18}H_{12}$
Diethyl phthalate: $C_{12}H_{14}O_4$
Methylene chloride: CH_2Cl_2
Naphthalene: $C_{10}H_8$
2-Nitrophenol: $C_6H_5NO_3$
Pentachlorophenol: C_6HCl_5O
PCB-1016: This pollutant is a mixture of mono-, di-, and trichloroisomers of the polychlorinated biphenyls. It has no single chemical formula.
Tetrachloroethylene: C_2Cl_4
Vinyl chloride: C_2H_3Cl

Note: Details on each of the PWPs are provided in the *Regulatory Chemicals Handbook* (see Problem REG.3).

REG.6   PCB-1016

One of the worst "actors" in the PWPs list is PCB-1016. Provide detailed information for this chemical.

Solution

(Drawn in part from Spero et al., *Regulatory Chemicals Handbook*, Marcel Dekker, New York, 2000.)
PCB-1016 is a mixture of mono-, di-, and trichloroisomers of the polychlorinated biphenyls (PCBs), 257.9 average molecular weight.

Gas/DOT Identification #: 12674-11-2/UN 2315.
Synonyms: Chlorodiphenyl, Aroclor 1016; PCPs, Polychlorinated biphenyls, Aroclors.
Physical Properties: Polychlorinated biphenyl contain ~16% chlorine; the solubility of PCB decreases with increasing chlorination (0.04–0.2 ppm); colorless oil; soluble in oils and organic solvents; odorless; boiling point

(BP) 325–356°C); density (DN) (1.33 g/mL at 25°C); specific gravity (SG) (1.4); vapor pressure, (VP) (4×10^{-4} torr at 25°C estimated).

Chemical Properties: Incompatible with strong oxidizers; generally nonflammable; chemically inert and stable to conditions of hydrolysis and oxidation in industrial use; freezing point (FP) (286°F).

Biological Properties: Aerobic degradation in semicontinuous activated sludge process; 30% degradation of < 1 mg/L concentration after 48 h incubation; partition coefficient between sediment and water: ~ 105; 15–10% degradation of 1 mg in 48 h, increased chlorine in molecule decreases degradation; catalytic dechlorination to biphenyl was achieved with 5% platinum or palladium on 60/80 mesh glass beads; 48% biodegraded at the end of 28 days at 5 ppm concentration in a static flask screening procedure using BOD dilution water, settled domestic wastewater innoculum, at 10 ppm, 13% biodegraded; can be detected in water by EPA Method 608 (gas chromatography) or EPA Method 625 (gas chromatography plus mass spectrometry).

Bioaccumulation: PCBs in pelagic organisms; a food chain interrelationship study; PCB concentration factor (wet weight); microplankton 170,000, macroplanktonic enphausiid (*Meganyctiphanes norvegica*) 50,000; camicorous decapod shrimp (*Sergestes arcticus*) 47,000; (*Pasiphaea sivado*) 20,000; myctophid fish (*Myctophus glaciale*) 6000; no biomagnification in this food chain, if whole organisms are considered.

Origin/Industry Sources/Uses: Prepared by the chlorination of biphenyl; used in the electrical industry in capacitors and transformers; used in the formulation of lubricating and cutting oils; pesticides; adhesives; plastics; inks; paints; sealants.

Exposure Routes: Inhalation of fume or vapor; percutaneous adsorption of liquid; ingestion; eye and skin contact; landfills containing PCB waste materials and products; incineration of municipal refuse and sewage sludge; waste transformer fluid disposal to open areas.

Regulatory Status: Criterion to protect freshwater aquatic life: 0.014 µg/L · 24 h avg.; criterion to protect saltwater aquatic life: 0.030 µg/L · 24 h avg.; criterion to protect human health: preferably 0.0; lifetime cancer risk of 1 in 100,000: 0.00079 ng/L; maximum contaminant level in drinking water: 0.5 µg/L (for PCBs as decachlorobiphenyl).

Probable Fate: Photolysis: too slow to be important, vapor-phase reaction with hydroxyl radicals has a half-life of 27.8 days to 3.1 months; oxidation: not important; hydrolysis: not important; volatilization: slow volatilization is the cause of global distribution of PCBs, but is inhibited by adsorption, may be significant over time, rapid volatilization from water in the absence of adsorption, half-life of 2–7 years in typical water bodies; sorption: PCBs are rapidly adsorbed onto solids, especially organic matter, and are often immobilized in sediments, but may reenter solution; biological processes: strong bioaccumulation, mono-, di-, and trichlorinated biphenyls are gradually biodegraded; biodegradation is probably the ultimate degradation process in water and soil.

REG.7   TYPES OF DISCHARGES

1. Answer the following questions related to types of discharges to surface waters:
   a. What are the two major categories of discharges into surface waters? Give an example of each category.
   b. Which of the two major categories would be easier to regulate and why?
2. Answer the following questions regarding the NPDES program:
   a. What does the acronym NPDES stand for?
   b. What are the three major parameters regulated under the NPDES program for municipal wastewater discharges?
   c. What are the maximum concentrations allowed for each of these parameters in NPDES permits?

Solution

1. The two major categories of discharges into surface waters are point sources, which are essentially any discharge that can be specifically located by finding the discharge pipe, and nonpoint sources. A few examples of point sources include:

   Industrial discharges

   Municipal wastewater discharges

   Municipal storm water discharges

   A few examples of nonpoint sources include:

   Rainfall runoff from large areas of land, as in agricultural areas, large fields, etc.

   Atmospheric deposition

   The easiest category of discharge source to regulate is the point source. The point source essentially has a physical structure that extends from the source of the discharge to the receiving body or surface water. Thus, there is one point or location that can be sampled to check for compliance with set discharge standards.

   The more difficult category of discharge source to regulate is the nonpoint source. This source of discharge covers a wide and varying area and a sample cannot be taken that is truly representative of a specific location. Thus, there are numerous variables that could impact the discharge sample that limits the enforcement of a set of discharge standards for this type of source.

2. NPDES stands for the *National Pollutant Discharge Elimination System*, a permit program established for each point source discharge in the United States under the Clean Water Act of 1972.

The three major parameters regulated by the NPDES permit program for municipal wastewater treatment plants and their regulated maximum concentrations are shown in the following table:

Effluent BOD$_5$ and TSS Concentrations for a Municipal Wastewater Treatment Plant as Defined by the NPDES Permit Program

	30-day Average Concentration	7-day Average Concentration
5-day BOD (BOD$_5$)	30 mg/L	45 mg/L
TSS[a]	30 mg/L	45 mg/L
pH[b]	6 to 9	6 to 9

[a] TSS = total suspended solids.
[b] pH = negative log$_{10}$ of hydrogen ion concentration in mole/liter (see Problem CHR.4, Chapter 30).

REG.8   TRACE CONCENTRATION

Some wastewater and water standards and regulations are based on a term defined as *parts per million*, ppm, or *parts per billion*, ppb. Define the two major classes of these terms and describe the interrelationship from a calculational point of view. Also convert 10 calcium parts per million parts of water on a mass basis to parts per million on a mole basis.

Solution

Water streams seldom consist of a single component. It may also contain two or more phases (a dissolved gas and/or suspended solids), or a mixture of one or more solutes. For mixtures of substances, it is convenient to express compositions in mole fractions or mass fractions. The following definitions are often used to represent the composition of component A in a mixture of components:

$$w_A = \frac{\text{Mass of A}}{\text{Total mass of water stream}} = \text{Mass fraction of A}$$

$$y_A = \frac{\text{Moles of A}}{\text{Total moles of water stream}} = \text{Mole fraction of A}$$

Trace quantities of substances in water streams are often expressed in parts per million (ppmw) or as parts per billion (ppbw) on a mass basis. These concentrations can also be provided on a mass per volume basis for liquids and on a mass per mass basis for solids. Gas concentrations are usually represented on a mole or volume basis (e.g., ppmm or ppmv). The following equations apply:

$$\text{ppmw} = 10^6 \, w_A$$

$$= 10^3 \, \text{ppbw}$$

$$\text{ppmv} = 10^6 \, y_A$$

$$= 10^3 \, \text{ppbv}$$

The two terms ppmw and ppmm are related through the molecular weight. To convert 10 ppmw Ca to ppmm, select a basis of 10^6 g of solution.

The mass fraction of Ca is first obtained by the following equation:

$$\text{Mass of Ca} = 10 \, \text{g}$$

$$\text{Moles Ca} = \frac{10 \, \text{g}}{40 \, \text{g/mol}} = 0.25 \, \text{mol}$$

$$\text{Moles H}_2\text{O} = \frac{10^6 \, \text{g} - 10 \, \text{g}}{18 \, \text{g/mol}} = 55{,}555 \, \text{mol}$$

$$\text{Mole fraction Ca} = y_{\text{Ca}} = \frac{0.25 \, \text{mol}}{0.25 \, \text{mol} + 55{,}555 \, \text{mol}} = 4.5 \times 10^{-6}$$

$$\text{ppmm of Ca} = 10^6 \, y_{\text{Ca}}$$

$$= (10^6)(4.5 \times 10^{-6})$$

$$= 4.5$$

30 Characteristics (CHR)

CHR.1 WASTEWATER CHARACTERISTICS

Answer the following three questions:

1. List the physical characteristics important in establishing the quality of municipal wastewaters.
2. The characteristics of a municipal wastewater can change drastically over time. Why is this true?
3. List the three categories of industrial wastewater treatment. Provide examples from each category.

Solution

1. Numerous wastewater physical characteristics include:

   Temperature

   Color

   Odor

   Turbidity

   Suspended solids

   Volatile suspended solids

2. Municipal wastewater characteristics are dependent on a number of factors that include solids, wastewater volume, and chemical content. The factors can change rapidly in response to daily and seasonal variations, precipitation, industrial discharges, and public habits. Combination of these factors can result in significant hourly fluctuations in wastewater characteristics.

3. Wastewater treatment categories include:

   Physical treatment: clarification, flotation

   Chemical treatment: coagulation/precipitation/flocculation, neutralization

   Biological treatment: aerobic suspended growth processes, fixed film processes, aerated lagoons, anaerobic treatment processes

These are treated in separate problem sets later in this part.

CHR.2   KEY WASTEWATER CHARACTERISTICS

List and describe some key wastewater characteristics.

Solution

Key characteristics include:

Suspended Solid (SS): A measure of solids that are suspended (not dissolved) in the wastewater. Suspended solids will lower the amount of light and oxygen in a body of wastewater. Over time, suspended solids will settle from the wastewater, blanketing the floor with a layer of sediment called *sludge.*

Biochemical Oxygen Demand (BOD): Primarily the level of organic content in a wastewater measured by the demand for oxygen that can be consumed by living organisms in the wastewater. Wastewater with high BOD content is characterized by low oxygen content and high biological activity.

Floating Materials: Materials seen on the surface of wastewater that indicate the presence of insoluble fats, oils, greases, and other immiscible materials such as wood, paper, plastics, etc.

Color: Reveals the presence of dissolved material in wastewater. Color in wastewater may originate from dyes, decaying organics, etc.

The three major biodegradable organics in wastewater are composed principally of proteins, carbohydrates, and fats. If discharged untreated to the environment, their biological stabilization can lead to the depletion of natural oxygen resources and to the development of septic conditions in rivers and other natural bodies of water.

Pathogens: Pathogenic organisms that can transmit communicable diseases via wastewater. Typical notified infectious diseases reported are cholera, typhoid, paratyphoid fever, salmonellosis, and shigellosis. *E. coli* and fecal coli are indicators of pathogens.

Nutrients: Carbon is the major nutrient source while nitrogen and phosphorus are secondary. When discharged to the receiving water, these nutrients can lead to the growth of undesirable aquatic life. When discharged in excessive amounts on land, they can also lead to the pollution of groundwater.

Priority Pollutants: Organic and inorganic compounds designated on the basis of their known or suspected carcinogenicity, mutagenicity, or high acute toxicity. Many of these compounds are found in wastewater.

Heavy Metals: Heavy metals are usually added to wastewater from municipal commercial and industrial activities and may have to be removed if the wastewater is to be reused, or discharged into a water body.

Floating solids and liquids include oils, greases, and other materials that float on the surface; if not treated, they not only make the river unsightly but also obstruct passage of light and clarity through the water.

Color contributed by textile and paper mills, tanneries, slaughterhouses, and other industries is an indicator of pollution. Compounds present in wastewaters absorb certain wavelength of light and reflect the remainder, a fact generally conceded to account for color development of streams. Color interferes with the transmission of sunlight into the stream and therefore lessens photosynthetic actions.

Regarding general definitions, municipal wastewater is composed of a mixture of dissolved and particulate organic and inorganic materials and infectious disease-causing bacteria. The total amount of each component accumulated in wastewater is referred to as the *mass loading* and is given units of pounds per day (lb/day). The concentration, given in pounds per gallon of water (lb/gal), of any individual component entering a wastewater treatment plant can change as a result of the activities that are producing this waste. The units used to express any concentration, lb/gal, can also be converted into other terms such as grams/liter, mg/L, ppm, or even μg/L, a term used frequently with very toxic substances. The concentration of each individual component while in the treatment plant is usually reduced significantly by the time it reaches the end of the plant prior to discharge.

Municipal wastewater normally contains approximately 99.9% water. The remaining materials (as described earlier) include suspended and dissolved organic and inorganic matter as well as microorganisms. These materials make up the physical, chemical, and biological qualities that are characteristic of municipal and industrial waters. Each of these three qualities are briefly described below.

Wastewater characteristics depend largely on the mass loading rates flowing from the various sources in the collection system. The flow in combined sewers is a composite of domestic and industrial wastewaters, infiltration into the sewer from cracks and leaks in the system, and flow from sanitary sewers. During wet weather, the addition of rainfall collected from the combined sewer system and the storm drain collection system (combined sewer overflow systems) can significantly change the characteristics of wastewater due to the increased flow carried by the sewer to the treatment plant. The peak flowrate can be two to three times the average dry (or sunny) weather flowrate and a portion of it may be by-passed to the receiving water. The mass loading rate into the plant also varies cyclically throughout the day with the usual peak occurring in the afternoon and the low in the early morning hours. The impact of flowrate is an important determining factor in the design and operation of wastewater treatment plant facilities. The records kept by the treatment plant should include the minimum, average, and maximum flow values (gallons per unit time) on a hourly, daily, weekly, and monthly basis for both wet and dry weather conditions. A moving 7-day daily average flow and mass loading rate entering the plant and at various locations throughout the plant can then be computed from the record. This intricate form of record keeping of all factors affecting a wastewater treatment plant must be considered to assess the wastewater flow and variations of wastewater strength in order to operate a facility correctly. The parameters used to indicate the total mass loading in the wastewater entering the treatment plant are the measurements of total suspended solids (TSS), and total dissolved solids (TDS)

(Holmes, Singh, and Theodore, *Handbook of Environmental Management and Technology*, 2nd ed., Wiley-Interscience, NYC, 2001). The parameters used to indicate the organic and inorganic chemical concentration in the wastewater are the measurements of the biochemical oxygen demand (BOD) and the chemical oxygen demand (COD). Both BOD and COD are discussed in more detail later in this section. Additionally, the total nutrients (carbon, nitrogen, and phosphorus), any toxic chemicals, and trace metals are also characterized for the plant design so that the treatment processes can successfully treat the varying waste loads.

A wastewater plant can handle a large volume of flow if its units are sufficiently designed. Unfortunately most wastewater plants are already in operation when a request comes to accept the flow of waste from some new industrial concern or expanding community.

Finally, other harmful constituents in industrial wastes can cause problems. Some problem areas and corresponding effects are:

1. Toxic metal ions that cause toxicity to biological oxidation.
2. Feathers that clog nozzles, overload digesters, and impede proper pump operation.
3. Rags that clog pumps and valves and interfere with proper operation.
4. Acids and alkalis that may corrode pipes, pumps, and treatment units, interfere with settling, upset the biological purification of sewage, release odors, and intensify color.
5. Flammables that cause fires and may lead to explosions.
6. Fat that clogs nozzles and pumps and overloads digesters.
7. Noxious gases that present a direct danger to workers.
8. Detergents that cause foaming.
9. Phenols and other toxic organic material.

CHR.3 UNITS CONVERSION

The suspended particulate concentration in an aqueous stream has been determined to be 27.6 mg/L. Convert this value to units of $\mu g/L$, g/L, lb/ft^3, and lb/gal.

Solution

Refer to Chapter 1.

$$C = 27.6 \frac{mg}{L}$$

$$= \left(27.6 \frac{mg}{L}\right)\left(10^3 \frac{\mu g}{mg}\right)$$

$$= 2.76 \times 10^4 \frac{\mu g}{L}$$

$$C = 27.6 \frac{\text{mg}}{\text{L}}$$

$$= \left(27.6 \frac{\text{mg}}{\text{L}}\right)\left(10^{-3} \frac{\text{g}}{\text{mg}}\right)$$

$$= 2.76 \times 10^{-2} \frac{\text{g}}{\text{L}}$$

$$C = 27.6 \frac{\text{mg}}{\text{L}}$$

$$= \left(27.6 \frac{\text{mg}}{\text{L}}\right)\left(\frac{1 \text{ lb}}{454,000 \text{ mg}}\right)\left(\frac{1000 \text{ L}}{35.3 \text{ ft}^3}\right)$$

$$= 1.72 \times 10^{-3} \frac{\text{lb}}{\text{ft}^3}$$

$$C = 27.6 \frac{\text{mg}}{\text{L}}$$

$$= \left(27.6 \frac{\text{mg}}{\text{L}}\right)\left(\frac{1 \text{ lb}}{454,000 \text{ mg}}\right)\left(\frac{1000 \text{ L}}{264 \text{ gal}}\right)$$

$$= 2.30 \times 10^{-4} \frac{\text{lb}}{\text{gal}}$$

CHR.4   pH CALCULATION

Calculate the hydrogen ion and the hydroxyl ion concentration of an aqueous solution if the pH of the solution is 1, 3, 5, 7, 8, 10, 12, and 14.

Solution

An important chemical property of an aqueous solution is its pH. The pH measures the acidity or basicity of the solution. In a neutral solution, such as pure water, the hydrogen (H^+) and hydroxyl (OH^-) ion concentrations are equal. At ordinary temperatures, this concentration is

$$C_{H^+} = C_{OH^-} = 10^{-7} \text{ g} \cdot \text{ion/L}$$

where C_{H^+} = hydrogen ion concentration
   C_{OH^-} = hydroxyl ion concentration

The unit g · ion stands for gram · ion, which represents an Avogadro number of ions. In all aqueous solutions, whether neutral, basic, or acidic, a chemical equilibrium or balance is established between these two concentrations, so that

$$K_{eq} = C_{H^+} \cdot C_{OH^-} = 10^{-14}$$

where K_{eq} is the equilibrium constant.

The numerical value for K_{eq} given above holds for room temperature and only when the concentrations are expressed in gram · ion per liter (g · ion/L). In acid solutions, C_{H^+} is > C_{OH^-}; in basic solutions, C_{OH^-} predominates.

The pH is a direct measure of the hydrogen ion concentration and is defined by

$$pH = -\log(C_{H^+})$$

Thus, an acidic solution is characterized by a pH below 7 (the lower the pH, the higher the acidity); a basic solution has a pH above 7; and, a neutral solution possesses a pH of 7.

Regarding the problem statement, for a pH of 1.0,

$$pH = -\log(C_{H^+})$$

$$C_{H^+} = 10^{-pH} = 10^{-1} = 0.1\,g \cdot ion/L$$

$$C_{H^+} \times C_{OH^-} = 10^{-14}$$

$$C_{OH^-} = \frac{10^{-14}}{C_{H^+}}$$

$$= 10^{-13}\,g \cdot ion/L$$

The remaining results are calculated in a similar fashion and are presented in the following table.

pH	C_{H^+} g · ion/L	C_{OH^-} g · ion/L
1	10^{-1}	10^{-13}
3	10^{-3}	10^{-11}
5	10^{-5}	10^{-9}
7	10^{-7}	10^{-7}
8	10^{-8}	10^{-6}
10	10^{-10}	10^{-4}
12	10^{-12}	10^{-2}
14	10^{-14}	1.0

It should be pointed out that the above equation employed is not the exact definition of pH but is a close approximation to it. Strictly speaking, the *activity* of the hydrogen ion, a_{H^+}, and not the ion concentration, C_{H^+}, belongs in the equation.

CHR.5 PROCESSES INVOLVING NITROGEN

Describe the nitrification, denitrification, and ammonia stripping processes.

Solution

Nitrification converts ammonia to the nitrate form, thus eliminating toxicity to fish and other aquatic life and reducing the nitrogenous oxygen demand. Ammonia is first oxidized to nitrite and then to nitrate by autotrophic bacteria. The reactions are

$$NH_3 + 3O_2 + biomass \rightarrow NO_2^- + H^+ + H_2O + more \ biomass$$

$$NO_2^- + \tfrac{1}{2}O_2 \rightarrow NO_3^- + biomass$$

The term "biomass" will be neglected in later reactions, since all biochemical reactions require and produce biomass. Temperature, pH, dissolved oxygen, and the ratio of BOD to total Kjeldahl nitrogen (TKN) are important factors in nitrification.

Nitrite and nitrate are reduced to gaseous nitrogen by a variety of facultative heterotrophs in an anoxic environment. An organic source, such as acetic acid, sewage, acetone, ethanol, methanol, or sugar is needed to act as hydrogen donor (oxygen acceptor) and to supply carbon for synthesis. Methanol is used, as it is frequently the least expensive. The basic reactions take the form:

$$3O_2 + 2CH_3OH \rightarrow 2CO_2 + 4H_2O$$

$$6NO_3^- + 5CH_3OH \rightarrow 3N_2 + 5CO_2 + 7H_2O + 6OH^-$$

$$2NO_2^- + CH_3OH \rightarrow N_2 + CO_2 + H_2O + 2OH^-$$

Biological phosphorus and nitrogen removal has received considerable attention in recent years. Basic benefits reported for biological nutrient removal includes monetary saving through reduced aeration capacity and the obviated expense for chemical treatment. Biological nutrient removal involves anaerobic and anoxic treatment of return sludge prior to discharge into the aeration basin. Based on the anaerobic, anoxic, and aerobic treatment sequence and internal recycling, several processes have been developed. Over 90% phosphorus and high nitrogen removal (by nitrification and denitrification) has been reported using biological means.

Ammonia gas can be removed from an alkaline solution by air stripping. This operation can be expressed in equation form as:

$$NH_4^+ + OH^- \rightarrow NH_3 \uparrow + H_2O$$

The basic equipment for an ammonia-stripping system includes chemical feed, a stripping tower, a pump and liquid spray system, a forced-air draft, and a recarbonation system. This process requires raising the pH of the wastewater to about 11, the formation of droplets in the stripping tower, and providing air–water contact and droplet agitation by countercurrent circulation of large quantities of air through the tower. Ammonia-stripping towers are simple to operate and can be very effective in ammonia removal, but the extent of their efficiency is dependent on the air temperature. As the air temperature decreases, the ammonia removal efficiency drops significantly. This process, therefore, is not recommended for use in a cold climate. A major operational disadvantage of stripping is the need for neutralization and the prevention of calcium carbonate scaling within the tower. Also, there is some concern over discharge of ammonia into the atmosphere.

The process has not found much use in the United States.

CHR.6   NITROGEN CONCENTRATION IN ACID

A wastewater contains 0.013% by mass of aminobenzoic acid. Calculate the percent by mass of nitrogen in the acid.

Solution

Although any basis may be chosen, it is convenient to select either 137 gmol (the approximate molecular weight) or 1.0 gmol of the acid. Choose 1 gmol of acid.

$$\text{Moles of acid} = 1 \, \text{gmol}$$

Molecular formula of aminobenzoic acid $= C_7H_7NO_2$

Moles $N = 1$
Moles $H = 7$
Moles $C = 7$
Moles $O = 2$

The corresponding mass of each component is

Mass $N = (1)(14) = 14 \, g$
Mass $H = (7)(1) = 7 \, g$
Mass $C = (7)(12) = 84 \, g$
Mass $O = (2)(16) = 32 \, g$

$$\text{Total} = 137 \, g$$
$$\text{MW of acid} = 137 \, g/mol$$
$$\text{Mass of acid} = (137 \, g/mol)(1 \, mol) = 137 \, g$$

The percent nitrogen in the acid is therefore

$$\%N_2 \text{ (by mass of acid)} = (14/137)\,100 = 10.2\%$$

CHR.7   NITROGEN CONCENTRATION IN WASTEWATER

With reference to Problem CHR.6, calculate the percent of nitrogen by mass in the wastewater.

Solution

The calculation for the percent nitrogen in the wastewater is slightly more complicated. Select a basis of 100 g of wastewater.

Mass of acid $= (0.00013)(100) = 0.013$ g
Mass of water $= 100$ g
Mass of nitrogen in acid $= (0.102)(0.013) = 0.001326$ g

Thus, the mass percent of nitrogen in the wastewater is

$$\text{Mass }\% = (0.001326\,\text{g}/100\,\text{g})\,100\%$$

$$= 0.001326\%$$

CHR.8   BOD AND COD

Discuss the difference between BOD and COD.

Solution

Biochemical oxygen demand and chemical oxygen demand are two important water quality parameters in wastewater engineering. The accurate measurement of these parameters is essential in the proper design of wastewater treatment systems and in the study of the transport and fate of contaminants in the aquatic environment.

Biochemical oxygen demand, or BOD, is the quantity of dissolved oxygen required to biochemically stabilize substrate materials in water. BOD is a measure of the oxygen demand of wastewater because it provides an approximate amount of oxygen needed in aerobic biological treatment. BOD is a measure of treated wastewater quality because its discharge into a natural receiving water exerts oxygen demand and is therefore a critical factor in the viability of the aquatic system's ecology. The U.S. Environmental Protection Agency (EPA) and state environmental regulatory agencies routinely require the monitoring of BOD in all municipal and industrial discharges.

BOD is generally measured in a three-step process. First, the sample is quantitatively diluted so that there is an appropriate concentration of oxidizable substrate and dissolved oxygen in the sample volume to be measured. For well-treated or "polished" wastewaters (5–30 mg/L BOD), dilution is small or may not be necessary at all. For raw, untreated wastewaters (100–400 mg/L BOD), a high level of dilution is necessary. Second, the sample is inocculated with microorganisms (seed), sealed, and incubated in a controlled temperature environment for a set period of time. During that time, dissolved oxygen (DO) and substrate are consumed through microbial activity. Third, DO is measured, typically once at the start of incubation and again at the end. The BOD is equal to the difference in DO concentration after it has been adjusted for the dilution factor. There is also typically an adjustment factor necessary to compensate for extra BOD introduced to the sample with the inocculating culture material (seed).

BOD is most commonly expressed as "five-day BOD" (BOD_5) and ultimate BOD (BODU). BOD_5 is the oxygen demand that is exerted after a standard period of 5 days of incubation and is the value required by regulatory agencies.

BODU represents the oxygen demand that would be exerted if incubation was allowed to occur long enough for virtually all the biologically oxidizable substrate to be consumed. Often, BODU is approximated by allowing incubation to occur for a standard period of 20 days, or longer.

Typically, BOD_5 is equal to approximately two-thirds of the BODU. Although this ratio is a good approximation for typical domestic wastewaters, it can also vary significantly depending on the nature of the wastewater source. Lower ratios are typical of industrial wastewaters with a lower relative composition of readily oxidizable substrate. For example, such a wastewater typically might contain higher colloidal or suspended solids substrates (such as petroleum refinery or paper and pulp mill effluent) or higher nitrogenous substrates (such as meat-processing effluents). Higher ratios are indicative of industrial wastewaters with a higher relative composition of readily oxidizable soluble substrate (such as brewery or carbohydrate food-processing wastewaters). Substrates such as these are often referred to by wastewater engineers as "jellybeans" because they are readily consumed during treatment by the process microorganisms.

BOD is often distinguished by the type of oxidizable substrate that is consumed during the incubation period. Carbonaceous BOD (CBOD) represents the oxygen demand exerted by the carbon-based components of the substrate. Nitrogenous BOD (NBOD) represents oxygen demand exerted by the process of nitrification, or the oxidation of ammonia to nitrite and nitrate. To measure CBOD only, an inhibitory chemical is added to the sample to stop the nitrification portion of oxidation from occurring during the incubation period. To measure NBOD, it is necessary to measure both total BOD (TBOD) and CBOD in samples that have incubated 5 or more days. NBOD is calculated as the difference between total BOD and CBOD.

Total BOD = CBOD + NBOD

TBOD (without inhibitor) − CBOD (with inhibitor) = NBOD

Because CBOD is the primary substrate consumed during the first 5 days of incubation, BOD_5 is approximately equal to $CBOD_5$, except in highly treated waters. The measurement of NBOD with unacclimated organisms typically requires longer incubation periods (such as 20 days) because nitrifying organisms are slower growing and do not start consuming oxygen to a measurable degree until well after carbonaceous oxidation has begun. This "delay" is generally considered to be approximately 8–10 days, in raw municipal wastewater.

Chemical oxygen demand (COD) is equal to the equivalent oxygen concentration required to chemically oxidize substrate materials in water. Because the test can be completed in a little over 2 h, it very quickly provides the concentration of the wastewater chemical oxygen demand. It also is an indicator of nonbiodegradable organics. To chemically oxidize all the substrate in wastewater, a strong chemical oxidant (such as dichromate in sulfuric acid) is used as the standard oxidant and a catalyst. The COD is reported as the dissolved oxygen concentration equivalent to the decrease in the acidic dichromate concentration. The level of oxidation provided by this reagent is generally sufficient to oxidize almost all of the oxidizable substrates in water. Because there are additional (albeit less commonly used) chemical oxidants other than acidic dichromate, which can serve as COD standard oxidants, COD measured using dichromate is often referred to as "dichromate COD."

COD is an important water quality parameter because for similar wastewaters it often correlates well with CBOD. COD often serves as a reliable measure of the CBOD once the wastewater being examined has been characterized. It is a faster, less expensive analysis than CBOD, and is therefore used where frequent, routine process control analyses are required. A wastewater that is well suited to biological treatment generally has a CBODU concentration approximately equal to the COD concentration. That is, the bulk of the substrate that can be oxidized ultimately by process microorganisms is also oxidized by a strong chemical oxidant. CBODU approximately equals COD for typical, primary-treated domestic sewage. Industrial wastewater, or domestic sewage with a significant industrial component, will often have CBODU/COD ratios less than 1 if they contain substrates that are chemically reactive, but biologically refractive. This type of wastewater is typical of several industries. Chemical oxygen demand is of direct significance in industrial treatment when there are relatively few biologically oxidizable components in the wastewater. COD then can be used to directly monitor those components.

Usually, regulatory agencies will not accept COD concentrations for permit limits, but COD/BOD relationships can be very useful in predicting effluent BOD_5 values from COD values. COD measurements provide immediate knowledge of a process efficiency, whereas BOD requires five days.

The theoretical chemical (or ultimate) oxygen demand, or ThCOD (BODU), can be calculated for a wastewater with relatively few oxidizable components. The procedure requires one to calculate the equivalent oxygen concentration necessary for the complete stoichiometric oxidation of each of the wastewater components to its corresponding highest oxidation state and sum them for the total wastewater ThCOD. Carbonaceous components of wastewater generally considered to be fully oxidized to carbon dioxide. Typically, a laboratory measured COD is very close or equal to ThCOD.

CHR.9   ULTIMATE OXYGEN DEMAND

A wastewater contains 250 mg/L benzoic acid, C_6H_5COOH. What is the ultimate oxygen demand (BODU) of this water? The molecular weight of benzoic acid is 122.

As described in the previous problem, the ultimate oxygen demand is a calculated oxygen demand that assumes oxidation of all species to their most highly oxidized stable form: CO_2, H_2O, etc.

Solution

The number of moles of benzoic acid in 1 liter is as follows:

$$n = 250/122$$

$$= 2.05 \, \text{mgmol}$$

Note that the units are milligram moles since the mass is given in milligrams.

Convert all of the species (in this case, only benzoic acid) to their (its) most highly oxidized stable form:

$$C_6H_5COOH + (\tfrac{15}{2})O_2 \;\rightarrow\; 7CO_2 + 3H_2O$$

The ultimate oxygen demand (BODU) for 2.05 mgmol of benzoic acid is

$$BODU = [(15/2)O_2/1 \, C_6H_5COOH](2.05 \, \text{mgmol} \, C_6H_5COOH)$$

$$= 15.4 \, \text{mgmol} \, O_2$$

$$= 493 \, \text{mg} \, O_2$$

The BODU (or ThCOD) for this wastewater is therefore 492 mg/L.

CHR.10   ThCOD OF A GLUCOSE SOLUTION

Calculate the ThCOD of a 100-mg/L solution of glucose.

Solution

First, balance the chemical equation for the reaction of glucose and oxygen:

$$C_6H_{12}O_6 + 6O_2 \rightarrow 6CO_2 + 6H_2O$$

Second, determine the oxygen/substrate stoichiometric ratio for the oxygen reaction:

$$\text{Ratio} = (6)(32)/[(1)(180)]$$
$$= 0.067 \text{ mg } O_2/\text{mg glucose}$$

Third, calculate the ThCOD for a 100-mg/L solution of glucose. This is equal to the product of the mass concentration of glucose and the stoichiometric ratio:

$$\text{ThCOD} = (100 \text{ mg glucose/L})(1.067 \text{ mg } O_2/\text{mg glucose})$$
$$= 106.7 \text{ mg } O_2/\text{L, or } 106.7 \text{ mg COD/L}$$

CHR.11   TOC OF A GLUCOSE SOLUTION

With reference to Problem CHR.10, calculate the total organic carbon, TOC.

Solution

As mentioned earlier, TOC is another important water quality parameter. TOC is equal to the total concentration of organic carbon in water and is reported as milligrams carbon/liter (mg C/L). It serves as a general indicator of the overall quantity of carbon-based matter contained in water. TOC frequently correlates to BOD and COD but is not necessarily always a reliable measure. If a repeatable empirical relationship is established between TOC and BOD or TOC and COD for a particular wastewater, TOC can be used as an estimate of the corresponding BOD or COD. The theoretical TOC (ThCOD) can be calculated stoichiometricaly in similar manner to the ThCOD.

First, establish the ratio of mg carbon/mg substrate. In the case of glucose, the ratio is as follows:

$$\text{Molar ratio} = 6 \text{ mol C}/1 \text{ mol } C_6H_{12}O_6$$
$$\text{Mass ratio} = (6)(12) \text{ g C}/(1)(180) \text{ of } C_6H_{12}O_6$$
$$= 0.40 \text{ mg C/mg glucose}$$

Second, calculate the ThTOC concentration for the particular wastewater. For a 100-mg glucose/L solution, the ThCOD is:

$$\text{ThCOD} = (100 \text{ mg glucose/L})(0.40 \text{ mg C/mg glucose})$$
$$= 40 \text{ mg C/L}$$
$$= 40 \text{ mg TOC/L}$$

CHR.12   BOD$_5$ DISCHARGE

A typical city of 55,000 people has a wastewater treatment discharge of 6.7 million gal/day (MGD) and a BOD$_5$ in the raw wastewater of 225 mg/L. What is the total discharge of BOD in lb/day? What is the BOD$_5$ discharge in lb/person · day?

Solution

The total BOD$_5$ produced per day is

$$\text{Total BOD}_5/\text{day} = (6.7\,\text{MGD})(225\,\text{mg/L})(8.34\,\text{lb L/MG mg})$$
$$= 12{,}570\,\text{lb BOD}_5/\text{day}$$

The BOD$_5$ discharge in lb/person-day is

$$\text{BOD, discharge} = \frac{12{,}570\,\text{lb BOD}_5/\text{day}}{55{,}000\,\text{people}}$$
$$= 0.23\ \frac{\text{lb BOD}_5}{\text{people} \cdot \text{day}}$$

31 Water Chemistry (WCH)

WCH.1 CONCENTRATION TERMS

Briefly describe each of the following concentration terms:

1. Weight or mole percent
2. Molar solutions
3. Molal solutions
4. Normal solutions

Solution

There are various ways of expressing the concentration of a solution containing a solute (dissolved) in a solvent, e.g., a pollutant in a water (wastewater) stream. Solutions may be either dilute or concentrated. When a solution contains a relatively small quantity of the solute, it is dilute; when the amount of solute is large, the solution is said to be concentrated. The precise quantity of solute contained in a given amount of solution is called the concentration. The latter may be designated in physical units as in weight or mole percent or it may be expressed in terms of chemical units as in molar and normal solutions.

1. *Weight or Mole Percent:* A weight or mass percent is a concentration in which the ratio of the weight of the solute to that of the solution is expressed in parts per 100. Thus, a 5% solution (by mass) of KCl contains 5 g of salt dissolved in 100 g of solution (or 5 g of salt in 95 g of water). Similarly, a 5% solution by mole contains 5 moles of salt per 100 moles of solution.

2. *Molar Solutions:* A molar solution contains 1 mol of the substance in a liter of solution. A mole of substance is its molecular weight expressed in grams. Thus, a one-molar solution of NaCl contains $23 + 35.5$ or 58.5 g NaCl in 1000 mL of solution. A 1 M (one molar) solution of $BaCl_2(H_2O)_2$ contains $137 + 71 + 36$ or 244 g $BaCl_2(H_2O)_2$ per liter. An ion also possesses some of the properties of a molecule in solution. In calculations, therefore, a mole of hydrogen ion, H^+, has a mass of 1 g. For the same reason a mole of Cl^- and a mole of $SO_4^=$ ions consist of 35.5 and 96 g, respectively. A 1 M solution of hydrogen ion therefore contains 1 g of hydrogen per liter. Similarly, a liter of a 1 M solution of $SO_4^=$ contains 96 g $SO_4^=$ ion.

The concentration of a solution bears no relation to the amount of solution. If the concentration of a solution is designated as 2 M, then every drop, every milliliter, every liter, or even every barrel of that solution has the same concentration of 2 M.

The number of moles of a solute contained in a given amount of solution is never equal to the molarity of that solution unless the volume should happen to be exactly one liter. The number of moles of a solute contained in any solution is equal to the molarity multiplied by the volume of the solution expressed in liters. In 1 mL of a 2-M NaCl solution there is present 0.001×2 mol or 0.002 mol of NaCl. It is to be noted that a 2-M solution may also be defined as one that contains 2 mmol per mL (1 mmol is 0.001 mol and a mL is 0.001 liter).

3. *Molal Solutions:* The molal concentration of a solution is defined as the number of gram-molecular-weights (moles) or gram-formula-weights dissolved in 1000 g of water. A 2 m (2 molal) solution of NaCl is composed of 116.9 grams or two gram-molecular weights of NaCl for every kg of water.

4. *Normal Solutions:* A normal solution contains 1 g equivalent of the solute in a liter of solution. The equivalent is discussed in problem WCH.2.

WCH.2   EQUIVALENT WEIGHT

Briefly describe the following three quantities:

1. Equivalent of an acid
2. Equivalent of a base
3. Equivalent of a salt

Solution

1. The equivalent of an acid can be defined as that weight containing 1.008 g of ionizable hydrogen. It is this hydrogen that is liberated when the acid is treated with a metallic element. Therefore, 1 mol of HCl, 0.5 mol of H_2SO_4, 1 mol of $HC_2H_3O_2$, and 0.5 mol of $H_2C_4H_4O_6$ are, respectively, the equivalents of these acids.

2. The active component of a base is the hydroxyl (OH^-) group. Since 17.008 g of OH^- are equivalent to 1.008 g of hydrogen, the equivalent of a base is that weight that contains 17.008 g of hydroxyl. Thus, 1 mol of NaOH, 0.5 mol of $Ca(OH)_2$ and 0.33 mol of $Fe(OH)_3$ are, respectively, the equivalents of these bases.

3. The components of a salt taking part in reactions are its ions. Hence, the equivalent of a salt is that quantity containing 1 g equivalent of the cation or anion, and this quantity is given by the moles of the salt divided by the total

valence of the cation or anion. Thus, 1 mol of NaCl, 1 mol of NH_4Cl, 0.5 mol of Ca_2SO_4 and 0.17 mol of $Ca_3(PO_4)_2$ are each the equivalents of these salts. The above rule applies only to simple normal salts.

Note: A water solution containing one equivalent of an acid, base, or salt in one liter of solution is a one normal (1 N) solution.

WCH.3   CONCENTRATION CONVERSION

Describe how to convert molarity to normality.

Solution

As described earlier, the molarity of a solution represents the number of q-moles of solute per liter of solution; the normality represents the number of gram equivalents of solute per liter of solution. Hence, the problem of changing molarity to normality consists merely in changing moles into gram equivalents. Thus, given a 2-M $CaCl_2$ solution, one may calculate its normality. A 1-M $CaCl_2$ solution contains 2 g equivalents per liter; hence a 2 M $CaCl_2$ solution should contain 4 equivalents per liter. Its concentration is therefore 4 N. Conversely, suppose one is required to calculate the molarity of a 4-N $CaCl_2$ solution. A 1-N solution of $CaCl_2$ contains 1 g equivalent of $CaCl_2$, which is equal to 0.5 mol calcium chloride; therefore a 4-N $CaCl_2$ solution should contain 4×0.5 mol or 2 mol $CaCl_2$, and hence the solution is 2 molar. Note that a solution's molarity is equal to its normality multiplied by a small integer such as 1, 2 or 3.

WCH.4   MOLARITY CALCULATIONS

Provide answers to the following four questions:

1. How many grams of KBr will be needed to make 300 mL of a 2-M solution?
2. How many milliliters of water must be added to 5 mL of a 12-M HCl solution to make a 3-M HCl solution?
3. How many milliliters of a 2-M HCl solution is necessary to neutralize 2 mL of a 0.5-M NaOH solution?
4. What is the molarity of a H_2SO_4 solution that contains 33.3% H_2SO_4 by weight and has a density of 1.25 g/mL?

Solution

1. 1 mol of KBr $= 39.1 + 79.9$
$$= 119.0\,g$$
   2 mol of KBr $= (2)(119.0)$
$$= 238.0\,g$$

   Thus, 1 liter of a 2-M KBr solution contains 238.0 g of KBr.

2. In all dilution problems involving liquid solutions, it is almost always assumed that the final volume is equal to the volume of the "initial" solution plus the volume of the other solutions (or water) added.

   In 1 mL of a 3-M solution there are 1/4 as many moles of solute as there are in 1 mL of a 12-M solution. Therefore, enough water must be added to the 12-M solution to make its final volume 4 times as great as it was originally. In this problem the final volume must be 4×5 or 20 mL. Therefore 15 mL of water must be added. In brief, the final volume must be 5×12 or 60 mL.

   The amount of water added is $5 \times (12 - 3)/3$ or 15 mL. In general, the volume of water added equals the initial volume of solution multiplied by $(C_2 - C_1)/C_1$, where C_2 is the solvent concentration and C_1 is the solute concentration.

3. Note that 1 mol of HCl neutralizes 1 mol of NaOH. The same number of moles of HCl are required as there are moles of NaOH in 3 mL of a 0.5-M NaOH solution.

   If the HCl solution were 0.5 M, instead of 2 M, equal volumes of each would be required. The HCl is 2 M or 4 times as concentrated as the NaOH; therefore, less is required. Specifically, 0.25 as much would be required as would be the case if the HCl were 0.5 M.

   The number of milliliters of HCl required is $3 \times 0.5/2$ or 0.75 mL.

4. One liter of the solution weighs 1000 mL $\times$ 1.25 g/mL or 1250 g.

   The number of grams of H_2SO_4 in 1 liter is $(0.3333)(1250)$ or 417 g.

   One mole of H_2SO_4 is 98.1 g. The number of moles in 1 liter is $417/98.1$ or 4.25 mol/L. The solution is therefore 4.25 M.

WCH.5   DEGREE OF IONIZATION

Calculate the hydrogen ion (H^+) concentration in a 0.1-M HCNO solution. What is the degree of ionization of cyanic acid in this same solution? The ionization constant, K_i, for the acid is 2×10^{-4}.

Solution

The ionization equation is

$$HCNO \rightarrow H^+ + CNO^-$$

The concentration (0.1 M) given for HCNO is that for the total HCNO in solution, both dissociated and undissociated. Set

$$(H^+) = X$$

Where (H^+) represents the hydrogen ion concentration in mol/L. (CNO^-) must also be X in this case, for as many CNO^- as H^+ ions are formed by the dissociation process. Thus,

$$HCNO = 0.1 - X$$

Substituting these values in the equilibrium expression gives

$$K_i = \frac{(H^+)(CNO^-)}{(HCNO)}$$

$$= \frac{X^2}{0.1 - X} = 2 \times 10^{-4}$$

Obviously, X is relatively small as compared with 0.1. Therefore, for all practical purposes

$$0.1 - X \approx 0.1$$

Then,

$$\frac{X^2}{0.1 - X} = \frac{X^2}{0.1} = 2 \times 10^{-4}$$
$$X^2 = 2 \times 10^{-5}$$

$$X = 4.5 \times 10^{-3}$$

and

$$H^+ = CNO^- = 0.0045 \text{ gmol/L}$$

From the value of X obtained one can readily see that neglecting X as compared with 0.1 was justified, since $0.1 - 0.0045$ equals 0.0955, which is close to 0.1. Solving the equation by the use of the quadratic equation leads to a value for X of

0.0044 gmol/L. This provides justification for the approximation method for the solution.

The reader should note that the degree of ionization is the fractional number of molecules dissociated, or the amount per liter of the dissociated weak electrolyte divided by the total concentration (both dissociated and undissociated). Since, in this particular example, the concentration of H^+ and CNO^- is 4.5×10^{-3} mol/L, the amount of the dissociated HCNO has this same value since 4.5×10^{-3} mol of HCNO gives 4.5×10^{-3} mol of H+ and 4.5×10^{-3} mol of CNO^- upon dissociation. For this case,

$$\text{Degree of dissociation} = \frac{4.5 \times 10^{-3}}{0.1}$$

$$= 4.5 \times 10^{-2}$$

$$= 0.045 \text{ or } 4.5\%$$

Also note that, if the acid contains more that one atom of hydrogen, e.g., H_2CO_3, there are two ionization constants, one for the removal of each H^+ ion. See Problems CHM.4 and CHM.5.

WCH.6   pH CALCULATION

Calculate the pH of a 0.01-M HCN solution if the ionic equilibrium constant, K(HCN), is 2.1×10^{-9}.

Solution

First calculate the (H+). The dissociation equation is as follows:

$$HCN \rightarrow H^+ + CN^-$$

Individual concentrations are

$$HCN = 0.01 - X$$

$$H^+ = X$$

$$CN^- = X$$

$$K(HCN) = \frac{(H^+)(CN^-)}{(HCN)} = \frac{X^2}{0.01 - X} = 2.1 \times 10^{-9}$$

Once again, neglecting X in the denominator.

$$\frac{X^2}{0.01} = 2.1 \times 10^{-9}$$

$$X = 4.6 \times 10^{-6} \text{ M} = (\text{H}^+)$$

$$\text{pH} = -\log(\text{H}^+)$$

$$\log(\text{H}^+) = \log(4.5 \times 10^{-6})$$

$$= -5.34$$

Therefore,

$$\text{pH} = -(-5.34)$$

$$= 5.34$$

WCH.7   SOLUBILITY PRODUCT

1. The solubility of $CaSO_4$ in water is an important consideration in flue gas desulfurizing. Calculate the solubility of $CaSO_4$ in grams per 100 ml if its solubility product constant, K_s, is 6.1×10^{-5}.
2. Calculate the solubility of $Mg(OH)_2$ in grams per liter if its solubility product constant, K_s, is 1.5×10^{-11}.

Solution

1. Set X equal to the number of moles of $CaSO_4$ in 1 liter of solution. Since the $CaSO_4$ that dissolves is completely dissociated, there will be X moles of Ca^{++} and X moles of $SO_4^=$.

$$\begin{array}{ccc} CaSO_4(\text{solid}) & \rightarrow & Ca^{++} + SO_4^= \\ 0 & & X \quad\;\; X \end{array}$$

Thus,

$$(Ca^{++})(SO_4^=) = X^2$$

$$X^2 = 6.1 \times 10^{-5}$$

$$X = 7.8 \times 10^{-3} \text{ gmol/L}$$

This is not only the concentration of the calcium ion and of the sulfate ion, but it also represents the concentration of the total amount of calcium sulfate in solution. The molecular weight of calcium sulfate is 136. The $CaSO_4$ concentration is therefore

$$(7.8 \times 10^{-3})(136) = 1.06 \text{ g CaSO}_4/\text{L}$$
$$= 0.106 \text{ g}/100 \text{ mL}$$

2. In this case, set

$$X = \text{number of moles of Mg (OH)}_2 \text{ dissolved}$$
$$X = \text{number of moles of Mg++}$$
$$2X = \text{number of moles of OH}^-$$
$$Mg^{++}(OH^-)^2 = X(2X)^2 = 4X^3 = 15 \times 10^{-12}$$
$$X^3 = 3.75 \times 10^{-12}$$

$$X = 1.55 \times 10^{-4} \text{ gmol/L}$$

The molecular weight of $Mg(OH)_2$ is 58.3. The $Mg(OH)_2$ concentration is therefore

$$(1.55 \times 10^{-4})(58.3) = 90.4 \times 10^{-4}$$
$$= 0.00904 \text{ g/L}$$

WCH.8 PRECIPITATION OF SULFATES

Metals in wastewater play an important role in terms of treatment. A solution contains 0.02 mol of Cd^{++} ion, 0.02 mol of Zn^{++} ion, and 1 mol of HCl per liter, and is saturated with H_2S at room temperature.

1. What is the concentration of the $S^=$ ion in this solution?
2. Will CdS precipitate?
3. Will ZnS precipitate?

The solubility constant for H_2S, K_s, is 1.1×10^{-23}.
The solubility constant for Cd, K_s, is 1×10^{-28}.
The solubility constant for Zn, K_s, is 1.2×10^{-23}.

Solution

The dissociation equation for H_2S is

$$H_2S \rightarrow 2H^+ + S^=$$

Assume that the solubility of H_2S in the solution is 0.1 M. Then,

$$\frac{(H^+)^2(S^=)}{(H_2S)} = \frac{(H^+)^2(S^=)}{0.1} = \frac{1.1 \times 10^{-23}}{0.1}$$
$$= 1.1 \times 10^{-22}$$

Thus,

$$(H^+)^2(S^=) = 1.1 \times 10^{-23}$$

If the (H^+) is 1M, then

$$(1)^2(S^=) = 1.1 \times 10^{-23}$$
$$(S^=) = 1.1 \times 10^{-23} \text{ M}$$

If precipitation of both sulfides takes place, then at equilibrium the reactions are

$$CdS_{(s)} = Cd^{++} + S^=$$
$$ZnS_{(s)} = Zn^{++} + S^=$$

The solubility product expressions are, respectively,

$$(Cd^{++})(S^=) = 1 \times 10^{-28}$$
$$(Zn^{++})(S^=) = 1.2 \times 10^{-23}$$

In the case of CdS, the ion product, $(0.02)(1.1 \times 10^{-28}) = 2.2 \times 10^{-25}$, is greater than the solubility product constant, so CdS precipitates. On the other hand, the ion product for ZnS, 2.2×10^{-25}, is less than the solubility product constant, so ZnS remains completely in solution.

WCH.9   PRECIPITATION OF AgCl

A given solution contains 0.01 mol of Cl^- ion and 0.07 mol of NH_3 per liter. If 0.01 mol of solid $AgNO_3$ is added to 1 liter of this solution, will AgCl precipitate?

The equilibrium constants, K, for AgCl and $Ag(NH_3)_2^+$ are 1.56×10^{-10} and 6.8×10^{-8}, respectively.

Solution

The solution of this problem involves two equilibria:

$$AgCl_{(s)} \rightarrow Ag^+ + Cl^-$$

$$Ag(NH_3)_2^+ \rightarrow Ag^+ + 2NH_3$$

The equilibrium expressions are

$$(Ag^+)(Cl^-) = 1.5 \times 10^{-10}$$

$$\frac{(Ag^+)(NH_3)^2}{\left(Ag(NH_3)_2^+\right)} = 6.8 \times 10^{-8}$$

Due to the great stability of the complex ion, one can first assume that 0.01 mol of this ion is formed from 0.01 mol of the Ag^+ ion. This process would consume 0.02 mol of NH_3; then 0.05 mol of NH_3 would be left in solution. Under these conditions, the concentration of the free Ag^+ ion in solution becomes

$$\frac{(Ag^+)(0.05)^2}{0.01} = 6.8 \times 10^{-8}$$

$$(Ag^+) = \frac{(6.8 \times 10^{-8})(0.01)}{2.5 \times 10^{-3}}$$

$$= 2.72 \times 10^{-7} \, gmol/L$$

Since 0.01 mol of the Cl^- ion is present per liter of solution, the product of the ion concentrations is $(2.72 \times 10^{-7})(0.01)$ or 2.72×10^{-9}. This value is greater than the solubility product constant; therefore, AgCl precipitates.

This problem could be solved in another manner. Calculate the amount of Ag^+ ion necessary to start the precipitation of AgCl when 0.01M Cl^- ion is present. This would be

$$(Ag^+) = \frac{1.56 \times 10^{-10}}{0.01} = 1.56 \times 10^{-8} \, M$$

With this amount of free Ag^+ ion in solution and assuming that $0.01 \, mol$ of $Ag(NH_3)_2^+$ is formed, one can then calculate the amount of free NH_3 that would be required to maintain these conditions. Then,

$$\frac{(1.56 \times 10^{-8})(NH_3)^2}{0.01} = 6.8 \times 10^{-8}$$

$$(NH_3)^2 = \frac{(0.01)(6.8 \times 10^{-8})}{1.56 \times 10^{-8}} = 4.36 \times 10^{-2}$$

$$(NH_3) = 2.1 \times 10^{-1}$$

$$= 0.21 \, M$$

This value for the amount of free ammonia necessary to maintain $0.01 \, mol$ of the complex in solution is much larger than the available ammonia; therefore, AgCl precipitates.

WCH.10   OXIDATION–REDUCTION TREATMENT

List the steps for an oxidation–reduction treatment process.

Solution

Contaminants in process wastewaters or remediable contaminated groundwaters often require a change in oxidation state to render them either less toxic or better suited to further treatment. Several examples of this include:

1. Cyanide from industrial manufacturing processes is reduced to ammonia.
2. Chlorine used for chemical disinfection of sewage is reduced to chloride.
3. Arsenite in contaminated groundwater is oxidized to arsenate.

The material process requirements for a specific treatment (chemical usage, residuals production, etc.) must usually be determined empirically through bench- and/or pilot-scale testing. However, it is preferable to understand the reaction mechanism and stoichiometry to best optimize the process. For example, an understanding of dechlorination chemistry permits greater control of the concentration of the residual reducing agent remaining in the effluent.

If the reaction is an oxidation–reduction reaction, the chemical equation must be balanced in a stepwise procedure that is presented below.

1. Major reactants must be identified as either oxidizing agents or reducing agents.
2. Appropriate half-reactions that include these agents must be identified from the electrochemical series tables for both oxidation and reduction.

3. Select a multiplier for the coefficients of each of the two half-reaction equations such that there is conservation of electron charge transfer (i.e., the number of electrons involved in oxidation is equal to the number of electrons involved in reduction).

4. Add the two half-reactions and simplify terms by reacting H^+ and OH^- terms and canceling H_2O terms.

5. Use the balanced chemical equation and process data to calculate chemical consumption and residuals production.

32 Physical Treatment (PHY)

PHY.1 ROLE OF PHYSICAL TREATMENT FOR WASTEWATER

Describe physical treatment in relation to the overall wastewater treatment process.

Solution

The overall wastewater treatment process consists of the following:

1. Wastewater treatment can be made up of roughly three (consecutive) steps with a preliminary process called pretreatment.
2. Pretreatment is the removal of stones, sand, grit, fat/grease, rags, plastics, etc. using mechanical processes such as screening, settling, or flotation.
3. Primary settling is the removal of suspended solids by passing wastewater through settling tanks and with accompanying skimming and removal of floatables.
4. In secondary treatment, a biological process is used where wastewater passes through tanks in which bacteria consume pollutants and transform them into carbon dioxide, water, and more biological cells. Following the biological process, settling is required, which is a physical treatment process.
5. Tertiary or advanced treatment includes nitrogen and phosphorus removal, disinfection by means of chlorination, ultraviolet (UV) radiation, or ozone treatment and filtration.

Note: This problem section will primarily address steps 2 and 3.

PHY.2 DESCRIPTION OF PHYSICAL TREATMENT DEVICES

Three physical treatment devices employed in wastewater treatment control include screens, grit chambers, and gravity sedimentation. Briefly describe each of these devices.

Solution

Wastewater physical treatment is directed toward removal of certain pollutants with the least effort. Three treatment steps are presented below.

1. *Screens.* Fine screens such as hydroscreens are used to remove moderate-size particles that are not easily compressed under fluid flow. Fine screens are normally used when the quantities of screened particles are large enough to justify the additional units. Mechanically cleaned fine screens have been used for separating large particles. These are used in industrial applications but not frequently for municipalities. Some industries as well use bar screens that catch large solids that could clog or damage pumps or equipment following the screens. Mechanically cleaned bar screens, with about 1-in. spacing, are used in most municipal plants. Manually cleaned bar screens are used in many small plants.

2. *Grit Chambers.* Industries and municipalities with sand or hard, inert heavy particles in their wastewaters use aerated grit chambers useful for the rapid separation of these inert particles. Aerated grit chambers are relatively small, with their total design volume based on 3-min retention at maximum flow. Diffused air is normally used to create the mixing pattern with the heavy, inert particles removed by centrifugal action and friction against the tank walls. The airflow rate is adjusted for the specific particles to be removed. It is important to provide for regular removal of floatable solids from the surface of the grit chambers; otherwise, nuisance conditions will be created. The settled grit is normally removed with a continuous screw and buried in a landfill. Smaller plants are frequently equipped with long channels for settling of the grit. Manual removal is required.

3. *Gravity Sedimentation.* Slowly settling particles are removed with gravity sedimentation tanks. For the most part, these tanks are designed on the basis of retention time, surface overflow rate, weir loading and minimum depth. A sedimentation tank can be rectangular or circular. The important factor affecting its removal efficiency is the hydraulic flow pattern through the tank. The flow energy contained in the incoming wastewater flow must be dissipated before the solids can settle. The wastewater flow must be distributed uniformly through the sedimentation volume for maximum settling efficiency. After the solids have settled, they should be collected without creating serious hydraulic currents that could adversely affect the sedimentation process. Effluent weirs are placed at the end of rectangular sedimentation tanks and around the periphery of circular sedimentation tanks to ensure uniform flow out of the tanks.

Once the solids have settled, they must be removed from the sedimentation tank floor by scraping or vacuum. Conventional sedimentation tanks have sludge hoppers to collect the concentrated sludge and to prevent removal of excess volumes of water with the settled solids. Design criteria for gravity sedimentation tanks normally provide for a 2-h retention time based on average flow, with longer retention periods used for light solids or inert solids that do not change during their retention in the tank. Care should be exercised that the sedimentation time is not too long; otherwise, the solids can compact, ferment, and possibly affect solids collection and removal and effluent quality.

PHY.3 CLARIFYING CENTRIFUGE

You have been requested by your company to study the feasibility of using a clarifying centrifuge to separate solid particles of specific gravity, 1.60, from water with a specific gravity of 1.00 and a viscosity of 1.20 cP. The pilot unit has a basket that has an inside diameter of 800 mm, is 400 mm in height, and operates at 1000 rpm (revolutions per minute). The thickness of the liquid layer in the centrifuge, i.e., the radial distance from the inside wall to the air–water surface, is to be 100 mm. As part of your study, you need to determine the time required for a particle 20.0 μm in diameter to settle through the 100 mm of water.

The equations of Problem FPD.4 (see Chapter 13) apply except that the gravitational acceleration, g, must be replaced by the centrifugal acceleration, $r\omega^2$, in consistent units.

Solution

To calculate the terminal settling velocity of the particle, the K value (see Problem FPD.4) must be used to determine the appropriate range of the fluid–particle dynamic laws. K is obtained from

$$K = d_p \left(\frac{r\omega^2 \rho (\rho_p - \rho)}{\mu^2} \right)^{1/3}$$

where K = dimensionless constant that determines the range of the fluid–particle
              dynamic laws
    r = radius or distance from centrifuge axis
    ω = angular velocity
    d_p = particle diameter
    ρ_p = particle density
    ρ = fluid (water) density
    μ = fluid (water) viscosity

When $K < 3.3$, Stokes' law applies; when $3.3 < K < 43.6$, the intermediate law applies; and when $43.6 < K < 2360$, Newton's law applies.

The inside radius of the centrifuge (or outside surface of the water), r_2, is

$$r_2 = (800/2)/1000 = 0.40 \, \text{m}$$

Since the thickness of the water is 100 mm or 0.10 m, the radius at the inside surface of the water, i.e., the distance from the axis to the air–water interface, r_1, is

$$r_1 = 0.40 - 0.10 = 0.30 \, \text{m}$$

The solid and water densities are

$$\rho_p = (1.60)(998) = 1597 \, \text{kg/m}^3$$

$$\rho = (1.0)(998) = 998 \, \text{kg/m}^3$$

The water viscosity and particle diameter in SI units are

$$\mu = (1.2)(0.001) = 1.2 \times 10^{-3} \, \text{kg/(m} \cdot \text{s)}$$

$$d_p = (20.0)(10^{-6}) = 2.0 \times 10^{-5} \, \text{m}$$

The angular velocity, ω, is

$$\omega = (2\pi)(1000)/60 = 104.7 \, \text{rad/s}$$

At r_2, the value of K is

$$K = (2.0 \times 10^{-5}) \left(\frac{(0.40)(104.7)^2(998)(1597 - 998)}{(1.2 \times 10^{-3})^2} \right)^{1/3}$$

$$= 2.44$$

Since $2.44 < 3.3$, Stokes' law applies. Note that, in a centrifugal field, K varies with radius. There is no need to check the value of K at r_1 (0.3 m) since K decreases as r decreases. Therefore, the value of K at 0.3 m has to be less than 2.44; Stokes' law applies everywhere from $r = 0.3$ to 0.4.

For the Stokes' law range, the terminal settling velocity (moving radially from the air–water interface to the centrifuge wall) as a function of r is given by

$$v = \frac{r\omega^2 d_p^2 \rho_p}{18\mu} = \frac{r(104.7)^2(2.0 \times 10^{-5})^2(1597)}{(18)(1.2 \times 10^{-3})}$$

$$= 0.324r$$

Since $v = dr/dt$,

$$dt = \frac{dr}{v} = \frac{dr}{0.324r}$$

Integrating from $r = 0.3$ to 0.4 yields

$$t = \left(\frac{1}{0.324} \right) \ln \left(\frac{0.40}{0.30} \right)$$

$$= 0.89 \, \text{s}$$

It takes 0.89 s for the particle to settle from the "top" (inside surface) of the water layer to the "bottom" (outside surface) in the centrifuge.

PHY.4   ETHANOL *Txy* DIAGRAM

The following information is requested as part of a wastewater treatment stripping operation for ethyl alcohol (EtOH). Generate a *Txy* diagram given the vapor–liquid equilibrium (VLE) data below. What is the vapor mole fraction of the ethanol if the liquid mixture contains 3.6 mol % ethanol?

T (°F)	x_{EtOH}	y_{EtOH}
414	0.000	0.000
378	0.072	0.390
367	0.124	0.470
358	0.238	0.545
356	0.261	0.557
351	0.397	0.612
349	0.520	0.661
345	0.676	0.738
343	0.750	0.812
342	0.862	0.925
340	1.000	1.000

Solution

First linearly interpolate between the first two x_{EtOH} and T data points to estimate the temperature of the liquid when $X_{EtOH} = 0.036$:

$$\frac{414 - 378}{0.0 - 0.072} = \frac{414 - T}{0.0 - 0.036}$$

$$T = 396°F$$

Next linearly interpolate between the first two T and y_{EtOH} data points to estimate the vapor fraction corresponding to the temperature, 396°F:

$$\frac{414 - 378}{0.0 - 0.390} = \frac{414 - 396}{0.0 - y_{EtOH}}$$

$$y_{EtOH} = 0.195$$

A graphical analysis of the interpolations is shown in Figure 88. Strictly speaking the results should be "smoothed-out" when graphed since the relationship is not

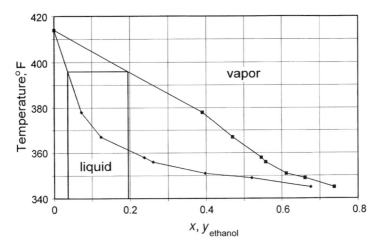

Figure 88. *Txy* diagram for ethanol.

linear at low values of x_{EtOH}. The actual ethanol vapor concentration could be significantly higher. The smoothed plot is shown in Figure 89.

The results from Figure 89 show that the vapor concentration, y_{EtOH}, is approximately 0.205.

PHY.5 AMMONIA REMOVAL FROM AIR BY SCRUBBING

Concerns regarding water discharges from an absorber have led to the need to perform the following calculation. Experimental data for an absorption system to be

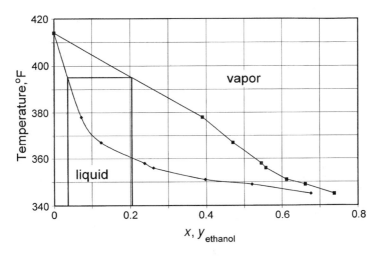

Figure 89. *Txy* diagram for ethanol ("Smooth" Plot).

used for scrubbing ammonia (NH_3) from air with water is provided. The water rate is 500 lb/min and the gas rate is 450 lb/min, both at 72°F.

The air to be scrubbed has 1.5 wt % NH_3 at 72°F and 1 atm pressure and is to be vented after 95% of the ammonia is removed. The inlet scrubber water is ammonia free.

1. Plot the equilibrium data in mole fraction units (x, y).
2. Perform the material balance and plot the operating line on the equilibrium plot.

NH_3 Partial Pressure (mmHg)	NH_3 Concentration (lb NH_3/100 lb H_2O)
3.4	0.5
7.4	1.0
9.1	1.2
12.0	1.6
15.3	2.0
19.4	2.5
23.5	3.0

Solution

The two key equations to obtain an $x–y$ graph are as follows:

$$y = p_{NH_3}/P$$

$$= p_{NH_3}/760$$

where p_{NH_3} is the partial pressure of the ammonia, and

$$x = \frac{\dfrac{w}{MW_{NH_3}}}{\dfrac{w}{MW_{NH_3}} + \dfrac{100 - w}{MW_{H_2O}}}$$

where w is the mass of NH_3 in H_2O.

Since the molecular weights of the NH_3 and H_2O are approximately the same, the equation above reduces to

$$x \cong \frac{w}{100}$$

For a partial pressure of 3.4 mm Hg, the gas mole fraction, y, is as follows:

$$y = p_{NH_3}/P$$
$$= 3.4/760$$
$$= 0.00447$$

Similarly,

NH$_3$ Partial Pressure (mmHg)	Mole fraction, y
3.4	0.00447
7.4	0.00973
9.1	0.01200
12.0	0.01580
15.3	0.02010
19.4	0.02550
23.5	0.03090

For an NH$_3$ concentration of 0.5 lb NH$_3$/100 lb H$_2$O, the liquid mole fraction, x, is

$$x = \frac{\dfrac{w}{MW_{NH_3}}}{\dfrac{w}{MW_{NH_3}} + \dfrac{100 - w}{MW_{H_2O}}}$$

$$= \frac{\dfrac{0.5}{17}}{\dfrac{0.5}{17} + \dfrac{100 - 0.5}{18}}$$

$$= 0.0053$$

Similarly,

NH$_3$ Concentration (lb NH$_3$/100 lb H$_2$O)	Mole fraction, x
3.4	0.0053
7.4	0.0106
9.1	0.0127
12.0	0.0169
15.3	0.0212
19.4	0.0265
23.5	0.0318

The water molar flowrate, L, and air molar flowrate, G, are

$$L = (500 \text{ lb/min})/(18 \text{ lb/lbmol})$$
$$= 27.78 \text{ lbmol/min}$$

$$G = (450 \text{ lb/min})/29 \text{ lb/lbmol})$$
$$= 15.52 \text{ lbmol/min}$$

The vapor concentrations can also be calculated:

$$y_1 = \frac{1.5/17}{1.5/17 + 98.5/29}$$
$$= 0.0253$$

$$y_2 = \frac{0.05\,(1.5/17)}{0.05\,(1.5/17) + (98.5/29)}$$
$$= 0.0013$$

From the problem statement x_2 is 0.0.

Applying a material balance across the column ($1 = \text{bottom}$, $2 = \text{top}$)

$$\frac{G}{L} = \frac{x_1 - x_2}{y_1 - y_2}$$

Rearranging gives

$$x_1 = \left(\frac{G}{L}\right)(y_1 - y_2) + x_2$$

Thus,

$$x_1 = \left(\frac{15.52}{27.78}\right)(0.0253 - 0.0013) + 0$$
$$= 0.0134$$

From Figure 90, the slope of the operating line (OL) is 1.79.

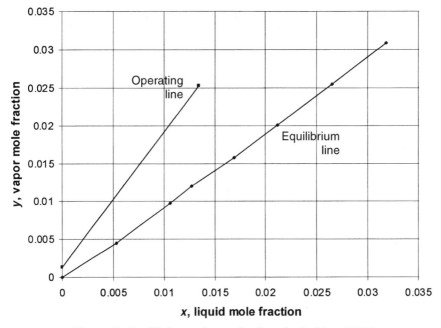

Figure 90. Equilibrium and operating lines for Problem PHY.5.

The above results can also be obtained analytically:

$$OL = \frac{0.025 - 0.0013}{0.0134 - 0.0}$$

$$= 1.77$$

PHY.6 STRIPPING OF ETHYLENE OXIDE

Following the absorption of ethylene oxide by water from a process stream, a 600-lbmol/h water stream (prior to discharge) is steam-stripped of the ethylene oxide as part of the regeneration step. For a feed ethylene oxide concentration of 0.5 mol %, determine the actual amount of steam required for stripping ethylene oxide to a concentration of 0.03 mol %.

The liquid flowrate is 600 lbmol/h and 40% excess ethylene oxide free steam is required for the separation in a packed column with 1-in. Raschig rings.

The system is at 30 psia and uses saturated steam at system conditions. Assume Henry's law applies and $y = 20x$ as the ethylene oxide equilibrium data for the system.

Solution

The reader is referred to the absorption problem provided in Part II for technical details on the calculations that follow.

For $x_2 = 0.005$, $y_2 = 0.10$ at the minimum liquid to gas ratio. Therefore,

$$\frac{G_{min}}{L_{min}} = \frac{x_{a1} - x_{a2}}{y_{a1} - y_{a2}}$$

$$= \frac{0.0003 - 0.005}{0 - 0.10}$$

$$= 0.047$$

For 40% excess

$$\frac{G}{L} = (0.047)(1.4)$$

$$= 0.0658$$

For stripping,

$$G_{min} = G_{actual}$$

Thus,

$$G_{actual} = (0.0658)(600 \text{ lbmol/h})(18 \text{ lb/lbmol})$$

$$= 710.64 \text{ lb/h of steam}$$

Refer to Figure 78 in Chapter 20. The x coordinate is

$$\left(\frac{L}{G}\right)\left(\frac{\rho}{\rho_L}\right)^{0.5} = \left(\frac{L_{min}}{G_{min}}\right)\left(\frac{\rho}{\rho_L}\right)^{0.5}$$

From steam tables, the boiling of steam at 30 psia is 250.34°F

$$\rho = \frac{P(\text{MW})}{RT} = \frac{(30 \text{ psia})(18 \text{ lb/lbmol})}{\left[10.73 \dfrac{\text{psia} \cdot \text{ft}^3}{(\text{lbmol} \cdot °R)}\right](250.34 + 459.67)°R}$$

$$= 0.0709 \frac{\text{lb}}{\text{ft}^3}$$

$$\rho_L = 62.4 \frac{\text{lb}}{\text{ft}^3}$$

$$\left(\frac{L}{G}\right)\left(\frac{\rho}{\rho_L}\right)^{0.5} = \left(\frac{1}{0.0658}\right)\left(\frac{0.0709}{62.4}\right)^{0.5} = 0.512$$

From Figure 78, the ordinate is 0.042. Thus,

$$0.042 = \frac{G^2 F \psi \mu^{0.2}}{\rho_L \rho g_c}$$

Note: G in the expression above is a mass flux rather than a mass flow rate and has units of $lb/ft^2 \cdot s$.

$$0.042 = \frac{G^2 (160)(1)(0.19\,cP)^{0.2}}{(0.07085)(62.4)\left[32.2 \dfrac{lb \cdot ft}{(lb_f \cdot s^2)}\right]}$$

Solving for G,

$$G = 0.0521 \frac{lb}{ft^2 \cdot s}$$

$$G_{actual} = (0.0521)(0.512) = 0.027 \frac{lb}{ft^2 \cdot s}$$

$$Area = \frac{Mass\ flowrate}{Mass\ velocity}$$

$$= \frac{\left(710.64 \dfrac{lb}{h}\right)}{\left(0.027 \dfrac{lb}{ft^2 \cdot s}\right)\left(3600 \dfrac{s}{h}\right)}$$

$$= 7.31\ ft^2$$

$$= \pi \frac{D^2}{4}$$

$$D = 3.05\ ft$$

PHY.7 CARBON ABSORPTION CALCULATION

A small municipal groundwater supply has become contaminated with a trace concentration of the wood preservative, pentachlorophenol (PCP). The average groundwater concentration of PCP is 7 mg/L and the average daily flow is 0.5 million gallons per day (MGD).

Calgon Filtrasorb 300 GAC (granular activated carbon) is being investigated for possible use in the treatment of the water supply. An investigation included laboratory scale testing using the Freundlich isotherm technique with an initial PCP concentration of 10 mg/L. The raw data resulting from the test are as follows:

GAC Dose (mg GAC/L)	Final Concentration (mg PCP/L)
2	9.34
5	8.41
10	7.05
20	4.92
50	1.76
100	0.48
200	0.10

Calculate the approximate time of breakthrough for a single-stage, 5000-lb canister of Filtrasorb. Assume that breakthrough occurs at an insignificantly short time before the bulk of the carbon is exhausted.

Solution

The capacity and intensity with which either GAC or PAC adsorbs a given solute has been modeled in several ways. The most commonly applied method is referred to as a *Freundlich isotherm*. The capacity of the activated carbon is defined as a function of the intrinsic characteristics of the carbon itself, the nature of the solute, and the final equilibrium concentration of the solute. The capacity is defined as:

$$X/M = K \, C_f^{(1/n)}$$

where X/M is the carbon capacity in milligram solute removed/gram of activated carbon, K may be viewed as the carbon capacity with a final equilibrium concentration of 1 mg solute/L, and $1/n$ is a function of the intensity of adsorption. It is essential to determine the characteristics, K and $1/n$, as a preliminary step in the evaluation of a particular activated carbon in a particular treatment application. To evaluate K and $1/n$, the mass of solute removed can be expressed as the difference between the original and final concentrations, where X is $C_0 - C_f$. The above equation can be rearranged into its linear, logarithmic form:

$$\ln(C_0 - C_f) - \ln M = \ln K + 1/n \, \ln C_f$$

The intercept of the line is $\log K$ and the slope is $1/n$. These parameters can be applied to process calculations in water treatment. In flow-through column applications, the final equilibrium concentration of the solute in the bulk of the carbon is approximately equal to the initial influent concentration.

Plot the given lab-scale data using both the normal and the logarithmic forms of the Freundlich equation (see Figures 91 and 92). Plot the normal form to check if the data approach an asymptote representing a maximum capacity. Plot the logarithmic form and check to determine how straight the line is.

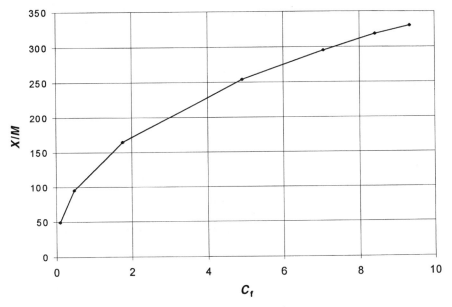

Figure 91. Freundlich equation.

Perform a linear regression on the logarithmic data (Figure 92) and determine the intercept (K) and the slope $(1/n)$.

A linear regression on the logarithmic data produces the equation:

$$y = 0.4194x + 4.8667$$

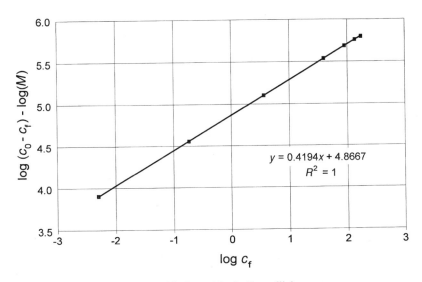

Figure 92. Logarithmic Freundlich.

Comparing this to the log form of the Freundlich equation shows that

$$\ln K = 4.8667$$

$$K = 130$$

and

$$1/n = 0.4194$$

$$n = 2.38$$

Use K and $1/n$ to calculate X/M for the equilibrium concentration. Because a flow-through column is intended for treatment, the equilibrium concentration is assumed to be approximately the influent concentration in the bulk of the carbon at breakthrough. Thus,

$$X/M = 130(7)^{0.42}$$

$$= 294 \, \text{mg PCP/g GAC}$$

Calculate breakthrough time based on plant parameters including average flow and influent PCP concentration.

$$t = \frac{(X/M)M_{\text{GAC}}}{qC_0}$$

where t = breakthrough time
M_{GAC} = mass of GAC
q = inlet flowrate
C_0 = inlet concentration of PCP

Noting that $1000 \, \text{L} = 264.17 \, \text{gal}$ and $1000 \, \text{g} = 2.2 \, \text{lb}$, the breakthrough time is given by

$$t = \frac{(294)(5000)}{(0.5 \times 10^6)(7)} \left(\frac{264.17}{1000}\right)\left(\frac{1000}{2.2}\right)$$

$$= 50.4 \, \text{days}$$

The reader should know that activated carbon is used in treatment by removing dilute concentrations of hazardous materials from ground and surface water supplies and wastewater discharge. Although activated carbon is capable of removing some metals, it is used primarily for the removal of volatile organic carbon (VOC) compounds from liquid streams. VOC-laden off-gases from processes such as air stripping and thermal oxidation are also treated with activated carbon.

VOCs and metals are removed from process streams by becoming adsorbed via Van der Waals forces within the pore volume of the carbon. When the activated carbon is prepared through the application of steam and high pressure, carbon is made a highly porous material with a very high surface area-to-mass ratio. Granular activated carbon (GAC) has a particle size that is generally >1 mm and typically is used in flow-through column applications for liquid and gas. Powdered activated carbon (PAC) is generally <1 mm and typically is used in batch treatments requiring clarification or filtration after contact. Activated carbon exhausted by VOC treatment can be recycled via thermal processes for reuse. However, carbon exhausted with metals or nonvolatile organics such as PCBs or pesticides must be properly disposed of.

PHY.8   FLOCCULATION CALCULATION

To determine the optimum chemical dose for flocculation, jar tests are used to estimate the amount of chemicals. In a jar test conducted for a given water treatment plant, 4-liter samples were poured into a series of jars. After the test, the jar dosed with 20 mL of alum solution containing 5 mg Al(III)/mL showed optimal results. If the amount of water to be treated is 90 MGD, calculate the pounds per day of alum [$Al_2(SO_4)_3 \cdot 14H_2O$, MW = 594.4 g/gmol] that should be added to the raw water.

Solution

As described earlier, flocculation is a physical process used to encourage small particles to aggregate into larger particles or floc. It is an essential component of most water treatment plants in which flocculation, sedimentation, and filtration processes are integrated to effectively remove suspended particles from water. Chemicals (such as alum, polyelectrolytes, etc.) are usually added to achieve agglomeration among small particles in water.

The dose administrated in the laboratory, neglecting dilution, is

$$\text{Dose} = \frac{[5 \text{ mg Al(III)/mL alum}](20 \text{ mL alum})(594.4 \text{ g/gmol alum})}{[27 \text{ g/gmol Al(III)}][2 \text{ gmol Al(III)/gmol alum}](4 \text{ L})}$$

$$= 275 \text{ mg alum/L}$$

Note that 1 gmol alum contains 2 gmol Al(III) ions.

The daily dosage applied at the plant is

$$\text{Alum dosage} = (90 \text{ MGD})(275 \text{ mg alum/L})[8.34 \text{ L} \cdot \text{lb/MG} \cdot \text{mg}]$$

$$= 206,415 \text{ lb/day}$$

Note: MG = 10^6 gal

PHY.9   PERMISSIBLE BOD AND TSS CONCENTRATION LEVELS

The following 5-day biochemical oxygen demand (BOD_5) and total suspended solids (TSS) data were collected from a clarifier at a local municipal wastewater treatment plant over a 7-day period. The National Pollutant Discharge Elimination System (NPDES) permit limitations for BOD_5 and TSS effluent concentrations from this wastewater treatment plant are 45 mg/L on a 7-day average. Based on this information, is the treatment plant within its NPDES permit limits?

Daily BOD_5 and TSS Effluent Concentration Data Collected Over a 7-Day Period at a Municipal Wastewater Treatment Plant

Day	BOD (mg/L)	TSS (mg/L)
1	45	20
2	79	100
3	64	50
4	50	42
5	30	33
6	25	25
7	21	15

Solution

The BOD_5 7-day average concentration based on the data tabulated in the problem statement is

$$(BOD_5)_7 = (45 + 79 + 64 + 50 + 30 + 25 + 21)/7$$
$$= 44.9 \, \text{mg/L}$$

The 7-day average concentration for TSS is

$$(TSS)_7 = (20 + 100 + 50 + 42 + 33 + 25 + 15)/7$$
$$= 40.7 \, \text{mg/L}$$

The wastewater treatment plant is still within its NPDES permit limit (but only marginally) of an average 7-day maximum concentration of 45 mg/L for both BOD_5 and TSS.

PHY.10   CONTAMINANT TRAVEL TIME IN SOIL

The velocity of a contaminant in groundwater is slowed by the presence of organic matter in the soil into which the contaminant will partition. Under equilibrium partitioning conditions, contaminant velocity is related to the pore water velocity by

$$v_c = v/R$$

where v_c = contaminant velocity, length/time
       v = pore water velocity, length/time
      R = retardation factor, dimensionless

For naphthalene in a particular aquifer, R has been found experimentally to equal 70. If the pore water velocity from a source to a well at a distance of 1000 m is 5×10^{-3} cm/s, what is the travel time of naphthalene to the well?

Solution

The pore water velocity, $v = 5 \times 10^{-3}$ cm/s, and the retardation factor, $R = 70$, are given in the problem statement. Thus, the velocity of the naphthalene is determined as follows:

$$v_c = \frac{v}{R}$$

$$= \frac{5 \times 10^{-3} \text{ cm/s}}{70}$$

$$= 7.14 \times 10^{-5} \text{ cm/s}$$

The naphthalene travel time is simply the distance divided by naphthalene's retarded velocity:

$$t = \frac{1000 \text{ m}}{7.14 \times 10^{-5} \text{ cm/s} \, (1 \text{ m}/100 \text{ cm})}$$

$$= 1.4 \times 10^9 \text{ s}$$

$$= 44.4 \text{ yr}$$

This long travel time is due to naphthalene's extremely low retarded velocity.

33 Biological Treatment (BIO)

BIO.1 AEROBIC VS. ANAEROBIC PROCESSES

Briefly describe the anaerobic process and explain its difference from an aerobic process.

Solution

The term *anaerobic digestion*, as opposed to *aerobic digestion*, is applied to a process in which organic material is decomposed biologically in an environment devoid of oxygen. Decomposition results from the activities of two major groups of bacteria. One group, called the acid formers, consists of facultative bacteria, which are also found in many aerobic environments and which, in a multistep anaerobic environment, convert carbohydrates, fats, and proteins to organic acids. The other group, the methane bacteria, converts the organic acids produced by the acid formers to methane and carbon dioxide. Some organic materials such as lignin are quite resistant to the activity of both groups and, hence, pass through the process relatively unaltered. Although cellulose is a relatively difficult material to degrade, it is treated due to the long detention time provided. The role of predator populations in anaerobic digestion is considered to be minor compared to that in aerobic processes. A detailed discussion of the biochemistry of anaerobic digestion is available in the literature.

BIO.2 DESIGN OF AEROBIC TREATMENT PROCESSES

Provide a qualitative overview of the design of aerobic biological wastewater treatment processes.

Solution

Through many years of designing and operating publicly owned treatment works (POTWs), typical values that reflect good wastewater treatment have been identified for most of the common biological treatment design parameters. Average values and typical ranges are often listed in wastewater engineering textbooks. However, "textbook" values for any given parameter are only applicable if the character of the wastewater being treated is also considered typical. A municipal wastewater might be considered typical if the common wastewater characterization parameters such as biochemical oxygen demand (BOD) and total suspend solids (TSS) fall within certain ranges (TSS, 100–350 mg/L; BOD_5, 110–300 mg/L). There must also be insignificant impact from industrial wastewaters that are inhibitory to biological treatability. If a wastewater can be characterized as typical, it may be possible to design a full-scale wastewater treatment system with a minimum of laboratory and pilot-scale treatability studies.

The cumulative wastewater engineering experience has evolved to the extent where the design of plants for aerobic suspended growth biological treatment systems can be based on previous design records and textbook values. Treatment processes typically begin with some form of bulk solids removal such as screening, grit removal, and/or primary sedimentation. Aerobic, suspended biological growth treatment disperses mechanically pressurized air supply within the treatment tank. The air is generally dispersed within the treatment tank "mixed liquor" (wastewater with a high concentration of process microorganisms) by means of either coarse bubble or fine bubble diffusers. The air serves two purposes. First, oxygen from air injected into the mixed liquor is dissolved and, therefore, is available to the microorganisms for the biological oxidation of substrate. Second, the injected air causes agitation of the liquid in the wastewater treatment tank and helps suspend and mix the microorganisms for efficient absorption and substrate oxidation.

BIO.3 TYPES OF BIOLOGICAL TREATMENT PROCESSES

Describe the various biological treatment processes employed for wastewater.

Solution

For many industrial wastewaters, one of the most important concerns has to do with those constituents that can exert an oxygen demand and have an impact on receiving waters. Most industries are discussed in terms of the toxics they discharge. They are also significant sources of biochemical oxygen demand (BOD) and chemical oxygen demand (COD). In many instances, the most effective and economical industrial treatment technology for removing this oxygen-demanding pollutant is biological treatment. Discussion of this problem will be limited to those processes frequently used in the industrial field: aerobic suspended growth processes (activated sludge), fixed film processes, and anerobic and aerated lagoons (stabilization ponds).

An aerobic suspended growth process is one in which the biological growth (microorganisms) are kept in suspension in a liquid medium and in contact with suspended, colloidal, and dissolved organic and inorganic materials. This biological process uses the metabolic reactions of the microorganisms to attain an acceptable effluent quality by removing those substances exerting an oxygen demand. Depending on the type and source of material in the raw wastewater stream, this process may be preceded by one or more other treatment technologies (i.e., clarification, oil and grease removal, etc.) to improve removal efficiencies.

In the suspended growth processes, wastewater enters a reactor basin where microorganisms are brought into contact with the organic components of the wastewater by mechanical mixing or air bubbles. This mixing not only maintains all material in suspension but also promotes transfer of oxygen to the wastewater, thus providing oxygen for sustaining the biological reactions in the basin. The organic matter in the wastewater serves as a carbon and energy source for microbial growth and is converted into microbial cell tissue and oxidized end products, mainly carbon dioxide and water. Contents of the reactor basin are referred to as *mixed liquor volatile suspended solids* (MLVSS) and consist mainly of microorganisms or biomass. *Mixed liquor total suspended solids* (MLTSS) consist of MLVSS plus inert and nonbiodegradable matter.

When the MLTSS are discharged from the reactor basin, a clarifier is used to separate them from the liquor. Concentrated microbial solids from the clarifier bottom are recycled back to the reactor basin to maintain a concentrated microbial population for the degradation of the wastewater. Because microorganisms are usually synthesized in the process, a means must be provided for "wasting" some of the microbial solids. Wasting of concentrated solids is usually accomplished from the settling basin, although wasting from the reactor basin is a less used alternative.

The first suspended growth process, now called the *conventional activated sludge process*, was developed to achieve carbonaceous BOD removal. However, since its inception, many modifications to the basic process have taken place. The variations in the activated sludge process are too numerous to be discussed in detail here. However, some of these are step aeration, contact stabilization, aerated lagoon, deep-tank aeration, and pure oxygen aeration.

Aerobic suspended growth systems can be adapted to treat a wide range of industrial wastewaters. The process can be easily expanded to accommodate increased flows. However, these systems do have some drawbacks. Suspended growth systems generally perform best under uniform hydraulic and pollutant loadings. For some industries, it is extremely difficult to maintain these conditions because of their manufacturing operations. Also, diurnal flows in industry and municipalities may cause effects on treatment quality. A common event is a "shock" loading of high-strength wastewater entering the treatment process, with the result being poor pollutant treatment. Also, certain toxic pollutants can kill microorganisms in the reactor basin, causing a loss of treatment from this part of the overall system. Operational skills and controls required to effectively operate an aerobic suspended growth system are higher than for most other biological processes. Finally, high energy costs are a part of the expense of providing mixing and oxygen in the process to sustain microbial growth.

Excluding those waste streams with very high concentrations of toxic or refractory pollutants, aerobic suspended growth systems can be used to treat any watewater containing biodegradable matter. Removal efficiencies in excess of 90% for carbonaceous BOD are often achieved through this process. For toxic organic wastes, pilot testing is necessary.

An aerobic attached growth process is one in which the biological growth products (microorganisms) are attached to some type of medium (i.e., rock, plastic sheets, plastic rings, etc.) and where either the wastewater trickles over the surface or the medium is rotated through the wastewater. The process is related to the aerobic suspended growth process in that both depend upon biochemical oxidation of organic matter in the wastewater to carbon dioxide, with a portion oxidized for energy to sustain and promote the growth of microorganisms. There are three general types of aerobic attached growth systems that are most frequently used: conventional trickling filters, roughing filters, and rotating biological contactors (RBCs).

There are several advantages for attached growth processes over other biological processes. First, microorganism growth can be easily reinstituted in the case of an accidental kill. Second, since oxygen is supplied naturally in most cases, the need for air- or oxygen-generating equipment is eliminated. This, along with a much simpler operation, lowers the requirements for highly skilled operational personnel. Both can result in substantial cost savings. The disadvantages are that attached growth treatment processes experience operating difficulty in cold climates. Enclosing the units for temperature protection can lead to other problems such as condensation. These units are also susceptible to clogging if dense media are used and/or high BOD and solids loadings are applied.

With proper design, aerobic attached growth processes can be used to treat any wastewater containing biodegradable matter. In general, the aerobic attached growth processes are not quite as efficient as the aerobic suspended growth processes in removing BOD suspended solids and toxic pollutants.

Aerobic lagoons are large, shallow earthen basins that are used for wastewater treatment by utilizing natural processes involving both algae and bacteria. The objective is microbial conversion of organic wastes into algae. Aerobic conditions prevail throughout the process with mechanical mixing. These lagoons have been used in rural areas and underdeveloped countries.

In aerobic photosynthesis, the oxygen produced by the algae through the process of photosynthesis is used by the bacteria in the biochemical oxidation and degradation of organic waste. Carbon dioxide, ammonia, phosphate, and other nutrients released in the biochemical oxidation reactions are, in turn, used by the algae, forming a cyclic-symbiotic relationship. Aerobic lagoons are used for treatment of weak industrial wastewater containing negligible amounts of toxic and/or nonbiodegradable substances.

Anaerobic lagoons are earthen ponds built with a small surface area and a deep liquid depth of 8–20 ft. These lagoons are usually anaerobic throughout their depth, except for an extremely shallow surface zone. Once greases form an impervious layer, completely anaerobic conditions develop. In a typical anaerobic lagoon, raw wastewater enters near the bottom of the lagoon (often at the center) and mixes with the active microbial mass in the sludge blanket, which is usually several feet deep. The discharge is located near one of the sides of the lagoon, submerged below the liquid surface. Excess undigested grease floats to the top, forming a heat-retaining and fairly airtight cover. Excess sludge is washed out with the effluent. Anaerobic lagoons are effective prior to aerobic treatment of high-strength organic wastewater that also contains a high concentration of solids. Under optimal operating conditions, BOD removal efficiencies of up to 85% are possible.

The advantage in using either aerobic or anaerobic lagoons are low cost (excluding land if not readily available), simplicity of operation, low operation and maintenance cost, and, when designed properly, high reliability. The disadvantages in using any lagoon process are high land requirements, possible odor emissions, and the potential for seepage of wastewater into groundwater, unless the lagoon is adequately lined. In addition, in most locales of the United States there are seasonal changes in both available light and temperature. Typically in the winter, biological activity decreases because of a reduction in temperature. *Note:* Lagoons are not frequently used in the United States but are used in some underdeveloped countries.

BIO.4   TREATMENT DESIGN PARAMETER CALCULATIONS

Provide typical design parameters (unless values have already been specified) and sample calculations for each of the following 10 sets of wastewater treatment design projects:

1. Mass loading rate of solids to the primary clarifier (at a flow of 30 million gallons per day (MGD) and a suspended solids concentration of 290 mg/L).
2. Mass loading rate of BOD_5 (240 mg/L) to the clarifier.
3. Mass production rate of solids in raw sludge (primary clarifier underflow) for 60% removal.
4. Hydraulic flow (underflow) of raw sludge from primary clarifier (1 lb H_2O/1.05 lb raw sludge; 6.5 lb dry solids/100 lb raw sludge).
5. Primary clarifier effluent (PE) hydraulic flow.
6. Production rate of solids in primary effluent.
7. Concentration of TSS in primary effluent.
8. Working volume of aeration basin.
9. Aeration basin width (20 ft deep; 1.5 ft width/depth).
10. Total aeration basin length.

Solution

1. Loading rate of solids $= \left(8.34 \dfrac{L \cdot lb}{MG \cdot mg}\right)\left(30 \dfrac{MG}{day}\right)\left(290 \dfrac{mg}{L}\right)$

   $= 72,600\, lb/\,day$

   Note: $MG = 10^6\, gal$

2. Loading rate of $BOD_5 = \left(8.34 \dfrac{L \cdot lb}{MG \cdot mg}\right)\left(30 \dfrac{MG}{day}\right)\left(240 \dfrac{mg}{L}\right)$

$= 60,000$ lb/day

3. Production rate $= \left(0.60 \dfrac{lb\ eff\ solids}{lb\ inf\ solids}\right)\left(72,600 \dfrac{lb\ inf\ solids}{day}\right)$

$= 43,600$ lb eff solids/day

4. $\left(43,600 \dfrac{lb\ dry\ solids}{day}\right)\left(\dfrac{1}{1.05} \dfrac{lb\ water}{lb\ raw\ sludge}\right)\left(\dfrac{100}{6.5} \dfrac{lb\ raw\ sludge}{lb\ dry\ solids}\right)$

$\times \left(454 \dfrac{g\ raw\ sludge}{lb\ water}\right)\left(\dfrac{1}{1000} \dfrac{L\ water}{g\ water}\right)\left(\dfrac{1.00}{3.78} \dfrac{gal\ raw\ sludge}{L\ water}\right)$

$= 76,700$ gal raw sludge/day

5. Hydraulic flow (PE) $= \left(30 \dfrac{MG}{day}\right) - \left(76,700 \dfrac{gal}{day}\right)\left(\dfrac{1}{10^6} \dfrac{MG}{gal}\right)$

$= 29.92$ MGD

6. Solid production rate in PE

$=$ Total solids $-$ Primary sludge solids

$= \left(72,600 \dfrac{lb\ dry\ wt}{day}\right) - \left(43,600 \dfrac{lb\ dry\ wt}{day}\right)$

$= 29,000$ lb dry wt/day

7. TSS conc. in PE $= \left(29,000 \dfrac{lb\ dry\ wt}{day}\right)\left(\dfrac{1}{30} \dfrac{day}{MG}\right)\left(\dfrac{1}{8.34} \dfrac{MG \cdot mg}{lb \cdot L}\right)$

$= 116$ mg solids/L primary eff

8. Volume of aeration basin

$= \left(30 \dfrac{MG}{day}\right)\left(10^6 \dfrac{gal}{MG}\right)\left(\dfrac{1}{24} \dfrac{day}{h}\right)\left(6 \dfrac{h}{tank\ volume}\right)\left(\dfrac{1}{7.48} \dfrac{ft^3}{gal}\right)$

$= 1,003,000$ ft^3/tank volume

9. Aeration basin width

$= (20\ ft\ basin\ depth)\left(1.50 \dfrac{ft\ basin\ width}{ft\ basin\ depth}\right)$

$= 30$ ft basin width

10. Total aeration basin length

$$= (1,003,000 \text{ ft}^3) \left(\frac{1}{20 \text{ ft basin depth}} \right) \left(\frac{1}{30 \text{ ft basin width}} \right)$$

$$= 1670 \text{ ft total length}$$

The number and length of basins will vary depending on land available, modification of the activated sludge process used, etc.

BIO.5 MICROBIAL REGROWTH

Microbial regrowth can be defined as an increase in viable microorganism concentrations in drinking water downstream of the point of disinfection after treatment. These microorganisms may be coliform bacteria, bacteria enumerated by the heterotrophic plate count (HPC bacteria), other bacteria, fungi, or yeasts. Regrowth of bacteria in drinking water can lead to numerous associated problems including multiplication of pathogenic bacteria such as *legionella pneumophila*, deterioration of taste, odor, and color of treated water, and intensified degradation of the water mains, particularly cast iron, by creating anaerobic conditions and reducing pH in a limited area. To obtain stable drinking water (i.e., to control regrowth), one needs to understand the sources of biological instability.

Explore the possible factors that affect regrowth in a water distribution system.

Solution

The following items can affect microbial regrowth in a drinking water distribution system:

Water quality (e.g., concentration of organic carbon and nutrients, temperature, disinfectant residual, etc.). Water quality certainly influences the potential for bacterial regrowth in the distribution system. The proliferation of microorganisms depends upon the presence of an adequate food supply and a favorable environment for their growth. Though the majority of organic compounds in water can be removed through treatment processes, low concentrations may remain that sufficiently stimulate bacterial growth. Some groups of bacteria (generally known as *oligotrophic bacteria*) are capable of surviving under low carbon and nutrient conditions. Examples of water quality parameters affecting microbial regrowth are temperature, pH, and dissolved oxygen. In general, disinfectant residual is used to suppress growth during storage and transportation of potable water.

Physical condition. The physical condition of distribution system pipes influences their tendency to foster biological regrowth. Distribution pipes that have tubercules or other surface irregularities commonly harbor microbial encrustations. Certain pipe materials and conditions can lead to a heavy accumulation of bacteria on their walls, a so-called biofilm. By attaching to the surface, microorganisms can be protected from washout and can exploit larger nutrient resources either accumulated at the surface or in the passing water. Moreover, attached bacteria appear to be less affected by disinfectants than those suspended in the disinfected liquid.

Flow conditions in the distribution system (e.g., detention time, flow velocity, wall shear stress, flow reversals, etc.). Long detention times and low flow velocities tend to encourage bacterial growth in the distribution system.

Water treatment processes. The efficiency of organic compound and nutrient removal depends upon the type of treatment processes utilized at a given plant. Generally, residual disinfectant in the water is required to prevent regrowth in the distribution system. Ozonation is known to increase bacterial nutrients by converting some of the nonbiodegradable organic matter into oxidized, degradable compounds, making it more available for uptake by microorganisms colonizing the distribution system.

Physical state of bacteria (type, concentration, and physiological state). It is evident that bacteria or other organisms must be present in a distribution system for biological regrowth to occur. The possible sources of bacteria in the water are:

1. Recovery of injured or dormant bacteria from disinfection
2. Cross-connections that allow back siphonage of contaminated water
3. Improperly protected distribution system storage
4. Line breaks and subsequent repair operations that fail to redisinfect the repaired lines effectively before placing them back into service

BIO.6 LAND APPLICATION OF WASTEWATER

Land treatment of industrial wastewater is a process in which wastewater is applied directly to the land. This type of treatment is most common for food-processing wastewater including meat, poultry, dairy, brewery, and winery wastes. The principal rationale of this practice is that the soil is a highly efficient biological treatment reactor, and food-processing wastewater is highly degradable. This treatment practice is usually carried out by distributing the wastewater through spray nozzles onto the land or letting the water run through irrigation channels.

Suppose that the rate of the wastewater flowing to a land application site is 178 gal/acre · min and the irrigated land area is 5.63 acres. The entire irrigation process lasts 7.5 h/day. If the wastewater has a BOD_5 concentration of 50 mg/L, what is the mass of BOD_5 remaining in the soil after the land treatment process is complete (assume a BOD_5 removal efficiency by the land treatment process is 95%)?

Solution

The amount of the wastewater flowing to the land is

$$\left(\frac{178 \text{ gal}}{\text{acre} \cdot \text{min}}\right)\left(\frac{60 \text{ min}}{\text{h}}\right)\left(\frac{7.5 \text{ h}}{\text{day}}\right)(5.63 \text{ acre}) = 451{,}000 \text{ gal/day}$$

The amount of BOD_5 remaining is

$$\left(\frac{451{,}000 \text{ gal}}{\text{day}}\right)(1.00 - 0.95)\left(\frac{50 \text{ mg}}{\text{L}}\right)\left(\frac{3.785 \text{ L}}{\text{gal}}\right) = 4{,}268{,}000 \text{ mg/day}$$

or

$$\left(\frac{4{,}268{,}000 \text{ mg}}{\text{day}}\right)\left(\frac{1 \text{ g}}{1000 \text{ mg}}\right)\left(\frac{454 \text{ g}}{\text{lb}}\right) = 9.4 \text{ lb/day}$$

BIO.7   MAXIMUM YIELD AND ENDOGENOUS DECAY COEFFICIENTS

Two of the primary parameters that are essential in the design of a biological wastewater treatment system are the maximum yield coefficient (Y) and the endogenous decay coefficient (k). The *yield coefficient* is defined as the ratio of the mass of cells produced to the mass of substrate consumed, as measured during any finite period of logarithmic growth. Logarithmic growth occurs when substrate is abundantly available to microorganisms, and cell growth is rapid with little effect on the mass of cells from endogenous respiration or death. Y is often reported in units of milligram of VSS (volatile suspended solids) produced/milligram BOD removed. The *endogenous decay coefficient* is a measure of the rate at which cell mass is being oxidized from the mixed liquor resulting due to low availability of substrate and starvation of the process microorganisms. It is usually reported in units of milligrams of VSS oxidized/milligram VSS of mixed liquor/day, or simply, 1/day.

The yield coefficient and the endogenous decay coefficient are measured for batch yield processes with a simple batch reactor or chemostat. The reactor is usually a 1- or 2-liter plastic or glass container that is temperature controlled by means of a water bath or other thermostatically controlled method or is run at room temperature. Wastewater and microorganisms are introduced through the opening at the top, and analytical samples can be removed through the top or drawn off through a tap near the container bottom. The mixed liquor is aerated via a forced-air supply and an air sparger at the container bottom.

In a typical lab-scale experiment to determine Y and k, a series of wastewater batch reactors are run at a controlled temperature (usually 20°C), or room temperature, until steady state is achieved. Several batch reactors are run at different organic loading rates. At steady state, the effluent, the substrate, and cell (VSS) concentrations are constant. Wasting of VSS daily is required to maintain a steady state concentration.

The relationship between substrate concentration and cell concentration is described by the following equation:

$$\frac{dX}{dt} = Y \frac{dS}{dt} - k X_v$$

where $dX/dt = $ mg MLVSS/(L·day) (MVLSS = mixed liquor volatile suspended solids)

$dS/dt = $ mg COD/(L·day)

$Y = $ mg MLVSS/mg COD

$k = $ decay coefficient; 1/day (b is frequently interchanged with k)

$X_v = $ mg MLVSS/L; average

The yield coefficient, Y, and the endogenous decay coefficient, k, are determined from the linear form of this equation:

$$(dX/dt)/X_v = Y (dS/dt)/X_v - k$$

where k (or b) has units of 1/day.

The slope of this equation is Y and the y intercept is $-k$.

A batch-fed, activated sludge pilot plant was developed for the treatment of an industrial waste. A total of six batch runs were made, each over a 24-h period and each with a different starting condition. The pilot plant study data (see table below) include initial and final concentrations of COD and VSS.

	COD (mg/L)		VSS (mg/L)	
Run Number	Initial S_i	Final S_f	Initial X_i	Final X_f
1	2370	70	8100	8300
2	7000	1800	4430	5960
3	4510	610	5150	5990
4	2920	120	6650	7350
5	6000	1400	5030	6470
6	3110	110	7230	7760

Compute the cell growth yield coefficient (Y) and cell decay rate (k) by using linear regression and by plotting the data.

Solution

Key calculations are presented in the tables that follow. The results appear graphically in Figure 93. As an example calculation, for the first run,

$$\left(\frac{dS}{dt}\right) = \frac{2370 - 70 \text{ mg/L}}{1 \text{ day}} = 2300 \text{ mg/(L} \cdot \text{day)}$$

$$\left(\frac{dX}{dt}\right) = \frac{8300 - 8100 \text{ mg/L}}{1 \text{ day}} = 200 \text{ mg/(L} \cdot \text{day)}$$

$$X_v(\text{average}) = \frac{8100 + 8300}{2} = 8200$$

$$\frac{dS/dt}{X_v} = \frac{2300}{8200} = 0.280$$

$$\frac{dX/dt}{X_v} = \frac{200}{8200} = 0.024$$

Determination of Yield (Y) and Endogenous Decay (k) Coefficients

	COD (mg/L)			VSS (mg/L)			
Run Number	Initial S_i	Final S_f	ΔS	Initial X_i	Final X_f	ΔX	Average X_v
1	2370	70	2300	8100	8300	200	8200
2	7000	1800	5200	4430	5960	1530	5195
3	4510	610	3900	5150	5990	840	5570
4	2920	120	2800	6650	7350	700	7000
5	6000	1400	4600	5030	6470	1440	5750
6	3110	110	3000	7230	7760	530	7495

Run Number	Unit Rate $(dS/dt)/X_v$	Unit Rate $(dX/dt)/X_v$	Regressed $(dX/dt)/X_v$
	0.000		−0.070
1	0.280	0.024	0.032
2	1.001	0.295	0.297
3	0.700	0.151	0.186
4	0.400	0.100	0.076
5	0.800	0.250	0.223
6	0.400	0.071	0.076

Based on a linear regression of the data, the yield coefficient, Y, and the endogenous decay coefficient, k, are found:

Slope $= Y = 0.367$ mg MLVSS/mg COD.

Intercept $= -0.070$ 1/day; k (or b) $= 0.070$ 1/day.

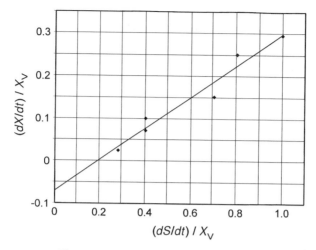

Figure 93. Linearization of growth curve 1.

BIO.8   WASTEWATER PILOT-SCALE STUDY—REQUIRED OXYGEN

A pilot-scale study of batch, aerobic, biological treatment has been conducted on an industrial wastewater from an organic chemical manufacturing process. Three reactors provide different levels of treatment, each starting with a different substrate concentration (as measured by COD in mg/L) and different concentration of biomass (as measured by MLVSS in mg/L). Oxygen uptake rates (OUR) and the remaining substrate COD are measured at intervals following the start of treatment. (*Note:* COD was measured in supernatant obtained following gravity settling of MLVSS.) Data from the three levels of treatment are presented in the table below.

Time Following Start of Treatment (h)	Effluent Substrate Concentration COD (mg/L)	Oxygen Uptake Rate OUR [mg/(L · min)]	Average Biomass Concentration MLVSS (mg/L)
		Reactor A	
0.0	245	1.20	1000
0.5	165	1.00	1000
2.0	90	0.58	1000
4.0	35	0.38	1000
		Reactor B	
0.0	265	1.55	4240
0.5	145	1.40	4240
2.0	40	0.95	4240
4.0	25	0.60	4240

(*continued*)

Time Following Start of Treatment (h)	Effluent Substrate Concentration COD (mg/L)	Oxygen Uptake Rate OUR [mg/(L · min)]	Average Biomass Concentration MLVSS (mg/L)
		Reactor C	
0.0	260	1.67	3000
0.5	90	1.50	3000
2.0	40	0.70	3000
4.0	30	0.47	3000

The solids production design equation once again is

$$(dX/dt) = Y(dS/dt) - bX_v$$

or in its linear form

$$(dX/dt)/X_v = Y(dS/dt)/X_v - b$$

where $dX/dt = $ mg MLVSS/L · day
$dS/dt = $ mg COD/L · day
$Y = $ mg MLVSS/mg COD
$b = 1/$day
$X_V = $ mg MLVSS/L average

The oxidation utilization design equation is given by

$$(dO/dt) = a'(dS/dt)/ + b'X_V$$

or in linear form

$$(dO/dt)/X_v = a'(dS/dt)/X_v + b'$$

where $dO/dt = $ mg $O_2/(L \cdot day)$
$dS/dt = $ mg COD$/(L \cdot day)$
$a' = $ mg $O_2/$mg COD
$b' = ($day$)^{-1}$
$X_v = $ mg MLVSS/L average

The relationship between solids production and oxygen utilization is

$$b' = 1.42b$$

and

$$a' + 1.42Y = 1$$

Based on the above data, compute the coefficients b and Y.

Solution

The calculation results are presented in tabular format below. In addition, the oxygen utilization results for each reactor are plotted in graphical form in Figure 94. Figure 95 allows one to obtain numerical values for both a' and b'.

Reactor	Time Following Start of Treatment (min)	Length of Treatment Interval (min)	Measured O$_2$ Uptake OUR [mg/(L · min)]	Average O$_2$ Uptake Rate for Interval [mg/(L · min)]	Oxygen Consumed During Time Interval (mg/L)	Total O$_2$ Consumed During 4-h Period (mg/L)	Total O$_2$ Consumed During 24-h Period (mg/L)
A	0	—	1.20	—	—		
A	30	30	1.00	1.1	33.0		
A	120	90	0.58	0.79	71.1		
A	240	120	0.38	0.48	57.6	162	970
B	0	—	1.55	—	—		
B	30	30	1.40	1.48	44.3		
B	120	90	0.95	1.18	105.8		
B	240	120	0.60	0.78	93.0	243	1458
C	0	—	1.67	—	—		
C	30	30	1.50	1.59	47.6		
C	120	90	0.70	1.10	99.0		
C	240	120	0.47	0.59	70.2	217	1301

Reactor	Average Biomass Conc. MLVSS (mg/L)	Initial COD (mg/L)	Final COD (mg/L)	Substrate Utilization per Day dS/dt (mg/L·day)	Oxygen Utilization per Day dO/dt (mg/L·day)	Normalized Substrate Utilization $(dS/dt)/X_v$ (1/day)	Normalized Oxygen Utilization $(dO/dt)/X_v$ (1/day)	Normalized Oxygen Utilization Regression $(dO/dt)/X_v$ (L/day)
						0.00		0.12
A	1000	245	35	1260	970	1.26	0.97	0.97
B	4240	265	25	1440	1458	0.34	0.34	0.35
C	3000	260	30	1380	1301	0.46	0.43	0.43

From the straight line in Figure 95,

$$a' = 0.677$$
$$b' = 0.118$$

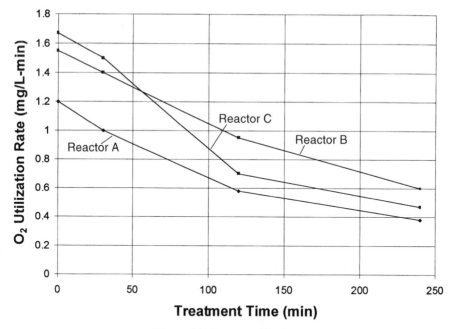

Figure 94. Oxygen utilization.

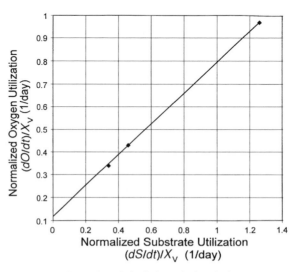

Figure 95. Calculation of a' and b'.

The oxygen utilization equation is therefore

$$(dO/dt) = 0.677(dS/dt) + 0.118X_V$$

Using the equations given in the problem statement, b and Y are calculated from a' and b':

$$b = k = 0.118/1.42$$

$$= 0.083 \text{ day}^{-1}$$

and

$$Y = (1.0 - 0.677)/1.42$$

$$= 0.23 \text{ mg MLVSS/mg COD}$$

BIO.9 ANALYSIS OF REACTOR LABORATORY DATA

A treatment system was designed to biologically remove a hazardous material from an industrial wastewater. To determine design parameters for the system, the following laboratory study was conducted.

Each of three aerated reactors was continuously fed a wastewater with a COD concentration of 1200 mg/L at the rate of 10 L/day until steady state was reached. Extensive data were collected over the next 4 days. Volumes fed to reactors 1, 2, and 3 were 4, 4, and 5 liters, respectively. Mixed liquor recycle was employed during treatment, but no solids were wasted from the system until the end of every 24-h period. Wasting at the end of the day adjusted the initial TVSS to 4000 mg/L. The collected data were used to produce the plot of $(dS/dt)/X_v$ vs. $(dX/dt)/X_v$ shown in Figure 96.

1. From the plot, compute the cell growth coefficient, Y, and the cell decay rate, b.
2. Using the design parameters listed below, calculate the solids retention time (SRT) (the average time the cells spend in the aeration tank):

Design Parameters

Sludge flowrate = 100,000 gpd
Aeration tank influent COD concentration = 1200 mg/L
Aeration tank effluent COD concentration = 60 mg/L
MLVSS = 4000 mg/L
% Volatiles = 80
Detention (retention) time = 7 h

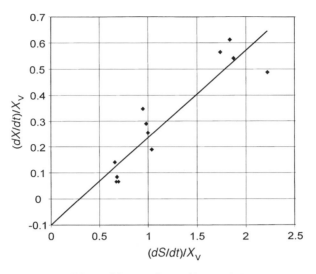

Figure 96. Data for Problem BIO.9.

Solution

1. From a linear regression, the intercept is -0.100 day^{-1}. The cell decay rate, b, is therefore approximately

$$b = 0.100 \text{ day}^{-1}$$

Using points $(0, -0.1)$ and $(2.0, 0.57)$, the slope (cell growth coefficient, Y) is given by

$$Y = \frac{0.57 - (-0.10)}{2.0 - 0} = 0.335$$

2. From the design data, the COD removed is

$$1200 - 60 = 1140 \text{ mg/L}$$

The COD removal rate, (dS/dt), in lb/day is

$$\left(\frac{dS}{dt}\right) = \left(1140 \frac{\text{mg}}{\text{L}}\right)\left(10^5 \frac{\text{gal}}{\text{day}}\right)\left(2.250 \times 10^{-6} \frac{\text{lb}}{\text{mg}}\right)\left(3.785 \frac{\text{L}}{\text{gal}}\right)$$

$$= 951.4 \frac{\text{lb}}{\text{day}}$$

The average MLVSS in the aeration tank, X_v, is

$$X_v = \left(4000\,\frac{mg}{L}\right)\left(\frac{7}{24}\,day\right)\left(10^5\,\frac{gal}{day}\right)\left(2.205 \times 10^{-6}\,\frac{lb}{mg}\right)\left(3.785\,\frac{L}{gal}\right)$$

$$= 973.7\,lb$$

The MLTSS (mixed liquor total suspended solids) in the aeration tank is therefore

$$973.7/0.8 = 1217\,lb$$

The rate of solids produced, dX/dt, is given by

$$\left(\frac{dX}{dt}\right) = Y\left(\frac{dS}{dt}\right) - bX_v$$

$$= (0.335)\left(951.4\,\frac{lb}{day}\right) - \left(\frac{0.10}{day}\right)(973.7\,lb)$$

$$= 221\,lb/day$$

The solids retention time (SRT) is then

$$SRT = \frac{\text{cells in systems}}{\text{rate of cells leaving systems}}$$

$$= \frac{973.7\,lb}{221\,lb/day}$$

$$= 4.41\,days$$

Note that, at steady state, the cells produced equals the cells leaving the system.

BIO.10   DESIGN OF A WASTEWATER TREATMENT PLANT

Design a wastewater treatment plant for the removal of carbonaceous and nitrogeneous BOD. It is desired to obtain at least 85% removal of CBOD and 90% oxidation of the influent ammonia nitrogen. The following data have been secured for the design:

Design flow $= 3.5\,MGD$

Aeration tank influent $BOD_5 = 186\,mg/L$

Aeration tank influent $NH_4^+N = 20\,mg/L$

SRT winter controls $= 12\,days$ at $14°C$

Sludge yield coefficient, $Y = 0.65$

Decay Coefficient, $b = 0.05\,(\text{day})^{-1}$

MLSS in aeration tank $= 2500\,\text{mg/L}$

MLVSS in aeration tank $= 2000\,\text{mg/L}$

In submitting the design, provide the following information:

1. The food-mass ratio, F/M
2. Aeration tank volume in gal and ft^3
3. Required aeration time in hours
4. Oxygen consumed in lb/day (employ 100% for peak flow, 4.2 mg O_2/mg NH_3–N_2, and 1.1 mgO_2/mg CBOD applied)
5. Nitrogen loading in lb $NH_4^+N/1000$ ft$^3 \cdot$ day and lb NH_4^+N/lb MLVSS $\cdot$ day

Solution

The design equation for F/M (BOD in MLSS) is derived from

$$\frac{dX}{dt} = Y\left(\frac{dS}{dt}\right) - bX$$

$$\frac{dX/dt}{X} = Y\frac{dS/dt}{X} - b$$

Substituting

$$\frac{dX/dt}{X} = \frac{1}{\text{SRT}}$$

and

$$\frac{dS/dt}{X} = \frac{F}{M}$$

into the equation above yields

$$\frac{1}{\text{SRT}} = Y\left(\frac{F}{M}\right) - b$$

This equation is equivalent (in rearranged form) to the earlier equations presented in this problem set. Substituting yields

$$\frac{1}{12} = 0.65\left(\frac{F}{M}\right) - 0.05$$

$$F/M = 0.205 \qquad F/M \text{ based on removal and MLVSS}$$

The ratio can be viewed as a loading factor that can be applied to the removed organics.

Based on the data provided,

$$\text{BOD} = (3.5)(186)(8.34)$$
$$= 5429 \text{ lb/day}$$

For 85% removal,

$$\text{BOD}_{\text{effluent}} = (0.15)(5429)$$
$$= 814 \text{ lb/day}$$

The BOD removed is then

$$\text{BOD}_{\text{removed}} = 5429 - 814$$
$$= 4615 \text{ lb/day}$$

Therefore, M is

$$M = 4615/0.205$$
$$= 22{,}510 \text{ lb MLVSS}$$

The aeration tank volume, V, can now be calculated:

$$V = \frac{(22{,}510)(10^6)}{(8.34)(2000)}$$
$$= 1.35 \times 10^6 \text{ gal}$$

The detention (or retention) time, t, is

$$t = \frac{V}{q} = \frac{(1.35)(24)}{3.5}$$
$$= 9.3 \text{ h}$$

The ammonia as nitrogen, $(NH_3-N)_{rem}$, removed is

$$(NH_3-N)_{rem} = (3.5)(20 - 2)(8.34)$$
$$= 525 \, lb/day$$

This is what leaves the effluent as nitrate. The oxygen required for the oxidation of NH_3-N to NO_3-N (nitrates–nitrogen) is

$$O_2 = (525)(4.2)$$
$$= 2207 \, lb/day$$

The oxygen required for the BOD removal is

$$(O_2)_{BOD} = (5429)(1.1)$$
$$= 5972 \, lb/day$$

The total oxygen consumed (required) is

$$(O_2)_{REQ} = 2207 + 5972$$
$$= 8200 \, lb/day$$

Adding 100% for peak flow leads to

$$(O_2)_{TOT} = (8200)(2)$$
$$= 16,400 \, lb/day$$

The MLSS wasted is

$$MLSS_W = MLSS_{under \, air}/SRT$$
$$= \frac{(1.35)(2500)(8.34)}{12}$$
$$= 2350 \, lb/day$$

The nitrogen loading is

$$N = \frac{(3.5)(8.34)(20)(1000)(7.48)}{(1.35)(10^6)}$$
$$= 3.2 \, lb \, NH_3-N/1000ft^3$$

BIO.11   SIZING OF AN AEROBIC DIGESTER

A municipality generates 1000 lb of solids daily. Size an aerobic digester to treat the solids. The following design parameters and information are provided:

Detention time, hydraulic, $= t_H = 20$ days
Detention time, solids, $t_S = 20$ days
Temperature $= 95°F$
Organic loading (OL) $= 0.2$ lbVS/(ft$^3 \cdot$ day)
Volatile solids (VS) $= 78\%$ of total solids
Percentage solids (TS) entering digester $= 4.4\%$
VS destruction $= 62\%$

Solution

Check the design based on the organic load and the hydraulic load. The volume based on the organic load V_{OL}, is

$$V_{OL} = (1000)(0.78)/(0.2)$$
$$= 3900 \text{ ft}^3$$

Based on the hydraulic load the volume, V_H, is

$$V_H = \frac{(1000)(20)}{(0.044)(8.33)(7.48)}$$
$$= 7300 \text{ ft}^3$$

Since $7300 > 3900$, the hydraulic detention time controls and the design volume is 7300 ft^3.

BIO.12   ANALYSIS OF AN ANAEROBIC CONTACT PROCESS

A 100,000-gal per day protein-contaminated wastewater with a COD of 4000 mg/L is generated by a meat-processing plant. This waste is to be treated by the anaerobic contact process using a loading of 0.15 lbCOD/ft$^3 \cdot$ day and a sludge detention time of 20 days. Assuming 80% efficiency, determine/provide the following:

1. A flow diagram of the process
2. Hydraulic detention time, hours
3. Daily solids accumulation, pounds (assume a 6% synthesis of solids from BOD)

4. Nitrogen and phosphorous requirements, pounds (10 and 1.5%, respectively, in solids)
5. Mixed liquor suspended solids concentration

Solution

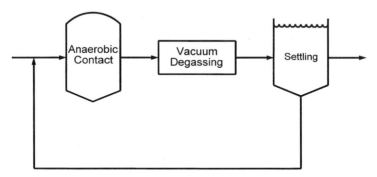

Figure 97. Flow diagram for Problem BIO.12.

A very simple flow diagram of the process is provided in Figure 97. The daily COD removed is

$$\text{COD} = (4000\,\text{mg/L})(0.1 \times 10^6\,\text{gal/day})(10^{-6}\,\text{L/mg})(8.34\,\text{lb/gal})$$
$$= 3336\,\text{lb COD/day}$$

The volume of the anaerobic contact tank is

$$V = (3336\,\text{lb COD/day})(7.48\,\text{gal/ft}^3)(\text{ft}^3 \cdot \text{day}/0.15\,\text{lb})$$
$$= 166,000\,\text{gal}$$

The hydraulic detention time, t_H, is then

$$t_\text{H} = (166,000\,\text{gal})\left(\frac{\text{day}}{0.1 \times 10^6\,\text{gal}}\right)$$
$$= 1.66\,\text{days}$$
$$= 39.8\,\text{h}$$

The solids growth (SG) may be calculated using the 6% synthesis figure and 80% efficiency:

$$SG = (3336)(0.8)(0.06)$$
$$= 160 \, \text{lb/day}$$

The SG represents the mass of suspended solids and is the active cell population. This concentration may be converted to mg/L as follows:

$$(1.0 \, \text{lb/gal}) = (454 \text{g/lb})(1000 \text{mg/g})(264 \text{gal/m}^3)(\text{m}^3/1000 \, \text{L})$$
$$= 120,000 \, \text{mg/L}$$

This number is equivalent to the conversion constant

$$\left(\frac{10^6 \, \text{mg/l}}{8.34 \, \text{lb/gal}} \right) \approx 120,000$$

The latter conversion constant (120,000) is most often used in industry. Thus,

$$MLVSS = \frac{(160)(20)}{166,000}(120,000)$$
$$= 2300 \, \text{mg/L}$$

Since 10% of the solids are nitrogen,

$$N = (160)(0.1)$$
$$= 16 \, \text{lb N/day}$$

Since 1.5% of solids is phosphorous,

$$P = (160)(0.015)$$
$$= 2.4 \, \text{lb P/day}$$

Since the wastewater is proteinaceous (rich in nitrogen and phosphorous) N and P will not be required.

34 Chemical Treatment (CHM)

CHM.1 CHEMICAL TREATMENT PROCESSES

Describe the two primary chemical treatment processes.

Solution

Coagulation–Precipitation The nature of an industrial wastewater is often such that conventional physical treatment methods will not provide an adequate level of treatment. Particularly, ordinary settling or flotation processes will not remove ultrafine colloidal particles and metal ions. In these instances, natural stabilizing forces (such as electrostatic repulsion and physical separation) predominate over the natural aggregating forces and mechanisms, namely, van der Waals forces and Brownian motion, which tend to cause particle contact. Therefore, to adequately treat such particles in industrial wastewaters, coagulation–precipitation may be warranted.

The first and most important part of this technology is coagulation, which involves two discrete steps. Rapid mixing is employed to ensure that the chemicals are thoroughly dispersed throughout the wastewater flow for uniform action. Next, the wastewater undergoes flocculation, which provides for particle contact at a slow mix so that the particles can agglomerate to a size large enough for removal. The second part of this technology involves precipitation, which is the separation of solids from liquid and thus can be performed in a unit similar to a clarifier.

Coagulation–precipitation is capable of removing pollutants such as biochemical oxygen demand (BOD), chemical oxygen demand (COD), and total suspended solids (TSS) from industrial wastewater. In addition, depending upon the specifics of the wastewater being treated, coagulation–precipitation can remove additional pollutants such as phosphorus, organic nitrogen and metals. This technology is attractive to industry because a high degree of clarification and toxic pollutant removal can be combined in one treatment process. A disadvantage of this process is the substantial quantity of sludge generated, which presents a sludge disposal problem and that all industrial wastewaters do not respond to this treatment.

Neutralization In virtually every type of manufacturing industry, chemicals play a major role. Whether they result from the raw materials or from the various processing agents used in the production operation, some residual compounds will

616

ultimately end up in a process wastewater. Thus, it can generally be expected that most industrial waste streams will deviate from the neutral state (i.e., will be acidic or basic in nature).

Highly acidic or basic wastewaters are undesirable for two reasons. First, they can adversely impact the aquatic life in receiving waters; second, they might significantly affect the performance of downstream biological treatment processes at the plant site or at a publicly owned treatment works (POTW). Therefore, to rectify these potential problems, one of the most fundamental treatment technologies, neutralization, is employed at industrial facilities. Neutralization involves adding an acid or a base to a wastewater to offset or neutralize the effects of its counterpart in the wastewater flow, i.e., adding acids to alkaline wastewaters and bases to acidic wastewaters.

The most important considerations in neutralization are a thorough understanding of the wastewater constituents so that the proper neutralizing chemicals are used, and proper monitoring to ensure that the required quantities of these chemicals are employed and that the effluent is in fact neutralized. For acid waste streams, lime, soda ash, and caustic soda are the most common base chemicals used in neutralization. In the case of alkaline waste streams, sulfuric, hydrochloric, and nitric acid are generally used. Some industries have operations that separate acid and alkaline waste streams. If properly controlled, these waste streams can be mixed to produce a neutralized wastewater with less or no additional neutralizing chemicals.

Neutralizing treats the pH level of a wastewater flow. Although most people do not think of pH as a pollutant, it is in fact designated by the Environmental Protection Agency (EPA) as such. Since many subsequent treatment processes are pH-dependent, neutralization can be considered as a preparatory step in the treatment of all pollutants.

Eliminating the adverse impacts on water quality and wastewater treatment system performance is not the only benefit of neutralization. Acidic or alkaline wastewaters can be very corrosive. Thus, by neutralizing its wastewaters, a plant can protect its treatment units and associated piping. The major disadvantage of neutralization is that the chemicals used in the treatment process are often themselves corrosive and can be dangerous.

CHM.2   TOLUENE CONCENTRATION DECAY

Determine the time, in hours, required for the concentration of toluene in a wastewater treatment process to be reduced to one-half of its initial values. Assume the first-order reaction velocity constant for toluene is 0.07/h. Also calculate the time for the toluene to be reduced to 99% of its initial value.

Solution

The decay (disappearance) of a chemical can often be described as a first-order function (see Chapter 10 for additional details):

$$\frac{dC}{dT} = -kC$$

where C = concentration at time t
  t = time
  k = first-order reaction rate constant

The integrated form of this equation is

$$\ln \frac{C_0}{C} = kt$$

where C_0 is the concentration at time zero.

When half of the initial material has decayed (reacted), C_0/C is equal to 2; the corresponding time is given by the following expression:

$$t_{1/2} = \frac{\ln(2)}{k}$$

$$= \frac{0.69}{0.07/h}$$

$$= 9.86 \ h$$

When 99% of the initial material has decayed, C_0/C is equal to 100 and

$$t = \frac{\ln(100)}{k}$$

$$= \frac{4.605}{0.07/h}$$

$$= 65.79 \ h$$

CHM.3   ALKALINE CHLORINATION OF A CYANIDE-BEARING WASTE

Industrial cyanide-bearing waste is to be treated in a batch process using alkaline chlorination. In this process, cyanide is reacted with chlorine under alkaline conditions to produce carbon dioxide and nitrogen as end products.

The cyanide holding tank contains $28 \ m^3$ with a cyanide concentration of $18 \ mg/L$. Assuming that the reaction proceeds stoichiometrically, answer the following questions:

1. How many pounds of chlorine are needed?
2. How long will the hypochlorinator have to operate if the hypochlorinator can deliver $900 \ L/day$ of chlorine?
3. How long should the caustic soda feed pump operate if the pump delivers $900 \ L/day$ of 10 wt % caustic soda solution?

Solution

1. The balanced equation can be written as follows:

$$2CN^- + 8OH^- \rightarrow 2CO_2 + N_2 + 4H_2O + 10e^-$$

$$5Cl_2 + 10e^- \rightarrow 10Cl^-$$

$$\overline{2CN^- + 8OH^- + 5Cl_2 \rightarrow 2CO_2 + N_2 + 10Cl^- + 4H_2O}$$

Therefore, 8 gmol of caustic and 5 gmol of chlorine are required to oxidize 2 gmol of cyanide to carbon dioxide and nitrogen gas.

Calculate the mass of cyanide to be treated:

$$m_{CN^-} = (28 \, m^3)(18 \, mg/L)(1 \, L/m^3)$$

$$= 504 \, g \; cyanide$$

Calculate the mass of chlorine needed to oxidize this amount of cyanide:

$$m_{Cl_2} = \frac{(504 \, g \; CN^-)(5 \, gmol \; Cl_2)}{(2 \, gmol \; CN^-)} \left(\frac{71 \, g/gmol \; Cl_2}{26 \, g/gmol \; CN^-} \right)$$

$$= 3441 \, g \; Cl_2$$

2. The time required for operation of the hypochlorinator is

$$t = \frac{(3441 \, g)(24 \, h/day)}{(900 \, L/day)(0.02)(1000 \, g/L)}$$

$$= 4.59 \, h$$

$$= 4 \, h, \, 35 \, min$$

3. The mass of caustic soda required is

$$m_{NaOH} = \frac{(504 \, g \; CN^-)(8 \, gmol \; NaOH)}{2 \, gmol \; CN^-} \left(\frac{40 \, g/gmol \; NaOH}{26 \, g/gmol \; CN^-} \right)$$

$$= 3102 \, g \; NaOH$$

The time for operation of the feed pump is

$$t = [(3102 \, g)(24 \, h/day)]/[(900 \, L/day)(0.10)(1000 \, g/L)]$$

$$= 0.83 \, h$$

$$= 50 \, min$$

CHM.4   pH OF WASTEWATER

An industrial wastewater has an alkalinity of 60 mg/L and a CO_2 content of 7 mg/L. Determine the pH. The first ionization constant of H_2CO_3, is

$$K_{i,1} = \frac{(H^+)(HCO_3^-)}{(H_2CO_3)}$$

$$= 3.98 \times 10^{-7}$$

Assume that the alkalinity is all in the HCO_3^- form. Note that the parentheses represent the concentration in mol/L.

Solution

"Alkalinity" is measured as equivalents of $CaCO_3$ which has a molecular weight of 100 and an equivalent weight of 50. For example, a water solution of 17 mg/L of OH^- (equivalent weight $= 17$) contains 10^{-3} equivalent weights of OH^-, which is equivalent to 10^{-3} equivalents of $CaCO_3$. The alkalinity of that solution is therefore $(50)(10^{-3})$ g/L or 50 mg/L.

For an alkalinity of 60 mg/L, the concentration of HCO_3^- is

$$(HCO_3^-) = \frac{60 \, mg/L}{50 \, mg/meq}$$

$$= 1.2 \, meq/L$$

$$= 1.2 \, mmol/L$$

The unit "meq" represents 10^{-3} equivalents.
For the CO_2,

$$(CO_2) = \frac{7 \, mg/L}{44 \, mg/mmol}$$

$$= 0.159 \, mmol/L$$

Note that $(CO_2) = (H_2CO_3)$.
  Substitution gives

$$3.98 \times 10^{-7} = \frac{(H^+)(1.2 \times 10^{-3})}{(0.159 \times 10^{-3}}$$

$$(H^+) = 5.27 \times 10^{-8}$$

The pH is given by

$$pH = -\log(H^+)$$
$$= 7.28$$

CHM.5 EFFECT OF HYDROLYSIS ON pH

With reference to CHM.4, the waste passes through a holding tank in 1 h (assume air contact is negligible), and the urea content is found to be close to zero coming out of the holding tank. Determine this pH. (Assume any CO_2 formed remains in solution.) The second ionization constant of H_2CO_3, i.e., $K_{i,2}$ for the reaction

$$HCO_3^- \rightarrow CO_3^{-2} + H^+$$

is 5.01×10^{-11}.

Solution

From the solution to Problem CHM.4, the initial concentrations of urea and $(NH_4)HCO_3$ are 0.300 and 1.200 mmol/L. (*Note:* This assumes that the alkalinity of 50 mg/L is due entirely to the HCO_3^- concentration.)

If the hydrolysis of urea is complete, the following reactions occur:

Reaction 1: $NH_2CONH_2 + H_2O \rightarrow 2NH_3 + CO_2$
Reaction 2: $NH_3 + CO_2 + H_2O \rightarrow (NH_4)HCO_3$
Reaction 3: $NH_3 + (NH_4)HCO_3 \rightarrow (NH_4)_2CO_3$

The concentrations of the reactants before and after each reaction are shown below. For reaction 1:

$$NH_2CONH_2 + H_2O \rightarrow 2NH_3 + CO_2$$

Start:	0.300		0	0.159
End:	0		0.600	0.459

For reaction 2:

$$NH_3 + CO_2 + H_2O \rightarrow (NH_4)HCO_3$$

Start:	0.600	0.459	1.200
End:	0.141	0	1.659

For reaction 3:

$$NH_3 + (NH_4)HCO_3 \rightarrow (NH_4)_2CO_3$$

Start:	0.141	1.659	0
End:	0	1.518	0.141

The pH may now be recalculated:

$$pH = pK_{i,2} + \log \frac{CO_3^{-2}}{HCO_3^-}$$

$$= -\log 5.0 \times 10^{-11} - \log \frac{0.141}{1.518} = 10.30 - 1.03$$

$$= 9.3$$

The final urea concentration is essentially zero because it degrades rapidly to form CO_2 and NH_3.

CHM.6   SOURCES OF Cr^{6+}

List the possible sources of highly toxic hexavalent chromium (Cr^{6+}) and methods to remove it from a wastewater stream.

Solution

Hexavalent chromium-bearing wastewater is produced in chromium electroplating, chromium conversion coating, etching with chromic acid, and in metal-finishing operations carried out on chromium as the base material.

Chromium wastes are commonly treated in a two-stage batch process. The primary stage is used to reduce the highly toxic hexavalent chromium to the less toxic trivalent chromium. There are several ways to reduce the hexavalent chrome to trivalent chrome including the use of sulfur dioxide, bisulfite, or ferrous sulfate. The trivalent chrome is then removed by hydroxide precipitation. Most processes use caustic soda (NaOH) to precipitate chromium hydroxide. Hydrated lime [$Ca(OH)_2$] may also be used. The chemistry of the reactions is described as follows:

Using SO_2 to convert Cr^{6+} to Cr^{3+}

$$SO_2 + H_2O \rightarrow H_2SO_3$$

$$3H_2SO_3 + 2H_2CrO_4 \rightarrow Cr_2(SO_4)_3 + 5H_2O$$

Using NaOH to precipitate,

$$6NaOH + Cr_2(SO_4)_3 \rightarrow 2Cr(OH)_3 + 3Na_2SO_4$$

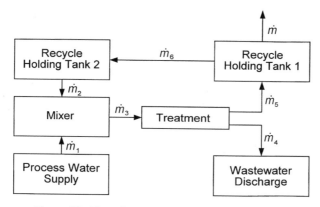

Figure 98. Flow diagram for Problem CHM.7, part 1.

CHM.7 MATERIAL BALANCES ON AN INDUSTRIAL WASTEWATER SYSTEM

Answer the following two-part problem.

1. Consider the flow diagram in Figure 98 for a wastewater treatment system. The following flowrate data are given:

$\dot{m}_1 = 1000 \, \text{lb/min}$
$\dot{m}_2 = 1000 \, \text{lb/min}$
$\dot{m}_4 = 200 \, \text{lb/min}$

Find the amount of water lost by evaporation in the operation.
2. Consider the system shown in Figure 99.

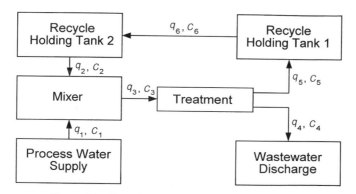

Figure 99. Flow diagram for Problem CHM.7, part 2.

The following volumetric flowrate and phosphate concentration data have been provided by the plant manager. Are the data correct and/or consistent?

$$q_1 = 1000 \text{ gal/day} \qquad C_1 = 4 \text{ ppm}$$
$$q_2 = 1000 \text{ gal/day} \qquad C_2 = 0 \text{ ppm}$$
$$q_3 = 2000 \text{ gal/day} \qquad C_3 = 2 \text{ ppm}$$
$$q_4 = 200 \text{ gal/day} \qquad C_4 = 20 \text{ ppm}$$
$$q_5 = 1800 \text{ gal/day} \qquad C_5 = 0 \text{ ppm}$$
$$q_6 = 1000 \text{ gal/day} \qquad C_6 = 0 \text{ ppm}$$

Solution

1. Apply a material balance around the treatment system to determine the value of $\dot{m}_5$. The value of $\dot{m}_5$ equals the number of gallons of water being turned into steam:

$$\dot{m}_3 = \dot{m}_4 + \dot{m}_5$$
$$\dot{m}_1 + \dot{m}_2 = \dot{m}_4 + \dot{m}_5$$
$$1000 + 1000 = 200 + \dot{m}_5$$
$$\dot{m}_5 = 1800 \text{ lb/min}$$

Similarly (for tank 2),

$$\dot{m}_6 = \dot{m}_2$$
$$\dot{m}_6 = 1000 \text{ lb/min}$$

Thus (for tank 1),

$$\dot{m}_6 = \dot{m}_5 + \dot{m}$$
$$1000 = 1800 - \dot{m}$$
$$\dot{m} = 800 \text{ lb/min}$$

One sees that 800 lb of water per minute are lost in the operation.

2. A componential balance around the mixer gives

$$C_1 q_1 + C_2 q_2 = C_3 q_3$$

$$\left(\frac{4}{120,000}\right)(1000) + \left(\frac{0}{120,000}\right)(1000) = \left(\frac{2}{120,000}\right)(2000)$$

$$4000 = 4000 \qquad \text{OK}$$

A balance around the treatment tank gives

$$C_3 q_3 = C_4 q_4 + C_5 q_5$$

$$\left(\frac{2}{120,000}\right)(2000) = \left(\frac{20}{120,000}\right)(200) + \left(\frac{0}{120,000}\right)(1800)$$

$$4000 = 4000 \qquad \text{OK}$$

A balance around hold tank 1 gives

$$C_5 q_5 = C_6 q_6$$

$$(0)(1800) = (0)(1000)$$

$$0 = 0 \qquad \text{OK}$$

A balance around hold tank 2 gives

$$C_2 q_2 = C_6 q_6$$

$$(0)(1000) = (0)(1000)$$

$$0 = 0 \qquad \text{OK}$$

CHM.8   ADSORPTION CALCULATIONS

Adsorption processes are often used as a follow-up to chemical wastewater treatment to remove organic reaction products that cause taste, odor, color, and toxicity problems. The equilibrium relationship between adsorbents (solid materials that adsorb organic matter, e.g., activated carbon) and adsorbates (substances that are bound to the adsorbates, e.g., benzene) may be simply expressed as

$$q = Kc^n$$

where q = amount of organic matter adsorbed per amount of adsorbent

c = concentration of organic matter in water

n = experimentally determined constant

K = equilibrium distribution constant

Based upon the data given in the table below that were obtained from a laboratory sorption experiment, a college student who worked for a company as an intern was asked to evaluate the K and n values for a certain type of activated carbon to be used to remove undesirable by-products formed during chemical treatment. What are the values of K and n that the intern should have generated?

Sorption Data Collected in Laboratory Experiments

c (mg/L)	q (μg/g)
50	118
100	316
200	894
300	1,640
400	2,530
650	5,240

Solution

The equilibrium relationship given in the problem statement can be linearized by taking the log of both sides of the equation. This yields the following equation:

$$\log(q) = \log(K) + n\log(c)$$

A plot of $\log(q)$ versus $\log(c)$ yields a straight line if this relationship can be used to represent the experimental data. The slope of this line is equal to n, while the

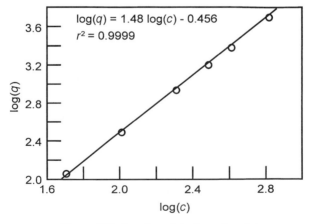

Figure 100. Log–log plot of sorption data.

intercept is $\log(K)$. The experimental data analyzed using the equation above are plotted in Figure 100, showing that the data fit this linearized isotherm quite well.
    The equation generated from a regression analysis indicates that

$$n = 1.48$$

$$\log(K) = -0.456 \quad \text{or} \quad K = 0.35$$

provided the units of c and q are mg/L and $\mu g/q$, respectively.

CHM.9   WATER SOFTENING BY ION EXCHANGE

A fixed-bed ion exchanger is to be used to soften 20,000 gal of water between regenerations. Hardness in the water averages 400 mg/L as $CaCO_3$. If the capacity to breakthrough of the exchange material is 25 kilograms (kg) (as $CaCO_3$)/ft^3, and the time interval between regenerations is 8 h, determine the dimensions of the exchanger.

Solution

The exchange capacity required is

$$C = \frac{(400)(20,000)}{(17.12)(1,000)}$$

$$= 467 \, \text{kg (as } CaCO_3)$$

The tentative volume of exchange material required is

$$V = \frac{467}{25} = 18.7 \, \text{ft}^3$$

Assuming a bed depth of 6 ft, the surface area will be

$$S = \frac{18.7}{3} = 6.23 \, \text{ft}^2$$

For a surface area of 6.23 ft^2, the superficial velocity of flow through the bed will be

$$v_s = \frac{20,000}{(6.23)(8)(7.48)(3600)} = 0.0149 \, \text{ft/s}$$

CHM.10   ULTRAVIOLET VS. CHLORINE DISINFECTION

A large wastewater treatment facility currently uses chlorine for disinfection. The average chlorine dosage is 6.0 mg/L at an average flowrate of 70 MGD. Chlorine (Cl_2) costs \$1.00/lb. Sulfur dioxide (SO_2) is used to dechlorinate the effluent before it is discharged (a requirement placed on the facility to protect the fish inhabiting the receiving water) and is consumed at an average dose of 2.0 mg/L. Its cost is \$1.20/lb.

Concerned about operating costs and the risks of Cl_2 usage, the utility has completed a pilot study showing that ultraviolet (UV) light disinfection could be used to replace Cl_2 and SO_2. The UV system will have a capital cost of \$12,000,000, will cost \$500,000 a year to operate with a 20-year service life. The utility will draw from bank savings accounts currently earning 8.0% simple interest per year to pay for the UV system's capital cost. Does the savings in operating costs justify the capital cost? Assume straight-line depreciation of the UV system, with \$0 salvage value at the end of its service life.

Additional Information

Conversion factor: A dose of $1 \text{ mg/L} = 8.34 \text{ lb}/10^6 \text{ gal}$ (MG)

Straight-line depreciation equation:

$$\text{Depreciation rate in \$/yr} = \frac{\text{Initial value} - \text{Salvage value}}{\text{Service life}}$$

Rate of return on investment (ROI) equation in %/yr:

$$\text{ROI} = \frac{\text{Scenario A operating costs} - \text{Scenario B operating costs}}{\text{Investment}}$$

where scenario A = estimated cost for continuing the status quo
     scenario B = estimated cost for the proposed alternative

Solution

Calculate the daily Cl_2 and SO_2 dosage rates as follows:

$$Cl_2 \text{ dosage rate} = (70 \text{ MGD})(6.0 \text{ mg/L})(8.34 \text{ L} \cdot \text{lb/MG} \cdot \text{mg})$$
$$= 3500 \text{ lb/day}$$

$$SO_2 \text{ dosage rate} = (70 \text{ MGD})(2.0 \text{ mg/L})(8.34 \text{ L} \cdot \text{lb/MG} \cdot \text{mg})$$
$$= 1170 \text{ lb/day}$$

These daily rates are then converted to annual rates:

$$\text{Annual Cl}_2 \text{ dosage rate} = (\text{daily Cl}_2 \text{ rate})(\text{day/yr})$$
$$= (3500 \text{ lb/day})(365 \text{ day/yr})$$
$$= 1{,}278{,}000 \text{ lb/yr}$$

$$\text{Annual SO}_2 \text{ dosage rate} = (\text{daily SO}_2 \text{ rate})(\text{day/yr})$$
$$= (1170 \text{ lb/day})(365 \text{ day/yr})$$
$$= 427{,}000 \text{ lb/yr}$$

The annual costs for Cl_2 and SO_2 addition are calculated as follows:

$$\text{Annual Cl}_2 \text{ cost} = (\text{annual Cl}_2 \text{ dosage rate})(\text{Cl}_2 \text{ cost in \$/lb})$$
$$= (1{,}278{,}000 \text{ lb/yr})(\$1.00/\text{lb})$$
$$= \$1{,}278{,}000/\text{yr}$$

$$\text{Annual SO}_2 \text{ cost} = (\text{annual SO}_2 \text{ dosage rate})(\text{SO}_2 \text{ cost in lb})$$
$$= (427{,}000 \text{ lb/yr})(\$1.20/\text{lb})$$
$$= \$512{,}000/\text{yr}$$

$$\text{Total operating cost} = \text{Annual Cl}_2 \text{ cost} + \text{Annual SO}_2 \text{ cost}$$
$$= \$1{,}278{,}000/\text{yr} + \$512{,}000/\text{yr}$$
$$= \$1{,}790{,}000/\text{yr}$$

The straight-line depreciation of the system is then calculated as follows:

$$\text{Depreciation} = (\text{initial value} - \text{salvage value})/\text{service life}$$
$$= (\$12{,}000{,}000 - \$0)/(20 \text{ yr})$$
$$= \$600{,}000/\text{yr}$$

The total annual cost of the UV system is then calculated as:

$$\text{Total annual cost} = \text{Straight-line depreciation} + \text{Operating cost}$$
$$= \$600{,}000/\text{yr} + \$500{,}000/\text{yr}$$
$$= \$1{,}100{,}000/\text{yr}$$

The annual cost savings for switching to the UV system is determined as follows:

Annual cost savings = Scenario A operating cost − Scenario B operating cost

where scenario A is the total annual Cl_2 cost and scenario B is the total annual UV cost.

$$\text{Annual cost savings} = \text{Annual } Cl_2 \text{ cost} - \text{Annual UV cost}$$
$$= \$1,790,000/\text{yr} - \$1,100,000/\text{yr}$$
$$= \$690,000/\text{yr}$$

The expected ROI on the investment in the UV system is determined as follows:

$$\text{ROI} = 100(\text{Annual cost savings})/(\text{investment})$$
$$= 100(\$690,000/\text{yr})/(2,000,000)$$
$$= 100(0.0575)$$
$$= 5.8\%$$

This is less than the 8.0% simple interest the utility could make from cash in its bank account (at the time of preparation of this problem). Thus, based on operating cost considerations alone, the utility should not invest in the UV system. However, other considerations may make UV more attractive. There may be hidden costs to continued Cl_2 and SO_2 usage, such as maintaining safety equipment and training, hazardous materials planning, worker and public concerns, stockholder support and, perhaps, capital costs to maintain or upgrade the Cl_2 and SO_2 storage and dosing systems. Additional calculations should be made that include these as well as potentially other hidden costs before a final decision is made.

35 Sludge Handling (SLU)

SLU.1 TYPES OF SLUDGE

Identify two types of sludges and three classes of sludge generated by wastewater treatment processes.

Solution

Primary settled or settleable sludge consists of thickened material that is generated in a primary sedimentation tank. Primary settled sludge is readily dewaterable and can be further thickened for wastewater volume reduction. The second type is the biological and chemical sludges produced in secondary and biological treatment processes. These sludges contain a significant level of biological mass that is more difficult to settle and dewater. A discussion of the three classes of sludge follows.

Primary Sludge   Primary sludge is generated during primary wastewater treatment, which removes those solids that settle out readily. Primary sludge contains 2–8% solids; usually thickening or other dewatering operations can easily reduce its water content.

Secondary Sludge   This sludge is often referred to as biological/process sludge because it is generated by secondary biological treatment processes, including activated sludge systems and attached growth systems such as trickling filters and RBCs (rotating biocontactors). Secondary sludge has a low solids content (0.5–2%) and is more difficult to thicken and dewater than primary sludge.

Tertiary Sludge   This is produced in advanced wastewater treatment processes, such as chemical precipitation and filtration. The characteristics of tertiary sludge depend on the wastewater treatment process that produced it. Chemical sludges result from treatment processes that add chemicals, such as lime, organic polymers, and aluminum and iron salts, to wastewater. Generally, lime or polymers improve the thickening and dewatering characteristics of a sludge, whereas iron or aluminum salts may reduce its dewatering and thickening capacity by producing very hydrous sludges that bind water. Chemicals are sometimes used in primary and secondary settling tanks for several of phosporus and poor settling floc.

SLU.2   WASTEWATER CHARACTERISTICS

Describe the wastewater characteristics of wastewater sludge.

Solution

Settled sludge is a by-product of wastewater treatment. It usually contains 95–99.5% water as well as solids and dissolved substances that were present in the wastewater or were added during wastewater treatment processing. Usually these wastewater solids are further dewatered to improve their characteristics prior to ultimate use/disposal.

The characteristics of a sludge depend on both the initial wastewater composition and the subsequent wastewater and sludge treatment processes used. Different treatment processes generate radically different types and volumes of sludge. For a given application, the characteristics of the sludge produced can vary annually, seasonally, or even daily because of variations in incoming wastewater composition and variations in the treatment processes. This variation is particularly pronounced in wastewater systems that receive a large proportion of industrial discharges.

The characteristics of a sludge can significantly affect its suitability for the various use/disposal options. Thus, when evaluating sludge use/disposal alternatives, one should first determine the amount and characteristics of its sludge and the degree of variation in these characteristics.

SLU.3   WASTEWATER SLUDGE MANAGEMENT

List and describe the three methods involved for wastewater sludge management.

Solution

Three methods are widely employed to use or dispose of wastewater sludge: land application, landfilling, and incineration. Their applicability depends on many factors, including the source, quantity and quality of wastewater sludge, geographic location of the community, hydrogeology of the region, land use, economics, public acceptance, and regulatory framework. Often, one must select and implement more than one sludge use/disposal option and must develop contingency and mitigation plans to ensure reliable capacity and operational flexibility.

Determining which of the use/disposal options is most suitable for a particular application is a multistage process. The first step is to define the needs, i.e., to determine the quantity and quality of sludge that must be handled and estimate future sludge loads based on growth projections. Alternative sludge use/disposal options that meet these needs and that comply with applicable environmental regulations must then be broadly defined. Unsuitable or noncompetitive alternatives must be weeded out in a preliminary evaluation based on readily available information. For example, seldom would a rural midwestern agricultural community elect to incinerate its wastewater sludge. Resources are then focused on a more detailed definition of the remaining alternatives and on their evaluation. Final selection of an option may require a detailed feasibility study.

One planning approach for tallying the important factors is a "system of criteria," which allows the proposed alternatives to be evaluated from different criteria including annualized costs compatibility with land, uses in close proximity to the proposed site (including public acceptance), recovery of resources (including energy), and reliability.

In this approach, the different criteria selected for evaluation are scored numerically to tally the comparative strengths and weaknesses of the various options. This overall approach is now described in the example below.

The FAT (Flesche and Theodore) Corporation has instituted a new program that is concerned with the management of solid waste. Three process/treatment options are under consideration:

A. land applications
B. landfill
C. incineration

Based on the data provided below, you have been asked—as the environmental engineer assigned to this project—to use the system of criteria method to determine which option is most attractive.

FAT Corporation has determined that annualized cost is the most important criterion, with a weight factor of 100 (from a maximum of 100). Other significant criteria include compatibility (weight of 80), recovery/reuse (weight of 70), and reliability (weight of 50). Options A, B, and C have also been assigned effectiveness factors. Option A is given a rating of 80 for annualized cost, 60 for compatibility, 40 for recovery/reuse and 20 for reliability. The corresponding effectiveness factors for B and C are 60, 30, 40, 20 and 30, 80, 50, 80, respectively.

In solving this example, first generate the overall rating (OR) for option A:

$$OR_A = (100)(80) + (80)(60) + (70)(40) + (50)(20)$$
$$= 16,600$$

Generate the overall rating for option B:

$$OR_B = (100)(60) + (80)(30) + (70)(40) + (50)(20)$$
$$= 12,200$$

Generate the overall rating for option C:

$$OR_C = (100)(30) + (80)(80) + (70)(50) + (50)(80)$$
$$= 16,900$$

From this screening, option C rates the highest with a score of 16,900. Option A's score is 16,600 and option B's score is 12,200. In this case, both option C and option A should be selected for further evaluation because their scores are high and close to each other. Consideration should also be given to combining the "best" features of options A and C.

The system of criteria is a valuable tool for demonstrating the objectivity of the planning to groups that may oppose project alternatives.

SLU.4 CALCULATIONS ON A SLUDGE DIGESTER

A wastewater sludge consisting of 4% solids and 96% water is to be digested for 30 days in a "standard-rate" digestion process. The solids initially are 70% volatile and 30% fixed, with corresponding specific gravities of 1.0 and 2.5, respectively. Based on past experience, the sludge withdrawn from the digester is expected to contain 3% solids. Estimate the required tank capacity if the digester is to treat 5000 lb/day of dry solids entering the system.

Assume the volatile solids reduction for 30 days retention time is 50%. The specific gravity (SG) of the solids can be estimated from the following equation:

$$SG = \frac{1}{x/sg_v + (1-x)/sg_f}$$

where sg_v and sg_f = specific gravities of the volatile and fixed solids
x = fraction of solids that are volatile

The volume, V, requirement for the digester can be estimated from

$$V = [V_t + \tfrac{2}{3}(V_0 - V_t)](t)$$

where V_0 and V_t are the volume occupied by a daily accretion to the sludge, initially and at the retention period, respectively.

Solution

For the undigested solids,

$$SG_U = \frac{1}{(0.7/1) + (0.3/2.5)}$$

$$= 1.220$$

For the digested solids,

$$SG_D = \frac{1}{(0.35/0.65)/1 + (0.30/0.65)/2.5}$$

$$= 1.383$$

The volumes of sludge initially and at the end of digestion are

$$V_0 = \frac{1000}{(1.220)(62.4)} + \frac{(1000)(1 - 0.04)}{(0.04)(62.4)}$$

$$= 397.2 \, \text{ft}^3$$

$$V_t = \frac{650}{(1.383)(62.4)} + \frac{(650)(1 - 0.03)}{(0.03)(62.4)}$$

$$= 344.3 \, \text{ft}^3$$

The digester volume can now be calculated:

$$V = [344.3 + (\tfrac{2}{3})(397.2 - 344.3)](30)$$

$$= 11,390 \, \text{ft}^3$$

The unit is rather large.

SLU.5   CALCULATION OF A LEACHATE RATE

Calculate how much leachate in gallons/hour would develop during the operation of the Fresh Kills landfill in Staten Island, New York. Assume the annual rainfall for the year in question is 45 in. and the affected surface area of the landfill is 1.0 square miles. Also assume that the leachate produced is 40% of the estimated rainfall.

Solution

To calculate the leachate, L, first calculate the effective height, H, of rainfill:

$$H = (0.40)(45)$$

$$= 18.0 \, \text{in./yr}$$

$$= 1.5 \, \text{ft}$$

The cross-sectional area, A, of the landfill is

$$A = 1.0\,\text{mi}^2$$
$$= (1.0)(5280)^2$$
$$= 2.79 \times 10^7\,\text{ft}^2$$

The leachate rate on a volume basis is then

$$L = (1.5)(2.79 \times 10^7\,\text{ft}^2)$$
$$= 4.18 \times 10^7\,\text{ft}^3/\text{yr}$$
$$= 3.13 \times 10^8\,\text{gal/yr}$$
$$= 8.57 \times 10^5\,\text{gal/day}$$

The leachate rate on a mass basis is

$$L = (4.18 \times 10^7)(62.4)$$
$$= 2.61 \times 10^9\,\text{lb/yr}$$
$$= 7.15 \times 10^6\,\text{lb/day}$$

Note: This result does not take into account evaporation, groundwater infiltration, or biological reactions.

SLU.6   BLENDING SLUDGE WITH LEAVES

Leaves, with a mass C/N ratio of 45, are to be blended with sludge from a wastewater treatment plant with a C/N ratio of 7.5. Determine the mass proportions of each component in order to achieve an optimum (for sludge management) blended C/N ratio in the 22–26 range. Assume that the following conditions apply:

Moisture content of solid sludge $= 65\%$
Moisture content of leaves $= 30\%$
Nitrogen content of solid sludge $= 5.4\%$
Nitrogen content of leaves $= 0.9\%$

Assume that the nonwater portion of the wastewater consists of solid sludge.

Solution

This requires two calculations. Assume

$$m_s = \text{mass of sludge}$$
$$m_l = \text{mass of leaves}$$

For the sludge

$$m_{H_2O} = 0.65m_s$$

$$\text{Solid sludge} = 0.35m_s$$
$$m_N = 0.054\,(1 - 0.65)m_s$$
$$= 0.0189m_s$$

$$m_C = (7.5)(0.054)(1 - 0.65)m_s$$
$$= 0.142m_s$$

The inerts, I, in the solid sludge is

$$m_l = (0.35 - 0.0189 - 0.142)m_s$$
$$= 0.149m_s$$

For the leaves:

$$m_{H_2O} = 0.3m_l$$
$$m_N = 0.009(1 - 0.3)m_l$$
$$= 0.0063m_l$$

$$m_C = (0.009)(45)(1 - 0.3)m_l$$
$$= 0.2835m_s$$
$$m_l = (0.7 - 0.0063 - 0.2835)m_l$$
$$= 0.410m_s$$

For C/N ratio of 22,

$$22 = \frac{0.142m_s + 0.410m_l}{0.0189m_s + 0.0063m_l}$$

Arbitrarily set $m_1 = 1$ and solve for m_s

$$m_s = 0.991$$

The mass ratio of sludge to leaves is approximately 1.0. Similarly for a ratio of 26, the mass ratio of sludge to leaves is 0.705 or approximately 0.7. Therefore, the mass ratio of sludge to leaves should be in the 0.7–1.0 range for optimum management/treatment of the sludge. Note that the leaves increase the nitrogen content of the waste, which in turn can improve the treatment by composting.

SLU.7   METHANE PRODUCTION FROM SLUDGE

Refer to Problem BIO.12 in Chapter 33. Determine the methane produced from the sludge digestion in ft^3/day and the corresponding energy value of the methane. The following data are to be employed:

Chemical oxygen demand (COD) removed/day $= 3330$ lb/day
Methane generation $= 0.162$ lb COD destroyed/ft^3 CH_4 produced
Methane recovery efficiency $= 80\%$
Heating value, methane $= 960$ Btu/ft^3
Energy value $= \$0.08$/kWh

Solution

The volumetric flowrate of methane produced, q, with 80% recovery is

$$q = (3330)(0.8)\left(\frac{1}{0.162}\right)$$

$$= 16{,}400 \ ft^3 \ CH_4/\, day$$

The total heating value of the methane, H, is

$$H = (16{,}400)(960)$$

$$= 15.7 \times 10^6 \ Btu/day$$

Convert Btu/day to kWh and multiply by $\$0.08$/kWh:

$$\text{Energy value} = (15.7 \times 10^6)\left(\frac{2.778 \times 10^{-7}}{9.84 \times 10^{-4}}\right)(0.08)$$

$$= \$368/\, day$$

SLU.8   LIME REQUIREMENT FOR SLUDGE STABILIZATION

A 10^6 gal/day (1.0 MGD) wastewater from a treatment plant contains 0.2 mg suspended solids (SS) per cubic meter of wastewater. The separated sludge from the plant consists of the SS. If 10% by weight of lime is required to stabilize the sludge treatment and 80% of the solids are captured, calculate the daily and annual lime requirements.

Solution

The sludge flowrate is

$$\dot{m}_S = (0.2)(10^6)(8.34)/(1000)$$
$$= 1668 \text{ lb/day}$$

The treated sludge is

$$\dot{m}_{TS} = (0.8)(1668)$$
$$= 1334 \text{ lb/day}$$

The lime requirement is therefore

$$\dot{m}_L = (0.1)(1334)$$
$$= 133.4 \text{ lb/day}$$

The annual requirement is

$$\dot{m}_L = (133.4)(365)$$
$$= 48,691 \text{ lb/yr}$$

36 Water Quality Analysis (WQA)

WQA.1 OVERVIEW OF WATER QUALITY

Answer the following questions.

1. Identify materials that can affect the chemical quality of a municipal waste-water.
2. What are the four types of microorganisms that can affect water quality?
3. Define *eutrophication*. What environmental factors affect the extent of eutrophication?
4. How has the use of mathematical models for predicting dispersion of compounds in water systems developed after 1900?
5. Identify three water systems, in the order of increasing complexity, that are commonly modeled for water quality analysis. What assumptions are made for each system?

Solution

1. Municipal wastewaters contain solid and dissolved substances including organic and inorganic compounds and dissolved gases, all of which affect the chemical quality of the water. Organic substances generally consist of proteins, carbohydrates, fats, and oils but may include pesticides, herbicides, phenolic compounds, polychlorinated biphenyls (PCBs), and dioxins. Inorganic substances will include chloride salts, bases, heavy metals, and inorganic chemicals. Dissolved gases often found in municipal wastewater include nitrogen, oxygen, carbon dioxide, hydrogen sulfide, ammonia, and methane.

2. The four types of microorganisms that can affect water quality are:

   Indicator bacteria

   Pathogenic bacteria

   Viruses

   Pathogenic protozoa

3. Eutrophication is the excessive growth of aquatic plants to levels that interfere with the desirable uses of the water body. Eutrophication is often associated with growth of algae and aquatic plants, water discoloration, odor discharge,

considerable oxygen reduction, and death of fish. Factors that affect the extent of eutrophication are:

Nutrient concentrations (nitrogen, phosphorous)

Solar radiation

Geometry of the water body

Water temperature

Water flow, velocity, and dispersion

Amount of existing phytoplankton

4. A chronology of the usage of models is:

   Precomputer age (before 1950): Focus concentrated on water quality. Little focus on environmental aspects of emissions.

   Transition period of 1950s: Improved data collection. Underdeveloped analysis of data.

   Early computer age: Simple mathematical models were developed.

   Current times: More complex models created and routinely used for water quality analyses.

5. Model water systems include:

   Rivers: Can be modeled in one, two and three dimensions and also with respect to time.

   Lakes: Closed system. Are subject to evaporative effects. Assumes poor mixing within the lake (temperature stratification with depth).

   Estuaries: Complex mass balance. Boundaries must be defined. Can often be defined as a steady-state condition. Assume time-averaged and distance-averaged conditions with respect to area, flow, and reaction rates.

WQA.2   WATER QUALITY ANALYSIS

Discuss the basic approach employed in water quality analysis (WQA).

Solution

The principal desirable uses of water are:

1. Water supply—municipal and industrial
2. Recreational—swimming, boating, and aesthetics
3. Fisheries—commercial and sport
4. Ecological balance

The basic objective of the field of water quality engineering is the determination of the environmental controls that must be instituted to achieve a specific environmental quality objective. The problem arises principally from the discharge of the residues of human and natural activities that result, in some way, in an interference of a desirable use of water. What constitutes a desirable use is, of course, a matter of considerable discussion and interaction between the social-political environment and the economic ability of a given region to live with or otherwise improve its water quality.

In general, the role of the water quality engineer and scientist is to analyze water quality problems by dividing the problem into its principal components:

1. Inputs, e.g., the discharge of residue into the environment from human and natural activities.
2. The reactions and physical transport (e.g., the chemical and biological transformations) and water movement that result in different levels of water quality at different locations in time in the aquatic ecosystem.
3. The output (e.g., the resulting concentration of a substance, such as dissolved oxygen or nutrients, at a particular location in the water body) during a particular time of the year or day.

The inputs are discharged into an ecological system such as a river, lake, estuary, or oceanic region. As a result of chemical, biological, and physical phenomena (such as bacterial biodegradation, chemical hydrolysis, and physical sedimentation), these inputs result in a specific concentration of the substance in the given water body. Concurrently, through various mechanisms of public hearings, legislation, and evaluation, a desirable water use is being considered or has been established for the particular region of the water body under study. Such a desirable water use is translated into public health and/or ecological standards, and such standards are then compared to the concentration of the substance resulting from the discharge of the residue. This desired versus actual comparison may result in the need for environmental engineering controls, if the actual or forecasted concentration is not equal to that desired. Environmental engineering controls are then instituted on the inputs to provide the necessary reduction to reach the desired concentration. The presentation of various environmental engineering control alternatives to reach the same objective has a central role in the decision-making process of water quality management.

There are several other issues that must be addressed to answer the basic WLA (waste load allocation) question, which is: "What is the permissible equitable discharge of residuals that will not exceed a water quality standard?" These questions include:

1. What does "permissible" mean? Is "permissible" meant in terms of maximum daily load, 7-day average load, or 30-day average load?
2. What does "exceed" mean? Does it mean "never" or 95% of the time and for all locations?
3. What are the design conditions to be used for the analysis?
4. How credible is the water quality model projection of expected responses due to the WLA, e.g., what is the "accuracy" of the model calculations and how should the level of the analysis be reflected, if at all, in the WLA?

From a water quality point of view, the basic relationship between waste load input and the resulting response is given by a mathematical model of the water system. The development and applications of such a water quality model in the specific context of a WLA involve a variety of considerations, including the specifications of parameters and model conditions.

The principal inputs can be divided into two broad categories: point sources and nonpoint sources. The point sources are those inputs that are considered to have a well-defined point of discharge, which, under most circumstances, is usually continuous. A discharge pipe or group of pipes can be located and identified with a particular discharger. The two principal point source groupings are: (a) municipal point sources that result in discharges of treated and partially treated sewage [with associated bacteria and organic matter, biochemical oxygen demand (BOD), nutrients, and toxic substances] and (b) industrial discharges that also result in the discharge of nutrients, BOD, and hazardous substances.

The principal nonpoint sources are agricultural, silviculture, atmospheric, urban and suburban runoff, and groundwater. In each case, the distinguishing feature of the nonpoint source is that the origin of the discharge is diffuse. That is, it is not possible to relate the discharge to a specific well-defined location.

Furthermore, the source may enter the given river or lake via overland runoff as in the case of agriculture or through the surface of the land and water as an atmospheric input. The urban and suburban runoff may enter the water body through a large number of smaller drainage pipes not specifically designed for the carriage of wastes but for the carriage of storm runoff. In some instances in urban runoff, the discharge may be a large pipe draining a similarly large area. Other nonpoint sources include pollution due to groundwater infiltration, drainage from abandoned mines and construction activities, and leaching from land disposal of solid wastes.

In addition to the fact that the nonpoint sources result from diffuse locations, point sources also tend to be transient in time, although not always. For example, agriculture, silviculture, and urban and suburban runoff tend to be transient, resulting from flows due to precipitation at various times of the year. Other inputs such as the atmospheric input and leaching of substances out of solid waste disposal sites are more or less continuous.

One of the more important aspects of water quality engineering is the determination of the input mass loading, e.g., the total mass of a material discharged per unit time into a specific body of water. The mass input depends on both the input flow and the input concentration and, for atmospheric inputs, includes the airborne deposition on an areawide basis.

In any given problem context, reductions in waste inputs may be required to mitigate existing water quality violations. The choice of reducing point or nonpoint sources will depend not only on the feasibility and economics of existing engineering controls but also on the relative magnitudes of the sources. Early identification of the major waste sources will allow concentration of available resources on the most significant inputs. To this end, published values of point and nonpoint sources may be used in a preliminary screening effort, although caution should be exercised because of the wide variability in such data. Data from areas similar to the study areas should be used as much as possible.

WQA.3 MODELING OF RIVERS AND STREAMS

Discuss the modeling of rivers and streams.

Solution

For anyone who has been thrilled by the excitement of canoeing on a river or quietly paddling on a stream, it is quite clear that the distinguishing feature of describing water quality in rivers and streams is the movement of the water, more or less rapidly, in a downstream direction. From a water quality engineering point of view, rivers have been studied more extensively and longer than other bodies of water, probably reflecting the fact that many people live close to or interact with rivers and streams. Hydrologically then, interest in rivers begins with the analysis of river flows.

The magnitude and duration of flows, coupled with the chemical quality of the waters, determine (to a considerable degree) the biological characteristics of the stream. The river is an extremely rich and diverse ecosystem, and any water quality analysis must recognize this diversity. The river system may therefore be considered from the physical, chemical, and biological perspective. The principal physical characteristics of rivers that are of interest include:

1. Geometry: width, depth
2. River slope, bed roughness, "tortuosity"
3. Velocity
4. Flow
5. Mixing characteristics (dispersion in the river)
6. River water temperature
7. Suspended solids and sediment transport

For river water quality management, the important chemical characteristics are:

1. Dissolved oxygen (DO) variations, including associated effects of oxidizable nitrogen on the DO regime
2. pH, acidity, alkalinity relationships in areas subjected to such discharges, for example, as drainage from abandoned mines
3. Total dissolved solids and chlorides in certain river systems, e.g., natural salt springs in the Arkansas–White–Red River basins
4. Chemicals that are potentially toxic

Biological characteristics of river systems that are of special significance in water quality studies are:

1. Bacteria, viruses, algae
2. Fish populations

3. Rooted aquatic plants
4. Biological slimes

As with all water quality analyses, the objective in river water quality engineering is to recognize and quantify, as much as possible, the various interactions between river hydrology, chemistry, and biology.

The study of river hydrology includes many factors of water movement in river systems, including precipitation, stream flow, droughts and floods, groundwater, and sediment transport. The most important aspects of river hydrology are the river flow, velocity, and geometry. Each of the characteristics are used in various ways in the water quality modeling of rivers. Measurements of river flows focus on those times when the flow is "low" due to the factor of dilution. If a discharge is running into a stream, then conditions will probably be most critical during the times when there is less water in the stream. The flow at a given point in a river will depend on:

1. Watershed characteristics such as the drainage area of the river or stream basin up to the given location
2. Geographical location of the basin
3. Slope of the river
4. Dams, reservoirs, or locks that may regulate flow
5. Flow diversions into or out of the river basin

The flow in the river can be obtained by several methods. A direct measurement of river velocity and cross-sectional area at a specific location can provide an estimate of the flow at that location and time. River velocities are measured either directly by current meters or indirectly by tracking the time for objects in the water to travel a given distance. Since the velocity of a river varies with width and depth due to frictional effects, the mean vertical velocity must be estimated.

With an estimate of the velocity at hand, a first approximation can be made to the time of travel between various points on the river. For example, the travel time, in days, to cover a given distance, in miles, can be estimated by knowing the velocity in miles per day. This relationship ignores dispersion or mixing in the river and any effects of "dead" zones such as deep holes or side channel coves.

With the flow and hydraulic properties of the river system defined and the estimates of these properties at hand, some approaches to describing the discharge of residual substances into rivers and streams can be examined. Such residuals may include discharges from waste treatment plants, from combined sewer overflows, or from agricultural and urban runoff.

The basic idea in describing the discharge of material into a river is to write a mass balance equation for various reaches of the river. Begin by examining the mass balance right at the point of discharge. The first key assumption is that the river is homogeneous with respect to water quality variables across the width (laterally) and with respect to depth (vertically).

Consideration of how to compute the distance from an outfall to complete mixing is a separate, more complicated topic. However, the order of magnitude of the distance from a single point source to the zone of complete mixing is obtained by knowing the average stream velocity, width, and depth.

The second key assumption to be made in this analysis of water quality in streams is that there is no mixing of water in the longitudinal downstream direction. Each element of water and its associated quality flows downstream in a unique and discrete fashion. There is no mixing of one parcel with another due to dispersion or velocity gradients. This condition is referred to as an *advective* system, a *plug flow* system, or *maximum gradient* system. A pulse discharge retains its identity, and any spreading of the pulse is assumed to be negligible. In actual streams and rivers, however, true plug flow is never really reached. Lateral and vertical velocity gradients, dead zones in the river (coves, deep holes, backwater regions) produce some mixing and retardation of the material in a stream. For many purposes, however, the nondispersive assumption is a good one.

The model discussed above (total or component) involves the application of the conservation law for mass on a rate basis. For the mass balance at the outfall, the principal equation is: Mass rate of substance upstream ($\dot{m}_{I,i}$) + mass rate added by outfall ($\dot{m}_{A,i}$) = mass rate of substance immediately downstream from outfall assuming complete mixing ($\dot{m}_{D,i}$) or

$$\dot{m}_{I,i} + \dot{m}_{A,i} = \dot{m}_{D,i}$$

A similar equation can also be written for the balance of the flows, that is, Flowrate upstream ($\dot{m}_I$) + flowrate added by outfall ($\dot{m}_A$) = flowrate immediately downstream from the outfall ($\dot{m}_D$) or

$$\dot{m}_I + \dot{m}_A = \dot{m}_D$$

The upstream conditions of flow and concentration are often known or can be measured, and typically some information is available on the effluent conditions. For example, the discharge may be a proposed industrial treatment plant, and an estimate is available of the flow and concentration of the waste substance to be expected upon completion of the plant. Interest then centers on estimating the concentration of the substance in the river at the outfall after mixing of the effluent (of the upstream) concentration. The downstream concentration is thus dependent on the upstream and downstream flows and the concentrations of the upstream and effluent inputs. If the upstream concentration of the substance is zero, then the downstream concentration is equivalent to the effluent concentration reduced by the ratio of influent flow to total river flow. This is a dilution effect. If the river flow is increased for a given mass discharge, this will result in a decreased concentration in the river due to the diluting effect of the increased river flow. These two relationships contain a considerable amount of information regarding water quality engineering controls, including reduction of the mass loading and/or upstream concentration of the substance (e.g., a toxic substance, bacteria, or BOD) at the outfall. Although the concentration at the outfall may be reduced due to increased river flow, the situation becomes more complex as one proceeds downstream from the point of discharge. Additional details are available in the classic work of Thomann and Mueller, *Principles of Surface Water Quality Modeling and Control*, Harper Collins, 1987.

WQA.4   OVERVIEW OF ENGINEERING CONTROLS

Provide an overview of the engineering controls employed in a water quality study/analysis.

Solution

There are several points at which the water quality in a system can be controlled. The initial concentration at the outfall can be controlled by:

1. Reducing the effluent concentration of the waste input by:

   Wastewater treatment

   Industrial in-plant process control and/or "housekeeping"

   Eliminating effluent constituents by pretreatment prior to discharge to municipal sewer systems or by different product manufacturing for an industry

2. Reducing the upstream concentration by upstream point and non-point-source controls

Reduction of effluent flow and/or augmentation of stream river flow reduces the initial concentration and hence may achieve water quality standards. Thus, the initial concentration can be controlled by the following:

3. Reducing the effluent volume by:

   Reduction in infiltration into municipal sewer systems

   Reduction of direct industrial discharge volumes into the sewer system

   Reduction, for industry, of waste volumes and concentrations through process modifications

   Increasing the upstream flow by low flow augmentation e.g., releases from upstream reservoir storage or from diversions from nearby bodies of water

However, these latter two controls on effluent flow and upstream river flow also may affect the downstream transport through the velocity. A reduction in concentration at the outfall due to increased dilution from low flow augmentation may result in an increase in the concentration downstream because of an increased velocity. A reverse situation may occur if the flow is reduced. The concentration profile also depends on the decay rate. Thus, a final general control point is to:

4. Increase the environmental, in-stream degradation rate of the substance.

The latter control can be accomplished by a redesign of the chemical to result in a more rapid breakdown of the chemical by the natural heterotrophic bacteria in the stream. Examples include the redesign of synthetic detergents to reduce foaming and downstream transport through increased biodegradation rate. Also, the thrust in contemporary manufacture of potentially toxic chemicals is to attempt as much as possible to increase biodegradation rates so that a chemical buildup or bioaccumulation does not occur.

The choice of the mix of the above controls involves issues of

1. The costs of the controls—locally, regionally, and nationally
2. The expected benefits of the resulting water quality in terms of water use
3. The technological bounds (e.g., available storage for low flow augmentation) on the control

WQA.5   NITROGEN DISCHARGE FROM A SEWAGE TREATMENT PLANT

A watershed has an area of 8 mi^2. Rainfall occurs on the average every 3 days at a rate of 0.06 mL/day and for a period of 5 h. Approximately 50% of the rain runoff reaches the sewers and contains an average total nitrogen concentration of 9.0 mg/L. In addition, the city wastewater treatment plant discharges 10 MGD with a total nitrogen concentration of 35 mg/L. Compare the total nitrogen discharge from runoff from the watershed with that of the city's sewage treatment plant.

Solution

First calculate the total nitrogen discharge, $\dot{m}_w$, from the treatment plant:

$$\dot{m}_w = (10)(35)(8.34)$$
$$= 2919 \text{ lb/day}$$

The volumetric flow of the runoff, q, is

$$q = (0.5)(0.06)(8)(5280)^2/(3600)(12)$$
$$= 155 \text{ ft}^3/\text{s}$$

The total nitrogen discharge, $\dot{m}_r$, from runoff is then

$$\dot{m}_r = \left(155\ \frac{\text{ft}^3}{\text{s}}\right)\left(9\ \frac{\text{mg}}{\text{L}}\right)\left(10^{-6}\ \frac{\text{L}}{\text{mg}}\right)\left(3600 \times 24\ \frac{\text{s}}{\text{day}}\right)\left(62.4\ \frac{\text{lb}}{\text{ft}^3}\right)$$
$$= 7521 \text{ lb/day}$$

During rain, the runoff is over 2.5 times that for the treatment plant.

WQA.6   SUSPENDED SOLIDS IN AN ESTUARY

The following suspended solids and daily volumetric average flows were obtained from an estuary.

Suspended Solids (mg/L)	Flowrate (cfs)
0.11	20
0.30	40
0.30	70
0.42	100
0.70	300
1.40	440
1.50	1000

Using the above data, determine the average suspended solids in the estuary during a 30-day period when the daily flowrates were 80, 200, 280 (10 days), 300 (10 days), 360, 400 (2 days), 500, 600 (2 days), 620, and 800 cfs.

Solution

Assume the suspended solids (SS) are related to the volume rate of flow of water (q) through the relationship:

$$SS = aq^b$$

A linear regression of the data leads to

$$SS = 0.0205q^{0.644}$$

where $SS = \text{mg/L}$
$q = \text{ft}^3/\text{s}$

The average SS mass load for each period is obtained by multiplying q by SS and converting to lb/day. For example, for the 10-day period with a flow of $300 \, \text{ft}^3/\text{s}$,

$$m_{SS} = (300)(0.0205)(300^{0.644})(5.4)(10)$$

$$= 13,080 \, \text{lb}$$

Similarly

q (ft^3/s)	m_{SS} (lb)
80	150
200	670
280	11,600
300	1,310
360	1,770
400	4,200
500	3,020
600	9,160
620	4,320
800	6,570
$\sum m_{SS} =$	53,540

The average daily flow of SS is therefore

$$m(\text{average}) = \frac{\sum m_{SS}}{30}$$
$$= \frac{53,540}{30}$$
$$= 1785 \text{ lb/day}$$

WQA.7   CHLORIDE CONCENTRATION CONTROL

An upstream flow of 25 cfs, with a background level of chlorides (nonreactive), of 30 mg/L is supplemented with

1. An industrial discharge (I) of 6.5 MGD with 1500 mg/L chloride
2. A municipal discharge (M) of 2 MGD with 500 mg/L chloride
3. A runoff (R) of 0.5 MGD with 100 mg/L chlorides
4. A downstream tributary (T) of 5 cfs with the same background chlorides concentration of 30 mg/L

To maintain a maximum chloride concentration of 250 mg/L at the water intake location, determine

1. The required industrial (I) reduction in flowrate, or
2. The required increase in the secondary tributary (T) flowrate, or
3. A combination of steps 1 and 2.

Solution

The system plus data are pictured in Figure 101. In Figure 101, subscripts I, M, R, T refer to industrial, municipal, runoff, and tributary, respectively.

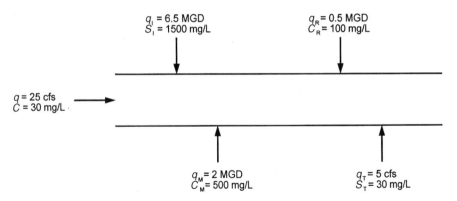

Figure 101. System and data for Problem WQA.7.

There are five input chlorine loads to the system.

$$\dot{m} = (25)(30)(5.4)$$
$$= 4050 \text{ lb/day}$$

$$\dot{m}_{\text{I}} = (6.5)(1500)(8.34)$$
$$= 81{,}315 \text{ lb/day}$$

$$\dot{m}_{\text{M}} = (2)(500)(8.34)$$
$$= 8340 \text{ lb/day}$$

$$\dot{m}_{\text{R}} = (0.5)(100)(8.34)$$
$$= 4170 \text{ lb/day}$$

$$\dot{m}_{\text{T}} = (5)(30)(5.4)$$
$$= 810 \text{ lb/day}$$

The allowable downstream chloride concentration is specified as 250 mg/L. Converting volume flows to ft^3/s leads to

$$q_{\text{I}} = (6.5)(1.547)$$
$$= 10 \text{ ft}^3/\text{s}$$

$$q_{\text{M}} = (2)(1.547)$$
$$= 3.1 \text{ ft}^3/\text{s}$$

$$q_{\text{R}} = (0.5)(1.547)$$
$$= 0.77 \text{ ft}^3/\text{s}$$

Assuming complete mixing at the downstream location, the allowable concentration in terms of the industrial loading, is given by

$$C = \frac{4050 + \dot{m}_1 + 8340 + 417 + 810}{25 + 10 + 3.1 + 0.77 + (5)(5.4)} = 250$$

Therefore,

$$\dot{m}_1 = 41{,}223 \text{ lb/day}$$

Therefore, a reduction of approximately 40,000 lb/day in the industrial loading must occur to achieve the required chloride concentration.

The increase in the tributary flow may be similarly calculated:

$$C = \frac{4050 + 81{,}315 + 8340 + 417 + (q_T)(30)(5.4)}{(25 + 10 + 310 + 0.77 + q_T)(5.4)} = 250$$

$$q_T = 275 \text{ ft}^3/\text{s}$$

As expected, the tributary flow must be significantly increased from 5 to 275 ft³/s.

Obviously, an infinite number of solutions are possible. One may simply select values of the industrial loading between 41,000 and 81,000 lb/day and proceed to calculate the corresponding (increased) value of q_T.

WQA.8   REACTING CHEMICAL CONCENTRATION

Consider the water system pictured in Figure 102. The upstream flowrate at point 1 and downstream flowrate at point 2 are 20 cfs and 28 cfs, respectively. If the decay (reaction) rate of the chemical can be described by a first-order reaction with a reaction velocity constant of 0.2 day^{-1}, determine the concentration profile of the chemical if both the upstream and infiltrating flows do not contain the chemical.

Assume the cross-sectional area available for flow is constant at 40 ft² and that the flow variation with distance can be a linear relationship. Perform the calculation for the following cases:

1. Use the inlet velocity and assume it to be constant. Neglect concentration variations arising due to the infiltration.
2. Use the average velocity (inlet, outlet) and assume it to be constant. Neglect concentration variations arising due to the infiltration.
3. Account for both velocity and concentration variation due to infiltration.

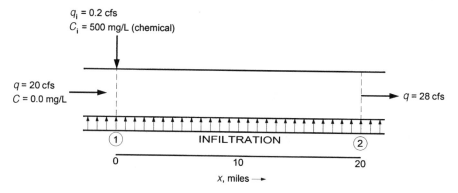

Figure 102. Water system for Problem WQA.8.

Solution

1. The inlet water velocity is

$$v = \frac{20.2}{40}$$

$$= 0.505 \,\text{ft/s}$$

$$= 8.28 \,\text{mpd (miles per day)}$$

The velocity at point 2 is

$$v = \frac{28}{40}$$

$$= 0.7 \,\text{ft/s}$$

$$= 11.5 \,\text{mpd}$$

Since the cross-sectional area for flow is constant, one may employ the "plug" flow model described in Chapter 10 with a constant inlet velocity. For this case,

$$\frac{dC}{dx} = -\frac{kC}{v}$$

with,

$$C_0 = \frac{(0.2)(500)}{20.2}$$

$$= 4.95 \,\text{mg/L}$$

Integrating the above,

$$C = C_0 e^{-k(x/v)}$$

$$= 4.95 e^{-0.2(x/8.28)}$$

$$= 4.95 e^{-0.0241x}$$

2. Using the average velocity of 9.89 mpd results in the following concentration profile equation:

$$C = 4.95 e^{-0.0202x}$$

3. Theodore (personal notes) has provided an approach to account for these two variations. Assume the increase in velocity (and volumetric flowrate) occurs linearly down the system. For this condition

$$v = 8.28(1 + 0.161x)$$

Since the volumetric flowrate correspondingly increases, a concentration reduction will occur due to its "dilution" effect. This accounting is given by dividing the concentration by the factor $(1 + 0.161x)$. The describing equation now becomes

$$\frac{dC}{dx} = -k\left(\frac{C}{1 + 0.161x}\right)\left(\frac{1}{1 + 0.161x}\right)$$

Separating the variables leads to

$$\frac{dC}{C} = -\frac{k\,dx}{(1 + 0.161x)^2}$$

Integrating gives

$$\ln \frac{C}{C_0} = k\left[-\frac{1}{(0.161)(1 + 0.161x)}\right]_0^x$$

$$C = C_0 \exp\left\{-k\left[\frac{1}{(0.161)(1 + 0.161x)} - \frac{1}{0.161}\right]\right\}$$

$$= (4.95)\exp\left(1.242 - \frac{1.242}{1 + 0.161x}\right)$$

Thomann and Mueller, *Principles of Surface Water Quality Modeling and Control*, Harper Collins, 1987, provides other approaches in solving this type of problem.

WQA.9 CALCULATION INVOLVING MCL AND RfD

The drinking water maximum contaminant level (MCL) set by the U.S. Environmental Protection Agency (EPA) for atrazine is 0.003 mg/L and its reference dose (RfD) is 3.5 mg/(kg · day). How many liters of water containing atrazine at its MCL would a person have to drink each day to exceed the RfD for this triazine herbicide?

Solution

As with most of these calculations, it is assumed that those exposed can be represented by a 70-kg individual. The volume rate, q, of drinking water at the MCL to reach the RfD for atrazine is (using the standard RfD equation)

$$q = [3.5 \, \text{mg}/(\text{kg} \cdot \text{day})](70 \, \text{kg})/(0.003 \, \text{mg/L})$$
$$= 81,700 \, \text{L/day}$$

This is a surprisingly high volume of water on a daily basis. This large volume indicates that there is considerable uncertainty (i.e., the product of the uncertainty factors is large) in estimating a reference dose for atrazine. However, based on the MCL and RfD, there appears to be no problem.

WQA.10 CANCER RISK CALCULATION

The odor perception threshold for benzene in water is 2 mg/L. The benzene drinking water risk is $8.3 \times 10^{-7}/(\mu g \cdot L)$. Calculate the potential benzene intake rate [mg benzene/(kg · day)] and the cumulative cancer risk from drinking water with benzene concentrations at half of its odor threshold for a 30-year exposure duration.

Use the following equation for estimating the benzene ingestion rate:

$$\text{Ingestion rate } [\text{mg}/(\text{kg} \cdot \text{day})] = \frac{(C)(I)(\text{EF})(\text{ED})}{(\text{BW})(\text{AT})}$$

where C = concentration, mg/L
I = water ingestion rate, 2 L/day
EF = exposure frequency, 350 day/yr
ED = exposure duration
BW = body weight, 70 kg
AT = averaging time, 70 yr

Solution

Using the ingestion rate equation given in the problem statement, the benzene ingestion rate (IR) can be determined:

$$IR = \frac{(1\,mg/L)(2\,L/day)(30\ yr)}{(70\,kg)(70\ yr)}$$
$$= 0.012\,mg/(kg \cdot day)$$

The cancer risk (CR) from this ingestion exposure at half of the odor threshold of 1 mg/L is calculated using the benzene unit risk of $8.3 \times 10^{-7}/(\mu g \cdot L)$

$$CR = (1000\ \mu g/L)[8.3 \times 10^{-7}/(\mu g \cdot L)]$$
$$= 8.3 \times 10^{-4}$$

A problem exists for this system since this risk is high relative to the widely accepted standard range of environmental risk of 1×10^{-6} to 1×10^{-4}.

PART VI
Pollution Prevention

Tara Fleck

37 Source Reduction (RED)

RED.1 REDUCTION TYPES

Define and discuss the differences among the following terms:

1. Pollution prevention
2. Pollution control
3. Waste minimization

Solution

1. *Pollution Prevention:* Pollution prevention refers to the reduction or prevention of pollutant generation at the source. This concept was first defined as "waste minimization," but waste minimization can refer to methods that reduce the volume of waste after it is generated. In contrast, pollution prevention implies prevention of waste before it is generated.

2. *Pollution Control:* Pollution control refers to "downstream" reduction of pollution, i.e., treatment of process streams after waste has been generated. Frequently, pollution control simply may involve the transfer pollutant from one medium to another (e.g., air pollutants to waste water).

3. *Waste Minimization:* Waste minimization was defined by the Environmental Protection Agency (EPA) in its 1986 report to Congress (EPA/530-SW-86-033) as: "The reduction, to the extent feasible, of hazardous waste that is generated or subsequently treated, stored, or disposed of. It includes any source reduction or recycling activity undertaken by a generator that results in either:

   a. The reduction of total volume or quantity of hazardous waste.
   b. The reduction of toxicity of hazardous waste.
   c. Or both (a) and (b), so long as such reduction is consistent with the goal of minimizing present and future threats to human health and the environment."

RED.2   SOURCE REDUCTION PROGRAMS

The major components of a company's source reduction program include the following:

1. Management commitment
2. Communication of the program to the rest of the company
3. Waste audits
4. Cost/benefit analysis
5. Implementation of the program
6. Follow-up

Describe each of these general principles necessary for a successful pollution prevention program involving source reduction.

Solution

1. *Management Commitment:* Gaining the approval and support of top management is vital to the success of the source reduction plan. It will be necessary to educate management about the pollution prevention program and its benefits through seminars and meetings.

2. *Communication of Program to the Rest of the Company:* Those in the best position to make suggestions as to where process improvements can be made are middle managers and employees with direct process line experience. In addition, a monetary incentive and corporate recognition of employees for practical source reduction ideas will also be effective.

3. *Waste Audits:* The company must identify the processes, the products, and the waste streams in which (hazardous) chemicals are used. Mass balances of specific (hazardous) chemicals will help to identify source reduction opportunities. Engineering interns could be very valuable in conducting such audits. An outside person can achieve significant progress in this area as well, due to the fact that he or she may be able to cut through some of the management and personnel barriers of the industry.

4. *Cost/Benefit Analysis:* Since any change or modification in the process requires additional capital, operation, and maintenance cost, a cost analysis must be included to help management make informed decisions. These factors may include cost avoidance, enhanced productivity, and decreased liability risks from the pollution prevention effort. Federal and state agencies have provided matching grants to small industries to implement source reduction programs.

5. *Implementation of Source Reduction Programs:* An impediment for the implementation of the source reduction program is a resistance to change by management and employees. People known to resist new ideas and changes

in the company must be included in the planning stages of the program. The CEO must be convinced of the merits of the program and must fully support the implementation of it.

6. *Follow-up:* Reduced energy costs, reduced raw materials, and reduced waste disposal fees must be tracked and communicated to the company personnel. This information will be very useful in the filing of the company's waste manifest and biennial reports on its source reduction efforts. In addition, this information will convince employees and management that pollution prevention programs, in general, make sense environmentally as well as economically.

The legislation enacted under the Clean Air Act Amendments of 1977 provided the foundation for EPA's controlled-trading program, the essential elements of which include:

1. Bubble policy [or bubble exemption under Prevention of Significant Deterioration (PSD)]. The bubble policy was under the state implementation plans (SIP) first. PSD does not allow bubbles; it only allows netting.
2. Offsets policy (under nonattainment).
3. Banking and brokerage (under nonattainment).

While these different policies vary broadly in form, their objective is essentially the same: to substitute flexible economic-incentive systems for the rigid, technology-based regulations that specify exactly how companies must comply. Although still not fully developed, these market mechanisms could make regulating easier for EPA and less burdensome and costly for industry.

Note: Additional details on pollution prevention, and source reduction in particular, are available in the text/reference book, *Pollution Prevention*, Dupont and co-workers (CRC Press, Boca Raton, FL, 2000).

RED.3   DEGREASING SOLVENT DISPOSAL

Consider a degreasing operation in a metal finishing process. Give an example of process modification that might be made to this part of the process that would represent source reduction.

Solution

Degreasing or solvent metal cleaning employs nonaqueous solvents to remove soils from the surface of metal articles that are to be electroplated, painted, repaired, inspected, assembled, or further machined. Metal work pieces are cleaned with organic solvents because water or detergent solutions exhibit a slow drying rate, electrical conductivity, high surface tension, a tendency to cause rusting, and a relatively low solubility for organic soils such as greases. A broad spectrum of organic solvents is available, such as petroleum distillates, chlorinated hydrocarbons, ketones, and alcohols. Although solvents may vary, there are basically three types of degreasers: cold cleaners, open-top vapor degreasers, and conveyorized degreasers. A description of these three degreasing processes follows.

Cold cleaners are the simplest, least expensive, and most common type of degreaser. They are used for the removal of oil-base impurities from metal parts in a batch-load procedure that can include spraying, brushing, flushing, and immersion. The cleaning solvent is generally at room temperature. Although it may be heated slightly, the solvent never reaches its boiling point. When parts are soaked to facilitate cleaning, it is not uncommon for the solvent to be agitated by pumps, compressed air, mechanical motion, or sound. There are several methods for materials handling in cold cleaning operations. Manual loading is used for simple, small-scale cleaning operations. Batch-loaded conveyorized systems are more efficient for complex, large-scale operations. Loading systems can be set to automatically lower, pause, and raise a workload. By dipping in a series of tanks, each with increasingly pure solvent or possibly a different solvent, a "cascade" cleaning system is established.

The *open-top vapor degreaser* cleans by condensing vaporized solvent on the surface of the metal parts. The soiled parts are batch loaded into the solvent vapor zone of the unit. Solvent vapors condense on the cooler surface of the metal parts until the temperature of the metal approaches the boiling point of the solvent. The condensing solvent dissolves oil and grease, washing the parts as it drips down into the tank. Sometimes the cleaning process is modified with spraying or dipping. To condense rising vapors and prevent solvent loss, the air layer or freeboard above the vapor zone is cooled by a series of condensing coils that ring the internal wall of the unit. Most vapor degreasers also have an external water jacket that cools the freeboard to prevent convection up hot degreaser walls. The freeboard protects the solvent vapor zone from disturbance caused by air movement around the equipment.

Conveyorized degreasers operate on the same principles as open-top degreasers; the only difference is the materials handling. In conveyorized cleaners, parts may be dipped but manual handling is mostly eliminated. In addition, conveyorized degreasers are almost always hooded or covered. There are many designs for conveyorized degreasers. These include monorail, cross-rod, vibra, ferris wheel, belt, and strip degreasers. Each conveying operation can be used with either cold or vaporized solvent. The first four designs listed above usually employ vaporized solvent. Conveyorized degreasers are used in a wide range of applications and are typically found in plants where there is enough production to provide a continuous stream of products to be degreased.

An example of a process modification is the substitution of a different, nontoxic or nonpolluting solvent. For example, a citric-acid-based solvent could replace a toxic solvent such as 1,1,1-trichloroethane (TCE).

RED.4   DEGREASING OPERATION

Perchloroethylene (PCE) is utilized in a degreasing operation and is lost from the process via evaporation from the degreasing tank. This degreasing process has an emission factor (estimated emission rate/unit measure of production) of 0.78 lb PCE released per lb PCE entering the degreasing operation. The PCE entering the degreaser is made up of recycled PCE from a solvent recovery operation plus a fresh PCE makeup. The solvent recovery system is 75% efficient, with the 25% reject going offsite for disposal. (Adopted from EPA 560/4-88-002, December 1987, "Eliminating Releases and Waste Treatment Efficiencies for the Toxic Chemical Release Inventory Form," Office of Pesticides and Toxic Substances, Washington DC).

1. Draw a flow diagram for the process.
2. Develop a mass balance around the degreaser.
3. Develop a mass balance around the solvent recovery system.
4. Develop a mass balance around the entire system.
5. Determine the mass of PCE emitted per pound of fresh PCE utilized.

Quantify the impact of the emission factor in the degreasing operation on the flowrates within the solvent recovery unit.

Solution

1. A flow diagram for the system is provided in Figure 103.

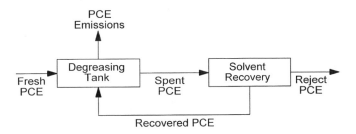

Figure 103. Flow diagram for Problem RED.4.

2. Assume a basis of 1 lb of fresh PCE feed. A PCE mass balance around the degreasing tank can now be written for the unit pictured in Figure 104.

$$\text{Input} = \text{Fresh} + \text{Recycled PCE}$$
$$= 1 \text{ lb} + X \text{ lb}$$

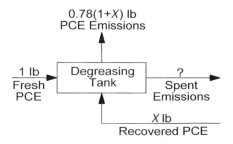

Figure 104. PCE mass balance around degreasing tank.

$$\text{Output} = \text{PCE emissions} + \text{Spent PCE}$$
$$= 0.78(1 + X) + \text{Spent PCE}$$

Equating the input with the output gives

$$\text{Spent PCE} = (1 - 0.78)(1 + X)$$
$$= 0.22(1 + X) \text{ lb PCE}$$

3. A PCE mass balance around the solvent recovery unit is shown in Figure 105.

$$\text{Input} = \text{Spent PCE}$$
$$= 0.22(1 + X) \text{ lb}$$

$$\text{Output} = \text{Recycle PCE} + \text{Reject PCE}$$

where

$$\text{Recycle PCE} = 75\% \text{ of spent PCE}$$
$$= 0.75[0.22(1 + X)] + \text{Reject PCE}$$

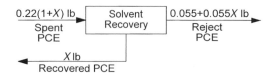

Figure 105. PCE mass balance around solvent recovery unit.

Since

$$\text{Input} = \text{Output}$$

$$0.22(1 + X) = 0.75[0.22(1 + X)] + \text{Reject PCE}$$

$$\text{Reject PCE} = (1 - 0.75)[0.22(1 + X)]$$

$$= 0.055 + 0.055X \text{ lb}$$

4. PCE mass balance around the entire system is shown in Figure 106.

$$\text{Input} = 1 \text{ lb PCE}$$

$$\text{Output} = \text{PCE emissions} + \text{Spent PCE}$$

$$= (0.78 + 0.78X) + (0.055 + 0.055X)$$

Since

$$\text{Input} = \text{Output}$$

$$1 \text{ lb PCE} = (0.78 + 0.78X) + (0.055 + 0.055X)$$

$$= 0.835 + 0.835X$$

$$X = 0.165/0.835$$

$$= 0.198 \approx 0.20 \text{ lb PCE}$$

5. From the flow diagram in Figure 106,

$$\text{PCE emissions} = 0.78 + 0.78X$$

$$= 0.78 + 0.78(0.20)$$

$$= 0.94 \text{ lb PCE emitted per lb fresh PCE}$$

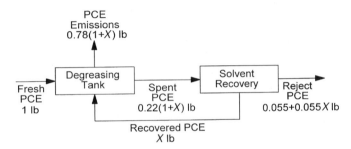

Figure 106. PCE mass balance around entire system.

If the emission factor were lower, the flowrates to the solvent recovery unit and the recycle stream would be higher. Additionally, there would be less PCE lost from the system. To determine the effect of the emissions factor on the system flow streams, the equations above were solved using three different emission factors: 0.78, 0.60, and 0.40. These results are summarized below.

Emission Factor	0.78	0.60	0.40
Fresh PCE	1.0	1.0	1.0
Recovered PCE	0.198	0.429	0.818
Spent PCE	0.263	0.571	1.091
PCE emissions	0.934	0.857	0.727
Reject PCE	0.066	0.142	0.273

The sum of the recovered and fresh PCE provides a measure of the degreasing capability of the system per kilogram feed. Notice that, as the emission factors decrease, this sum goes up significantly.

RED.5   EMISSION FACTORS

Surface coating entails the deposition of a solid film on a surface through the application of a coating material such as paint, lacquer, or varnish. Surface-coating operations are significant volatile organic compound (VOC) emission sources. Most coatings contain VOCs, which evaporate during the coating application and curing processes, rather than becoming part of the dry film.

The use of the following expressions will help solve many of the coating calculations.

1. Emission factor (EF) in reasonable available control technology (RACT) units of lb VOC/gal H_2O.

$$EF = \frac{vs\rho_v}{1 - ws}$$

where EF = emission factor, lb VOC/gal H_2O
   v = volume fraction of organic volatiles in solvent
   w = volume fraction of H_2O in solvent
   s = volume fraction of solvent in the paint
   ρ_v = density of organic volatiles, lb/gal

2. Emission factor on a solids basis (lb VOC/gal solids).

$$EF' = vs\rho_v/(1 - s)$$

where EF' is the emission factor, lb VOC/gal solids.

3. Percent emission reduction:

$$\% \text{ reduction} = [(EF'_{orig} - EF'_{rep})/EF'_{orig}]100$$

where EF'_{orig} = emission factor on a solids basis for original coating
EF'_{rep} = emission factor on a solids basis for replacement coating

Using the equations above, determine the emission factors EF and EF' for an organic solvent-borne coating that contains 40% organic solvent having a density of 7.36 lb/gal. Also determine the emission factors EF and EF' for a waterborne coating containing 65% solvent with 80% of the solvent being water. The density of the organic portion of the solvent is 7.36 lb/gal. Finally, calculate the percent reduction in volatile organic emissions achieved by switching from the solvent-borne to the waterborne coating.

Solution

The emission factor, EF, for the organic solvent-borne coating in lb/gal H_2O is first calculated:

$$EF = vs\rho_v/(1 - ws)$$
$$= (1)(0.40)(7.36)(1 - 0)$$
$$= 2.94 \text{ lb/gal } H_2O$$

The emission factor, EF', for the organic solvent-borne coating in lb/gal solids is

$$EF' = vs\rho_v/(1 - s)$$
$$= (1)(0.40)(7.36)(1 - 0.40)$$
$$= 4.91 \text{ lb/gal solids}$$

Similarly, the emission factor, EF, for the waterborne coating in lb/gal H_2O is

$$EF = (0.2)(0.65)(7.36)/[1 - (0.80)(0.65)]$$
$$= 1.99 \text{ lb/gal } H_2O$$

and the emission factor, EF', for the waterborne coating in lb/gal solids is

$$EF' = (0.20)(0.65)(7.36)/(1 - 0.65)$$
$$= 2.73 \text{ lb/gal solids}$$

The percent reduction achieved in VOC emissions by switching from the solvent-borne to the waterborne coating is then

$$\% \text{ reduction} = 100(EF'_{orig} - EF'_{rep})/EF'_{orig}$$
$$= 100(4.91 - 2.73)/4.91$$
$$= 44.4\%$$

The reader should note that the same volume of coating solids must be deposited on an object to coat it to a desired film thickness regardless of the type of coating or volatile organic compound content of the coating used. Solids make the film. Volatiles (VOC, water, and nonphotochemically reactive solvents) evaporate. For example, 4 gal of a 25 vol % solids coating must be used to get 1 gal of coating solids. However, only 2 gal of a 50 vol % solids coating must be used to get 1 gal of coating solids. This means that twice as much "work" can be done with a gallon of 50 vol % solids coating than with a gallon of 25 vol % solids coating, i.e., twice as many gallons of 25% solids coating are needed than gallons of 50% solids coating to do the same job.

RED.6   FURNITURE COATING

Metal furniture coating involves the application of prime and top coatings to any piece of metal furniture or any metal part that will be assembled with other metal, wood, fabric, plastic, or glass parts to form a furniture piece. Most metal furniture is finished with a single coat. However, some furniture pieces require a prime coat application. The prime coat is applied by electrostatic spraying, conventional spraying, dipping, or flowcoating techniques. The topcoat of a single coat is also applied by spraying, dipping, or flow-coating techniques.

EPA's *Control Technology Guide* (CTG) document recommends a single emission limit for metal furniture coating applications. This limit is based on the use of low organic solvent-borne coatings. These coatings include waterborne, high solids, electrodeposition, and powder coatings. The limit can also be met by the installation of add-on control equipment. To comply with the limit, 80% of the VOC solvent vapor emitted from the application of conventional coatings must be destroyed (using an afterburner) or recovered (using a carbon adsorption unit).

An existing metal furniture coating operation uses a coating with a VOC content of 4.5 lb/gal of coating minus water (lb/gal H_2O). The process utilizes a new minibell automatic electrostatic spray system recently purchased as part of a pollution prevention program that has a transfer efficiency of 80%. The allowable VOC content of the coating is 3.0 lb VOC/gal H_2O. The reasonable available control technology (RACT) transfer efficiency is 30%. Does the operation meet the equivalency requirements and is it operating in compliance with air quality regulations? If in compliance, how much credit could be given for a bubble or offset? The density of the VOC (solvent) is 7.36 lb/gal.

Solution

Calculate the lb VOC/gal solids for the coating presently in use employing the equation provided in Problem RED.5.

$$EF' = 4.5/[1 - (4.5/7.36)]$$
$$= 11.6 \text{ lb VOC/gal solids}$$

The lb VOC/gal solids for the RACT coating is similarly calculated:

$$EF' = 3.0/[1 - (3.0/7.36)$$
$$= 5.06 \text{ lb VOC/gal solids}$$

The lb VOC/gal solids applied for the coating presently in use is

$$VOC = EF'/0.8$$
$$= 11.6/0.8$$
$$= 14.5 \text{ lb VOC/gal solids applied}$$

The lb VOC/gal solids applied for the RACT coating is

$$VOC = EF'/0.3$$
$$= 5.06/0.3$$
$$= 16.9 \text{ lb VOC/gal solids applied}$$

Thus, the coating line is in compliance.
The credit available is

$$\text{Credit} = \text{RACT coating} - \text{Coating in use}$$
$$= 16.9 - 14.5$$
$$= 2.4 \text{ lb VOC/gal solids applied}$$

When spray guns are used to apply coatings, much of the coating material either bounces off the surface being coated or misses it altogether. Transfer efficiency (TE) is the ratio of the amount of coating solids deposited on the coated part to the amount of coating solids used. Regardless of the TE, all of the VOCs in the dispensed coating are emitted whether or not the coating actually reaches and adheres to the surface. Consequently, improved TE can reduce VOC emissions because less coating material is used.

RED.7 AUTOMOBILE APPLICATIONS

An automobile coating facility utilizes a prime coat and topcoat operation. The prime coat has a VOC content of 2.75 lb VOC/(gal H_2O) at 100% transfer efficiency using a dip coating operation. The top coat has a VOC content of 4.25 lb VOC/(gal H_2O), is equipped with minibell electrostatic sprays, and operates with a transfer efficiency of 80%. The topcoat has an allowable VOC content of 2.84 lb VOC/(gal H_2O). The state implementation plans (SIP) assumed that the transfer efficiency is 30%. The prime coat has an allowable VOC content of 1.92 lb VOC/(gal H_2O). The prime coat transfer efficiency is 100%. The plant processes 50 cars per hour with a paint usage of 1.5 gal/car for prime coat and 1.0 gal/car for topcoat. Since the prime coat is not in compliance, the company has requested a bubble determination. As a regulatory official, you are asked to determine if the company has a viable bubble. Assume the density of the VOC is 7.36 lb/gal; in addition, the paint contains no water. Also indicate source reduction measures that can be taken to reduce emissions.

Solution

The actual emissions for the prime coat, e_p, in lb VOC/h is

$$e_p = \frac{2.75 \text{ lb VOC}}{\text{gal } H_2O} (50 \text{ car/h})(1.5 \text{ gal/car})$$
$$= 206.3 \text{ lb/h}$$

Similarly, the actual emissions for the topcoat, e_t, in lb VOC/h is

$$e_t = \frac{4.25 \text{ lb VOC}}{\text{gal } H_2O} (50 \text{ car/h})(1.0 \text{ gal/car})$$
$$= 212.5 \text{ lb/h}$$

The total actual emissions becomes

$$e_{tot} = e_p + e_t$$
$$= 206.3 + 212.5$$
$$= 418.8 \text{ lb/h}$$

The emission factor on a solids basis, EF', for the prime coat (p) is

$$EF'_p = 1.92/[1 - (1.92/7.36)]$$
$$= 2.60 \text{ lb VOC/gal solids}$$

Similarly, the emission factor on a solids basis for the topcoat (t) is

$$EF'_t = 2.84/[1 - (2.84/7.36)]$$
$$= 4.62 \text{ lb VOC/gal solids}$$

The fraction of solids in each coating, f, may now be calculated:

$$f_p = 1 - 2.75/7.36$$
$$= 0.626$$

$$f_t = 1 - 4.25/7.36$$
$$= 0.423$$

The allowable emissions, E, in lb VOC/h for the prime coat is

$$E_p = (EF'_p)(f_p)(1.5 \text{ gal/car})(50 \text{ car/h})/1.0$$
$$= (2.60)(0.626)(75)$$
$$= 122.0 \text{ lb/h}$$

Similarly, the allowable emissions for the topcoat is

$$E_t = (EF'_t)(f_t)(1.0 \text{ gal/car})(50 \text{ car/h})(0.8)/0.3$$
$$= (4.62)(0.423)(133.3)$$
$$= 260.5 \text{ lb/h}$$

Finally, the total allowable emissions in lb VOC/h becomes

$$E = E_p + E_t$$
$$= 122.0 + 260.5$$
$$= 382.5 \text{ lb/h}$$

Since 418.72 is greater than 382.5, the company does not have a viable bubble.

The bubble concept was formally proposed as EPA policy on January 18, 1979, with the final policy statement being issued on December 11, 1979. The bubble policy allows a company to find the most efficient way to control the emissions of a plant as a whole rather than by meeting individual point-source requirements. If it is found less expensive to tighten controls of a pollutant at one point and relax controls at another, this would be possible as long as the total pollution from the plant would not exceed the sum of the current limits on individual point sources of pollution in the plant. Properly applied, this approach would promote greater economic efficiency and increased technological innovation.

There are some restrictions, however, in applying the bubble concept:

1. The bubble may only be used for pollutants in an area where the state implementation plan has an approved schedule to meet air quality standards for that pollutant.
2. The alternatives used must ensure that air quality standards will be met.
3. Emissions must be quantifiable, and trades among them must be even. Each emission point must have a specific emission limit, and that limit must be tied to enforceable testing techniques.
4. Only pollutants of the same type may be traded, that is, particulates for particulates, hydrocarbons for hydrocarbons, etc.
5. Control of hazardous pollutants cannot be relaxed through trades with less toxic pollutants.
6. Development of the bubble plan cannot delay enforcement of federal and state requirements.

Some additional considerations must be noted:

1. The bubble may cover more than one plant within the same area.
2. In some circumstances, states may consider trading open dust emissions for particulates (although EPA warns that this type of trading will be difficult).
3. EPA may approve compliance-date extensions in special cases. For example, a source may obtain a delay in a compliance schedule to install a scrubber if such a delay would have been permissible without the bubble.

Source reduction measures that can be taken to reduce emissions include using coatings with lower VOC content.

RED.8 EMISSIONS CALCULATION—VOLUME BASIS

A major chemical company has recently developed a new paint for can coating that contains a replacement organic solvent (MW = 136) with a density of 9.24 lb/gal. The solids are 63% of the coating material by volume. Show that this paint cannot meet RACT emission limits of 2.79 lb VOC/gal of coating material less water (gal H_2O).

Although the paint does not meet RACT requirements, the company argues that this new coating material is replacing a paint that is emitting a higher volume (ppm) of VOCs to the atmosphere. As a regulatory official, you are requested to act on their request for a variance and comment on their analysis from a pollution prevention perspective. The density and molecular weight of the previous solvent were 7.36 lb/gal and 68, respectively.

Solution

The emission of the new coating material (CH) in lb VOC per gallon of coating material is

$$EF_{new} = \frac{vs\rho_v}{1 - ws}$$

$$= \frac{1(1 - 0.63)(9.24)}{1 - 0}$$

$$= 3.42 \text{ lb VOC/gal CM}$$

Thus, the emission of the new coating material is less than the RACT emission limit of 2.79 lb VOC/gal CM.

The emission of the previous solvent (pre) in lb VOC/gal CM is

$$EF_{pre} = (1 - 0.63)(7.36)$$

$$= 2.72 \text{ lb VOC/gal CM}$$

Thus, the previous solvent was in compliance.

The ratio of the volumes emitted may now be calculated:

$$V_{new}/V_{pre} = \frac{\rho_{new}/MW_{new}}{\rho_{pre}/MW_{pre}}$$

$$= \frac{9.24/136}{7.36/68}$$

$$= 0.628$$

The volume emitted by the new coating material is lower since the above ratio is less than 1. This simply means that less of the new coating solvent is being emitted on a volume basis.

The example presented above is a real-world application drawn from the files of a major chemical company located in the Northeast.

RED.9   ASPHALT EMULSION

An automobile manufacturer that has a past history of applying sound pollution prevention procedures to its processes has proposed to construct a plant in a nonattainment area in Pennsylvania. The required offset is to be obtained by replacing emulsion asphalt by a waterborne mix. The area has traditionally used 200,000 lb/month of medium setting emulsion asphalt. Calculate the maximum annual offset (20%) if this conversion is adopted.

State of Pennsylvania regulations for cutback asphalt are given below. Note that a 20% offset indicates that the ratio of actual emission reductions to new emissions is equal to or greater than 1.2 to 1.

VOC regulations vary from state to state. The following is an excerpt from the regulatory provision provided by the state of Pennsylvania regarding VOC usage in cutback asphalt paving:

1. After April 30, 1980, and before May 1, 1982, no person may cause, allow, or permit the mixing, storage, use, or application of cutback asphalt for paving operations except when:

   a. long-life stockpile storage is necessary;

   b. the use or application between October 31 and April 30 is necessary; or

   c. the cutback asphalt is to be used solely as a penetrating prime coat, a dust palliative, a tack coat, a precoating of aggregate, or a protective coating for concrete.

2. After April 30, 1982, no persons may cause, allow, or permit the use or application of cutback asphalt for paving operations except when:

   a. long-life stockpiles are necessary;

   b. the use or application between October 31 and April 30 is necessary; or

   c. the cutback asphalt is used solely as a tack coat, a penetrating prime coat, a dust palliative, or precoating of aggregate.

3. After April 30, 1982 emulsion asphalts may not contain more than the maximum percentage of solvent as shown in the following table:

Emulsion Grade	Type	% Solvent, Max.
E-1	Rapid Setting	0
E-2	Rapid Setting (Anionic)	0
E-3	Rapid Setting (Cationic)	3
E-4	Medium Setting	12
E-5	Medium Setting	12
E-6	Slow Setting (Soft Residue)	0
E-8	Slow Setting (Hard Residue)	0
E-10	Medium Setting (High Float)	7
E-11	High Float	7
E-12	Medium Setting (Cationic)	8

Source: The provisions of Section 129.64 amended April 21, 1981, effective June 20, 1981, 11 Pa. B. 2118.

Solution

The average amount of medium setting emulsion asphalt used by the automobile manufacturer in pounds/year, $\dot{m}$, is

$$\dot{m} = (200{,}000\ \text{lb/month})(12\text{months/yr})$$

$$= 2{,}400{,}000\ \text{lb/yr}$$

The amount of solvent used per year, $\dot{m}_s$, is

$$\dot{m}_s = 0.12\dot{m}$$

$$= (0.12)(2{,}400{,}000)$$

$$= 288{,}000\ \text{lb solvent/yr}$$

It will be the responsibility of industry to suggest alternative control approaches and demonstrate satisfactorily that the proposal is equivalent in pollution reduction, enforceability, and environmental impact to existing individual process standards.

Offsets were the EPA's first application of the concept that one source could meet its environmental protection applications by getting another source to assume additional control actions. In nonattainment areas, pollution from a proposed new source, even one that controls its emissions to the lowest possible level, would aggravate existing violations of ambient air quality standards and trigger the statutory prohibition. The offsets policy provided these new sources with an alternative. The source could proceed with the construction plans, provided that:

1. The source would control emissions to the lowest achievable level.
2. Other sources owned by the applicant were in compliance or on an approved compliance schedule.
3. Existing sources were persuaded to reduce emissions by an amount at least equal to the pollution that the new source would add.

The maximum annual offset assuming 75% of the solvent is vaporized is therefore:

$$\text{Offset} = 0.75\dot{m}_s/1.2$$

$$= (0.75)(288{,}000)/1.2$$

$$= 180{,}000\ \text{lb/yr}$$

The above calculation can be rejected assuming 100% of the solvent is vaporized:

$$\text{Offset} = \dot{m}_s/1.2$$

$$= 288{,}000/1.2$$

$$= 240{,}000\ \text{lb/yr}$$

Note that cutback asphalt contains VOCs, but the waterborne mix has no VOCs. The above result indicates that, if this conversion is adopted (using the 75% vaporization adopted in the solution), a new source could operate that emits a maximum of 180,000 lb/yr.

RED.10   PROCESS AND EQUIPMENT MODIFICATIONS

A nickel electroplating line uses a dip-rinse tank to remove excess plating metals from the parts. Currently, a single tank is used that requires R gal/h of fresh rinse water to clean F parts/h (see Figure 107). Assume the cleaning is governed by the following equilibrium relation:

$$\lambda = \frac{f_i}{r_i} = \frac{\text{ounces of metal residue/part}}{\text{ounces of metal residue/gal bath}}$$

1. Calculate the reduction in rinse water flowrate (a pollution prevention measure) if a two-stage countercurrent rinse tank is used (as compared to the single-stage unit), and 99% of the residue must be removed. Assume the drag-out volume is negligible.
2. Has the total metal content of the exit rinse water been altered? Discuss implications for further wastewater treatment/reuse.

Solution

Using the flow diagram provided in Figure 107, a material balance for the residue may be written:

$$f_{in}F + r_{in}R = f_1F + r_1R$$

This equation may be rearranged in terms of λ. For the single-stage operation, set $i = 1$:

$$R/F = (f_{in} - f_1)/r_1 = \lambda(f_{in} - f_1)/f_1$$

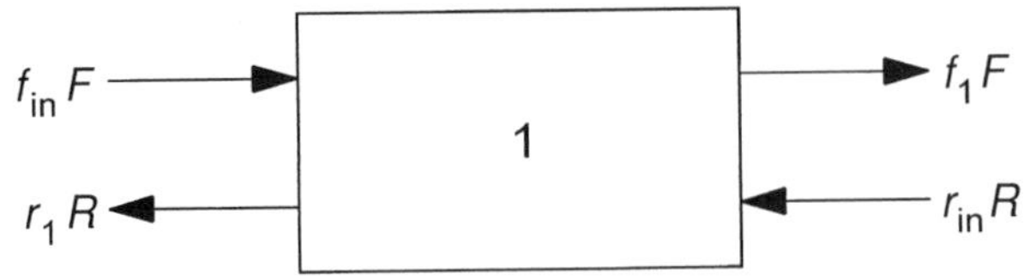

Figure 107. Flow diagram for one-stage operation.

The fraction of residue removed, x, is

$$x = (f_{in} - f_1)/f_{in}$$

Now R/F may be expressed in terms of λ and x:

$$R/F = \lambda[x/(1 - x)]$$

The flow diagram for a two-stage (stage 1, stage 2) countercurrent operation is provided in Figure 108. Material balances on the residue for each stage are

$$\text{stage 1: } f_{in}F + r_2R = f_1F + r_1R$$
$$\text{stage 2: } f_1F + r_{in}R = f_2F + r_2R$$

Each of these equations may be solved for R/F in terms of λ and f_i using the defining equation for λ:

$$\text{stage 1: } R/F = \lambda(f_{in} - f_1)/(f_1 - f_2)$$
$$\text{stage 2: } R/F = \lambda(f_1 - f_2)/f_2$$

If one substitutes x into the mass balance expressions,

$$\text{stage 1: } \frac{R}{F} = \frac{f_{in} - f_1}{f_1 - f_{in}(1 - x)}$$

$$\text{stage 2: } \frac{R}{F} = \frac{f_1 - f_{in}(1 - x)}{f_{in}(1 - x)}$$

The right-hand sides (RHS) of the stage 1 and stage 2 equations may be set equal to each other:

$$\frac{f_{in} - f_1}{f_1 - f_{in}(1 - x)} = \frac{f_1 - f_{in}(1 - x)}{f_{in}(1 - x)}$$

These equations may be rearranged to obtain a quadratic equation in f_1:

$$f_1^2 - [(1 - x)f_{in}]f_1 - [x(1 - x)f_{in}^2] = 0$$

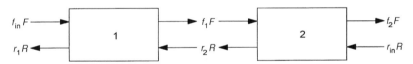

Figure 108. Flow diagram for 2-stage countercurrent operation.

Solving for f_1 yields

$$f_1 = (f_{in}(1-x) \pm \{[-f_{in}(1-x)]^2 - 4(1)x(1-x)f_{in}^2\}^{0.5})/2$$

This can be rewritten as:

$$f_1 = (f_{in}/2)\{1 - x \pm [(3x+1)(1-x)]^{0.5}\}$$

The following expression results if $(1-x)$ is factored out:

$$f_1 = \left(\frac{f_{in}(1-x)}{2}\right)\left[1 \pm \left(\frac{3x+1}{1-x}\right)^{0.5}\right]$$

Only the "+" term is physically reasonable; therefore,

$$f_1 = \left(\frac{f_{in}(1-x)}{2}\right)\left[1 + \left(\frac{3x+1}{1-x}\right)^{0.5}\right]$$

The term f_1 may be substituted into the stage 2 mass balance equation:

$$\frac{R}{F} = \lambda\left\{\left(\frac{f_{in}(1-x)}{2}\right)\left[1 + \left(\frac{3x+1}{1-x}\right)^{0.5}\right] - f_2\right\}\Big/f_2$$

However, f_2 can be expressed in terms of x and f_{in}:

$$f_2 = f_{in}(1-x)$$

This can be substituted into the above equation to yield

$$\frac{R}{F} = \frac{\lambda\left\{\left(\frac{f_{in}(1-x)}{2}\right)\left[1 + \left(\frac{3x+1}{1-x}\right)^{0.5}\right] - f_{in}(1-x)\right\}}{f_{in}(1-x)}$$

which reduces to

$$\frac{R}{F} = \left(\frac{\lambda}{2}\right)\left[\left(\frac{3x+1}{1-x}\right)^{0.5} - 1\right]$$

The rinse water requirements for both a single- and a two-stage countercurrent unit becomes.

$$\text{Single-stage:} \quad R = \lambda\left(\frac{0.99}{1-0.99}\right)F$$

$$= 99.0\,\lambda F$$

$$\text{Two-stage:} \quad R = \left(\frac{\lambda}{2}\right)\left\{\left[\frac{3(0.99)+1}{1-0.99}\right]^{0.5} - 1\right\}F$$

$$= 9.46\,\lambda F$$

The rinse water flowrate reduction is therefore

$$\frac{99.0\,\lambda F - 9.46\,\lambda F}{(99.0\,\lambda F)} = 0.904 = 90.4\%\ \text{reduction}$$

For fixed residue removal, the total mass of metals in the rinse water will be the same; with reduced water duty, the metals concentration increases. A smaller water volume now needs to be processed (or sewered). Alternatively, if the metals concentration is sufficiently high, it can be returned to the plating bath for reuse or can be recovered.

38 Recycle/Reuse (RCY)

RCY.1   RECYCLING

Describe why waste recycle is a viable option in a pollution control strategy.

Solution

Recycling or reuse can take two forms: preconsumer and postconsumer applications. Preconsumer recycling involves raw materials, products, and by-products that have not reached a consumer for an intended end use but are typically reused within an original process. Postconsumer recycled materials are those that have served their intended end use by a business, consumer, or institutional source and have been separated from municipal solid waste for the purpose of recycling.

Recycling techniques allow waste materials to be used for a beneficial purpose. A material is recycled if it is used, reused, or reclaimed. Recycling through use and/or reuse involves returning waste material either to the original process as a substitute for an input material or to another process as an input material. Recycling through reclamation is the processing of waste for recovery of a valuable material or for regeneration. Recycling of wastes can provide a very cost-effective waste management alternative. This option can help eliminate waste disposal costs, reduce raw material costs, and provide income from salable waste.

Recycling is the second most preferred option in the pollution prevention hierarchy and as such should be considered only when all source reduction options (see Chapter 37) have been investigated and implemented. Reducing the amount of waste generated at the source will often be more cost effective than recycling since waste is primarily lost raw material or product, which requires time and money to recover. It is important to note that recycling can increase a generator's risk or liability as a result of the associated handling and management of the materials involved. The measure of effectiveness in recycling is dependent upon the ability to separate any recoverable waste from other process waste that is not recoverable.

RCY.2   RECYCLING SELECTION

In the Environmental Protection Agency (EPA) hierarchy of pollution prevention, recycling is listed as a viable technique. Glass, food, and beverage containers are excellent candidates for recycling as these can be used many times. However, there are some glass items that should not be recycled.

1. Suggest five containers that should not be recycled and why.
2. Does a similar problem exist for plastics, i.e., limiting their ability and desirability to be recycled?

Solution

1. Some glass products contain metal, paint, organics, and other contaminants that render them unsuitable for food and/or beverage containers before use in food and beverage packaging, and some glass products may be amenable to current sterilization techniques. Five glass items that should not be recycled and the reasons why are summarized below.
   a. Light bulbs: Contain filaments and other metals.
   b. Dishes: Contain paint.
   c. Windshields: Composite containing metal and organics.
   d. Plate glass: May contain metals and organics.
   e. Ceramic glass: Contains inorganic salts and metals.
2. A similar problem also exists with plastics. Most plastics contain plasticizers, fillers, and solvents. Some plastics also contain residual monomers. Plastic beverage containers may also contain metals and organic compounds in the inks and dyes used for their design.

RCY.3   ONSITE AND OFFSITE RECYCLING

The use of solvent recovery has increased due to the increased utilization of industrial solvents. To prevent pollution and minimize waste, industry has to choose between onsite recycle/reuse and offsite recycle/reuse. Briefly discuss the advantages and the disadvantages of each technique.

Solution

Advantages of onsite recycle/reuse include:

Less (or no) waste leaves the facility
Owner control of purity of reclaimed solvent
Reduced reporting
Lower liability
Possible lower unit cost of reclaimed solvent

Disadvantages of onsite recycle/reuse include:

Capital investment for equipment
Workers liability: health, fires, explosion, leaks, spills and other risks
Operator training
Additional operating costs

Advantages of offsite recycle/reuse include:

Since several wastes are received at the off-site facility, this will improve the economics for processing.
Outlets for reclaimed chemicals.
Fractionation capabilities.
Technical support.
Ability to blend by-products to obtain the most cost-effective method.

Disadvantages of offsite recycle/reuse include:

No control of the fate or what is to be done with the waste.
Considerable waste leaves the facility.

Reuse involves finding a beneficial purpose for a recovered waste in a different process. Three factors to consider when determining the potential for reuse are:

1. The chemical composition of the waste and its effect on the reuse process.
2. Whether the economic value of the reused waste justifies modifying a process in order to accommodate it.
3. The extent of availability and consistency of the waste to be reused.

If an insufficient amount of waste is generated onsite to make an in-plant recovery system cost effective, or if the recovered material cannot be reused onsite, offsite recovery is preferable. Some materials commonly reprocessed offsite are oils, solvents, electroplating sludges and process baths, scrap metal, and lead-acid batteries. The cost of offsite recycling is dependent upon the purity of the waste and the market for the recovered material.

RCY.4   RECOVERY OF POTASSIUM NITRATE

Crystallization is the process of forming a solid phase from solution. It is employed heavily as a separation process in the inorganic chemistry industry, particularly where salts are recovered from aqueous media. In the production of organic chemicals, crystallization is also used to recover product, to refine intermediate chemicals, and to remove undesired salts. The feed to a crystallization system consists of a solution from which solute is crystallized through one or more of a variety of processes. The solids are normally separated from the crystallizer liquid, washed, and discharged to downstream equipment for further processing.

Potassium nitrate is obtained from an aqueous solution of 20% KNO_3. During the process, the aqueous solution of potassium nitrate is evaporated, leaving an outlet stream with a concentration of 50% KNO_3, which then enters a crystallization unit where the outlet product is 96% KNO_3 (anhydrous crystals) and 4% water. A residual aqueous solution that contains 0.55 g of KNO_3 per gram of water also leaves the crystallization unit and is mixed with the fresh solution of KNO_3 at the evaporator inlet. Such recycling provides a means of preventing pollution by minimizing the loss of valuable raw materials. The flow diagram in Figure 109 depicts the process.

What is the mass flowrate, M, in kilograms/hour of the recycled material (R) when the feedstock flowrate is 5000 kg/h?

Solution

Using Figure 109, an overall material balance on KNO_3 leads to:

$$0.96C = 0.2(5000)$$

This equation may be solved for the rate of product leaving the crystallizer:

$$C = 1000/0.96$$
$$= 1042 \, \text{kg/h}$$

An overall mass balance around the crystallizer may also be written:

$$M = R + C$$
$$= R + 1042$$

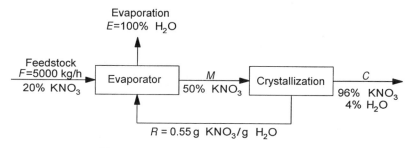

Figure 109. Flow diagram for Problem RCY.4.

Since the mass concentration of the KNO_3 in R is $(0.55/1.55)$, the equation for the KNO_3 mass balance around the crystallizer is

$$0.5M = 0.96(1042) + (0.55/1.55)R$$
$$= 1000 + 0.355R$$

These two equations may be solved simultaneously for the amount of KNO_3 leaving the evaporator in terms of the amount of recycle. Solving the second equation for M yields

$$M = 2000 + 0.710R$$

Therefore,

$$2000 + 0.710R = R + 1042$$

and the new mass flow rate of the recycle stream is

$$R = 3301 \text{ kg/h}$$

RCY.5 BYPASSING STREAMS

A liquid stream contaminated with a pollutant is being cleansed with a control device. If the liquid has 600 ppm (parts per million) of pollutant, and it is permissible to have 50 ppm of this pollutant in the discharge stream, what fraction of the liquid can bypass the control device?

Solution

Using a basis of 1 lb of liquid fed to the control device, the flow diagram in Figure 110 applies.

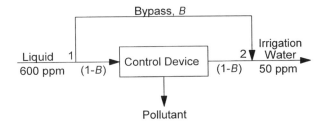

Figure 110. Flow diagram for Problem RCY.5.

Note that:

$$B = \text{fraction of liquid bypassed}$$

$$1 - B = \text{fraction of liquid treated}$$

Performing a pollutant balance around point 2 in Figure 110 yields

$$(1 - B)(0) + 600B = (50)(1.0)$$

Solving gives

$$B = 0.0833$$

Note that in some operations a process does a more complete job than is required. For example, if moist air is passed through a fresh silica gel dryer, the air will leave the system almost bone dry. If it were desirable to have air containing some moisture, one would have to reintroduce water vapor into the air. This would be a wasteful process compared to bypassing the proper amount of original moist air. In general, a finished product is made only as good as it has to be to meet competition and/or satisfy the user. "Product quality giveaway" is costly and is often minimized by bypassing. This was illustrated in the problem above.

RCY.6   UNREACTED ETHYLENE RECYCLE

Ethylene oxide is produced by the catalytic oxidation of ethylene with oxygen, $C_2H_4 + \frac{1}{2}O_2 \rightarrow (C_2H_2)_2O$. The feed to the catalyst bed is a 10:1 (volume) ratio of air to ethylene, and conversion of ethylene is 23% per pass. The ethylene oxide is removed from the reaction products selectively. The unreacted ethylene must be recycled.

What are the recycle ratio (in mol recycled/mol or total feed), and the total inlet and outlet gas compositions?

Solution

Select as a basis 1.0 lbmol of ethylene feed. The following table can then be constructed. Calculational details are left as an optional exercise.

Compound	Into Reactor					Out of Reactor				
	Moles	C	H	O	N	Moles	C	H	O	N
C_2H_4	1	2	4	—	—	0.77	1.54	3.08	—	—
N_2	7.9	—	—	—	15.8	7.9	—	—	—	15.8
O_2	2.1	—	—	4.2	—	1.99	—	—	3.97	—
CH_2CH_2O	0	—	—	—	—	0.23	0.46	0.92	0.23	—
Total	11.0	2	4	4.2	15.8	10.89	2.00	4.00	4.20	15.8

Thus,

   Ethylene recycle ratio: $0.77/1 = 0.77$
   Inlet composition: 9.09% C_2H_4, 71.8% N_2, 19.1% O_2
   Outlet composition: 7.08% C_2H_4, 72.5% N_2, 18.3% O_2, 2.11% CH_2CH_2O

   Recycle problems can become more complicated since there can be unknowns in the overall stoichiometry of the process and "local" unknowns involving the reactor and its feed streams. These problems are solved by using two material balances: one an overall balance of the material entering and leaving the reactor complex; the other a once-through, or local, balance about the reactor itself. The latter balance includes recycle streams; the former does not.

RCY.7   UNCONVERTED ETHANE RECYCLE

Ethane is converted to ethylene, a valuable monomer, via pyrolysis. The one-pass conversion efficiency of ethane ranges from 40 to 70% depending on operating conditions of a particular facility but can be easily set to a desired value by controlling the temperature of the reaction. If ethane is not recycled within the plant, it must be sent to a flare and be wasted. The flow schematic for the process is shown in Figure 111.

1. For a fresh ethane feed rate of 100,000 kg/h, evaluate the impact of one-pass ethane conversion efficiency on the plant recycle rate and hydraulic loading by carrying out a material balance on the process and tabulating recycle flows for 40, 50, 60, and 70% conversion efficiency. For each calculation assume that no ethane leaves the product recovery unit, i.e., $E_3 = 0$.
2. Ethylene yield ($Y =$ mass ethylene/mass ethane converted) is related to one-pass ethane conversion (C, %) by the following expression:

$$Y = 0.85 + (40 - C)/300$$

Given the following additional data, what one-pass conversion efficiency should be recommended and why?

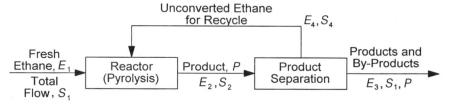

Figure 111. Flow diagram for Problem RCY.7.

Solution

For a fresh ethane feed rate of 100,000 kg/h, the impact of one-pass ethane conversion efficiency on the plant recycle rate and hydraulic loading is determined through a material balance on the process. Material balance calculations were made for recycling flows of 40, 50, 60, and 70% conversion efficiency as described below.

Material balances are performed around each node, a point where the streams enter and leave. The total flow material balance around each node is

$$S_1 + S_4 - S_2 = 0 \text{ (for the reactor)}$$

$$S_2 - S_4 + S_3 = 0 \text{ (for the recovery section)}$$

Now the material balance is conducted for the components. There are two components in this problem, ethane (E) and products (P). For any stream,

$$E + P = S$$

Thus,

$$E_1 + E_4 - E_2 = \text{ethane converted}; \ P_2 - P_1 - P_4 = 0 \text{ (for the reactor)}$$
$$E_2 - E_4 - E_3 = 0; \ P_2 = P_4 + P_3 \text{ (for the separation section)}$$

From the problem statement, it is also known that

$$S_1 = E_1 = S_3 = 100,000 \text{ kg/h}$$

and

$$S_4 = E_4$$

In addition, $E_2 = (1 - C/100)(100,000 + E_4)$ where C is the one-pass conversion efficiency.

Making these substitutions and rewriting the original four equations, the following result may be obtained:

$$S_1 + S_4 - S_2 = 0 \text{ (original)}$$
$$100,100 + E_4 - S_2 = 0 \tag{1}$$

$$E_1 + E_4 - E_2 = \text{ethane converted (original)}$$
$$E_4 = E_2 + (\text{ethane converted} - E_1)$$
$$E_4 = E_2 \tag{2}$$
$$E_4 = (1 - C/100)(100,100 + E_4)$$

$$E_2 - E_4 - E_3 = 0 \tag{3}$$

For a given conversion efficiency, C, E_4 is calculated from Equation (2). Then S_2 is determined from Equation (1), and finally E_3 is determined from Equation (3).

The solutions for 40, 50, 60, and 70% conversion efficiency are summarized in the table below.

Stream Flowrates (1000 kg/h)	Conversions			
	40%	50%	60%	70%
E_1	100	100	100	100
E_2	150	100	66.7	42.9
E_3	0	0	0	0
E_4	150	100	66.7	42.9
S_1	100	100	100	100
S_2	250	200	166.7	142.9
S_3	100	100	100	100
S_4	150	100	66.7	42.9

Note that P_1, P_2, P_3 and P_4 can be calculated from the equation

$$P_i = S_i - E_i; \quad i = 1, 2, 3, 4$$

With the ethylene yield expression given in the problem statement, the net yield and ratio of total flow from the reactor, S_2, to net ethylene yield are calculated below as a function of ethylene conversion efficiency.

Conversions	40%	50%	60%	70%
Net ethylene yield (1000 kg/h)	85.0	81.7	78.3	75.0
S_2/Net ethylene yield	2.94	2.45	2.13	1.91

Although the lowest conversion produces the least raw material waste ($P_2 = $ Net yield of ethylene), it also requires the highest flowrate of streams; this will result in higher utility costs and higher capital cost. These must be balanced as part of an economic analysis.

RCY.8   DEGREASER EMISSION REDUCTION

Metal surfaces are often cleaned using organic solvents in an open-top degreasing tank. One of the widely used solvents for such operations is 1,1,1-trichloroethane (TCE). TCE belongs to a group of highly stable chemicals known as ozone depleters. Figure 112 depicts a typical degreasing operation. The emission factor for the process shown is estimated to be 0.6 lb/lb of TCE entering the degreaser. The

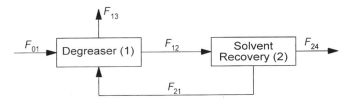

Figure 112. Schematic of a typical degreasing operation.

solvent from the degreaser is sent to a solvent recovery unit where 80% of the solvent is recovered and 20% of the solvent is disposed with the sludge.

1. To ascertain the feasibility of the installation of a vapor recovery system (see Figure 113), determine the amount of TCE vented to the atmosphere per pound of fresh TCE used.
2. If the vapor recovery system is 90% efficient, determine the fraction of trichloroethane lost to the atmosphere and the fraction going with the sludge.

Solution

Assume a basis of 1 kg for F_{01}. The mass balance equations around the two units for the process without the vapor recovery unit is

$$1 + F_{21} = F_{12} + F_{13} \tag{1}$$

$$F_{12} = F_{21} + F_{24} \tag{2}$$

The equation for the amount of TCE emissions in terms of the amount of TCE entering the degreaser is

$$F_{13} = 0.6 \, (1 + F_{21}) \tag{3}$$

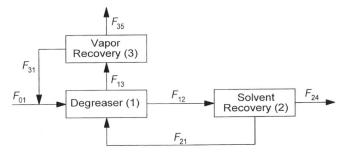

Figure 113. Schematic of degreasing operation with vapor recovery unit.

In addition, the amount of TCE recovered in terms of the amount entering solvent recovery is

$$F_{21} = 0.8F_{12} \tag{4}$$

To calculate the amount of solvent recovered, note that there are four equations [(1)–(4)] and four unknowns (F_{21}, F_{12}, F_{13}, and F_{24}). First, rearrange Equation (4) in terms of F_{12}:

$$F_{12} = 1.25F_{21} \tag{5}$$

Substituting this result into Equation (1) leads to Equation (6):

$$F_{21} = 1.25F_{21} + F_{13} - 1$$
$$0.25F_{21} + F_{13} - 1 = 0$$

Simultaneously solving Equations (3) and (6) gives the following result for F_{21}:

$$F_{21} = 0.471 \text{ kg/kg fresh solvent}$$

The amount of TCE emissions per kilogram of fresh TCE is then:

$$(0.25)(0.471) + F_{13} - 1 = 0$$
$$F_{13} = 0.882 \text{ kg/kg fresh solvent}$$

The amount of TCE in the sludge is

$$F_{12} = 1.25F_{21}$$
$$= (1.25)(0.471)$$
$$= 0.588$$

Therefore,

$$F_{24} = F_{12} - F_{21}$$
$$= 0.588 - 0.471$$
$$= 0.118 \text{ kg/kg fresh solvent}$$

Proceed to the vapor recovery unit calculation and write the mass balance equations around the three units for the process:

$$1 + F_{21} + F_{31} = F_{12} + F_{13} \tag{7}$$

$$F_{12} = F_{21} + F_{24} \tag{8}$$

$$F_{13} = F_{31} + F_{35} \tag{9}$$

An equation for the amount of TCE emissions in terms of the amount of TCE entering the degreaser can also be written:

$$F_{13} = 0.6(1 + F_{21} + F_{31}) \tag{10}$$

The equation for the amount of TCE recovered in terms of the amount entering solvent recovery is

$$F_{21} = 0.8F_{12} \tag{11}$$

Further, the equation for the amount of TCE lost to the atmosphere is

$$F_{35} = 0.1F_{13} \tag{12}$$

Equation (11) may be rearranged to give

$$F_{12} = 1.25F_{21} \tag{13}$$

This may be substituted into Equation (7):

$$1 + F_{21} + F_{31} = 1.25F_{21} + F_{13} \tag{14}$$

Arrange Equations (10) and (14) in terms of F_{31} and set them equal to each other:

$$0.25F_{21} + F_{13} - 1 = F_{13}/0.6 - 1 - F_{21}$$
$$1.25F_{21} = F_{13}$$
$$F_{21} = 0.533F_{13}$$

Now, write the equation for the amount of TCE recovered in vapor recovery in terms of the amount of emissions from the degreaser. Use Equations (9) and (12):

$$F_{13} = F_{31} + 0.1F_{13}$$
$$F_{31} = 0.9F_{13}$$

To calculate the amount of TCE emissions from the degreaser, use Equation (10):

$$F_{13} = 0.6(1 + 0.533F_{13} + 0.9F_{13})$$
$$= 4.286 \text{ kg/kg fresh solvent}$$

To calculate the amount of TCE lost to the atmosphere, note that

$$F_{35} = 0.1F_{13}$$
$$= (0.1)(4.286)$$
$$= 0.429 \text{ kg/kg fresh solvent}$$

Further, the amount of TCE lost in the sludge is

$$F_{21} = 1.6F_{13}/3$$
$$= (1.6)(4.286)/3$$
$$= 2.286 \text{ kg/kg fresh solvent}$$

The amount of solvent recovered can now be calculated:

$$F_{12} = F_{21} + F_{24}$$
$$= 1.25F_{21}$$
$$F_{24} = F_{21}(1.25 - 1)$$
$$= (0.25)(2.286)$$
$$= 0.571 \text{ kg/kg fresh solvent}$$

The amount of solvent lost to the atmosphere and the amount of fresh feed per solvent disposed of in the sludge for the system without vapor recovery is therefore

$$F_{13}/F_{24} = 0.882/0.118$$
$$= 7.5 \text{ kg lost/kg waste solvent}$$

$$F_{01}/F_{24} = 1/0.118$$
$$= 8.5 \text{ kg fresh solvent/kg waste solvent}$$

The amount of solvent lost to the atmosphere and the amount of fresh feed per solvent disposed of in the sludge for the system with vapor recovery is

$$F_{35}/F_{24} = 0.429/0.571$$

$$= 0.75 \, \text{kg lost/kg waste solvent}$$

$$F_{01}/F_{24} = 1/0.571$$

$$= 1.75 \, \text{kg fresh solvent/kg waste solvent}$$

By employing the vapor recovery system, the amount of solvent lost due to evaporation and fresh solvent requirements are considerably reduced.

39 Treatment (TRT)

TRT.1   GAS RELEASE TREATMENT

Discuss why the "doughboys" in World War I employed gas masks to prevent problems with poisonous gas releases.

Solution

The gas masks caused all breathing air being drawn in to pass through a canister filled with activated carbon, which is a highly porous granular or pelleted form of carbon. The activated carbon readily adsorbed the organic molecules of poisonous gases but did not (essentially) adsorb oxygen and nitrogen, which passed through the canister freely.

Adsorbers are one type of several control devices available for treating gaseous wastes/pollutants. A gas mask can be viewed as a bench-scale version of an industrial adsorber.

TRT.2   ODOR TREATMENT

Explain why a large open bottle of a liquid waste with a finite vapor pressure ultimately fills the room with the odor of that waste.

Solution

Through the process of mass transfer the waste evaporates from the open bottle due to its vapor pressure. Then it diffuses through the air in the room from locations of high concentrations, e.g., at the mouth of the open bottle, to locations of lower concentrations, e.g., at the far ends of the room. Diffusion will continue throughout the room. Given the sensitivity of the human nose, plus the nature of the waste evaporated, the odor of waste would then be detected throughout the room.

694

TRT.3   ONSITE AND OFFSITE TREATMENT

The use of solvent recovery has increased due to the increased utilization of industrial solvents. To prevent pollution and minimize waste, industry has often chosen between onsite and offsite treatment for pollution prevention purposes. Briefly discuss the advantages and disadvantages of both. As one might expect, the solution is, in many respects, similar to that for problem RCY. 3.

Solution

Some advantages of onsite treatment are:

1. Less (or no) waste leaves the facility.
2. Owner control of purity of reclaimed solvent.
3. Reduced reporting.
4. Lower liability.
5. Possible lower unit cost offsite treatment.

Some disadvantages of onsite treatment include:

1. Capital investment for equipment.
2. Workers liability; health, fires, explosions, leaks, spills, and other risks.
3. Operator training.
4. Additional operating costs.

The advantages of offsite treatment are:

1. Since several wastes are received at the offsite facility, the costs for processing would improve.
2. Outlets for reclaimed chemicals.
3. Fractionation capabilities.
4. Technical support.
5. Ability to blend by-products to obtain the most cost-efficient method.

The disadvantages of offsite treatment include:

1. There is no control of the fate or what is to be done with the waste.
2. Considerable amounts of waste leave the facility.
3. Transportation costs may be expensive.
4. Liability concerns associated with transport.

Treatment and disposal practices are viewed as low-priority options or simply not considered as part of the pollution prevention hierarchy. However, the four pollution prevention options—source reduction, recycling, treatment, and ultimate disposal—

are required in a total systems approach to pollution prevention, thus constituting part of the pollution prevention hierarchy.

TRT.4   TREATMENT METHODS

Discuss the advantages and disadvantages of chemical, physical, and biological treatment. Also discuss the differences among these three treatment methods.

Solution

Chemical treatment involves the waste to be treated undergoing a chemical reaction to convert it to a less hazardous substance. Examples of chemical treatment are oxidation, incineration, and neutralization. An advantage of chemical treatment is that the waste may be converted into a substance that can be released into the environment as is. Also, chemical treatment may produce a usable end product. Disadvantages include possible need for an energy source, as well as reactants. Some reactions may also require extensive process control to be effective.

Physical treatment involves transferring the waste from one form to another or concentrating the waste. Examples of physical treatment are sedimentation and air stripping. One advantage is that the processes are often quite simple. A disadvantage is that physical treatment only transfers the waste from one state to another or one concentration to another.

Biological treatment involves the use of microorganisms to break down organic wastes to less offensive materials. Examples of biological treatment include activated sludge and anaerobic digestion. Advantages of biological treatment are that the waste is fully treated (i.e., changed into another substance), and that once started, the biological growths are often self-sustaining. Disadvantages are that some substances in high concentrations (i.e., toluene) cannot be treated and can actually kill the microbiology. If there is a major loss of microbes, restarting the process can take a long period of time.

Each of these treatment methods has its own strengths and weaknesses. Often a combination can be used to totally treat a waste stream. A physical process may be used to concentrate a waste before being sent to a chemical process or may be used to transfer the waste from the air to water before being sent to an activated sludge plant. Generally, biological treatment is effective for nontoxic organic wastes or very low concentration toxic wastes. Chemical treatment works well for almost all situations but may include high capital and operating costs. Physical treatment may remove all of the waste from the main stream, but frequently this is not the case and further treatment is required.

TRT.5   GASEOUS POLLUTION TREATMENT

The four main types of gas-phase air pollution treatment control are

1. Absorption (packed or plate columns)

2. Adsorption (with activated carbon or alumina silica gel)
3. Incineration (flaring, thermal, or catalytic oxidation)
4. Condensation (contact or surface condensers)

Which of these control devices should be used to reduce H_2S emissions from the drilling of a gas well with a discharge containing a mixture of natural gas with hydrogen sulfide?

Solution

Absorption is a mass transfer operation in which a gas is dissolved in a liquid. A contaminant (pollutant exhaust stream) contacts a liquid and the contaminant diffuses (is transported) from the gas phase into the liquid phase. The absorption rate is enhanced by (1) high diffusion, (2) high solubility of the contaminant, (3) large liquid-gas contact area, and (4) good mixing between liquid and gas phases (turbulence). The liquid most often used for absorption is water because it is inexpensive, is readily available, and can dissolve a number of contaminants. Reagents can be added to the absorbing water to increase the removal efficiency of the system. Certain reagents merely increase the solubility of the contaminant in the water. Other reagents chemically react with the contaminant after it is absorbed. In reactive scrubbing the absorption rate is higher, so in some cases a smaller, economical system can be used. However, the reactions can form precipitates that could cause plugging problems in the absorber or in associated equipment.

   Adsorption is a mass transfer process in which a gaseous contaminant adheres to the surface of a solid. Adsorption can be classified as physical or chemical. In physical adsorption, a gas molecule adheres to the surface of the solid due to an imbalance of natural forces (electron distribution). In *chemisorption*, once the gas molecule adheres to the surface, it reacts chemically with it. The major distinction is that physical adsorption is readily reversible whereas chemisorption is not. All solids physically adsorb gases to some extent. Certain solids, called *adsorbents*, have a high attraction for specific gases; they also have a large surface area that provides a high capacity for gas capture. By far the most important adsorbent for air pollution control is activated carbon. Because of its unique surface properties, activated carbon will preferentially adsorb hydrocarbon vapors and odorous organic compounds from an airstream. Most other adsorbents (molecular sieves, silica gel, and activated aluminas) will preferentially adsorb water vapor, which may render them incapable of removing other contaminants. Adsorption can be a very useful removal technique since it is capable of removing very small quantities (a few parts per million) of vapor from an airstream. The vapors are not destroyed; instead they are stored on the adsorbent surface until they can be condensed and recycled or burned as an ultimate disposal technique.

   Combustion (or *incineration*) is defined as a rapid, high-temperature gas-phase oxidation. Simply, the containment (a carbon–hydrogen substance) is burned with air and converted to carbon dioxide and water vapor. The operation of any combustion source is governed by the three T's of combustion: temperature, turbulence, and time. For complete combustion to occur, each contaminant molecule must come in contact (turbulence) with oxygen at a sufficient temperature, while being maintained at this temperature for an adequate time. These three variables are dependent on each other. For example, if a higher temperature is used, less mixing of the contaminant and combustion air or shorter residence time may be required. If adequate turbulence cannot be provided, a higher temperature or longer residence time may be employed for complete combustion. Combustion devices can be categorized as flares, thermal incinerators, or catalytic incinerators.

Condensation is a process in which the volatile gases are removed from the contaminant stream and changed into a liquid. Condensation is usually achieved by reducing the temperature of a vapor mixture until the partial pressure of the condensable component equals its vapor pressure. Condensation requires low temperatures to liquefy most pure contaminant vapors. Condensation is affected by the composition of the contaminant gas stream. The presence of additional gases that do not condense at the same conditions—such as air—hinders condensation. Condensers are classified as being either contact condensers or surface condensers. Contact condensers cool the vapor by spraying liquid directly on the vapor stream. These devices resemble a simple spray scrubber. Surface condensers are normally shell-and-tube heat exchangers. Coolant flows through the tubes, while vapor is passed over and condenses on the outside of the tubes. In general, contact condensers are more flexible, simpler, and less expensive than surface condensers. However, surface condensers require much less water and produce nearly 20 times less wastewater that must be treated than do contact condensers. Surface condensers also have an advantage in that they can directly recover valuable contaminant vapors.

Based on the above brief description of the various treatment methods, it appears that only a flare or an adsorption tower would be technically feasible for H_2S control. However, none of these control devices would likely be cost effective during the drilling process. Some cost-effective precautionary measures that can be taken to reduce H_2S releases include:

1. Safety training in the hazards of H_2S inhalation
2. Provisions of safety equipment for all personnel
3. Use of safety and first-line defense blow-out prevention equipment
4. Downhole chemical treatment to neutralize or absorb the H_2S in the well bore

TRT.6   ESTIMATING INCINERATOR ADIABATIC FLAME TEMPERATURE

Incineration is one of the major types of treatment. An important parameter in incineration is the operating temperature. The operating temperature in an incinerator is a function of many variables. For most hazardous waste incinerators, the operating temperature is calculated by determining the flame temperature under adiabatic or near-adiabatic conditions. From a calculational point of view, the flame temperature has a strong dependence on the excess air requirement and the heating value of the combined waste–fuel mixture. The Theodore–Reynolds equation shown below (and presented earlier in Chapter 27, Problem HWI. 6) can be used to estimate the average temperature in the incinerator in lieu of using a rigorous model that may require extensive experimental data and physical/chemical properties.

$$T = 60 + \frac{NHV}{(0.325)[1 + (1 + EA)(7.5 \times 10^{-4})(NHV)]}$$

where T = temperature, °F
 NHV = net heating value of the inlet mixture, Btu/lb
 EA = excess air on a fractional basis

The value of EA may be estimated by

$$EA = \frac{0.95Y}{(21 - Y)}$$

where Y is the dry mol % O_2 in the combustion (incinerated) gas.

Additional details on these equations (including their derivations) can be found in the book by Santoleri, Reynolds, and Theodore titled *Introduction to Hazardous Waste Incineration* (Wiley–Interscience, New York, 2000).

Estimate the theoretical flame temperature of a hazardous waste mixture containing 25% cellulose, 35% motor oil, 15% water (vapor), and 25% inerts, by mass. Assume 5% radiant heat losses. The flue gas contains 11.8% CO_2, 13 ppm CO, and 10.4% O_2 (dry basis) by volume.

NHV of cellulose = 14,000 Btu/lb

NHV of motor oil = 25,000 Btu/lb

NHV of water = 0 Btu/lb

NHV of inerts (effective) = −1000 Btu/lb

Solution

The NHV for the mixture is obtained by multiplying the component mass fractions by their respective NHVs and taking the sum of the products. Thus,

$$NHV = \sum_{i=1}^{n} (NHV)_i w_i$$

Substituting,

$$NHV = 0.25(14{,}000\,\text{Btu/lb}) + 0.35(25{,}000\,\text{Btu/lb})$$
$$+ 0.15(0.0\,\text{Btu/lb}) + 0.25(-1000\,\text{Btu/lb})$$
$$= 12{,}000\,\text{Btu/lb}$$

The excess air employed is obtained from the equation provided in the problem statement.

$$EA = \frac{0.95Y}{21 - Y}$$
$$= \frac{0.95(10.4)}{21 - 10.4}$$
$$= 0.932$$

The flame temperature is estimated using the Theodore–Reynolds equation provided in the problem statement:

$$T = 60 + \frac{NHV}{(0.325)[1 + (1 + EA)(7.5 \times 10^{-4})(NHV)]}$$
$$= 60 + \frac{12{,}000}{(0.325)[1 + (1 + 0.932)(7.5 \times 10^{-4})(12{,}000)]}$$
$$= 2068°F$$

The reader should also note that this equation is sensitive to the value assigned to the coefficient which represents the average heat capacity of the flue gas, in this case, 0.325. This is a function of both the temperature (T) and excess air fraction (EA) and also depends on the flue products since the heat capacities of air and CO_2 are about half that of H_2O. In addition, the 7.5×10^{-4} term may vary slightly with the composition of the waste–fuel mixture incinerated. The overall relationship between operating temperature and composition is therefore rather complex, and its prediction not necessarily as straightforward as shown here.

Finally, a key design specification for any reactor, including incinerators, is the operating temperature. Materials of construction must withstand the operating temperature without experiencing any damage for safe and efficient plant operation. Also, the temperature affects the degree of destruction of waste, which is very important. In addition, the volumetric flowrate is a strong function of temperature as shown by the ideal gas law and plays an important role in properly sizing an incinerator.

TRT.7   COLD CLEANER DEGREASERS

A cold cleaner degreaser presently operates at a temperature of 20°C. At this temperature the vapor pressure of the solvent is 19 mm Hg.

1. Calculate the partial pressure of the solvent at the air–solvent interface.
2. Calculate the parts per million of the solvent in the gas at the interface.
3. A vent is employed as a treatment measure to control emissions into the workplace. If the vent air flowrate is 220 ft³/min (measured at 20°C), calculate the *maximum* emission rate of the solvent in pounds/hour.
4. Comment on the results of part 3.
5. Design a cold cleaner degreaser that has an opening of 20 ft² that contains some simple pollution prevention measures.

Solution

1. The partial pressure of the solvent, p_s, at the interface is equal to the vapor pressure, p'_s.

$$p_s = p'_s$$
$$= 19 \text{ mm Hg}$$

2. The ppm of the solvent at the interface is

$$\text{ppm}_s = (p_s/P)(10^6)$$
$$= (19/760)(10^6)$$
$$= 25{,}000 \text{ ppm}$$

This corresponds to a mole (or volume) fraction of 0.025.
3. The maximum emission rate of the solvent can now be calculated:

$$\dot{m}_s = (220 \text{ ft}^3/\text{min})(60 \text{ min/h})(0.025)(0.075 \text{ lb/ft}^3)$$
$$= 24.75 \text{ lb/h}$$

4. This is a fairly large emission rate.
5. Regarding the design, one state's regulations regarding cold cleaner degreasers follows.

Cold cleaning degreasers that have a degreaser opening that is greater than 10 ft² (0.93 m²) shall:

1. Be equipped with
   a. A cover to prevent evaporation of solvent during periods of nonuse
   b. Equipment for draining cleaned parts
   c. A permanent, conspicuous label summarizing the operating requirements
2. Be operated in accordance with the following requirements:
   a. Do not dispose of waste solvent or transfer it to another party, such that greater than 20% of the waste solvent (by weight) can evaporate into the atmosphere; store waste solvent only in covered containers;
   b. Close degreaser cover whenever not handling parts in the cleaner.
   c. Drain cleaned parts for at least 15 s or until dripping ceases.

Thus, the design must include (at a minimum):

1. A cover to prevent evaporation when the unit is not in use
2. Equipment for draining cleaned parts
3. A permanent, conspicuous label summarizing the operating requirements

Treatment with one of the gaseous control devices discussed in Problem TRT.5 should be included if emissions are a concern.

TRT.8   RECOVERY OF GASOLINE VAPORS FROM TANK TRUCKS

The *Code of Federal Regulations*, 40 CFR Part 60, Subpart xx, gives standards of performance for bulk gasoline terminals. This standard describes collection and processing of vapor displaced from tank trucks being filled. The emissions to the atmosphere are not to exceed 35 mg of total organic compounds/liter of liquid gasoline loaded. The tank transfer connections are to be vapor tight, which implies that the vapor processing system (unspecified) must meet the 35-mg limit. The tank vapor is air saturated with gasoline; the temperature may be assumed to be 75°F. Assume also that one liter of vapor will be displaced for each liter of liquid gasoline transferred. The following additional data apply:

Gasoline molecular weight at 75°F = 70 g/gmol
Vapor pressure of liquid gasoline at 75°F = 6.5 psia
Specific gravity of condensed vapor = 0.62

1. Calculate the partial pressure of gasoline vapor at the 35-mg/L limit.
2. What total pressure would be required for this emission level to be obtained by condensation at 75°F?
3. What is the fraction of gasoline recovered by the vapor recovery system that reduces the gasoline concentration in the vapor from saturation at 75°F to 35 mg/L?

4. Describe some additional pollution prevention measures that can reduce emissions.

Solution

1. First convert the vapor concentration of 35 mg/L to mole fraction and partial pressure:

$$y_3 = \frac{(35\,\text{mg/L})(22.4\,\text{L/gmol})(535/492)}{(1000\,\text{mg/g})(70\,\text{g/gmol})}$$

$$= 0.0122$$

In addition,

$$p = y_3 P$$

$$= (0.0122)(1\,\text{atm})$$

$$= 0.0122\,\text{atm}$$

2. Since the vapor pressure of gasoline at 75°F is 6.5 psia, the total pressure required for condensation to occur at 75°F is

$$P = p'/y_3$$

$$= 6.5/0.0122$$

$$= 533\,\text{psia} = 36.2\,\text{atm}$$

3. Details regarding the fraction of the gasoline recovered follow. The mole fraction of the original vapor is

$$y_1 = 6.5/14.7$$

$$= 0.442$$

Assuming as a basis 1 gmol of original vapor, a total mass balance around the tank/vapor recovery system can be written:

$$n_1 = n_2 + n_3$$

where $n_1 = 1$ gmol of original vapor
$n_2 =$ gmol of gasoline removed
$n_3 =$ gmol of residual vapor stream

A gasoline balance around the tank truck is

$$y_1 n_1 = y_2 n_2 + y_3 n_3$$

where $y_1 = 0.442 = $ mole fraction of original vapor
$y_2 = 1.0 = $ mole fraction of recovered stream
$y_3 = 0.0122 = $ mole fraction of residual stream

The above total mass and gasoline balances can be solved simultaneously for n_2 and n_3

$$1.0 = n_2 + n_3$$
$$0.442 = n_2 + 0.0122 n_3$$
$$0.442 = n_2 + 0.0122(1 - n_2)$$
$$0.442 = 0.0122 + 0.9878 n_2$$
$$n_2 = 0.435 \text{ gmol}$$

The term n_3 may now be calculated:

$$1.0 = 0.435 + n_3$$
$$n_3 = 0.565 \text{ gmol}$$

Therefore, the recovery is

$$R = y_2 n_2 / y_1 n_1$$
$$= 0.435/0.442$$
$$= 0.984 = 98.4\%$$

4. Some additional measures that can reduce emissions include:
   a. Inspect all connections for leaks.
   b. Recycle condensed gasoline vapors back to the tank.
   c. Bottom load tank trucks to reduce splashing.

A "terminal" is defined by the Environmental Protection Agency's (EPA's) *Control Technology Guideline* (CTG) document as a gasoline distribution facility that has a daily gasoline throughput greater than 76,000 L/day (20,000 gal/day). The daily gasoline throughput at a typical size terminal is 950,000 L/day (250,000 gal/day). Only the loading of gasoline tank trucks is covered by the CTG document.

Tank trucks are filled at loading racks. These facilities contain the equipment necessary to meter and to deliver the gasoline from the storage tanks to the tank trucks. The equipment includes pumps, piping, valves, fittings, meters, and loading arm assemblies. The loading rack may also be a ground-level facility for bottom loading of tank trucks.

The CTG document recommends a single volatile organic compound (VOC) emission limit for loading of gasoline tank trucks at terminals. To comply with the limit, an active vapor control system must be installed to condense, absorb, adsorb, or incinerate the VOC vapors. Control equipment to fulfill this requirement is presently commercially available. However, a good maintenance and inspection program must minimize leakage from the vapor control system and tank trucks.

40 Ultimate Disposal (ULT)

ULT.1 ULTIMATE DISPOSAL

Briefly describe ultimate disposal.

Solution

Ultimate disposal is described by many as the final process in the treatment and management of wastes. The term *ultimate disposal methods* was coined by the U.S. Environmental Protection Agency (EPA) and originally assigned to the four processes listed below.

1. Landfilling
2. Landfarming
3. Ocean dumping
4. Deep-well injection

More recently, one of the co-authors of this book (L. Theodore) has added atmospheric dispersion to this list. The top four practices are addressed in the next problem (ULT.2). Dispersion is addressed in Problem ULT.3.

The authors disagree with the nomenclature described above, preferring that *ultimate disposal* be reserved for a process in which wastes are chemically or biologically rendered innocuous or a process that has been modified so that such wastes are no longer generated. Before disposal, most wastes undergo various treatments previously described (biological, chemical, and/or physical) to concentrate, detoxify, and reduce the volume of wastes. Yet this has not always been the case. Before the early 1970s, most wastes were haphazardly landfilled or ocean dumped with little concern for the environmental effects of these practices.

ULT.2 DISPOSAL METHODS

Disposal is defined in Section 1004 of the Resource Conservation and Recovery Act (RCRA) Subtitle A as

> the storage, deposit, injection, dumping, spilling, leaking, or placing of any
> solid waste or hazardous waste into or in any land or water so that such solid

waste or hazardous waste or any constituent thereof may not enter the environment or be emitted into the air or discharged into any waters including ground waters.

The ultimate disposition of residuals from treatment processes depends on similar considerations as those connected with the selection of a particular treatment method. Specifically, disposal is based on

1. Federal, state, and local environmental regulations
2. Potential environmental hazards
3. Liabilities and risks
4. Geography
5. Demography

The four ultimate disposal methods defined by the EPA are landfarming, deep-well injection, landfilling, and ocean dumping. Landfarming, used for organic wastes, relies on nutrients in the soil to convert wastes into nonhazardous materials that may enrich the soil. The latter three methods concentrate on containing, not converting, wastes and may be used on various waste types. Briefly describe each of these four disposal methods.

Solution

Landfarming, consisting of disposing of wastes in the upper layer of the soil, has been widely used for many years in the petroleum industry for disposing of oily petroleum wastes. It is a biological treatment method in which nutrients within the soil convert the hazardous waste into nonhazardous materials.

Deep-well injection is used to transfer liquid wastes far beneath the ground. The injection wells must be placed as far away as possible from drinking water sources. The type of waste to be injected will affect the depth of the wells.

Landfilling has been a widely used technique for disposing of both inorganic and organic wastes, usually in the form of sludges, and municipal waste. Basically, landfilling involves either area fill or trench fill.

The practice of *ocean dumping* involves transporting waste material out to sea, usually by barge, and then releasing it either directly or within the containers. Despite the relative simplicity, the long-term consequences associated with ocean dumping are complex and not yet well understood.

ULT.3   OTHER DISPOSAL OPTIONS

Although the term *ultimate disposal* is a misnomer, Theodore has included another disposal option available to industry (see Problem ULT.1). This fifth disposal option is defined as *atmospheric dispersion*. Briefly describe this other option.

Solution

It is rarely possible to treat/control all pollutants; the discharge to the atmosphere from a process of one of these control units may contain trace or significant quantities of the pollutant. "Solution to pollution by dilution" offers the engineer another ultimate disposal method, particularly with odors and difficult to remove pollutants. The concentration of the pollutant on the ground that results from the dilution is primarily important in the economical and safe design of stacks. It is also used in the development of urban air pollution sources, in accidental discharges to a population complex, and in planning the location of future sources of pollutants.

Control by dilution encompasses the subjects of both atmospheric dispersion and stack design. Stacks are expensive items of construction, and there is need for a sound basis for determining stack heights and diameters. The motion in the atmosphere primarily dictates the movement and subsequent dilution of the pollutant after discharge from a stack. The stack control system must provide ground-level concentrations that will be acceptable and not seriously affect receptors (humans, livestock, vegetation, materials of construction, etc.).

ULT.4   PACKAGING POLLUTION PREVENTION

A key to identifying and applying pollution prevention principles is to understand the total life cycle of a product or material. From an ultimate disposal view, identify and discuss potential opportunities for pollution prevention for a juice pack of orange juice. Note that both the packaging and content are components of this popular product.

Solution

Life-cycle analysis (LCA) has developed over the past 20 years to provide decision makers with analytical tools that attempt to accurately and comprehensively account for the environmental consequences and benefits of competing projects. LCA is a procedure to identify and evaluate "cradle-to-grave" natural resource requirements and environmental releases associated with processes, products, packaging, and services. LCA concepts can be particularly useful in ensuring that identified pollution prevention opportunities are not causing unwanted secondary impacts by shifting burdens to other places within the life cycle of a product or process. LCA is an evolving tool undergoing continued development. Nevertheless, LCA concepts can be useful in gaining a broader understanding of the true environmental effects of current practices and of proposed pollution prevention opportunities.

It is important to note that LCA is a tool to evaluate all environmental effects of a product or process throughout its entire life cycle. This includes identifying and quantifying energy and materials used and wastes released to the environment, assessing their environmental impact, and evaluating opportunities for improvement. LCA can also be used in various ways to evaluate alternatives including in-process analysis, material selection, product evaluation, product comparison, and policy-making. The method is appropriate not only for process engineering evaluations but also for use by a wide range of facility personnel ranging from materials acquisition staff to new product design staff to staff involved in investment evaluation.

Regarding this problem, an important, but perhaps not obvious, environmental aspect of this product is that juice pack technology has deployed environmentally beneficial technology through such features as:

1. Replacement of the relatively heavy and "odd-shaped" juice bottles with lightweight, easily stacked containers, resulting in much more efficient transportation and reductions in associated environmental impacts or transport.

2. Eliminating the need for refrigeration of juices during transport and storage, thereby saving energy, the need for coolants, etc.

Factors associated with the life-cycle analysis of the orange juice include:

1. Where and how were the oranges grown? Did the orange grower employ sustainable agricultural practices, e.g., natural pesticides, appropriate fertilizers, etc.?

2. Were oranges transported and processed into juice using environmentally responsible methods?

3. Were process residues, e.g., pulp, seeds, skins, etc., beneficially used?

Factors associated with the life-cycle analysis of the packaging include:

1. Was recycled paper or virgin pulp from trees used to make the container? Were trees from an old-growth forest or tree farm used in the production process?

2. Did the paper mill operate in an environmentally responsible manner?

3. What materials were used to manufacture the inks used to print the package? Was ink produced and was printing performed with environmental responsibility?

4. Was aluminum produced from recycled aluminum, or was virgin aluminum used and produced via bauxite ore, alumina, smelting, etc.? (If the source was not recycled aluminum, the environmental impacts of aluminum production must be identified.)

5. What type of plastic was used, and where did it originate? (To be comprehensive in this analysis, one should identify the environmental impacts of plastic manufacturing beginning with the extraction of crude oil.)

ULT.5 SOLID WASTE DISPOSAL

Assume that a waxed paper cup has a mass of 10 g, consisting of 8.0 g of paper and 2.0 g of wax. Each paper cup is crushed to a volume of $50 \, cm^3$ before disposal in a landfill. How much mass (kg) and crushed volume (m^3) of paper cups will have been kept out of the landfill over the 1000-use life of a single, reusable plastic cup by an individual?

Solution

The total mass of paper cups thrown away, m, is

$$m = 10\,g/cup\ (1000\,cups) = 10,000\,g$$

Converting to kilograms,

$$m = 10,000\,g/(1000\,g/kg) = 10\,kg$$

The total crushed volume of paper cups thrown away, V, is

$$V = (50\,cm^3/cup)(1000\,cups) = 50,000\,cm^3$$

Converting to cubic meters,

$$V = 50,000\,cm^3/(1,000,000\,cm^3/m^3) = 0.050\,m^3$$

This essentially represents the volume savings in a landfill due to the multicup use by an individual. If one million individuals adjusted to this reusable process, the volume savings (VS) would be

$$VS = (10^6)(V)$$
$$= (10^6)(0.050)$$
$$= 5.0 \times 10^4\,m^3$$
$$= (50,000)(ft^3/0.28317\,cm^3)$$
$$= 1,766,000\,ft^3$$

This can obviously have a very significant effect on a landfill.

ULT.6 WASTE COMPOSITION

The national average composition (weight percent) of materials discarded in municipal solid waste (MSW) is as follows:

Paper 41%
Glass 8%
Plastic 9%
Other 7%
Yard waste 18%
Metal 9%
Food waste 8%

An environmentally conscious city in Florida has recently decided to do something about the amount of MSW the city has been dumping into its landfill. The first and logical target for waste reduction is the paper waste, which accounts for more than 41 wt% of the total waste generated. After an aggressive campaign drive, the amount of the paper going into the landfill decreased by 25%. What is the composition (weight percent) of landfill discards now?

Solution

Assume as a basis 100 lb of discards in the landfill before the paper waste reduction campaign began. The paper waste was therefore 41 lb of the original total.

Because of the campaign, the paper waste is reduced to 30.75 lb (25% decrease). Thus,

$$\text{Total weight of discards} = 30.75 + 18 + 8 + 9 + 9 + 8 + 7$$
$$= 89.75 \text{ lb}$$

The new weight percent averages are

$$\text{Paper waste} = 30.75/89.75$$
$$= 34.3 \text{ wt } \%$$

$$\text{Yard waste} = 18/89.75$$
$$= 20.1 \text{ wt } \%$$

$$\text{Glass} = 8/89.75$$
$$= 8.9 \text{ wt } \%$$

$$\text{Metals} = 9/89.75$$
$$= 10.0 \text{ wt } \%$$

$$\text{Plastic} = 9/89.75$$
$$= 10.0 \text{ wt } \%$$

$$\text{Food waste} = 8/89.75$$
$$= 8.9 \text{ wt } \%$$

$$\text{Other} = 7/89.75$$
$$= 7.8 \text{ wt } \%$$

ULT.7   WASTE IDENTIFICATION

Many plastic containers today are stamped with symbols as an aid to recycling. Identify the source of plastic listed below and give examples of containers that are usually produced from each type of material.

1. PET
2. HDPE
3. V
4. LDPE
5. PP
6. PS

Solution

Symbol	Type of Plastic	Examples of Container
PET	Polyethylene terephthalate	Beverage bottles, frozen food, boil-in-bag pouches, microwave food trays
HDPE	High-density polyethylene	Milk jugs, trash bags, detergent bottles, bleach bottles, aspirin bottles
V	Vinyl	Cooking oil bottles, meat packaging
LDPE	Low-density polyethylene	Grocery store produce bags, bread bags, food wraps, squeeze bottles
PP	Polypropylene	Yogurt containers, shampoo bottles, straws, syrup bottles, margarine tubs
PS	Polystyrene (Styrofoam)	Hot-beverage cups, fast-food clamshell containers, egg cartons, meat trays

ULT.8   ENERGY SAVING AND DISPOSAL

As part of an ultimate disposal project, an engineer has been asked to demonstrate energy savings by employing reusable plastic cups as opposed to disposable waxed paper cups. Assume that an estimated 400,000 joules (J) of energy input during the life of one paper cup is required for its manufacture from bulk paper, transport, use, and disposal. There are approximately 142,000,000 J in a gallon of oil. How much oil will have been used to make 1000 waxed paper cups?

Solution

To solve this problem, the energy expended for the paper cups, E, is first calculated:

$$E = 400,000 \, \text{J/cup} \, (1000 \, \text{cups})$$

$$= 4.0 \times 10^8 \, \text{J}$$

To calculate the quantity of oil used, E is divided by the energy content in a gallon of oil as follows:

$$V = E/\text{Energy per gallon of oil}$$

$$= 4.0 \times 10^8 \, \text{J}/(142{,}000{,}000 \, \text{J/gal})$$

$$= 2.817 \, \text{gal}$$

The total quantity of oil (TQO) is

$$\text{TQO} = (2.817 \, \text{gal})(128 \, \text{fluid ounces/gal})$$

$$= 361 \, \text{fluid ounces}$$

ULT.9   FLY ASH DISPOSAL

A 6% ash coal is the source of fossil fuel at a local utility. The average coal feed rate to the boiler is approximately 28,000 lb/h. Estimate the amount of ash that must be disposed of for the following percentages of ash in the coal that "flies," i.e., leaves the boiler with the gas.

a. 0%
b. 25%
c. 50%
d. 75%
e. 100%

The ash that does not "fly" exits the bottom of the boiler as a solid waste that must be disposed of in a controlled landfill. Assume steady-state round-the-clock operation. Comment on pollution prevention measures that can be instituted to reduce the emission of fly ash from the boiler.

Solution

The rate of ash feed, $\dot{m}_f$, to the boiler is

$$\dot{m}_f = (0.06)(28{,}000)$$

$$= 1680 \, \text{lb/h}$$

The rate of ash feed to the boiler in pounds/week is

$$\dot{m}_f = (1680 \, \text{lb/h})(24 \, \text{h/day})(7 \, \text{days/week})$$

$$= 282{,}200 \, \text{lb/week}$$

The ash collected in the boiler bottoms, $\dot{m}_0$, in pounds/week if 0% of the ash flies is

$$\dot{m}_0 = (1 - 0)(282{,}200)$$
$$= 282{,}200 \, \text{lb/week}$$

If 25% of the ash flies, the ash collected in the boiler bottoms in pounds/week is

$$\dot{m}_{25} = (1 - 0.25)(282{,}200)$$
$$= 211{,}700 \, \text{lb/week}$$

If 50% of the ash flies,

$$\dot{m}_{50} = (1 - 0.5)(282{,}200)$$
$$= 141{,}100 \, \text{lb/week}$$

If 75% of the ash flies,

$$\dot{m}_{75} = (1 - 0.75)(282{,}200)$$
$$= 70{,}600 \, \text{lb/week}$$

Finally, if 100% of the ash flies,

$$\dot{m}_{100} = (1 - 1)(282{,}200)$$
$$= 0 \, \text{lb/week}$$

The following pollution prevention measures are recommended:

1. Use an alternative fuel rather than coal.
2. Air pollution control equipment can be utilized.
3. Investigate if the ash can be used elsewhere.
4. Carefully consider ultimate disposal systems for the collected ash.

The reader is left the exercise of calculating the rate of ash discharged from the boiler in the flue gas. An electrostatic precipitator (ESP) or baghouse (fabric filter) is normally used to treat the flue gas to reduce the quantity of ash discharged to the atmosphere. Particulate and/or ash collection efficiencies rarely are below 99% by mass; most operate in excess of 99.5% efficiency.

ULT.10   DEEP-WELL INJECTION IN SALT BEDS

A large, deep cavern (formed from a salt dome) located north of Houston, Texas, has been proposed as an ultimate disposal site for both solid hazardous and municipal wastes. Preliminary geological studies indicate that there is little chance that the wastes and any corresponding leachates will penetrate the cavern walls and contaminate adjacent soil and aquifers. A risk assessment analysis was also conducted during the preliminary study and the results indicate that there was a greater than 99% probability that no hazardous and/or toxic material would "meander" beyond the cavern walls during the next 25 years.

The company preparing the permit application for the Texas Water Pollution Board has provided the following data and information:

Approximate total volume of cavern $= 0.78 \text{ mi}^3$

Approximate volume of cavern available for solid waste depository $= 75\%$ of total volume

Proposed maximum waste feed rate to cavern $= 20{,}000 \text{ lb/day}$

Feed rate schedule $= 6 \text{ days/week}$

Average bulk density of waste $= 30 \text{ lb/ft}^3$

Based on the above data, estimate the minimum amount of time it will take to fill the volume of the cavern available for the waste deposition.

Note: The proposed operation could extend well beyond the 25 years upon which the risk assessment analysis was based. The decision whether to grant the permit is somewhat subjective since there is a finite, though extremely low, probability that the cavern walls will be penetrated. Another, but more detailed and exhaustive, risk analysis study should be considered.

Solution

The volume of the cavern, V, in cubic miles available for the solid waste is

$$V = (0.75)(0.78)$$
$$= 0.585 \text{ mi}^3$$

This volume can be converted to cubic feet:

$$V = (0.585 \text{ mi}^3)(5280 \text{ ft/mi})^3$$
$$= 8.61 \times 10^{10} \text{ ft}^3$$

The daily volume rate of solids deposited within the cavern in cubic feet/day, q, is

$$q = (20,000\,\text{lb/day})/(30\,\text{lb/ft}^3)$$
$$= 667\,\text{ft}^3/\text{day}$$

The solids volume rate can now be converted to cubic feet/year:

$$q = (667\,\text{ft}^3/\text{day})(6\,\text{days/week})(52\,\text{weeks/yr})$$
$$= 208,000\,\text{ft}^3/\text{yr}$$

The time it will take to fill the cavern is therefore

$$t = V/q$$
$$= 8.61 \times 10^{10}/208,000$$
$$= 414,000\,\text{yr}$$

As described in an earlier problem, deep-well injection is an ultimate disposal method that transfers liquid wastes far underground and away from freshwater sources. Like landfarming, this disposal process has been used for many years by the petroleum industry. It is also used to dispose of saltwater in oil fields. When the method first came into use, the injected brine would often eventually contaminate groundwater and freshwater sands because the site was poorly chosen. The process has since been improved, and laws such as the Safe Drinking Water Act of 1974 ensure that sites for potential wells are better surveyed.

Many factors are considered in the selection of a deep-well injection site. For example, the rock formation surrounding the disposal zone must be strong but permeable enough to absorb liquid wastes, and the site must be far enough from drinking water sources to prevent contamination. Once a site is selected, it must be tested by drilling a pilot well. The performance data from the pilot well, besides testing permeability and water quality, also aid in the design of the final well and in determining the proper injection rate.

Finally, the type of waste injected into the well is a determinant in how deep the injection will be made. The more toxic the waste, the farther down the disposal zone must usually be. Disposal zones have been classified into five different types:

1. *Zone of Rapid Circulation.* This designation describes the area that runs from the soil surface to only a few hundred feet below. Waste is not injected into this zone.
2. *Zone of Delayed Circulation.* This zone contains circulating fresh water and may be used for certain wastewaters if properly monitored. The water circulation is slow enough so that residence times of a few decades to a few centuries can be achieved for the waste.

3. *Subzone of Lethargic Flow.* The liquid flowing in this zone is very slow moving and saline. More concentrated wastes are injected here.

4. *Stagnant Subzone.* The liquid contained in this region is hydrodynamically trapped, and the zone is generally several thousand feet below the soil surface. Highly toxic wastes may be injected here if the zone can properly accept and keep the waste.

5. *Dry Subzone.* Salt beds fall in this classification. This zone does not contain water and is nearly impermeable. Because there is a possibility that liquid movement could occur through hydrofractures, this zone must be monitored. Otherwise, waste injected here would be isolated from any water sources.

More recently, salt beds have been considered for storing nuclear and solid wastes. One possibility is to inject water to remove some of the salt and thereby form an underground cavern. The cavern could then be filled with the waste, plus (possibly) a solidifying agent, and then cement sealed.

41 Energy Conservation (ENC)

ENC.1 POWER GENERATION AND ITS IMPACT

Even with an aggressive energy conservation program, the growing population will continue to demand increasing amounts of electricity. Identify and describe the environmental impacts, both positive and negative, of the two means of power generation: coal-fired steam boilers and nuclear power.

Solution

The following answer itemizes the positive and negative aspects of each of the two energy generation methods from the standpoint of their impact on the environment.

A contemporary coal-fired boiler and electric generation facility requires three primary raw materials, coal (the energy source), water (for steam, cooling, and probably emissions control), and limestone (for emissions control of SO_2). Therefore, the potential impacts of raw materials suppliers and waste management, as well as the potential impacts of coal combustion, must be considered. Some negative impacts usually include:

1. Air pollution caused by SO_2, NO_x, particulate matter, and CO_2 (global warming)
2. Water pollution from boiler operations (thermal pollution), surface or groundwater contamination from mining of coal and limestone
3. Land pollution from mining wastes and disposal of scrubber sludge, i.e., calcium sulfate

Some positive impacts can include:

1. Producing huge amounts of electricity at one location where highly efficient environmental controls are cost effective
2. Producing a potentially useful waste/by-product in the form of calcium sulfate
3. Producing potentially useful surplus heat, e.g., hot water, low-pressure steam, etc.

718

The principal raw materials for a nuclear power facility are uranium and water (for cooling). The potential impacts of nuclear fission must be considered, as well as the potential impacts of uranium mining and processing. Some negative impacts may include:

1. Accidental release of radiation to the environment
2. Thermal pollution of the cooling water supply
3. Voluminous uranium mining and processing wastes, since only a very small percentage of uranium bearing ore is beneficially used
4. Difficult and costly storage and disposal of spent nuclear fuel, with a potentially continuous, indefinite threat to the environment

Some positive impacts may include:

1. Producing huge amounts of electricity at one facility, although highly toxic waste volumes are relatively small.
2. Virtually contaminant-free stack emissions if the plant is operating properly. No particulate emissions, heavy metals from fuel combustion, etc., are generated from nuclear power.
3. No waste materials generated in the treatment of gas streams, so that the impact of nuclear power plants to the land are minimal when operated properly.

ENC.2   POWER GENERATION

Identify and describe both the positive and negative impacts of the two alternative means of power generation: photovoltaic solar panels and hydroelectric dams.

Solution

The following answer itemizes the positive and negative aspects of each of the two energy generation methods from the standpoint of their impact on the environment.

The principal materials used in photovoltaic solar panel power generation are the photovoltaic cells and water (for cooling). The potential impacts of solar cell power generation must be considered, as well as the potential impacts of the manufacturing of the cells themselves, and any batteries used to store energy during low solar radiation periods. Some negative impacts may include:

1. Generation of small amounts of hazardous waste in the manufacture of the solar cells
2. Requirement for large land areas to provide adequate collector space to meet energy demands

3. Generation of hazardous waste from mining and production of materials in batteries used to store energy for later delivery to the public when solar energy is not available

Some positive impacts may include:

1. Producing small amounts of electricity at many separate facilities and providing it to the power grid or storing it when the generating unit does not use it
2. Completely contaminant-free emissions during the energy generation process
3. No waste materials generated in the treatment of gas streams, water streams, etc.

The principal material required for a hydroelectric power facility is water (as energy supply and cooling). The potential impacts of hydroelectric power generation are primarily related to the impact of the physical structure, i.e., the dam, that is required for storage and management of the water used to power the turbines, and its effect on the local ecosystem in the river that has been converted to an electric power facility. Some negative impacts may include:

1. Disruption of the aquatic environment in the area where the facility is constructed
2. Disruption of the economy affected by the disruption of the ecosystem, and the consequent environmental and socio/political implications
3. Disruption of the aquatic environment upstream and downstream of the facility due to its impact on spawning, changing downstream water temperatures, etc.
4. Thermal pollution of the cooling water supply

Some positive impacts may include:

1. Producing large amounts of electricity at one facility
2. Virtually contaminant-free operations
3. No waste materials produced in the generation of power

ENC.3   CONVERSION EFFICIENCY

In 1900 it took about 20,000 Btu fuel input to produce 1 kWh of electricity. Estimate the efficiency of conversion and compare it with a typical value for today's power industry.

Solution

The solution to this problem is based, in part, on unit conversions. From standard conversion tables, 1 kWh is equivalent to 3412 Btu. Since only 1 kWh was being produced in 1900 from 20,000 Btu, the energy requirement (ER) was

$$ER = (20,0000 \text{ Btu})/(3412 \text{ Btu/kWh})$$
$$= 5.86 \text{ kWh}$$

Since only 1 kWh was being produced, the efficiency (E) of energy conversion is

$$E = \frac{\text{Actual energy produced}}{\text{Energy producation potential}}$$
$$= \frac{1 \text{ kWh}}{5.86 \text{ kWh}}$$
$$= 0.171$$
$$= 17.1\%$$

Today's energy conversion efficiency has improved over this value from a century ago (thankfully!), but not as much as many would like. Typical values range from 30 to 35% efficiency, or approximately 100% better than before. However, as one can see, there is significant room for improvement.

ENC.4   ENERGY LOADS

The James David University runs it own coal-fired power plant, consuming Utah bituminous coal with an energy content (in the combustion literature, energy content is defined as the lower heating value, LHV) of 25,000 kJ/kg. The coal contains, on average, 1.0 wt % sulfur and 1.2 wt % ash (based on the total mass of the coal). The power plant is 35% efficient (meaning that 35% of the energy in the coal is actually converted to electrical energy), and is operated at a 2.0-MW average daily electrical load (ADL).

Assume that the coal is completely burned during combustion, and also that the power plant captures 99% of the ash and 70% of the sulfur dioxide produced during combustion. After a U.S. Environmental Protection Agency (EPA) Green Lights energy audit, James David found that it could install energy-efficient lighting and reduce its average daily electrical generating needs by 25%.

Using the information given above, calculate the average reduction in electrical load and the new average daily load for the power plant.

Solution

With 25% reduction in electrical load resulting from the implementation of energy conservation measures, the new electrical load will be 75% of the old electrical load.

For a 2.0-MW power plant, the new average daily load (ADL) will be

$$\text{New ADL} = (2.0\ \text{MW})\,(0.75)$$
$$= 1.5\ \text{MW}$$

The average reduction (AR) in electrical load becomes

$$\text{AR} = (\text{new ADL}) - (\text{old ADL})$$
$$= 2.0\ \text{MW} - 1.5\ \text{MW}$$
$$= 0.5\ \text{MW}$$

ENC.5   ENERGY-SAVING REDUCTION

Refer to Problem ENC.4. Using the efficiency of the power plant, the heating value of the coal, and the results from Problem ENC.4, calculate the daily reduction in the quantity of coal (kg/day) consumed by the university's power plant and the daily reduction in the quantity of ash (kg/day) produced when the university implements this energy-saving lighting program.

Solution

First, use the available data to calculate the thermal energy input to the plant. Recall that

$$1\ \text{W} = 1\ \text{J/s}$$
$$1\ \text{kJ} = 1.0 \times 10^3\ \text{J}$$
$$1\ \text{MW} = 1.0 \times 10^6\ \text{W}$$

Further, the energy contained in 1 J can raise the temperature of 1 g of water by 25°C.

Consider the system before energy conservation. By definition:

$$\text{Thermal energy input} = \frac{\text{Electrical output}}{\text{Fractional thermal efficiency}}$$

where fractional thermal efficiency is expressed as a decimal.

For a 35% efficient power plant, the fractional thermal efficiency is

$$35\%/100\% = 0.35$$

Therefore, the thermal energy input (TEI) is given by

$$TEI = (2.00 \times 10^6 \text{ J/s})/(0.35)$$
$$= 5.71 \times 10^6 \text{ J/s}$$

After energy conservation, the thermal energy input is

$$TEI \text{ (new)} = (1.50 \times 10^6 \text{ J/s})/(0.35)$$
$$= 4.29 \times 10^6 \text{ J/s}$$

Next, the energy content of the coal [called the lower heating value (LHV)] is used to calculate the mass of coal required for the thermal energy input.

Before energy conservation, the mass flowrate of coal (MC) is

$$MC = \frac{\text{Thermal energy input}}{\text{LHV of coal}}$$
$$= \frac{5.71 \times 10^6 \text{ J/s}}{25.0 \times 10^6 \text{ J/kg}}$$
$$= 0.228 \text{ kg/s}$$

After energy conservation:

$$MC = \frac{4.29 \times 10^6 \text{ J/s}}{25.0 \times 10^6 \text{ J/kg}}$$
$$= 0.172 \text{ kg/s}$$

Therefore, the reduction in mass rate of coal used is

$$\text{Before mass rate} - \text{After mass rate} = 0.228 \text{ kg/s} - 0.172 \text{ kg/s}$$
$$= 0.056 \text{ kg/s}$$

Converting the above to a daily reduction:

$$\text{Mass coal/day} = (0.056 \text{ kg/s}) (86,400 \text{ s/day})$$
$$= 4840 \text{ kg/day}$$

To obtain the fractional particulate capture efficiency, E, of the power plant, divide the percentage of ash captured by 100:

$$E = 99/100$$
$$= 0.99 \text{ kg ash captured}/1.00 \text{ kg ash in coal}$$

To obtain the mass fraction of ash (MFA) in the coal, divide the percent ash by 100:

$$MFA = 1.2/100$$
$$= 0.012 \text{ kg ash/kg coal}$$

To calculate the daily reduction (DR) in captured ash produced by the power plant, multiply the mass of coal per day by the mass fraction of ash in the coal, and multiply the result by the fractional capture efficiency of the power plant. Thus,

$$DR = (\text{mass coal/day}) \text{ (mass ash/mass coal) (capture efficiency)}$$
$$= \left(4840\frac{\text{kg coal}}{\text{day}}\right)\left(0.012\frac{\text{kg ash}}{\text{kg coal}}\right)\left(0.99\frac{\text{kg ash captured}}{\text{kg ash}}\right)$$
$$= 57.5 \text{ kg ash captured/day}$$

Note that the 4840 kg coal is the reduction in the amount of coal needed per day and that the 57.5 kg/day result is the reduction in the amount of ash that is captured per day.

ENC.6   COAL-FIRED POWER PLANT

A power plant generates 2.76×10^5 megawatt hours (MWh) of electricity per year by burning 1.66×10^5 tons of a low-grade coal containing 3.2% sulfur and 15.4% ash. The ratio of fly ash to bottom ash is 0.65 and the plant's particulate collection efficiency is 85%. Determine the amount of each pollutant (particulates, NO_4 and SO_2) emitted per year in pounds/kilowatt-hour for the coal-fired power plant.

The following pollutant emission equations, which apply to coal-fired plants, have been adapted from Ronald White's *The Price of Power Update—Electric Utilities and the Environment*, Council on Economic priorities, 1977.

Particulates:	$e_p = (a)\,(b)\,(c)[1 - (p/100)] \times 10^{-2}$
Sulfur Dioxide:	$e_{SO_2} = (1.90)\,(c)\,(s) \times 10^{-2}$
Nitrous Oxides:	$e_{NO_x} = (15)\,(c) \times 10^{-3}$

where e = emissions in tons/year
  a = mass percent ash content of coal burned
  b = ratio of fly ash to bottom ash
  c = coal consumption in tons/year
  p = percent particulate collection efficiency of the precipitator, baghouse, and/or mechanical collector
  s = average annual mass percent coal sulfur content

Solution

The amount of each pollutant emitted per year in pounds/kilowatt-hour is
Particulates:

$$e_p = (15.4)\,(0.65)\,(166{,}000 \text{ tons/yr})\,[1 - (85/100)] \times 10^{-2}$$
$$= 2492 \text{ tons/yr}$$
$$= 4.98 \times 10^6 \text{ lb/yr}$$

Converting,

$$e_p = \frac{4.98 \times 10^6 \text{ lb/yr}}{276{,}000 \text{ MWh/yr}}$$
$$= 18.0 \text{ lb/MWh}$$
$$= 0.0180 \text{ lb/kWh}$$

Sulfur dioxide:

$$e_{SO_2} = (1.90)\,(166{,}000 \text{ tons/yr})\,(3.2) \times 10^{-2}$$
$$= 10{,}100 \text{ tons/yr}$$
$$= 20.2 \times 10^6 \text{ lb/yr}$$

Converting again,

$$e_{SO_2} = \frac{20.2 \times 10^6 \text{ lb/yr}}{276{,}000 \text{ MWh/yr}}$$
$$= 73.1 \text{ lb/MWh}$$
$$= 0.0731 \text{ lb/kWh}$$

Nitrogen oxides:

$$e_{SO_x} = (15)\,(166{,}000 \text{ tons/yr}) \times 10^{-3}$$
$$= 2490 \text{ tons/yr}$$
$$= 4.98 \times 10^6 \text{ lb/yr}$$

Converting again,

$$e_{SO_x} = \frac{4.98 \times 10^6 \text{ lb/yr}}{276{,}000 \text{ MWh/yr}}$$
$$= 18.0 \text{ lb/MWh}$$
$$= 0.0180 \text{ lb/kWh}$$

ENC.7   REDUCTION IN POLLUTION PRODUCTION

The Jones family illuminates each of the six rooms in their apartment with one 100-W incandescent light per room. Each light is used for 5 h each day. Refer to Problem ENC.6 to determine the reduction in the amount of each pollutant (particulates, NO_x and SO_2) achieved each year by utilizing 34-W fluorescent bulbs instead of incandescent bulbs.

Solution

The number of killowatt-hours utilized by the Jones family each year is

$$(1 \text{ light/room}) (6 \text{ rooms}) (5 \text{ h/day}) (0.1 \text{ kW/light}) = 3 \text{ kWh/day}$$
$$= 1095 \text{ kWh/yr}$$

Therefore, the amount of pollution generated by the production of the required energy is
  Particulates:

$$(0.0181 \text{ lb/kWh}) (1095 \text{ kWh}) = 19.7 \text{ lb}$$

Sulfur dioxide:

$$(0.0731 \text{ lb/kWh}) (1095 \text{ kWh}) = 80.0 \text{ lb}$$

Nitrogen oxides:

$$(0.0180 \text{ lb/kWh}) (1095 \text{ kWh}) = 19.8 \text{ lb}$$

The amount of pollution generated per year by using 34-W fluorescent lights can also be calculated:

$$(1 \text{ light/room}) (6 \text{ rooms}) (5 \text{ h/day}) (0.04 \text{ kW/light}) = 1.2 \text{ kWh/day}$$
$$= 438 \text{ kWh/yr}$$

The pollutants generated is
  Particulates:

$$(0.0180 \text{ lb/kWh}) (438 \text{ kWh}) = 7.88 \text{ lb}$$

Sulfur dioxide:

$$(0.0731 \text{ lb/kWh}) (438 \text{ kWh}) = 32.0 \text{ lb}$$

Nitrogen oxides:

$$(0.0180 \text{ lb/kWh}) (438 \text{ kWh}) = 7.88 \text{ lb}$$

Thus, the reduction in pollution production realized by utilizing the fluorescent bulbs is
  Particulates:

$$19.7 - 7.88 = 11.8 \text{ lb}$$

Sulfur dioxide:

$$80.0 - 32.0 = 48.0 \text{ lb}$$

Nitrogen oxides:

$$19.8 - 7.88 = 11.9 \text{ lb}$$

The past several years have seen an increased social awareness of the impact of lifestyle on the environment. As a result, many are becoming more eager to do what can be done to prevent pollution and protect the environment. The purpose of the this problem is to demonstrate the effect of energy use on air pollution as well as the pollution reduction achieved by the implementation of more efficient lighting (a 34-W fluorescent light provides as much light as a 100-W incandescent bulb). According to Consolidated Edison of New York, indoor lighting accounts for approximately 32% of the annual electricity usage for the typical commercial customer.

ENC.8   RETROFITTING OFFICE LIGHTING SYSTEMS

The Aldo Leone Corporation's 100,000 ft^2 office center contains 1253 standard four-lamp fluorescent fixtures consuming an average of 174 W per fixture or about 43.5 W per lamp. These lamps operate for an average of 16 h a day, 6 days a week, thus accounting for a yearly operational time of 4992 h. Also, it is estimated that the local coal-fired electric power plant emits 0.0175 lb of sulfur dioxide, 0.00824 lb of nitrous oxides, and 2.25 lb of carbon dioxide per kilowatt-hour generated.

   To effectively decrease the amount of pollutants resulting from the consumption of electricity in the building, the most effective method of pollution prevention, i.e., source reduction, is to be implemented. An effective guideline for properly retro-fitting a new lighting system is to keep the watts per square foot to a maximum of 1.5.

   1. Calculate the present electricity consumption (watts per square foot), the new energy consumption, and the number and new wattage of lamps to be installed.

2. Determine the total present lighting load, the new lighting load, and the load reduction (in kilowatts) as well as the present annual load, the new annual load, and the annual load reduction (in kilowatt-hours).
3. Calculate the effects of this energy consumption reduction on the amount of pollutants emitted from the local power plant.

Solution

Regarding the current present load, calculate the total number of lights:

$$N = (1253 \text{ units}) (4 \text{ lights/units})$$
$$= 5012 \text{ lights}$$

The present lighting load, P, in watts and kilowatts is

$$P = (1553 \text{ units}) (174 \text{ W/units})$$
$$= 218,000 \text{ W}$$
$$= 218 \text{ kW}$$

The present annual load (PA) in kilowatt-hours is

$$\text{PA} = (218 \text{ kW}) (16 \text{ h/day}) (6 \text{ days/week}) (52 \text{ weeks/yr})$$
$$= 1.088 \times 10^6 \text{ kWh}$$

The annual pollution contribution of SO_2, CO_2, and NO_x can now be calculated:

SO_2: $(0.0175 \text{ lb/kWh}) (1.088 \times 10^6 \text{ kWh}) = 19,040 \text{ lb/yr}$
CO_2: $(2.25 \text{ lb/kWh}) (1.088 \times 10^6 \text{ kWh}) = 2.45 \times 10^6 \text{ lb/yr}$
NO_x: $(0.00824 \text{ lb/kWh}) (1.088 \times 10^6 \text{ kWh}) = 8970 \text{ lb/yr}$

The present watts per square foot is

$$(218,000 \text{ W})/(100,000 \text{ ft}^2) = 2.18 \text{ W/ft}^2$$

For retrofitting the lighting system, calculate the number of lights removed:

$$(5012 \text{ lights}) (0.2) = 1002 \text{ lights} = 250 \text{ units}$$

The new lighting load in watts and kilowatts becomes

$$P = (1253 - 250)\,(106)$$
$$= 106{,}000 \text{ W}$$
$$= 106 \text{ kWh}$$

Thus, the new annual load in kilowatt-hours is:

$$P_a = (106 \text{ kW})\,(16 \text{ h/day})\,(6 \text{ days/week})\,(52 \text{ weeks/yr})$$
$$= 529{,}000 \text{ kWh}$$

Calculate the new watts per square foot:

$$106{,}000 \text{ W}/100{,}000 \text{ ft}^2 = 1.06 \text{ W/ft}^2$$

Therefore, the load reduction (PR) in kilowatts and kilowatt-hours/year is

$$PR = 218 - 106$$
$$= 112 \text{ kW}$$

Converting,

$$PR = (112 \text{ kW})\,(16 \text{ h/day})\,(6 \text{ days/week})\,(52 \text{ weeks/yr})$$
$$= 559{,}000 \text{ kWh/yr}$$

Calculate the new amounts of pollutants emitted in pounds/year:

SO_2: $(0.0175 \text{ lb/kWh})\,(531{,}000 \text{ kWh}) = 9290 \text{ lb/yr}$
CO_2: $(2.25 \text{ lb/kWh})\,(531{,}000 \text{ kWh}) = 1.19 \times 10^6 \text{ lb/yr}$
NO_x: $(0.00824 \text{ lb/kWh})\,(531{,}000 \text{ kWh}) = 4370 \text{ lb/yr}$

Therefore, the amount of pollutants reduced by the new lighting system is

SO_2: $19{,}040 - 9290 = 9800 \text{ lb/yr}$
CO_2: $2.45 \times 10^6 - 1.19 \times 10^6 = 1.26 \times 10^6 \text{ lb/yr}$
NO_x: $8970 - 4370 = 4600 \text{ lb/yr}$

The process of removing and replacing lighting is an effective method of source reduction, saving energy and ultimately reducing a large amount of air pollution from power plants.

ENC.9   ALTERNATE ENERGY SOURCES

A lake is located at the top of a mountain. A power plant has been constructed at the bottom of the mountain. The potential energy of the water traveling downhill can be used to spin turbines and generate electricity. This is the operating mode in the daytime during peak electrical demand. At night, when demand is reduced, the water is pumped back up the mountain. The operation is shown in Figure 114.

Using the method of power "production" described above, determine how much power (watts) is generated by the lake located at an elevation of 3000 ft above the power plant. The flowrate of water is 500,000 gpm. The turbine efficiency is 30%. Neglect friction effects.

Note: This programmed-instructional problem is a modified and edited version (with permission) of an illustrative example prepared by Marie Gillman, a graduate mechanical engineering student at Manhattan College.

Solution

First, convert height and flowrate to SI units in order to solve for the power in watts:

$$(3000 \text{ ft}) \, (0.3048 \text{ m/ft}) = 914.4 \text{ m}$$
$$(500,000 \text{ gal/min}) \, (0.00378 \text{ m}^3/\text{gal}) = 1890 \text{ m}^3/\text{min}$$

The mass flow rate of the water in kilograms/second is

$$\frac{(1890 \text{ m}^3/\text{min}) \, (1000 \text{ kg/m}^3)}{60 \text{ s/min}} = 31,500 \text{ kg/s}$$

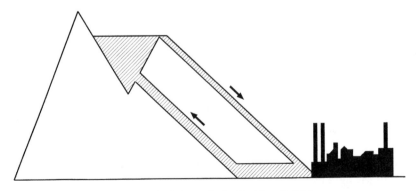

Figure 114. Schematic for Problem ENC.9

The drop in potential energy, ΔPE, of the water flow is given by

$$\Delta PE = \frac{mgh}{g_c}$$

Substituting yields

$$\Delta PE = (31{,}500 \text{ kg/s}) (9.8 \text{ m/s}^2) (914.4 \text{ m})$$
$$= 2.82 \times 10^8 \text{ kg} \cdot \text{m/s}^3$$
$$= 2.82 \times 10^8 \text{ N/S}$$
$$= 282 \text{ MW}$$

Note that $g_c = 1$ in the SI system of units.

Assuming that the potential energy decrease is entirely converted to energy input to the turbine, the actual power output is

$$P = (0.30) (282) = 84.7 \text{ MW}$$

This is enough power for a small town. No pollution is generated because no fossil fuel is required. The initial construction expense would be quite high, but the long-term cost of producing electricity would probably be very economical.

ENC.10   COMBUSTION OF COAL

The New Jersey Power Company PSE&G has determined that for every 10,000 kW it generates, it must burn approximately one ton of coal. The coal used at this particular facility is Illinois #6, a form of bituminous coal with the approximate chemical formula $C_{100}H_{85}S_{2.1}N_{1.5}O_{9.5}$. The company is embarking on a new public relations project to show consumers the impact of wasted energy on the environment. As a PSE&G engineer, you have been asked to calculate the amount of CO_2 and SO_2 discharged into the atmosphere for every kilowatt of energy produced.

Solution

Calculate the molecular weight of the coal. The atomic weights of the components are

$$C = 12.00 \text{ lb/lbmol}$$
$$H = 1.000 \text{ lb/lbmol}$$
$$S = 32.06 \text{ lb/lbmol}$$
$$N = 14.00 \text{ lb/lbmol}$$
$$O = 16.00 \text{ lb/lbmol}$$

The molecular weight is determined from the molecular formula:

$$MW = (100)\,(12.00) + (85)(1.000) + (2.1)(32.06) + (1.5)(14.00) + (9.5)(16.00)$$
$$= 1525 \text{ lbmol}$$

Assuming complete combustion, the stoichiometric equation for the combustion of the coal can now be written:

$$C_{100}H_{85}S_{2.1}N_{1.5}O_{9.5} + 120.1O_2 \rightarrow 100CO_2 + 42.5H_2O + 2.1SO_2 + 1.5NO_2$$

Calculate the lbmol of coal per ton of coal and then determine the amount of products generated in lbmol per ton of coal combusted.

Coal:	$(2000 \text{ lb})/(1525) = 1.311 \text{ lbmol}$
CO_2:	$(1.311)(100) = 131.1 \text{ lbmol}$
H_2O:	$(1.311)(42.5) = 55.7 \text{ lbmol}$
SO_2:	$(1.311)(2.1) = 2.75 \text{ lbmol}$
NO_2:	$(1.311)(1.5) = 1.97 \text{ lbmol}$

Calculate the amount of products generated in pounds:

CO_2:	$(131.1 \text{ lbmol})/(44.00 \text{ lb/lbmol}) = 5770 \text{ lb}$
H_2O:	$(55.73 \text{ lbmol})/(18.00 \text{ lb/lbmol}) = 1003 \text{ lb}$
SO_2:	$(2.754 \text{ lbmol})/(64.06 \text{ lb/lbmol}) = 176 \text{ lb}$
NO_2:	$(1.967 \text{ lbmol})/(46.00 \text{ lb/lbmol}) = 91 \text{ lb}$

The amount of CO_2 and SO_2 generated per kilowatt can now be calculated:

CO_2:	$5770/10,000 = 0.577 \text{ lb/kW}$
SO_2:	$176/10,000 = 0.0176 \text{ lb/kW}$
NO_2:	$91/10,000 = 0.0091 \text{ lb/kW}$

Note: NO_2 plays an important role in smog formation.

42 DOMESTIC APPLICATIONS (DOM)

DOM.1 COMMON DOMESTIC WASTES

The following toxic compounds are often found in a domestic household. What are their structural and molecular formulas?

1. Tetrachloroethene
2. Formaldehyde
3. Carbon tetrachloride

Solution

The structural formulas are:

1. Tetrachloroethene:

$$\begin{array}{c}
\text{Cl} \qquad\qquad \text{Cl} \\
\diagdown \qquad \diagup \\
\text{C}=\text{C} \\
\diagup \qquad \diagdown \\
\text{Cl} \qquad\qquad \text{Cl}
\end{array}$$

2. Formaldehyde:

$$\begin{array}{c}
\text{O} \\
\parallel \\
\text{H}-\text{C}-\text{H}
\end{array}$$

3. Carbon tetrachloride:

$$\begin{array}{c}
\text{Cl} \\
| \\
\text{Cl}-\text{C}-\text{Cl} \\
| \\
\text{Cl}
\end{array}$$

The molecular formulas are:

1. C_2Cl_4, tetrachloroethene
2. CH_2O, formaldehyde
3. CCl_4, carbon tetrachloride

DOM.2   PROMOTING LOCAL POLLUTION PREVENTION

Discuss activities a local government can undertake to promote pollution prevention.

Solution

Local governments can carry out the following types of activities to promote pollution prevention:

1. Educational programs to raise awareness in both local businesses and the community of the need to conserve resources and reduce waste and pollution
2. Technical assistance programs that provide onsite help to businesses, local citizens, and organizations to reduce pollution at the source
3. Regulatory programs that promote prevention at the domestic level through mechanisms such as permits, codes, and licenses
4. Procurement policies regarding government purchase of recycled products, products designed to be recycled, and reusable products

DOM.3   DOMESTIC WASTE MEASURES

Discuss pollution prevention measures that can be instituted at home and at the office.

Solution

Pollution prevention measures that can be performed at home:

1. Purchase products with the least amount of packaging.
2. Keep hazardous materials out of reach of children.
3. Install water-flow restriction devices on sink faucets and showerheads.

Pollution prevention measures that can be performed in an office environment:

1. Pass on verbal memos when written correspondence is not required.
2. Know building evacuation procedures.
3. Do not waste utilities simply because one is not paying for them.
4. Take public transportation to the office.

DOM.4   WASTE MINIMIZATION ASSESSMENT PROCEDURE

As a rule, point-source pollution sites are easily identified (large chemical plants, sewage treatment, military bases, etc.) and can be dealt with more or less satisfactorily by means of regulation and/or public pressure. In contrast, non-point-source pollution (homes, restaurants, dry cleaning establishments, etc.) are far more numerous, widely distributed, and therefore more difficult to identify and regulate. It is particularly important that these smaller sources of pollution adopt the desired "pollution prevention" attitude. Small colleges are one example of an intermediate source of pollution in which several departments and offices on campus may use, store, and dispose of pollutants quite independently of one another.

The four major elements of a waste minimization assessment procedure are:

1. Planning and organization
2. Assessment
3. Feasibility analysis
4. Implementation

The assessment element of this procedure is to select and to prioritize assessment targets. From the list of campus sites below, select the five most likely sources of pollution in a typical small college. Prioritize choices from 1 to 5 with 1 being most likely. Cite the likely items contributing to the waste at these sites. Check the suggested answers before continuing.

English/History Dept.	Cafeteria
Library	Chemistry Dept.
Biology Dept.	Maintenance/Shop
Dormitories	Psychology Dept.
Physics Dept.	Art/Theater

A further part of the assessment phase is to generate options for dealing with generated wastes. Use the list above (or the suggested answers) and discuss the various options available for minimizing major hazardous waste sources on a small college campus.

Solution

The five most likely sources of pollution are detailed below.

1. General chemistry labs: mostly inorganic wastes, heavy metals, iodine, xylene, naphtha, P-dichlorobenzene, mercury, chromium, lead, and carbon tetrachloride. Organic chemistry labs: mostly organic solvents, some organic solids and chromium salts.
2. Biology Department labs: formaldehyde, mercury salts, solvents, and a small amount of osmium tetroxide are generated as wastes.

3. Art Department: acids, waste paint, and solvents are generated from silkscreen and printing processes.

4. Psychology department: some waste solvents are generated.

5. Maintenance/Shop Department: generates waste oils, paints, etc. from vehicle and/or equipment maintenance.

The options available for managing the wastes in the five above source are provided below.

1. To the extent possible, all chemistry labs should convert to microscale experiments. This is the single most productive way to avoid the generation of significant wastes and to minimize student exposure. Experiments using toxic substances can often be conducted with alternative reagents [bleach instead of chromium (VI) for oxidation, for example].

2. Biology Department options are similar to the chemistry labs.

3. In the Art Department switch to water-base paints where possible. Use adequate ventilation to protect students.

4. Psychology Department options are similar to the chemistry labs.

5. The Maintenance/Shop Department can switch to water-base paints, avoid spills, and recycle oil.

Something to ponder is whether the cafeteria should have been included in the above answers. A good reference (if available) for this open-ended problem is *Guides to Pollution Prevention—Research and Educational Institutions*, EPA/625/7-90/010, Office of Pollution Prevention, Washington, DC, 1990.

DOM.5 WATER APPLICATION

A suburban town council projects that there will be approximately 500 new homes built in and around the central town within the next 10 years. Currently, the town's wastewater treatment facility has the capacity to handle 1.2 MGD (million gallons per day) of wastewater from homes (the current daily flow of wastewater is 1.0 MGD). It is also projected that each new home will use 375 gallons per day of water, or a total of 0.1875 MGD for the 500 new homes. Unfortunately, the facility manager claims that the facility can handle no more than 0.16 MGD additional wastewater before the end of 12 years (at which time the new 2.0 MGD facility will have been completed).

In an attempt to solve the above problem, the owner of a plumbing supply outlet in the town suggests that each new home be fitted with his company's new Airflush toilets, which he claims use considerably less water per flush. Projections also predict that each new home will have 2.5 toilets. If the average number of flushes per toilet is 5/day, and the Airflush toilet saves 4.5 gallons per flush over the conventional toilet, will the plumber's proposition work? Wastewater discharges from other domestic sources are to be neglected in this analysis.

Solution

The gallons of water saved per toilet per day is

$$\dot{V}_{toi} = (4.5\,\text{gal/flush})(5\,\text{flushes/day})$$
$$= 22.5\,\text{gpd/toilet}$$

The water saved per home per day is

$$\dot{V}_{home} = (22.5\,\text{gpd/toilet})(2.5\,\text{toilets/home})$$
$$= 56.25\,\text{gpd/home}$$

The total amount of water saved per day with the new toilets can now be calculated:

$$\dot{V}_{tot} = (56.25\,\text{gpd/home})(500\,\text{homes})$$
$$= 28,125\,\text{gpd}$$

The new amount of water sent to the waste treatment facility is

$$\dot{V}_{treat} = 187,500\,\text{gpd} - 28,125\,\text{gpd}$$
$$= 159,400\,\text{gpd}$$
$$= 0.159\,\text{MGD}$$

Thus, the air-assisted toilets would conserve enough water to allow the current housing construction project to continue as planned.

DOM.6   VOLUME REDUCTION DUE TO HOME COMPACTORS

The following table provides a weekly volume estimate for some common household solid wastes:

TABLE 1

Waste	Weight (lb)	Density (lb/ft³)	Volume (ft³)
Food	10	18.0	0.6
Paper	50	5.0	10.0
Cardboard	10	6.5	1.5
Plastic	2	4.0	0.5
Textile	1	4.0	0.3
Leather	2	10.0	0.2
Garden trimmings	10	6.5	1.5
Wood	5	15.0	0.3
Nonferrous metal	2	10.0	0.2
Ferrous metal	4	20.0	0.2
Dirt, ash, bricks, etc.	1	30.0	0.03

Note: It is assumed that the household recycles glass and aluminum cans.

Estimate the percent volume reduction that would be achieved in the solid waste collected in the community if all households installed garbage compactors that compacted the solid wastes to a density of $20\,lb/ft^3$. Note that garden trimmings, wood, ferrous metals, dirt, ash, and brick are usually not placed in trash compactors.

Solution

The total weight of all solid wastes is

$$W_{tot} = 10 + 50 + 10 + 2 + 1 + 2 + 10 + 5 + 2 + 4 + 1$$
$$= 97\,lb$$

The total weight, excluding items not placed in the compactor is

$$W_{tcom} = 10 + 50 + 10 + 2 + 1 + 2 + 2$$
$$= 77\,lb$$

The total volume of all solid wastes is therefore

$$V_{tot} = 0.6 + 10.0 + 1.5 + 0.5 + 0.3 + 0.2 + 1.5 + 0.3 + 0.2 + 0.2 + 0.03$$
$$= 15.33\,ft^3$$

The total volume, excluding items not placed in the compactor is

$$V_{tcom} = 0.6 + 10.0 + 1.5 + 0.5 + 0.3 + 0.2 + 0.2$$
$$= 13.30\,ft^3$$

The volume of the compacted waste becomes

$$V_{com} = 77/20$$
$$= 3.85\,ft^3$$

The percent volume reduction (VR) for the solid waste deposited into the compactor is therefore

$$VR = (100)(13.30 - 3.85)/(13.30)$$
$$= 71.1\%$$

The overall volume reduction achieved with the household compactor (including garden trimmings, wood, ferrous metals, dirt, ashes, and brick) is then

$$V_{t,fin} = (15.33 - 13.30) + 3.85$$
$$= 5.88 \text{ ft}^3$$

The percent volume reduction (VR$'$) for the solid waste deposited into the compactor if garden trimmings, wood, ferrous metals, dirt, ashes, and brick are included is therefore

$$VR = (100)(15.33 - 5.88)/(15.33)$$
$$= 61.6\%$$

If one compares the overall volume reduction to the volume reduction excluding items not placed in the compactor, the effectiveness of household compactors is reduced. But there is still substantial volume reduction and if an entire community used household compactors, the results could be substantial.

DOM.7   RAPID CURING CUTBACK ASPHALT

A local asphalt company has provided the State Air Pollution Agency with its records, indicating that 40,000 lb of rapid curing cutback asphalt (containing 45% diluent by volume) is annually employed. Assume the density of naphtha (the diluent) and asphalt cement to be 5.9 lb/gal and 9.3 lb/gal, respectively. Regarding the complaints of local residents, the state's pollution prevention agency has asked the company to determine the annual volatile organic compound (VOC) emissions. Field data suggest that 95% of the diluent evaporates.

Solution

A conservation of mass equation relating the total asphalt mass with the volume of naphtha (x) and volume of asphalt (y) yields

$$40,000 \text{ lb} = (x \text{ gal})(5.9 \text{ lb/gal}) + (y \text{ gal})(9.3 \text{ lb/gal})$$

Using a volume balance, the volume fraction of diluent can be expressed in terms of x and y.

$$0.45 = x/(x+y)$$

These two equations can be solved simultaneously for x and y in gallons.

$$x = 2312 \, \text{gal}$$

$$y = 2821 \, \text{gal}$$

The amount of diluent employed by the company in pounds/year is then

$$x = (2312)(5.9)$$
$$= 13{,}640 \, \text{lb/yr}$$

Therefore, the VOC emissions rate is

$$\text{VOC} = \frac{(0.95)(13{,}640 \, \text{lb/yr})}{(365 \, \text{day/yr})(24 \, \text{h/day})}$$
$$= 1.48 \, \text{lb/h}$$
$$= 12{,}960 \, \text{lb/yr}$$

DOM.8 REFUELING AUTOMOBILES

The average gasoline tank in an automobile has a 14-gal capacity. Every time the gas tank is filled, the vapor space in the tank is displaced to the environment. Since all forms of hydrocarbons in the atmosphere contribute to the formation of ozone and need to be controlled, this problem attempts to quantify some of these emissions.

Assume the automobile tank vapor space, the air, and the gasoline supply is all at 20°C. The vapor space is saturated with gasoline. The vapor-phase mole fraction of gasoline under these conditions is approximately 0.4. The lost vapor has a molecular weight of about 70 g/gmol and a liquid specific gravity of 0.62.

1. Calculate the amount of gasoline (in gallons of liquid) that is lost to the air during a 10-gal fill.
2. How much is lost annually from 50 million cars filled once each week with 10 gal of gasoline.

Solution

The vapor volume in m^3/kgmol is

$$\frac{V}{n} = \frac{RT}{P}$$

$$= \frac{(8.314)\,(293)}{101.3}$$

$$= 24.05\,\text{m}^3/\text{kgmol}$$

The amount of gasoline vapor in the tank in kgmol is

$$n = \frac{(0.4)\,(10\,\text{gal})}{(264.1\,\text{gal/m}^3)\,(24.05\,\text{m}^3/\text{kgmol})}$$

$$= 6.298 \times 10^{-4}\,\text{kgmol}$$

The liquid volume of the gasoline vapor in the tank in gallons is

$$V_1 = \frac{(6.298 \times 10^{-4}\,\text{kgmol})\,(70\,\text{kg/kgmol})\,(264.1\,\text{gal/m}^3)}{(620\,\text{kg/m}^3)}$$

$$= 0.01878\,\text{gal}$$

The gasoline loss in gallons per car per year can be calculated:

$$\text{Lost} = (0.01878\,\text{gal/fill})(52\,\text{fills/yr})$$

$$= 0.976\,\text{gal/(car} \cdot \text{yr)}$$

The estimated annual loss (AL) arising because of the vapor displaced during filling is

$$\text{AL} = [0.976\,\text{gal/(car} \cdot \text{yr)}](50,000,000\,\text{cars})$$

$$= 4.88 \times 10^7\,\text{gal/yr}$$

PART VII
Health, Safety, and Accident Management

Contributing Author: Mary Keane

43 Toxicology (TOX)

TOX.1 TOXICOLOGY

Discuss the science of toxicology.

Solution

Toxicology is the science dealing with the effects, conditions, and detection of toxic substances or poisons. Six primary factors affect human response to toxic substances or poisons. These are detailed below:

1. The chemical itself: Some chemicals produce immediate and dramatic biological effects, whereas others produce no observable effects or produce delayed effects.
2. The type of contact: Certain chemicals appear harmless after one type of contact (e.g., skin) but may have serious effects when contacted in another way (e.g., lungs).
3. The amount (dose) of a chemical: The dose of a chemical exposure depends upon how much of the substance is physically contacted.
4. Individual sensitivity: Humans vary in their response to chemical substance exposure. Some types of responses that different persons may experience at a certain dose are serious illness, mild symptoms, or no noticeable effect. Different responses may also occur in the same person at different exposures.
5. Interaction with other chemicals: Toxic chemicals in combination can produce different biological responses than the responses observed when exposure is to one chemical alone.
6. Duration of exposure: Some chemicals produce symptoms only after one exposure (acute), some only after exposure over a long period of time (chronic), and some may produce effects from both kinds of exposure.

TOX.2 ROUTES OF EXPOSURE

Briefly discuss the various routes by which a chemical can enter the body.

Solution

To protect the body from hazardous chemicals, one must know the route of entry into the body. All chemical forms may be inhaled. After a chemical is inhaled into the mouth, it may be ingested, absorbed into the bloodstream, or remain in the lungs. Various types of personal protective equipment (PPE) such as dust masks and respirators prevent hazardous chemicals from entering the body through inhalation. Ingestion of chemicals can also be prevented by observing basic housekeeping rules, such as maintaining separate areas for eating and chemical use or storage, washing hands before handling food products, and removing gloves when handling food products. Wearing gloves and protective clothing prevent hazardous chemicals from entering the body through skin absorption.

After a chemical has entered the body, the body may break it down or metabolize it, the body may excrete it, or the chemical may remain deposited in the body.

The route of entry of a chemical is often determined by the physical form of the chemical. Physical chemical forms and the routes of entry are summarized in the following table.

Chemical Form	Principal Danger
Solids and fumes	Inhalation, ingestion, and skin absorption
Dusts and gases	Inhalation into lungs
Liquids, vapors and mists	Inhalation of vapors and skin absorption

TOX.3 TOXICOLOGY TERMINOLOGY

Describe the following toxicology terms:

1. Threshold limit value (TLV)
2. Immediately dangerous to life and health (IDLH)
3. Lethal dose (LD)
4. Effective dose (ED)
5. Toxic dose (TD)
6. Lethal concentration (LC)

Solution

The concept of *threshold* is used to assess the toxicity of noncarcinogenic chemical substances. The dose–effect relationship is generally characterized by a threshold below which no effects can be observed. However, the threshold value for a toxic substance cannot be identified precisely. Instead, it can only be bracketed based on analyses of data from animal tests in which other parameters are used to evaluate the hazard.

The *threshold limit value* (TLV) is the maximum limit of the amount of a chemical to which a human can be exposed without experiencing toxic effects. The TLV is categorized into the TLV-TWA, TLV-STEL, and TLV-C, where -TWA, -STEL, and -C represent *time-weighted average, short-term exposure limit,* and *ceiling*, respectively. A TWA can be the average concentration over any period of time. Most often, however, a TWA is the average concentration of a chemical that workers can typically be exposed to during a 40-h week and a normal 8-h day without showing any toxic effects. A STEL is a 15 min time-weighted average exposure. Excursions to the STEL should be at least 60 min apart, no longer than 15 min in duration, and should not be repeated more than four times per day. Since the excursions are calculated into an 8 h TWA, the exposure must be limited to avoid exceeding the TWA. Ceiling values, C, exist for substances whose exposure results in a rapid and particular type of response. It is used where TWA (with its allowable excursions) would not be appropriate. The American Conference of Governmental Industrial Hygienists (ACGIH) and the Occupational Safety and Health Administration (OSHA) state that a ceiling value should not be exceeded even instantaneously. The National Institute for Occupational Safety and Health (NIOSH) also uses ceiling values. However, its ceiling values are similar to a STEL.

Similar exposure limits employed are the *permissible exposure limit* (PEL) and the *recommended exposure limit* (REL). PELs are extracted from the TLVs and other standards including standards for benzene and 13 carcinogens. Since OSHA is a regulatory agency, its PELs are legally enforceable standards and apply to all private industries and federal agencies. RELs are used in developing OSHA standards, but there are many that have not been adopted and are in the same status as the exposure guidelines of ACGIH.

The IDLH (immediately dangerous to life and health) is the maximum concentration of a substance to which a human can be exposed for 30 min without experiencing irreversible health effects.

Dosages of a chemical can be described as a lethal dose (LD), effective dose (ED), or toxic dose (TD). The LD50 or LD_{50} is a common parameter used in toxicology. It represents the dose at which 50% of a test population would die when exposed to a chemical at that dose. Similarly, the lethal concentration (LC) is the concentration of a substance in air that will cause death.

TOX.4   TOXICITY FACTORS

Define and compare the following pairs of parameters used in toxicology.

1. NOEL and NOAEL
2. LOEL and LOAEL
3. ADI and RfD

Solution

The parameters NOEL and NOAEL, LOEL and LOEAL, and ADI and RfD are used to establish thresholds.

The NOEL (*no observed effect level*) is the highest dose of the toxic substance that will not cause an effect. The NOAEL (*no observed adverse effect level*) is the highest dose of the toxic substance that will not cause an adverse effect.

The LOEL (*lowest observed effect level*) is the lowest dose of the toxic substance tested that shows effects. The LOAEL (*lowest observed adverse effect level*) is the lowest dose of the toxic substance tested that shows adverse effects. The LOEL and LOAEL give no indication of individual variation in susceptibility.

The ADI (*acceptable daily intake*) is the level of daily intake of a particular substance that will not produce an adverse effect. The RfD (*reference dose*) is an estimate of the daily exposure level for the human population. The RfD development follows a stricter procedure than that followed for the ADI. This sometimes results in a lower value for the ADI. The ADI approach is used extensively by the Food and Drug Administration (FDA) and the World Health Organization (WHO), while the RfD is a contemporary replacement for the ADI used by the Environmental Protection Agency (EPA).

When using data for LOELs, LOAELs, NOELs, or NOAELs, it is important to be aware of their limitations. Statistical uncertainty exists in the determination of these parameters due to the limited number of animals used in the studies to determine the values. In addition, any toxic effect might be used for the NOAEL and LOAEL so long as it is the most sensitive toxic effect and considered likely to occur in humans.

TOX.5   OSHA AND NIOSH

The *Occupational Safety and Health Act* (OSHA) enforces basic duties that must be carried out by employers. Discuss these basic duties. Also, state the major roles of the *National Institutes of Safety and Health* (NIOSH) and the *Occupational Safety and Health Administration* (OSHA).

Solution

Employers are bound by OSHA to provide each employee with a working environment free of recognized hazards that cause or have the potential to cause physical harm or death. Employers must have proper instrumentation for the evaluation of test data provided by an expert in the area of toxicology and industrial hygiene. This instrumentation must be obtained because the presence of health hazards cannot be evaluated by visual inspection. This data collection effort provides the employer with substantial evidence to disprove invalid complaints by employees alleging a hazardous working situation. This law also gives employers the right to take full disciplinary action against those employees who violate safe working practices in the workplace.

NIOSH recommends standards for industrial exposure that OSHA uses in its regulations. OSHA has the power to enforce all safety and health regulations and standards recommended by NIOSH.

TOX.6 TOXICOLOGY DETERMINATION

Calculating toxicological effects attributable to an environmental contaminant often begins with the following simple equation:

$$\text{Health risk} = (\text{Human exposure})\,(\text{Potency of contaminant})$$

Identify several factors that may lead to an "uncertain" calculation of the actual risk to human health of a particular chemical.

Solution

Uncertainties associated with determining the "human exposure" term include the following:

1. Limited knowledge of source characteristics, e.g., how much of a contaminant is released and for how long the release has occurred
2. Difficulty in describing and calculating the "fate and transport" of the contaminant as it travels from the release point in the environment to the exposed population, i.e., the receptor (sometimes referred to as pathway analysis)
3. Mobility of the exposed individual or population, thereby constantly changing the individual's exposure

Uncertainties associated with determining the "potency of contaminant" term include the following:

1. Potency factors are often based on animal toxicity studies and then applied to humans.
2. Variable effects on exposed humans due to differences in age, sex, health condition, etc.
3. Extrapolation from measured high-dose effects to determine low-dose effects.

TOX.7 TOXIC CHEMICAL EXPOSURE

Hazardous waste sites provide an easy opportunity for individuals to be exposed to some of the most dangerous poisons known. Federal standards (29CFR1910) require protection for workers involved in hazardous waste remediation. Protection must be provided through engineering controls, work practices, and/or personal protective equipment (PPE).

Hazardous waste remediation workers and industrial/manufacturing workers have a greater risk of toxic chemical overexposure and/or poisoning than the general public due to the nature of their work. Each of the chemicals listed in this problem is commonly used in industry. Toluene and methylene chloride are common solvents and act as narcotics at high concentrations. Vinyl chloride is used in the plastics industry, and Aroclor 1260 [polychlorinated biphenyl (PCB)] is used in electrical capacitors and transformers. Both are suspected carcinogens. Pentachlorophenol finds application as an insecticide and wood preservative. It causes damage to the lungs, liver, and kidneys and also causes contact dermatitis.

New York City obtains the majority of its drinking water from several reservoirs in the Catskill region of New York State. At an abandoned site approximately $\frac{1}{4}$ mile from one such reservoir, several dozen leaking barrels of toxic chemicals have been discovered in a buried trench.

Given the following limited information for five of these chemicals, discuss each compound in relative terms of its potential hazards to (a) hazardous waste remediation workers involved in sampling and cleanup of the site and (b) ground and surface water supplies.

Assume the barrels have corroded and have been leaking slowly over several years. The site consists of porous soil, is at a slightly higher elevation than the reservoir, and is situated over a groundwater source.

Data for the five chemicals are provided in the following table:

Chemical	Form	Water Solubility (mg/L)	Henry's Law Constant [atm · m³/(gmol)]	Density (g/mL)
Methylene chloride	Liquid	6900	0.003	1.32
Vinyl chloride	Liquid	1.1	2.40	0.91
Toluene	Liquid	535.0	0.0059	0.87
Pentachlorophenol	Crystal	14.0	N/A	1.98
Aroclor 1260 (PCB)	Resin	0.0027	0.0071	>1

Solution

Methylene chloride, vinyl chloride, and toluene will be present in liquid form in the trench. All would also be expected to be present in the vapor state, with vinyl chloride being the most volatile due to its high Henry's law constant. The principal dangers to site remediation workers, therefore, would be from inhalation of vapors and skin absorption of contacted liquids. Pentachlorophenol and PCBs will be present in solid and semisolid forms. Minimal vapors of these compounds would be expected. The principal dangers would be from inhalation of dusts and fumes, ingestion of solids, and skin absorption due to solids contact. In the absence of engineering controls and work practices, the necessary worker PPE would include, at a minimum, respirators and chemical protective clothing.

The higher a compound's water solubility, the more easily the compound will disperse within a water source. Methylene chloride would be expected to spread rapidly upon reaching groundwater and would be most likely to find its way into the reservoir. Toluene will dissolve more readily than the remaining compounds. Thus, it can disperse to a greater extent. Vinyl chloride, pentachlorophenol, and Aroclor 1260 have low or very low water solubilities. These compounds would not travel large distances in stagnant water.

The compounds with relatively low Henry's law constants (methylene chloride, toluene, pentachlorophenol, and Aroclor 1260) have the lowest potential volatility. They would not evaporate readily and would therefore present a higher risk to the ground and surface waters. Vinyl chloride has the highest relative volatility and would present the least risk to ground and surface water.

Methylene chloride, pentachlorophenol, and Aroclor 1260 all have densities greater than that of water, i.e., greater than 1.0 g/mL. As such, these compounds would be expected to sink to the bottom of the ground and surface water bodies. Vinyl chloride and toluene have densities less than that of water and would be expected to float on the ground and surface water surfaces. Chemicals sinking to and settling on the bottom of a water body generally present a lesser risk to humans and animal life unless disturbed. Chemicals that float on water surfaces are more likely to contact human and animal life because they can be carried large distances by winds and currents.

TOX.8   THRESHOLD LIMIT VALUES

The dynamic seal for a control valve suddenly starts leaking toluene-2,4-diisocynate (TDI) vapor at a rate of $40 \, \text{cm}^3/\text{h}$ into a 12 ft × 12 ft × 8 ft high room. The air in the room is uniformly mixed by a ceiling fan. The background TDI vapor concentration is 1.0 ppb. Air temperature and pressure are 77°F and 1 atm, respectively. Calculate the ppm value of leaking TDI vapor if its vapor pressure is 35 mm Hg at 25°C. Also, calculate the number of minutes after the leak starts that a person sleeping on the job would be at risk of being exposed to TDI vapor with respect to the STEL, $\frac{1}{4}$ LEL, and TLV.

TDI exposure limit values are as follows:

1. Short-term exposure limit (STEL) = 0.02 ppm
2. 25% of lower exposure limit ($\frac{1}{4}$ LEL) = 0.325 v/v
3. Threshold limit value (TLV) = 0.005 ppm

Solution

The maximum TDI vapor concentration, C_{max}, in ppm based on its vapor pressure at the leak is

$$C_{max} = (35 \, \text{mm Hg}/760 \, \text{mm Hg})(10^6)$$
$$= 46,100 \, \text{ppm}$$

The TDI vapor concentration in the room, C (ppm), as a function of time, t (min), is calculated as follows:

$$C \text{ (ppm)} = 0.001 + \frac{\left(40 \, \frac{cm^3}{h}\right)\left(\frac{h}{60 \, min}\right)\left(\frac{m^3}{10^6 \, cm^3}\right)(t)(46,100 \text{ ppm})}{(12 \, ft)(12 \, ft)(8 \, ft)\left(\frac{0.02832 \, m^3}{ft^3}\right)}$$

$$= 0.001 + 0.94 \times 10^{-3} t$$

Note that t is the time from the start of the leak, and 0.001 ppm is the background concentration in the room.

The time, t, to reach the STEL of 0.02 ppm is obtained by using the mean value theorem,

$$\int_{t_1}^{t_2} C \, dt = C_{ave}(t_2 - t_1)$$

where C_{ave} = average concentration or STEL (ppm)
$(t_2 - t_1)$ = range of the time-weighted value (min)

Integrating the equation yields the following solution:

$$\int_{t_1}^{t_2} (0.001 + 0.94 \times 10^{-3} t) \, dt = 0.02(15)$$

$$\left(0.001t + \frac{0.94 \times 10^{-3} t^2}{2}\right)\Big|_{t_1}^{t_2} = 0.02(15)$$

$$(0.001)(t_2 - t_1) + (0.47 \times 10^{-3})(t_2^2 - t_1^2) = 0.30$$

$$(0.001)(15) + (0.47 \times 10^{-3})(t_2 - t_1)(t_2 + t_1) = 0.30$$

$$(0.001)(15) + (0.47 \times 10^{-3})(15)(t_2 + t_1) = 0.30$$

Substituting $t_2 = 15 + t_1$,

$$0.015 + (0.47 \times 10^{-3})(15)(15 + 2t_1) = 0.30$$

$$t_1 = 12.71 \text{ min}$$

The total time to reach the STEL of 0.02 ppm on a TWA basis is $15 + t_1$, or $t_2 = 27.7 \text{ min}$.

The time in minutes, t, to reach $\frac{1}{4}$ LEL is given by

$$325,000 \ (ppm) = 0.001 \ (ppm) + 0.94 \times 10^{-3}t$$
$$t = 3,460,000 \ \text{min} \ (6.6 \ \text{yr})$$

The average concentration in terms of time may be expressed as follows:

$$C_{ave} = \frac{(0.001) + (0.001 + 0.94 \times 10^{-3}t)}{2}$$

Thus, for an 8-h averaging period,

$$C_{ave} = \frac{(0.001) + (0.001 + (0.94 \times 10^{-3})(480 \ \text{min})}{2}$$
$$= 0.227 \ \text{ppm}$$

Therefore, the TLV is exceeded by the mean TDI vapor concentration in the room over the 8-h averaging period by a factor of 45.4.

TOX.9 IDLH AND LETHAL LEVEL

As described earlier, the *immediately dangerous to life and health* (IDLH) level is the maximum concentration of a substance to which one can be exposed for 30 min without irreversible health effects or death. A lethal level is the concentration at which death is almost certain to occur. The IDLH values were determined by the *National Institute for Occupational Safety and Health* (NIOSH) for the purpose of respirator selection. Respirators provide protection against the inhalation of toxic or harmful materials and may be necessary in certain hazardous situations.

Carbon dioxide is not normally considered to be a threat to human health. It is exhaled by humans and is found in the atmosphere at about 3000 parts per million (ppm). However, at high concentrations it can be a hazard and may cause headaches, dizziness, increased heart rate, asphyxiation, convulsions, or coma.

Two large bottles of flammable solvent were ignited by an undetermined ignition source after being knocked over and broken by a janitor while cleaning a 10 ft $\times$ 10 ft $\times$ 10 ft research laboratory. The laboratory ventilator was shut off and the fire was fought with a 10-lb CO_2 fire extinguisher. As the burning solvent had covered much of the floor area, the fire extinguisher was completely emptied in extinguishing the fire.

The IDLH level for CO_2 set by NIOSH is 50,000 ppm. At that level, vomiting, dizziness, disorientation, and breathing difficulties occur after a 30 min exposure. At a 10% level (100,000 ppm), death can occur after a few minutes even if the oxygen in the atmosphere would otherwise support life.

Calculate the concentration of CO_2 in the room after the fire extinguisher is emptied. Does it exceed the IDLH value? Assume that the gas mixture in the room is uniformly mixed, that the temperature in the room is 30°C (warmed by the fire above normal room temperature of 20°C), and that the ambient pressure is 1 atm.

Solution

First, calculate the number of moles of CO_2, n_{CO_2}, discharged by the fire extinguisher:

$$n_{CO_2} = (10 \text{ lb } CO_2)(454 \text{ g/lb})/(44 \text{ g/gmol } CO_2)$$

$$= 103 \text{ gmol of } CO_2$$

The volume of the room, V, is

$$V = (10 \text{ ft})(10 \text{ ft})(10 \text{ ft})(28.3 \text{ L/ft}^3)$$

$$= 28,300 \text{ liters}$$

The ideal gas law is used to calculate the total number of moles of gas in the room, n:

$$n = \frac{PV}{RT}$$

$$= \frac{(1 \text{ atm})(28,300 \text{ L})}{\left(0.08206 \dfrac{\text{atm} \cdot \text{L}}{\text{gmol} \cdot \text{K}}\right)(303 \text{ K})}$$

$$= 1138 \text{ gmol gas}$$

The concentration or mole fraction of CO_2 in the room, x_{CO_2}, may now be calculated:

$$x_{CO_2} = (\text{gmol } CO_2)/(\text{gmol gas})$$

$$= (103 \text{ gmol } CO_2)/(1138 \text{ gmol gas})$$

$$= 0.0905$$

$$= 9.05\%$$

The IDLH level is 5.0% and the lethal level is 10.0%. Therefore, the level in the room of 9.05% does exceed the IDLH level for CO_2. It is also dangerously close to the lethal level. The person extinguishing the fire is in great danger and should take appropriate safety measures.

If a dangerous level is present, consideration must be given to using protective equipment such as a respirator. Respirators protect the individual from harmful materials in the air. Air-purifying respirators will clean the air but will not protect users against an oxygen-deficient atmosphere. Thus, air-purifying respirators are not used in IDLH applications. The only respirators that are recommended for fighting fires are self-contained breathing apparatuses with full facepieces. Recommendations for the selection of the proper respirator are based on the most restrictive of the occupational exposure limits.

If a toxic material is dispersed into the air, the engineer/scientist must know how high its concentration can be without causing danger to people. The Occupational Safety and Health Administration (OSHA) has set concentration levels for many substances. This concentration level is called the *permissible exposure limit* (PEL). The PEL is synonymous in most application with the TLV-TWA (threshold limit value–time-weighted average).

Most toxicity studies are performed by using test animals. Humans obviously cannot be exposed to lethal concentrations of toxic materials to determine toxicity. Please note that there are some differences in chemical tolerance levels for humans and animals. The differences include metabolism and other factors. Thus, toxicity tests are not always easy to interpret. However, the results of animal toxicity tests are used to guide the selection of acceptable exposure limits for humans.

TOX.10   ACCIDENTAL VAPOR EMISSION

Many industrial chemicals are toxic or flammable, or sometimes both. Regardless of whether the chemical is toxic or flammable, it can present a danger to plant operators and the public if it is released from its container. Substantial efforts are taken to assure that toxic or flammable materials are not spilled or released from containment. There is always a chance, however, that such materials might be released. Therefore, provisions must be made to protect the plant operators and anyone who lives or works in the vicinity.

A certain poorly ventilated chemical storage room ($10\,\text{ft} \times 20\,\text{ft} \times 8\,\text{ft}$) has a ceiling fan but no air conditioner. The air in the room is at $51\,°\text{F}$ and $1.0\,\text{atm}$ pressure. Inside this room, a 1-lb bottle of iron (III) sulfide (Fe_2S_3) sits next to a bottle of sulfuric acid containing 1 lb H_2SO_4 in water. An earthquake (or perhaps the elbow of a passing technician) sends the bottles on the shelf crashing to the floor where the bottles break, and their contents mix and react to form iron (III) sulfate [$Fe_2(SO_3)$] and hydrogen sulfide (H_2S).

1. Calculate the maximum H_2S concentration that could be reached in the room assuming rapid mixing by the ceiling fan with no addition of outside air (poor ventilation). Compare your result with the TLV (10 ppm) and IDLH (300 ppm) levels for H_2S.

2. Later, when exhausted from the room, the H_2S mixes with outside air. What will be the final volume of the H_2S cloud when the concentration finally reaches the TLV?

Solution

1. Balance the chemical equation:

amount	Fe_2S_3	$+$	$3H_2SO_4$	$\rightarrow$	$Fe_2(SO_4)_3$	$+$	$3H_2S$
before	1 lb		1 lb		0		0
reaction	0.0048 lbmol		0.010 lbmol		0		0

The molecular weights of Fe_2S_3 and H_2SO_4 are 208 and 98, respectively.

The terms *limiting reactant* and *excess reactant* refer to the actual number of moles present in relation to the stoichiometric proportion required for the reaction to proceed to completion. See Problem STC.3 in Chapter 5. From the stoichiometry of the reaction, 3 lbmol of H_2SO_4 are required to react with each lbmol of Fe_2S_3. The sulfuric acid is the limiting reactant and the iron (III) sulfide is the excess reactant. In other words, 0.0144 lbmol of H_2SO_4 is required to react with each 0.0048 lbmol of Fe_2S_3, or 0.030 lbmol of Fe_2S_3 is required to react with 0.010 lbmol of H_2SO_4.

Calculate the moles of H_2S generated, n_{H_2S}:

$$n_{H_2S} = (0.010 \text{ lbmol } H_2SO_4)(3H_2S/3H_2SO_4)$$

$$= 0.010 \text{ lbmol}$$

Next, convert the moles to mass:

$$m_{H_2S} = (0.010 \text{ lbmol } H_2S)(34 \text{ lb/lbmol } H_2S)$$

$$= 0.34 \text{ lb}$$

The final H_2S concentration in the room in ppm, C_{H_2S}, can now be calculated. At 32°F and 1 atm, one lbmol of an ideal gas occupies 359 ft^3; at 51°F, one lbmol occupies

$$V = 359\left(\frac{460 + 51}{460 + 32}\right) = 373 \text{ ft}^3$$

Therefore,

$$C_{H_2S} = \frac{(0.34 \text{ lb})\left(\dfrac{373 \text{ ft}^3}{\text{lbmol air}}\right)\left(\dfrac{\text{lbmol air}}{29 \text{ lb}}\right)(10^6)}{1600 \text{ ft}^3}$$

$$= 2733 \text{ ppm}$$

This concentration of H_2S far exceeds the TLV (10 ppm) as well as the IDLH (300 ppm).

2. Calculate the volume of H_2S in the room at a room concentration of 10 ppm. Determine the dilution factor required to decrease the H_2S concentration to the TVL.

$$\text{Dilution factor} = 2733\,\text{ppm}/10\,\text{ppm} = 273.3$$

Assuming that the outside air is also at 51°C and 1.0 atm, calculate the total gas volume required to reach the TVL.

$$V_{cloud} = (273.3)(1600\,\text{ft}^3)$$
$$= 437,000\,\text{ft}^3$$

Note that this volume includes the gas still in the room.

44 Health Risk Analysis (HRA)

HRA.1 HEALTH RISK ASSESSMENT

List and describe the four major steps in a health risk assessment.

Solution

The four major steps in a health risk assessment are hazard identification, dose–response assessment, exposure assessment, and risk characterization.

A *hazard* is defined as a toxic agent or a set of conditions that has the potential to cause adverse effects to human health or the environment. *Hazard identification* is a process that determines the potential human health effects that could result from exposure to a hazard. This process requires a review of the scientific literature. The literature could include information published by the Environmental Protection Agency (EPA), federal or state agencies, and health organizations.

Dose–response, or *toxicity, assessment* is the determination of how different levels of exposure to a hazard or pollutant affect the likelihood or severity of health effects. Responses/effects can vary widely since all chemicals and contaminants vary in their capacity to cause adverse effects. The dose–response relationship can be evaluated for either carcinogenic or noncarcinogenic substances.

Exposure assessment is the determination of the magnitude of exposure, frequency of exposure, duration of exposure, and routes of exposure by contaminants to human populations and ecosystems. There are three components to this step. The first is the identification of contaminants being released. The second is an estimation of the amounts of contaminants released from all sources or the source of concern. Third, there is an estimation of the concentration of contaminants.

Finally, in *risk characterization*, toxicology and exposure data/information are combined to obtain a qualitative or quantitative expression of risk.

An expanded treatment of and comparative analysis with hazard risk assessment is provided in Problem HZA.1 in Chapter 45.

HRA.2 RISK ASSESSMENT TERMINOLOGY

Define the following terms:

758

1. Risk management
2. Risk communication
3. Risk estimation
4. Risk perception
5. Comparative risk assessment

Solution

1. *Risk management* is an evaluation of various options to reduce the risk to the exposed population. Risk management usually follows a risk assessment. Specific actions that may be involved in risk management include consideration of engineering constraints, regulatory issues, social issues, political issues, and economic issues.

2. *Risk communication* is the part of the risk management process that includes exchanging risk information among individuals, groups, and government agencies. The major challenge in this phase of the risk management process is transferring information from the experienced expert to the nonexperienced, but greatly concerned, public.

3. *Risk estimation* is based on the nature and extent of the source, the chain of events, pathways, and processes that connect the cause to the effects. It is also based on the relationship between the characteristics of the impact (dose) and the type of effects (response).

4. *Risk perception* describes an individual's intuitive judgment of the risk. Risk perception is not often in agreement with the actual level of risk.

5. *Comparative risk assessment* is the comparison of potential risks associated with a variety of activities and situations so that a specific action can be placed in perspective with other risks. An attempt is often made, for example, to compare an individual's risk of death or cancer from exposure to a hazardous waste site with that associated with traveling in an automobile or eating a peanut butter sandwich (both of these latter events have relatively high risks but are perceived by the public to have a relatively low risk when compared to the risk of a hazardous waste site).

HRA.3   RISK MANAGEMENT

To apply risk assessments to large groups of individuals, certain assumptions are usually made about an "average" person's attributes. List the average or standard values used for:

1. Body weight
2. Daily drinking water intake
3. Amount of air breathed per day

4. Expected life span
5. Dermal contact area

Solution

1. Average body weight is 70 kg for an adult and 10 kg for a child.
2. The average daily drinking water intake is 2 liters for an adult and 1 liter for a child.
3. The average amount of air breathed per day is 20 m³ for an adult and 10 m³ for a child.
4. The average expected life span is 70 years.
5. The average dermal contact area is 1000 cm² for an adult and 300 cm² for a child.

HRA.4   REFERENCE DOSE

Describe and illustrate the process of correlating a reference dose (RfD) with a schematic of a dose–response curve. Label both axes and the critical points on the curve.

Solution

A reference dose (RfD) estimates the lifetime dose that does not pose a significant risk to the human population. This estimate may have an uncertainty of one order of magnitude or more. The RfD is determined by dividing the *no observed adverse effect level* (NOAEL) dose of a substance by the product of the uncertainty and modifying factors as shown in the following equation

$$RfD = \frac{NOAEL}{(UF)(MF)}$$

The uncertainty factor (UF) is usually represented as a multiple of 10 to account for variation in the exposed population (to protect sensitive subpopulations), uncertainties in extrapolating from animals to humans, uncertainties resulting from the use of subchronic data instead of data obtained from chronic studies, and uncertainties resulting from the use of the lowest observable adverse effect level (LOAEL) instead of the NOAEL. The modifying factor (MF) reflects qualitative professional judgment of additional uncertainties in the data.

The schematic of the dose–response curve shown in Figure 115 illustrates that the value of the reference dose is less than the value of the NOAEL by a safety factor.

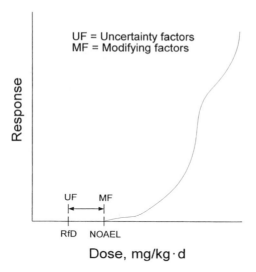

Figure 115. Dose–response curve.

HRA.5   CARCINOGENS

Determine the action level in $\mu g/m^3$ for an 80-kg person with a life expectancy of 70 years exposed to benzene over a 15-year period. The acceptable risk is one incident of cancer per one million persons, or 10^{-6}. Assume a breathing rate of 15 m^3/day and an absorption factor of 75% (0.75). The potency factor for benzene is $1.80 \, mg/(kg \cdot day)^{-1}$.

The following equation has been used in health risk assessment studies for carcinogens:

$$C_m = \frac{RWL}{PIA(\text{ED})}$$

where C_m = action level, i.e., the concentration of carcinogen above which remedial action should be taken

R = acceptable risk or probability of contracting cancer
W = body weight
L = assumed lifetime
P = potency factor
I = intake rate
A = absorption factor, the fraction of carcinogen absorbed by the human body
ED = exposure duration

Solution

Using the equation provided in the problem statement, the action level for the carcinogen benzene for an 80-kg person with a life expectancy of 70 years and an exposure duration of 15 years is

$$C_m = \frac{(10^{-6})(80\,\text{kg})(70\text{ yr})}{\left(1.80\,\dfrac{\text{kg}\cdot\text{d}}{\text{mg}}\right)\left(15\,\dfrac{\text{m}^3}{\text{d}}\right)(0.75)(15\text{ yr})\left(\dfrac{\text{mg}}{1000\,\mu\text{g}}\right)}$$

$$= 0.0184\;\mu\text{g/m}^3$$

The risk for this person would be classified as unacceptable if the exposure exceeds $0.0184\,\mu\text{g/m}^3$ in a 15-year period.

HRA.6   CHEMICAL EXPOSURE IN A LABORATORY

The American Conference of Government and Industrial Hygienist (ACGIH), a private organization of professionals working in the field of industrial hygiene, publishes *threshold limit values* (TLVs) annually for a large number of chemicals. As described earlier, the TLV is the concentration to which workers can be exposed for 40 h a week for 20 years without ill effects. The National Institute for Occupational Safety and Health (NIOSH), a government agency, published *immediately danger-ous to life or health* (IDLH) levels for many chemicals; the IDLH level represents a maximum concentration from which one could escape within 30 min without irreversible health effects.

Many accidents occur in a laboratory environment. This is because a laboratory setting has a great potential for accidental exposure to toxic chemicals, among other dangers. Therefore, laboratory workers must be especially cautious and prepared to deal with exposures should they occur. It is essential to know what chemicals and what quantities are being used and stored in the laboratory. These chemicals should be used and stored properly according to established guidelines and regulations. A Material Safety Data Sheet (MSDS) describes a chemical and dangers associated with the chemical. MSDSs should be readily available in chemical use and storage areas. In the event of an exposure, safe actions must be taken to minimize injury to health and life.

Over a period of 1 year, 1 gal of each of the four common chemicals shown below has been spilled in an unventilated 20-ft × 20-ft × 8-ft laboratory room at 68°F and 1 atm pressure on separate occasions. The information and data are provided in the following table:

Chemical	TLV (ppm)	IDLH (ppm)	MW (g/gmol)	Evaporation Rate (mL/min)	Density (g/mL)
Bromine	0.1	10	160	5.8	3.119
Pyridine	5	3600	79	0.60	0.982
n-Hexane	50	5000	86	4.1	0.660
Methyl acrylate	10	1000	86	2.3	0.954

1. Calculate how many milliliters of each chemical would have to evaporate in order to reach the TLV concentration and how much would have to evaporate to reach the IDLH level.
2. Determine how long it would take each chemical to reach the TLV concentration and how long it would take to reach the IDLH level.
3. Calculate the air flowrate in ft^3/min that would be needed to ventilate the laboratory to keep the chemical concentrations below their respective TLVs.
4. Comment on the hazards of the bromine spill.
5. Comment on the need for ventilation and controls in working with the three chemicals.

Solution

As the solution methodology is identical for each compound, calculations are shown for bromine, and answers are provided for all compounds.

First, use the ideal gas law to find the number of gmol of gas, n_{gas} in the laboratory at 1 atm and $68°F = 293$ K:

$$V_{lab} = (20 \text{ ft})(20 \text{ ft})(8 \text{ ft})(28.3 \text{ L/ft}^3)$$
$$= 90,600 \text{ L}$$

$$n_{gas} = \frac{(1 \text{ atm})(90,600 \text{ L})}{\left(0.0821\dfrac{\text{atm} \cdot \text{L}}{\text{gmol} \cdot \text{K}}\right)(293 \text{ K})}$$
$$= 3760 \text{ gmol}$$

1. Determine the volume of liquid of each chemical that must evaporate to bring the concentration to the TLV with the equation

$$\text{mL to reach TLV} = \frac{[\text{TLV(ppm)}][n_{gas}(\text{gmol})][\text{MW(g/gmol)}]}{\rho(\text{g/mL})}$$

For Br$_2$,

$$= \frac{(0.1/10^6)(3760 \text{ gmol})(160 \text{ g/gmol})}{3.119 \text{ g/mL}}$$

$$= 0.0193 \text{ mL Br}_2$$

Determine the volume of liquid of each chemical that must evaporate to bring the concentration to the IDLH with the equation

$$\text{mL to reach IDLH} = \frac{[\text{IDLH(ppm)}][n_{gas}(\text{gmol})][\text{MW(g/gmol)}]}{\rho(\text{g/mL})}$$

For Br$_2$,

$$= \frac{(10/10^6)(3760 \text{ gmol})(160 \text{ g/gmol})}{3.119 \text{ g/mL}}$$

$$= 1.93 \text{ mL Br}_2$$

The calculated values for the other chemicals are shown in the following table:

Chemical	TLV (mL)	IDLH (mL)
Pyridine	1.51	1089
n-Hexane	3.39	339
Methyl acrylate	24.5	2450

2. Determine the time to reach the TLV concentration of each chemical with the equation

$$\text{Time to reach TLV} = \frac{\text{TLV volume evaporated (mL)}}{\text{Rate of evaporation (mL/min)}}$$

For Br$_2$,

$$= (0.0193 \text{ mL})/(5.8 \text{ mL/min})$$

$$= 0.0033 \text{ min}$$

Determine the time to reach the IDLH concentration of each chemical with the equation

$$\text{Time to reach IDLH} = \frac{\text{IDLH volume evaporated (mL)}}{\text{Rate of evaporation (mL/min)}}$$

For Br_2,

$$= (1.93 \text{ mL}/5.8 \text{ mL/min})$$

$$= 0.33 \text{ min}$$

The calculated values for the other chemicals are shown in the following table:

Chemical	TLV (min)	IDLH (min)
Pyridine	2.5	1815
n-Hexane	1.47	147
Methyl acrylate	5.98	598

3. Determine the flowrate of air, q, needed to remain below the TLV level of each chemical in the laboratory with the equation

$$q(\text{ft}^3/\text{min}) = \frac{[\text{Rate of evaporation (mL/min)}][\rho(\text{g/mL})](R)(T)}{[\text{MW(g/gmol)}](P)[\text{TLV(ppm)}]}$$

For Br_2,

$$= \frac{(5.8 \text{ mL/min})(3.119 \text{ g/mL})\left(0.0821 \dfrac{\text{L} \cdot \text{atm}}{\text{mole} \cdot \text{K}}\right)(293 \text{ K})(0.0353 \text{ ft}^3/\text{L})}{(160 \text{ g/gmol})(1 \text{ atm})((0.1/10^6)\text{ppm})}$$

$$= 960,100 \text{ ft}^3/\text{min}$$

The calculated values for the other chemicals are shown in the following table:

Chemical	Flowrate (ft³/min)
Pyridine	1267
n-Hexane	2167
Methyl acrylate	535

Note that the volume of gas in the laboratory is not included in this calculation. In practice, the calculated flowrates would be multiplied by a safety factor of 3–10. The numerical value of this safety factor depends on the layout of the laboratory, the nature of the work being done in it, the position of the air vents, etc.

4. A bromine spill would cause bromine vapor concentrations in the unventilated laboratory to reach IDLH values in less than 1 min, an instantaneously dangerous situation. Even if the laboratory were ventilated, the flowrate necessary to keep the concentration of the atmosphere below the TLV cannot be achieved by practical means. Procedures for the prevention of such spills must be carefully developed, and materials necessary to neutralize a bromine spill must be readily available to minimize the danger posed by the use of this chemical.

5. Although the time to reach the IDLH levels of pyridine, *n*-hexane, and methyl acrylate are substantially longer than bromine, the time to reach the TLV of all of these chemicals in the unventilated laboratory is only a few minutes. For each of these chemicals, the concentration in the laboratory could be kept below the TLV with reasonable air flowrates, even when increased by an appropriate safety factor. A spill or routine repeated use of these chemicals in an unventilated laboratory would definitely result in a significant health risk and should be avoided. If a large safety factor is necessary, requiring air flowrates that are impractical, other measures to reduce the concentration in the air, such as localized hoods, should be considered.

The bromine spill results in an instantaneously dangerous situation. Ventilation rates necessary to keep the concentration of the atmosphere below the TLV during a bromine spill are difficult to achieve. For the other chemicals, it is practical to achieve the required ventilation rate; however, reliable systems must be provided to maintain the required air flow.

Laboratory safety is absolutely essential for any organization. The protection of human health is foremost when dealing with toxic chemicals. Proper protective clothing and equipment should be used by all who may be exposed to these chemicals. Correct storage and handling procedures should always be employed in the laboratory. Emergency plans should be developed and made aware to all involved in case of an accidental exposure.

HRA.7   VENTILATION

Ventilation is an important method of reducing the level of toxic airborne contaminants in the process environment. Ventilation includes "general (dilution) ventilation" and "local exhaust (vent) ventilation." General ventilation involves dilution of air and hence the term "dilution ventilation." Local exhaust ventilation is a method of removing contaminants before they enter the workplace air. Local ventilation is typically achieved by employing a hood that covers the specific area of contamination.

The Occupational Safety and Health Administration (OSHA) has set the permissible exposure level (PEL) of vinyl chloride (VC) at 1.0 part per million (ppm) as a maximum time-weighted average (TWA) for an 8-h workday. The PEL was set at 1.0 ppm because vinyl chloride is a suspected carcinogen. Thus, if vinyl chloride escapes into the air, the concentration of vinyl chloride must be maintained at or below 1.0 ppm. The major source of VC escape into the workplace air in typical process conditions is fugitive emissions from pipe connections such as valves, flanges, and pump seals.

The vinyl chloride fugitive emission rate in a process was estimated to be 10 g/min by a series of bag tests conducted for the major pieces of connections (i.e., flanges and valves) and pump seals. Determine the flowrate of air (25°C) necessary to maintain the PEL level of 1.0 ppm by dilution ventilation. Correct for incomplete mixing by employing a safety factor of 10. Also consider partially enclosing the process and using local exhaust ventilation. Assume that the process can be carried out in a hood with an opening of 30 in. wide by 25 in. high with a face velocity greater than 100 ft/min to ensure high capture efficiency. What will be the flowrate of air required for local exhaust ventilation? Which ventilation method seems better?

Solution

Convert the mass flowrate of the vinyl chloride (VC) to volumetric flowrate, q, in cm^3/min and acfm. First, use the ideal gas law to calculate the density.

$$\rho = \frac{P(MW)}{RT}$$

$$= \frac{(1\,atm)(78\,g/gmol)}{\left(82.06\,\dfrac{cm^3 \cdot atm}{mol \cdot K}\right)(298\,K)}$$

$$= 0.00319\,g/cm^3$$

$$q = (mass\ flowrate)/(density)$$

$$= (10\,g/min)/(0.00319\,g/cm^3)$$

$$= 3135\ cm^3/min$$

$$= 0.1107\,acfm$$

Calculate the air flowrate in acfm, q_{air}, required to meet 1 ppm PEL with the equation

$$q_{air} = (0.1107\,acfm)/10^{-6}$$

$$= 1.107 \times 10^5\,acfm$$

Apply the safety factor to calculate the actual air flowrate for dilution ventilation:

$$q_{air,dil} = (10)(1.107 \times 10^5 \, \text{acfm})$$
$$= 1.107 \times 10^6 \, \text{acfm}$$

Now consider the local exhaust ventilation by first calculating the face area of the hood, A, in square feet:

$$A = (\text{Height}) (\text{Width})$$
$$= (30 \, \text{in.})(25 \, \text{in.})(\text{ft}^2/144 \, \text{in.}^2)$$
$$= 5.21 \, \text{ft}^2$$

The air flowrate in acfm, $q_{air,exh}$, required for a face velocity of 100 ft/min is then

$$q_{air,exh} = (5.21 \, \text{ft}^2)/(100 \, \text{ft/min})$$
$$= 521 \, \text{acfm}$$

Since the air flowrate for dilution ventilation is approximately 2000 times higher than the local ventilation air flowrate requirement, and considering the high cost of large blowers to handle high air flowrates, the local ventilation method is the better method for this case.

HRA.8   RESPIRATORS

Briefly describe the role respirators can play in health risk management.

Solution

Respirators provide protection against inhaling harmful materials. Different types of respirators may be used depending on the level of protection desired. For example, supplied-air respirators (e.g., a self-contained breathing apparatus) may be required in situations where the presence of highly toxic substances is known or suspected and/or in confined spaces where it is likely that toxic vapors may accumulate. On the other hand, a full-face or half-face air-purifying respirator may be used in situations where measured air concentrations of identified substances will be reduced by the respirator below the substance's threshold limit value (TLV) and the concentration is within the service limit of the respirator (i.e., that provided by the canister).

Air-purifying respirators contain cartridges (or canisters) that contain an adsorbent, such as charcoal, to adsorb the toxic vapor and thus purify the breathing air. Different cartridges can be attached to the respirator depending on the nature of the contaminant. For example, a cartridge for particulates will contain a filter rather than charcoal. The charcoal in a cartridge acts like a fixed-bed adsorber. The performance of any charcoal cartridge may be evaluated by treating it as a fixed-bed adsorber.

HRA.9   PERFORMANCE OF A CARBON CARTRIDGE RESPIRATOR

A respirator cartridge contains 80 g of the above blend of charcoal, and tests have shown that breakthrough (when A starts to be emitted from the cartridge) will occur when 80% of the charcoal is saturated. How long will this cartridge be effective if the ambient concentration of substance A is 700 ppm and the temperature is 30°C? Assume that the breathing rate of a normal person is 45 L/min (45,000 cm^3/min).

For a particular blend of carbon cartridge (for organic vapor), the adsorption potential (equilibrium concentration) for substance A can be expressed by the following equation:

$$\log_{10}(C_A) = -0.11\left[\left(\frac{T}{V}\right)\log_{10}\left(\frac{p'_A}{p_A}\right)\right] + 2.076$$

where C_A = amount of A adsorbed in charcoal, cm^3 liq/100 g charcoal at partial pressure. p_A

T = temperature, K

V = molar volume of A (as liquid) at the normal boiling point, 100 cm^3/gmol for substance A

p'_A = saturation (vapor pressure) of A

p_A = partial pressure of A

The vapor pressure of substance A at 30°C is 60 mm Hg and the liquid density is 1.12 g/cm^3. The molecular weight is 110.

Solution

First, calculate the partial pressure of substance A, p_A, in units of mm Hg:

$$p_A = (700 \times 10^{-6})(760 \text{ mm Hg})$$
$$= 0.532 \text{ mm Hg}$$

The ratio of vapor pressure to the partial pressure, p'_A/p_A, is then

$$p'_A/p_A = (60 \text{ mm Hg})/(0.532 \text{ mm Hg})$$
$$= 112.8$$

The amount of A adsorbed per 100 g of charcoal in $cm^3/100\,g$ charcoal is

$$\log_{10}(C_A) = -0.11\left[\left(\frac{T}{V}\right)\log_{10}\left(\frac{p'_A}{p_A}\right)\right] + 2.076$$

$$= -0.11\left[\left(\frac{30 + 273K}{100\,cm^3/gmol}\right)\log_{10}(112.8)\right] + 2.076$$

$$= 1.392$$

$$C_A = 24.65\,cm^3/100\,g$$

The mass of substance A adsorbed in the cartridge in grams at 80% saturation (i.e., at breakthrough) is

$$m_A = (0.8)\left(\frac{24.65\,cm^3\,of\,A}{100\,g\,charcoal}\right)(80\,g\,charcoal)(1.12\,g\,A/cm^3)$$

$$= 17.67\,g\,of\,A$$

Next, calculate the volumetric flowrate of A inhaled through the cartridge from the vapor pressure and breathing rate:

$$q_{air} = (700 \times 10^{-6})(45,000\,cm^3/min)$$

$$= 31.5\,cm^3/min$$

The intake mass flowrate of A is

$$\dot{m}_A = \frac{Pq(MW)}{RT}$$

$$= \frac{(1\,atm)(31.5\,cm^3/min)(110\,g/gmol)}{\left(82.06\,\dfrac{cm^3 \cdot atm}{mol \cdot K}\right)(303\,K)}$$

$$= 0.139\,g/min$$

Finally, the time for breakthrough in minutes is

$$t = (17.67\,g)/(0.139\,g/min)$$

$$= 127\,min$$

The typical economic concentration limit of organic vapor cartridges is approximately 1000 ppm. If the ambient or local concentration is above 1000 ppm, other methods of personal protection are recommended. It is also important to select the appropriate type of adsorbent since some toxic chemicals are not readily adsorbed by charcoal. For example, hydrogen cyanide is not well adsorbed by charcoal. Since the *immediately dangerous to life and health* (IDLH) level of hydrogen cyanide is 50 ppm and the odor threshold is greater than 50 ppm, by the time the worker smells the vapor, it will be too late to avoid death. Therefore, it is crucial to select a cartridge adsorbent that is specifically designed for hydrogen cyanide. Examples of such special cartridges are the cartridges for chlorine gas and pesticide vapors. Typically, cartridges are color coded so that they are easily distinguished (e.g., black for common organic vapors).

HRA.10   RISK CHARACTERIZATION

Reynolds, Theodore, and Morris Associates have been contracted by the EPA to provide two separate benzene lifetime risk estimates for the population in the United States. The two calculations are to be based on

1. One tenth (0.1) of the OSHA PEL (permissible exposure limit)
2. An ambient benzene concentration (ABC) of $9.5 \times 10^{-3}\,mg/m^3$

The following data are provided by the EPA:

OSHA PEL $= 1.0\,ppmv$
Breathing rate $= 20\,m^3/day$
Percent absorption (PEL) $= 75\%$
Percent absorption (ABC) $= 50\%$
Dose–response relationship $= \dfrac{0.028kg \cdot day \cdot probability}{mg \cdot lifetime}$
Exposure duration $= 70$ years
Average weight of individual $= 70\,kg$
Exposed population $= 200{,}000{,}000$
Ambient conditions $= 25°C$, 1 atm

Solution

The calculation to characterize the risk to a population exposed to an event or chemical agent can be divided into the four steps described here. The units correspond to a gaseous chemical emission to the atmosphere posing a risk to a population.

1. Exposure (E)

$$E\left(\frac{mg \cdot day}{m^3}\right) = \left[\text{Pollutant concentration}\left(\frac{mg}{m^3}\right)\right][\text{Exposure duration (day)}]$$

This represents the total exposure over the days in question. The days can often be the lifetime of an individual (25,550 days). Thus,

$$E = (PC)(ED)$$

2. Dose (D)

$$D\left(\frac{mg}{kg \cdot day}\right) = \left[E\left(\frac{mg \cdot day}{m^3}\right)\right]\left[\text{Receptor dose factor }\left(\frac{m^3}{kg \cdot day^2}\right)\right]$$

This represents the average mass of pollutant intake per unit mass of receptor on a daily basis. The receptor dose factor (RDF) is given by

$$RDF = \frac{\left[\text{"Contract rate"}\left(\frac{m^3}{day}\right)\right]\left[\text{Intake fraction }\left(\frac{\%}{100}\right)\right]}{[\text{Average receptor weight (kg)}][\text{Exposure duration (day)}]}$$

Thus,

$$D = E(RDF)$$

3. Lifetime individual risk (LIR)

$$LIR\left(\frac{\text{Probability}}{\text{Lifetime}}\right) = \left[D\left(\frac{mg}{kg \cdot day}\right)\right]$$
$$\times \left[\text{Dose–response relationship }\left(\frac{kg \cdot day \cdot probability}{mg \cdot lifetime}\right)\right]$$

This represents an individual's risk over a lifetime. Thus,

$$LIR = D(DRR)$$

4. Risk to exposed population (REP)

$$REP\left(\frac{\text{individuals}}{yr}\right) = \frac{\left[LIR\left(\frac{\text{probability}}{\text{lifetime}}\right)\right][\text{Exposed population (individuals)}]}{\text{Years per lifetime}\left(\frac{yr}{\text{lifetime}}\right)}$$

This provides a reasonable estimate of the number of individuals or cases in an exposed population per year. Thus,

$$REP = (LIR)(EP)/(YPL)$$

The years per lifetime (YPL) is normally 70-yr/lifetime.

Solution

Convert the OSHA PEL for benzene to one tenth its value with units of mg/m^3 and compare with the ambient benzene concentration (ABC).

$$(0.1)(PEL) = 0.1(1.0\,ppmv)$$
$$PEL = 0.1\,ppmv$$
$$= \frac{(0.1/10^6)(78\,g/gmol)(1000\,mg/g)}{0.245\,m^3/gmol}$$
$$= 0.318\,mg/m^3$$

This value is approximately 30 times greater than the ABC.
Calculate the exposure (E) in $mg \cdot day/m^3$ for both the PEL and ABC:

$$ED = (365\,day/yr)(70\,yr)$$
$$= 25{,}550\,day$$
$$E(PEL) = (PC)(ED)$$
$$= (0.318\,mg/m^3)(25{,}550\,day)$$
$$= 8.13 \times 10^3\,mg \cdot day/m^3$$
$$E(ABC) = (PC)(ED)$$
$$= (9.5 \times 10^{-3}\,mg/m^3)(25{,}550\,day)$$
$$= 243\,mg \cdot day/m^3$$

Calculate the receptor dose factor (RDF) in $m^3/(kg \cdot day^2)$ for both cases:

$$RDF(PEL) = \frac{(20\,m^3/day)(0.75)}{(70\,kg)(25{,}550\,day)}$$
$$= 8.39 \times 10^{-6}\,m^3/(kg \cdot day^2)$$
$$RDF(ABC) = \frac{(20\,m^3/day)(0.50)}{(70\,kg)(25{,}550\,day)}$$
$$= 5.59 \times 10^{-6}\,m^3(kg \cdot day^2)$$

Calculate the dose (D) in mg/(kg· day) for both cases:

$$D = E(\text{RDF})$$

$$D(\text{PEL}) = \left(8.13 \times 10^3 \frac{\text{mg} \cdot \text{day}}{\text{m}^3}\right)\left(8.39 \times 10^{-6} \frac{\text{m}^3}{\text{kg} \cdot \text{day}^2}\right)$$

$$= 0.0682 \, \text{mg}/(\text{kg} \cdot \text{day})$$

$$D(\text{ABC}) = \left(243 \frac{\text{mg} \cdot \text{day}}{\text{m}^3}\right)\left(5.59 \times 10^{-6} \frac{\text{m}^3}{\text{kg} \cdot \text{day}^2}\right)$$

$$= 0.00136 \, \text{mg}/(\text{kg} \cdot \text{day})$$

Estimate the lifetime individual risk (LIR) in units of probability/lifetime to those exposed to benzene for both cases:

$$\text{LIR} = D(\text{DRR})$$

$$\text{LIR(PEL)} = \left(0.0682 \frac{\text{mg}}{\text{kg} \cdot \text{day}}\right)\left(0.028 \frac{\text{kg} \cdot \text{day} \cdot \text{probability}}{\text{mg} \cdot \text{lifetime}}\right)$$

$$= 1.91 \times 10^{-3} \, \text{probability}/\text{lifetime}$$

$$\text{LIR(ABC)} = \left(0.00136 \frac{\text{mg}}{\text{kg} \cdot \text{day}}\right)\left(0.028 \frac{\text{kg} \cdot \text{day} \cdot \text{probability}}{\text{mg} \cdot \text{lifetime}}\right)$$

$$= 3.80 \times 10^{-5} \, \text{probability}/\text{lifetime}$$

The annual risk to the exposed population in number of individuals per year for both cases may also be calculated:

$$\text{REP} = (\text{LIR})(\text{EP})/(\text{YPL})$$

$$\text{REP(PEL)} = \frac{\left(1.91 \times 10^{-3} \dfrac{\text{probability}}{\text{lifetime}}\right)(2.00 \times 10^8 \text{ individuals})}{70 \dfrac{\text{yr}}{\text{lifetime}}}$$

$$= 5460 \, \text{individuals}/\text{yr}$$

$$\text{REP(ABC)} = \frac{\left(3.80 \times 10^{-5} \dfrac{\text{probability}}{\text{lifetime}}\right)(2.00 \times 10^8 \text{ individuals})}{70 \dfrac{\text{yr}}{\text{lifetime}}}$$

$$= 109 \, \text{individuals}/\text{yr}$$

As expected, the above results are significantly higher for the PEL case. The 50-fold factor difference can be primarily attributed to the pollutant concentration (PC) employed in both calculations; the OSHA PEL is based, to an extent, on workplace conditions rather than ambient conditions in the atmosphere, which can be orders of magnitude lower. Regarding an individual's lifetime risk, the "workman's" risk is approximately 1 in 1000 or 10^{-3} over his/her lifetime, as compared to approximately 40 in a million or 40×10^{-6} for an average citizen.

Perhaps the most important consideration in these calculations is the value of the dose–response relationship provided by the EPA. Determining a numerical value for this factor is critical to the calculation; however, a significant degree of uncertainty exists with these probability values provided by researchers in the field.

45 Hazard Risk Analysis (HZA)

HZA.1 HAZARD RISK ASSESSMENT

Discuss the key features of hazard risk assessment. Also discuss the differences between hazard and health risk assessment.

Solution

Hazard risk assessment, among other things, involves the identification of a potential hazard. Hazard risk assessment can also include the process of assessing the likelihood and severity of a hazardous material release from a particular location, developing designs and procedures to minimize the chance of a release, and developing plans to respond to a release. Proper planning, containment, prevention, and emergency response, when used together, minimize both the likelihood and risk of exposure to an accidental chemical release. Combining the probability of a hazard occurring and the consequences of a hazard occurring produces a risk characterization value.

As discussed in the previous problem set, the four major steps in a health risk assessment are hazard identification, dose–response assessment, exposure assessment, and risk characterization. A *health risk assessment* initially involves the identification of human health effects attributed to exposure to a chemical, usually on a continuous basis. A *dose–response assessment* determines how different levels of exposure to a hazard or pollutant affect the likelihood or severity of the health effects. An *exposure assessment* determines the extent of human exposure. These are combined to provide a *risk characterization value.*

Regarding risk differences, hazard risk assessment is used for any hazard, particularly accidents. There are hazards practically everywhere, including chemical plants, factories, and homes. Health risk assessment deals specifically with the health effects that can result from continuous (usually) human exposure to chemicals, whereas hazard risk assessment deals with acute (short-term) exposures.

Both risk assessment approaches are very valuable because they thoroughly examine a potential hazard. They explore all possible situations and their probabilities and consequences. Also, they provide a risk characterization value that can be used to compare different hazards. This is very important in relating a hazard to the risk that it presents.

HZA.2 PREPARING FOR EMERGENCIES

Answer the following questions.

1. Briefly explain SARA. Explain what the acronym SARA represents, and identify the other name by which this legislation is known. State why and when this legislation was passed.
2. List at least three specific types of chemical emergencies that might occur in a large metropolitan area.
3. Briefly describe at least five specific features of an emergency preparedness plan that would be put in place to respond to a major accidental release of a volatile hazardous chemical.

Solution

1. Title III of the *Superfund Amendments and Reauthorization Act* (SARA) is a freestanding piece of legislation that is also known as the *Emergency Planning and Community Right-to-Know Act* of 1986. This act requires county, state, and federal government agencies to work with industries to develop "community right-to-know" reporting on hazardous chemicals. Industries are required to report to local emergency planning agencies regarding the quantities and locations of hazardous chemicals stored at their facilities. Government agencies and industry then develop emergency preparedness plans to handle chemical emergencies at these industrial storage and production facilities.
2. Examples of three types of chemical emergencies that could occur in a large metropolitan area are:
   a. A spill from a railroad tanker car containing a hazardous chemical resulting in the release of a gas cloud
   b. An explosion and fire at a petrochemical plant
   c. A corrosive acid leak from storage drums into a nearby river or lake
   These examples provide a sample of the types of industrial accidents that are common in most major and even smaller, industrialized American cities. Readers are encouraged to review local and national newspapers for current examples of such chemical accidents and emergencies.
3. Five typical features of an emergency preparedness plan are:
   a. Evacuation plans and routes for nearby dwellings, schools, offices, and industries
   b. Access routes and predicted response times for emergency teams
   c. Ability to predict the trajectory and concentration of airborne or waterborne toxic releases
   d. Ability to provide emergency services to the site, e.g., adequate municipal water flowrate and pressure to extinguish a major petrochemical fire
   e. Plans for mobilization of area medical personnel to treat casualties associated with major airborne accidental chemical releases

Additional components of emergency management plans are:

a. Methods for communicating risk and evacuation information, etc., to the general public during an emergency
b. Identification of roles and responsibilities of participating entities responding to and controlling hazards
c. Mechanisms for testing and updating emergency management plans through mock drills and exercises
d. Mechanisms for training professional and voluntary response personnel to make sure the plan and services provided by response personnel are up to date
e. Methods for reviewing the performance of personnel involved in real and/or "practice" emergencies in order to update and improve planning and response procedures.

HZA.3   PLANT AND PROCESS SAFETY

Answer the following questions.

1. List the types of process events that can result in a plant accident and discuss the various kinds of equipment failure that can occur in a process plant. Also list several design approaches that should be employed to minimize accidents in a process plant.
2. Discuss the various pieces of protective equipment that may be found at a chemical plant.

Solution

1. The types of process events that can result in a plant accident are:

   Abnormal temperatures
   Abnormal pressures
   Material flow stoppage

   Equipment failures that can occur in a process plant may be described within the major equipment categories of reactors, heat exchangers, vessels, mass transfer unit operations, pipes and valves, and pumps. The failures associated with these categories are discussed below:

Reactors

a. Runaway reaction: During a runaway reaction, heat is not adequately removed from the system. As a result, excessive temperature and/or pressure buildup in the system.
b. Metallic creep failure: Metallic creep failure occurs when temperature limits are exceeded.

c. Deactivation of catalyst: When a catalyst becomes deactivated due to age or contamination, a dramatic change in temperature and or pressure may occur.

Heat Exchangers

a. Fouling: Heat transfer is reduced when deposited materials accumulate in the heat exchanger and foul the heat exchanger.
b. Tube rupture: Tube rupture may be caused by fouling, which leads to high fluid velocities, tube vibration, corrosion, or erosion.
c. Leakage: Leakage is usually caused by corrosion of baffles in the heat exchanger.

Vessels

a. Pressure/temperature excursions: High pressure and temperature in a vessel may result in tank rupture, especially when the condition repeats itself.
b. Stability: Vessels, tanks, and phase separators are designed according to specifications. Factors such as fabrication, construction, and corrosion must be considered when selecting a vessel for a particular service.

Mass Transfer Unit Operations

a. Compatibility: The materials used to construct supports, plates, and valves must be compatible with the service for which the device is intended.
b. Reactivity: Some metals may react with certain materials or chemicals.

Pipes and Valves

a. Insufficient or ineffective supports
b. Poor weld quality
c. Temperature stress
d. Overpressure
e. Dead ends
f. Material compatibility

Pumps

a. Gland failure
b. Dead-heading: Dead-heading a pump is a condition where the pump is allowed to run against a closed valve. When this occurs, the temperature rise may lead to seal damage and leakage of process fluid.

Plant safety and design may be divided into four major areas. These areas are plant equipment and layout, controls, emergency safety devices, and safety factors and backup. These four areas are discussed below:

Plant and Equipment Layout

a. Adequate water supplies, electrical power sources, and roadways should be available at the plant location.

b. Population centers should be avoided.

c. Climate and natural hazards should be evaluated.

d. Equipment should be adequately spaced. When possible, the equipment should be located in a grid-type pattern.

e. Offices, laboratories, cafeteria, storage, loading, and transportation facilities should all be located on the periphery of the plant site.

f. Fire protection equipment should be readily available at all locations within the plant.

Controls

a. All instruments and controls should be of the fail-safe type.

b. Instruments should be made of corrosion-resistant material.

c. Instruments should be easily accessible for inspection and maintenance.

d. Audible and visible alarms should be provided so operators can determine what type of emergency is occurring.

e. Separate indicators should be used for each critical hazard point of the operation.

Emergency Safety Devices

a. All vessels that may be subject to changes in internal pressure should be equipped with pressure and vacuum relief valves.

b. A relief piping system should be sized for the maximum flow resulting from simultaneous relief of safety valves activated by a common failure mode.

c. Relief piping should be in a horizontal plane to avoid the trapping of liquids.

Safety Factors and Backup

a. Proper design and safety factors should be used to size all equipment and structural components.

b. Backup systems should be investigated for critical operations in the process.

2. Various types of protective equipment found at a chemical plant include:

Relief Devices and Collection Systems

a. Safety valves

b. Rupture disks

c. Piping networks

Treatment Systems

a. Liquid disengagement
b. Gas quenching
c. Scrubbing

Disposal Systems

a. Atmospheric venting
b. Flaring (elevated and ground)
c. Recovery and containment: The recovery and containment of materials includes returning the material to the process, containment of the material external to the process, and flare gas recovery.

HZA.4   SERIES AND PARALLEL SYSTEMS

Determine the reliability of the electrical system shown in Figure 116 using the reliabilities shown beneath each component.

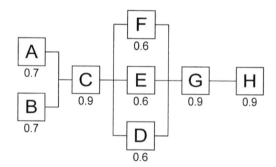

Figure 116. Diagram for Problem HZA.4.

Solution

Many systems consisting of several components can be classified as series, parallel, or a combination of both. A series system is one in which the entire system fails to operate if any one of its components fails to operate. If such a system consists of n components that function independently, then the reliability of the system is the product of the reliabilities of the individual components. If R_s denotes the reliability of a series system and R_i denotes the reliability of the ith component. $(i = 1, 2, \ldots, n)$, then the formula for R_s, the reliability of a series system, is as follows:

$$R_s = R_1 R_2 \ldots R_n = \prod_i R_i$$

A parallel system is one that fails to operate only if all of its components fail to operate. If R_i is the reliability of the ith component, then $1 - R_i$ is the probability that the ith component fails. Assuming all n components function independently, the probability that all n components fail is $(1 - R_1)(1 - R_2) \ldots (1 - R_n)$. Subtracting this product from unity yields the following formula for R_p, the fractional reliability of a parallel system,

$$R_p = 1 - (1 - R_1)(1 - R_2) \ldots (1 - R_n) = 1 - \prod_i (1 - R_1)$$

The reliability formulas for series and parallel systems can be used to obtain the reliability of a system that combines features of series and parallel systems.

The reliability of the parallel subsystem consisting of components A and B in the problem statement is obtained by applying the equation for a parallel system:

$$R_p = 1 - (1 - 0.7)(1 - 0.7)$$
$$= 0.91$$

The reliability of the parallel subsystem consisting of components D, E, and F is

$$R_p = 1 - (1 - 0.6)(1 - 0.6)(1 - 0.6)$$
$$= 0.936$$

The reliability of the entire system is obtained by applying the equation for a series system:

$$R_p = (0.91)(0.9)(0.936)(0.9)(0.9)$$
$$= 0.621$$

HZA.5   PROBABILITY DISTRIBUTION FUNCTION

The difference between the magnitude of a large earthquake, as measured on the Richter scale, and the threshold value of 3.25, is a random variable X having the following probability distribution function (PDF):

$$f(x) = 1.7 \exp(-1.7x) \qquad x > 0$$
$$= 0 \qquad \text{elsewhere}$$

Find the probability that X will have a value between 2 and 6, i.e., $P(2 < X < 6)$, and find the variance of X.

Solution

The probability distribution of a random variable concerns the distribution of probability over the range of the random variable. The distribution of probability is specified by the probability distribution function (PDF). The random variable may be discrete or continuous. Special PDFs finding application in risk analysis are considered in later problems. The PDF of a continuous random variable X has the following properties:

1. $\int_a^b f(x) \, dx = P(a < X < b)$
2. $f(x) \geq 0$
3. $\int_{-\infty}^{\infty} f(x) \, dx = 1$

where $P(a < X < b) =$ probability assigned to an outcome or an event corresponding to the number x in the range of X between a and b

$f(x) =$ PDF of the continuous random variable X

Property 1 indicates that the PDF of a continuous random variable generates probability by integration of the PDF over the interval whose probability is required. When this interval contracts to a single value, the integral over the interval becomes zero. Therefore, the probability associated with any particular value of a continuous random variable is zero. Consequently, if X is continuous,

$$P(a \leq X \leq b) = P(a < X \leq b)$$
$$= P(a < X < b)$$

Property 2 restricts the values of $f(x)$ to non-negative numbers. Property 3 follows from the fact that

$$P(-\infty < X < \infty) = 1$$

The PDF can also provide information on the average value of X, the expected value of X, $E(X)$, and the variance of X, σ^2.

By definition,

$$\sigma^2 = E(X^2) - [E(X)]^2$$

where

$$E(X) = \int_{-\infty}^{\infty} xf(x)\, dx$$

$$E(X^2) = \int_{-\infty}^{\infty} x^2 f(x)\, dx$$

Note that the lower limit of the integral can begin at zero rather than negative infinity since $f(x) = 0$ at $x \leqslant 0$. The expected value of X, $E(X)$, is also called "the mean of X" and is often designated by the special symbol μ. The variance is calculated from

$$\sigma^2 = E(X^2) - [E(X)]^2$$

Since the range is between 2 and 6, the equation for $f(x)$ for $x > 0$ applies:

$$P(2 < X < 6) = \int_{2}^{6} f(x)\, dx$$

$$= \int_{2}^{6} 1.7 e^{-1.7x}\, dx$$

Since

$$\int e^{ax}\, dx = \frac{1}{a} e^{ax}$$

$$P(2 < X < 6) = \left(\frac{1.7}{-1.7}\right) e^{-1.7x}\Big|_{2}^{6}$$

$$= -[e^{-1.7(6)} - e^{-1.7(2)}]$$

$$= e^{-3.4} - e^{-10.2}$$

$$= 0.0333$$

By definition,

$$E(X) = \int_{0}^{\infty} x(1.7 e^{-1.7x})\, dx$$

Since

$$\int xe^{ax}\,dx = \left(\frac{ax-1}{a^2}\right)e^{ax}$$

$$E(X) = 1.7\left[\frac{-1.7x-1}{(-1.7)^2}\right]e^{-1.7x}\Big|_0^\infty$$

$$= \left[\frac{1.7}{(-1.7)^2}\right]\{[-1.7(\infty)-1](e^{-1.7(\infty)}) - [-1.7(0)-1](e^{-1.7(0)})\}$$

$$= \left(\frac{1}{1.7}\right)(0+1)$$

$$= 0.5882$$

By definition,

$$E(X^2) = \int_0^\infty x^2(1.7e^{-1.7x})\,dx$$

Since

$$\int x^2 e^{ax}\,dx = \left(\frac{1}{a}\right)(x^2 e^{ax}) - \left(\frac{2}{a}\right)\int xe^{ax}\,dx$$

$$E(X^2) = 1.7\left(\frac{1}{-1.7}\right)(x^2 e^{-1.7x})\Big|_0^\infty - \left(\frac{2}{-1.7}\right)(0.5882)$$

$$= \left(\frac{1.7}{-1.7}\right)[(\infty)^2 e^{-1.7(\infty)} - (0)^2 e^{-1.7(0)}] - \left(\frac{2}{-1.7}\right)(0.5882)$$

$$= -[0-0] - \left(\frac{2}{-1.7}\right)(0.5882)$$

$$= 0.6920$$

The variance, σ^2, is then

$$\sigma^2 = E(X^2) - [E(X)]^2$$

$$= 0.6920 - (0.5882)^2$$

$$= 0.3460$$

HZA.6   BINOMIAL DISTRIBUTION

A coolant sprinkler system in a reactor has 20 independent spray components, each of which fails with probability of 0.1. The coolant system is considered to "fail" only if 4 or more of the sprays fail. What is the probability that the sprinkler system fails?

Solution

Several probability distributions figure prominently in reliability calculations. The binomial distribution is one of them. Consider n independent performances of a random experiment with mutually exclusive outcomes that can be classified "success" and "failure." These outcomes do not necessarily have the ordinary connotation of success or failure. Assume that P, the probability of success on any performance of the random experiment, is constant. Let $q = 1 - P$ be the probability of failure. The probability distribution of X, the number of successes in n performances of the random experiment, is a binomial distribution with probability distribution function (PDF) specified by

$$f(x) = \frac{P^x q^{n-x} n!}{x!(n-x)!} \qquad x = 0, 1, 2, \ldots, n$$

where $f(x) =$ probability of x successes in n performances
$\qquad n =$ number of independent performances of a random experiment

The binomial distribution can therefore be used to calculate the reliability of a redundant system. A redundant system consisting of n identical components is a system that fails only if more than r components fail. Typical examples include single-usage equipment such as missile engines, short-life batteries, and flash bulbs that are required to operate for one time period and not to be reused.

Assume that the n components of the spray system are independent with respect to failure, and that the reliability of each is $1 - P$. One may associate "success" with the failure of a component. Then X, the number of failures, has a binomial PDF and the reliability of the random system is

$$P(X \leqslant r) = \sum_{x=0}^{r} \frac{P^x q^{n-x} n!}{x!(n-x)!}$$

For this problem, X has a binomial distribution with $n = 20$ and $P = 0.10$; the probability that the system fails is given by

$$P(X \geqslant 4) = \sum_{x=4}^{20} \frac{(0.10^x)(0.90^{20-x})20!}{x!(20-x)!}$$

$$= 1 - P(X \leqslant 3)$$

This simplifies the problem:

$$P(X \geqslant 4) = 1 - \sum_{x=0}^{3} \frac{(0.10^x)(0.90^{20-x})20!}{x!(20-x)!}$$

$$= 0.13$$

HZA.7 WEIBULL DISTRIBUTION

The life in hours of a component in an electrostatic precipitator (ESP) is a random variable having a Weibull distribution with $\alpha = 0.025$ and $\beta = 0.50$. What is the average life of the component? What is the probability that the component will last more than 4000 h?

Solution

Frequently, the failure rate of equipment exhibits three stages: a break-in stage with a declining failure rate, a useful life stage characterized by a fairly constant failure rate, and a wearout period characterized by an increasing failure rate. An example of this type of failure rate is the death rate of humans. A failure rate curve exhibiting these three phases is called a "bathtub curve." The Weibull distribution provides a mathematical model of all three stages of the bathtub curve. The probability distribution function (PDF) is given by

$$f(t) = \alpha \beta t^{\beta-1} \exp\left(-\int_0^t \alpha \beta t^{\beta-1}\right) dt$$

$$= \alpha \beta t^{\beta-1} \exp(-\alpha t^\beta) \qquad t > 0, \ \alpha > 0, \ \beta > 0$$

where $\alpha, \beta = $ constants
$\qquad t = $ time

Let T denote the life in hours of the electronic component in the ESP. The PDF of T is obtained by applying the above formula and substituting $\alpha = 0.025$ and $\beta = 0.50$ to yield

$$f(t) = (0.025)(0.50)t^{(0.50-1)} \exp(-0.025t^{0.50}) \qquad t > 0$$

$$= (0.0125)t^{(-0.50)} \exp(-0.025t^{0.50}) \qquad t > 0$$

By definition, the average value of T is given by

$$E(T) = \int_0^\infty t f(t) \, dt$$

$$= \int_0^\infty t(0.0125)t^{(-0.50)} \exp(-0.025t^{0.50}) \, dt$$

$$= 3200 \text{ h}$$

The probability that the component will last more than 4000 h is given by

$$P(T > 4000) = \int_{4000}^\infty f(t) \, dt$$

$$= \int_{4000}^\infty (0.0125)t^{(-0.50)} \exp(-0.025t^{0.50}) \, dt$$

$$= 0.2057$$

HZA.8   PROBABILITY OF A THREAD DEFECT

The measurement of the pitch diameter of the thread of a fitting is normally distributed with mean 0.4008 in. and standard deviation 0.0004 in. The specifications are given as 0.40000 ± 0.0010 in. What is the probability that a "defect" will occur?

Solution

When a variable, T, has a normal distribution, its probability distribution function (PDF) is given by

$$f(T) = \left(\frac{1}{(2\pi)^{1/2}\sigma}\right) \exp\left[-0.5\left(\frac{T-\mu}{\sigma}\right)^2\right] \qquad -\infty < t < \infty$$

where μ = mean value of T
  σ = standard deviation of T

Thus, if T is normally distributed with mean μ and standard deviation σ, then the random variable, $(T - \mu)/\sigma$, is also normally distributed with mean 0 and standard deviation 1. The term $(T - \mu)/\sigma$ is called a *standard normal variable* (designated by Z) and the graph of its PDF is called a standard normal curve. The literature provides a tabulation of areas under a standard normal curve to the right of Z_0 for non-negative values of Z_0. From these tables, probabilities about a standard normal variable, Z, can be determined. (See Chapter 52 for additional details.)

To calculate the probability of a defect occurring, the probability of meeting the specification must be calculated.

First determine the standard normal variable, Z:

$$Z = (X - \mu)/\sigma$$
$$= (X - 0.4008)/0.0004$$

Determine the lower and upper limits of the probability of meeting specification:

$$\text{Lower limit (LL)} = 0.4000 - 0.0010 = 0.3990$$
$$\text{Upper limit (UL)} = 0.4000 + 0.0010 = 0.4010$$

Determine the probability of meeting specification, P_s, from the area under the standard normal curve between the lower and upper limits:

$$P_s = P\left(\frac{LL - \mu}{\sigma} < Z < \frac{UL - \mu}{\sigma}\right)$$
$$= P\left(\frac{0.3990 - 0.4008}{0.0004} < Z < \frac{0.4010 - 0.4008}{0.0004}\right)$$
$$= P\left(-4.5 < Z < 0.5\right)$$

From standard normal table (see Problem NUC.9, Chapter 24).

$$P_s = 0.5 + 0.191 = 0.691$$

The probability of a defect occurring, P_d, is therefore

$$P_d = 1 - P_s$$
$$= 1 - 0.691 = 0.309$$

HZA.9   CODED QUALITY OF BAG FABRIC

Let X denote the coded quality of bag fabric used in a particular utility baghouse. Assume that X is normally distributed with mean 10 and standard deviation 2. Find c such that $P(|X - 10| < c) = 0.90$. Also find k such that $P(X > k) = 0.90$.

Solution

If X is normally distributed with mean 10 and standard deviation 2, then $(X - 10)/2$ is a standard normal variable:

$$P(|X - 10| < c) = P(-c < (X - 10) < c)$$

$$= P\left(\frac{-c}{2} < \frac{X - 10}{2} < \frac{c}{2}\right)$$

$$= P\left(\frac{-c}{2} < Z < \frac{c}{2}\right)$$

Since $P(|X - 10| < c) = 0.90$, the value of c must be such that $P(0 < Z < c/2) = 0.45$. Therefore (from the standard normal table provided in Problem NUC.9, Chapter 24)

$$c/2 = 1.645$$

$$c = 3.29$$

To determine k, note that

$$P(X > k) = P\left(Z > \frac{k - 10}{2}\right)$$

$$= 0.90$$

Therefore (from the table),

$$(k - 10)/2 = -1.28$$

$$k = 7.44$$

HZA.10   MONTE CARLO SIMULATION

According to state regulations, three thermometers (A, B, C) are positioned near the outlet of an afterburner. Assume that the individual thermometer component lifetimes are normally distributed with means and standard deviations given in the following table:

Thermometer	A	B	C
Mean (weeks)	100	90	80
Standard deviation (weeks)	30	20	10

Using the following random numbers from 0 to 1, simulate the lifetime (time to thermometer failure) of the temperature recording system and estimate its mean and standard deviation. The lifetime is defined as the time (in weeks) for one of the thermometers to "fail."

For A		For B		For C	
0.52	0.01	0.77	0.67	0.14	0.90
0.80	0.50	0.54	0.31	0.39	0.28
0.45	0.29	0.96	0.34	0.06	0.51
0.68	0.34	0.02	0.00	0.86	0.56
0.59	0.46	0.73	0.48	0.87	0.82

Solution

Monte Carlo simulation is a procedure for mimicking observations on a random variable that permits verification of results that would ordinarily require difficult mathematical calculations or extensive experimentation. The method normally uses computer programs called random number generators. A random number is a number selected from the interval (0,1) in such a way that the probabilities that the number comes from any two subintervals of equal length are equal. For example, the probability the number is in the subinterval (0.1, 0.3) is the same as the probability that the number is in the subinterval (0.5, 0.7). Thus, random numbers are observations on a random variable X having a uniform distribution on the interval (0,1). This means that the PDF of X is specified by

$$f(x) = 1 \qquad 0 < x < 1$$

$$0 \qquad \text{elsewhere}$$

The above PDF assigns equal probability to subintervals of equal length in the interval (0,1). Using random number generators, Monte Carlo simulation can generate observed values of a random variable having any specified PDF. For example, to generate observed values of T, the time to failure, when T is assumed to have a PDF specified by $f(t)$, first use the random number generator to generate a value of X between 0 and 1. The solution is an observed value of the random variable T having a PDF specified by $f(t)$.

Let T_A, T_B, and T_C denote the lifetimes of thermometer components A, B, and C, respectively. Let T_S denote the lifetime of the system.

The random number generated is the cumulative probability, and the cumulative probability is the area under the standard normal distribution curve. Since the standard normal distribution curve is symmetrical, the negative values of Z and the corresponding area are found by symmetry. For example, as described in the two previous problems,

$$P(Z < -1.54) = 0.062$$

$$P(Z > 1.54) = 0.062$$

$$P(0 < Z < 1.54) = 0.5 - P(Z > 1.54)$$

$$= 0.5 - 0.062$$

$$= 0.438$$

Note that the lifetime or time to failure of each component, T, is calculated using the equation

$$T = \mu + \sigma Z$$

where μ = mean
σ = standard deviation
Z = standard normal variable

First determine the values of the standard normal variable, Z, for component A using the 10 random numbers given in the problem statement and a standard normal table. Then calculate the lifetime of thermometer component A, T_A, using the equation for T.

Random No.	Z (from standard normal table)	$T_A = 100 + 30\,(Z)$
0.52	0.05	102
0.80	0.84	125
0.45	−0.13	96
0.68	0.47	114
0.59	0.23	107
0.01	−2.33	30
0.50	0.00	100
0.29	−0.55	84
0.34	−0.41	88
0.46	−0.10	97

Next, determine the values of the standard normal variable and the lifetime of the thermometer component for component B.

Random No.	Z (from standard normal table)	$T_B = 90 + 20\,(Z)$
0.77	0.74	105
0.54	0.10	92
0.96	1.75	125
0.02	−2.05	49
0.73	0.61	102
0.67	0.44	99
0.31	−0.50	80
0.34	−0.41	82
0.00	−3.90	12
0.48	−0.05	89

Determine the values of the standard normal variable and the lifetime of the thermometer component for component C.

Random No.	Z (from standard normal table)	$T_C = 80 + 10\,(Z)$
0.14	-1.08	69
0.39	-0.28	77
0.06	-1.56	64
0.86	1.08	91
0.87	1.13	91
0.90	1.28	93
0.28	-0.58	74
0.51	0.03	80
0.56	0.15	81
0.82	0.92	89

For each random value of each component, determine the system lifetime, T_S. Since this is a series system, the system lifetime is limited by the component with the minimum lifetime.

T_A	T_B	T_C	T_S
102	105	69	69
125	92	77	77
96	125	64	64
114	49	49	49
107	102	91	91
30	99	30	30
100	80	74	74
84	82	80	80
88	12	12	12
97	89	89	89
Total			635

The mean value, μ, of T_S is

$$\mu = 635/10 = 63.5 \text{ years}$$

Calculate the standard deviation, σ, of T_S using the equation

$$\sigma^2 = \frac{1}{n}\sum(T_S - \mu)^2$$

where n is 10, the number in the population.

T_S	$(T_S - \mu)^2$
69	30.25
77	182.25
64	0.25
49	210.25
91	756.25
30	1122.25
74	110.25
80	272.25
12	2652.25
89	650.25
Total	5987

Therefore,

$$\sigma = (5987/10)^{0.5} = 24.5 \text{ years}$$

Monte Carlo simulation is an extremely powerful tool available to the scientist/ engineer that can be used to solve multivariable systems, ordinary and partial differential equations, numerical integrations, etc.

HZA.11   EVENT TREE ANALYSIS

If a building fire occurs, a smoke alarm sounds with probability 0.9. The sprinkler system functions with probability 0.7 whether or not the smoke alarm sounds. The consequences are minor fire damage (alarm sounds, sprinkler works), moderate fire damage with few injuries (alarm sounds, sprinkler fails), moderate fire damage with many injuries (alarm fails, sprinkler works), and major fire damage with many injuries (alarm fails, sprinkler fails). Construct an event tree and indicate the probabilities for each of the four consequences.

Solution

An event tree provides a diagrammatic representation of event sequences that begin with a so-called initiating event and terminate in one or more undesirable consequences. In contrast to a fault tree (considered in the next problem), which works backward from an undesirable consequence to possible causes, an event tree works forward from the initiating event to possible undesirable consequences. The initiating event may be equipment failure, human error, power failure, or some other event that has the potential for adversely affecting the environment or an ongoing process and/or equipment.

Note that for each branch in an event tree, the sum of probabilities must equal 1.0. Note again that an event tree includes the following: (1) works forward from the initial event, or an event that has the potential for adversely affecting an ongoing process, and ends at one or more undesirable consequences; (2) is used to represent

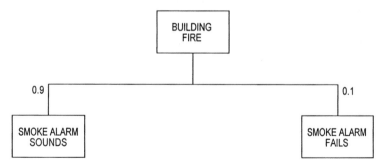

Figure 117. Event tree with first set of consequences.

the possible steps leading to a failure, or accident; (3) uses a series of branches that relate the proper operation and/or failure of a system with the ultimate consequences; (4) is a quick identification of the various hazards that could result from a single initial event; (5) is beneficial in examining the possibilities and consequences of a failure; (6) usually does not quantify (although it can) the potential of the event occurring; and (7) can be incomplete if all the initial events are not identified.

Thus, the use of event trees is sometimes limiting for hazard analysis because it usually does not quantify the potential of the event occurring. It may also be incomplete if all the initial occurrences are not identified. Its use is beneficial in examining, rather than evaluating, the possibilities and consequences of a failure. For this reason, a fault tree analysis should supplement this model to establish the probabilities of the event tree branches. This topic is introduced in the next problem.

The first consequence(s) of the building fire and the probabilities of the first consequence(s) are shown in Figure 117.

The second consequence(s) of the building fire and the probabilities of the consequence(s) are shown in Figure 118.

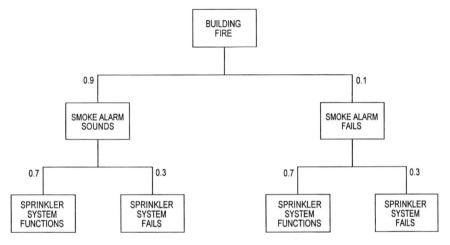

Figure 118. Event tree with second set of consequences.

The final consequences and the probabilities of minor fire damage, moderate fire damage with few injuries, moderate fire damage with many injuries, and major fire damage with many injuries are shown in Figure 119.

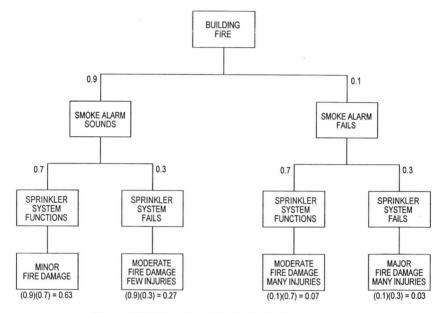

Figure 119. Event tree with final set of consequences.

HZA.12   FAULT TREE ANALYSIS

A runaway chemical reaction can occur if coolers fail (A) or there is a bad chemical batch (B). Coolers fail only if both cooler 1 fails (C) and cooler 2 fails (D). A bad chemical batch occurs if there is a wrong mix (E) or there is a process upset (F). A wrong mix occurs only if there is an operator error (G) and instrument failure (H).

1. Construct a fault tree.
2. If the following annual probabilities are provided by the plant engineer, calculate the probability of a runaway chemical reaction occurring in a year's time given the following probabilities:

$$P(C) = 0.05$$

$$P(D) = 0.08$$

$$P(F) = 0.06$$

$$P(G) = 0.03$$

$$P(H) = 0.01$$

Solution

Fault tree analysis seeks to relate the occurrence of an undesired event to one or more antecedent events. The undesired event is called the "top event" and the antecedent events are called "basic events." The top event may be, and usually is, related to the basic events via certain intermediate events. The fault tree diagram exhibits the casual chain linking the basic events to the intermediate events and the latter to the top event. In this chain, the logical connection between events is illustrated by so-called *logic gates*. The principal logic gates are the AND gate, symbolized on the fault tree by ⌂, and the OR gate, symbolized by ⌂.

The reader should note that a fault tree includes the following: (1) works backward from an undesirable event or ultimate consequence to the possible causes and failures, (2) relates the occurrence of an undesired event to one or more preceding events, (3) "chain links" basic events to intermediate events that are in turn connected to the top event, (4) is used in the calculation of the probability of the top event, (5) is based on the most likely or credible events that lead to a particular failure or accident, and (6) analysis includes human error as well as equipment failure.

Begin with the top event for this problem shown in Figure 120.

```
┌─────────────┐
│  RUNAWAY    │
│  CHEMICAL   │
│ REACTION (T)│
└─────────────┘
```

Figure 120. Top event of fault tree.

Generate the first branch of the fault tree, applying the logic gates (see Figure 121).

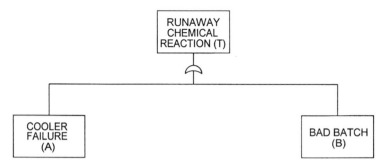

Figure 121. Fault tree with first branch.

Generate the second branch of the fault tree, applying the logic gates as shown in Figure 122.

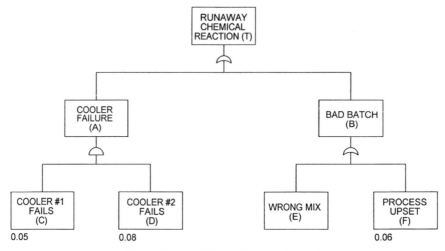

Figure 122. Fault tree with second branch.

Generate the third branch of the fault tree, applying the logic gates as shown in Figure 123.

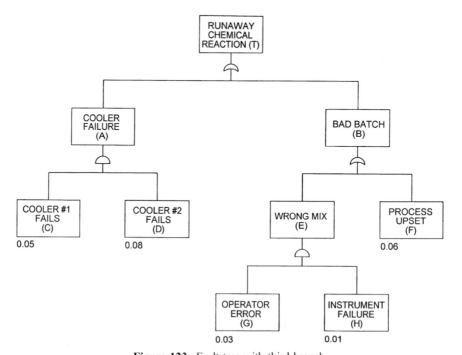

Figure 123. Fault tree with third branch.

Finally, calculate the probability that the runaway reaction will occur:

$$P = (0.5)(0.08) + (0.01)(0.03) + 0.06$$
$$= 0.064$$

Note that the process upset, F, is the major contributor to the probability.

46 Hazard Risk Assessment (HZR)

HZR.1 DILUTION WITH NITROGEN

Solvent vapor emissions from a coating operation are diluted with air to bring the solvent concentrations below the lower explosive level. As a pollution prevention measure you are asked to consider nitrogen as an alternative for air to transport the solvent vapor. What are the advantages of the nitrogen system over the air system?

Solution

Nitrogen-based systems are much safer from an explosion point of view than conventional air systems. A nitrogen system can be built as a self-contained closed system. This type of system minimizes solvent vapor discharge to the atmosphere. In addition, the solvent vapor may exist in the nitrogen without being an explosion hazard at a much higher concentration than in air. Recovery by condensation would be more efficient at the higher concentration and could result in energy savings.

HZR.2 FLAMMABILITY LIMITS

A gas mixture of methane, ethane, and pentane entering an adsorber has an upper flammability limit of 12.5% and a lower flammability limit of 2.85%. Given a methane concentration of 30%, calculate the concentrations of the other two components of the gas mixture. Flammability limits for methane, ethane, and propane at various concentrations are given in the following table.

Component	Lower Flammability Limit (vol %)	Upper Flammability Limit (vol %)
Methane	4.6	14.2
Ethane	3.5	15.1
Pentane	1.4	7.8

The lower flammability limit (LFL) and upper flammability limit (UFL) of a multicomponent system with n components are calculated using these equations.

$$\text{Lower flammability limit, LFL} = \frac{1}{\dfrac{y_1}{LFL_1} + \dfrac{y_2}{LFL_2} + \cdots \dfrac{y_n}{LFL_n}}$$

$$\text{Upper flammability limit, UFL} = \frac{1}{\dfrac{y_1}{UFL_1} + \dfrac{y_2}{UFL_2} + \cdots \dfrac{y_n}{UFL_n}}$$

where $y_i = $ volume % (or mole %) of component, $i(i = 1, 2, \ldots n)$.

Solution

The lower flammability limit of the mixture in percent can be calculated using the equation for the LFL of a multicomponent system:

$$LFL = \frac{1}{\dfrac{y_1}{LFL_1} + \dfrac{y_2}{LFL_2} + \cdots \dfrac{y_n}{LFL_n}}$$

$$0.0285 = \frac{1}{\dfrac{30}{4.6} + \dfrac{y_2}{3.5} + \dfrac{y_3}{1.4}}$$

The upper flammability limit of the mixture in percent can be calculated using the equation for the UFL of a multicomponent system:

$$UFL = \frac{1}{\dfrac{y_1}{UFL_1} + \dfrac{y_2}{UFL_2} + \cdots \dfrac{y_n}{UFL_n}}$$

$$0.125 = \frac{1}{\dfrac{30}{14.2} + \dfrac{y_2}{15.1} + \dfrac{y_3}{7.8}}$$

There are two equations and two unknowns. One may solve the lower flammability equation in terms of y_2 and use it to solve the upper flammability equation for y_3:

$$2.113 + \frac{y_2}{15.1} + \frac{y_3}{7.8} = 8$$

$$y_2 = 88.90 - 1.94 y_3$$

Substituting into the lower flammability equation leads to

$$0.0285 = \cfrac{1}{\cfrac{30}{4.6} + \left(\cfrac{88.90 - 1.94y_3}{3.5}\right) + \cfrac{y_3}{1.4}}$$

$$= \frac{1}{31.92 + 0.16y_3}$$

$$y_3 = 19.8\% = 0.198$$

$$y_2 = 50.2\% = 0.502$$

Note that the LEL and UEL, where E means explosive, are terms equivalent to LFL and UFL that are employed in industry and the literature.

The reader should also note that the width of the flammability limits (the range between the lower and upper flammability limits) is a function of ignition energy, ignition pressure, ignition temperature, inert gas concentration, and relative humidity of the mixture.

The width of the flammability range becomes wider when the ignition energy is higher, moving the UFL to a higher concentration. The higher the initial pressure at which the ignition source is activated, the wider the width of the flammability range. The higher the temperature at the moment of ignition, the easier the reaction will propagate, making the width of the flammability range wider. The flammability limits can also be changed significantly by changing the oxygen content or adding an inert gas to the gas/air mixture. The heat capacity of the inert or diluent gas plays a role since a diluent with a higher heat capacity acts as a better heat sink. For example, carbon dioxide is a better diluent than nitrogen because carbon dioxide has a higher heat capacity. The width of the flammability range is wider for a drier (i.e., lower moisture content) mixture. Finally, the *flash point* of a flammable liquid is defined as the temperature at which the vapor pressure of the liquid is the same as the vapor pressure that corresponds to the lower flammability concentration.

HZR.3 HYDROCARBON EMISSIONS FROM A REACTOR

A reactor is located in a relatively large laboratory with a volume of $1100\,m^3$ at $22°C$ and 1 atm. The reactor can emit as much as 0.75 gmol of hydrocarbon (HC) into the room if a safety valve ruptures. A hydrocarbon mole fraction in the air greater than 425 parts per billion (ppb) constitutes a health and safety hazard.

Suppose the reactor valve ruptures and the maximum amount of HC is released instantaneously. Assume the air flow in the room is sufficient to cause the room to behave as a continuously stirred tank reactor (CSTR), i.e., the air composition is spatially uniform. Calculate the ppb of hydrocarbon in the room. Is there a health risk? From a treatment point of view, what can be done to decrease the environmental hazard or to improve the safety of the reactor?

Solution

Calculate the total number of gmols of air in the room, n_{air}. Assuming that air is an ideal gas, 1 gmol of air occupies 22.4 liters ($0.0224 \, m^3$) at standard temperature and pressure (273 K, 1 atm). Since the room temperature is not 273 K,

$$n_{air} = (1100 \, m^3)\left(\frac{1 \, gmol}{0.0224 \, STP \, m^3}\right)\left(\frac{273 \, K}{295 \, K}\right)$$

$$= 45,445 \, gmols$$

Note: STP m^3 indicates the volume (in m^3) the gas would have at standard temperature and pressure.

The mole fraction of hydrocarbon in the room, x_{HC}, is

$$x_{HC} = \frac{0.75 \, gmol \, HC}{45,445 \, gmol \, air + 0.75 \, gmol \, HC} = 16.5 \, ppm = 16,500 \, ppb$$

Since 16,500 ppb $\gg$ 850 ppb, the hazard presents a health risk.

To implement safety measures, the potential rupture area should be vented directly into a hood or a duct to capture any leakage in the event of a rupture. Another alternative is input substitution, a source reduction measure. Input substitution is the replacement of the material in the reactor with material with a lower vapor pressure.

HZR.4 HAZARD OPERABILITY STUDY

List and discuss the key guide words that are employed in a hazard operability (HAZOP) study.

Solution

A HAZOP study is a very useful technique that may lead to a more reliable and safer process. Whether it is applied at preliminary design stages or to the detailed layout of an existing plant, its benefits can be invaluable. It reduces the possibility of accidents for the process involved, improves on-stream availability of the process, can lead to a better understanding of the process and possible malfunctions, and provides training for the evaluation of any process. Finally, it is also a way of optimizing a process and providing a reliable and cost-effective system.

Generally, HAZOP focuses on a major piece of equipment, although a lesser piece of equipment such as a pump or a valve may be chosen depending upon the nature of the materials being handled and the operating conditions. Once an intended operation is defined, a list of possible deviations from the intended operation is developed. The degrees of deviation from normal operation are conveyed by a set of guide words, some of which are listed in the following table:

Guide Words	Meaning	Examples
NO or NOT	No part of the intention is achieved but nothing else happens.	No flow, no agitation, no reaction.
MORE and LESS	Quantitative increase(s) or decrease(s) to the intended activity.	More flow, higher pressure, lower temperature, less time.
AS WELL AS	All of the intention is achieved but some additional activity occurs.	Additional component contaminants, extra phase.
PART OF	Only part of the intention is achieved, part is not.	Component omitted, part of multiple destinations omitted.
REVERSE	The opposite of the intention occurs.	Reverse flow, reverse order of addition.
OTHER THAN	No part of the intention is achieved. Something different happens.	Wrong component, startup and shutdown problems, utility failure.

The purpose of these guide words is to develop the thought process and encourage discussion that is related to any potential deviation(s) in the system. Upon recognizing a possible deviation, the possible cause(s) and consequence(s) can be determined.

HZR.5   HAZOP OF A BOILER SYSTEM

Consider a small industrial boiler system generating low-pressure saturated steam. The boiler drum receives preheated feedwater that is required to maintain the water level in the drum between 30 and 40% of its total height. There is a visual level indicator (LI) on the drum in addition to a level indicator controller (LIC) that is connected to a level indicator control valve (LICV) in the feedwater line. Natural gas fuel is fired below the boiler drum. The fuel line contains a pressure control valve (PCV) that is connected to a pressure controller (PC) located in the steam line in and leaving the drum.

Perform a HAZOP study on the system with the intention of providing the required amount of correctly treated feedwater to the boiler drum to maintain the prescribed water level.

Solution

The overall HAZOP method is summarized in the following steps:

1. Define objectives.
2. Define plant limits.

3. Appoint and train a team.
4. Obtain complete preparative work.
5. Conduct examination meetings in order to
   a. Select a manageable portion of the process.
   b. Review the flowsheet and operating instructions.
   c. Agree on how the process is intended to operate.
   d. State and record the intention.
   e. Search for possible ways to deviate from the intention, utilizing the guide words.
   f. Determine possible causes for the deviation.
   g. Determine possible consequences of the deviation.
   h. Recommend action(s) to be taken.
6. Issue meeting report.
7. Follow up on recommendations.

After the serious hazards have been identified with a HAZOP study or some other type of qualitative approach, a quantitative examination should be performed. Hazard quantification or hazard analysis (HAZAN) involves the estimation of the expected frequencies or probabilities of events with adverse or potentially adverse consequences. It logically ties together historical occurrences, experience, and imagination. To analyze the sequence of events that lead to an accident or failure, event and fault trees are used to represent the possible failure sequences.

A line diagram of the system is provided in Figure 124.

Apply the first guide word NO to the system and list the deviation(s), cause(s), and consequence(s).

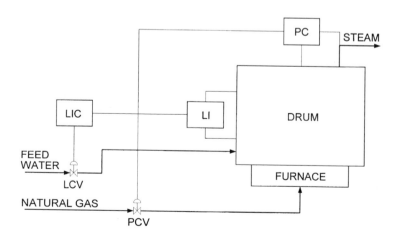

Figure 124. Line diagram for Problem HZR.5.

Deviations	Causes	Consequences
NO flow of feedwater.	LICV closed.	Loss of level in drum and explosion of drum by flame impingement on dry shell if flame continues.
NO level in drum.	Loss of feedwater pressure. Massive leak.	Same as above.
NO fuel flow.	Fuel line closed.	Flame extinguished. No steam generation.

Apply the second guide word MORE to the system and list the deviation(s), cause(s), and consequence(s).

Deviations	Causes	Consequences
MORE level in drum than 40%.	Level control fault (failure).	Excessive entrainment in steam.

Apply the guide word LESS to the system.

Deviations	Causes	Consequences
LESS level in drum than 30%.	Similar to NO level.	Same as NO level.

Apply the guide word AS WELL AS to the system.

Deviations	Causes	Consequences
Contaminants AS WELL AS feedwater.	Water treatment fault.	Fouling of boiler. Corrosive steam.

Apply the guide word PART OF to the system.

Deviations	Causes	Consequences
PART OF feedwater (treatment chemical) omitted.	Water treatment fault.	Same as above.

Apply the guide word REVERSE to the system.

Deviations	Causes	Consequences
REVERSE flow in feedwater line.	Loss of feedwater pressure.	Steam in feedwater system.

Apply the guide word OTHER THAN to the system.

Deviations	Causes	Consequences
Unplanned shutdown (OTHER THAN normal operation).	Utility failure.	Does LICV fail closed or open? Same consequences as NO level in drum or MORE level in drum.
OTHER THAN feedwater	Not possible	

HZR.6 THE FAR CONCEPT

The fatal accident rate (FAR) is the number of fatal accidents per 1000 workers in a working lifetime (10^8 h). A responsible chemical company typically employs a FAR equal to 2 for chemical process risks such as fires, toxic releases, or spillage of corrosive chemicals.

Identify potential problem areas that may develop for a company if acceptable FAR numbers are exceeded.

Solution

Potential problems that may develop for a company within the community if acceptable FAR numbers are exceeded include:

1. Adverse publicity by media
2. Adverse community relations
3. Decreased public trust in company

Potential problems that may develop for a company concerning legal and regulatory issues if acceptable FAR numbers are exceeded include:

1. Legal action against the company by those affected
2. Potential notices of violations by appropriate regulatory agencies, i.e., National Institute of Occupational Safety and Health (NIOSH), Occupational Safety and Health Administration (OSHA), Environmental Protection Agency (EPA), etc.

3. Advisory actions regarding permit compliance

Potential problems that may develop for a company concerning employees if acceptable FAR numbers are exceeded include:

1. Employee safety concerns and discontent
2. Increased employee turnover
3. Decreased productivity

Potential problems that may develop for a company concerning economics and finances if acceptable FAR numbers are exceeded include:

1. Increased insurance costs
2. Decreased profits due to customer dissatisfaction
3. Decreased profits due to decreased employee productivity
4. Fines for violating regulations

If acceptable FAR numbers are not maintained within a company, a lack of concern for health and safety becomes apparent.

HZR.7   CALCULATING THE FAR

As described earlier, a HAZOP (*hazard and operability study*) is a technique used to identify process hazards and obstacles to the efficient operation of industrial facilities. It provides a systematic approach to quantifying hazard potentials. A HAZAN (*hazard analysis*) is a technique for quantitative evaluation after a serious hazard has been identified or after an accident has occurred.

You have been hired as a consultant to an administrator who has a limited budget for the mitigation of hazards in a certain chemical plant. The plant employs two kinds of workers: day employees who work one 8-h shift daily, and shift employees who rotate through three 8-h shifts each day. A HAZOP-HAZAN report reveals two kinds of accidents are possible during plant operation.

Accidents of the first kind result in the death of one day employee per incident and occur with a frequency of 2.92×10^{-5} accidents per year. Accidents of the second kind result in the death of 100 shift workers per incident and occur with a frequency of 8.76×10^{-7} accidents per year.

1. Calculate the *fatal accident rate* (FAR) for the first kind of accident. The FAR (see Problem HZR.6) is a measure of the risk associated with an accident or event in units of number of deaths/1000 worker lifetimes (10^8 h).
2. Calculate the FAR for the second kind of accident.

3. What considerations would influence your recommendation on the allocation of funds to reduce these hazards?

Risk assessment describes the challenges to environmental decision-making today. Included in these challenges are a national tradition of narrowly focusing on separate environmental problems, the danger of deadlock on crucial environmental issues, and a divergence of public and scientific views on what are the most risky environmental problems.

Solution

The FAR for the first type of accident is

$$\mathrm{FAR} = \left(\frac{1\ \text{fatality}}{\text{accident}}\right)\left(2.92 \times 10^{-5} \frac{\text{accidents}}{\text{yr}}\right)\left(\frac{1\ \text{yr}}{365\ \text{day}}\right)\left(\frac{1\ \text{day}}{8\ \text{h}}\right)\left(\frac{10^8\ \text{h}}{1000\ \text{lifetimes}}\right)$$

$$= 1\ \text{fatality}/1000\ \text{worker lifetimes}$$

The FAR for the second type of accident is

$$\mathrm{FAR} = \left(\frac{100\ \text{fatalities}}{\text{accident}}\right)\left(8.76 \times 10^{-7} \frac{\text{accidents}}{\text{yr}}\right)\left(\frac{1\ \text{yr}}{365\ \text{day}}\right)$$

$$\times \left(\frac{1\ \text{day}}{24\ \text{h}}\right)\left(\frac{10^8\ \text{h}}{1000\ \text{lifetimes}}\right)$$

$$= 1\ \text{fatality}/1000\ \text{worker lifetimes}$$

While any answer to the question on the allocation of funds is likely to be incomplete, the following recommendation is suggested for further discussion.

The FARs for the two types of accidents are equal, and it is likely that over long periods of time an equal number of deaths may be expected from both types of accidents. This does not mean that equal consequences to the company will result from the two types of accidents. Accidents of the first type involve a low (but, perhaps, steady) loss of life. Accidents of the second type, however, are sure to attract more attention in the media. Adverse public relations are nearly certain, as well as unfavorable attention from legislators and other public officials.

Also, accidents of the second type have a catastrophic effect on production. Not only will the entire facility be demolished (in all likelihood), but a large fraction of the pool of trained personnel will be lost all at once. Who will train replacement personnel if everyone is lost in the disaster? Similar concerns make disruption in the community much greater for accidents of the second type. Some of these considerations can be factored into decision-making as direct economic losses that increase the burden to both the company and the community for accidents of the second type.

One view (one that is the only acceptable view to many people) is to give priority to the prevention of both types of accidents.

HZR.8   OVERPRESSURE

Calculate the peak overpressure of a 50-lb trinitrotoluene (TNT) explosion at a distance 200 ft from the ignition, if the peak overpressure at 1000 ft is 0.10 psi when 150 lb of TNT is detonated.

Solution

An explosion pressure, P_{ex}, is the pressure in excess of the initial pressure at which the explosive mixture is ignited. The rate of pressure rise is represented by dP/dt. This is the change in pressure with respect to time. The rate of pressure rise is a measure of the speed of the flame propagation; therefore, it is also a measure of the violence of the explosion. Typical values of maximum explosion pressures in a closed vessel range from 7 to 8 bar. The rate of pressure rise can vary considerably with the flammable gas. The influence of vessel volume on the maximum rate of pressure rise for a given flammable gas is characterized by the cubic law:

$$(dP/dt)_{max} V^{1/3} = K_G$$

where V = vessel volume, m^3
$\quad K_G$ = constant, bar · m/s
$(dP/dt)_{max}$ = maximum pressure rise rate, bar/s

Another method of estimating the effects of explosions in process plants is represented by the following equation developed from data on the high explosive, TNT:

$$\frac{P_1^o}{P_2^o} = \left(\frac{m_1}{m_2}\right)^{n/3} \left(\frac{r_2}{r_1}\right)^n$$

where P_1^o, P_2^o = peak overpressure
$\quad m_1, m_2$ = mass of explosive
$\quad r_1, r_2$ = distance
$\quad n$ = 1.6 for overpressures of 1–10 psi
$\quad$ = 2.3 for overpressures of 10–100 psi

The peak overpressure can be calculated from this second equation by substituting the values given in the problem statement and assuming the lowest value of n. This yields

$$\frac{P_1^o}{0.1\,\text{psi}} = \left(\frac{50\,\text{lb}}{150\,\text{lb}}\right)^{1.6/3} \left(\frac{1000\,\text{ft}}{200}\right)^{1.6}$$

$$P_1^o = 0.73\,\text{psi}$$

HZR.9   FLUID EXPANSION ENERGY

Calculate the fluid expansion energy for an isothermal expansion of a cylindrical vessel at 550°C and initial and final pressures of 147 and 450 psi, respectively.

Solution

Fluid expansion energy in an explosion, W_{ex}, can be determined from the equation

$$W_{ex} = RT \ln\left(\frac{P_1}{P_2}\right) \qquad \text{(for isothermal expansion of an ideal gas)}$$

The energy released in an explosive expansion of a liquid is given by the equation

$$W_{ex} = \tfrac{1}{2}\beta P^2 V$$

where β = bulk modulus of liquid
    V = volume of vessel

The fluid expansion energy is calculated from the first equation for isothermal expansion:

$$W_{ex} = RT \ln\left(\frac{P_1}{P_2}\right)$$

where $T = 550°C = 823\,K$
    $P_1 = 147\,psi$
    $P_2 = 450\,psi$
    $R = 8.314\,J/(gmol \cdot K)$

$$W_{ex} = \left(8.314\,\frac{J}{gmol \cdot K}\right)(823\,K) \ln\left(\frac{147\,psi}{450\,psi}\right)$$

$$= -7655\,J/gmol$$

$$= -1830\,cal/gmol$$

Note: In this book, work is positive if it is done *on* the system and negative if it is done *by* the system.

HZR.10   WEIBULL DISTRIBUTION

The life of an automobile seal has a Weibull distribution with failure rate $Z(t) = 1/t^{1/2}$, where t is measured in years. What is the probability that the seal will last at least 4 years?

Solution

As described in Problem HZA.7, the failure rate of equipment frequently exhibits three stages: a break-in stage with a declining failure rate, a useful life stage characterized by a fairly constant failure rate, and a wearout period characterized by an increasing failure rate. Many industrial parts and components follow this path. A failure rate curve exhibiting these three phases is called a *bathtub curve*.

The Weibull distribution provides a mathematical model of all three stages of the bathtub curve. An assumption about failure rate that reflects all three stages of the bathtub stage is

$$Z(t) = \alpha\beta t^{\beta-1} \qquad t > 0$$

where α, β = constants
$\qquad t$ = time

The probability distribution function (PDF) of T, time to failure, is

$$f(t) = \alpha\beta t^{\beta-1} \exp\left(-\int_0^t \alpha\beta t^{\beta-1}\, dt\right)$$
$$= \alpha\beta t^{\beta-1} \exp(-\alpha t^\beta) \qquad t > 0, \qquad \alpha > 0, \qquad \beta > 0$$

This integral may be evaluated by realizing that

$$\int e^v\, dv = e^v$$

Replacing v by $-2t^{1/2}$ yields

$$dv = -\frac{1}{t^{1/2}}\, dt$$

or

$$dt = -t^{1/2}dv = \frac{v}{2}\, dv$$

Adjusting the limits,

$$t = 4 \qquad v = -4$$
$$t = \infty \qquad v = -\infty$$

Thus,

$$P(T > 4) = \int_{-4}^{-\infty} \left(\frac{-2}{v}\right) e^v \left(\frac{v}{2}\right) dv$$

$$= -\int_{-4}^{-\infty} e^v \, dv$$

$$= -e^v \big|_{-4}^{-\infty}$$

$$= -(0 - e^{-4})$$

$$= e^{-4}$$

$$= 0.0183$$

HZR.11 INSTANTANEOUS "PUFF" MODEL

A rather significant amount of data and information is available for sources that emit continuously to the atmosphere. See Chapter 48 for more details. Unfortunately, little is available on instantaneous or "puff" sources. Other than computer models that are not suitable for classroom and/or illustrative example calculations, only Turner's *Workbook of Atmospheric Dispersion Estimates*, USEPA Publication No. AP-26, Research Triangle Park, NC, 1970 provides an equation that may be used for estimation purposes. Cases of instantaneous releases, as from an explosion, or short-term releases on the order of seconds, are also and often of practical concern.

To determine concentrations at any position downwind, one must consider both the time interval after the time of release and diffusion in the downwind direction, as well as lateral and vertical diffusion. Of considerable importance, but very difficult to determine, is the path or trajectory of the puff. This is most important if concentrations are to be determined at specific points. Determining the trajectory is of less importance if knowledge of the magnitude of the concentration for particular downwind distances or travel times is required without the need to know exactly at what points these concentrations occur.

An equation that may be used for estimates of concentration downwind from a release from height, H, is

$$C(x, y, 0, H) = \left[\frac{2m_T}{(2\pi)^{1.5}\sigma_x\sigma_y\sigma_z}\right]\left\{\exp\left[-0.5\left(\frac{x-ut}{\sigma_x}\right)^2\right]\right\}$$

$$\times \left\{\exp\left[-0.5\left(\frac{H}{\sigma_z}\right)^2\right]\right\}\left\{\exp\left[-0.5\left(\frac{y}{\sigma_y}\right)^2\right]\right\}$$

where m_T = total mass of the release
           u = wind speed

t = time after the release
x = distance in the x direction
y = distance in the y direction
$\sigma_x, \sigma_y, \sigma_z$ = dispersion efficients in the x, y and z direction, respectively.

The dispersion coefficients above are not necessarily those evaluated with respect to the dispersion of a continuous source at a fixed point in space. (See Chapter 48 for more details.) This equation can be simplified for centerline concentrations and ground-level emissions by setting $y = 0$ and $H = 0$, respectively. The dispersion coefficients in the above equation refer to dispersion statistics following the motion of the expanding puff. The σ_x is the standard deviation of the concentration distribution in the puff in the downwind direction, and t is the time after release. Note that there is essentially no dilution in the downwind direction by wind speed. The speed of the wind mainly serves to give the downwind position of the center of the puff, as shown by examination of the exponential term involving σ_x. In general, one should expect the σ_x value to be about the same as σ_y.

Unless another model is available for treating instantaneous sources, it is recommended that the above equation be employed. The use of appropriate values of σ for this equation is not clear cut. As a first approximation, the reader may consider employing the values of provided in Chapter 48.

A 20-m high tank in a plant containing a toxic gas suddenly explodes. The explosion causes an emission of 400 g/s for 3 min. A school is located 200 m west and 50 m south of the plant. If the wind velocity is 3.5 m/s from the east, how many seconds after the explosion will the concentration reach a maximum in the school? Humans will be adversely affected if the concentration of the gas is greater than 1.0 µg/L. Is there any impact on the students in the school? Assume that stability category D applies (see Problem DSP.4 in Chapter 48 for a description of stability categories.)

Solution

First determine the total amount of toxic gas released, m_T:

$$m_T = (400\ \text{g/s})(3\ \text{min})(60\ \text{s/min})$$
$$= 72{,}000\ \text{g}$$

The values of the three standard deviations are obtained from Figures 125 and 126 in Chapter 48:

$\sigma_x = \sigma_y = 16\ \text{m}$ at downwind distance $= 200\ \text{m}$ and stability category D
$\sigma_z = 8.5\ \text{m}$ at downwind distance $= 200\ \text{m}$ and stability category D

The time at which the maximum concentration will occur at the school is

$$t = \text{(downward distance)}/\text{(wind speed)}$$
$$= (200\,\text{m})/(3.5\,\text{m/s})$$
$$= 57.1\,\text{s}$$

The maximum concentration at the school can now be calculated using the equation given in the problem statement:

$$C(x, y, 0, H) = \left[\frac{2m_T}{(2\pi)^{1.5}\sigma_x\sigma_y\sigma_z}\right]\left\{\exp\left[-0.5\left(\frac{x-ut}{\sigma_x}\right)^2\right]\right\}$$
$$\left\{\exp\left[-0.5\left(\frac{H}{\sigma_z}\right)^2\right]\right\}\left\{\exp\left[-0.5\left(\frac{y}{\sigma_y}\right)^2\right]\right\}$$

Note that for the maximum concentration, $x = ut$ and the first exponential term at the maximum concentration becomes 1.0 (unity).
  Substituting yields

$$C(200, 50, 0, 20) = \left[\frac{2(72{,}000)}{(2\pi)^{1.5}(16)(16)(8.5)}\right][1]\left\{\exp\left[-0.5\left(\frac{20}{8.5}\right)^2\right]\right\}$$
$$\times\left\{\exp\left[-0.5\left(\frac{50}{16}\right)^2\right]\right\}$$
$$= 2.00 \times 10^{-3}\,\text{g/m}^3$$

The concentration can also be expressed in units of micrograms/liter:

$$C = (2.00 \times 10^{-3}\,\text{g/m}^3)(1 \times 10^6\,\mu\text{g/g})/(1000\,\text{L/m}^3) = 2.00\,\mu\text{g/L}$$

Since the calculated maximum concentration, $2.00\,\mu\text{g/L}$ which is greater than $1.0\,\mu\text{g/L}$, there is an impact on students in the school.

HZR.12 INSTANTANEOUS "PUFF" SOURCES

EPA considers it a health hazard if a particular chemical gas has a ground-level concentration (GLC) greater than $1 \times 10^{-9}\,\text{g/m}^3$. An explosion in a chemical plant releases 1000 kg of this gas as a "puff" from ground level conditions. A residential community is located 4000 m downwind and 1000 m crosswind from the source of the explosion. If the wind speed is 6 m/s and the stability category is "unstable," how much time do the residents have to evacuate the town?

Suggested values of σ_y and σ_z for quasi-instantaneous sources are given in the following table.

	$x = 100\,m$		$x = 4\,km$	
	σ_y	σ_z	σ_y	σ_z
Unstable	10.	15.	300	220
Neutral	4.	3.8	120	50
Very stable	1.3	0.75	35	7

Solution

From the table, determine the values of the dispersion coefficients, σ_y and σ_z.

$$\sigma_y = 300\,m$$

$$\sigma_z = 220\,m$$

Also, assign a value to the dispersion coefficient, σ_x:

$$\sigma_x = \sigma_y = 300\,m$$

Since the explosion occurs at ground level, $H = 0$:

$$C(x, y, 0, H) = \left[\frac{2m_T}{(2\pi)^{1.5}\sigma_x\sigma_y\sigma_z}\right]\left\{\exp\left[-0.5\left(\frac{x - ut}{\sigma_x}\right)^2\right]\right\}$$
$$\times \left\{\exp\left[-0.5\left(\frac{H}{\sigma_z}\right)^2\right]\right\}\left\{\exp\left[-0.5\left(\frac{y}{\sigma_y}\right)^2\right]\right\}$$

Set the left-hand side (LHS) of the instantaneous puff equation to the maximum GLC allowed and solve for the time, t:

$$1 \times 10^{-9} = \left[\frac{2(1000)}{(2\pi)^{1.5}(300)(300)(220)}\right]\left\{\exp\left[-0.5\left(\frac{4000 - 6t}{300}\right)^2\right]\right\}[1]$$
$$\times \left\{\exp\left[-0.5\left(\frac{1000}{300}\right)^2\right]\right\}$$

$$t = 442\,s$$

$$= 7.4\,min$$

47 Industrial Applications (IAP)

IAP.1 SELECTING A PLANT SITE

List several guidelines that should be followed when selecting a "safe" site for a plant.

Solution

The following guidelines should be followed when selecting a "safe" site for a plant:

1. *Topography.* A fairly level site is needed to contain spills and prevent spills from migrating and creating more of a hazard. Firm soil above water level is recommended.

2. *Utilities and water supply.* The water supply must be adequate for fire protection and cooling. The source of electricity should be reliable to prevent unplanned shutdowns.

3. *Roadways.* Roadways should allow access to the site by emergency vehicles such as ambulances and fire engines in the event of an emergency.

4. *Neighboring communities and plants.* Population density and proximity to the plant should be considered for the initial site and in anticipation of a possible future expansion.

5. *Waste disposal.* Waste disposal systems containing flammable, corrosive, or toxic materials should be a minimum distance of 250 ft from plant equipment.

6. *Climate and natural hazards.* Lightning arrestors should be installed to reduce/eliminate ignition sources in flammable areas. Storm drainage systems should be maintained.

7. *Emergency services.* Emergency services should be readily available, well trained, and appropriately equipped.

IAP.2 PLANT LAYOUT

Outline the key safety factors that should be considered in the layout of a new plant.

Solution

The key safety factors that should be considered in the layout of a new plant are:

1. Site selection
2. Water supply
3. Utilities
4. Offices and ancillary equipment
5. Storage and loading areas
6. Process equipment
7. Safety equipment
8. Access in and out of the plant

IAP.3   EXHAUST VENTILATION

Discuss transport air velocities in exhaust ventilation systems.

Solution

When designing a local exhaust ventilation system for a process that generates dust particles, it is important to consider the minimum air velocity. The minimum air velocity is the velocity required to prevent settling of dust particles in the air ducts. The minimum velocity is a function of dust particle size and particle density. Listed in the table below are the minimum air velocities recommended for the transport of various types of particulate contaminants.

Types of Particulates	Examples	Recommended Velocity (ft/min)
Very fine light dust	Cotton lint, wood flour	2000–2500
Dry dusts and powders	Fine rubber dust, cotton dust, light shavings	2500–3000
Average industrial dust, general foundry dust	Sawdust, grinding dust, limestone dust	3500–4000
Heavy dusts	Metal turnings, sand blast dust, lead dust	4000–4500
Heavy or moist dust (very heavy dust)	Lead dust with small chips, moist cement dust, quick lime dust	4500 and up

From *Industrial Ventilation — A Manual of Recommended Practices*, 19th Ed., ACGIH, Cincinnati, 1986.

For vapors, gases, smokes, and fumes, any economic air velocity is adequate. Typically, it is recommended that an air cleaner be located upstream of the blower or fan prior to discharge; this is usually required for air recirculation systems.

IAP.4   DESIGN OF EXHAUST VENTILATION SYSTEMS

An iron foundry has four workstations that are connected to a single duct. Each workstation has a hood that transports 3000 acfm of air flow. The duct length is 400 ft, and the pressure loss at the hood entrance is 0.5 in. of water. There is also a cyclone air cleaner that creates 3.5 in. H_2O pressure drop. Determine the diameter of the duct to ensure adequate transport of the dust. Also determine the power required for a combined blower/motor efficiency of 40%.

Solution

Select the minimum air velocity, v_{air}, required for general foundry dust (see Problem IAP.3):

$$v_{air} = 4000\,\text{ft/min}$$
$$= 66.67\,\text{ft/s}$$

Since there are four sources, the total air flow, q_{air}, required in acfm is

$$q_{air} = (3000\,\text{acfm})(4)$$
$$= 12,000\,\text{acfm}$$

Thus, the required cross-sectional area, A, in square feet is

$$A = q_{air}/v_{air}$$
$$= (12,000\,\text{acfm})/(4000\,\text{ft/min})$$
$$= 3\,\text{ft}^2$$

The duct diameter, D, can now be calculated from the equation for the area of a circle:

$$D = \left(\frac{4A}{\pi}\right)^{1/2}$$
$$= \left(\frac{4(3\,\text{ft}^2)}{3.14}\right)^{1/2}$$
$$= 1.95\,\text{ft}$$
$$= 24\,\text{in.}$$

To calculate power requirements, the pressure drop across the system needs to be calculated first.

Calculate the Reynolds number for the 24-in. duct using the velocity of the air calculated earlier.

$$N_{Re} = \frac{Dv\rho}{\mu}$$

$$= \frac{(24 \text{ in.})(\text{ft}/12 \text{ in.})(66.67 \text{ ft/s})(0.075 \text{ lb/ft}^3)}{1.21 \times 10^{-5} \text{ lb/(ft} \cdot \text{s})}$$

$$= 8.08 \times 10^5$$

The Fanning friction factor may be approximated as 0.003 since $N_{Re} > 20{,}000$. See Chapter 6 for additional details. Using Bernoulli's equation, the pressure drop in the duct is

$$\Delta P_{duct} = \frac{4fLv^2\rho}{2g_cD}$$

$$= \frac{4(0.003)(400 \text{ ft})(66.67 \text{ ft/s})^2(0.075 \text{ lb/ft}^3)}{2\left(32.2\dfrac{\text{lb} \cdot \text{ft}}{\text{lb}_f \cdot \text{s}^2}\right)(2 \text{ ft})}$$

$$= 12.4 \text{ lb}_f/\text{ft}^2$$

The total system pressure drop is generated by summing all contributory effects:

$$\Delta P_{tot} = \Delta P_{duct} + \Delta P_{hood} + \Delta P_{cyc}$$

$$= 12.4\frac{\text{lb}_f}{\text{ft}^2} + (0.5 \text{ in. } H_2O)\left(\frac{5.2 \text{ lb}_f/\text{ft}^2}{\text{in. } H_2O}\right) + (3.5 \text{ in. } H_2O)\left(\frac{5.2 \text{ lb}_f/\text{ft}^2}{\text{in. } H_2O}\right)$$

$$= 33.2 \text{ lb}_f/\text{ft}^2$$

Calculate the horsepower with the following equation:

$$HP = \frac{\Delta P q_{air}}{\eta}$$

Substituting and using consistent units yields

$$= \frac{(33.2 \text{ lb}_f/\text{ft}^2)(12{,}000 \text{ ft}^3/\text{min})(\text{min}/60 \text{ s})}{(550 \text{ ft} \cdot \text{lb}_f/\text{s})(0.4)}$$

$$= 30.2 \text{ hp}$$

The reader should note that there exists a phenomena called "drag effect" which actually reduces the pressure drop when particulates are present in the gaseous flow.

The pressure drop reduction is said to occur due to the ball-bearing effect of the particulates. The magnitude of pressure drop reduction is a function of particle size and particle characteristics (e.g., density). However, some researchers argue that it is an unsteady-state phenomenon on that cannot be relied on to occur consistently.

IAP.5 RUPTURE DISKS AND RELIEF VALVES

Discuss the major differences between rupture disks and relief valves.

Solution

Overpressure protection is one of the major requirements to prevent accidents from occurring in modern chemical operations. The two common devices used for overpressure protection are rupture disks and relief valves. Relief valves are usually used for process protection and rupture disks for vessel protection.

The relief system is designed to relieve the pressure in equipment if the process gets out of control. To avoid an explosion, the relief system should be designed to relieve the pressure at a rate determined by the heat transfer to the vessel during a fire.

The following table provides a comparison of relief valves and rupture disks.

Relief Valve	Rupture Disk
May leak	Zero leakage
Reusable	Need to replace
Requires maintenance	Low maintenance
Responds to pressure slowly	Responds to pressure quickly
Set pressure (adjustable)	Burst pressure (nonadjustable)
More expensive	Less expensive

IAP.6 TANK PRESSURE

Concern has arisen regarding an explosive mixture of liquid n-hexane and stoichiometric air in a closed process tank. As a health, safety, and accident prevention consultant, you have been hired to perform the following calculations for the industrial complex.

1. If the initial temperature and pressure of the mixture are 27°C and 3 psig, respectively, calculate the final pressure if combustion occurs isothermally.
2. Repeat the calculation with an initial pressure equal to 5 psia.
3. Outline how to calculate the final pressure if the combustion occurs adiabatically. Also indicate what additional data are required.

4. Outline how to calculate the final pressure if the combustion occurs adiabatically with 100% excess air.

Solution

1. The balanced stoichiometric equation for the combustion of liquid n-hexane is written first.

$$C_6H_{14}(l) + 9.5O_2 + 35.74N_2 \rightarrow 6CO_2 + 7H_2O + 35.74N_2$$

Assume a basis of 1.0 mol of liquid n-hexane. On the left-hand side of the equation, the initial number of gaseous moles, n_i, is 45.24. Note that this neglects the n-hexane. On the right hand side, the final number of moles, n_f, is 48.74.

Apply the ideal gas law at constant temperature between initial and final conditions to calculate the final pressure. Note that the units of temperature and pressure must be absolute when using the ideal gas law:

$$P_f = (n_f/n_i)P_i$$
$$= (48.74 \, \text{mol}/45.24 \, \text{mol})/(17.7 \, \text{psia})$$
$$= 19.07 \, \text{psia}$$

2. If $P_i = 5 \, \text{psia}$,

$$P_f = (48.74 \, \text{mol}/45.24 \, \text{mol})/(5 \, \text{psia})$$
$$= 5.39 \, \text{psia}$$

3. A thermochemical calculation is required, details of which are provided in Problem IAP.10. Data required include the heat of combustion and the heat capacities for CO_2, H_2O, and N_2.
4. Follow the same procedure as step 3, except use the stoichiometric equation with 100% excess air:

$$C_6H_{14}(l) + 19O_2 + 71.44N_2 \rightarrow 6CO_2 + 7H_2O + 71.48N_2 + 9.5O_2$$

IAP.7   EXPLOSION AND FIRE HAZARDS

A laboratory technician needs to attach a regulator valve assembly to a compressed gas cylinder of hydrogen, but the only regulator available is already in use on a cylinder of oxygen. Moreover, the fittings are not compatible, so she cannot simply switch the regulator to the hydrogen tank. However, she knows that the hydrogen cylinder is nearly empty, and she believes that she could use some heavy rubber tubing and some good screw-type clamps to hook the regulator from the oxygen tank to the hydrogen tank.

Assume the internal volume of the regulator valve assembly is $0.15 \, \text{ft}^3$ and contains oxygen gas at 1.0 atm (14.7 psia) when it is disconnected from the oxygen tank. Also assume that the regulator valve assembly is transferred rapidly from the oxygen tank so that none of the oxygen is lost. When the valve of the hydrogen tank is opened and hydrogen flows into the regulator valve assembly, the pressure gauge of the regulator climbs to 30.3 psig. After taking a break for lunch, the technician opens the needle valve and a small amount of gas flows out of the regulator valve assembly where it is ignited by some ignition source. Assume that the pressure rise gas constant, K_G, for hydrogen is 550 bar·m/s.

Verify that combustion is possible for the gas mixture in the regulator valve assembly described above. Calculate the maximum rate at which the pressure will rise inside the valve. Knowing that the regulator flame velocity for hydrogen flames can be as low as 11 ft/s, estimate the time it takes for the flame to move from one end of the valve assembly to the other, a distance of 6 in. Would the risk associated with this misuse of a regulator valve assembly be the same if the contents of the two tanks were reversed?

Solution

The balanced stoichiometric equation for the combustion of hydrogen is

$$2H_2 + O_2 \rightarrow 2H_2O$$

The partial pressure of hydrogen is the pressure reading on the regulator attached to the hydrogen tank:

$$p_{H_2} = (P_{gauge} - P_{atm})$$
$$= 45 \, \text{psi} - 14.7 \, \text{psi}$$
$$= 30.3 \, \text{psi}$$

The molar ratio of the two gases is equivalent to the ratio of their partial pressures:

$$\text{Ratio} = (\text{gmol } H_2)/(\text{gmol } O_2)$$
$$= (30.3 \, \text{psi})/(14.7 \, \text{psi})$$
$$= 2.1$$

Since the gas mixture is approximately at the stoichiometric ratio, the possibility of combustion in enhanced.

The equation for the maximum change in pressure with time, $(dP/dt)_{max}$ is

$$\left(\frac{dP}{dt}\right)_{max} = \frac{K_G}{V^{1/3}}$$

$$= \frac{550\,(\text{bar}\cdot\text{m/s})}{\left[(0.15\,\text{ft}^3)\left(\dfrac{\text{m}^3}{35.3\,\text{ft}}\right)\right]^{1/3}}$$

$$= 3400\,\text{bar/s}$$

The time, t, for the flame to move through the valve assembly is given by the valve assembly distance, d, divided by the regulator flame velocity, v:

$$t = d/v$$

$$= (0.5\,\text{ft})/(11\,\text{ft/s})$$

$$= 0.046\,\text{s}$$

The pressure rise or change in pressure inside the valve assembly is then

$$\Delta P = t(dP/dt)_{max}$$

$$= (3400\,\text{bar/s})(0.046\,\text{s})$$

$$= 156\,\text{bar}$$

The risk of putting a hydrogen valve on an oxygen cylinder is less than the risk of the reverse situation. If the explosion in the valve should result in a leak from the cylinder, the contents of the hydrogen cylinder will burn in air, while those of the oxygen cylinder will not.

IAP.8   REDUCTION OF EMISSIONS

In attempting to quantify the present and future risk associated with the greenhouse effect, the U.S. Environmental Protection Agency (EPA) has hired Theodore-Keane Associates to estimate the reduction of these harmful emissions to the atmosphere from automobile combustion through the use of a new gasoline additive, Wrieden FA, developed by Wrieden Enterprises, a small technology-oriented company based in Long Island, New York. EPA plans to use the results not only in its national inventory emission data bank but also in a revised meteorological study of the effect of these reductions on the greenhouse effect. In addition, EPA is considering mandating the use of the Wrieden FA additive in all gasoline-powered vehicles. Because of the timeliness and the importance of the project, EPA has requested that both Dr. Theodore and Ms. Keane be assigned to the project team. Specific information required includes the reduction of fuel usage, the mass reduction of hydrocarbon emissions to the atmosphere, the mass reduction of carbon dioxide to the atmosphere, and the consumer savings on an annual basis.

EPA has provided the following test data:

Current gasoline vehicle consumption rate $= 7.1 \times 10^6$ bbl/day

Percent increase in gasoline economy with the Wrieden FA $= 4.5\%$

Percent hydrocarbon emission reduction with the Wrieden FA $= 11\%$

The following assumptions apply:

1. The chemical formula for gasoline is C_8H_{18}.
2. Combustion of gasoline in a standard engine is approximately 99% complete.
3. One gallon of the Wrieden FA is required to treat 1500 gal of fuel; the cost is 1 cent/gal of fuel usage.
4. The specific gravity of gasoline is 0.8.
5. The price of gasoline is \$1.88/gal.

Solution

First determine the gasoline usage in gal/yr and lbmol/yr. There are 42 U.S. gal/bbl and 7.48 gal/ft^3. The molecular weight of the gasoline may be obtained from its chemical formula, C_8H_{18}. The specific gravity of the gasoline is specified in the problem.

$$\text{Gasoline usage} = \left(7.1 \times 10^6 \, \frac{\text{bbl}}{\text{d}}\right)\left(\frac{365 \, \text{d}}{\text{yr}}\right)\left(\frac{42 \, \text{gal}}{\text{bbl}}\right)$$

$$= 1.09 \times 10^{11} \, \text{gal/yr}$$

The molecular weight of C_8H_{18} is 114:

$$\text{Gasoline usage} = \left(1.09 \times 10^{11} \, \frac{\text{gal}}{\text{yr}}\right)(0.8)\left(62.4 \, \frac{\text{lb}}{\text{ft}^3}\right)\left(\frac{\text{ft}^3}{7.48 \, \text{gal}}\right)\left(\frac{\text{lbmol}}{114 \, \text{lb}}\right)$$

$$= 6.38 \times 10^9 \, \text{lbmol/yr}$$

Calculate the fuel usage reduction in gal/yr and lbmol/yr with the Wrieden FA additive. The reduction on a percent basis is 4.5%:

$$\text{Fuel usage reduction} = (1.09 \times 10^{11} \, \text{gal/yr})(0.045)$$

$$= 4.9 \times 10^9 \, \text{gal/yr}$$

In units of lbmol/yr,

$$\text{Fuel usage reduction} = \left(4.9 \times 10^9 \frac{\text{gal}}{\text{yr}}\right)(0.8)\left(62.4 \frac{\text{lb}}{\text{ft}^3}\right)\left(\frac{\text{ft}^3}{7.48 \text{ gal}}\right)\left(\frac{\text{lbmol}}{114 \text{ lb}}\right)$$

$$= 2.87 \times 10^8 \text{ lbmol/yr}$$

Calculate the volume of gasoline (gal/yr) that is discharged to the atmosphere because of incomplete combustion. The gasoline combustion efficiency is given as 99%; the unburned gasoline is therefore 1%.

$$\text{Volume of gasoline discharged} = (1.09 \times 10^{11} \text{ gal/yr})(0.01)$$

$$= 1.09 \times 10^9 \text{ gal/yr}$$

Calculate the mass of the gasoline that is discharged to the atmosphere (lb/yr) because of incomplete combustion:

$$\text{Mass of gasoline discharged} = \left(1.09 \times 10^9 \frac{\text{gal}}{\text{yr}}\right)(0.8)\left(62.4 \frac{\text{lb}}{\text{ft}^3}\right)\left(\frac{\text{ft}^3}{7.48 \text{ gal}}\right)$$

$$= 7.27 \times 10^9 \text{ lb/yr}$$

The stoichiometric equation for the combustion of gasoline, C_8H_{18}, and oxygen, O_2, is now written assuming complete combustion to CO_2 and H_2O:

$$C_8H_{18} + 12.5O_2 \rightarrow 8CO_2 + 9H_2O$$

The reduction in CO_2 generated can now be calculated neglecting the amount of gasoline that is not combusted.

From the stoichiometric equation, for every 1 lbmol of C_8H_{18} that is combusted, 8 lbmol of CO_2 are generated:

$$\text{molar rate of } CO_2 \text{ reduction} = \left(\frac{8 \text{ lbmol } CO_2}{1 \text{ lbmol } C_{18}H_{18}}\right)\left(2.87 \times 10^8 \frac{\text{lbmol } C_{18}H_{18}}{\text{yr}}\right)$$

$$= 2.30 \times 10^9 \text{ lbmol/yr}$$

$$\text{new rate of } CO_2 \text{ reduction} = (2.30 \times 10^9 \text{ lbmol/yr})(44 \text{ lb/lbmol})$$

$$= 1.01 \times 10^{11} \text{ lb}$$

The savings due to fuel usage reduction through the use of the Wrieden FA is

$$\text{Annual savings} = (4.9 \times 10^9 \text{ gal/yr})(1.88 \text{ \$/gal})$$

$$= 9.21 \times 10^9 \text{\$/yr}$$

The cost of using the Wrieden FA is

$$\text{Annual cost} = (1.09 \times 10^{11} \text{ gal/yr})(0.01 \text{ \$/gal})$$
$$= 1.09 \times 10^{9} \text{ \$/yr}$$

Therefore, the annual savings is

$$\text{Annual savings} = 9.21 \times 10^{9} \text{ \$/yr} - 1.09 \times 10^{9} \text{ \$/yr}$$
$$= 8.12 \times 10^{9} \text{ \$/yr}$$
$$\cong \$8 \text{ billion per year}$$

IAP.9   ACCIDENTAL/EMERGENCY DISCHARGE INTO A LAKE/RESERVOIR

Thomann and Mueller (*Principles of Surface Water Quality Modelling and Control*, Harper & Row, 1987) have provided simple, easy-to-use equations that can be employed to describe the concentrations of species in different bodies of water for (a large number of) various conditions. For the case of a steady continuous discharge of a pollutant species into a lake or reservoir undergoing an irreversible first-order reaction, they have shown that the concentration of the pollutant can be described by the following equation

$$C = \frac{(\dot{m})\left\{ 1 - \exp\left[-\left(\frac{1}{t_{\text{d}}} + k \right)t \right] \right\}}{q(1 + kt_{\text{d}})}$$

where C = concentration of pollutant at time t
$\dot{m}$ = mass flowrate of pollutant discharge
q = net volume flowrate through lake or reservoir
t_{d} = lake retention time = V/q
V = lake or reservoir volume
k = reaction velocity constant, $(\text{time})^{-1}$

The concentration at a time θ following termination of the discharge may be calculated using the following equation:

$$C = C_{\text{o}}\left\{ 1 - \exp\left[-\left(\frac{1}{t_{\text{d}}} + k \right)\theta \right] \right\}$$

where C_{o} is the concentration that would be achieved if the pollution discharge continued indefinitely.

A near core meltdown at a nuclear power plant brought about the implementation of an emergency response procedure. Part of the response plan resulted in the steady discharge of a radioactive effluent into a nearby reservoir. The mass flowrate of the discharge was 120,000 lb/h with a radioactive waste concentration of 10^6 pico-curies/L (pCi/L) over an 11-h period. (One gram of radium undergoes 3.7×10^{10} nuclear disintegrations in 1 s. This number of disintegrations is known as a curie, Ci, which is the unit used to measure nuclear activity). The reservoir volume and the net throughput volumetric flowrate are approximately (annual average) 3.6×10^8 ft³ and 200 ft³/s, respectively. If the waste decays in a first order manner with a decay constant of 0.23 (h)$^{-1}$, determine the following:

1. The equilibrium concentration associated with the steady waste discharge
2. The maximum concentration
3. The time after termination of the waste discharge for the concentration to reach an acceptable level of 10 pCi/L.

Solution

The describing equation for the equilibrium concentration is obtained using the equation provided in the problem statement:

$$C = \frac{(\dot{m})\left\{1 - \exp\left[-\left(\frac{1}{t_d} + k\right)t\right]\right\}}{q(1 + kt_d)}$$

Setting $t = \infty$ leads to

$$C_{eq} = \frac{\dot{m}}{q(1 + kt_d)}$$

Values may now be substituted into this equation. Assume that the effluent has the properties of water. Also, note that $t_d = V/q$:

$$\dot{m} = \left(120{,}000\,\frac{\text{lb}}{\text{h}}\right)\left(\frac{1}{62.4\,\frac{\text{lb}}{\text{ft}^3}}\right)\left(\frac{\text{L}}{0.0353\,\text{ft}^3}\right)\left(\frac{10^6\,\text{pCi}}{\text{L}}\right)$$

$$= 5.45 \times 10^{10}\,\text{pCi/h}$$

$$q = (200\,\text{ft}^3/\text{s})(3600\,\text{s/h})$$

$$= 7.2 \times 10^5\,\text{ft}^3/\text{h}$$

$$t_d = (3.6 \times 10^8\,\text{ft}^3)/(7.2 \times 10^5\,\text{ft}^3/\text{h})$$

$$= 500\,\text{h}$$

$$k = 0.23\,\text{h}^{-1}$$

Substituting the values of $\dot{m}$, q, t_d, and k into the equation yields

$$C_{eq} = \frac{\dot{m}}{q(1 + kt_d)}$$

$$= \frac{5.45 \times 10^{10} \dfrac{pCi}{h}}{\left(7.2 \times 10^5 \dfrac{ft^3}{h}\right)\left[1 + \left(0.23\dfrac{1}{h}\right)(500\ h)\right]}$$

$$= 652\ pCi/ft^3$$

$$= 652\ pCi/ft^3(0.0353\ ft^3/L)$$

$$= 23.02\ pCi/L$$

The maximum concentration, C_{max}, is achieved when the discharge is stopped, i.e., when $t = 11$ h. This is obtained from the following equation:

$$C_{max} = \frac{(\dot{m})\left\{1 - \exp\left[-\left(\dfrac{1}{t_d} + k\right)t\right]\right\}}{q(1 + kt_d)}$$

Substituting yields

$$C_{max} = C_{eq}\left\{1 - \exp\left[-\left(\frac{1}{t_d + k}\right)t\right]\right\}$$

$$= 23.02\frac{pCi}{L}\left\{1 - \exp\left[-\left(\frac{1}{500\ h} + 0.23\frac{1}{h}\right)(11\ h)\right]\right\}$$

$$= 21.23\ pCi/L$$

The time, θ, after the termination of the waste discharge, that the concentration will reach an acceptable level is given by

$$C = C_{max}\left\{1 - \exp\left[-\left(\frac{1}{t_d} + k\right)t\right]\right\}$$

$$10\frac{pCi}{L} = 21.23\frac{pCi}{L}\left\{1 - \exp\left[-\left(\frac{1}{500\ h} + 0.23\frac{1}{h}\right)\theta\right]\right\}$$

$$\frac{10}{21.23} = 1 - \exp(-0.232\theta)$$

$$0.529 = \exp(-0.232\theta)$$

$$\ln(0.529) = -0.232\theta$$

$$\theta = -0.638/-0.232$$

$$= 2.75\ h$$

Since the waste is discharged for 11 h, the time from the start of the discharge that the concentration will reach the acceptable level is

$$t = \theta + 11$$
$$= 2.75 + 11$$
$$= 13.75 \text{ h}$$

In reality, a concentration profile will exist in the body of water. An average or mean concentration may not be realistic over this short period of time since complete mixing will probably not occur.

Thomann and Mueller's work (*Principles of Surface Water Quality Modelling and Control*, Harper & Row, 1987) is an exceptionally informative and well-written text/reference book that is presented in a clear, concise manner. A significant amount of information is provided for describing the fate of pollutant species in oceans, rivers, lakes, etc.

IAP.10   THERMODYNAMICS

Knockout drums and seal drums are used in refineries and other chemical plants, usually to remove condensable vapors in process streams. They are operated at low pressures (0–5 psig) and may contain flammable mixtures of vapor and air. Although the chance of explosion is remote since there is usually no source of ignition, the American Petroleum Institute (API) recommends a design pressure of 50 psig. This design pressure is based on the peak explosion pressure of 7–8 times the operating pressure. In addition, a safety factor of 4 for this low-pressure operation is recommended by the American Society of Mechanical Engineers (ASME). Therefore, a knockout drum with a design pressure of 50 psig should not rupture until the pressure is equal to or greater than 200 psig.

Determine if the 50 psig design pressure suggested by API will contain an explosive mixture of air and *n*-pentane with an initial temperature of 25°C and 5 psig internal pressure. Assume a stoichiometric concentration of pentane in air and that the combustion reaction proceeds to completion. The internal energy of reaction of *n*-pentane at 25°C is 777.46 kcal/gmol. The heat capacity (C_p) constants (at constant pressure) for the flue gas components are given in the following table.

	a	b	c	d
N_2	6.529	0.149×10^{-2}	-0.0227×10^{-5}	0
CO_2	5.316	1.429×10^{-2}	-0.8362×10^{-5}	1.784×10^{-9}
H_2O	6.970	0.345×10^{-2}	-0.0483×10^{-5}	0

Note that $C_p = a + bT + cT^2 + dT^3$, with units for C_p of cal/(gmol·K) and for T, of degrees kelvin, K.

Solution

The pressure rise due to the explosion in a knockout drum can be determined using basic thermodynamic principles. Assume no heat loss. An energy balance on the drum contents yields

$$\Delta U_r(\text{at } T_{\text{in}}) + \int_{T_{\text{in}}}^{T_{\text{out}}} \sum (n_i C_{v,i})_{\text{out}} \, dT = 0$$

where ΔU_r = internal energy of reaction, cal/gmol
  T_{in} = initial temperature, K
  T_{out} = final temperature, K
  n_i = number of moles of the ith compound
  $C_{v,i}$ = constant volume molar heat capacity of the ith compound, cal/(gmol·K)
  i = component in flue or product gas

Assuming a basis of 1 gmol of n-pentane, $\Delta U_r = -777.46$ kcal/mol (see problem statement). The equation above for a batch process is similar to that for the theoretical adiabatic flame temperature (TAFT) employed with flow incinerators (see Problem THR.9 in Chapter 9). The first term of the energy balance, i.e., the change in internal energy of reaction, is somewhat similar to and approximately given by the enthalpy change of reaction (sometimes referred as the net heating value, NHV). The second term of the balance, the internal energy increase as the reaction products are raised in temperature from T_{in} to T_{out}, is determined as follows.
    The heat capacity at constant volume can be expressed by

$$C_v = C_p - R$$

where C_p = heat capacity at constant pressure, cal/(gmol·K)
  R = ideal gas law constant, cal/(gmol·K)

The heat capacity at constant pressure, in turn, can be expressed by the following equation given in the problem statement.

$$C_p = a + bT + cT^2 + dT^3$$

where a, b, c, d = constants
  T = temperature, K

The combustion reaction equation for 1 mol of pentane with stoichiometric air is

$$C_5H_{12} + 8O_2 + 30.1N_2 \rightarrow 5CO_2 + 6H_2O + 30.1N_2$$

Note that the number of moles of nitrogen, n_{N_2}, is present in proportion to the number of moles of oxygen, n_{O_2}, in the air:

$$n_{N_2} = (0.79/0.21)n_{O_2}$$
$$= (0.79/0.21)(8)$$
$$= 30.1 \, \text{mol of nitrogen}$$

The number of moles present before and after the reaction are therefore

$$n_{\text{in}} = \sum n_{\text{LHS}}$$
$$= n_{C_5H_{12}} + n_{O_2} + n_{N_2}$$
$$= (1) + (8) + (30.1)$$
$$= 39.1 \, \text{mol}$$
$$n_{\text{out}} = \sum n_{\text{RHS}}$$
$$= n_{CO_2} + n_{H_2O} + n_{N_2}$$
$$= (5) + (6) + (30.1)$$
$$= 41.1 \, \text{mol}$$

The constant volume heat capacity can be expressed as a function of the constant pressure heat capacity:

$$C_v = C_p - R$$
$$R = 1.987 \, \text{cal/(gmol} \cdot \text{K)}$$

	a	$a - R$	b	c	d	n_i
N_2	6.529	4.542	0.149×10^{-2}	-0.0227×10^{-5}	0	30.1
CO_2	5.316	3.329	1.429×10^{-2}	-0.8362×10^{-5}	1.784×10^{-9}	5
H_2O	6.970	4.983	0.345×10^{-2}	-0.0483×10^{-5}	0	6

The term $\sum(n_i, C_{v,i})_{\text{out}}$ can now be calculated:

$$
\begin{aligned}
\sum(n_i, C_{v,i})_{\text{out}} = {} & [(4.542)(30.1) + (3.329)(5) + (4.983)(6)] + [(0.149 \times 10^{-2})(30.1) \\
& + (1.429 \times 10^{-2})(5) + (0.345 \times 10^{-2})(6)]T \\
& + [(-0.0227 \times 10^{-5})(30.1) + (-0.8362 \times 10^{-5})(5) \\
& + (-0.0483 \times 10^{-5})(6)]T^2 + [(1.784 \times 10^{-9})(5)]T^3
\end{aligned}
$$

The second term of the energy balance, i.e., the change in internal energy due to the temperature rise of the products, can be obtained by integrating the above equation with respect to T from T_{in} (298 K) to T_{out}.

$$\int_{T_{in}}^{T_{out}} \sum(n_i, C_{v,i})_{out} \, dT = [(4.542)(30.1) + (3.329)(5) + 4.983)(6)][(T_{out} - 298)]$$

$$+ [(0.149 \times 10^{-2})(30.1) + (1.429 \times 10^{-2})(5)$$
$$+ (0.345 \times 10^{-2})(6)][(T_{out})^2 - (298)^2]/2$$
$$+ (-0.0227 \times 10^{-5})(30.1) + (-0.8362 \times 10^{-5})(5)$$
$$+ (-0.0483 \times 10^{-5})(6)][(T_{out})^3 - (298)^3]/3$$
$$+ [(1.784 \times 10^{-9})(5)][(T_{out})^4 - (298)^4]/4$$

Substituting into the energy balance given

$$0 = -777.76 \times 10^3 + [(4.542)(30.1) + (3.329)(5) + (4.983)(6)][(T_{out} - 298)]$$
$$+ [(0.149 \times 10^{-2})(30.1) + (1.429 \times 10^{-2})(5) + (0.345 \times 10^{-2})(6)]$$
$$\times [(T_{out})^2 - (298)^2]/2 + [(-0.0227 \times 10^{-5})(30.1) + (-0.8362 \times 10^{-5})(5)$$
$$+ (-0.0483 \times 10^{-5})(6)][(T_{out})^3 - (298)^3]/3 + [(1.784 \times 10^{-9})(5)]$$
$$\times [(T_{out})^4 - (298)^4]/4$$

By trial-and-error calculation, $T_{out} = 2870$ K.

The final pressure in the vessel is obtained by applying the ideal gas law:

$$PV = nRT$$

$$\frac{P_f}{n_f T_f} = \frac{P_i}{n_i T_i}$$

$$P_f = (14.7 + 5)\,\text{psia}\left(\frac{41.1\,\text{mol}}{39.1\,\text{mol}}\right)\left(\frac{2870\,\text{K}}{298\,\text{K}}\right)$$

$$= 199.4\,\text{psia}$$

$$= 184.7\,\text{psig}$$

Since the final pressure of 184.7 psig is less than the burst pressure of 200 psig, the vessel will withstand the explosion.

PART VIII
Other Topics

Roberto Diaz, Udomlug Siriphonlai

48 Dispersion (DSP)

DSP.1   ADVANTAGES AND DISADVANTAGES

Discuss the advantages and disadvantages of employing atmospheric dispersion as an ultimate disposal method of waste gases.

Solution

The advantages are listed below:

1. Dispersion of the waste gases leads to the dilution of the pollutants in the atmosphere. Self-purification mechanisms of atmospheric air also assists the process.
2. Tall stacks emit gas into the upper layer of the atmosphere and lower the ground concentration of the pollutants.
3. The method is commonly used, cheap and easily applicable.
4. By selecting the proper location of stacks through the use of different models for dispersion, it is possible to significantly reduce the concentration of waste gases in the atmosphere.

The disadvantages are:

1. Any particulate matter contained in the dispersed gases have a tendency to settle down to the ground level.
2. The location of the industrial source may prohibit dispersion as an option.
3. Plume rise can significantly vary with ambient temperature, stability conditions, molecular weight, and exit velocity of the stack gases.
4. The models of atmospheric dispersion are rarely accurate. They should only be used for estimation and comparative analysis.

DSP.2   GROUND SURFACE DEPOSITION

An incinerator stack is emitting fly ash at the rate of 2 tons per hour. Natural processes are capable of removing these particles from the affected ground surface at a steady rate, provided no more than 0.02% of the ground is covered by them per hour. Assume the particles are, on the average, spheres of radius 10^{-4} ft and have an average density of $120\,\text{lb/ft}^3$. The wind speed is $5\,\text{mph}$.

If L is the distance through which an average particle is carried by the wind, the particles will settle out uniformly over a wedge-shaped area the central angle of which is $20°$, at distances ranging from $0.5L$ to $20L$. Determine the minimum stack height H required to prevent ground level accumulation. Assume monolayer deposition, with the ground area covered by a particle equal to its cross-sectional area [i.e. $\pi(d_p)^2/4$].

Solution

Select $1.0\,\text{h}$ as a basis for the calculations:

Volume of one particle $= \frac{4}{3}\pi r^3$

Mass (weight) of one particle $= \rho_p V = (120)(4.19 \times 10^{-12})$
$$= 5.027 \times 10^{-10}\,\text{lb}$$

Number of particles emitted per hour $= (2)(2000)/(5.027 \times 10^{-10})$
$$= 7.958 \times 10^{12}$$

Area covered by particles per hour $=$ (number)(area) $=$ (number)$(\pi r^2) =$ $(7.958 \times 10^{12})(\pi)(10^{-4})^2 = 2.5 \times 10^5\,\text{ft}^2$

Set a (maximum deposition distance) $= 20L$ and b (minimum deposition distance) $= 0.5L$.

Area of deposit $= \pi(a^2 - b^2)(20/360) = \pi[(20L)^2 - (0.5L)^2](20/360)$
$$= 69.8L^2$$

0.02% of deposited area $= (2 \times 10^{-4})(69.8)L^2 = 1.396 \times 10^{-2}L^2$

Substitution yields

$$1.39 \times 10^{-2}L^2 = 2.5 \times 10^5$$

$$L^2 = 1.79 \times 10^7$$

$$L = 4230\,\text{ft}$$

The Stokes' law region applies for the settling particles. Therefore,

$$v_s = \frac{d_p^2 g \rho_p}{18\mu}$$

$$= \frac{(2 \times 10^{-4})^2(32.2)(120)}{(18)(1.23 \times 10^{-5})}$$

$$= 0.698\,\text{ft/s}$$

The residence time t is given by

$$t = L/u = H/v_s$$

$$\frac{H}{L} = \frac{0.698}{(5.0)(5280/3600)} = 0.0952$$

Since $L = 4230$ ft,

$$H = (0.0952)(4230) = 403 \text{ ft}$$

The minimum height of the stack is approximately 400 ft.

DSP.3   PLUME RISE

If a waste source emits a gas with a buoyancy flux of $50 \text{ m}^4/\text{s}^3$, and the wind averages 4 m/s, find the plume rise at a distance of 750 m downward from a stack that is 50 m high under unstable atmospheric conditions. Use the equation proposed by Briggs.

Solution

Several plume rise equations are available. Briggs used the following equations to calculate the plume rise:

$$\Delta h = 1.6 F^{1/3} u^{-1} x^{2/3} \qquad x < x_f$$

$$= 1.6 F^{1/3} u^{-1} x_f^{2/3} \qquad \text{if } x \geq x_f$$

$$x^* = 14 F^{5/8} \qquad \text{when } F < 55 \text{ m}^4/\text{s}^3$$

$$= 34 F^{2/5} \qquad \text{when } F \geq 55 \text{ m}^4/\text{s}^3$$

$$x_f = 3.5 x^*$$

where Δh = plume rise, m
     F = buoyancy flux, $\text{m}^4/\text{s}^3 = 3.7 \times 10^{-5} \dot{Q}_H$
     u = wind speed, m/s
     x^* = downward distance, m
     x_f = distance of transition from first stage of rise to the second stage of rise, m
     $\dot{Q}_H$ = heat emission rate, kcal/s

If the term $\dot{Q}_H$ is not available, the term F may be estimated by

$$F = (g/\pi)q(T_s - T)/T_s$$

where $g = $ gravity term 9.8 m/s^2
    $q = $ stack gas volumetric flowrate, m^3/s (actual conditions)
$T_S, T = $ stack gas and ambient air temperature, K, respectively

Calculate x_f to determine which plume equation applies.

$$x^* = 14F^{5/8} \qquad \text{since } F \text{ is less than } 55 \text{ m}^4/\text{s}^3$$

$$= (14)(50)^{5/8}$$

$$= 161.43 \text{ m}$$

$$x_f = 3.5x^*$$

$$= (3.5)(161.43)$$

$$= 565.0 \text{ m}$$

The plume rise is therefore

$$h = 1.6F^{1/3}u^{-1}x_f^{2/3} \qquad \text{since } x \geqslant x_f = (1.6)(50)^{1/3}(4)^{-1}(565)^{2/3} = 101 \text{ m}$$

Many more plume rise equations may be found in the literature. The Environmental Protection Agency (EPA) is mandated to use Brigg's equations to calculate plume rise. In past years, industry has often chosen to use the Holland or Davidson–Bryant equation. The Holland equation is

$$\Delta h = d_s(v_s/u)[1.5 + 2.68 \times 10^{-3}P(\Delta T/T_s)/d_s]$$

where $d_s = $ inside stack diameter, m
    $v_s = $ stack exit velocity, m/s
    $u = $ wind speed, m/s
    $P = $ atmospheric pressure, mbar
$T_s, T = $ stack gas and ambient temperature, respectively, K
    $\Delta T = T_s - T$
    $\Delta h = $ plume rise, m

The Davidson–Bryant equation is

$$\Delta h = d_s(v_s/u)^{1.4}[1.0 + (T_s + T)/T_s]$$

The reader should also note that the "plume rise" may be negative in some instances due to surrounding structures, topography, etc.

DSP.4   POWER PLANT EMISSION

A power plant burns 12 tons of 2.5% sulfur content coal per hour. The effective stack height is 120 m and the wind speed is 2 m/s. At one hour before sunrise, the sky is clear. A dispersion study requires information on the approximate distance of the maximum concentration under these conditions. (*Hint:* Calculate concentrations for downward distances of 0.1, 1.0, 5, 10, 20, 25, 30, 50 and 70 km.)

Solution

The coordinate system used in making atmospheric dispersion estimates of gaseous pollutants, as suggested by Pasquill and modified by Gifford, is described below. The origin is at ground level or beneath the point of emission, with the x axis extending horizontally in the direction of the mean wind. The y axis is in the horizontal plane perpendicular to the x axis, and the z axis extends vertically. The plume travels along or parallel to the x axis (in the mean wind direction). The concentration, C, of gas or aerosol at (x, y, z) from a continuous source with an effective height, H_e, is given by:

$$C(x, y, z, H_e) = (\dot{m}/2\pi\sigma_y\sigma_z u)[e^{-(1/2)(y/\sigma_y)^2}][e^{-(1/2)((z-H_e)/\sigma_z)^2} + e^{-(1/2)((z+H_e)/\sigma_z)^2}]$$

where H_e = effective height of emission (sum of the physical stack height, H_s and
       the plume rise, Δh), m
   u = mean wind speed affecting the plume, m/s
   $\dot{m}$ = emission rate of pollutants, g/s
σ_y, σ_z = dispersion coefficients or stability parameters, m
   C = concentration of gas, g/m^3
x, y, z = coordinates, m

The assumptions made in the development of the above equation are: (1) the plume spread has a Gaussian (normal) distribution in both the horizontal and vertical planes, with standard deviations of plume concentration distribution in the horizontal and vertical directions of σ_y, and σ_z, respectively; (2) uniform emission rate of pollutants, $\dot{m}$; (3) total reflection of the plume at ground $z = 0$ conditions; and (4) the plume moves downstream (horizontally in the x direction) with mean wind spead, u. Although any consistent set of units may be used, the cgs system is preferred.
    For concentrations calculated at ground level ($z = 0$), the equation simplifies to

$$C(x, y, 0, H_e) = (\dot{m}/2\pi\sigma_y\sigma_z u)[e^{-(1/2)(y/\sigma_y)^2}][e^{-(1/2)(H_e/\sigma_z)^2}]$$

If the concentration is to be calculated along the centerline of the plume ($y = 0$), further simplification gives

$$C(x, 0, 0, H_e) = (\dot{m}/\pi\sigma_y\sigma_z u)[e^{-(1/2)(H_e/\sigma_z)^2}]$$

In the case of a ground-level source with no effective plume rise ($H_e = 0$), the equation reduces to

$$C(x, 0, 0, 0) = (\dot{m}/\pi\sigma_y\sigma_z u)$$

It is important to note that the two dispersion coefficients are the product of a long history of field experiments, empirical judgments, and extrapolations of the data from those experiments. There are few knowledgeable practitioners in the dispersion modeling field who would dispute that the coefficients could easily have an inherent uncertainty of $\pm25\%$.

The six applicable stability categories for these coefficients are shown in the following table:

Surface Wind Speed at 10 m (m/s)	Day			Night	
	Incoming Solar Radiation			Thinly Overcase or $>4/8$ Low Cloud	$<3/8$ Cloud
	Strong	Moderate	Slight		
2	A	A–B	B	—	—
2–3	A–B	B	C	E	F
3–5	B	B–C	C	D	E
5–6	C	C–D	D	D	D
6	C	D	D	D	D

Note that A, B, C refer to daytime with unstable conditions; D refers to overcast or neutral conditions at night or during the day; E and F refer to night time stable conditions and are based on the amount of cloud cover. "Strong" incoming solar radiation corresponds to a solar altitude greater than $60°$ with clear skies (e.g., sunny midday in midsummer); "slight" insolation (rate of radiation from the sun received per unit of Earth's surface) corresponds to a solar altitude from $15°$ to $35°$ with clear skies (e.g., sunny midday in midwinter). For the A–B, B–C, and C–D stability categories, one should use the average of the A and B values, B and C values, and C and D values, respectively. Figures 125 and 126 provide the variation of σ_y and σ_z with stability categories and distances.

First calculate the sulfur dioxide emission rate, $\dot{m}$, in grams/second. The molecular weights of S and SO_2 are 32 and 64, respectively.

$$\dot{m} = (12)(0.025)(64/32)$$
$$= 0.60 \text{ ton/h}$$
$$= (0.60)(2000)(454)/(3600)$$
$$= 151.3 \text{ g/s}$$

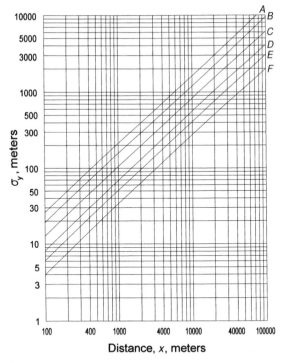

Figure 125. Dispersion coefficients, *y* direction.

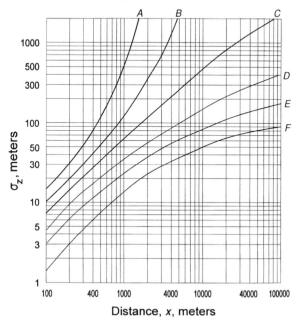

Figure 126. Dispersion coefficients, *z* direction.

Since it is night time, the sky is clear, and the wind speed is 2 m/s. Therefore, the stability category is F (see table above).

Determine the values of σ_y and σ_z for each downward distance for stability category F from Figures 125 and 126.

x (km)	σ_y (m)	σ_z (m)
0.1	4	2.3
1.0	34	14.0
5	148	34.0
10	275	46.5
20	510	60.0
25	610	64.0
30	720	68.0
50	1120	78.0
70	1500	84.0

Since the maximum groundlevel ($z = 0$) concentration along the x axis is desired and the maximum concentration occurs along the centerline ($y = 0$) of the plume,

$$C(x, 0, 0, H_e) = (\dot{m}/\pi\sigma_y\sigma_z u)[e^{-1/2(H_e/\sigma_z)^2}]$$

The concentrations of sulfur dioxide at the downward distances are obtained by completing the table below.

x (km)	C (g/m^3)
0.1	0 (infinitely small)
1.0	5.61×10^{-18}
5	9.41×10^{-6}
10	6.72×10^{-5}
20	1.06×10^{-4}
25	1.06×10^{-4}
30	1.03×10^{-4}
50	8.41×10^{-5}
70	7.03×10^{-5}

For example, at 1.0 km,

$$C(1.0, 0, 0, H_e) = [(151.3)/(\pi)(34)(14)(2)][e^{-1/2(120/14)^2}]$$
$$= 5.61 \times 10^{-18} \text{ g/m}^3$$

From the table above, the maximum concentration is approximately 1.06×10^{-4} g/m^3 and the corresponding location is between 20 and 25 km.

DSP.5   MAXIMUM GROUND-LEVEL CONCENTRATION

Determine the required recovery of an air pollution control device knowing that the maximum concentration along the ground-level centerline of the plume must not exceed $1.05 \times 10^{-6} \, \text{g/m}^3$ at a downward distance of 50 km. The wind speed is 6 m/s and the effective height of emission is 150 m. The concentration was measured during the day, when incoming solar radiation was very intense. The incoming rate of pollutant to the control device is known to be 4189 g/s.

Solution

Based on the problem conditions, the stability category is C. From Figures 125 and 126, at $x = 50 \, \text{km}$, $\sigma_y = 3300 \, \text{m}$, and $\sigma_z = 1500 \, \text{m}$:

$$C(50 \, \text{km}, 0, 0, 150 \, \text{m}) = 1.05 \times 10^{-6} \, \text{g/m}^3$$

Using the Pasquill–Gifford equation,

$$C(x, y, z, H_e) = (\dot{m}/2\pi\sigma_y\sigma_z u)[e^{-(1/2)(y/\sigma_y)^2}][e^{-(1/2)((z-H_e)/\sigma_z)^2} + e^{-(1/2)((z+H_e)/\sigma_z)^2}]$$

Solving for $\dot{m}$ yields

$$\dot{m} = 94.44 \, \text{g/s}$$

Thus, the collection efficiency required is

$$E(\text{required}) = (4189 - 98.44)/4189$$
$$= 0.9765 = 97.65\%$$

DSP.6   OBSERVATION TOWER CONCENTRATION

What is the concentration of pollutants at the top of a recently constructed ranger's observation tower that is 2 km north and 100 m west of a stack emitting pollutants at a rate of 8.0 g/s. The stack is known to have a centerline ground zero concentration of $3 \, \text{mg/m}^3$ at 1.0 km. The volumetric flowrate of the stack gas is 40,000 scfm, the diameter of the stack is 6.0 ft, the temperature of the emission is 300°F, the wind is out of the south at a rate of 10.0 miles/h, and the height of the ranger's tower is 12 m. Assume Holland's equation to apply, both the standard and ambient temperature to be 68°F (298 K), and the stability category to be D.

Solution

A top-view schematic of the system is provided below in Figure 127. Pertinent data follow:

$$q_s = 40,000 \, \text{scfm}$$

$$u = 10 \, \text{mph} = 14.67 \, \text{ft/s} = 4.47 \, \text{m/s}$$

$$T_s = 300°\text{F} = 422 \, \text{K}$$

$$T = 68°\text{F} = 298 \, \text{K}$$

$$P = 1 \, \text{atm} = 1013.25 \, \text{mbar}$$

$$d_s = (\text{stack diameter}) = 6 \, \text{ft} = 1.8288 \, \text{m}$$

Key calculations for the actual volumetric flowrate (q_a), stack area (A_s), stack velocity (v_s), and plume rise (Δh) are listed below.

$$q_a = 40,000(300 + 460)/528 = 57,575.8 \, \text{acfm}$$

$$A_s = \pi(6)^2/4 = 28.234 \, \text{ft}^2$$

$$v_s = q_a/A_s = 57,575.8/28.234 = 33.94 \, \text{ft/s} = 10.35 \, \text{m/s}$$

$$\Delta h = [v_s(d_s/u)][1.5 + 0.00268P(T_s - T)(d_s/T)]$$

$$= [(10.35)(1.83)/(4.47)][1.5 + (0.00268)(1013.28)[(432 - 298)/298](1.83)]$$

$$= 15.1 \, \text{m}$$

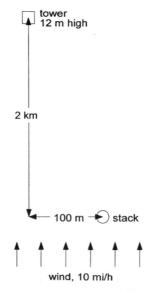

Figure 127. Schematic for Problem DSP.6.

The Pasquill–Gifford equation is used to estimate the effective stack height, H_e, noting that $\dot{m}$ is 8 g/s.

From Figures 125 and 126, at $x = 1$ km, $\sigma_y = 70$ m, and $\sigma_2 = 31$ m. Thus,

$$C(x, 0, 0, \bar{H}_e) = \left(\frac{\dot{m}}{\pi\sigma_y\sigma_z u}\right)e^{-(1/2)(\bar{H}_e/\sigma_z)^2}$$

$$3.0 \times 10^{-6} = \left[\frac{8}{(\pi)(70)(31)(4.47)}\right]e^{-(1/2)(\bar{H}_e/31)^2}$$

$$\bar{H}_e = 92.7\,\text{m}$$

Therefore, the stack height is

$$H_e = 92.7 - 15.1 = 77.6m = 255\,\text{ft}$$

Find $C(2000, 100, 12, 92.7) = C(x, y, z, H_e)$.

$$C = \left(\frac{\dot{m}}{2\pi\sigma_y\sigma_2 u}\right)(e^{-(1/2)(y/\sigma_y)^2})[e^{-(1/2)(z-\bar{H}_e/\sigma_z)^2} + e^{-(1/2)(z+\bar{H}_e/\sigma_z)^2}]$$

$$\sigma_y = 130\,\text{m} \qquad \sigma_z = 50\,\text{m}$$

$$C = \left[\frac{8.0}{(2\pi)(130)(50)(4.47)}\right](e^{-(1/2)(100/130)^2})[e^{-(1/2)(10-92.7/50)^2} + e^{-(1/2)(10+92.7/50)^2}]$$

$$= 4.38 \times 10^{-5}(0.7439)(0.2719 + 0.11165)$$

$$1.25 \times 10^{-5}\,\text{g/m}^3 = 12.5\,\mu\text{g/m}^3$$

DSP.7   LINE SOURCE APPLICATION

A six-story hospital building is located 300 m east and downwind from an expressway. The expressway runs north–south and the wind is from the west at 4 m/s. It is 5:30 in the afternoon on an overcast day. The measured traffic flow is 8000 vehicles per hour during this rush hour and the average vehicle, traveling at an average speed of 40 mph, is expected to emit 0.02 g/s of total hydrocarbons. Concentrations at the hospital are required as part of a risk assessment study. How much lower, in percent, will the hydrocarbon concentration be on top of the building (where the elderly patients are housed) as compared with the concentration estimated at ground level? Assume a standard floor to be 3.5 m in height.

Solution

Line source applications are generally confined to roadways and streets along which there are well-defined movements of motor vehicles. For these types of line sources, data are required on the width of the roadway and its center strip, the types and amounts per unit time per unit length [g/(s · m)] of pollutant emissions, the number of lanes, the emissions from each lane, and the height of emissions. In some situations, e.g., a traffic jam at a tollbooth, or a series of industries located along a river, or heavy traffic along a stright stretch of highway, the pollution problem may be modeled as a continuous emitting infinite line source. Concentrations downwind of a continuous emitting infinite line source, when the wind direction is normal to the line, can be calculated from

$$C(x, y, 0, H_e) = [2q/(2\pi)^{1/2}\sigma_z u](e^{-(1/2)(H_e/\sigma_z)^2})$$

where q = source strength per unit distance, g/(s · m)
$\quad H_e$ = effective stack or discharge height, m
$\quad u$ = wind speed, m/s
$\quad \sigma_z$ = vertical dispersion coefficients, m

Note that the horizontal dispersion coefficient, σ_y, does not appear in this equation since it is assumed that lateral-dispersion from one segment of the line is compensated by dispersion in the opposite direction from adjacent segments. Also, y does not appear since the concentration at a given x is the same for any value of y.

Since it is early evening and the wind speed is 4 m/s, the stability category is D. The number of vehicles per meter, n, is

$$n = \frac{8000\,\text{vehicles/m}}{(40\,\text{mi/h})(1609\,\text{m/mi})}$$

$$= 0.1243\,\text{vehicles/m}$$

Calculate the source strength per unit length, q, from n and the emission rate per vehicle:

$$q = (n)(0.02)$$

$$= (0.1243)(0.02)$$

$$= 2.49 \times 10^{-3}\,\text{g HC}/(\text{s} \cdot \text{m})$$

The vertical dispersion coefficient, σ_z, for stability category D from Figure 126 is

$$\sigma_z = 12\,\text{m at}\,x = 300\,\text{m}$$

The height of the building, H_b, is

$$H_b = (6)(3.5)$$
$$= 21\,\text{m}$$

The concentration 300 m downwind at ground level conditions is then

$$C(x, y, 0, H_e) = (2q/(2\pi)^{1/2}\sigma_z u)(e^{-1/2(H_e/\sigma_z)^2})$$
$$C(300, 0, 0, 0) = [(2)(2.49 \times 10^{-3})/(2\pi)^{1/2}(12)(4)](1)$$
$$= 4.14 \times 10^{-5}\,\text{g/m}^3$$

Note that the exponential term becomes 1.0 if $H_e = 0$.

The concentration at 300 m downwind at the top of the building may also be calculated.

$$C(300, 0, 0, H_b) = (2q/(2\pi)^{1/2}\sigma_z u)(e^{-1/2(H_b/\sigma_z)^2})$$
$$C(300, 0, 0, 21) = [(2)(2.49 \times 10^{-3})/(2\pi)^{1/2}(12)(4)](e^{-1/2(21/12)^2})$$
$$= 8.95 \times 10^{-6}\,\text{g/m}^3$$

Determine how much lower the concentration, in percent, will be at the top of the building:

$$\% = 100\,(\text{ground conc.} - \text{top conc.})/(\text{ground conc.})$$
$$= 100(4.14 \times 10^{-5} - 8.95 \times 10^{-6})/(4.14 \times 10^{-5})$$
$$= 78.4\%$$

Concentrations from infinite line sources, when the wind is not perpendicular to the line, can also be approximated. If the angle between the wind direction and the line source is θ

$$C(x, y, 0, H_e) = [2q/\sin(\theta)(2\pi)^{1/2}\sigma_z u](e^{-(1/2)(H_b/\sigma_z)^2})$$

This equation should not be used when θ is less than 45°.

When the continuously emitting line source is reasonably short in length or "finite," one can account for the edge effects caused by the two ends of the source. If the line source is perpendicular to the wind direction, then it is convenient to define the x axis in the direction of the wind and also passing through the sampling point downwind. The ends of the line source are then at two positions in the crosswind direction, y_1 and y_2, where y_1 is less than y_2. The concentration along the x axis at ground level is

$$C(x, y, 0, H_e) = [2q/(2\pi)^{1/2}\sigma_z u](e^{-(1/2)(H_b/\sigma_z)^2}) \int_{p_1}^{p_2} [(1/(2\pi)^{1/2})e^{(-1/2)p^2}]\,dp$$

where $p_1 = y_1/\sigma_y$
$\qquad p_2 = y_2/\sigma y$

Once the limits of integration are established, the value of the integral may be determined from standard tables of integrals.

DSP.8   AREA SOURCE APPLICATION

An inventory of emissions has been made in an urban location by square areas, 1524 m on a side, as part of an emergency planning and response study. The emissions from two such adjacent areas (source areas B and D) are estimated to be 6.0 g/s for each area. The effective stack height of the sources within each area is approximately 20 m. The wind is 2.5 m/s on a thinly overcast night and perpendicular to the two adjacent square areas (i.e., source areas B and D are aligned perpendicular to the wind). What is the percentage contribution of emissions from the two sources to the center point of the square area (point A) immediately downwind of the source area D?

Solution

Area sources include the multitude of minor sources with individually small emissions that are impractical to consider as separate point or line sources. Area sources are typically treated as a grid network of square areas, with pollutant emissions distributed uniformly within each grid square. Area source information required are types and amounts of pollutant emissions, the physical size of the area over which emissions are prorated and representative stack height(s) for the area.

In dealing with dispersion of pollutants in areas having large numbers of sources, e.g., as in fugitive dust from (coal) piles, a large number of automobiles in a parking lot, or a multistack situation, there may be too many sources to consider each source inividually. Often an approximation can be made by combining all of the emissions in a given area and treating this area as a source having an initial horizontal standard deviation, σ_{y0}. A virtual distance, x_y, that will give this standard deviation can be found. This virtual distance is the distance that will yield the appropriate value for σ_y from Figure 125 provided in Problem DSP.4. Values of x_y will vary with stability. The equation for point sources may then be used, determining σ_y as a function of $x + x_y$. This procedure effectively treats the area source as a cross-wind line source with a normal distribution across the area source. Finally, the initial standard deviation for a square area source can be approximated by $\sigma_{y0} = s/4.3$, where s is the length of a side of the area.

Since it is a thinly overcast night and the wind speed is 2.5 m/s, stability category E applies. Calculate the initial standard deviation, σ_{y0}.

$$\sigma_{y0} = s/4.3$$
$$= 1524/4.3$$
$$= 354\,\text{m}$$

Obtain the virtual point source distance, x_y using σ_{y0} of 354 m from Figure 125.

$$x_y = 8.4\,\text{km for } \sigma_{y0} = 354\,\text{m and stability category E}$$

The total distance, x_t, is therefore

$$x_t = 8.4 + 1.524$$
$$= 9.9\,\text{km}$$

Also obtain σ_y, which corresponds to an x_t of 9.9 km, from Figure 125.

$$\sigma_y = 410\,\text{m}$$

Since the area source can be assumed to be well mixed, σ_z may be obtained at $x = 1524$ m from Figure 126.

$$\sigma_z = 26\,\text{m}$$

Calculate the concentration at point A because of point D (directly downward):

$$C(1524, 0, 0, H) = (\dot{m}/\pi\sigma_y\sigma_z u)(e^{-(1/2)(y/\sigma_y)^2})(e^{-(1/2)(H_e/\sigma_z)^2})$$
$$= \{(6.0)/[(\pi)(410)(26)(2.5)]\}(e^{-(1/2)(20/26)^2})$$
$$= 5.33 \times 10^{-5}\,\text{g/m}^3$$

Calculate the concentration at point A (directly downward from point D).

$$C(1524, 1524, 0, H) = (\dot{m}/\pi\sigma_y\sigma_z u)(e^{-(1/2)(H_e/\sigma_z)^2})$$
$$= \{(6.0)/[(\pi)(410)(26)(2.5)]\}(e^{-(1/2)(1524/410)^2})(e^{-(1/2)(20/26)^2})$$
$$= 5.33 \times 10^{-8}\,\text{g/m}^3$$

Finally, calculate the ratio of the contribution to point A from points B and D.

$$\text{Ratio} = 5.33 \times 10^{-8}/5.33 \times 10^{-5}$$

$$0.001$$

$$\text{Percent contribution of } B = (0.001)(100)$$

$$= 0.1\%$$

DSP.9   STACK DESIGN RECOMMENDATIONS

List some of the key rules of thumb of stack design.

Solution

As experience in designing stacks has accumulated over the years, several guidelines have evolved.

1. Stack heights should be at least 2.5 times the height of any surrounding buildings or obstacles so that significant turbulence is not introduced by these factors.
2. The stack gas exit velocity should be greater than 60 ft/s so that stack gases will escape the turbulent wake of the stack. In many cases, it is good practice to have the gas exit velocity on the order of 90 or 100 ft/s.
3. A stack located on a building should be set in a position that will assure that the exhaust escapes the wakes of nearby structures.
4. Gases from the stacks with diameters less than 5 ft and heights less than 200 ft will hit the ground part of the time, and the ground concentration will be excessive. In this case, the plume becomes unpredictable.
5. The maximum ground concentration of stack gases subjected to atmospheric dispersion occurs about 5–10 effective stack heights downwind from the point of emission.
6. When stack gases are subjected to atmospheric diffusion and building turbulence is not a factor, ground-level concentrations on the order of 0.001–1% of the stack concentration are possible for a properly designed stack.
7. Ground concentrations can be reduced by the use of higher stacks; the ground concentration varies inversely as the square of the effective stack height.
8. Average concentrations of a contaminant downwind from a stack are directly proportional to the discharge rate; an increase in the discharge rate by a given factor increases ground-level concentrations at all points by the same factor.
9. In general, increasing the dilution of stack gases by the addition of excess air in the stack does not affect ground-level concentrations appreciably. Practical

stack dilutions are usually insignificant in comparison to the later atmospheric dilution by plume diffusion. Addition of diluting will increase the
effective stack height, however, by increasing the stack exit velocity. This
effect may be important at low wind speeds. On the other hand if the stack
temperature is decreased appreciably by the dilution, the effective stack
height may be reduced.

10. Stack dilution has an appreciable effect on the concentration in the plume
    close to the stack.

These 10 guidelines represent the basic design elements of a stack "pollution
control" system. An engineering approach suggests that each element be evaluated
independently and as part of the whole control system. However, the engineering
design and evaluation must be an integrated part of the complete pollution control
program.

DSP.10   RISK OF METAL EMISSION

Surface deposition and soil concentrations of particulate metals released by a
hazardous waste incinerator are to be determined as part of an exposure assessment
being conducted. To estimate the settling velocity of the metal particulate material,
the following equation for terminal velocity, v_t is to be used:

$$v_t = \frac{d_p^2 \rho_p g}{18\mu}$$

where d_p = particle diameter, cm
  ρ_p = density of air, g/cm^3
  g = gravitational acceleration, 980.7 cm/s^2
  μ = air viscosity, g/(cm · s)

Assume that the incinerator emits the state's legal maximum (0.08 gr/dscf) particulate concentration as metals, and that it operates at 50% excess air with a stack gas
flow of 10,000 dscf. Also assume that the particles are emitted from an effective
height of 200 ft and that they have a density of 2.5 g/cm^3.

Use the Gaussian dispersion equation for the estimation of downwind ambient
particulate concentrations. To account for particle settling, a "titled" plume model
may be used, in which H is decreased to account for the vertical settling motion of
the particles over a travel distance x.

The following expressions may be used for the estimation of dispersion
coefficients σ_y and σ_z for stability class D (neutral stability), with x in meters:

$$\sigma_y = 0.32x^{0.78}$$

$$\sigma_z = 0.22x^{0.78}$$

Using the data presented above, determine the following for particles of 0.5-, 5-, and 50-μm diameter under conditions of neutral stability with a 5-mph wind:

1. The emission rate in lb/h and g/s
2. The settling velocity of the particles in m/s
3. The distance at which the centerline of the titled plume would touch the ground surface in miles
4. The ambient concentration at the distance determined in part 3 in $\mu g/m^3$
5. The surface flux due to settling at 0.25 mile downwind of the stack in $g/(m^2 \cdot s)$
6. Assuming that the meteorological conditions described above apply 5% of the time, that the facility operates for 30 years, and that the deposited particles become evenly mixed in the top 10 cm of the soil, calculate the metal concentration in the upper soil in ppm. Use a soil density of $200\,kg/m^3$.
7. The annual probability of an individual dying a carcinogenic death from metal deposition over a 30-year period. The analysis should be based on the following:

   a. The probability that a food crop will be planted on the contaminated site in any given year is 0.25.
   b. The probability that the crop planted will produce a marketable yield is 0.80.
   c. The probability that the marketable yield is contaminated; this will vary with metal soil concentration as follows:

Metal Soil Concentration (ppmw)	Probability of Contamination (independent of particle size)
0.013	0.0001
1.3	0.001
185	0.10

The probability of human consumption of a fatal amount of contaminated crop yield is 0.005.

Solution

1. The emission rate is

$$\dot m = \left(\frac{0.08\,gr}{dscf}\right)\left(\frac{1\,lb}{7000\,gr}\right)\left(\frac{10,000\,dscf}{min}\right)\left(\frac{60\,min}{h}\right)$$

$$= 6.86\,lb/h$$

$$= 0.865\,g/s$$

2. At 18°C,

$$v_t = \frac{d_p^2 \rho_p g}{18\mu} = \frac{d_p^2(2.5\,\text{g/cm}^3)(980.7\,\text{cm/s}^2)}{(18)[1.827 \times 10^{-4}\,\text{g/(cm} \cdot \text{s})]}$$

$$= 7.455 \times 10^5 d_p^2 \,\text{cm/s} \quad (d_p \equiv \text{cm})$$

$$= 7.455 \times 10^{-3} d_p^2 \,\text{cm/s} \quad (dp \equiv \mu\text{m})$$

3. The settling time, t, is given by

$$t = (1 - v_t(200\,\text{ft})(0.305\,\text{m/ft})$$

where t is in s and v_t is in m/s. The horizontal distance traveled, x, is

$$x = (5\,\text{mi/h})(1\,\text{h/3600 s})t$$

where x is in mi and t is in s.

Settling velocity and travel distance variations with particle diameter presented below.

Particle diameter	Settling velocity (m/s)	Travel distance (mi)
0.5	1.86×10^{-5}	4460
5.0	1.86×10^{-3}	44.6
50	1.86×10^{-1}	0.446

4. The Pasquill–Gifford model is applicable for $x < 50$ km. Therefore, calculations need only be performed for 50-μm particles where the distance traveled is 0.446 miles or 710 m. For this condition:

$$\sigma_y = 0.32(719)^{0.78}$$

$$\sigma_z = 0.22(719)^{0.78}$$

Based on the problem statement, $H = 0, y = 0$, and $u = 5\,\text{mph} = 2.2\,\text{m/s}$. Therefore

$$C(719m, 0, 0, 0) = \frac{0.865}{(2)(\pi)(2.2)(54)(37)}$$

$$= 3.13 \times 10^{-5}\,\text{g/m}^3$$

$$= 31.3\,\mu\text{g/m}^3$$

5. For surface deposition, D,

$$D = (C)(v)$$

For this condition, $y = 0$, and $H = 200\,\text{ft} = 61\,\text{m}$.

For 0.5-µm and 5.0-µm particles (where the particulate settling can be neglected) at a distance of 0.25 miles or 403 m,

$$\sigma_y = (0.32)(403)^{0.78} = 34\,\text{m}$$

$$\sigma_z = (0.22)(403)^{0.78} = 24\,\text{m}$$

Thus,

$$C(719\text{m}, 0, 0, 0) = \left(\frac{0.865}{(2)(\pi)(2.2)(34)(24)}\right)(e^{-1/2(61/24)^2})$$

$$= 3.03 \times 10^{-6}\,\text{g/m}^3$$

$$= 3.03\,\mu\text{g/m}^3$$

A settling correction, h_y, must be included for the 50 µm particles. At 0.25 miles,

$$h_y = 61[(1 - (0.25/0.44)] = 26.3\,\text{m}$$

$$C(403\text{m}, 0, 0, 0) = \left(\frac{0.865}{(2)(\pi)(2.2)(34)(24)}\right)(e^{-1/2(26.3/24)^2})$$

$$= 4.21 \times 10^{-6}\,\text{g/m}^3$$

$$= 4.21\,\mu\text{g/m}^3$$

Thus,

$$D(0.5\,\mu\text{m}) = (3.03)(1.86 \times 10^{-5}) = 5.63 \times 10^{-11}\,\text{g/(m}^2 \cdot \text{s})$$

$$D(5.0\,\mu\text{m}) = (3.03)(1.86 \times 10^{-3}) = 5.63 \times 10^{-9}\,\text{g/(m}^2 \cdot \text{s})$$

$$D(50.0\,\mu\text{m}) = (42.1)(1.86 \times 10^{-1}) = 7.83 \times 10^{-6}\,\text{g/(m}^2 \cdot \text{s})$$

6. The mass, m, of 0.5-µm particles deposited in a 1.0-m^2 area during a 30-year period is

$$m_{0.5\mu\text{m}} = (1.0\,\text{m}^2)(5.63 \times 10^{-11}\,\text{g/(m}^2 \cdot \text{s})](30\text{ yr})(8760\,\text{h/yr})(3600\,\text{s/h})(0.05)$$

$$= 0.0027\,\text{g}$$

The particles are deposited in the upper 10 cm of soil with a mass

$$m_{0.5\mu m} = (1.0\,\text{m}^2)[2.0 \times 10^6\,\text{g}/(\text{m}^2 \cdot \text{s})](0.1\,\text{m}) = 2.0 \times 10^5\,\text{g}$$

The mass fraction of metals (particulates) is

$$C_{0.5\mu m} = 0.0026/2.0 \times 10^5 = 1.3 \times 10^{-9} = 0.013\,\text{ppmw}$$

In a similar manner,

$$C_{5.0\mu m} = 1.3\,\text{ppmw}$$

$$C_{50.0\mu m} = 185\,\text{ppmw}$$

7. The metal soil concentrations provided in the problem statement correspond to those calculated in part 6. (The probability of contamination would otherwise be obtained by linear interpolation.) The annual probability of an individual dying a carcinogenic death from metal deposition over a 30-year period is therefore given by the product of the probabilites provided in (a), (b), and (c), in part 7 of the problem statement.

$$P_{0.5\mu m} = (0.25)(0.80)(0.0001)(0.005) = 10^{-7}$$

$$P_{5.0\mu m} = (0.25)(0.80)(0.001)(0.005) = 10^{-6}$$

$$P_{50.0\mu m} = (0.25)(0.80)(0.10)(0.005) = 10^{-4}$$

49 Noise Pollution (NOP)

NOP.1 DEFINITIONS

Describe the general subject of noise pollution.

Solution

An estimated 20 million Americans are exposed to noise that poses a threat to their hearing. Everyone at some time or another has experienced the effects of noise pollution. Many people are unaware that the sounds that cause them so much annoyance may also be affecting their hearing. Hearing loss is one of the most serious health threats as a result of noise pollution.

Noise pollution is traditionally not placed among the top environmental problems facing the nation; however, it is one of the more frequently encountered sources of pollution in everyday life. Noise pollution can be defined simply by combining the meaning of both environmental terms, *noise* and *pollution*. Noise is typically defined as unwanted sound, and pollution is generally defined as the presence of matter or energy whose nature, location, or quantity produces undesired environmental effects. Noise is typically thought of as a nuisance rather than a source of pollution. This is due in part because noise does not leave a visible impact on the environment as do other sources of pollution.

The damage done by the traditional pollution of our air and water is widely recognized. The evidence is right before one's eyes—in contaminated water, oil spills, and dying fish, as well as in smog that burns the eyes and sears the lungs. However, noise is a more subtle pollutant. Aside from sonic booms that can break windows, noise usually leaves no visible evidence, although it also can pose a hazard to our health and well-being. Approximately 15 million Americans are exposed to noise that poses a threat to their hearing on the job. Another 15 million are exposed to dangerous noise levels without knowing it from trucks, airplanes, motorcycles, stereos, lawnmowers, and kitchen appliances.

NOP.2 EFFECT OF SOUND ON HEARING

Describe sound and its effects on hearing.

Solution

Sound travels in waves through the air like waves through water. The higher the wave, the greater its power. The greater the number of waves a sound has, the greater is its frequency or pitch.

The strength of sound or sound level is measured in decibels (dB). The decibel scale ranges from 0, which is regarded as the threshold of hearing for normal, healthy ears, to 194, which is regarded as the theoretical maximum for pure tones. Because the decibel scale, like the pH scale, is logarithmic, 20 dB is 100 times louder than 0; 30 dB is 1000 times louder; 40 dB is 10,000 times louder, etc. Thus at high levels, even a small reduction in level values can make a significant difference in noise intensity.

The frequency is measured in hertz (Hz) (cycles per second) and can be described as the rate of vibration. The faster the movement, the higher the frequency of the sound pressure waves created. The human ear does not hear all frequencies. Our normal hearing ranges from 20 to 20,000 Hz, or, roughly, from the lowest note on a great pipe organ to the highest note on a violin.

The human ear also does not hear all sounds equally. Very low and very high notes sound more faint to one's ear than do 1000-Hz sounds of equal strength. This is the way ears function. The human voice in conversation covers a median range of 300–4000 Hz. The musical scale ranges from 30 to 4000 Hz. Noise in these ranges sound much louder than do very low or very high pitched noises of equal strength.

Because hearing also varies widely among individuals, what may seem loud to one person may not be to another. Although loudness is a personal judgment, precise measurement of sound is made possible by use of the decibel scale. This scale measures sound pressure or energy according to international standards. This table compares some common sounds and shows how they rank in potential harm. Note that 70 dB is the point at which noise begins to harm hearing. To the ear, each 10-dB increase seems twice as loud.

NOP.3   SOUND PRESSURE LEVEL

Calculate the sound pressure level (SPL) (often denoted as L_p) in decibels (dB) if the sound pressure, p, is

1. 2.0×10^{-4} Pa
2. 20.0×10^{-4} Pa
3. 200.0×10^{-4} Pa

Solution

The sound pressure level in decibels (dB) is defined by the following equation:

$$SPL = 20 \log(p/p_o)$$

where SPL = sound pressure, dB
      p = sound pressure (rms), μPa
      p_o = reference pressure, 20 Pa

The reference pressure, 20 μPa, is the threshold of human hearing. It is also the pressure equivalent of 10^{-12} W/m^2, 2.0×10^{-5} Pa, 2.0×10^{-5} N/m^2, 2.9×10^{-9} psi, and 2.0×10^{-4} dyn/cm.

Since the value of effective sound pressure p_{rms} increases about 10^6 times from a pin drop to a thunderclap, and since the human hearing sensation seems to vary in a logarithmic way, logarithms are used in measurement of sound.

For 2.0×10^{-4} Pa,

$$\text{SPL} = 20 \log(2.0 \times 10^{-4}/2.0 \times 10^{-5})$$

$$= 20 \, \text{dB}$$

Similarly, for 20.0×10^{-4} and 200.0×10^{-4} Pa,

$$\text{SPL}(20.0 \times 10^{-4}) = 40 \, \text{dB}$$

$$\text{SPL}(200.0 \times 10^{-4}) = 60 \, \text{dB}$$

NOP.4 PERMISSIBLE NOISE EXPOSURES

List the permissible noise exposures of the Occupational Safety and Health Administration (OSHA) and the Environmental Protection Agency (EPA).

Solution

Both OSHA and the EPA have provided information on permissible noise exposures. Protection from noise is required when sound levels exceed those provided below. These are measured on the A scale at a slow response on a standard sound-level meter (except for certain alarms, etc.) as provided by OSHA.

Permissible Noise Exposures

Duration per Day (h)	Sound Level, dB Slow Response
8	90
6	92
4	95
3	97
2	100
1.5	102
1	105
0.5	110
0.25 or less	115

NOP.5   CLASSIFICATION OF SOUNDS

List various sounds. Also provide information on the associated noise level (in dB) and the effects.

Solution

Information on sounds, noise levels and effects are tabulated below.

Sound Levels and Human Response

Common Sounds	Noise Level (dB)	Effect
Carrier deck jet operations	140	Painfully loud
Air raid siren	130	
Jet takeoff (200 ft)	120	Maximum vocal effort
Thunderclap		
Discoteques		
Auto horn (3 ft)		
Stereo	Up to 120	
Pile drivers	110	
Garbage truck	100	
Chain saw		
Gasoline riding mower	90–95	
Food disposal (grinder)	67–93	
Gasoline power mower	87–92	
Heavy truck (50 ft)	90	Very annoying
City traffic		Hearing damage (8 h)
Home shop tools	85	
Vacuum cleaner	62–85	
Dishwasher	54–85	
Electric lawn edger	81	
Alarm clock (2 ft)	80	Annoying
Hair dryer		
Washing machine	47–78	
Electric shaver	75	
Sewing machine	64–74	
Noisy restaurant	70	Telephone use difficult
Freeway traffic		
Man's voice (3 ft)		
Floor fan	38–70	
Air-conditioning unit (20 ft)	60	Intrusive
Clothes dryer	55	
Light auto traffic (100 ft)	50	Quiet

(*continued*)

Living room	40	
Bedroom		
Quiet office		
Refrigerator		
Library	30	Very quiet
Soft whisper (15 ft)		
Broadcasting studio	20	
	10	Just audible
	0	Hearing begins

NOP.6   EMPLOYEE EXPOSURE

An employee is exposed at these following noise levels and periods:

110 dB for 0.25 h

100 dB for 0.5 h

90 dB for 1.5 h

Is the noise exposure within permissible limits?

Solution

If the variations in noise level involve maxima at intervals of 1 s or less, it is to be considered continuous. When the daily noise exposure is composed of two or more periods of noise exposure of different levels, their combined effect should be considered, rather than the individual effect of each. Exposure to different levels for various periods of time are to be computed according to the following formula:

$$F_e = (T_1/L_1) + (T_2/L_2) + \cdots + (T_n/L_n)$$

where F_e is the equivalent noise exposure factor; T is the period of noise exposure at any essentially constant level; and L is the duration of the permissible noise exposure at the constant level. If the value of F_e is greater than 1, the exposure exceeds permissible levels.

$$F_e = (0.25/0.5) + (0.5/2) + \cdots + (1.5/8)$$
$$= 0.500 + 0.25 + 0.188$$
$$= 0.938$$

Since the value of F_e is not greater than 1, the exposure is within permissible limits. Also note that exposure to impulsive or impact noise should not exceed 140-dB peak sound-pressure level.

NOP.7 VIOLATION OF OSHA RULES

A worker spends 3 h exposed to a noise level at 90 dB and 2.5 h at a level of 94 dB. Two hours are spent in a relatively quiet control room. Is this in violation of OSHA rules? If there is a violation, suggest a solution?

Solution

Using the OSHA information provided in Problem NOP.4, the following table can be generated.

dB	Time (h)	Permissible (OSHA) Time (h)
90	3	8
94	2.5	4

Using the equation presented in the previous Problem (NOP.6)

$$F_C = 3/8 + 2.5/4 = 1.0$$

Thus, the result is at the OSHA limit. Correction could include decreasing the time of exposure for either the 90-dB sound level or the 94-dB level, or both.

NOP.8 FAN NOISE

Calculate the sound level in decibels of a vane-axial fan operating at the following conditions:

Gas flowrate $= 100{,}000$ acfm
Fan speed $= 500$ rpm
Pressure differential $= 2.0$ in. H_2O
Number of blades $= 10$

Solution

Sources of sound transmission from a fan can include all or some of the following:

- Inlet and outlet joints
- Fan casing
- Ductwork
- Vibrations

A fan's sound level may be calculated using the following relationship:

$$FSPL = 10 \log q + 20 \log p$$

where FSPL = fan sound power level
　　　q = volumetric flowrate, acfm
　　　p = pressure differential, in. of H_2O

FSPL can now be computed from the above equation:

$$FSPL = 10 \log(100,000) + 20 \log(2.0)$$
$$= 10(5.0) + 20(0.30)$$
$$= 56 \, dB$$

NOP.9   TRANSFORMER SOUND PRESSURE

A 5000 kVA, 480-V transformer is approximately 6 ft wide by 10 ft high in size and is located inside a room with approximately 1200 ft^2 of wall area. Estimate the operating sound pressure in decibels at 6.0 ft from the transformer.

An equation for estimating a transformer's sound pressure level (TSPL) indoors is:

$$TSPL = TSPL(0) + 10 \log(A_T/4\pi r^2 + 4A_T/R)$$

where TSPL(0) = reference sound level, dB
　　　A_T = area of the transformer, ft^2
　　　R = room constant, ft^2
　　　r = distance to center of transformer, ft

The term R is given by

$$R = S\alpha/(1 - \alpha)$$

where S = room area
　　　α = absorption coefficient

For this unit and system,

$$\alpha = 0.01$$
$$TSPL(0) = 80 \, dB$$

Solution

The transformer area is

$$A_T = (2)(6)(10) + (10)(5) + (2)(10)(10) = 370 \, ft^2$$

Replacing R by $S\alpha/(1 - \alpha)$ in the describing equation leads to

$$\begin{aligned}
TSPL &= TSPL(0) + 10 \log[A_T/4\pi r^2 + 4A_T(1 - \alpha)/S\alpha] \\
&= 80 + 10 \log\{[(370/[(4\pi)(50)^2] + (4)(0.99)(370)/[(0.01)(1200)]\} \\
&= 100.9 \, dB
\end{aligned}$$

The outdoor sound level can be assumed to be given by TSPL(0), i.e.,

$$TSPL(\text{outdoors}) = TSPL(0) = 80 \, dB$$

NOP.10   ABATEMENT METHODS

Discuss various noise abatement methods and techniques.

Solution

Noise abatement measures are under the jurisdiction of local government, except for occupational noise abatement efforts. It is impossible for an active person to avoid exposure to potentially harmful sound levels in today's mechanized world. Therefore, hearing specialists now recommend that individuals get into the habit of wearing protectors to reduce the annoying effects of noise. Muffs worn over the ears and inserts worn in the ears are two basic types of hearing protectors. Since ear canals are rarely the same size, inserts should be separately fitted for each ear. Protective ear muffs should be adjustable to provide a good seal around the ear, proper tension of the cups against the head, and comfort. Both types of protectors are well worth the small inconvenience they cause the wearer, and they are available at most sport stores and drugstores. Hearing protectors are recommended at work, and during recreational and home activities such as target shooting and hunting, power tool use, lawnmowing, and snowmobile riding.

The main industrial control approaches include use of sound absorption, enclosures, barriers, and vibration isolation and damping. These methods are briefly described below.

1. Most porous materials absorb sound, and those materials specially made for this purpose include porous foams and fiberglass. However, ordinary materials such as carpets and drapes are also effective, and can be used in building spaces to reduce reverberant sound buildup and noise.

2. Enclosures can be used to partially or completely enclose machines (machine enclosures) or to enclose operators of machines (personnel enclosures). The first approach may be regarded as *path* control and the second as *receiver* control.

3. Barriers are used to shield personnel from sound sources. The effectiveness of a barrier depends not only on the effective height of the barrier but also how far the receiver point is into the sound shadow.

4. The vibration isolation of machine sources from their supports can be particularly useful in reducing the noise produced, especially if the machine is small compared with a large flexible support or enclosure that can act as a sounding board and radiate the sound. Soft metal springs or elastomeric isolators are often used as isolators.

5. Damping materials can also be effective at reducing noise when applied properly to structures if their vibration is resonant in nature. Damping materials that are viscous applied with thicknesses two or three times that of the vibrating metal panel are particularly effective.

50 Economics (ECO)

ECO.1 CHROME PLATING OPERATION

The rinse water from a chrome plating operation has an average total chromium concentration (as Cr) of 20 mg/L; 450,000 gal/yr of this rinsewater is generated as a process wastewater. What is the value of the chromium ($/yr) as chromic acid ($H_2CrO_4$) if it is recovered from the rinse water. Chromic acid costs $3.75/lb.

Solution

The value of the chromic acid is equal to the product of the concentration, flowrate, cost, and stoichiometric ratio. Since one mole of H_2CrO_4 produces one mole of Cr, the stoichiometric ratio between the two is

$$\frac{1 \text{ mol } H_2CrO_4}{1 \text{ mol Cr}}$$

Therefore,

$$\text{Value of } H_2CrO_4 = \left(\frac{20 \text{ mg}}{L}\right)\left(\frac{450,000 \text{ gal}}{\text{yr}}\right)\left(\frac{\$3.75}{\text{lb}}\right)\left(\frac{1 \text{ mol } H_2CrO_4}{1 \text{ mol Cr}}\right)\left(\frac{3.785L}{\text{gal}}\right)$$
$$\times \left(\frac{1 \text{ lb}}{453.6 \text{ mg}}\right)$$
$$= \$281.62/\text{yr}$$

ECO.2 METHODS OF ANALYSIS

1. Define the straight-line method of analysis that is employed in calculating depreciation allowances.
2. Define the double-declining balance (DDB) method of analysis.
3. Define the sum-of-the-year's digits (SYD) method of analysis.
4. Compare the three methods (1–3) above.

Solution

The straight-line rate of depreciation is a constant equal to $1/r$, where r is the life of the facility for tax purposes. Thus, if the life of the plant is 10 yr, the straight-line rate of depreciation is 0.1. This rate, applied over each of the 10 yr, will result in a depreciation reserve equal to the initial investment.

A declining balance rate is obtained by first computing the straight-line rate and then applying some multiple of that rate to each year's unrecovered cost rather than to the original investment. Under the double-declining balance method, twice the straight-line rate (see Problem ECO.12) is applied to each year's remaining unrecovered cost. Thus, if the life of a facility is 10 years, the straight-line rate will be 0.1, and the first year's double-declining balance will be 0.2. If the original investment is I, the depreciation allowance the first year will be $0.2I$. For the second year, it will be $0.16I$, or 0.2 of the unrecovered cost of $0.8I$. The depreciation allowances for the remaining years are calculated in a similar manner until the tenth year has been completed. Since this method involves taking a fraction of an unrecovered cost each year, it will never result in the complete recovery of the investment. To overcome this objection, the U.S. Internal Revenue Service allows the taxpayer to shift from the DDB depreciation method to the straight-line method any time after the start of the project.

The rate of depreciation for the sum-of-the-year's digits method is a fraction. The numerator of this fraction is the remaining useful life of the property at the beginning of the tax year, while the denominator is the sum of the individual digits corresponding to the total years of life of the project. Thus, with a project life, r, of 10 years, the sum of the year's digits will be $10 + 9 + 8 + 7 + 6 + 5 + 4 + 3 + 2 + 1 = 55$. The depreciation rate the first year will be $10/55 = 0.182$. If the initial cost of the facility is I, the depreciation for the first year will be $0.182I$, $9/55 = 0.164I$ for the second year, and so on until the last year. The SYD method will recover 100% of the investment at the end of r years. A shift from SYD to straight-line depreciation cannot be made once the SYD method has been started.

A tabular summary of the results of depreciation according to the straight-line, double-declining, and sum-of-the-year's digits methods are shown below:

Year	Straight-Line	Double-Declining	Sum-of-the Year's Digits
0	1.000	1.000	1.000
1	0.900	0.800	0.818
2	0.800	0.640	0.655
3	0.700	0.512	0.510
4	0.600	0.410	0.383
5	0.500	0.328	0.274
6	0.400	0.262	0.183
7	0.300	0.210	0.110
8	0.200	0.168	0.056
9	0.100	0.134	0.018
10	0.000	0.108	0.000

ECO.3   PAYOUT TIME

When considering the cost of environmental control systems, it is often important to know how long it will take to recover the cost of an investment. The time required to recover this cost is defined as the *payout time*.

A plant manager spends $10,000 for new scrubbing packing for a tower that strips toxics out of a water stream. The manager decides to depreciate the equipment at $1430/yr (7-yr straight-line method depreciation) and estimates that the equipment will generate $1500/yr in annual profit. Define the commonly accepted formula for payout time and calculate the payout time for this system.

Solution

The payout time is calculated as the fixed capital investment divided by the sum of the annual profit plus the annual depreciation.

$$\text{Payout time} = \frac{\text{Fixed capital investment}}{\text{Annual profit} + \text{Annual depreciation}}$$

When the formula is applied to the data presented in the problem statement, the following result is obtained:

$$\text{Payout time} = \frac{\$10,000}{\$1500 + \$1430}$$

$$= 3.44 \text{ yr}$$

ECO.4   RETURN OF INVESTMENT

A plant manager must often decide if it is truly worthwhile to invest in new equipment. The fundamental question is, "Will the company earn more money per year by investing in this equipment than it would if the money was invested in an interest-bearing financial account?" The calculation that is used to make this decision is called the *percent rate of return on investment* (ROI) and is carried out as follows:

$$\text{Percent ROI} = \frac{\text{Annual profit}}{\text{Initial investment cost}} \times 100$$

Calculate the percent rate of return on investment for the plant manager's $10,000 investment described in the previous problem.

Solution

When the percent rate of return on investment formula is applied to the data presented in the problem statement for Problem ECO.4, the following result is obtained:

$$\text{Percent ROI} = \frac{\$1500/\text{yr}}{\$10,000} \times 100$$

$$= 15\%/\text{yr}$$

Since this rate is higher than could be earned with many financial investments, the plant manager would be advised to make the investment.

ECO.5   DISCOUNTED CASH FLOW

Discounted cash flow is a measure of profitability and is based on the amount of investment that is unreturned every year, taking into consideration the time-value of money (interest, I). Consider a plant manager's $10,000 investment in a control device that has a 3-yr lifetime. Carry out the discounted cash flow calculation for this new unit assuming that the investment returns $5000/yr for 3 yr, and that the investment has $0 salvage value at the end of the 3-yr period.

Solution

After 3 years, the compounded cash flow, CCF, will be as follows:

$$\text{CCF} = (\$5000)(1 + i)^2 + (\$5000)(1 + i) + \$5000$$

where i is the interest rate. The compounded cash flow for the plant manager's investment is

$$\text{CCF} = (10,000)(1 + i)^3$$

Combining these two equations yields

$$(\$5000)(1 + i)^2 + (\$5000)(1 + i) + \$5000 = (10,000)(1 + i)^3$$

Solving this equation for i by trial and error yields the following results:

$$i = 0.234$$
$$\text{CCF} = \$18,790$$

Thus, the investment has an effective rate of return of 23.4%.

ECO.6   CAPITAL RECOVERY FACTOR

The annual operation costs of an outdated environmental control device is $75,000. Under a proposed emission reduction plan, the installation of a new processing system will require an initial cost of $150,000 and an annual operating cost of $15,000 for the first 5 years. Determine the annualized cost for the new processing system by assuming the system has only 5 years (n) operational life. The interest rate (i) is 7%. The capital recovery factor (CRF) or annual payment of a capital investment can be calculated as follows:

$$\text{CRF} = \left(\frac{A}{P}\right)_{i,n} = \frac{i(1+1)^n}{(1+i)^n - 1}$$

where A is the annual cost and P is the present worth.
    Compare the costs for both the outdated and proposed operations.

Solution

The annualized cost for a new process is determined based on the following input data:

Capital cost $= \$150,000$
Interest, $i = 7\%$
Term, $n = 5\,\text{yr}$

For $i = 0.07$ and $n = 5$, the CRF is

$$\text{CRF} = \frac{0.07(1+0.07)^5}{(1+0.07)^5 - 1}$$

$$= 0.2439$$

The total annualized cost for the new process is then

$$\text{Annualized cost} = \text{installation cost} + \text{operation cost}$$

$$= (0.2439)(\$150,000) + \$15,000 = \$51,585$$

Since this cost is lower than the annual cost of $75,000 for the old process, the proposed plan should be substituted.

ECO.7   NET PRESENT WORTH

In the adsorption process used to remove hazardous air pollutants from gas streams, there are two options that can be considered for media (adsorbent) regeneration. These are dry regeneration and wet regeneration. The initial capital cost of dry regeneration is $250,000 while that of wet regeneration is $100,000. Wet regeneration has an additional yearly operating cost of $20,000 as compared to the dry process. Both systems would require an additional $50,000 expenditure at the end of year 5 for maintenance and overhauling. Using the concept of net present worth (NPW), determine the following:

1. If the interest rate, i, is 10%, how many years must the two systems operate to yield an option 1 NPW that is less than that of option 2?
2. If the lifetime of these systems is 10 years, what is the minimum interest rate that would yield a lower option 1 NPW than option 2 over the entire lifetime of the system?

The following expressions can be used to determine the NPW for single or uniform series payments over a time period, n:

Single-payment present worth factor:

$$\left(\frac{P}{F}\right)_{i,n} = \frac{1}{(1+i)^n}$$

where P is the present worth and F is the worth after n years at annual interest rate, i.

Uniform series present worth factor (the inverse of the CRF):

$$\left(\frac{P}{A}\right)_{i,n} = \frac{(1+i)^n - 1}{i(1+i)^n}$$

where A is the annual cost or payment.

Solution

1. The NPW of option 1, the dry regeneration option, is

$$NPW(1) = \$250{,}000 + \$50{,}000(P/F)_{0.1,5}$$

The NPW of option 2, the wet regeneration option, is

$$\$100{,}000 + \$50{,}000(P/F)_{0.1,5} + \$20{,}000(P/A)_{0.1,n}$$

Equating both NPW expressions yields

$$\$250{,}000 + \$50{,}000(P/F)_{1.0,5} = \$100{,}000 + \$50{,}000(P/F)_{0.1,5}$$
$$+ \$20{,}000(P/A)_{0.1,n}$$

Simplifying yields

$$\$150,000 = \$20,000(P/A)_{0.1,n}$$

$$\left(\frac{P}{A}\right)_{0.1,n} = 7.5 = \frac{(1+0.1)^n - 1}{(0.1)(1+0.1)^n}$$

By trial and error, the following value results:

$$n = 14.55 \qquad P/A = 7.50$$

Therefore, $n = 14.55$ years.

2. Substituting $n = 10$ into the expression for the uniform series present worth factor,

$$\left(\frac{P}{A}\right)_{i,10} = \frac{\$150,000}{\$20,000} = 7.5 = \frac{(1+i)^{10} - 1}{I(1+i)^{10}}$$

By trial and error, the following results:

$$i = 5.6\% \qquad P/A = 7.50$$

Therefore, $i = 5.6\%$.

Note that both problems can be completed using built-in iteration and convergence routines on a spreadsheet. NPW tables can also be used to converge on the final solution.

ECO.8   STRAIGHT-LINE DEPRECIATION

Three different control devices are available for the removal of a toxic contaminant from a stream. The service life is 10 years for each device. Their capital and annual operating costs are as follows:

Device	Initial Cost	Annual Operating Cost	Salvage Value in Year 10
A	$300,000	$50,000	0
B	$400,000	$35,000	0
C	$450,000	$25,000	0

Which is the most economical unit? Employ a straight-line depreciation method of analysis.

Solution

To select the most economical control device, a comparison can be performed among the three units based on the total annualized cost (TAC). The following table can be used to simplify these calculations:

Unit	A	B	C
Capital investment	$300,000	$400,000	$450,000
Depreciation	$30,000	$40,000	$45,000
Operating costs	$50,000	$35,000	$25,000
Total annual costs	$80,000	$75,000	$70,000

A comparison among units A, B, and C indicates that unit C has the lowest TAC and should be selected as the most economical unit of the three being evaluated. A similar result would be obtained if a return on investment (ROI) method of analysis were employed.

ECO.9   ENVIRONMENTAL CONTROL OPTIONS

A baghouse is needed at a coal-fired power plant for a design operating period of 20 years. If the unit fails at anytime (bag meltdown), a 45% (of the initial cost) reinvestment cost will result. Two companies submit bids for this particulate control device with the following cost and operating characteristics data:

Company	Initial Cost	Time to Failure (yr)	Annual Cost	Salvage Value in Year 20
A	$15,000	10	$1,300	0
B	$22,000	45	$1,100	0

If each alternative has an annual probability of failure that is inversely proportional to its time to failure, choose the best option for particulate control for this facility assuming an annual interest rate of 4%.

Solution

The annual probability of failure of the baghouse provided by Company A is $1/10 = 0.10$, while that of Company B is $1/45 = 0.022$. With these results, the estimated annual cost of each of the baghouse options can be determined as follows:

Annual cost = Annualized capital cost + Reinvestment cost + Annual O&M cost

$\quad$ = $P(\text{CRF}) + P(\%\text{Reinvestment Cost})(\text{Annual probability of failure})$

$\quad$ + A

The value of the CRF for 20 years at 4% is 0.0736 (see Problem ECO.6). For Company A, an estimated annual cost is

$$\text{Annual cost} = \$15,000\left(\frac{A}{P}\right)_{0.04,20} + \$15,000(0.45)(0.10) + \$1,300$$

$$= \$15,000(0.0736) + \$675 + \$1,300$$

$$= \$3,079/\text{yr}$$

For Company B, an estimated annual cost is

$$\text{Annual cost} = \$22,000\left(\frac{A}{P}\right)_{0.04,20} + \$22,000(0.45)(0.022) + \$1,100$$

$$= \$22,000(0.0736) + \$220 + \$1,100$$

$$= \$2,919/\text{yr}$$

From this present worth evaluation, Company B's bid for baghouse installation should be selected.

ECO.10   BREAK-EVEN POINT

From an economic point of view, the break-even point of a process operation is defined as that condition when the costs (C) exactly balance the income (I). The profit (P) is therefore $P = I - C$. At break-even, the profit is zero.

The cost and income (in dollars) for a particular operation are given by the following equations:

$$I = \$60,000 + 0.021N$$

$$C = \$78,000 + 0.008N$$

where N is the yearly production of the item being manufactured.
Calculate the break-even point for this operation.

Solution

Write the equation relating C to I. Note that at break-even operation, $P = 0$.

$$I = C$$

Substitute for C and I in terms of N:

$$\$60,000 + 0.021N = \$78,000 + 0.008N$$

Solving for N at the break-even point:

$$N = 1{,}384{,}600$$

Calculating the cost at the break-even point:

$$
\begin{aligned}
C &= \$78{,}000 + 0.008N \\
&= \$78{,}000 + 0.008(1{,}384{,}600) \\
&= \$89{,}077
\end{aligned}
$$

The reader should note that as N decreases below 1,384,600 items, P is negative (there is a cost). Higher values of N lead to profits.

ECO.11   RECOVERED DUST

A process emits 50,000 acfm of gas containing a dust (it may be considered ash and/or metal) at a loading of 2.0 gr/ft^3. A particulate control device is employed for particle capture and the dust captured from the unit is worth $0.03/lb of dust. Experimental data have shown that the collection efficiency, E, is related to the system pressure drop, ΔP, by the formula:

$$E = \frac{\Delta P}{\Delta P + 15.0}$$

where $E =$ fractional collection efficiency
   $\Delta P =$ pressure drop, lb_f/ft^2

If the overall fan is 55% efficient (overall) and electric power costs $0.18/kWh, at what collection efficiency is the cost of power equal to the value of the recovered material? What is the pressure drop in inches of water at this condition?

Solution

The value of the recovered material (RV) may be expressed in terms of the collection efficiency E, the volumetric flowrate q, the inlet dust loading w, and the value of the dust (DV):

$$RV = (q)(w)(DV)(E)$$

Substituting yields

$$RV = \left(\frac{50{,}000\ \text{ft}^3}{\text{min}}\right)\left(\frac{2.0\ \text{gr}}{\text{ft}^3}\right)\left(\frac{1\ \text{lb}}{7000\ \text{gr}}\right)\left(\frac{0.03\$}{\text{lb}}\right)(E) = 0.429E\$/\text{min}$$

The recovered value can be expressed in terms of pressure drop, i.e., replace E by ΔP:

$$RV = \frac{(0.429)(\Delta P)}{\Delta P + 15.0} \, \$/min$$

The cost of power (CP) in terms of ΔP, q, the cost of electricity (CE) and the fan efficiency, E_f, is

$$CP = (q)(\Delta P)(CE)/(E_f)$$

Substitution yields

$$CP = \left(\frac{50,000 \, \text{ft}^3}{\text{min}}\right)\left(\frac{\Delta P \, \text{lb}_f}{\text{ft}^2}\right)\left(\frac{0.18\$}{\text{kWh}}\right)\left(\frac{1 \, \text{min} \cdot \text{kW}}{44,200 \text{ft} \cdot \text{lb}_f}\right)\left(\frac{1}{0.55}\right)\left(\frac{1\text{h}}{60 \, \text{min}}\right)$$

$$= 0.006 \Delta P \$/min$$

The pressure drop at which the cost of power is equal to the value of the recovered material is found by equating RV with CP:

$$RV = CP$$

$$\Delta P = 66.5 \, \text{lb}_f/\text{ft}^2$$

$$= 12.8 \, \text{in} \, H_2O$$

Figure 128 shows the variation of RV, CP, and profit with pressure drop.

The collection efficiency corresponding to the above calculated ΔP is

$$E = \frac{\Delta P}{\Delta P + 15.0}$$

$$= \frac{66.5}{66.5 + 15.0}$$

$$= 0.82$$

$$= 82.0\%$$

The reader should note that operating below this efficiency (or the corresponding pressure drop) will produce a profit; operating above this value leads to a loss.

The operating condition for maximum profit can be estimated from the graph. Calculating this value is left as an exercise for the reader. [*Hint*: Set the first derivative of the profit (i.e., RV − CP) with respect to ΔP equal to zero. The answer is $13.9 \, \text{lb}_f/\text{ft}^2$.]

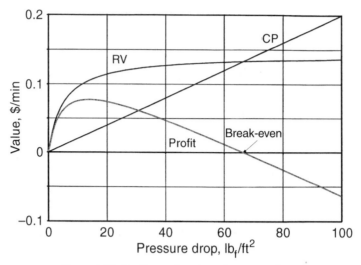

Figure 128. Profit as a function of pressure drop.

ECO.12   DISCOUNTED CASH FLOW APPLICATION

Two small commercial incineration facility designs are under consideration. The first design involves a liquid injection incinerator and the second a rotary kiln incinerator. For the liquid injection system, the total capital cost (TCC) is $2.5 million, the annual operating costs (AOC) are $1.2 million, and the annual revenue generated from the facility (R) is $3.6 million. For the rotary kiln system, TOC, AOC, and R are $3.5, 1.4, and 5.3 million, respectively. Using straight-line depreciation and the discounted cash flow method, which design is more attractive? Assume a 10-yr facility lifetime and a 2-yr construction period. Note that the solution involves the calculation of the rate of return for each of the two proposals.

Usually, an after-tax rate of return on the initial investment of at least 30% is desirable. The method used to arrive at a rate of return is discussed below. An annual after-tax cash flow can be computed as the annual revenues (R) less the annual operating costs (AOC) and less total income taxes (IT). Total income taxes can be estimated at 50% (this number could be lower subject to the passage of the new tax laws) of taxable income (TI).

$$IT = 0.5(TI)$$

The taxable income is obtained by subtracting the AOC and the depreciation of the plant (D) from the revenues generated (R) or

$$TI = R - AOC - D$$

As indicated, straight-line depreciation is assumed; that is, the plant will depreciate uniformly over the life of the plant. For a 10-yr lifetime, the facility will depreciate 10% each year.

$$D = 0.1 \text{ (TCC)}$$

The annual after-cash flow (A) is then

$$A = R - \text{AOC} - \text{IT}$$

This procedure involves a trial-and-error solution. There are both positive and negative cash flows. The positive cash flows consist of A and the recoverable working capital in year 10. Both should be discounted backward to time $= 0$, the year the facility begins operation. The negative cash flows consist of the TCC and the initial working capital (WC). In actuality, the TCC is assumed to be spent evenly over the 2-yr construction period. Therefore, one-half of this flow is adjusted forward from after the first construction year (time $= -1$ yr) to the year the facility begins operating (time $= 0$). The other half, plus the WC, is assumed to be expended at time $= 0$. Forward adjustment of the 50% TCC is accomplished by multiplying by an economic parameter (defined earlier) known as the *single-payment compound amount factor F/P*, given by

$$\frac{F}{P} = (1 + i)^m$$

where $i =$ rate of return (fraction)
$\quad m =$ the number of years (in this case, 1 yr)

For the positive cash flows, the annual after-tax cash flow (A) is discounted backward by using a parameter known as the uniform series present worth factor (P/A). This factor is dependent on both interest rate (rate of return) and the lifetime of the facility and was defined earlier by

$$\frac{P}{A} = \frac{(1 + i)^n - 1}{I(1 + i)^n}$$

where n is the lifetime of the facility (in this case, 10 yr)

The recoverable working capital at year 10 is discounted backward by multiplying WC by the single payment present worth factor (P/F), which was also defined earlier.

$$\frac{P}{F} = \frac{1}{(1 + i)^n}$$

where n is the lifetime of the facility (in this case, 10 yr).

The positive and negative cash flows are now equated and the value of i, the rate of return, may be determined by trial and error from the equation below.

$$\text{Term } 1 = \text{Term } 2 - \text{Term } 3 + \text{Term } 4$$

where Term $1 = \dfrac{(1+i)^{10} - 1}{i(1+i)^{10}} A$; worth at year $= 0$ of annual after-tax cash flows

Term $2 = \dfrac{1}{(1+i)^{10}} WC$; worth at year $= 0$ of recoverable WC after 10 yr

Term $3 = (WC + \frac{1}{2}TCC)$; assumed expenditures at year $= 0$

Term $4 = \frac{1}{2}(TCC)(1+i)^{1}$; worth at year $= 0$ of assumed expenditures at year $= -1$

Solution

For the liquid injection system, calculate, D, WC, TI, IT, and A. The depreciation is

$$D = 0.1(TCC)$$
$$= (0.1)(\$2,500,000)$$
$$= \$250,000$$

The WC is set at 10% of the TCC.

$$WC = 0.1(TCC)$$
$$= (0.1)(\$2,500,000)$$
$$= \$250,000$$

In addition,

$$TI = R - AOC - D$$
$$= \$3,600,000 - \$1,200,000 - \$250,000$$
$$= \$2,150,000$$

and

$$IT = (0.5)TI$$
$$= (0.5)(\$2,150,000)$$
$$= \$1,075,000$$

The after-tax cash flow is calculated using

$$A = R - \text{AOC} - \text{IT}$$
$$= \$3,600,000 - \$1,200,000 - \$1,075,000$$
$$= \$1,325,000$$

The rate of return, i, for the liquid injection unit is also calculated.
  The rate of return can be computed by solving the equation below:

$$\left[\frac{(1+i)^{10} - 1}{I(1+i)^{10}}\right]A + \left[\frac{1}{(1+i)^{10}}\right]\text{WC} = \text{WC} + (0.5)\text{TCC} + (0.5)\text{TCC}(1+i)^1$$

or

$$\left[\frac{(1+i)^{10} - 1}{I(1_i)^{10}}\right](1.325 \times 10^6) + \left[\frac{1}{(1+i)^{10}}\right](0.250 \times 10^6)$$
$$= (0.250 \times 10^6) + (0.5)(1.250 \times 10^6) + (0.5)(1.250 \times 10^6)(1+i)^1$$

By trial and error,

$$i = 39.6\%$$

For the rotary kiln system,

$$\text{WC} = D$$
$$= (0.1)(\$3,500,000)$$
$$= \$350,000$$
$$\text{TI} = \$5,300,000 - \$1,400,000 - \$350,000$$
$$= \$3,550,000$$
$$\text{IT} = (0.5)(\$3,550,000)$$
$$= \$1,775,000$$

The annual after-tax cash flow is

$$A = \$5,300,000 - \$1,400,000 - \$1,775,000$$
$$= \$2,125,000$$

The rate of return equation for the rotary kiln unit becomes

$$\left[\frac{(1+i)^{10} - 1}{i(1+i)^{10}} \right] (2.125 \times 10^6) + \left[\frac{1}{(1+i)^{10}} \right] (0.3650 \times 10^6)$$

$$= (0.350 \times 10^6 + (0.5)(1.750 \times 10^6) + (0.5)(1.750 \times 10^6)(1+i)^1$$

By trial and error,

$$i = 44.8\%$$

Hence, by the discounted cash flow method, the rate of return on the initial capital investment is approximately 5% greater for the rotary kiln system than the liquid injection system. From a purely financial standpoint, the rotary kiln system is the more attractive.

51 Ethics* (ETH)

ETH. 1 NO MORE LATE NIGHTS

Joshua works in the environmental division of the ABC Asbestos Testing Company. His position is senior laboratory analyst. He is in charge of analyzing transmission electron microscope (TEM) airborne asbestos samples. In the past few weeks a client has been sending him samples that have been failing, or coming up positive, for asbestos. When the results of the samples get back to the client, that client has to clean up the site again and take new samples. He then has to send the new samples back to the lab for reanalysis. The new samples have been coming to the lab at approximately 10:30 p.m. for immediate turnaround; Joshua is paged by the client to return at night to the lab and analyze the new samples. It takes a minimum of 3 h to prepare and read the TEM samples, so Joshua goes home at around one in the morning.

After 2 weeks, Joshua's sleep patterns are disrupted. He cannot take much more of this, and decides to pass the next set of samples so that he can catch up on some sleep. The next day the client calls him on the phone.

"Joshua, please."

"Speaking," mutters Joshua reluctantly because he recognizes the voice on the other end of the line.

"Guess what?" says the client. "I have another set of TEMs that need immediate turnaround time because this building has to reopen tomorrow."

"Sure thing. Those samples will be done in a jiffy," exclaims Joshua.

"I'm surprised to see that you are so enthusiastic about reading my samples, Josh. I figured that you would be fed up by now; you've had to come back at night seven of the past ten days to analyze my resamples," says the client.

---

*The problems presented in this chapter have been primarily drawn (with permission) from the John Wiley & Sons 1998 text *Engineering and Environmental Ethics: A Case Study Approach*, by John R. Wilcox and Louis Theodore. The authors used case studies to discuss ethical issues; both believe the case study method is one of the best ways to engage students and professionals in the discussion of moral challenges facing the engineering and scientific communities. Each of the 12 problems/case studies that follows contains a problem statement and data, and a solution that is highlighted by questions for discussion with an accompanying "answer."

"When will they be here?"

"They should be there around six o'clock," says the client confidently.

"Well, all right, then . . . talk to you later," says Joshua.

"Damn it!" yells Joshua after slamming down the receiver. "I was planning to leave early today. It's a nice day outside, and I wanted to work on my car so I can get it ready for the winter."

Then he thinks to himself, "I guess I'll have to work on my car another day."

Six-thirty rolls around and the samples haven't arrived yet. Finally at seven o'clock the client brings the samples to Joshua.

"Sorry, Josh, the field technicians didn't finish sampling until just now. Here you go," the client says as he rushes to his car to get home for the big game on TV.

At this point Joshua tells himself, "Forget it. I was going to analyze the samples and come back tonight if necessary, but I'm fed up. No more late nights!"

Joshua decides to forgo reading the samples this time. He calls the client from his couch at home 3 hours later to disguise the fact that he didn't really analyze the samples.

"Hey, aren't I lucky today! The samples are negative for asbestos," says Joshua without even a quiver in his voice.

The client answers, "Great. Talk to you tomorrow."

Solution

- What are the facts in this case?
- Why does Joshua want to pass the next set of samples that come in from the client?
- Why does he pass the late set of samples for the client?
- What risks does Joshua take by not analyzing the samples?
- What will happen to the people in the building if the samples are actually positive for asbestos?
- What will happen to Joshua if the client finds out that he did not read the samples?
- What can the client do to Joshua to make sure he will ever pass another set of samples without reading them again?

Joshua is wrong to lie to the client. Once a liar, can the client ever trust him in the future? This assumes the client finds out. Is that possible? The liar always faces that "danger." One lie then leads to another. Moreover, in this case Joshua may get away with the lie. Finding this action successful, it will make it easier to lie in the future, increasing the risk of being caught with the attendant lack of trust, loss of business, and criminal prosecution because of the seriousness of lying about possible asbestos contamination.

ETH. 2   WHAT THEY DON'T KNOW, WON'T HURT THEM

Jake is a geologist working on a current site evaluation and report for a project that his company has been hired as a subcontractor to conduct. The scope of the project involves transforming a solid waste dump into a park through habitat creation and land redevelopment. This project has been deemed essential to the vitality of Jake's company if they are to stay in business and continue to subcontract to the much wealthier firm on other future projects.

Jake has been informed that his evaluation and report should be limited to the current state of the land as it applies to the goal of creating a park. While conducting his research, Jake comes across a report as a reference; he reads it out of curiosity, and discovers that a section of the land belonging to the current project site was once accidentally contaminated with toxic waste by Jake's own firm. The toxic waste was promptly cleaned and the Environmental Protection Agency (EPA) designated the area to be safe. Jake informs his supervisor, Marie, of his findings and asks whether this information should be included in his report for the contracting firm.

> "No, why would you want to include that information in your report?" Marie asks. "That happened fifteen years ago and was completely taken care of. Besides, your report is supposed to be limited to the current state of the site; isn't the land fine?"
>
> "Yes, Marie, the EPA recently checked the land and was delighted by our plan to turn the site into a park."
>
> "Well, then, you have your answer. If the land is fine, why would you want to cause any alarm that might upset the contracting firm? As you already know, Jake, our company needs the work that this firm is providing us, or we'll all be out of a job."
>
> "You're right. I was just afraid that the firm might find out about the report later and wonder why we hadn't told them," Jake mentions.
>
> "That is the only copy of the report we have, so just don't show them," Marie states.
>
> "I guess you're right," mutters Jake.
>
> "Trust me, do what I say and none of us will have to worry about looking for a job tomorrow," Marie says.

Solution

- What are the facts in this case?
- What is the problem Jake is facing?
- Is it an ethical problem? How so, or why not?
- What are the risks Jake and the company might encounter if he reveals the findings of the report?

- What are the risks Jake and the company might encounter if he *doesn't* reveal the findings of the report?
- Is Marie's advice beneficial or harmful to Jake?
- What final action do you believe Jake will take?

Jake cannot follow Marie's directive. Marie has put money before public and professional responsibility. If the contracting firm is not alerted to the documentation on the section of land in question, Jake's company may well face a lawsuit. Given the environmental nature of the case, big bucks may be involved. It is in Jake's and Marie's self-interest to come clean.

ETH. 3 BEST-KEPT SECRET IN TOWN

Karen is an environmental management coordinator for EPA working on a current evaluation and report for the agency on the aftereffects of transforming solid waste landfills into productive designs for the community. The scope of the project is to take past data on a number of regional landfills that have been constructed and to coordinate it with data on the current usage (if any) of those properties. The purpose is to see if the government can justify the cost of spending so much money on remediating landfills.

Karen has been informed that the evaluation and report should be limited to the state of New Jersey. While conducting her research on past data, Karen comes across a remedial design report on a solid waste landfill that was written 20 years ago. Karen does not plan to use the report in her evaluation, but she reads it out of curiosity to learn how it might pertain to her project. In her reading, Karen discovers that a contractor employed by EPA had made some evaluation mistakes in the remedy chosen to protect the environment.

Karen informs a fellow coordinator, Peter, of her findings and asks whether she should inform the emergency response division about these findings.

> "Yes, you should inform someone about your findings because recontamination of the site may cause damage to the surrounding environment," Peter exclaims. "Twenty years of uncertain protection may have caused serious problems in the area."
>
> "You're right, but I was afraid that the agency might be just as liable as the contractor. I wonder why we've not told the public of the problem," Karen replies.
>
> "You need to inform someone fast so that they can secure the site before something happens," says Peter.
>
> "Right again," says Karen.

Solution

- What are the facts in this case?

- What is the problem Karen is facing?
- What risks is Karen taking if she does not tell anyone?
- What would take place is she reveals her findings?
- What final action do you believe Karen will take?

Reading the facts, it looks as if a honest mistake was made by the contractor employed by EPA. Karen and Peter are both correct about the need to inform someone about the error. However, "should" sounds too weak. Karen "must" report the mistakes to EPA and secure the site ASAP.

ETH. 4   CAMPAIGN TRAIL

It is election time again, and this year's campaign has been a heated one. Up for reelection is Mayor Martha Vineyard. Besides a number of other claims, she has been promoting herself as the environmentally conscious candidate. She has repeatedly pointed to the island's improved harbor and remarked about the number of permanent beach closures that were recently reversed.

"With Mayor Vineyard in office, you went to a clean beach this past summer. My team got rid of the floatable problem," she boasts. "You and your families were able to eat that lobster without worrying about pathogen contamination and to go clamming Sunday morning. I am the candidate who cares about the future of this island environmentally, and I proved it with the reopening of Beach A after ten years of contamination problems and Beach B after fifteen years."

The truth is that there have been major improvements in the water quality but not because of Mayor Vineyard specifically, if at all. The island has been working for 30 years to correct this problem. Federal regulation prompted the abatement of many combined sewer overflows as well as illegal sewer connections. The island improved the wastewater treatment facilities to stop primary effluent escaping the system. All of these continued while Mayor Vineyard was in office, but the relevant legislation was passed by prior administrations. If fact, Mayor Vineyard pulled some of the funding for research and diverted it to her limo service.

Linda is an environmental engineer who works for the island's Department of Environmental Protection on the pathogen contamination project, which she has led for the past 10 years. She is not really sure who to vote for, but she knows that Mayor Vineyard's boasts are not the whole truth. Two weeks before election time she is contacted by Mayor Vineyard's public relations director.

"Hey, Linda, we need you to do us a bit of a favor down here. Bill C., the newspaper reporter, would like to talk to you about the improvements in the harbor and the mayor's role in them."

"Well, the measures to effect these changes were approved before the mayor came into office, and I'm not sure she's really had a role in the improved water quality of the harbor," Linda states.

"Linda, the mayor's office is asking a favor from you," the PR director answers, his voice a bit strained. "The fact remains that those beaches opened while she was in office, so she's going to take the credit. If you cannot help us out with this matter, we will remember that when the election is over. You should remember you are still an employee of the island, and Mayor Vineyard is up in the polls!"

Solution

- What are the facts in this case?
- What is Mayor Vineyard implying in her campaign?
- What were the harbor's water quality improvements?
- What is the truth behind the improvements? Did Mayor Vineyard play an important role?
- What is the public relations director asking of Linda?
- What is the implication if Linda does not support the mayor's position?
- Is what the public relations director asks of Linda an ethical task, if we disregard his implied threat?
- What should Linda do? Come up with more than one option.
- Determine Linda's *best* option and develop her course of action.

Being a professional entails rights and responsibilities. Linda is now faced with a particular responsibility. In no uncertain terms, she has been told that her position depends on putting a positive spin on the mayor's environmental record, when, in fact, the mayor can make no claim to an environmental record except for ribbon cutting based on the work of others. The reader knows, but Linda does not know, that the mayor actually misappropriated environmental funds. What Linda does know is that credit should be given where credit is due, and she must ask herself whether her job is worth an outright lie.

ETH. 5 NO DUMPING ALLOWED

Mike is a technician in a wastewater treatment facility. He has recently been put in charge of disposing all of the sludge that is generated by the facility. Although he is not familiar with the legal aspects of this process, Mike calls both state and federal agencies and discovers that he has two options. He can either sell the sludge to farmers for use as fertilizer or dispose of it in landfills. He is also informed that the dumping of sludge into the ocean is no longer legal.

After weeks of sludge disposal, Mike has discovered that it is very hard to sell sludge on the open market, since most farmers already have contracts with waste-water facilities to buy their sludge. Mike has also discovered that landfills charge on a per ton basis for the dumping of sludge; due to the large amount of sludge produced by the facility, it is very costly.

Although Mike has not been on the job long, he realizes that if he continues to spend money at the same rate for the disposal of sludge, he will grossly surpass the facilities budget. Mike, not knowing how to deal with this dilemma, decides to contact the previous technician, Joe, who was in charge of the disposal of sludge.

"Well, Mike I can solve your problem for you. All you have to do is dump your sludge into the ocean."

"Into the ocean? That's illegal. They stopped doing that about five years ago," replies Mike.

"Some of the facilities stopped. However, we weren't one of them," states Joe. "That law claims that the dumping of sludge affects the biota of the ocean, and that's not true. Sludge has been dumped into the ocean for years."

"Does the boss know about this?" asks Mike.

"Of course not. And don't tell him. If he knew, I might lose my pension," replies Joe. "I'll give you the number to call, and the people there will take care of you."

"I don't know if I should do that," states Mike.

"If you don't dump the sludge my way, you will spend much more than your allotted budget, and that will not look good for you," Joe comments.

Solution

- What are the facts in this case?
- What problem is Mike facing?
- If Mike explains the situation to his boss, what reactions do you think the boss will have?
- What about the previous technician's comments? What type of mind frame does he have?
- If you were faced with the dilemma Mike has, what would you do?

If Mike takes Joe's suggestion, he is breaking the law and is failing in his engineering responsibility to the public. Joe claims the sludge has no effect on the ocean biota. Where is the evidence for his assertion? If, in fact, Joe is correct, then Mike should bring the evidence to his boss as well as EPA. Could what Joe is advocating be broadcast on the front page of the local paper? Hardly.

ETH. 6   MEDICINE VS. THE ENVIRONMENT

Carol is facing a medical dilemma not likely to be solved during her lifetime:

As a member of the Audubon Society and the Nature Conservancy, she has actively embraced the conservation movement and led children on nature hikes.

But Carol, now 65 years old, is also battling ovarian cancer and wants an experimental drug called Taxol. The drug, derived from the rare Pacific yew tree, has achieved a 30–40% response rate in advanced cases of ovarian cancer. Medical researchers say the only way to produce enough Taxol needed for treatment and research would be to chop down hundreds of thousands of yews, which are a refuge for the spotted owl.

"I want to be treated," Carol says, quite apologetically. "However, I don't want to see the forests destroyed."

The conservationists say that they are concerned about saving the owl and the yew trees as part of their overall mission to preserve the diversity and integrity of threatened forests. Conservationists, of course, deny that they are inhumane. They also say that medical researchers and a major drug company that manufactures Taxol are not moving quickly enough in pursuing other options for making the drug. After all, the environmentalists strongly believe that the ancient forests that gave us the yew may provide us with answers to other medical problems we have not thought about yet. "Are people entitled to destroy all of nature for their own selfish purposes?" they ask.

Clearly, this is part of the ultimate confrontation between medicine and the environment.

Solution

- What are the facts in this case?
- The environmentalists insist they have patients' long-term interests. Do you agree?
- Balancing finite resources with finite lives is the challenge now facing medical researchers. Does one concern outweigh the other?
- What are the possible courses of action the researcher could take?

There are a number of competing values in this case: a cure for Carol and conservation of a scarce species of tree and owl. Are the medical researchers dragging their heels because yews are available? Can that question be answered with reliability? Probably not and the conservationists should not accuse them of doing so without some evidence. If there is no evidence of such foot dragging, then the environmentalists' credibility is damaged. These conservationists are certainly correct that the yew should not disappear because other medical problems may be solved through their use. Is there a "win–win" solution? Certainly Carol's finite life is more valuable than finite resources. At the time, it is hard to believe that "hundreds of thousands of yews" need to be cut down as medical researchers claim. The latter's credibility is at stake just as much as the environmentalist's is at stake. Arbitration and scientific integrity should provide a solution. Adversarialism and hyperbole will not provide one.

ETH. 7   LIKELY STROKE CIGARETTE COMPANY

Introduction

It may seem surprising to many, given what we now know about the dangers of tobacco products, that manufacturers of tobacco products thoroughly test their products, not just for quality control, but also for such reasons as cancer research and litigation protection. Additives are usually quite benign if inhaled separately, but in the cone of intense heat at the tip of a cigarette, reactions may take place between additives and the 10,000-odd natural plant products in the tobacco matrix. Cigarette paper may also be a source of toxins when it is burned; for instance, the carcinogen benzo[*a*]pyrene, a polycyclic aromatic hydrocarbon, is a combustion product of tobacco and paper. The carcinogenic *N*-nitrosamines are naturally present, and some are formed during combustion from nicotine and amino acids that are naturally present.

Part One

Product development researchers at Likely Stroke, Inc. (LS) would like to introduce to their cigarettes a new flavor additive that smells like mint; they feel it would greatly enhance the quality of the smoke, and very little additive is required for olfactory detection. Joan, who works in research and development, is asked to do a preliminary inhalation toxicity study on mice using varying levels of the additive in an already marketed product. Several weeks later, the cigarettes are spiked with different amounts of the new additive and testing begins on four groups of 50 mice. The mice, each in an individual chamber, are given smoke at a rate commensurate with their body weight, roughly equivalent to 30 cigarettes/day under standard smoking conditions (the smoke machine drags 35 µL for 1.5 s/puff, once per minute).

   The study is carried out for a duration of 10 weeks, during which time 4 of the 200 mice in the study expire, and 6 develop precancerous lesions in the mouth and throat. No trend exists in the mortality rate, which is in fact typical among mice. The cancerous lesions occur in all groups, but the two groups with the two highest amounts of the flavor additive each produce 2 mice with cancer; the 2 groups with lower amounts of additive, each with one mouse with cancer. This would seemingly point to a trend; however, even if all 4 groups had experienced the same conditions, there is a significant chance this type of distribution would have occurred anyway.

   The conclusion is that there is little indication that the additive makes the product more carcinogenic. If there had been more cancers in proportion to the amount of additive, this study would have been continued.

Solution to Part One

- What are the facts in this case?
- Does the additive pose an increased risk of cancer?
- Can this study be considered conclusive?

Joan cannot conclude that this study has proven anything. She should get out of the tobacco production field. Cigarette manufacturers are under a very dark cloud of lying, hypocrisy, and financial catastrophe. Even thinking about new additives to "enhance the quality of the smoke" is equivalent to rearranging the deck chairs on the *Titanic*.

Note: Only key questions are provided for the remaining parts of this problem. Any open-ended answers can be given by the reader.

Part Two

Meanwhile, Tom in corporate analytical has been given the routine task of analyzing the smoke from the cigarettes of the same batch used in the animal testing. The amounts of tar and nicotine are all well within the normal ranges, as are the usual nitrosamines. However, there are changes in the relative amounts of some of the nitrosamines, and these changes correspond to the amount of additive.

It is necessary to give some background into how this sort of testing is done. Cigarettes are smoked on a machine under the industry standard conditions mentioned in part one of this case study, and the smoke condensate is collected on a filter in the smoke machine. The filter is then placed in solvent, and the chemicals to be tested are extracted by the solvent. Most of the solvent is then removed, and an internal standard chemical is added. The sample is then diluted to a specified volume.

Looking for specific chemicals in cigarette smoke condensate is tantamount to looking for a "needle in a haystack" because there are so many similar compounds that are extremely difficult to separate cleanly. Looking for one kind in particular requires special instrumentation. To test for nitrosamines, a few microliters of the sample are injected into a gas chromatograph. This device consists of a thin column packed with fine-grained silica powder (easy and cheap to make). The tube has a heated injector (to vaporize the sample in the gas phase) at one end, where the sample is injected, and a detector at the other end, which gives a response linearly proportional to the amount of that type of chemical present. The column is run through an oven and is kept hot so that all of the compounds in the sample will remain in the vapor phase and all will make it out the other end eventually. Chemicals are separated in the column (mostly in order of increasing boiling point) so that they emerge from the column one or a few at a time; conditions are adjusted so that the best or cleanest separation takes place among the chemicals being analyzed. The result is a chromatogram, which is a chart of detector response versus time, with chemical detection in the form of "peaks," and with certain chemicals emerging at specific times relative to the internal standard.

The choice of detector is critical to the specific types of chemicals to be analyzed. Some units will detect everything and the output will resemble noise; some will detect only certain types of chemicals. The latter type is preferred, and in the case of nitrosamines, the thermal energy analyzer/detector is the method of choice. Because smoke condensate is filthy, the cheaper, packed column is invariably used. Generally it gives adequate separation of the nitrosamines.

There are five to seven kinds of nitrosamines normally present in tobacco smoke condensate. The internal standard is a nitrosamine that does not occur in tobacco or its smoke, but is used to confirm the emergence time and quantify the amounts of the nitrosamines in the samples. The curious effect of the additive is that one of these peaks, corresponding to a particularly toxic cancer-causing nitrosamine called NNK, decreases in size, roughly in proportion to the amount of additive in the cigarette. An increase in the size of a peak corresponding to a less cancerous nitrosamine called NAT is also observed. This is a promising result. Could it be an additive that improves the flavor of the product and makes it less harmful?

Solution to Part Two

- What are the facts in this case?
- Because tobacco is a natural product, the content of each type of nitrosamine varies with each crop or even each plant in a crop. Is such variation relevant to this case?

Part Three

A meeting has been scheduled between the head of product development and the research staff to discuss the results of the various studies under way on the new additive. Tom and Joan present their findings, which basically show that there is little conclusive evidence of increased cancer or any other kind of risk resulting directly from the new additive, pending further study.

Tom is truly an expert at chromatography, and in his presentation he discusses the fact that along with the increase in the size of the NAT peak, another effect is noticed: The height/width of the peak is somewhat lower than it should be for a compound emerging from the column in only 15 minutes. (Peak broadening, as it is called, increases proportionally with the time it takes for a chemical to emerge from the column). This is slightly disturbing because it means that there may in fact be two compounds emerging from the column at very nearly the same time. Also, this effect occurs only in the samples containing the additive. If this is the case, then a new type of nitrosamine may be present, and it should be investigated. Unfortunately, separating NAT from another possible compound with a very similar emergence time from a packed column is impossible, and it may not be worth ruining an expensive capillary column to investigate another compound that may not even be present. Joe, the head of the division, has concluded that further study is not necessary, and LS will go ahead and use the additive.

Solution to Part Three

- What are the facts at this point?
- Since very little evidence exists at this point of significant risk, is it worth investigating this new compound from a management standpoint?

Part Four

Although Tom is very busy, he decides to run samples from the smoke tests, starting with a nonadditive sample followed by one taken from the test containing the most additive using an old capillary column about to be discarded. The result is that the sample from the additive cigarettes indeed contains two peaks (although the peaks still overlap somewhat), whereas the one for NAT shows only one peak. This means that there is indeed a new compound formed; the column held up only long enough to inject two more samples, two intermediate ones, thereby roughly establishing a trend.

Solution to Part Four

- What are the facts at this point?
- Should Tom inform Joe about the confirmation of the presence of a new nitrosamine?
- Because capillary columns are so expensive, most independent labs are not likely to repeat this experiment; this new chemical would probably not be detected by them. From a legal standpoint, it is worth determining what this new chemical may be and investigating its properties?

Part Five

Tom has logged his finding in his lab notebook and decided to tell Joe about it when Joe returns from vacation in Vegas. Meanwhile, he is very exicited about his find, so he chooses to tell one other person: his friend Sue, a synthetic organic chemist from another department with whom he frequently has lunch. (It should be noted that in most corporations, particularly those for whom public relations may involve sensitive issues, there are strict rules about discussion of work matters outside the supervisor/subordinate/group relationship.) Tom asks Sue to have lunch with him at a restaurant not usually frequented by employees of LS.

She concludes that there may indeed be a reaction occurring between the additive and one or more components within the tobacco matrix and that the new compound must have a very similar structure to NAT. Because she has synthesized NAT in the past, she decides to check her notes to see what side products may have been formed when she last made it, what similar N-nitrosamine chemicals may exist, and how they could be made if not already available.

A week later, a very depressed Joe is back from his vacation ("Viva Lost Wages," as the saying goes). Tom decides to give him some space for a while. Meanwhile, Sue has come up with three or four possible identities for the new chemical, and two can be easily synthesized. A third possibility may be that the same chemical has an impurity or side product that occurs in small amounts in the synthesis of NNK but gets removed in purification. Tom injects a solution containing only purified NAT and the unpurified NNK in a capillary column and finds that one impurity in the NAT solution has the same emergence time as the mystery chemical.

Solution to Part Five

- What are the new facts at this point?
- Exact emergence times in gas chromatography are strong but not conclusive confirmation of the identity of a compound. Should Tom wait until he has tested the other possibilities before informing Joe about his findings?

Part Six

Tom has decided to wait before telling Joe, so he tests the other possible chemicals. None of them have the same emergence time; he concludes that the mystery chemical is most likely the same as the impurity from the laboratory synthesis of NNK. Sue has recently characterized this to be the nitrosamine NNAL. It does not normally occur in tobacco or tobacco smoke, and its properties have not been studied.

Tom discloses his findings to Joe and gets a lecture about responsibility and proper conduct in the workplace. Joe is pretending to be angry that any time was spent on this study after he had given the final order on this additive, which has been very successful in capturing more market share for LS. But, Joe's real concerns are that Tom has unwittingly created more work for him and that this compound's possible adverse effects could cause LS to yank the product; this would not bode well for Joe's expected promotion. Now the identity of this chemical must be confirmed and toxicity and carcinogenicity data collected.

Joe realizes that he is being unfair to Tom and that he should really be glad to have such a diligent employee, so after apologizing to Tom for the unnecessary lecture on corporate conduct, he explains to Tom that this study will be continued by the department of toxicology at some later date and that Tom should now forget about the matter.

Solution to Part Six

- What are the facts at this point?
- Should Tom now let the matter drop?

Part Seven

Tom chalks his discovery up to a nice bit of chemical investigation and turns his attention to other analyses. Meanwhile, the toxicology department performs the usual battery of tests and confirms the identity of the chemical. NNAL has also been determined to be nearly 50 times as carcinogenic as NNK. This information is kept secret. Ten years later an independent research foundation reaches similar conclusions and publishes the findings in the *Journal of Carcinogenesis*.

Solution to Part Seven

- What are the facts now?
- Should LS voluntarily yank the product because of the presence of NNAL in the smoke?
- There is little empirical data that this flavored cigarette is responsible for more deaths and illnesses than any other similar product now on the market. What would you do if you were the decision maker at LS?

Part Eight

A consumer watchdog group and a state government have filed a lawsuit to recover medical costs from LS on the basis of the finding that NNAL is a cancerous component unique to this product. LS at first denies having any knowledge the NNAL exists in their product, buying some time to craft a defense. The CEO of LS has decided that it can tie up the courts for a few more years by claiming that LS itself will carry out the same study, and then it will simply dispute the findings of the research foundation.

Solution to Part Eight

- Is this ethical behavior on the part of LS?
- This may not be unusual behavior for an industry that has such an enormous profit margin. Besides, tobacco products contain numerous carcinogens, many equally as potent as NNAL, so why should it be unique in determining the fate of the product, which has been a huge success worldwide for LS?

ETH. 8   DECISION FROM THE HEART

Bill is a senior engineer in the biomedical division of a major corporation and the head of a research department that specializes in the construction of artificial organs. About a year ago, an artificial heart was created and tested in a human patient; unfortunately, the patient survived for only 9 months. However, it was a huge step in the fight against heart disease, and it brought worldwide recognition to his company and himself.

Bill is extremely proud of the progress he and his staff have made. He remembers how much work it took to convince the board of directors to support him financially in such an experimental field. The risks of failure were immense. But he and they know that a successful intervention could double the company's net profit.

One day, Mary, his top research assistant, reports to him that a problem has been detected in the tricuspid valve of the artificial heart model. With further testing, it is discovered that the rate at which this valve allows blood to pass tends to slow down after 8 months of continuous usage. The coroner's report states that the patient's death was due to the body's rejection of the artificial heart. However, it is very likely that the patient's death was brought on by this flaw in the artificial heart.

Bill becomes extremely worried. If he tells his superiors this piece of information, there is a high possibility that the project will be terminated. And, if this knowledge becomes public, not only will it bring humiliation to the company and probably cause dismissal, but also the company will be highly susceptible to a million-dollar lawsuit by the patient's family. If Bill decides to withhold this information, a new model could be created with the flaw corrected, without anyone knowing.

Bill decides to ask for advice from two of his dearest friends. He first asks Bob, a fellow chemical engineer and someone who understands the technical aspects of the project.

"You have no choice, Bill," replies Bob. "You made a mistake and now have to suffer the consequences. Also, if you withhold this information and it is discovered later, the situation becomes ten times worse."

Bill then phones his sister, Sheila, and explains the situation to her.

"It's a tough call, Bill," replies Sheila. "Ordinarily, I would say let the truth be known, but now that your wife's rare cancer has spread, you cannot afford to lose your medical benefits or Helen will be denied treatment."

Solution

- What are the facts in this case?
- What ethical dilemma has Bill encountered?
- How much of a factor is Helen's medical condition?
- Is Bill being selfish if he does not report the flaw in the artificial heart?
- If Bill's wife did not have cancer, would the decision be any easier?
- Should Bill's loyalty to the company play a major role?
- What do you think Bill will do?

Bill is head of research, not production. Finding a flaw is built into his type of work. There is no way the company will reap profits if he does not reveal the flaw. Given the nature of his work in an experimental field, why can't he give the board of directors a progress report and then build a new model? If the board rejects this proposal and wants to bury the flaw, he should quit in protest; if they hush the problem up and market the artificial heart, he should go public.

ETH. 9   HAIRMAGIC COMPANY

Joe Murray, a chemical engineer at the HairMagic Company, is in charge of the production of a new hair coloring product, which is expected to be ready for mass production in a month. Joe has been reviewing the results of the hair coloring tests conducted on volunteers, which show that when the product was used on dark-haired men and women, it changed the color of the hair with an accuracy of 99.2%. However, on fair and light-colored hair types, it had an accuracy of only 89.3%. Joe realizes that the product needs to be refined further and that some substitutions should be made in its chemical makeup.

Joe decides to go to his manager and explain that production will have to be delayed. His manager, on the other hand, feels that production should be initiated and not waste any more money or time.

Ethically, Joe feels strongly that this is wrong: The refining needs special attention, and production should not even be considered when the product is not yet perfected. Joe's only alternative is to go above his manager and try to convince upper management to delay production. However, doing this could anger his manager and could also lead to Joe's dismissal. But it might also have the positive effect of making him appear to be confident and conscientious and could also serve to publicize his abilities as an engineer.

Solution

- What are the facts in this case?
- What issues are involved?
- What are the choices Joe must make?
- What could be the consequences of each decision he could make?

To a great extent, the consequences depend on the "truth-in-advertising" claims that the company makes. Will the labeling indicate the two levels of accuracy or will it say "100% accurate." If the latter, then the company is misleading the public. There will be complaints, requests for refunds, and word-of-mouth negative advertising. Worse still, all the company needs is an exposé in the media or from *Consumer Reports*. Joe's manager is very short-sighted himself or else he reflects the short-term thinking of upper managment.

ETH. 10   NEW KID ON THE BLOCK

Lou is a recent college graduate who works for a local construction company. His job is to obtain a sample from any concrete used on the job and to test it for various properties. Basically, Lou is required to make sure that the concrete contractor has not cut any corners with the product. The job has been fairly simple so far, even boring at times, which probably explains why Lou is looking for a more challenging and better-paying position.

At his present assignment, Lou is waiting to obtain his next sample when he is approached by one of the concrete truck drivers, who offers Lou $500 if he will lie and say the concrete from his truck passes inspection. The driver tries to convince Lou that there is no way you can tell the difference between his concrete and the concrete specified by the engineer. The older man further says that this sort of thing is done all the time and is one of the big perks of being a concrete inspector. Lou asks the driver if the structure will still be safe, and the driver assures him that it will.

Lou wonders if he should take the money, figuring he can always lie and say that he tested the concrete and it passes his inspection. He knows that these structures are always overdesigned anyway, so no one would be in jeopardy. By taking the extra money, Lou feels he can make up for the small salary he is earning with his company. How else is he going to pay off his college loans?

Solution

- What are the facts in this case?
- What ethical problem is Lou facing?
- Do you think Lou's financial situation justifies making a decision in favor of corruption?
- Is the truck driver's advice logical?

Lou would be violating his public trust if he took what is commonly called a "bribe." His rationalizing about his low salary and his college loans is naive thinking. Lou is the point man in this case. The buck will stop at his desk if there are any problems, especially if there is a building collapse and lives are lost. Even if the problem is less serious, Lou will still be held responsible. His professional life is in jeopardy and, given today's legal climate, jail time will be a real possibility. Lou's dilemma is the classic one since there is always a window of opportunity for persons in positions such as his.

ETH. 11   INTELLECTUAL PROPERTY

Mike has been working for a large defense company for 5 years and has been thinking of looking for another, higher-paying job. The department he works for makes radios for the military; he writes the software that controls digital signal processing (DSP) chips in the units. Mike also works on data encryption and searches for better ways of making transmissions more secure so enemy forces cannot decode the signals.

Just recently Mike has thought of a better way of encrypting transmissions. His idea has a wide variety of applications in the telecommunications field and in cellular phones. Mike knows that according to his employment contract, any inventions he comes up with while working for the company are the intellectual property of the company. Mike would receive only a dollar from the company for his patent (due to legal formalities), and the company could stand to make millions.

Mike knows that he can leave the company in 6 months when his employment contract expires, and he will be able to patent his idea and market it to Bell Atlantic, Sprint, or NYNEX. But Mike knows this will be unethical and unlawful. He knows that the company may sue him, saying the invention was discovered while he was still in its employ and therefore belongs to the company. Mike does not know what to do.

Solution

- What are the facts in this case?
- What is the ethical dilemma?
- Does Mike deserve more than one dollar for his invention?
- What do you think Mike should do?

Surely Mike knew from day one that the company owned inventions he came up with while working for it. But in this case, is it an invention or just an idea, a hunch without any research? Does the company own all ideas that Mike thinks of during his employment. Well, it depends in the first instance on the wording of Mike's contract. The company may claim ownership of his ideas even if not realized in experimental work or patents. After all, the company may assert that Mike would never have had the idea if it were not for the type of work he was doing and for which the company paid him. Futhermore, Mike's idea is directly related to the company's own work, and it may well turn out to be a question of national security.

ETH. 12   SOME SMOKE

Curts Inc. is a company specializing in power plants. Recently, its business has taken a jump forward, and the company has started obtaining contracts faster than it can hire and train individuals.

John is a mechanical engineer for Curts Inc. Because of the recent dirth of employees, he has been juggling two or more projects at the same time. One project that he neglected for a while is a contract with the local utility company to modify an existing burner to use a more economical fuel and still produce the same amount of energy.

The project is near the end, and he just got started on his assignment of giving a detailed report about the fuel to be used. Initial information was gathered by a team of engineers who concluded the fuel will be easy to obtain if imported from South America and that it will produce the same amount of heat. John finds out that the fuel, which is a very crude (tarlike) oil, is suspected of being carcinogenic when burned, according to a study done on this type of fuel. When he brings this to the attention of his peers, they accuse him of trying to destroy all of their work at the last minute.

John takes a moment to think over his options: Should he destroy the whole team's work or forget about that *one* study and finish the project?

Solution

- What are the facts in this case?
- What is the ethical dilemma John faces?

- Does John have more than two options
- Which course of action would you recommend to John?

John is right to note that there is one study at issue. Are there others? How valid is that one study? John must do more homework, but he cannot disregard the issue at hand. As an engineer, he is responsible for public safety. Obtaining expert opinion on the one study and searching for other analyses are two steps he must take.

If it turns out that this is the only study, then John and his team must get advice from expert researchers in the field as to the next step. The company itself, as well as the team John is working with, must realize that the project has not received the attention it deserves and that there would be serious consequences. They cannot blame John for being a poor team player. To ignore such a potentially serious problem is to flirt with potential loss of life and bankrupting lawsuits if that one study proves to be accurate.

52 Statistics (STT)

STT.1 REPRESENTATIVE SAMPLING

Discuss the problems associated with the need to obtain representative samples for statistical analysis.

Solution

One of the main problems with sampling is the need to obtain samples that are representative of the population. This usually requires both taking a large enough sample and doing it randomly. To sample randomly is to select from a population such that each sample drawn has an equal chance of being chosen. This requires including the whole population when selecting and removing all possible bias(es) from the selection process. It also requires sampling from the population in such a manner that each removal of a sample is taken into account on a subsequent sample analysis.

As an example of what can happen in sampling, consider this notorious failure: A magazine took a telephone survey of eligible voters, and on the basis of this survey incorrectly predicted that Landon would beat Roosevelt in the 1936 U.S. presidential election. The problem: The sample included only those voters who had telephones and was not representative because voters without telephones had no chance of being included. An analogy of this sampling error in a process application would be to draw samples only during the weekday shift and ignore the weekend period entirely.

It is also important to see how a sample compares with the overall population in terms of a measurable characteristic. If the sample matches the population on this characteristic, it is often valid to conclude that it matches the population on others. However, this assumption is not always valid, so that it is a weakness in the concept of sampling.

There are so many kinds of populations and samples in chemical/environmental applications that it may not be possible to guarantee a representative sample by any single method. The reader is cautioned on this matter and it is recommended that care be exercised.

STT.2   PCB ANALYSIS

Consider the set of data below, which represents polychlorinate biphenyl (PCB) levels in a contaminated water stream for a given hour for 25 days. As a first step in summarizing the data, you are requested to form a frequency table, a frequency polygon, a cumulative frequency table, and a cumulative frequency distribution curve.

Days	PCB concentration (ppb)
1	53
2	72
3	59
4	45
5	44
6	85
7	77
8	56
9	157
10	83
11	120
12	81
13	35
14	63
15	48
16	180
17	94
18	110
19	51
20	47
21	55
22	43
23	28
24	38
25	26

Solution

Data are usually unmanageable in the form in which they are collected. In this section, the graphical techniques of summarizing such data so that meaningful information can be extracted from it is considered. Basically, there are two kinds of variables to which data can be assigned: continuous variables and discrete variables. A continuous variable is one that can assume any value in some interval of values. Examples of continuous variables are weight, volume, length, time, and temperature. Most environmental data are taken from continuous variables. Discrete variables, on the other hand, are those variables the possible values of which are integers. Therefore, they involve counting rather than measuring. Examples of discrete variables are the number of sample stations, number of people in a room, and number of times a control standard is violated.

Since any measuring device is of limited accuracy, measurements in real life are actually discrete in nature rather than continuous, but this should not keep one from regarding such variables as continuous. When a weight is recorded as 165 lb, it is assumed that the actual weight is somewhere between 164.5 and 165.5 lb.

A frequency table of the above data is first constructed.

In constructing the frequency table (shown below), it can be seen that the data have been divided into 11 class intervals with each interval being 15 units in length. The choice of dividing the data into 11 intervals was purely arbitrary. However, in dealing with data, it is a rule of thumb to choose the length of the class interval such that 8–15 intervals will include all of the data under consideration. Deriving the frequency column involves nothing more than counting the number of values in each interval. From observation of the frequency table, one can now see the data taking form. The values appear to be clustered between 25 and 85 ppb. In fact, nearly 80% are in this interval.

As a further step, one can graph the information in the frequency table. One way of doing this would be to plot the frequency midpoint of the class interval. The solid line connecting the points of Figure 129 forms a frequency polygon.

Another method of graphing the information would be by constructing a histogram as shown in Figure 130. The histogram is a two-dimensional graph in which the length of the class interval is taken into consideration. The histogram can be a very useful tool in statistics, especially if one converts the given frequency scale to a relative scale so that the sum of all the ordinates equals one. This is also shown in Figure 130. Thus, each ordinate value is derived by dividing the original value by the number of observations in the sample, in this case 25. The advantage in constructing a histogram like this is that one can read probabilities from it, if one can assume a scale on the abscissa such that a given value will fall in any one

Frequency Table

Class Interval (ppb)	Frequency of Occurence
25–40	4
40–55	7
55–70	4
70–85	4
85–100	2
100–115	1
115–130	1
130–145	0
145–160	1

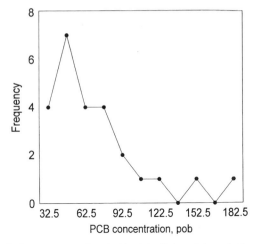

Figure 129. Pollution concentration (midpoint of class interval) frequency polygon.

interval is the area under the curve in that interval. For example, the probability that a value will fall between 55 and 70 is equal to its associated interval's portion of the total area of intervals, which is 0.16.

From the frequency table and histogram discussed above, one can also construct a cumulative frequency table and graph (Figure 131). These are shown below.

The cumulative frequency table gives the number of observations less than a given value. Probabilities can be read from the cumulative frequency curve (Figure 131). For example, to find the probability that a value will be less than 85, one should read the curve at the point $x = 85$ and read across to the value 0.74 on the y axis.

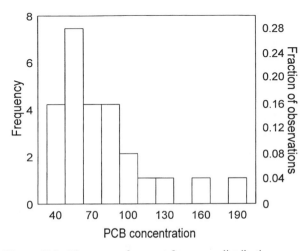

Figure 130. Histogram of percent frequency distribution curve.

Cumulative Frequency Table

PCP Level	Cumulative Frequency	Fractional Cumulative Frequency of Concentrations Less Than Stated Value
≤ 40	4	0.16
≤ 55	11	0.44
≤ 70	15	0.60
≤ 85	19	0.76
≤ 100	21	0.84
≤ 115	22	0.88
≤ 130	23	0.92
≤ 145	23	0.92
≤ 160	24	0.96
≤ 175	24	0.96
≤ 190	25	1.00

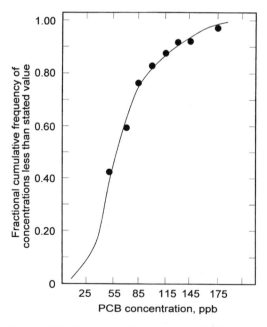

Figure 131. Cumulative frequency distribution curve.

STT.3 DISCRETE VARIABLES ANALYSIS

Using the raw data provided in Problem STT.2, generate an arithmetic-probability plot and a log-probability plot.

Solution

In drawing a histogram for a set of data, one is representing the distribution of the data. Different sets of data will vary in relation to one another and, consequently, their histograms will be different. Basically, there are three characteristics that will distinguish the distributions of different sets of data. These are central location, dispersion, and skewness. These are characterized in Figure 132. Curves A and B have the same central location, but B is more dispersed. However, both A and B are symmetrical and are, therefore, not skewed. Curve C is skewed to the right and has a different central location than A and B. Mathematical measures of central location and dispersion are discussed later in this problem set.

In most statistical work, data that closely approximate a particular symmetrical curve, called the normal curve, are required. Both curves A and B in Figure 132 are examples of normal curves. In dealing with skewed curves, such as C in the same figure, transforming the data in some way so that a symmetrical curve resembling the normal curve results is desirable. Referring to the frequency table of the data used earlier in Problem STT.2, it can be seen that for this set of data the distribution is skewed in the opposite direction as depicted by curve C in Figure 132.

One of the most successful ways of possibly generating a symmetrical distribution from a skewed distribution is by expressing the original data in terms of logarithms. The logarithms of the original data are given in the table below. By arbitrarily dividing the logarithmic data into nine class intervals, each of 0.1 unit in length, one can prepare the logarithmic frequency table as shown below. As one can see, a frequency plot versus the logarithmic scale would more closely approximate a symmetrical curve than would the arithmetic plot.

Probability graph paper is used in the analysis of cumulative frequency curves; for example, graph paper can be used as a rough test of whether the arithmetic or the logarithmic scale best approximates a normal distribution. The scale, arithmetic or

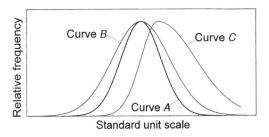

Figure 132. Sample distributions.

Logarithmic Transformation

Day	Pollutant Conc. X	$\log_{10} X$
1	53	1.724
2	72	1.857
3	59	1.771
4	45	1.653
5	44	1.644
6	85	1.929
7	77	1.887
8	56	1.748
9	157	2.196
10	83	1.919
11	120	2.079
12	81	1.909
13	35	1.544
14	63	1.799
15	48	1.681
16	180	2.255
17	94	1.973
18	110	2.041
19	51	1.708
20	47	1.672
21	55	1.740
22	43	1.634
23	28	1.447
24	38	1.580
25	26	1.415

Logarithmic Frequency

Class Interval	Tally	Frequency	Cumulative Frequency
1.4–1.5	11	2	2
1.5–1.6	11	2	4
1.6–1.7	11111	5	9
1.7–1.8	111111	6	15
1.8–1.9	11	2	17
1.9–2.0	1111	4	21
2.0–2.1	11	2	23
2.1–2.2	1	1	24
2.2–2.3	1	1	25

logarithmic, in which the cumulative frequency distribution of the data is more nearly a straight line is the one providing the better approximation to a normal distribution. By plotting the cumulative distribution curve of the data above on the two scales, it can be seen that the logarithmic scale yields the better fit (Figure 133).

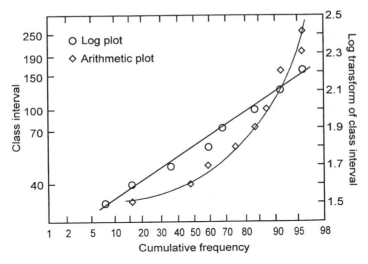

Figure 133. Normalized data plot vs. nontransformed data.

These probability plots can be used to estimate the mean and standard deviation of the data. The estimate of the mean, as will be shown in a later problem, is the 50th percentile point, and the estimation of the standard deviation is the distance from the 50th percentile to the (approximately) 16th or 84th percentile.

STT.4   CHEMILUMINESCENT NO$_x$ ANALYSIS

The following data were collected during a calibration of a chemiluminescent NO$_x$ analyzer.

$x = $ Concentration NO$_x$ (ppm)	0.05	0.10	0.20	0.30	0.45
$y = $ Instrument response (volts)	1.20	2.15	3.90	6.20	9.80

Obtain a linear regression equation that describes the relationship between x and y.

Solution

If the relationship between two variables is significant, a linear regression line, or line of "best fit," may be drawn to represent the data. Algebraically, a straight line has the following form:

$$y = mx + b$$

where $y =$ variable plotted on the ordinate
$x =$ variable plotted on abscissa
$b = y$ intercept
$m =$ slope $=$ change in y/change in $x = \Delta y/\Delta x$

Linearly regressing the data minimizes the sum of the squares of the vertical distances between all data points and the straight line. The constants m and b for the best-fit line can be determined using the following equations:

$$m = \frac{\sum xy - (\sum x)(\sum y)/n}{\sum x^2 - (\sum x)^2/n}$$

where n is the number of observations.

Values for m and b for the best-fit line can be calculated from $\sum x$, $\sum y$, $\sum x^2$, $\sum xy$, n, $\bar{y}$, and $\bar{x}$, where

$$\bar{y} = \sum y/n$$
$$\bar{x} = \sum x/n$$

Substituting the data yields

$$\sum x = 0.05 + 0.10 + 0.20 + 0.30 + 0.45 = 1.1$$
$$\sum y = 1.20 + 2.15 + 3.90 + 6.20 + 9.80 = 23.25$$
$$\sum x^2 = (0.05)^2 + (0.10)^2 + (0.20)^2 + (0.30)^2 + (0.45)^2 = 0.345$$
$$\sum xy = (0.05)(1.20) + (0.10)(2.15) + (0.20)(3.90) + (0.30)(6.20)$$
$$+ (0.45)(9.80) = 7.33$$

Thus, since $n = 5$

$$\bar{x} = \frac{\sum x}{n} = \frac{1.1}{5} = 0.22$$

$$\bar{y} = \frac{\sum y}{n} = \frac{23.25}{5} = 4.65$$

$$m = \frac{7.33 - (1.1)(23.25)/5}{0.345 - (1.1)^2/5} = \frac{2.22}{0.103} = 21.6$$

$$b = 4.65 - (21.6)(0.22) = -0.102$$

The equation for this calibration curve would be

$$y = 21.6x - 0.102$$

or

$$\text{Volts} = 21.6\,(\text{ppm}) - 0.102$$

This equation may be rearranged to yield

$$\text{ppm} = (\text{volts} + 0.102)/21.6$$

STT.5   MEDIAN, MEAN, AND STANDARD DEVIATION

The average weekly wastewater temperatures ($^\circ$C) for six consecutive weeks are

$$22, \quad 10, \quad 8, \quad 15, \quad 13, \quad 18$$

Find the median, the arithmetic mean, the geometric mean, and the standard deviation.

Solution

One basic way of summarizing data is by the computation of a central value. The most commonly used central value statistic is the arithmetic average, or the mean. This statistic is particularly useful when applied to a set of data having a fairly symmetrical distribution. The mean is an efficient statistic in that it summarizes all the data in the set and because each piece of data is taken into account in its computation. The formula for computing the mean is

$$\bar{X} = \frac{X_1 + X_2 + X_3 + \cdots + X_n}{n} = \frac{\sum\limits_{i=1}^{n} X_i}{n}$$

where $\bar{X} = $ arithmetic mean
$X_i = $ any individual measurement
$n = $ total number of observations
$X_1, X_2, X_3 \ldots = $ measurements 1, 2, and 3, respectively

The *arithmetic mean* is not a perfect measure of the true central value of a given data set. Arithmetic means can overemphasize the importance of one or two extreme data points. Many measurements of a normally distributed data set will have an arithmetic mean that closely approximates the true central value.

When a distribution of data is asymmetrical, it is sometimes desirable to compute a different measure of central value. This second measure, known as the *median*, is simply the middle value of a distribution, or the quantity above which half the data lie and below which the other half lie. If n data points are listed in their order of magnitude, the median is the $[(n + 1)/2]$th value.

If the number of data is even, then the numerical value of the median is the value midway between the two data nearest the middle. The median, being a positional value, is less influenced by extreme values in a distribution than the mean. However, the median alone is usually not a good measure of central tendency. To obtain the median, the data provided must first be arranged in order of magnitude, such as:

$$8, \quad 10, \quad 13, \quad 15, \quad 18, \quad 22$$

Thus, the median is 14, or the value halfway between 13 and 15 since this data set has an even number of measurements.

Another measure of central tendency used in specialized applications is the *geometric mean*, $\overline{X_G}$. The geometric mean can be calculated using the following equation:

$$\overline{X_G} = \sqrt[n]{(X_1)(X_2)\ldots(X_n)}$$

For the above wastewater temperatures (substituting T for X),

$$\overline{T_G} = [(8)(10)(13)(15)(18)(22)]^{1/6} = 13.54°C$$

where the arithmetic mean, $\bar{T}$, is

$$\bar{T} = (8 + 10 + 13 + 15 + 18 + 22)/6$$
$$= 14.33°C$$

The most commonly used measure of dispersion, or variability, of sets of data is the *standard deviation*, σ. Its defining formula is given by the expession:

$$\sigma = \sqrt{\frac{\sum(X_i - \bar{X})^2}{n-1}}$$

where σ = standard deviation (always positive)
X_i = value of the ith data point
$\bar{X}$ = mean of the data sample
n = number of observations

The expression $(X_i - \bar{X})$ shows that the deviation of each piece of data from the mean is taken into account by the standard deviation. Although the defining formula for the standard deviation gives insight into its meaning, the following algebraically equivalent formula makes computation much easier (now applied to the temperature, T):

$$\sigma = \sqrt{\frac{\sum(T_i - \bar{T})^2}{n-1}} = \sqrt{\frac{n\sum T_i^2 - (\sum T_i)^2}{n(n-1)}}$$

The standard deviation may be calculated for the data at hand:

$$\sum T_i^2 = (8)^2 + (10)^2 + (13)^2 + (15)^2 + (18)^2 + (22)^2 = 1366$$
$$(\sum T_i)^2 = (8 + 10 + 13 + 15 + 18 + 22)^2 = 7396$$

Thus,

$$\sigma = \sqrt{\frac{6(1366) - 7396}{(6)(5)}} = 5.16°C$$

SST.6 NORMAL DISTRIBUTION

Describe the normal distribution and briefly discuss its importance.

Solution

One reason the normal distribution is so important is that a number of natural phenomena are normally distributed or closely approximate it. In fact, many experiments, when repeated a large number of times, will approach the normal distribution curve. In its pure form, the normal curve is a continuous, symmetrical, smooth curve shaped like the one shown in Figure 134. Naturally, a finite distribution of discrete data can only approximate this curve.

The normal curve has the following definite relations to the descriptive measures of a distribution. The normal distribution curve is symmetrical; therefore, both the mean and the median are always to be found in the middle of the curve. Recall that, in general, the mean and median of an asymmetrical distribution do not coincide. The normal curve ranges along the x axis from minus infinity to plus infinity. Therefore, the range of a nomal distribution is infinite. The standard deviation, σ, becomes a most meaningful measure when related to the normal curve. A total of 68.2% of the area lying under a normal curve is included by the part ranging from one standard deviation below to one standard deviation above the mean. A total of 95.4% lies -2 to $+2$ standard deviations from the mean (see Figure 135). By using tables found in standard statistics texts and handbooks, one can determine the area lying under any part of the normal curve.

These areas under the normal distribution curve can be given probability interpretations. For example, if an experiment yields a nearly normal distribution

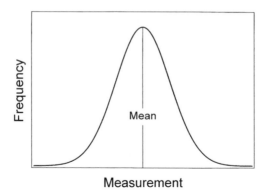

Figure 134. Gaussian distribution curve ("normal curve").

with a mean equal to 30 and a standard deviation of 10, one can expect about 68% of a large number of experimental results to range from 20 to 40, so that the probability of any particular experimental result having a value between 20 and 40 is about 0.68.

Applying the properties of the normal curve to the testing of data and/or readings, one can determine whether a change in the conditions being measured is shown or whether only chance fluctuations in the readings are represented.

For a well-established set of data, a frequently used set of control limits is ±3 standard deviations. Thus, these limits can be used to determine whether the conditions under which the original data were taken have changed. Since the limits of three standard deviations on either side of the mean include 99.7% of the area under the normal curve, it is very unlikely that a reading outside these limits is due to the conditions producing the criterion set of data. The purpose of this technique is to separate the purely chance fluctuations from other causes of variation. For example, if a long series of observations of an environmental measurement yield a mean of 50 and a standard deviation of 10, then control limits can be set up as 50

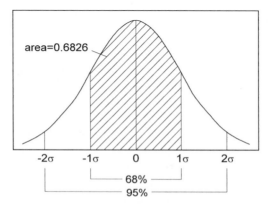

Figure 135. Characteristics of the Gaussian distribution.

± 30, i.e., ± 3 standard deviations, or from 20 to 80. A value above 80 would therefore suggest that the underlying conditions have changed and that a large number of similar observations at this time would yield a distribution of results with a mean different (larger) than 50.

STT.7 TOXIC WASTEWATER

The part-per-million concentration of a particular toxic in a wastewater stream is known to be normally distributed with mean, $m = 100$ and a standard deviation $\sigma = 2.0$. Calculate the probability that the toxic concentration, C, is between 98 and 104.

Solution

If C is normally distributed with mean m and standard deviation σ, then the random variable $(C - m)/\sigma$ is normally distributed with mean 0 and standard deviation 1. The term $(C - m)/\sigma$ is called a "standard normal variable," and the graph of its probability distribution function (PDF) is called a "standard normal curve." (The probability distribution of a random variable concerns the distribution of probability over the range of the random variable. The distribution of probability is specified by the probability distribution function, or PDF.) Referring to Figure 136, areas under a standard normal curve to the right of z_0 for non-negative values of z_0 can be found in any standard statistics book. Such an area is referred to as a "right hand tail" (see Figure 85). Probabilities about a standard normal variable Z can be determined from the tables in Problem NUC.9 in Chapter 24. For example,

$$P(Z > 1.54) = 0.062$$

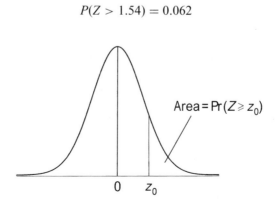

Figure 136. Standard normal, cumulative probability curve.

Since C is normally distributed with mean $m = 100$ and standard deviation $\sigma = 2$, then $(C - 100)/2$ is a standard normal variable and

$$P(98 < C < 104) = P\{-1 < [(C - 100)/2] < 2\} = P(-1 < Z < 2)$$

From tabulated or graphical values,

$$P(98 < C < 104) = 0.341 + 0.477 = 0.818 = 81.8\%$$

SST.8   TOXIC ASH

The regulatory specification on a toxic in a solid waste ash calls for a level of 1.0 ppm or less. Earlier observations of the concentration of the ash indicates a normal distribution with a mean of 0.60 ppm and a standard deviation of 0.20 ppm. Estimate the probability that ash will exceed the regulatory limit.

Solution

This problem requires the calculation of $P(C > 1.0)$. Normalizing the variable C,

$$P\{[(C - 0.6)/0.2] > [(1.0 - 0.6)/0.20]\}$$
$$P(Z > 2.0)$$

From the standard normal table (see Problem NUC.9 in Chapter 24),

$$P(Z > 2.0) = 0.0228$$
$$= 2.28\%$$

For this situation, the area to the right of the 2.0 is 2.28% of the total area. This represents the probability that ash will exceed the regulatory limit of 1.0 ppm.

Thus, one can show that for any random variable X that is normally distributed with mean, m, and standard deviation, s, (see Problem STT.7),

$$P(m - \sigma < X < m + \sigma) = [-1 < (X - m)/\sigma < 1]$$
$$= P(-1 < Z < 1) = 0.68$$
$$P(m - 2\sigma < X < m + 2\sigma) = [-2 < (X - m)/\sigma < 2]$$
$$= P(-2 < Z < 2) = 0.95$$
$$P(m - 3\sigma < X < m + 3\sigma) = [-3 < (X - m)/\sigma < 3]$$
$$= P(-3 < Z < 3) = 0.99$$

53 Indoor Air Quality

IAQ.1   RADON EXPOSURE

What steps can be taken to reduce the level of indoor radon accumulation?

Solution

One means of reducing radon exposure is to minimize radon entry by sealing off areas that may be a possible source of radon (i.e., exposed soil, cracks in the foundation, etc.). Proper ventilation, either natural or mechanical, can be effective in drawing radon from poorly ventilated areas. A third option is the use of mechanical devices such as a filter to remove airborne particles with attached radon particles.

IAQ.2   MONITORING TECHNIQUES

Describe the different techniques for monitoring indoor air quality.

Solution

Methods for monitoring indoor air quality differ in cost, sensitivity, accuracy, and precision. Monitoring methods can be divided into several categories. One category is defined by location (fixed, portable, or personal monitors). Another category is defined by the method by which the sample was taken, either using a pump or passive diffusion/permeation. A third category is defined by the length of time during which the sample was taken (grab, continuous, or time averaged).

IAQ.3   CHLORINE CONCENTRATION

Indoor concentrations of (toxic) pollutants are usually reported using two classes of units:

1. Mass of pollutant per volume of air, i.e., mg/m^3, $\mu g/m^3$, ng/m^3, etc. or
2. Parts of pollutant per parts of air (by volume), i.e, ppmv, ppbv, etc.

917

To compare data collected at different conditions, actual concentrations are often converted to standard temperature and pressure (STP). According to the Environmental Protection Agency (EPA), STP conditions for atmospheric or ambient sampling are usually 25°C and 1 atm.

The concentration of chlorine vapor is measured to be $15 \, mg/m^3$ at a pressure of 600 mm Hg and at a temperature of 10°C.

1. Convert the concentration units to ppmv.
2. Calculate the concentration in units of mg/m^3 at STP.

Solution

1. Choosing a basis of $1 \, m^3$ of air, the number of moles of Cl_2 (MW = 71) is

$$n_{Cl_2} = 0.015 \, g/(71 \, g/gmol) = 2.11 \times 10^{-4} \, gmol$$

The volume contribution of chlorine to the total volume, V_{Cl_2}, often referred to as the *pure component volume*, is

$$V_{Cl_2} = \frac{(2.11 \times 10^{-4} \, gmol)(0.082 \, atm \cdot L/gmol \cdot K)(10 + 273 \, K)}{(600/760) \, atm}$$

$$= 0.00620 \, L = 6.20 \, mL$$

Since there are $10^6 \, mL$ in a m^3, the concentration, C, in ppmv can be expressed as ML/m^3 or $mL/10^6 \, mL$. Thus,

$$C = 6.20 \, mL/m^3 = 6.20 \, ppmv$$

2. Applying the ideal gas law to adjust the air volume ($1 \, m^3$) to STP,

$$V = (1 \, m^3)\left(\frac{273 \, K}{283 \, K}\right)\left(\frac{600 \, mm}{760 \, mm}\right)$$

$$= 0.762 \, m^3$$

The concentration, C, in mg/m^3, is then

$$C = \frac{15 \, mg}{0.762 \, m^3}$$

$$= 19.7 \, mg/m^3$$

IAQ.4   WORKPLACE EXPOSURE

Exposure to indoor contaminants in a workplace can be reduced by proper ventilation. Ventilation can be provided either by dilution ventilation or by a local exhaust system. In dilution ventilation, air is brought into the work area to dilute the contaminant sufficiently to minimize its concentration and subsequently reduce worker exposure. In a local exhaust system, the contaminant itself is removed from the source through hoods. Discuss why a local exhaust system is often preferred to a dilution ventilation system.

Solution

A local exhaust is generally preferred over a dilution ventilation system for health hazard control because a local exhaust system removes the contaminants directly from the source, whereas dilution ventilation merely mixes the contaminant with uncontaminated air to reduce the contaminant concentration. Dilution ventilation may be acceptable when the contaminant concentration has a low toxicity or a threshold limit value (TLV) > 100 ppm, and the rate of contaminant emission is constant and low enough that the quantity of required dilution air is not prohibitively large. However, dilution ventilation generally is not acceptable when the TLV is less than 100 ppm.

In determining the quantity of dilution air required, one must also consider the mixing characteristics of the work area in addition to the quantity (mass or volume) of contaminant to be diluted. Thus, the amount of air required in a dilution ventilation system is much higher than the amount required in a local exhaust system. In addition, if the replacement air requires heating or cooling to maintain an acceptable workplace temperature, then the operating cost of a dilution ventilation system may further exceed the cost of a local exhaust system.

The amount of dilution air required in a dilution ventilation system can be estimated using the following expression:

$$q = K(q_c/C_a)$$

where q = dilution air flowrate
K = dimensionless mixing factor
q_c = flowrate of pure contaminant vapor
C_a = acceptable contaminant concentration

For more details, see Problem IAQ.6.

IAQ.5   VENTILATION SYSTEMS

List and briefly describe the major components of an industrial ventilation system that is employed for indoor air quality management.

Solution

The major components of an industrial ventilation system include the following:

1. Exhaust hood
2. Ductwork
3. Contaminant control devices
4. Exhaust fan
5. Exhaust vent of stack

Several types of hoods are available. One must select the appropriate hood for a specific operation to effectively remove contaminants from a work area and transport them into the ductwork. The three major types of hoods are capturing hoods, receiving hoods, and enclosure hoods. The ductwork must be sized such that the contaminant is transported without being deposited within the duct. Adequate velocity must be maintained in the duct to accomplish this. Selecting a control device that is appropriate for the contaminant removal is important to meet certain pollution control removal efficiency requirements. The exhaust fan is the workhorse of the ventilation system. The fan must provide the volumetric flow at the required static pressure and must be capable of handling contaminated air characteristics such as dustiness, corrosivity, and moisture in the airstream. Properly venting the exhaust out of the building is equally necessary to avoid contaminant recirculation into the air intake or into the building through the other openings. Such problems can be minimized by properly locating the vent pipe in relation to the aerodynamic characteristics of the building. Thus, a well-designed ventilation system can often provide the necessary health protection to the workers in an effective manner.

IAQ. 6   DILUTION VENTILATION

Estimate the dilution ventilation required in an indoor work area where a toluene-containing adhesive is used at a rate of 3 gal/8-h workday. Assume that the specific gravity of toluene (C_7H_8) is 0.87, that the adhesive contains 40 vol% toluene, and that 100% of the toluene is evaporated into the room air at 20°C. The plant manager has specified that the concentration of toluene must not exceed 80% of its threshold limit value (TLV) of 100 ppm.

The following equation can be used to estimate the dilution air requirement:

$$q = K(q_c/C_a)$$

where q = dilution air flowrate
   K = dimensionless mixing factor that accounts for less than complete mixing characteristics of the contaminant in the room, the contaminant toxicity level, and the number of potentially exposed workers. Usually, the value of K varies from 3 to 10, where 10 is used under

poor mixing conditions and when the contaminant is relatively toxic (TLV < 100 ppm).

q_c = volumetric flowrate of pure contaminant vapor, c

C_a = acceptable contaminant concentration in the room, volume or mole fraction (ppm × 10^{-6})

Solution

The dilution air can be estimated from

$$q = K(V/C_a)$$

Since the TLV for toluene is 100 ppm and C_a is 80% of the TLV,

$$C_a = [0.80(100)] \times 10^{-6} = 80 \times 10^{-6} \quad \text{(volume fraction)}$$

The mass flowrate of toluene is

$$\dot{m}_{tol} = \left(\frac{3 \text{ gal}_{adhesive}}{8 \text{ h}}\right)\left(0.4 \frac{\text{gal}_{toluene}}{1 \text{ gal}_{adhesive}}\right)\left[\frac{(0.87)(8.34 \text{ lb})}{1 \text{ gal}_{toluene}}\right]$$

$$= 1.09 \text{ lb/h}$$

$$= \left(\frac{1.09 \text{ lb}}{1 \text{ h}}\right)\left(\frac{454 \text{ g}}{1 \text{ lb}}\right)\left(\frac{1 \text{ h}}{60 \text{ min}}\right)$$

$$= 8.24 \text{ g/min}$$

Since the molecular weight of toluene is 92,

$$\dot{n}_{tol} = 8.24/92$$

$$= 0.0896 \text{ gmol/min}$$

The resultant toluene vapor volumetric flowrate, q_{tol}, is calculated directly from the ideal gas law:

$$q_{tol} = \frac{(0.0896 \text{ gmol/min})[0.08206 \text{ atm} \cdot \text{L}/(\text{gmol} \cdot \text{K})](293 \text{ K})}{1 \text{ atm}}$$

$$2.15 \text{ L/min}$$

Therefore, the required diluent volumetric flowrate is

$$q = \frac{(5)(2.15\,\text{L/min})}{80 \times 10^{-6}}$$

$$= 134{,}375\,\text{L/min}$$

$$= \left(134{,}375\,\frac{\text{L}}{\text{min}}\right)\left(\frac{1\,\text{ft}^3}{28.36\,\text{L}}\right)$$

$$= 4748\,\text{ft}^3/\text{min}$$

54 ISO 14000 (ISO)

ISO.1 ISO GOALS

Define the acronym ISO. Also discuss some of the purposes and goals of ISO.

Solution

The acronym ISO stands for *International Organization for Standardization*. It is a worldwide program that was founded in 1947 to promote the development of international manufacturing, trade, and communication standards. ISO membership includes over 100 countries. The American National Standards Institute (ANSI) is the U.S. counterpart to ISO and is the U.S. representative to ISO.

ISO essentially receives input from government, industry, and other interested parties before developing a standard. All standards developed by ISO are voluntary; thus, there are no legal requirements to force countries to adopt them. However, countries and industries often adopt ISO standards as requirements for doing and maintaining business.

ISO develops standards in all industries except those related to electrical and electronic engineering. Standards in these areas are developed by the Geneva-based International Electrotechnical Commission (IEC), which has more than 40 member countries, including the United States.

The purpose and goal of ISO is to improve the climate for international trade by "leveling the playing field." The concept is that by encouraging uniform practices around the world, barriers to trade will be reduced. If the management processes of companies in any country could be compared more readily with the management processes of companies in any other country, then international trade would be made simpler.

ISO.2 ISO 14000 COMPONENTS

Briefly describe ISO 14000.

Solution

ISO 14000 is the *International Organization for Standardization's* standard for developing and maintaining quality environmental management systems. ISO 14000 describes in considerable detail what a company must do without prescribing how it must or can be accomplished. When completed, ISO 14000 will be comprised of approximately 20 components. It will be sufficiently specific so that it will be possible to audit companies for their conformance with the standard.

Examples of the components of the ISO 14000 environmental management systems are:

1. Environmental management principles
2. Environmental labeling
3. Environmental performance evaluation
4. Life cycle assessment
5. Principles of environmental auditing
6. Terms and definitions

ISO.3   ISO 14000 DEFINITIONS

Describe the relationship and interrelationship between ISO 14000 and regulatory compliance.

Solution

As described earlier, ISO 14000 is a voluntary standard for environmental management systems. It does NOT require compliance with the regulations of the country in which the company is located. In some countries, it is possible that regulations may be more stringent than the standard. It seems likely, however, that in some countries achieving certification of adherence to the standard would improve the quality of environmental practices in that country. If, as expected, many countries adopt laws that require imported products to have been produced by companies certified to be adhering to ISO 14000, then environmental practices will almost certainly be improved worldwide.

ISO.4   EMS COMPONENTS

List the five major components of ISO 14001 EMS (environmental management system).

Solution

The five major components are:

1. Commitment and policy
2. Planning

3. Implementation
4. Measurement, monitoring and evaluation
5. Review and improvement

These must be implemented by the company in order to pass the audit requirement for certification. The environmental management system (EMS) of ISO 14001 is part of the general management system that includes organizational structure, planning activities, responsibilities, practices, procedures, processes, and resources for developing, implementing, achieving, reviewing, and maintaining the environmental policy of an organization. It is a structured process for the achievement of continual improvement related to environmental matters. The facility has the flexibility to define its boundaries and may choose to carry out this standard with respect to the entire organization or to focus the EMS on specific operating units or activities of the organization.

The EMS enables an organization to identify the significant environmental impacts that may have arisen or that may arise from the organization's past, existing, or planned activities, products, or services. It helps the organization to identify relevant environmental, legislative, and regulatory requirements that may be imposed on it. Finally, the EMS helps in planning, monitoring, auditing, corrective action, and review activities to assure compliance with established policy and allows a company to be proactive in terms of meeting anticipated new standards and compliance objectives.

ISO.5   ADVANTAGES AND DISADVANTAGES

List and briefly discuss the advantages and disadvantages of the ISO 14000 series of standards.

Solution

Advantages

1. The ISO 14000 standards provide industry with a structure for managing their environmental problems, which presumably will lead to better environmental performance.
2. It facilitates trade and minimizes trade barriers by harmonization of different national standards. As a consequence, multiple inspections, certifications, and other conflicting requirements could be reduced.
3. It expands possible market opportunities.
4. In developing countries, ISO 14000 can be used as a way to enhance regulatory systems that are either nonexistent or weak in their environmental performance requirements.
5. A number of potential cost savings can be expected, including:

> Increased overall operating efficiency
>
> Minimized liability claims and risk
>
> Improved compliance record (avoided fines and penalties)
>
> Lower insurance rates

Disadvantages

1. Implementation of ISO 14000 standards can be a tedious and expensive process.
2. ISO 14000 standards can indirectly create a technical trade barrier to both small businesses and developing countries due to limited knowledge and resources (e.g., complexity of the process and high cost of implementation, lack of registration and accreditation infrastructure, etc.)
3. ISO 14000 standards are voluntary. However, some countries may make ISO 14000 standards a regulatory requirement that can potentially lead to a trade barrier for foreign countries who cannot comply with the standards.
4. Certification/registration issues, including

   > The role of self-declaration versus third-party auditing
   >
   > Accreditation of the registrars
   >
   > Competence of ISO 14000 auditors
   >
   > Harmonization and worldwide recognition of ISO 14000 registration

ISO.6   AUDIT PLAN

What must be included in an audit plan?

Solution

Auditing a facility for certification involves several steps. Proper planning and management are very essential for effective auditing. The (lead) auditor must prepare an audit plan to ensure a smooth audit process. The audit plan must, in general, remain flexible so that any changes to the audit that are found necessary during the actual audit process can be made without compromising the audit.

An audit plan must include the following 10 items:

1. A stated scope and objectives for the audit. This includes the reason for conducting the audit, the information required, and the expectation of the audit.
2. Specification of the place, the facility, the date of the audit, and the number of days required to perform the audit.

3. Identification of high-priority items of the facility's and/or organization's EMS.

4. Identification of key personnel who will be involved in the auditing process.

5. Identification of standards and procedures (ISO 14001) that will be used to determine the conformance of various EMS elements.

6. Identification of audit team members including their special skills, experience, and audit background.

7. Specification of opening and closing meeting times.

8. Specification of confidentiality requirements during the audit process.

9. Specification of the format of the audit report, the language, distribution requirement, and the expected date of issue of the final report.

10. Identification of safety and related issues associated with entry and inspection of various portions of the facility, along with other equipment required to conduct an effective and efficient audit.

55 Measurements (MEA)

MEA.1 SOURCE TESTING OBJECTIVES

Discuss the need and objectives of source testing (monitoring).

Solution

An accurate quantitative analysis of the discharge of pollutants from a process must be determined prior to the design and/or selection of control equipment. If the unit is properly engineered, utilizing the emission data as input to the control device and the code requirements as maximum effluent limitations, most particulate pollutants can be successfully controlled by one or a combination of the methods to be discussed later.

The objective of source testing is to obtain data representative of the process being sampled. The steps followed in obtaining representative data from a large source are:

1. Obtaining a measurement that reflects the time magnitude of the characteristic being measured at the location where the measurement is made
2. Taking a number of measurements in such a manner that the data obtained from these measurements are representative of the source

Sampling is the keystone of source analysis. Sampling methods and tools vary in their complexity according to the specific task; therefore, a degree of both technical knowledge and common sense is needed to design a sampling function. Sampling is done to measure quantities or concentrations of pollutants in effluent streams, to measure the efficiency of an environmental control device, to guide the designer of control equipment and facilities, and/or to appraise contamination from a process or source. A complete measurement requires determination of the concentration and contaminant characteristics, as well as the associated flow. Most statutory limitations require mass rates of emission; both concentration and volumetric flowrate data are, therefore, required.

MEA.2 PITOT TUBE LOCATIONS

Determine the pitot tube locations in an 8-ft inner diameter (ID) circular pipe for a 16-point traverse.

Solution

The selection of a sampling site and the number of sampling points required are based on attempts to obtain representative samples. To accomplish this, the sampling site should be at least 8 duct diameters downstream and 2 diameters upstream from any bend, expansion, contraction, valve, fitting, or visible flame. For a rectangular cross section, the equivalent diameter, D_{eq}, is determined from:

$$D_{eq} = \text{flow own-sector/wetted perimieter} = 2(\text{Length} \times \text{Width})/(\text{Length} + \text{Width})$$

Once the sampling location is chosen, the cross section is laid out in a number of equal areas, the center of each being the point where the measurement is to be taken. For rectangular ducts, the cross section is divided into equal areas of the same shape, and the traverse points are located at the center of each equal area, as shown in Figure 137. For circular ducts of radius R, the cross section is divided into equal annular areas, and the traverse points are located at the centroids of each area, also shown in Figure 137.

An example showing sampling locations for an 8-ft ID circular stack (16-point traverse with equal annular areas) is presented below.

Traverse Point Number on Diameter	Distance from Inside Wall to Traverse Point (ft)
1	0.016(8) = 0.128
2	0.049(8) = 0.392
3	0.085(8) = 0.680
4	0.125(8) = 1.000
5	0.169(8) = 1.352
6	0.220(8) = 1.760
7	0.283(8) = 2.264
8	0.375(8) = 3.000
9	0.625(8) = 5.000
10	0.717(8) = 5.736
11	0.780(8) = 6.240
12	0.831(8) = 6.648
13	0.875(8) = 7.000
14	0.915(8) = 7.320
15	0.951(8) = 7.608
16	0.984(8) = 7.872

MEA.3   HIGH VOLUME SAMPLING

Determine the particle mass concentration from the following high-volume sampling and analysis data:

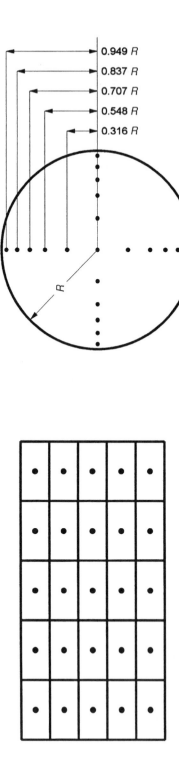

Figure 137. (*a*) Rectangular stack (10-point traverse) and (*b*) circular stack (measure at center of equal areas).

Mass of filter
    Before   3.182 g
    After   3.455 g
Flowrate
    Start   1.70 STP m^3/min
    Finish   1.41 STP m^3/min
Time
    Start   Midnight 8/11/00
    Finish   Midnight 8/12/00

Note: STP m^3/min is the volumetric flow rate corrected to standard temperature and pressure.

Solution

The high-volume sampler has become the most widely used tool for monitoring particulate matter air pollution. It is a low-cost, portable, easily maintained, and reasonably precise sampling device. Improvements in sampler performance have resulted from incorporation of automatic timers, flowrate recorders, and size separation devices into the basic system. Although hi-vols are the accepted standard in particulate matter monitoring, they inherently lack the ability to provide "real-time" particulate matter measurements.

For the problem at hand:

$$\text{Average sampling standard flowrate} = (1.70 + 1.41)/2$$

$$= 1.555 \text{ STP m}^3/\text{min}$$

$$\text{Sampled volume} = (1.555 \text{ STP m}^3/\text{min})(24\text{h})(60 \text{ min}/\text{h})$$

$$= 2239 \text{ STP m}^3$$

$$\text{Mass of particles collected} = 3.455 - 3.182$$

$$= 0.273 \text{ g}$$

$$\text{Particulate matter concentration} = (0.273 \text{ g}/2239 \text{ STD m}^3) \times 10^6 \text{ µg/g}$$

$$= 122 \text{ µg/STP m}^3$$

MEA.4   ACTIVATED CARBON REQUIREMENT

A sample gas stream containing sulfur dioxide at 15°C is to be contacted with an adsorbent while the sample collection is performed. The SO_2 concentration is known to be approximately 10 ppmv (10 µL/L). The adsorbent to be used is activated carbon. The sampling rate is 200 L/min and is to be maintained for 24 h. What quantity of activated carbon is required to remove all of the SO_2 for the entire length of the sampling period? At $15°C$, the carbon will adsorb approximately $380\,cm^3$ SO_2/g carbon.

Solution

The total volume, V, of sampled air can be calculated as follows:

$$V = (200\,L/min)(60\,min/h)(24\,h)$$
$$= 288{,}000\,L$$

The total amount of SO_2, V_{SO_2} that must be adsorbed from the sample is

$$V_{SO_2} = (288{,}000\,L\ of\ air)(10\,\mu L/L\ of\ air)$$
$$= 2{,}880{,}000\,\mu L\ of\ SO_2$$
$$= (2.88 \times 10^6\,\mu L)(10^{-3}\ cm^3/\mu L)$$
$$= 2880\ cm^3\ SO_2$$

Since the activated carbon will adsorb $380\,cm^3$ of SO_2 per gram of adsorbent, the required carbon, m_C, is

$$m_C = (2880\ cm^3\ SO_2)(1\ g\ carbon)/(380\ cm^3\ SO_2)$$
$$= 7.6\ g\ carbon$$

Thus, approximately, $7.6\,g$ of carbon would be needed to effectively remove the SO_2 from the gas sample.

MEA.5   CONTINUOUS EMISSION MONITORING

Continuous emission monitoring (CEM) systems are required to be installed in facilities specified by the Environmental Protection Agency (EPA) Standard of Performance for New Stationary Sources and by other federal and state regulations. The systems are used to continuously monitor the effectiveness of air pollution control equipment/systems and to determine if source compliance standards are being met.

A new CEM analyzer has been designed to measure NH_3. The current readings in the small table (next page) were obtained with three specially prepared calibration gase. Current readings of 18 and 4.4 mA were obtained from the analyzer at the inlet and outlet, respectively, of an absorber control device. What was the operating efficiency of the absorber?

ppmv	Current (ma)
0	4.0
40	8.0
80	16.0
100	24.0

Solution

First estimate the inlet ppm of the NH_3 gas by linearly interpolating between the 80- and 100-mA readings:

$$\text{ppm}_{\text{in}} = 80 + (100 - 80)\left(\frac{18 - 16}{24 - 16}\right)$$

$$= 85 \text{ ppm}$$

In a similar manner, the outlet ppm of the NH_3 gas is determined to be

$$\text{ppm}_{\text{out}} = 0 + (40 - 0)\left(\frac{4.4 - 4.0}{8.0 - 4.0}\right)$$

$$= 4.0 \text{ ppm}$$

Therefore, the control efficiency for the NH_3 gas is

$$\text{Efficiency} = (85 - 4.0)/85$$

$$= 0.953$$

$$= 95.3\%$$

CEM systems have been developed to monitor pollutant gases, such as SO_2 and NO, and the so-called diluent gases, CO_2 and O_2, present in the exhaust gas streams of combustion sources. Systems have also been developed to monitor flue-gas opacity. A *system* is defined as the total equipment required for the determination of flue gas opacity, a gas concentration, or the emission rate. A CEM system is normally composed of a sample interface, the pollutant and diluent analyzers, and a data recording subsystem. The system is used to generate emission data that are representative of the total emissions from the facility.

For more accurate results, one should apply Beer–Lambert's law to calculate the gas concentration:

$$I/I_0 = e^{-aC}$$

The term I/I_0 is the ratio of the current reading at concentration C divided by the reading at concentration zero, and a is a constant. A plot of I/I_0 vs. C on a semilog plot should result in a straight line.

MEA.6   HAZARDOUS WASTE INCINERATOR APPLICATION

A hazardous waste incinerator operates at a chemical plant to treat a liquid production waste stream. Your manager requests that in preparation for a trial burn, you institute a procedure for compliance with emission regulations. The waste stream feed rate is 800 lb/h and the stack gas flowrate is 22,760 scfm (60°F, 1 atm). The incinerator is to operate at a destruction and removal efficiency (DRE) of 99.995% for hexachlorobenzene.

The analytical laboratory informs you that the detection limit of hexachlorobenzene is $10\,\mu g/L$. If the sample is concentrated in 25 mL of solvent, what is the minimum volume of flue gas that must be collected to detect 99.995% DRE? The feed stream initially contains 1% (by mass hexachlorobenzene). The sample can be collected at 1 L/min (standard conditions).

Solution

Determine the mass flowrate of hexachlorobenzene:

$$\dot{m} = (0.01)(8000\,\text{lb/h})$$
$$= 80\,\text{lb/h}$$

Determine the principal organic hazardous constituents (POHC) flowrate in the stack for a 99.995% DRE.

$$\text{DRE} = \left(\frac{\dot{m}_{\text{in}} - \dot{m}_{\text{out}}}{\dot{m}_{\text{in}}}\right)(100)$$

$$= 0.99995 = 1 - \frac{\dot{m}_{\text{out}}}{80}$$

Solving for $\dot{m}_{\text{out}}$

$$\dot{m}_{\text{out}} = 0.004\,\text{lb/h}$$
$$= \left(\frac{0.004\,\text{lb}}{\text{h}}\right)\left(\frac{454 \times 10^6\,\mu g}{\text{lb}}\right)\left(\frac{1\,\text{h}}{60\,\text{min}}\right)$$
$$= 30,300\,\mu g/\text{min}$$

Determine the POHC concentration in the stack gas with units of μg/scf:

$$\text{POHC} = \frac{30{,}300\,\mu g/min}{22{,}760\,scf/min}$$

$$= 1.33\,\mu g/scf$$

Determine the sample gas volume required in scfm and the time required for sample collection in minutes:

$$\text{Volume} = \left(\frac{10\,\mu g}{L}\right)\left(\frac{1L}{1000\,mL}\right)(25\,mL)\left(\frac{1\,scf}{1.33\,\mu g}\right)$$

$$= 0.188\,scf$$

$$\text{Time} = (0.188\,scf)\left(\frac{28.3\,L}{ft^3}\right)\left(\frac{1\,min}{L}\right)$$

$$= 5.3\,min$$

MEA.7   TRIAL BURN

A trial burn is being designed for a liquid waste feed containing three principal organic hazardous compounds (POHCs) that have volatile organics sampling train (VOST) protocol requirements for recovery in the optimum ranges shown below:

POHC			
Mass Recovered	Carbon Tetrachloride (CCl_4) (ng)	Trichloroethylene (TCE) (ng)	Dichloromethane (DCM) (ng)
Minimum	200	250	500
Maximum	4000	5000	5000

Samples are collected at a rate of 1.0 L/min for 20 min.

1. Given an incinerator stack gas flowrate of 1000 dscm/min, what is the minimum feed rate for each POHC (kg/h) necessary, with a safety factor of 10, to demonstrate 99.99% DRE for each POHC?
2. What would be the impact on the success of the sampling effort for each POHC if one used the calculated feed rates, but the incinerator achieved only 99.98% DRE?

Solution

1. Calculate the sample volume.

$$(1.0\,\text{L/min})(20\,\text{min}) = 20\,\text{L of stack gas}$$

Calculate the minimum concentrations for detection in the stack gas:

$$\text{CCL}_4 \quad (200\,\text{ng})/(20\,\text{L}) = 10\,\text{ng/L} = 10\,\mu\text{g/m}^3$$

$$\text{TCE} \quad (250\,\text{ng})/20\,\text{L}) = 12.5\,\text{mg/L} = 12.5\,\mu\text{g/m}^3$$

$$\text{DCM} \quad (500\,\text{ng})/(20\,\text{L}) = 25\,\text{ng/L} = 25\,\mu\text{g/m}^3$$

Calculate the POHC mass flowrate out in mg/min.

$$\text{CCl}_4 \quad \dot{m}_{\text{out}} = (10\,\mu\text{g/m}^3)(1000\,\text{dscm/min}) = 10{,}000\,\mu\text{g/min}$$
$$= 10\,\text{mg/min}$$

$$\text{TCE} \quad \dot{m}_{\text{out}} = (12.5\,\mu\text{g/m}^3)(1000\,\text{dscm/min}) = 12{,}500\,\mu\text{g/min}$$
$$= 12.5\,\text{mg/min}$$

$$\text{DCM} \quad \dot{m}_{\text{out}} = (25\,\mu\text{g/m}^3)(1000\,\text{dscm/min}) = 25{,}000\,\mu\text{g/min}$$
$$= 25\,\text{mg/min}$$

Calculate the mass flowrate of each of the above components entering the incinerator in kg/h.

$$\text{CCl}_4 \quad \dot{m}_{\text{in}} = (10\,\text{mg/min})(1 - 0.9999) = 100{,}000\,\text{mg/min}$$
$$= 6\,\text{kg/h}$$

$$\text{TCE} \quad \dot{m}_{\text{in}} = (12.5\,\text{mg/min})(1 - 0.9999) = 125{,}000\,\text{mg/min}$$
$$= 7.5\,\text{kg/h}$$

$$\text{DCM} \quad \dot{m}_{\text{in}} = (25\,\text{mg/min})(1 - 0.9999) = 250{,}000\,\text{mg/min}$$
$$= 15\,\text{kg/h}$$

2. Reevaluate the above results using a safety factor of 10 and recalculate the stack gas mass flowrate in mg/min for a fractional DRE of 0.9998 and the flue gas concentrations with units of mg/L.

Mass flowrates:

$$\text{CCL}_4 \quad \dot{m}_{\text{in}} = (6\,\text{kg/h})(10) = 60\,\text{kg/h}$$

$$\text{TCE} \quad \dot{m}_{\text{in}} = (7.5\,\text{kg/h})(10) = 75\,\text{kg/h}$$

$$\text{DCM} \quad \dot{m}_{\text{in}} = (15\,\text{kg/h})(10) = 150\,\text{kg/h}$$

$$\text{CCl}_4 \quad \dot{m}_{\text{out}} = (1 - 0.9998)(60\,\text{kg/h}) = 120\,\text{mg/min}$$

$$\text{TCE} \quad \dot{m}_{\text{out}} = (1 - 0.9998)(75\,\text{kg/h}) = 250\,\text{mg/min}$$

$$\text{DCM} \quad \dot{m}_{\text{out}} = (1 - 0.9998)(150\,\text{kg/h}) = 500\,\text{mg/min}$$

Concentrations:

$$\text{CCl}_4 \quad (120\,\text{mg/min})/(1000\,\text{dscm/min}) = 120\,\text{ng/L}$$

$$\text{TCE} \quad (250\,\text{mg/min})/(1000\,\text{dscm/min}) = 250\,\text{ng/L}$$

$$\text{DCM} \quad (500\,\text{mg/min})/(1000\,\text{dscm/min}) = 500\,\text{ng/L}$$

Loadings:

$$\text{CCl}_4 \quad (120\,\text{ng/L})(20\,\text{L}) = 2400\,\text{ng}$$

$$\text{TCE} \quad (250\,\text{ng/L})(20\,\text{L}) = 5000\,\text{ng}$$

$$\text{DCM} \quad (500\,\text{ng/L})(20\,\text{L}) = 10{,}000\,\text{ng}$$

Comparing the loadings to the maximum limits,

	Loading (ng)	Maximum Limit (ng)
CCL$_4$	2400	4000
TCE	5000	5000
DCM	10,000	5000

Since the DCM loading is twice the maximum recovery limit, breakthrough for the DCM will be unacceptable. The loading for TCE is at the maximum level and therefore breakthrough will also be a concern. The CCl$_4$ is safely below the maximum loading level.

In real operations, the system should not run at the maximum detection limit. Sufficient backup devices must be employed to ensure breakthrough does not occur. Usually at least one extra sorbent tube will be used as an extra safeguard to handle potential overflow/breakthrough and in case there is total failure in one tube. Breakthroughs can occur when there is a process upset causing DRE to decrease.

MEA.8   ANDERSON 2000 SAMPLER

Given Andersen 2000 sampler data from an oil-fired boiler, you have been requested
to plot a cumulative distribution curve on log-probability paper and determine the
mean particle diameter and geometric standard deviation of the fly ash. Pertinent
data are provided below.

Sampler volumetric flowrate, $q = 0.5$ cfm

Anderson 2000 Sampler Data		
Plate Number	Tare Weight (g)	Final Weight (g)
0	20.48484	20.48628
1	21.38338	21.38394
2	21.92025	21.92066
3	21.55775	21.55817
4	11.40815	11.40854
5	11.61862	11.61961
6	11.76540	11.76664
7	20.99617	20.99737
Backup filter	0.20810	0.21156

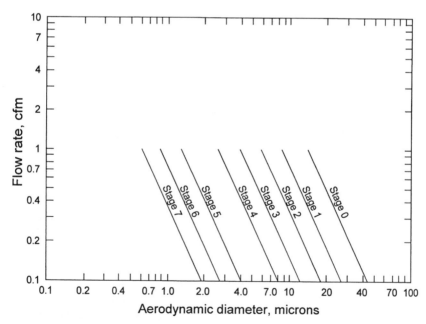

Figure 138. Aerodynamic diameter vs. Flowrate through Anderson Sampler for an impaction
efficiency of 95%.

See also Figure 138 for aerodynamic diameter vs. flowrate data for an Anderson sampler.

Solution

The table below provides the net weight, percent of total weight, and cumulative percent for each plate. A sample calculation (for plate 0) follows:

$$\text{Net weight} = \text{Final weight} - \text{Tare weight}$$
$$= 20.48628 - 20.48484$$
$$= 1.44 \times 10^{-3} \, \text{g}$$
$$= 1.44 \, \text{mg}$$

$$\text{Percent of total wt.} = (\text{net wt.}/\text{total net wt.}) \, (100)$$
$$= (1.44/10.11) \, (100\%)$$
$$= 14.2\%$$

Anderson 2000 Sampler Data

Plate Number	Net Weight (mg)	Percent of Total Weight
0	1.44	14.2
1	0.56	5.5
2	0.41	4.1
3	0.42	4.2
4	0.39	3.9
5	0.99	9.8
6	1.24	12.3
7	1.20	11.9
Backup filter	3.46	34.2
Total	10.11	100.0

Calculate the cumulative percent for each plate. Again for plate 0,

$$\text{Cumulative} \% = 100 - 14.2$$
$$= 85.8\%$$

For plate 1,

$$\text{Cumulative \%} = 100 - (14.2 + 5.5)$$

$$= 80.3$$

The table below shows the cumulative percent for each plate.

Plate No.	Cumulative Percent
0	85.8
1	80.3
2	76.2
3	72.0
4	68.1
5	58.3
6	46.0
7	34.1
Backup filter	—

Using the Andersen graph shown in Figure 138, determine the 95% aerodynamic diameter at $q = 0.5$ cfm for each plate (stage).

The cumulative distribution curve is provided on log-probability coordinates in Figure 139.

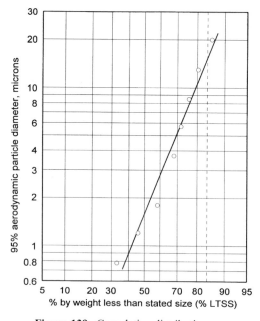

Figure 139. Cumulative distribution curve.

Plate No.	95% Aerodynamic Diameter (μm)
0	20.0
1	13.0
2	8.5
3	5.7
4	3.7
5	1.8
6	1.2
7	0.78

The mean particle diameter is the particle diameter corresponding to a cumulative percent of 50%.

$$\text{Mean particle diameter} = Y_{50} = 1.6\,\mu\text{m}$$

The distribution appears to approach log-normal behavior. The particle diameter at a cumulative percent of 84.13 is

$$Y_{84.13} = 15^{+}\,\mu\text{m}$$

Therefore, the geometric standard deviation is

$$\sigma_G = Y_{84.13}/Y_{50}$$
$$= 15/1.6$$
$$= 9.4$$

Index